Probability and Statistics
for Engineers and Scientists

Ronald E. Walpole

LATE PROFESSOR OF MATHEMATICS AND STATISTICS, ROANOKE COLLEGE

Raymond H. Myers

PROFESSOR OF STATISTICS, VIRGINIA POLYTECHNIC INSTITUTE AND STATE UNIVERSITY

Probability and Statistics for Engineers and Scientists

FIFTH EDITION

MACMILLAN PUBLISHING COMPANY

NEW YORK

Maxwell Macmillan Canada

TORONTO

Maxwell Macmillan International

NEW YORK   OXFORD   SINGAPORE   SYDNEY

Editor: Robert W. Pirtle
Production Supervisor: Elaine W. Wetterau
Production Manager: Paul Smolenski
Text Designer: Robert Freese
Cover Designer: Robert Freese
Illustrations: York Graphic Services, Inc.

This book was set in Times Roman and Bodoni by York Graphic Services, Inc., printed and bound by R. R. Donnelley & Sons Company–Crawfordsville. The cover was printed by R. R. Donnelley & Sons Company–Crawfordsville.

Macmillan Publishing Company
866 Third Avenue, New York, New York 10022

Macmillan Publishing Company is part
of the Maxwell Communication Group of Companies.

Maxwell Macmillan Canada, Inc.
1200 Eglinton Avenue East
Suite 200
Don Mills, Ontario M3C 3N1

Library of Congress Cataloging in Publication Data

Walpole, Ronald E.
    Probability and statistics for engineers and scientists / Ronald
E. Walpole, Raymond H. Myers.—5th ed.
        p.   cm.
    Includes bibliographical references and index.
    ISBN 0-02-424201-2
    1. Engineering—Statistical methods.   I. Myers, Raymond H.
II. Title.
TA340.W35   1993
519'.02462—dc20
                                                        92-16390
                                                          CIP
Printing: 1   2   3   4   5   6   7   8   Year: 3   4   5   6   7   8   9   0   1   2

Preface

The first four editions of *Probability and Statistics For Engineers and Scientists* were designed to provide an introductory probability and statistic textbook for students who major in engineering, mathematics, computer science, statistics, or any of the natural sciences. The heavy use of the high-speed computer by data analysts and recent emphasis on statistical methods in quality improvement necessitated the need to make important changes in the fourth and now in the fifth edition. One major change in the fifth edition centers around the strategic use of annotated computer printouts for data analysis examples and case studies in which real-life data sets are treated with extensive analysis, including graphical, diagnostic, and formal approaches. Graphical approaches to data analysis are emphasized even more than in the fourth edition. In the experimental design area the standard two-level factorial and fractional factorial material is enhanced with methodology that deals with quality improvement. Methods advocated by Genichi Taguchi are discussed. Alternative methods, including the construction of variance models, are illustrated.

Although most of the new features in the text deal with data analytic concepts, it is still our intention to strike the proper balance between statistical foundation and applications. The backbone of the book features basic probability theory and statistical inference. The important strength remains in the use of numerous exercises and examples. Again, real-life data sets are featured and answers to exercises appear in the back of the book. Both examples and exercises have been "freshened up" and expanded in several chapters to produce larger and more relevant data sets for the student.

As in the past, the prerequisites involve a course in differential and integral calculus through partial differentiation and multiple integration. A course in linear algebra and matrices is helpful for the regression chapter, although it is not necessary.

The important changes that enhance the text are as follows:

1. New exercises that emphasize additional modern real life applications and real-life data sets are introduced.
2. A new introductory chapter gives the students an overview of the need for dealing with variability in scientific problems. The essence of statistical inference and sampling is discussed from the point of view of the scientist or engineer. In addition, an overview of statistical graphics is given.
3. Complete case studies are presented in material dealing with two-sample inference, one- and two-factor analysis of variance, multifactor analysis of variance, and the analysis of 2^k factorial experiments.
4. Annotated computer printouts are used to illustrate results from examples and case studies.
5. New and continued emphasis on special graphical methods is featured. They are used in examples and case studies. Quantile plots, normal quantile-quantile plots, and normal probability plots are emphasized. Normal probability plots of ''effects'' are introduced as diagnostics in the case of nonreplicated 2^k factorial experiments. Residual plotting is highlighted in both simple and multiple linear regression.
6. New and important attention is given to the motivation of the usefulness of the t and F distributions at the point in which they are introduced. They are introduced in the chapter that precedes chapters on estimation and hypothesis testing. Connective tissue is provided so the student can anticipate the need for these two distributions in what follows. This was done as a response to an excellent suggestion from a reviewer.
7. Additional motivation is given for the use of tolerance intervals. They appear, as before, in the material on estimation. A clear distinction is made between confidence intervals on the mean and tolerance intervals on future observations.
8. The chapter on 2^k factorial and fractional factorial experiments contains new information on the analysis of unreplicated experiments. Pooling and diagnostic normal probability plotting of effects are covered and illustrated with a case study.
9. Quality improvement methods by Genichi Taguchi are discussed and illustrated with an example. The methodology is then critiqued. Simultaneous modeling of mean and variability is also discussed as an alternative to the Taguchi approach.

Once again the book is designed for either a one- or a two-semester course. A one-semester course containing treatment in probability and statistical inference is now accommodated in the material in Chapters 1 through 10. However, much flexibility exists in these chapters. Following the new introduction, Chapters 2, 3, and 4 cover basic probability and random variables. Chapters 5 and 6 treat specific distributions, their applications, and relationships between distributions. Chapter 5 deals with discrete distributions, including binomial and multinomial, hypergeometric, negative binomial, geometric, and Poisson. Chapter 6 discusses the normal (including normal approximation to the binomial), gamma, exponential, chi square, and other useful distributions. It is quite possible that for a more theoretical course, an instructor may choose not to use all of the material in those two chapters. On the other hand, a course for engineers would certainly benefit from the discussion of application areas of these important distributions and the use of the exercises. Chapter 6 of the fourth edition contained a rather large amount of material on somewhat disjoint topics. This material has been partitioned into two chapters: they are chapters 7 and 8. Chapter 7 is a short chapter entitled "Functions of Random Variables." It is the most mathematical topic in the book. Included is the use of moment generating functions and techniques for computing the distribution of functions of random variables. Some instructors will choose to eliminate this material, but an instructor who is more inclined to emphasize mathematical statistics may want to include it.

Chapter 8 contains three important and related topics: random sampling, data description, and fundamental sampling distributions. As we mentioned earlier, considerable attention is paid to data displays and graphics. In addition, a foundation was put in place for the use of quantile and normal probability plotting in applications forthcoming in later chapters. The distribution of $\bar{X}$ and S^2 are discussed and motivation is given for their use in the two chapters that follow. In this same regard, the t and F distributions are introduced and motivation for their use in interval estimation and hypothesis testing is given.

Chapters 9 and 10 contain material on one- and two-sample estimation and hypothesis testing, respectively. New exercises and examples are included and the student is exposed to many real-life data sets. The illustrations involve the use of graphical comparisons as well as formal inference. Annotated computer printout is a part of the illustration. Much flexibility remains with the use of Chapter 9. The instructor may wish to exclude one or more of the sections on maximum likelihood estimation, Bayes estimation, and decision theory. A certain amount of flexibility also appears in Chapter 10. Normal theory tests involving the t, F, and χ^2 distributions are given. Tests involving the use of categorical data are also presented. Chapters 11 and 12 contain simple linear regression and multiple regression, respectively. As in the past, single-number model selection criteria, stepwise procedures, diagnostic residual plots, and a discussion of transformations are given.

Chapter 13 contains a presentation of the one-factor problem. Analysis of variance for testing means is introduced. Blocking, latin squares, and the random effects model are included. An extensive case study illustrates the union of statistical graphics and formal inference. Chapter 14 contains the general factorial experiment, with graphics and annotated computer printout highlighted once again. Chapter 15 contains the aforementioned material on 2^k factorial and fractional factorial experiments, including blocking and confounding. Modern quality improvement methods are highlighted through examples and case studies, and the methodology of Taguchi is discussed. Chapter 16 is dedicated to nonparametric procedures and Chapter 17 is a chapter on statistical quality control.

We feel as if Chapters 11–17 provide sufficient material for a second semester. In some cases, entire chapters may very well be excluded. Each chapter contains sufficient flexibility and will certainly allow omission of sections without the loss of continuity. All answers to questions at the end of sections appear in the back of the book. Solutions to review exercises appear in the Instruction Solutions Supplement.

I would like to acknowledge those who contributed to the preparation of this, the fifth edition. I am grateful once again to my wife Sharon, who proofread everything. She is considerably better at it than I am. I would also like to thank Cindy Link, who typed the manuscript. Again, many thanks to the Statistical Consulting Center at Virginia Tech from which many real-life data sets were obtained and used as exercises and examples.

I am indebted to many reviewers for their helpful suggestions. They are Glen Bauer, Lawrence Technological University; Sol Blumenthal, Ohio State University; Arup Bose, Purdue University; Michael Chamberlain, United States Naval Academy; Dennis Lin, University of Tennessee at Knoxville; Kunliang Lu, University of Nebraska at Lincoln; Piotr Mikulski, University of Maryland; Giovanni Parmigiani, Duke University; and Ron Westman, Lawrence Institute of Technology.

We are indebted to the literary executor of the late Sir Ronald A. Fisher, F.R.S., Cambridge, and to Oliver & Boyd Ltd., Edinburgh, for their permission to reprint a table from their book *Statistical Methods for Research Workers*; to Professor E. S. Pearson and the Biometrika trustees for permission to reprint in abridged form Tables 8 and 18 from *Biometrika Tables for Statisticians*, Vol. I; to Oliver & Boyd Ltd. for permission to reproduce tables from their book *Design and Analysis of Industrial Experiments* by O. L. Davies; to the McGraw-Hill Book Company for permission to reproduce Tables A-25d and A-25e from their book *Introduction to Statistical Analysis* by W. J. Dixon and R. J. Massey, Jr; to C. Eisenhart, M. W. Hastay, and W. A. Wallis for permission to reproduce two tables from their book *Techniques of Statistical Analysis*. We wish also to express our appreciation for permission to reproduce tables from the *Annals of Mathematical Statistics*, from the *Bulletin of the Educational Research at Indiana University*, from a publication by the American Cyanamid Company, from *Biometrics*, from *Biometrika*, Vol. 38, and from the *Journal of the American Statistical Association*.

R. H. M.

Contents

3. Random Variables and Probability Distributions ————— 49

4. Mathematical Expectation ————————— 83

5. Some Discrete Probability Distributions ————— 113

6. Some Continuous Probability Distributions ————— 141

7. Functions of Random Variables ————————— 179

11. Simple Linear Regression and Correlation ——————— 365

12. Multiple Linear Regression ————————————— 413

13. One-Factor Experiments: General ———————————— 463

14. Factorial Experiments ───────────────── 533

15.2 2^k Factorial Experiments and Fractions ─────────── 567

Probability and Statistics
for Engineers and Scientists

Introduction to Statistics and Data Analysis

1.1 Overview

In the decade of the 1980s and thus far in the 1990s an inordinate amount of attention has been focused on improvement of quality in American industry. Much has been said and written about the Japanese "industrial miracle" which began in the middle of the twentieth century. They were able to succeed where we and other countries have failed—namely, to create an atmosphere that allows the production of high-quality products. Much of the success of the Japanese has been attributed to the use of *statistical methods* and statistical thinking among management personnel.

The use of statistical methods in manufacturing, development of food products, computer software, pharmaceuticals, and many other areas involves the gathering of information or **scientific data**. Of course, the gathering of data is nothing new. It has been done for well over a thousand years. Data have been collected, summarized, reported, and stored for perusal. However, there is a profound distinction between collection of scientific information and **inferential statistics**. It is the latter that has received rightful attention in recent decades. Inferential statistics has produced an enormous number of analytical tools that allow the engineer or scientist to better understand the systems that generate the data. This reflects the true nature of the science that we call inferential statistics, namely that of using techniques that allow us to go beyond merely reporting data, but rather, allows the drawing of conclusions (or inferences) about the scientific system. Statisticians make use of fundamental laws of probability and statistical inference to draw conclusions about scientific systems. Information is gathered in the form of **samples** or collections of **observations**. The process of sampling is introduced in Chapter 2 and the discussion continues throughout the entire book.

Samples are collected from **populations** that are collections of all individuals or individual items of a particular type. At times a population signifies a scientific system. For example, a manufacturer of computer boards may wish to eliminate defects. A sampling process may involve collecting information on 50 computer boards sampled randomly from the process. Here, the population is all computer boards manufactured by the firm over a specific period of time. In a drug experiment, a sample of patients is taken and each is given a specific drug to reduce blood pressure. The interest is focused on drawing conclusions about the population of those who suffer from hypertension. If an improvement is made in the computer board process and a second sample of boards is collected, any conclusions drawn regarding the effectiveness of the change in process should extend to the entire population of computer boards produced under the "improved process."

Often, it is very important to collect scientific data in a systematic way, with planning being high on the agenda. At times the planning is, by necessity, quite limited. An engineer may need to study the effect of process conditions, temperature, humidity, amount of a particular ingredient, and so on, on product output. He or she can systematically move these **factors** to whatever levels are suggested according to whatever prescription or **experimental design** is desired. However, a forest scientist who is interested in a study of factors that influence wood density in a certain kind of tree cannot necessarily design an experiment. In this case it may require an **observational study** in which data are collected in the field but **factor levels** could not be preselected. Both of these types of studies lend themselves to methods of statistical inference. In the former, the quality of the inferences will depend on proper planning of the experiment. In the latter, the scientist is at the mercy of what can be gathered. For example, it is sad if an agronomist is interested in studying the effect of rainfall on plant yield and the data are gathered during a drought.

One should gain an insight into the importance of statistical thinking by managers and the use of statistical inference by scientific personnel. Research scientists gain much from scientific data. Data provide understanding of scientific phenomenae. Product and process engineers learn more in their off-line efforts to improve the process. They also gain valuable insight by gathering production data (on-line monitoring) on a regular basis. This allows for determination of necessary modifications in order to keep the process at a desired level of quality.

1.2 The Role of Probability

In this book, Chapters 2 to 6 deal with fundamental notions of probability. A thorough grounding in these concepts allows the reader to have a better understanding of statistical inference. Without some formalism in probability the student cannot appreciate the true interpretation of data analysis through modern statistical methods. It is quite natural to study probability prior to studying statistical inference. Elements of probability allow us to quantify the strength or "confidence" in our conclusions.

EXAMPLE 1.1 Suppose that an engineer encounters data from a manufacturing process in which 100 items are sampled and 10 are found to be defective. It is expected and anticipated that occasionally there will be defective items. However, it has been determined that in the long run, the company can only tolerate 5% defective in the process. Now, the elements of probability allow the engineer to determine how conclusive the sample information is regarding the nature of the process. Suppose we learn that *if the process is acceptable,* that is, if it does produce items 5% of which are defective, there is a probability of 0.00001 of obtaining 10 or more defective items in a random sample of 100 items from the process. This small probability suggests that the process does, indeed, have a long-run percent defective that exceeds 5%. In other words, under the condition of an acceptable process, the sample information obtained would almost never occur. Clearly, though, it would occur with a much higher probability if the process defective rate exceeded 5% by a significant amount.

From this example it becomes clear that the elements of probability aid in the translation of sample information into something conclusive or inconclusive about the scientific system. The example that follows provides a second illustration.

EXAMPLE 1.2 Often the nature of the scientific study will dictate the role that probability and deductive reasoning play in statistical inference. Exercise 6 at the end of Section 9.8 provides data associated with a study conducted at the Virginia Polytechnic Institute and State University on the development of a relationship between the roots of trees and the action of a fungus. Minerals are transferred from the fungus to the trees and sugars from the trees to the fungus. Two samples of 10 northern red oak seedlings are planted in a greenhouse, one containing seedlings treated with nitrogen and one containing no nitrogen. All other environmental conditions are held constant. All seedlings contain the fungus *Pisolithus tinctorus.* More details are supplied in Chapter 9. The stem weights in grams were recorded after the end of 140 days. The data are as follows:

No Nitrogen	Nitrogen
0.32	0.26
0.53	0.43
0.28	0.47
0.37	0.49
0.47	0.52
0.43	0.75
0.36	0.79
0.42	0.86
0.38	0.62
0.43	0.46

It is instructive to plot the data as shown in Figure 1.1. The 0 values represent the "with nitrogen" data and the × values represent the "without nitrogen" data. Now, the purpose of this experiment is to determine whether or not the use of nitrogen has an influence on the growth of the roots. Notice that the general appearance of the data might suggest to the reader that, on the average, the use of nitrogen increases the stem weight. Four nitrogen observations are considerably larger than any of the no-nitrogen observations. Most of the no-nitrogen observations appear to be below the center of the data. The appearance of the data set would seem to indicate that nitrogen is effective. But how can this be quantified? How can all of the apparent visual evidence be summarized in some sense? As in the preceding example, the fundamentals of probability can be used. The conclusions may be summarized in a probability statement or p-value. The issue revolves around the "probability that data like these could be observed" *given that nitrogen has no effect,* in other words, given that both samples were generated from the same population. Suppose that this probability is small, say 0.03. This would certainly be strong evidence that the use of nitrogen does indeed influence (apparently increase) average stem weight.

We have given two examples in which the elements of probability have provided a summary that the scientist or engineer can use as evidence on which to build a decision. The bridge between the data and the conclusion is, of course, based on foundations of statistical inference, distribution theory, and sampling distributions discussed and illustrated in Chapters 7 to 11. It is instructive at this point to give some attention to measures of sample location and variability. Both exploratory or intuitive data analysis and formal statistical inference depend on these measures.

1.3 Measures of Location: The Sample Mean

Location measures in a data set are designed to give the analyst some quantitative measure of where the data center is in a sample. In Example 1.2 it certainly appeared as if the center of the nitrogen sample clearly exceeds that of the no-nitrogen sample. One obvious and very useful measure is the **sample mean**. The mean is

FIGURE 1.1   Stem weight data.

simply a numerical average. Suppose that the observations in a sample are given by $x_1, x_2, \ldots, x_n$. The sample mean is given by

$$\bar{x} = \sum_{i=1}^{n} \frac{x_i}{n} = \frac{x_1 + x_2 + \cdots + x_n}{n}.$$

There are other measures of central tendency that will be discussed in future chapters. In the "two-sample data set" of Example 1.3 the two sample means are

$$\bar{x}(\text{no nitrogen}) = 0.40 \text{ gram}$$

and

$$\bar{x}(\text{nitrogen}) = 0.57 \text{ gram}.$$

1.4 Measures of Variability

Sample variability plays an important role in data analysis. Process and product variability is a fact of life in engineering and scientific systems. In fact, the control or reduction of process variability is often the source of major difficulty. Variability in population values and sample data is a fact of life. Larger variability among sample observations can often "wash out" any effects that the engineer may try to detect. Measures of location in a sample do not provide a proper summary of the nature of a data set. For instance, in Example 1.3 we cannot conclude that the use of nitrogen enhances growth without taking sample variability into account. For example, contrast the two data sets below. Each contains two samples and the difference in the means is roughly the same for the two samples.

Data set A: × × × × × × 0 × × 0 0 × × × 0 0 0 0 0 0 0 0

$$\bar{x}_\times \qquad\qquad\qquad \bar{x}_0$$

Data set B: × × × × × × × × × × × × 0 0 0 0 0 0 0 0 0 0 0

$$\bar{x}_\times \qquad\qquad\qquad \bar{x}_0$$

Data set B seems to provide a much sharper contrast between the two populations from which the samples were taken. This is not the case with data set A. If the purpose of such an experiment is to detect distinction between the two populations, the task is accomplished in the case of data set B. However, in data set A the large variability within the two samples creates difficulty. In fact, it is not clear that there is a distinction between the two populations.

Just as there are many measures of central tendency or location there are many measures of spread or variability. Perhaps the simplest one is the **sample range**

$X_{max} - X_{min}$. The range can be very useful and is discussed at length in Chapter 17 on *statistical quality control*. The sample measure of spread that is used most often is the **sample standard deviation**. We again let $x_1, x_2, \ldots, x_n$ denote sample values; the sample standard deviation is given by

$$s = \sqrt{\sum_{i=1}^{n} (x_i - \bar{x})^2/(n - 1)}.$$

The quantity $s^2 = \sum_{i=1}^{n} (x_i - \bar{x})^2/(n - 1)$ is called a **sample variance**. It should be clear to the reader that the sample standard deviation is, in fact, a measure of variability. The quantity $n - 1$ is often called the **degrees of freedom associated with the variance** estimate. In this simple example the degrees of freedom depict the number of independent pieces of information available for computing variability. For example, suppose that we wish to compute the sample variance and standard deviation of the data set (5, 17, 6, 4). The sample average is $\bar{x} = 8$. The computation of the variance involves

$$(5 - 8)^2 + (17 - 8)^2 + (6 - 8)^2 + (4 - 8)^2 = (-3)^2 + (9)^2 + (-2)^2 + (-4)^2.$$

The quantities inside parentheses sum to zero. In fact, in general $\sum_{i=1}^{n} (x_i - \bar{x}) = 0$.

Then the computation of a sample variance does not involve n **independent squared deviations** from the mean $\bar{x}$. Thus there are $n - 1$ degrees of freedom rather than n degrees of freedom for computing a sample variance.

1.5 Discrete and Continuous Data

Statistical inference through the analysis of observational studies or designed experiments is common in many scientific areas. The data gathered may be **discrete** or **continuous**, depending on the area of application. For example, a chemical engineer may be interested in conducting an experiment that will lead to conditions where yield is maximized. Here, of course, the yield may be in percent, or grams/pound, measured on a continuum. On the other hand, a toxocologist in a biomedical study conducting a combination drug experiment may encounter data that are binary in nature (i.e., the patient either responds or not). In these two distinct situations the probability theory that allows us to draw statistical inferences is quite different.

1.6 Statistical Modeling, Scientific Inspection, and Graphical Diagnostics

Quite often the end result of a statistical analysis is the estimation of parameters of a **postulated model**. This is often quite natural for scientists and engineers since they often deal in modeling. A statistical model is not deterministic but, rather, must entail some probabilistic aspects. A model form is often the foundation of **assumptions** that are made by the analyst. For example, in our Example 1.3 the scientist may wish to draw some level of distinction between the "nitrogen" and "no nitrogen" population through the sample information. The analysis may require a certain model for the data, for example, that the two samples come from **normal** or **Gaussian distributions**. See Chapter 6 for a discussion of a normal distribution.

At times the model postulated may take on a somewhat more complicated form. Consider, for example, a textile manufacturer who designs an experiment in which cloth specimens are being produced that contain various percentages of cotton. Consider the following data:

Cotton Percentage	Tensile Strength
15	7, 7, 9, 8, 10
20	19, 20, 21, 20, 22
25	21, 21, 17, 19, 20
30	8, 7, 8, 9, 10

Five cloth specimens were manufactured for each of the four cotton percentages. Now, in this case both the model for the experiment and the type of analysis used should take into account the goal of the experiment and important input from the textile scientist. Some simple graphics can shed important light on the clear distinction between the samples. See Figure 1.2; the sample means and variability are depicted nicely in the data plot. One possible goal of this experiment is simply to determine which cotton percentages are truly distinct from the others. In other words, as in the case of the nitrogen/no nitrogen data, for which cotton percentages are there clear distinctions between the populations, or more specifically, between the population means? In this case, perhaps a reasonable model is that each sample comes from a normal distribution. Here the goal is very much like that of the nitrogen/no nitrogen data except that more samples are involved. The formalism of

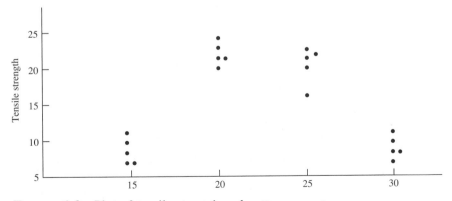

FIGURE 1.2 Plot of tensile strength and cotton percentages.

the analysis involves notions of hypothesis testing discussed in Chapter 10. Incidentally, this formality is perhaps not necessary in light of the diagnostic plot. But does this describe the real goal of the experiment and hence the proper approach to data analysis? It is likely that the scientist anticipates the existence of a *maximum population mean tensile strength* in the range of cotton concentration in the experiment. Here the analysis of the data should revolve around a different type of model, one that postulates a type of structure relating the population mean tensile strength to the cotton concentration. In other words, a model may be written

$$\mu_{t,c} = \beta_0 + \beta_1 C + \beta_2 C^2,$$

where μ_t is the population mean tensile strength, which varies with the amount of cotton in the product, C. The implication of this model is that for a fixed cotton level, there is a population of tensile strength measurements and the population mean is $\mu_{t,c}$. This type of model, called a **regression model**, is discussed in Chapters 11 and 12. The functional form is chosen by the scientist. At times the data analysis may suggest that the model be changed. Then the data analyst ''entertains'' a model that may be altered after some analysis is done. The use of an empirical model is accompanied by **estimation theory**, in which β_0, β_1, and β_2 are estimated by the data. Further, statistical inference can then be used to determine model adequacy.

Several points become evident from the two data illustrations here: (1) the type of model used to describe the data often depends on the goal of the experiment; and (2) the structure of the model should take advantage of nonstatistical scientific input. A selection of a model represents a **fundamental assumption** upon which the resulting statistical inference is based. It will become apparent throughout the book how important graphics can be. Often, plots can illustrate information that allows the results of the formal statistical inference to be better communicated to the scientist or engineer. At times, plots or **exploratory data analysis** can teach the analyst something not retrieved from the formal analysis. Almost any formal analysis requires assumptions that evolve from the model of the data. Graphics can nicely highlight **violation of assumptions** that would otherwise go unnoticed. Throughout the book, graphics are used and illustrated extensively to supplement formal data analysis.

2

Probability

2.1 Sample Space

In the study of statistics we are basically concerned with the presentation and interpretation of **chance outcomes** that occur in a planned study or scientific investigation. For example, we may record the number of accidents that occur monthly at the intersection of Driftwood Lane and Royal Oak Drive, hoping to justify the installation of a traffic light; we might classify items coming off an assembly line as "defective" or "nondefective"; or we may be interested in the volume of gas released in a chemical reaction when the concentration of an acid is varied. Hence the statistician is often dealing either with **experimental data**, representing **counts** or **measurements**, or perhaps with **categorical data** that can be classified according to some criterion.

We shall refer to any recording of information, whether it be numerical or categorical, as an **observation**. Thus the numbers 2, 0, 1, and 2, representing the number of accidents that occurred for each month from January through April during the past year at the intersection of Driftwood Lane and Royal Oak Drive, constitute a set of observations. Similarly, the categorical data N, D, N, N, and D, representing the items found to be defective or nondefective when five items are inspected, are recorded as observations.

Statisticians use the word *experiment* to describe any process that generates a set of data. A very simple example of a statistical experiment might be the tossing of a coin, In this experiment there are only two possible outcomes, heads or tails. Another experiment might be the launching of a missile and observing the velocity at specified times. The opinions of voters concerning a new sales tax can also be considered as observations of an experiment. We are particularly interested in the observations obtained by repeating the experiment several times. In most cases the outcomes will depend on chance and, therefore, cannot be predicted with certainty. If a chemist runs an analysis several times under the same conditions, he will obtain different measurements, indicating an element of chance in the experimental procedure. Even when a coin is tossed repeatedly, we cannot be certain that a given toss will result in a head. However, we know the entire set of possibilities for each toss.

DEFINITION 2.1   *The set of all possible outcomes of a statistical experiment is called the* **sample space** *and is represented by the symbol S.*   ∎

Each outcome in a sample space is called an **element** or a **member** of the sample space or simply a **sample point**. If the sample space has a finite number of elements, we may *list* the members separated by commas and enclosed in braces. Thus the sample space S, of possible outcomes when a coin is tossed, may be written

$$S = \{H, T\},$$

where H and T correspond to "heads" and "tails," respectively.

EXAMPLE 2.1   Consider the experiment of tossing a die. If we are interested in the number that shows on the top face, the sample space would be

$$S_1 = \{1, 2, 3, 4, 5, 6\}.$$

If we are interested only in whether the number is even or odd, the sample space is simply

$$S_2 = \{\text{even, odd}\}.$$

Example 2.1 illustrates the fact that more than one sample space can be used to describe the outcomes of an experiment. In this case S_1 provides more information than S_2. If we know which element in S_1 occurs, we can tell which outcome in S_2 occurs; however, a knowledge of what happens in S_2 is of no help in determining which element in S_1 occurs. In general, it is desirable to use a sample space that gives the most information concerning the outcomes of the experiment.

In some experiments it will be helpful to list the elements of the sample space systematically by means of a **tree diagram**.

EXAMPLE 2.2   An experiment consists of flipping a coin and then flipping it a second time if a head occurs. If a tail occurs on the first flip, then a die is tossed once. To list the elements of the sample space providing the most information, we construct the tree diagram of Figure 2.1. Now, the various paths along the branches of the tree give the distinct sample points. Starting with the top left branch and moving to the right along the first path, we get the sample point HH, indicating the possibility that heads occurs on two successive flips of the coin. Likewise, the sample point $T3$ indicates the

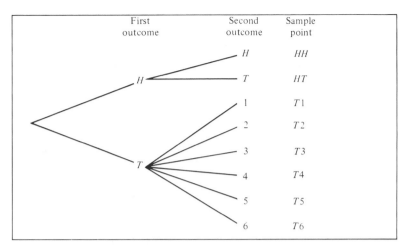

First outcome	Second outcome	Sample point
	H	HH
H	T	HT
	1	T1
	2	T2
	3	T3
T	4	T4
	5	T5
	6	T6

FIGURE 2.1   Tree diagram for Example 2.2.

possibility that the coin will show a tail followed by a 3 on the toss of the die. By proceeding along all paths, we see that the sample space is

$$S = \{HH, HT, T1, T2, T3, T4, T5, T6\}.$$

EXAMPLE 2.3   Suppose that three items are selected at random from a manufacturing process. Each item is inspected and classified defective, D, or nondefective, N. To list the elements of the sample space providing the most information, we construct the tree diagram of Figure 2.2. Now, the various paths along the branches of the tree give the distinct sample points. Starting with the first path, we get the sample point DDD, indicating the possibility that all three items inspected are defective. As we proceed along the other paths, we see that the sample space is

$$S = \{DDD, DDN, DND, DNN, NDD, NDN, NND, NNN\}.$$

Sample spaces with a large or infinite number of sample points are best described by a *statement* or *rule*. For example, if the possible outcomes of an experiment are the set of cities in the world with a population over 1 million, our sample space is written

$$S = \{x \mid x \text{ is a city with a population over 1 million}\},$$

which reads "S is the set of all x such that x is a city with a population over 1 million." The vertical bar is read "such that." Similarly, if S is the set of all points (x, y) on the boundary or the interior of a circle of radius 2 with center at the origin, we write

$$S = \{(x, y) \mid x^2 + y^2 \le 4\}.$$

Whether we describe the sample space by the rule method or by listing the elements will depend on the specific problem at hand. The rule method has practical advantages, particularly in the many experiments where a listing becomes a very tedious chore.

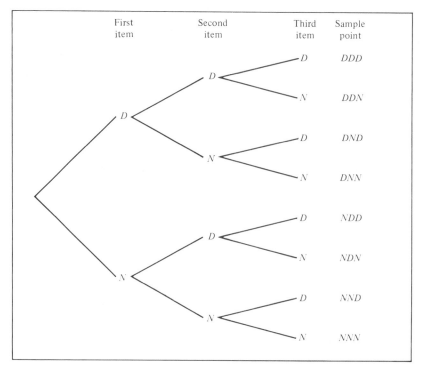

First item	Second item	Third item	Sample point

FIGURE 2.2   Tree diagram for Example 2.3.

2.2 Events

In any given experiment we may be interested in the occurrence of certain **events** rather than in the outcome of a specific element in the sample space. For instance, we might be interested in the event *A* that the outcome when a die is tossed is divisible by 3. This will occur if the outcome is an element of the subset $A = \{3, 6\}$ of the sample space S_1 in Example 2.1. As a further illustration, we might be interested in the event *B* that the number of defectives is greater than 1 in Example 2.3. This will occur if the outcome is an element of the subset $B = \{DDN, DND, NDD, DDD\}$ of the sample space *S*.

To each event we assign a collection of sample points, which constitute a subset of the sample space. This subset represents all of the elements for which the event is true.

DEFINITION 2.2   *An* **event** *is a subset of a sample space.*   ■

EXAMPLE 2.4   Given the sample space $S = \{t \mid t \geq 0\}$, where *t* is the life in years of a certain electronic component, then the event *A* that the component fails before the end of the fifth year is the subset $A = \{t \mid 0 \leq t < 5\}$.

It is conceivable that an event may be a subset that includes the entire sample space S, or a subset of S called the null set and denoted by the symbol $\varnothing$, which contains no elements at all. For instance, if we let A be the event of detecting a microscopic organism by the naked eye in a biological experiment, then $A = \varnothing$. Also, if $B = \{x \mid x$ is an even factor of 7$\}$, then B must be the null set, since the only possible factors of 7 are the odd numbers 1 and 7.

Consider an experiment in which the smoking habits of the employees of some manufacturing firm are recorded. A possible sample space might classify an individual as a nonsmoker, a light smoker, a moderate smoker, or a heavy smoker. Let the subset of smokers be some event. Then all the nonsmokers correspond to a different event, also a subset of S, which is called the **complement** of the set of smokers.

DEFINITION 2.3   *The **complement** of an event A with respect to S is the subset of all elements of S that are not in A. We denote the complement of A by the symbol A'.* ◼

EXAMPLE 2.5   Let R be the event that a red card is selected from an ordinary deck of 52 playing cards, and let S be the entire deck. Then R' is the event that the card selected from the deck is not a red but a black card.

EXAMPLE 2.6   Consider the sample space $S = \{$book, catalyst, cigarette, precipitate, engineer, rivet$\}$. Let $A = \{$catalyst, rivet, book, cigarette$\}$. Then $A' = \{$precipitate, engineer$\}$.

We now consider certain operations with events that will result in the formation of new events. These new events will be subsets of the same sample space as the given events. Suppose that A and B are two events associated with an experiment. In other words, A and B are subsets of the same sample space S. For example, in the tossing of a die we might let A be the event that an even number occurs and B the event that a number greater than 3 shows. Then the subsets $A = \{2, 4, 6\}$ and $B = \{4, 5, 6\}$ are subsets of the same sample space $S = \{1, 2, 3, 4, 5, 6\}$. Note that *both* A and B will occur on a given toss if the outcome is an element of the subset $\{4, 6\}$, which is just the intersection of A and B.

DEFINITION 2.4   *The **intersection** of two events A and B, denoted by the symbol $A \cap B$, is the event containing all elements that are common to A and B.* ◼

EXAMPLE 2.7   Let P be the event that a person selected at random while dining at a popular cafeteria is a taxpayer, and let Q be the event that the person is over 65 years of age. Then the event $P \cap Q$ is the set of all taxpayers in the cafeteria who are over 65 years of age.

EXAMPLE 2.8   Let $M = \{a, e, i, o, u\}$ and $N = \{r, s, t\}$; then it follows that $M \cap N = \varnothing$. That is, M and N have no elements in common and, therefore, cannot both occur simultaneously.

In certain statistical experiments it is by no means unusual to define two events A and B that cannot both occur simultaneously. The events A and B are then said to be **mutually exclusive.** Stated more formally, we have the following definition:

DEFINITION 2.5   *Two events A and B are **mutually exclusive** or disjoint if $A \cap B = \varnothing$, that is, if A and B have no elements in common.* ∎

EXAMPLE 2.9   A cable television company offers programs on eight different channels, three of which are affiliated with ABC, two with NBC, and one with CBS. The other two are an educational channel and the ESPN sports channel. Suppose that a person subscribing to this service turns on a television set without first selecting the channel. Let A be the event that the program belongs to the NBC network and B the event that it belongs to the CBS network. Since a television program cannot belong to more than one network, the events A and B have no programs in common. Therefore, the intersection $A \cap B$ contains no programs, and consequently the events A and B are mutually exclusive.

Often one is interested in the occurrence of at least one of two events associated with an experiment. Thus, in the die-tossing experiment, if $A = \{2, 4, 6\}$ and $B = \{4, 5, 6\}$, we might be interested in either A or B occurring, or both A and B occurring. Such an event, called the **union** of A and B, will occur if the outcome is an element of the subset $\{2, 4, 5, 6\}$.

DEFINITION 2.6   *The **union** of the two events A and B, denoted by the symbol $A \cup B$, is the event containing all the elements that belong to A or B or both.* ∎

EXAMPLE 2.10   Let $A = \{a, b, c\}$ and $B = \{b, c, d, e\}$; then $A \cup B = \{a, b, c, d, e\}$.

EXAMPLE 2.11   Let P be the event that an employee selected at random from an oil drilling company smokes cigarettes. Let Q be the event that the employee selected drinks alcoholic beverages. Then the event $P \cup Q$ is the set of all employees who either drink or smoke, or who do both.

EXAMPLE 2.12   If $M = \{x \mid 3 < x < 9\}$ and $N = \{y \mid 5 < y < 12\}$, then $M \cup N = \{z \mid 3 < z < 12\}$.

The relationship between events and the corresponding sample space can be illustrated graphically by means of **Venn diagrams**. In a Venn diagram we let the sample space be a rectangle and represent events by circles drawn inside the rectangle. Thus, in Figure 2.3, we see that

$$A \cap B = \text{regions 1 and 2,}$$
$$B \cap C = \text{regions 1 and 3,}$$
$$A \cup C = \text{regions 1, 2, 3, 4, 5, and 7,}$$
$$B' \cap A = \text{regions 4 and 7,}$$
$$A \cap B \cap C = \text{region 1,}$$
$$(A \cup B) \cap C' = \text{regions 2, 6, and 7,}$$

and so forth. In Figure 2.4 we see that events A, B, and C are all subsets of the sample space S. It is also clear that event B is a subset of event A; event $B \cap C$ has

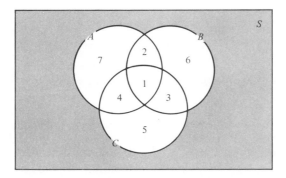

FIGURE 2.3 Events represented by various regions.

no elements and hence B and C are mutually exclusive; event $A \cap C$ has at least one element; and event $A \cup B = A$. Figure 2.4 might, therefore, depict a situation in which we select a card at random from an ordinary deck of 52 playing cards and observe whether the following events occur:

A: the card is red,

B: the card is the jack, queen, or king of diamonds,

C: the card is an ace.

Clearly, the event $A \cap C$ consists only of the 2 red aces.

Several results that follow from the foregoing definitions, which may easily be verified by means of Venn diagrams, are as follows:

1. $A \cap \varnothing = \varnothing$.
2. $A \cup \varnothing = A$.
3. $A \cap A' = \varnothing$.
4. $A \cup A' = S$.
5. $S' = \varnothing$.
6. $\varnothing' = S$.
7. $(A')' = A$.
8. $(A \cap B)' = A' \cup B'$.
9. $(A \cup B)' = A' \cap B'$.

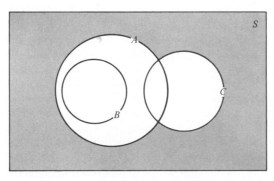

FIGURE 2.4 Events of the sample space S.

Exercises

1. List the elements of each of the following sample spaces:
  (a) the set of integers between 1 and 50 divisible by 8;
  (b) the set $S = \{x \mid x^2 + 4x - 5 = 0\}$;
  (c) the set of outcomes when a coin is tossed until a tail or three heads appear;
  (d) the set $S = \{x \mid x$ is a continent$\}$;
  (e) the set $S = \{x \mid 2x - 4 \geq 0$ and $x < 1\}$.

2. Use the rule method to describe the sample space S consisting of all points in the first quadrant inside a circle of radius 3 with center at the origin.

3. Which of the following events are equal?
  (a) $A = \{1, 3\}$.
  (b) $B = \{x \mid x$ is a number on a die$\}$.
  (c) $C = \{x \mid x^2 - 4x + 3 = 0\}$.
  (d) $D = \{x \mid x$ is the number of heads when six coins are tossed$\}$.

4. An experiment involves tossing a pair of dice, 1 green and 1 red, and recording the numbers that come up. If x equals the outcome on the green die and y the outcome on the red die, describe the sample space S
  (a) by listing the elements (x, y);
  (b) by using the rule method.

5. An experiment consists of tossing a die and then flipping a coin once if the number on the die is even. If the number on the die is odd, the coin is flipped twice. Using the notation $4H$, for example, to denote the event that the die comes up 4 and then the coin comes up heads, and $3HT$ to denote the event that the die comes up 3 followed by a head and then a tail on the coin, construct a tree diagram to show the 18 elements of the sample space S.

6. Two jurors are selected from 4 alternates to serve at a murder trial. Using the notation A_1A_3, for example, to denote the simple event that alternates 1 and 3 are selected, list the 6 elements of the sample space S.

7. Four students are selected at random from a chemistry class and classified as male or female. List the elements of the sample space S_1 using the letter M for "male" and F for "female." Define a second sample space, S_2, where the elements represent the number of females selected.

8. For the sample space of Exercise 4,
  (a) list the elements corresponding to the event A that the sum is greater than 8;
  (b) list the elements corresponding to the event B that a 2 occurs on either die;
  (c) list the elements corresponding to the event C that a number greater than 4 comes up on the green die;
  (d) list the elements corresponding to the event $A \cap C$;
  (e) list the elements corresponding to the event $A \cap B$;
  (f) list the elements corresponding to the event $B \cap C$;
  (g) construct a Venn diagram to illustrate the intersections and unions of the events A, B, and C.

9. For the sample space of Exercise 5,
  (a) list the elements corresponding to the event A that a number less than 3 occurs on the die;
  (b) list the elements corresponding to the event B that 2 tails occur;
  (c) list the elements corresponding to the event A';
  (d) list the elements corresponding to the event $A' \cap B$;
  (e) list the elements corresponding to the event $A \cup B$.

10. An experiment consists of asking 3 women at random if they wash their dishes with brand X detergent.
  (a) List the elements of a sample space S using the letter Y for "yes" and N for "no."
  (b) List the elements of S corresponding to event E that at least 2 of the women use brand X.
  (c) Define an event that has as its elements the points $\{YYY, NYY, YYN, NYN\}$.

11. The résumés of 2 male applicants for a college teaching position in psychology are placed in the same file as the résumés of 2 female applicants. Two positions become available and the first, at the rank of assistant professor, is filled by selecting 1 of the 4 applicants at random. The second position, at the rank of instructor, is then filled by selecting at random one of the remaining 3 applicants. Using the notation M_2F_1, for example, to denote the simple event that the first position is filled by the second male applicant and the second position is then filled by the first female applicant,

(a) list the elements of a sample space S;

(b) list the elements of S corresponding to event A that the position of assistant professor is filled by a male applicant;

(c) list the elements of S corresponding to event B that exactly 1 of the 2 positions was filled by a male applicant;

(d) list the elements of S corresponding to event C that neither position was filled by a male applicant;

(e) list the elements of S corresponding to the event $A \cap B$;

(f) list the elements of S corresponding to the event $A \cup C$;

(g) construct a Venn diagram to illustrate the intersections and unions of the events A, B, and C.

12. A developer from Saudi Arabia has decided to invest large sums of money in real estate. Four states, Virginia, New York, Connecticut, and Massachusetts, are being considered for the construction of hotels, motels, and condominiums, all of which will be located either directly on the beach or at resorts in the mountains. Using the notation Cmb, for example, to denote the simple event that the developer selects Connecticut as the place to build a motel on a beach, construct a tree diagram to show the 24 elements of the sample space.

13. Construct a Venn diagram to illustrate the possible intersections and unions for the following events relative to the sample space S consisting of all students at Roanoke College:

J: a student is a junior,

M: a student is a mathematics major,

W: a student is a woman.

14. If $S = \{0, 1, 2, 3, 4, 5, 6, 7, 8, 9\}$ and $A = \{0, 2, 4, 6, 8\}$, $B = \{1, 3, 5, 7, 9\}$, $C = \{2, 3, 4, 5\}$, and $D = \{1, 6, 7\}$, list the elements of the sets corresponding to the following events:

(a) $A \cup C$;   (b) $A \cap B$;

(c) C';   (d) $(C' \cap D) \cup B$;

(e) $(S \cap C)'$;   (f) $A \cap C \cap D'$.

15. Consider the sample space

$S = \{$copper, sodium, nitrogen, potassium,
uranium, oxygen, zinc$\}$

and the events

$A = \{$copper, sodium, zinc$\}$

$B = \{$sodium, nitrogen, potassium$\}$

$C = \{$oxygen$\}$.

List the elements of the sets corresponding to the following events:

(a) A';   (b) $A \cup C$;

(c) $(A \cap B') \cup C'$;   (d) $B' \cap C'$;

(e) $A \cap B \cap C$;   (f) $(A' \cup B') \cap (A' \cap C)$.

16. If $S = \{x \mid 0 < x < 12\}$, $M = \{x \mid 1 < x < 9\}$, and $N = \{x \mid 0 < x < 5\}$, find

(a) $M \cup N$;

(b) $M \cap N$;

(c) $M' \cap N'$.

17. Let A, B, and C be events relative to the sample space S. Using Venn diagrams, shade the areas representing the following events:

(a) $(A \cap B)'$;

(b) $(A \cup B)'$;

(c) $(A \cap C) \cup B$.

18. Which of the following pairs of events are mutually exclusive?

(a) A golfer scoring the lowest 18-hole round in a 72-hole tournament and losing the tournament.

(b) A poker player getting a flush (all cards in the same suit) and 3 of a kind on the same 5-card hand.

(c) A mother giving birth to a baby girl and a set of twin daughters on the same day.

(d) A chess player losing the last game and winning the match.

19. Suppose that a family is leaving on a summer vacation in their camper and that M is the event that they will experience mechanical problems, T is the event that they will receive a ticket for committing a traffic violation, and V is the event that they will arrive at a campsite with no vacancies. Referring to the Venn diagram of Figure 2.5 on page 18, state in words the events represented by the following regions:

(a) Region 5.

(b) Region 3.

(c) Regions 1 and 2 together.

(d) Regions 4 and 7 together.

(e) Regions 3, 6, 7, and 8 together.

20. Referring to Exercise 19 and the Venn diagram of Figure 2.5, list the numbers of the regions that represent the following events:

(a) The family will experience no mechanical problems and commit no traffic violation but will find a campsite with no vacancies.

(b) The family will experience both mechanical problems and trouble in locating a campsite with a vacancy, but will not receive a ticket for a traffic violation.

(c) The family will either have mechanical trouble or find a campsite with no vacancies but will not receive a ticket for committing a traffic violation.

(d) The family will not arrive at a campsite with no vacancies.

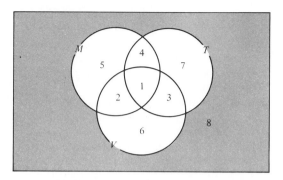

FIGURE 2.5 Venn diagram for Exercise 19.

2.3 Counting Sample Points

One of the problems that the statistician must consider and attempt to evaluate is the element of chance associated with the occurrence of certain events when an experiment is performed. These problems belong in the field of probability, a subject to be introduced in Section 2.4. In many cases we shall be able to solve a probability problem by counting the number of points in the sample space without actually listing each element. The fundamental principle of counting often referred to as the **multiplication rule**, is stated as follows:

THEOREM 2.1   *If an operation can be performed in n_1 ways, and if for each of these a second operation can be performed in n_2 ways, then the two operations can be performed together in $n_1 n_2$ ways.*   ■

EXAMPLE 2.13   How many sample points are in the sample space when a pair of dice is thrown once?

SOLUTION
The first die can land in any one of $n_1 = 6$ ways. For each of these 6 ways the second die can also land in $n_2 = 6$ ways. Therefore, the pair of dice can land in

$$n_1 n_2 = (6)(6) = 36$$

possible ways.

EXAMPLE 2.14   A developer of a new subdivision offers prospective home buyers a choice of Tudor, rustic, colonial, and traditional exterior styling in ranch, two-story, and split-level floor plans. In how many different ways can a buyer order one of these homes?

SOLUTION

Since $n_1 = 4$ and $n_2 = 3$ a buyer must choose from

$$n_1 n_2 = (4)(3) = 12$$

possible homes.

The answers to the two preceding examples can be verified by constructing tree diagrams and counting the various paths along the branches. For instance, in Example 2.14 there will be $n_1 = 4$ branches corresponding to the different exterior styles, and then there will be $n_2 = 3$ branches extending from each of these 4 branches to represent the different floor plans. This tree diagram yields the $n_1 n_2 = 12$ choices of homes given by the paths along the branches as illustrated in Figure 2.6.

The multiplication rule of Theorem 2.1 may be extended to cover any number of operations. Suppose, for instance, that a customer wishes to install a Trimline telephone and can choose from $n_1 = 10$ decorator colors, which we shall assume are available in any of $n_2 = 3$ optional cord lengths with $n_3 = 2$ types of dialing, namely, rotary or Touch-Tone. These three classifications result in

$$n_1 n_2 n_3 = (10)(3)(2) = 60$$

different ways for a customer to order one of these phones. The **generalized multiplication rule** covering k operations is stated in the following theorem.

THEOREM 2.2   *If a operation can be performed in n_1 ways, and if for each of these a second operation can be performed in n_2 ways, and for each of the first two a third operation can be performed in n_3 ways, and so forth, then the sequence of k operations can be performed in $n_1 n_2, \ldots , n_k$ ways.*   ■

EXAMPLE 2.15   How many lunches consisting of a soup, sandwich, dessert, and a drink are possible if we can select from 4 soups, 3 kinds of sandwiches, 5 desserts, and 4 drinks?

SOLUTION

Since $n_1 = 4$, $n_2 = 3$, $n_3 = 5$, and $n_4 = 4$, there are

$$n_1 \times n_2 \times n_3 \times n_4 = 4 \times 3 \times 5 \times 4 = 240$$

different ways to choose a lunch.

EXAMPLE 2.16   How many even three-digit numbers can be formed from the digits 1, 2, 5, 6, and 9 if each digit can be used only once?

SOLUTION

Since the number must be even, we have only $n_1 = 2$ choices for the units position. For each of these we have $n_2 = 4$ choices for the hundreds position and then $n_3 = 3$ choices for the tens position. Therefore, we can form a total of

$$n_1 n_2 n_3 = (2)(4)(3) = 24$$

even three-digit numbers.

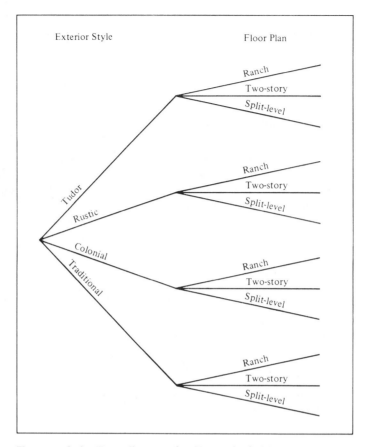

FIGURE 2.6   Tree diagram for Example 2.14.

Frequently, we are interested in a sample space that contains as elements all possible orders or arrangements of a group of objects. For example, we might want to know how many different arrangements are possible for sitting 6 people around a table, or we might ask how many different orders are possible for drawing 2 lottery tickets from a total of 20. The different arrangements are called **permutations**.

DEFINITION 2.7   *A* **permutation** *is an arrangement of all or part of a set of objects.*   ■

Consider the three letters *a*, *b*, and *c*. The possible permutations are *abc*, *acb*, *bac*, *bca*, *cab*, and *cba*. Thus we see that there are 6 distinct arrangements. Using Theorem 2.2 we could arrive at the answer 6 without actually listing the different orders. There are $n_1 = 3$ choices for the first position, then $n_2 = 2$ for the second, leaving only $n_3 = 1$ choice for the last position, giving a total of $n_1 n_2 n_3 = (3)(2)(1) = 6$ permutations. In general, *n* distinct objects can be arranged in $n(n - 1)(n - 2) \cdots (3)(2)(1)$ ways. We represent this product by the symbol *n*!, which is read ''*n* factorial.'' Three objects can be arranged in 3! = (3)(2)(1) = 6 ways. By definition, 1! = 1 and 0! = 1.

THEOREM 2.3   *The number of permutations of n distinct objects is n!.*   ∎

The number of permutations of the four letters a, b, c, and d will be $4! = 24$. Let us now consider the number of permutations that are possible by taking the four letters two at a time. These would be ab, ac, ad, ba, ca, da, bc, cb, bd, db, cd, dc. Using Theorem 2.1 again, we have two positions to fill with $n_1 = 4$ choices for the first and then $n_2 = 3$ choices for the second for a total of $n_1 n_2 = (4)(3) = 12$ permutations. In general, n distinct objects taken r at a time can be arranged in $n(n-1)(n-2) \cdots (n-r+1)$ ways. We represent this product by the symbol ${}_n P_r = n!/(n-r)!$

THEOREM 2.4   *The number of permutations of n distinct objects taken r at a time is*

$$ {}_n P_r = \frac{n!}{(n-r)!}. $$ ∎

EXAMPLE 2.17   Two lottery tickets are drawn from 20 for first and second prizes. Find the number of sample points in the space S.

SOLUTION
The total number of sample points is

$$ {}_{20} P_2 = \frac{20!}{18!} = (20)(19) = 380. $$

EXAMPLE 2.18   How many ways can a local chapter of the American Chemical Society schedule 3 speakers for 3 different meetings if they are all available on any of 5 possible dates?

SOLUTION
The total number of possible schedules is

$$ {}_5 P_3 = \frac{5!}{2!} = (5)(4)(3) = 60. $$

Permutations that occur by arranging objects in a circle are called **circular permutations**. Two circular permutations are not considered different unless corresponding objects in the two arrangements are preceded or followed by a different object we proceed in a clockwise direction. For example, if 4 people are playing bridge, we do not have a new permutation if they all move one position in a clockwise direction. By considering one person in a fixed position and arranging the other three in 3! ways, we find that there are 6 distinct arrangements for the bridge game.

THEOREM 2.5   *The number of permutations of n distinct objects arranged in a circle is $(n-1)!$.*
∎

So far we have considered permutations of distinct objects. That is, all the objects were completely different or distinguishable. Obviously, if the letters b and c are both equal to x, then the six permutations of the letters a, b, c become axx, axx, xax,

xax, *xxa*, and *xxa*, of which only three are distinct. Therefore, with three letters, two being the same, we have 3!/2! = 3 distinct permutations. With four different letters *a*, *b*, *c*, and *d* we have 24 distinct permutations. If we let *a* = *b* = *x* and *c* = *d* = *y*, we can list only the following: *xxyy*, *xyxy*, *yxxy*, *yyxx*, *xyyx*, and *yxyx*. Thus we have 4!/(2!2!) = 6 distinct permutations.

THEOREM 2.6	*The number of distinct permutations of n things of which* n_1 *are of one kind,* n_2 *of a second kind,* . . . , n_k *of a kth kind is* $$\frac{n!}{n_1! \, n_2! \cdots n_k!}.$$ ■
EXAMPLE 2.19	How many different ways can 3 red, 4 yellow, and 2 blue bulbs be arranged in a string of Christmas tree lights with 9 sockets?

SOLUTION
The total number of distinct arrangements is

$$\frac{9!}{3! \, 4! \, 2!} = 1260.$$

Often we are concerned with the number of ways of partitioning a set of *n* objects into *r* subsets called **cells**. A partition has been achieved if the intersection of every possible pair of the *r* subsets is the empty set $\varnothing$ and if the union of all subsets gives the original set. The order of the elements within a cell is of no importance. Consider the set $\{a, e, i, o, u\}$. The possible partitions into two cells in which the first cell contains 4 elements and the second cell 1 element are $\{(a, e, i, o), (u)\}$, $\{(a, i, o, u), (e)\}$, $\{(e, i, o, u), (a)\}$, $\{(a, e, o, u), (i)\}$, and $\{(a, e, i, u), (o)\}$. We see that there are 5 such ways to partition a set of 4 elements into two subsets or cells containing 4 elements in the first cell and 1 element in the second.

The number of partitions for this illustration is denoted by the symbol

$$\binom{5}{4, \, 1} = \frac{5!}{4! \, 1!} = 5,$$

where the top number represents the total number of elements and the bottom numbers represent the number of elements going into each cell. We state this more generally in the following theorem.

THEOREM 2.7	*The number of ways of partitioning a set of n objects into r cells with* n_1 *elements in the first cell,* n_2 *elements in the second, and so forth, is* $$\binom{n}{n_1, \, n_2, \, \ldots, \, n_r} = \frac{n!}{n_1! \, n_2! \cdots n_r!},$$ *where* $n_1 + n_2 + \cdots + n_r = n$. ■
EXAMPLE 2.20	In how many ways can seven scientists be assigned to one triple and two double hotel rooms?

SOLUTION

The total number of possible partitions would be

$$\binom{7}{3, 2, 2} = \frac{7!}{3!\ 2!\ 2!} = 210.$$

In many problems we are interested in the number of ways of *selecting r* objects from *n* without regard to order. These selections are called **combinations**. A combination is actually a partition with two cells, the one cell containing the *r* objects selected and the other cell containing the $(n - r)$ objects that are left.

The number of such combinations, denoted by $\binom{n}{r, n - r}$, is usually shortened to $\binom{n}{r}$, since the number of elements in the second cell must be $n - r$.

THEOREM 2.8   *The number of combinations of n distinct objects taken r at a time is*

$$\binom{n}{r} = \frac{n!}{r!(n - r)!}. \qquad \blacksquare$$

EXAMPLE 2.21   From 4 chemists and 3 physicists find the number of committees that can be formed consisting of 2 chemists and 1 physicist.

SOLUTION

The number of ways of selecting 2 chemists from 4 is

$$\binom{4}{2} = \frac{4!}{2!\ 2!} = 6.$$

The number of ways of selecting 1 physicist from 3 is

$$\binom{3}{1} = \frac{3!}{1!\ 2!} = 3.$$

Using the multiplication rule of Theorem 1.1 with $n_1 = 6$ and $n_2 = 3$, we can form

$$n_1 n_2 = (6)(3) = 18$$

committees with 2 chemists and 1 physicist.

Exercises

1. Registrants at a large convention are offered 6 sightseeing tours on each of 3 days. In how many ways can a person arrange to go on a sightseeing tour plannced by this convention?

2. In a medical study patients are classified in 8 ways according to whether they have blood type AB^+, AB^-, A^+, A^-, B^+, B^-, O^+, or O^-, and also according to whether their blood pressure is low, normal, or high. Find the number of ways in which a patient can be classified.

3. If an experiment consists of throwing a die and then drawing a letter at random from the English alphabet, how many points are in the sample space?

4. Students at a private liberal arts college are classified as being freshmen, sophomores, juniors, or seniors, and also according to whether they are male or female. Find the total number of possible classifications for the students of this college.

5. A certain shoe comes in 5 different styles with each style available in 4 distinct colors. If the store wishes to display pairs of these shoes showing all of its various styles and colors, how many different pairs would the store have on display?

6. In a California study, Dean Lester Breslow and Dr. James Enstrom of the University of California at Los Angeles' School of Public Health concluded that by following 7 simple health rules a man's life can be extended by 11 years on the average and a woman's life by seven years. These 7 rules are: no smoking, regular exercise, use alcohol moderately, get 7 to 8 hours of sleep, maintain proper weight, eat breakfast, and do not eat between meals. In how many ways can a person adopt 5 of these rules to follow
(a) if the person presently violates all 7 rules?
(b) if the person never drinks and always eats breakfast?

7. A developer of a new subdivision offers a prospective home buyer a choice of 4 designs, 3 different heating systems, a garage or carport, and a patio or screened porch. How many different plans are available to this buyer?

8. A drug for the relief of asthma can be purchased from 5 different manufacturers in liquid, tablet, or capsule form, all of which come in regular and extra strength. In how many different ways can a doctor prescribe the drug for a patient suffering from asthma?

9. In a fuel economy study, each of 3 race cars is tested using 5 different brands of gasoline at 7 test sites located in different regions of the country. If 2 drivers are used in the study, and test runs are made once under each distinct set of conditions, how many test runs are needed?

10. In how many different ways can a true–false test consisting of 9 questions be answered?

11. If a multiple-choice test consists of 5 questions each with 4 possible answers of which only 1 is correct,
(a) in how many different ways can a student check off one answer to each question?
(b) in how many ways can a student check off one answer to each question and get all the answers wrong?

12. (a) How many distinct permutations can be made from the letters of the word *columns*?
(b) How many of these permutations start with the letter *m*?

13. A witness to a hit-and-run accident told the police that the license number contained the letters RLH followed by three digits, the first of which was a five. If the witness cannot recall the last two digits, but is certain that all three digits are different, find the maximum number of automobile registrations that the police may have to check.

14. (a) In how many ways can 6 people be lined up to get on a bus?
(b) If a certain 3 persons insist on following each other, how many ways are possible?
(c) If a certain 2 persons refuse to follow each other, how many ways are possible?

15. A contractor wishes to build 9 houses, each different in design. In how many ways can he place these houses on a street if 6 lots are on one side of the street and 3 lots are on the opposite side?

16. (a) How many three-digit numbers can be formed from the digits 0, 1, 2, 3, 4, 5, and 6, if each digit can be used only once?
(b) How many of these are odd numbers?
(c) How many are greater than 330?

17. In how many ways can 4 boys and 5 girls sit in a row if the boys and girls must alternate?

18. Four married couples have bought 8 seats in a row for a concert. In how many different ways can they be seated
(a) with no restrictions?
(b) if each couple is to sit together?
(c) if all the men sit together to the right of all the women?

19. In a regional spelling bee, the 8 finalists consist of 3 boys and 5 girls. Find the number of sample points in the space S for the number of possible orders at the conclusion of the contest for
(a) all 8 finalists;
(b) the first 3 positions.

20. In how many ways can 5 starting positions on a basketball team be filled with 8 men who can play any of the positions?

21. Find the number of ways in which 6 teachers can be assigned to 4 sections of an introductory psychology course if no teacher is assigned to more than one section.

22. Three lottery tickets for first, second, and third prizes are drawn from a group of 40 tickets. Find the number of sample points in S for awarding the three prizes if each contestant holds only one ticket.

23. In how many ways can 5 different trees be planted in a circle?

24. In how many ways can a caravan of 8 covered wagons from Arizona be arranged in a circle?

25. How many distinct permutations can be made from the letters of the word *infinity*?

26. In how many ways can 3 oaks, 4 pines, and 2 maples be arranged along a property line if one does not distinguish between trees of the same kind?

27. A college plays 12 football games during a season. In how many ways can the team end the season with 7 wins, 3 losses, and 2 ties?

28. Nine people are going on a skiing trip in 3 cars that will hold 2, 4, and 5 passengers, respectively. In how many ways is it possible to transport the 9 people to the ski lodge using all cars?

29. How many ways are there to select 3 candidates form 8 equally qualified recent graduates for openings in an accounting firm?

2.4 Probability of an Event

Perhaps it was man's unquenchable thirst for gambling that led to the early development of probability theory. In an effort to increase their winnings, gamblers called upon mathematicians to provide optimum strategies for various games of chance. Some of the mathematicians providing these strategies were Pascal, Leibniz, Fermat, and James Bernoulli. As a result of this early development of probability theory, statistical inference, with all its predictions and generalizations, has branched out far beyond games of chance to encompass many other fields associated with chance occurrences, such as politics, business, weather forecasting, and scientific research. For these predictions and generalizations to be reasonably accurate, an understanding of basic probability theory is essential.

What do we mean when we make the statements "John will probably win the tennis match," "I have a fifty-fifty chance of getting an even number when a die is tossed," "I am not likely to win at bingo tonight," or "Most of our graduating class will likely be married within 3 years"? In each case we are expressing an outcome of which we are not certain, but owing to past information or from an understanding of the structure of the experiment, we have some degree of confidence in the validity of the statement.

Throughout the remainder of this chapter we consider only those experiments for which the sample space contains a finite number of elements. The likelihood of the occurrence of an event resulting from such a statistical experiment is evaluated by means of a set of real numbers called **weights** or **probabilities** ranging from 0 to 1. To every point in the sample space we assign a probability such that the sum of all probabilities is 1. If we have reason to believe that a certain sample point is quite likely to occur when the experiment is conducted, the probability assigned should be close to 1. On the other hand, a probability closer to zero is assigned to a sample point that is not likely to occur. In many experiments, such as tossing a coin or a die, all the sample points have the same chance of occurring and are assigned equal probabilities. For points outside the sample space, that is, for simple events that cannot possibly occur, we assign a probability of zero.

To find the probability of an event A, we sum all the probabilities assigned to the sample points in A. This sum is called the **probability of A** and is denoted by $P(A)$.

DEFINITION 2.8   *The **probability of an event** A is the sum of the weights of all sample points in A. Therefore,*

$$0 \le P(A) \le 1, \qquad P(\varnothing) = 0, \qquad and \qquad P(S) = 1 \qquad \blacksquare$$

EXAMPLE 2.22   A coin is tossed twice. What is the probability that at least one head occurs?

SOLUTION
The sample space for this experiment is

$$S = \{HH, HT, TH, TT\}.$$

If the coin is balanced, each of these outcomes would be equally likely to occur. Therefore, we assign a probability of w to each sample point. Then $4w = 1$ or $w = 1/4$. If A represents the event of at least one head occurring, then

$$A = \{HH, HT, TH\}$$

and

$$P(A) = \tfrac{1}{4} + \tfrac{1}{4} + \tfrac{1}{4} = \tfrac{3}{4}.$$

EXAMPLE 2.23   A die is loaded in such a way that an even number is twice as likely to occur as an odd number. If E is the event that a number less than 4 occurs on a single toss of the die, find $P(E)$.

SOLUTION
The sample space is $S = \{1, 2, 3, 4, 5, 6\}$. We assign a probability of w to each odd number and a probability of $2w$ to each even number. Since the sum of the probabilities must be 1, we have $9w = 1$ or $w = 1/9$. Hence probabilities of 1/9 and 2/9 are assigned to each odd and even number, respectively. Therefore,

$$E = \{1, 2, 3\}$$

and

$$P(E) = \tfrac{1}{9} + \tfrac{2}{9} + \tfrac{1}{9} = \tfrac{4}{9}.$$

EXAMPLE 2.24   In Example 2.23 let A be the event that an even number turns up and let B be the event that a number divisible by 3 occurs. Find $P(A \cup B)$ and $P(A \cap B)$.

SOLUTION
For the events $A = \{2, 4, 6\}$ and $B = \{3, 6\}$ we have $A \cup B = \{2, 3, 4, 6\}$ and $A \cap B = \{6\}$. By assigning a probability of 1/9 to each odd number and 2/9 to each even number, we have

$$P(A \cup B) = \tfrac{2}{9} + \tfrac{1}{9} + \tfrac{2}{9} + \tfrac{2}{9} = \tfrac{7}{9}$$

and

$$P(A \cap B) = \tfrac{2}{9}.$$

If the sample space for an experiment contains N elements, all of which are equally likely to occur, we assign a probability equal to $1/N$ to each of the N points. The probability of any event A containing n of these N sample points is then the ratio of the number of elements in A to the number of elements in S.

THEOREM 2.9   *If an experiment can result in any one of N different equally likely outcomes, and if exactly n of these outcomes correspond to event A, then the probability of event A is*

$$P(A) = \frac{n}{N}.$$ ∎

EXAMPLE 2.25   A mixture of candies contains 6 mints, 4 toffees, and 3 chocolates. If a person makes a random selection of one of these candies, find the probability of getting (a) a mint, or (b) a toffee or a chocolate.

SOLUTION
Let M, T, and C represent the events that the person selects, respectively, a mint, toffee, or chocolate candy. The total number of candies is 13, all of which are equally likely to be selected.

(a) Since 6 of the 13 candies are mints, the probability of event M, selecting a mint at random, is

$$P(M) = \tfrac{6}{13}.$$

(b) Since 7 of the 13 candies are toffees or chocolates, it follows that

$$P(T \cup C) = \tfrac{7}{13}.$$

EXAMPLE 2.26   In a poker hand consisting of 5 cards, find the probability of holding 2 aces and 3 jacks.

SOLUTION
The number of ways of being dealt 2 aces from 4 is

$$\binom{4}{2} = \frac{4!}{2!\,2!} = 6$$

and the number of ways of being dealt 3 jacks from 4 is

$$\binom{4}{3} = \frac{4!}{3!\,1!} = 4$$

By the multiplication rule of Theorem 2.1, there are $n = (6)(4) = 24$ hands with 2 aces and 3 jacks. The total number of 5-card poker hands, all of which are equally likely, is

$$N = \binom{52}{5} = \frac{52!}{5!\,47!} = 2{,}598{,}960.$$

Therefore, the probability of event C of getting 2 aces and 3 jacks in a 5-card poker hand is

$$P(C) = \frac{24}{2,598,960} = 0.9 \times 10^{-5}.$$

If the outcomes of an experiment are not equally likely to occur, the probabilities must be assigned on the basis of prior knowledge or experimental evidence. For example, if a coin is not balanced, we could estimate the probabilities of heads and tails by tossing the coin a large number of times and recording the outcomes. According to the **relative frequency** definition of probability, the true probabilities would be the fractions of heads and tails that occur in the long run.

To find a numerical value that represents adequately the probability of winning at tennis, we must depend on our past performance at the game as well as that of our opponent and to some extent in our belief in being able to win. Similarly, to find the probability that a horse will win a race, we must arrive at a probability based on the previous records of all the horses entered in the race as well as the records of the jockeys riding the horses. Intuition would undoubtedly also play a part in determining the size of the bet that we might be willing to wager. The use of intuition, personal beliefs, and other indirect information in arriving at probabilities is referred to as the **subjective** definition of probability.

In most of the applications of probability in this book the relative frequency interpretation of probability is the operative one. Its foundation is the statistical experiment rather than subjectivity. It is best viewed as the **limiting relative frequency**. As a result, many applications of probability in science and engineering must be based on experiments that can be repeated. Less objective notions of probability are encountered when we assign probabilities based on prior information and opinions. As an example, "There is a good chance that the Lions will lose the Super Bowl." When opinions and prior information differ from individual to individual, subjective probability becomes the relevant tool.

2.5 Additive Rules

Often it is easier to calculate the probability of some event from known probabilities of other events. This may well be true if the event in question can be represented as the union of two other events or as the complement of some event. Several important laws that frequently simplify the computation of probabilities follow. The first, called the **additive rule**, applies to unions of events.

THEOREM 2.10   *If A and B are any two events, then*

$$P(A \cup B) = P(A) + P(B) - P(A \cap B). \qquad \blacksquare$$

PROOF.   Consider the Venn diagram in Figure 2.7. The $P(A \cup B)$ is the sum of the probabilities of the sample points in $A \cup B$. Now $P(A) + P(B)$ is the sum of all the

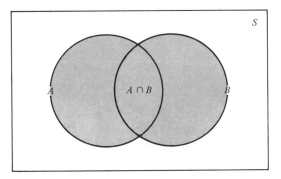

FIGURE 2.7 Additive rule of probability.

probabilities in A plus the sum of all the probabilities in B. Therefore, we have added the probabilities in $(A \cap B)$ twice. Since these probabilities add up to give $P(A \cap B)$, we must subtract this probability once to obtain the sum of the probabilities in $A \cup B$, which is $P(A \cup B)$.

COROLLARY 1   *If A and B are mutually exclusive, then*

$$P(A \cup B) = P(A) + P(B).$$ ■

Corollary 1 is an immediate result of Theorem 2.10, since if A and B are mutually exclusive, $A \cap B = \varnothing$ and then $P(A \cap B) = P(\varnothing) = 0$. In general, we write

---

COROLLARY 2   *If A_1, A_2, A_3, . . . , A_n are mutually exclusive, then*

$$P(A_1 \cup A_2 \cup \cdots \cup A_n) = P(A_1) + P(A_2) + \cdots + P(A_n).$$ ■

COROLLARY 3   *If A_1, A_2, . . . , A_n is a partition of a sample space S, then*

$$\begin{aligned} P(A_1 \cup A_2 \cup \cdots \cup A_n) &= P(A_1) + P(A_2) + \cdots + P(A_n) \\ &= P(S) \\ &= 1. \end{aligned}$$ ■

As one might expect, Theorem 2.10 extends in an analogous fashion.

THEOREM 2.11   *For three events A, B, and C*

$$P(A \cup B \cup C) = P(A) + P(B) + P(C) - P(A \cap B) - P(A \cap C) - P(B \cap C) + P(A \cap B \cap C).$$ ■

EXAMPLE 2.27   The probability that Paula passes mathematics is 2/3, and the probability that she passes English is 4/9. If the probability of passing both courses is 1/4, what is the probability that Paula will pass at least one of these courses?

SOLUTION
If M is the event "passing mathematics" and E the event "passing English," then by the additive rule we have

$$P(M \cup E) = P(M) + P(E) - P(M \cap E)$$

$$= \tfrac{2}{3} + \tfrac{4}{9} - \tfrac{1}{4}$$

$$= \tfrac{31}{36}.$$

EXAMPLE 2.28   What is the probability of getting a total of 7 or 11 when a pair of dice are tossed?

SOLUTION
Let A be the event that 7 occurs and B the event that 11 comes up. Now a total of 7 occurs for 6 of the 36 sample points and a total of 11 occurs for only 2 of the sample points. Since all sample points are equally likely, we have $P(A) = 1/6$ and $P(B) = 1/18$. The events A and B are mutually exclusive, since a total of 7 and 11 cannot both occur on the same toss. Therefore,

$$P(A \cup B) = P(A) + P(B)$$

$$= \tfrac{1}{6} + \tfrac{1}{18}$$

$$= \tfrac{2}{9}.$$

This result could also have been obtained by counting the total number of points for the event $A \cup B$, namely 8, and writing

$$P(A \cup B) = \frac{n}{N} = \frac{8}{36} = \frac{2}{9}. \qquad \blacksquare$$

Theorem 2.10 and its three corollaries should help the reader gain more insight into probability and its interpretation. Corollaries 1 and 2 suggest the very intuitive result dealing with the probability of occurrence of at least one of a number of events, no two of which can occur simultaneously. The probability that at least one occurs is the sum of the probabilities of occurrence of the individual events. The third corollary simply states that the highest value of a probability (unity) is assigned to the entire sample space S.

EXAMPLE 2.29   If the probabilities are, respectively, 0.09, 0.15, 0.21, and 0.23 that a person purchasing a new automobile will choose the color green, white, red, or blue, what is the probability that a given buyer will purchase a new automobile that comes in one of those colors?

SOLUTION
Let G, W, R, and B be the events that a buyer selects, respectively, a green, white, red, or blue automobile. Since these four events are mutually exclusive, the probability is

$$P(G \cup W \cup R \cup B) = P(G) + P(W) + P(R) + P(B)$$

$$= 0.09 + 0.15 + 0.21 + 0.23$$

$$= 0.68.$$

Often it is more difficult to calculate the probability that an event occurs than it is to calculate the probability that the event does not occur. Should this be the case for some event A, we simply find $P(A')$ first and then using Theorem 2.11, find $P(A)$ by subtraction.

THEOREM 2.12   *If A and A' are complementary events, then*

$$P(A) + P(A') = 1.$$ ■

PROOF. Since $A \cup A' = S$ and the sets A and A' are disjoint, then

$$1 = P(S)$$
$$= P(A \cup A')$$
$$= P(A) + P(A').$$

EXAMPLE 2.30   If the probabilities that an automobile mechanic will service 3, 4, 5, 6, 7, or 8 or more cars on any given workday are, respectively, 0.12, 0.19, 0.28, 0.24, 0.10, and 0.07, what is the probability that he will service at least 5 cars on his next day at work?

SOLUTION
Let E be the event that at least 5 cars are serviced. Now, $P(E) = 1 - P(E')$, where E' is the event that fewer than 5 cars are serviced. Since $P(E') = 0.12 + 0.19 = 0.31$, it follows from Theorem 2.12 that

$$P(E) = 1 - 0.31 = 0.69.$$

Exercises

1. Find the errors in each of the following statements:
   (a) The probabilities that an automobile salesperson will sell 0, 1, 2, or 3 cars on any given day in February are, respectively, 0.19, 0.38, 0.29, and 0.15.
   (b) The probability that it will rain tomorrow is 0.40 and the probability that it will not rain tomorrow is 0.52.
   (c) The probabilities that a printer will make 0, 1, 2, 3, or 4 or more mistakes in printing a document are, respectively, 0.19, 0.34, −0.25, 0.43, and 0.29.
   (d) On a single draw from a deck of playing cards the probability of selecting a heart is 1/4, the probability of selecting a black card is 1/2, and the probability of selecting both a heart and a black card is 1/8.

2. Assuming that all elements of S in Exercise 8 on page 16 are equally likely to occur, find
   (a) the probability of event A;
   (b) the probability of event C;
   (c) the probability of event $A \cap C$.

3. A box contains 500 envelopes of which 75 contain $100 in cash, 150 contain $25, and 275 contain $10. An envelope may be purchased for $25. What is the sample space for the different amounts of money? Assign probabilities to the sample points and then find the probability that the first envelope purchased contains less than $100.

4. Referring to the important health practices advocated by the California study in Exercise 6 on page 24, suppose that in a senior college class of 500 students it is found that 210 smoke, 258 drink alcoholic beverages, 216 eat between meals, 122 smoke and drink alcoholic beverages, 83 eat between meals and drink alcoholic beverages, 97 smoke and eat between meals, and 52 engage in all three of these bad health practices. If a member of this senior class is selected at random, find the probability that the student
   (a) smokes but does not drink alcoholic beverages;

(b) eats between meals and drinks alcoholic beverages but does not smoke;

(c) neither smokes nor eats between meals.

5. The probability that an American industry will locate in Munich is 0.7, the probability that it will locate in Brussels is 0.4, and the probability that it will locate in either Munich or Brussels or both is 0.8. What is the probability that the industry will locate
   (a) in both cities?
   (b) in neither city?

6. From past experiences a stockbroker believes that under present economic conditions a customer will invest in tax-free bonds with a probability of 0.6, will invest in mutual funds with a probability of 0.3, and will invest in both tax-free bonds and mutual funds with a probability of 0.15. At this time, find the probability that a customer will invest
   (a) in either tax-free bonds or mutual funds;
   (b) in neither tax-free bonds nor mutual funds.

7. If a letter is chosen at random from the English alphabet, find the probability that the letter
   (a) is a vowel exclusive of y;
   (b) is listed somewhere ahead of the letter j;
   (c) is listed somewhere after the letter g.

8. If a permutation of the word *white* is selected at random, find the probability that the permutation
   (a) begins with a consonant;
   (b) ends with a vowel;
   (c) has the consonant and vowels alternating.

9. If each coded item in a catalog begins with 3 distinct letters followed by 4 distinct nonzero digits, find the probability of randomly selecting one of these coded items with the first letter a vowel and the last digit even.

10. A pair of dice is tossed. Find the probability of getting
    (a) a total of 8;
    (b) at most a total of 5.

11. Two cards are drawn in succession from a deck without replacement. What is the probability that both cards are greater than 2 and less than 8?

12. If 3 books are picked at random from a shelf containing 5 novels, 3 books of poems, and a dictionary, what is the probability that
    (a) the dictionary is selected?
    (b) 2 novels and 1 book of poems are selected?

13. In a poker hand consisting of 5 cards, find the probability of holding
    (a) 3 aces;
    (b) 4 hearts and 1 club.

14. In a game of *Yahtzee*, where 5 dice are tossed simultaneously, find the probability of getting 4 of a kind.

15. In a high school graduating class of 100 students, 54 studied mathematics, 69 studied history, and 35 studied both mathematics and history. If one of these students is selected at random, find the probability that
    (a) the student took mathematics or history;
    (b) the student did not take either of these subjects;
    (c) the student took history but not mathematics.

2.6 Conditional Probability

The probability of an event B occurring when it is known that some event A has occurred is called a **conditional probability** and is denoted by $P(B|A)$. The symbol $P(B|A)$ is usually read "the probability that B occurs given that A occurs" or simply "the probability of B, given A."

Consider the event B of getting a perfect square when a die is tossed. The die is constructed so that the even numbers are twice as likely to occur as the odd numbers. Based on the sample space $S = \{1, 2, 3, 4, 5, 6\}$, with probabilities of 1/9 and 2/9 assigned, respectively, to the odd and even numbers, the probability of B occurring is 1/3. Now suppose that it is known that the toss of the die resulted in a number greater than 3. We are now dealing with a reduced sample space $A = \{4, 5, 6\}$, which is a subset of S. To find the probability that B occurs, relative to the space A, we must first assign new probabilities to the elements of A proportional to their original probabilities such that their sum is 1. Assigning a probability of w

to the odd number in A and a probability of $2w$ to the two even numbers, we have $5w = 1$ or $w = 1/5$. Relative to the space A, we find that B contains the single element 4. Denoting this event by the symbol $B|A$, we write $B|A = \{4\}$, and hence

$$P(B|A) = \tfrac{2}{5}.$$

This example illustrates that events may have different probabilities when considered relative to different sample spaces.

We can also write

$$P(B|A) = \frac{2}{5} = \frac{2/9}{5/9} = \frac{P(A \cap B)}{P(A)},$$

where $P(A \cap B)$ and $P(A)$ are found from the original sample space S. In other words, a conditional probability relative to a subspace A of S may be calculated directly from the probabilities assigned to the elements of the original sample space S.

DEFINITION 2.9   *The* **conditional probability** *of B, given A, denoted by $P(B|A)$, is defined by*

$$P(B|A) = \frac{P(A \cap B)}{P(A)} \qquad if \qquad P(A) > 0. \qquad \blacksquare$$

As an additional illustration, suppose that our sample space S is the population of adults in a small town who have completed the requirements for a college degree. We shall categorize them according to sex and employment status:

	Employed	Unemployed	Total
Male	460	40	500
Female	140	260	400
Total	600	300	900

One of these individuals is to be selected at random for a tour throughout the country to publicize the advantages of establishing new industries in the town. We shall be concerned with the following events:

M: a man is chosen,

E: the one chosen is employed.

Using the reduced sample space E, we find that

$$P(M|E) = \tfrac{460}{600} = \tfrac{23}{30}.$$

Let $n(A)$ denote the number of elements in any set A. Using this notation, we can write

$$P(M|E) = \frac{n(E \cap M)}{n(E)} = \frac{n(E \cap M)/n(S)}{n(E)/n(S)} = \frac{P(E \cap M)}{P(E)},$$

where $P(E \cap M)$ and $P(E)$ are found from the original sample space S. To verify this result, note that

$$P(E) = \tfrac{600}{900} = \tfrac{2}{3},$$

and

$$P(E \cap M) = \tfrac{460}{900} = \tfrac{23}{45}.$$

Hence

$$P(M|E) = \frac{23/45}{2/3} = \frac{23}{30},$$

as before.

EXAMPLE 2.31   The probability that a regularly scheduled flight departs on time is $P(D) = 0.83$; the probability that it arrives on time is $P(A) = 0.82$; and the probability that it departs and arrives on time is $P(D \cap A) = 0.78$. Find the probability that a plane (a) arrives on time given that it departed on time, and (b) departed on time given that it has arrived on time.

SOLUTION
(a) The probability that a plane arrives on time given that it departed on time is

$$P(A|D) = \frac{P(D \cap A)}{P(D)}$$

$$= \frac{0.78}{0.83} = 0.94.$$

(b) The probability that a plane departed on time given that it has arrived on time is

$$P(D|A) = \frac{P(D \cap A)}{P(A)}$$

$$= \frac{0.78}{0.82} = 0.95.$$

In the die-tossing experiment discussed on page 32 we note that $P(B|A) = 2/5$ while $P(B) = 1/3$. That is, $P(B|A) \neq P(B)$, indicating that B *depends* on A. Now consider an experiment in which 2 cards are drawn in succession from an ordinary deck, with replacement. The events are defined as

A: the first card is an ace,

B: the second card is a spade.

Since the first card is replaced, our sample space for both the first and second draws consists of 52 cards, containing 4 aces and 13 spades. Hence

$$P(B|A) = \tfrac{13}{52} = \tfrac{1}{4}$$

and

$$P(B) = \tfrac{13}{52} = \tfrac{1}{4}.$$

That is, $P(B|A) = P(B)$. When this is true, the events A and B are said to be **independent**.

The notion of conditional probability provides the capability of reevaluating the idea of probability of an event in light of additional information, that is, when it is known that another event has occurred. The probability $P(A|B)$ is an "updating" of $P(A)$ based on the knowledge that event B has occurred. In Example 2.31 it was important to know the probability that the flight arrives on time. One is given the information that the flight did not depart on time. Armed with this additional information, the more pertinent probability is $P(A|D')$, that is, the probability that it arrives on time, given that it did not depart on time. In many situations the conclusions drawn from observing the more important conditional probability change the picture entirely. In this example the computation of $P(A|D')$ is given by

$$P(A|D') = \frac{P(A \cap D')}{P(D')} = \frac{0.82 - 0.78}{0.17}$$

$$= 0.24.$$

As a result, the probability of an on-time arrival is diminished severely in the presence of the extra information.

Independent Events

Although conditional probability allows for an alteration of the probability of an event in the light of additional material, it also enables us to understand better the very important concept of **independence** or, in the present context, independent events. In the airport illustration, $P(D|A)$ differed from $P(D)$. This suggests that the occurrence of A influenced D, and this is certainly expected in this illustration. However, consider the situation in which we have events A and B and

$$P(A|B) = P(A).$$

In other words, the occurrence of B had no impact on the odds of occurrence of A. Here the occurrence of A is *independent* of the occurrence of B.

The importance of the concept of independence cannot be overemphasized. It plays a vital role in material in virtually all the chapters in this book and in all areas of application of statistics.

DEFINITION 2.10   *Two events A and B are* **independent** *if and only if*

$$P(B|A) = P(B)$$

and

$$P(A|B) = P(A).$$

Otherwise, A and B are **dependent**.   ■

The condition $P(B|A) = P(B)$ implies that $P(A|B) = P(A)$, and conversely. For the card-drawing experiments, where we showed that $P(B|A) = P(B) = 1/4$, we also can see that $P(A|B) = P(A) = 1/13$.

2.7 Multiplicative Rules

Multiplying the formula in Definition 2.9 by $P(A)$, we obtain the following important **multiplicative rule,** which enables us to calculate the probability that two events will both occur.

THEOREM 2.13   *If in an experiment the events A and B can both occur, then*

$$P(A \cap B) = P(A)P(B|A).$$ ∎

Thus the probability that both A and B occur is equal to the probability that A occurs multiplied by the probability that B occurs, given that A occurs. Since the events $A \cap B$ and $B \cap A$ are equivalent, it follows from Theorem 2.13 that we can also write

$$P(A \cap B) = P(B \cap A) = P(B)P(A|B).$$

In other words, it does not matter which event is referred to as A and which event is referred to as B.

EXAMPLE 2.32   Suppose that we have a fuse box containing 20 fuses, of which 5 are defective. If 2 fuses are selected at random and removed from the box in succession without replacing the first, what is the probability that both fuses are defective?

SOLUTION
We shall let A be the event that the first fuse is defective and B the event that the second fuse is defective; then we interpret $A \cap B$ as the event that A occurs, and then B occurs after A has occurred. The probability of first removing a defective fuse is 1/4; then the probability of removing a second defective fuse from the remaining 4 is 4/19. Hence

$$P(A \cap B) = (\tfrac{1}{4})(\tfrac{4}{19}) = \tfrac{1}{19}.$$

EXAMPLE 2.33   One bag contains 4 white balls and 3 black balls, and a second bag contains 3 white balls and 5 black balls. One ball is drawn from the first bag and placed unseen in the second bag. What is the probability that a ball now drawn from the second bag is black?

SOLUTION
Let B_1, B_2, and W_1 represent, respectively, the drawing of a black ball from bag 1, a black ball from bag 2, and a white ball from bag 1. We are interested in the union of the mutually exclusive events $B_1 \cap B_2$ and $W_1 \cap B_2$. The various possibilities and their probabilities are illustrated in Figure 2.8. Now

$$P[(B_1 \cap B_2) \text{ or } (W_1 \cap B_2)] = P(B_1 \cap B_2) + P(W_1 \cap B_2)$$

$$= P(B_1)P(B_2|B_1) + P(W_1)P(B_2|W_1)$$

$$= (\tfrac{3}{7})(\tfrac{6}{9}) + (\tfrac{4}{7})(\tfrac{5}{9}) = \tfrac{38}{63}.$$

If, in Example 2.32, the first fuse is replaced and the fuses thoroughly rearranged before the second is removed, then the probability of a defective fuse on the second

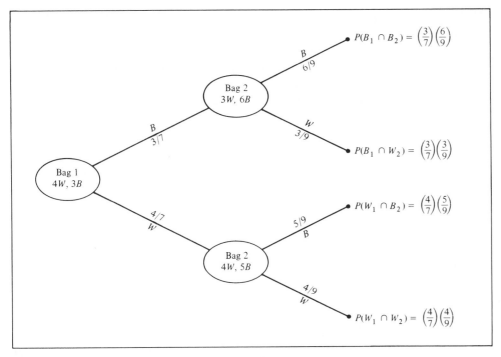

FIGURE 2.8   Tree diagram for Example 2.33.

selection is still 1/4; that is, $P(B|A) = P(B)$ and the events A and B are independent. When this is true, we can substitute $P(B)$ for $P(B|A)$ in Theorem 2.13 to obtain the following **special multiplicative rule**.

THEOREM 2.14   *Two events A and B are independent if and only if*

$$P(A \cap B) = P(A)P(B).$$ ■

Therefore, to obtain the probability that two independent events will both occur, we simply find the product of their individual probabilities.

EXAMPLE 2.34   A small town has one fire engine and one ambulance available for emergencies. The probability that the fire engine is available when needed is 0.98, and the probability that the ambulance is available when called is 0.92. In the event of an injury resulting from a burning building, find the probability that both the ambulance and the fire engine will be available.

SOLUTION
Let A and B represent the respective events that the fire engine and the ambulance are available. Then

$$P(A \cap B) = P(A)P(B)$$
$$= (0.98)(0.92)$$
$$= 0.9016.$$

EXAMPLE 2.35   A pair of dice is thrown twice. What is the probability of getting totals of 7 and 11?

SOLUTION
Let A_1, A_2, B_1, and B_2 be the respective independent events that a 7 occurs on the first throw, a 7 occurs on the second throw, an 11 occurs on the first throw, and an 11 occurs on the second throw. We are interested in the probability of the union of the mutually exclusive events $A_1 \cap B_2$ and $B_1 \cap A_2$. Therefore,

$$P[(A_1 \cap B_2) \cup (B_1 \cap A_2)] = P(A_1 \cap B_2) + P(B_1 \cap A_2)$$
$$= P(A_1)P(B_2) + P(B_1)P(A_2)$$
$$= (\tfrac{1}{6})(\tfrac{1}{18}) + (\tfrac{1}{18})(\tfrac{1}{6})$$
$$= \tfrac{1}{54}.$$

Theorems 2.13 and 2.14 may be generalized to cover any number of events, as stated in the following theorem.

THEOREM 2.15   *If, in an experiment, the events A_1, A_2, A_3, . . . , A_k can occur, then*

$$P(A_1 \cap A_2 \cap A_3 \cap \cdots \cap A_k) = P(A_1)P(A_2|A_1)P(A_3|A_1 \cap A_2) \cdots P(A_k|A_1 \cap A_2 \cap \cdots \cap A_{k-1}).$$

If the events A_1, A_2, A_3, . . . , A_k are independent, then

$$P(A_1 \cap A_2 \cap A_3 \cap \cdots \cap A_k) = P(A_1)P(A_2)P(A_3) \cdots P(A_k). \quad \blacksquare$$

EXAMPLE 2.36   Three cards are drawn in succession, without replacement, from an ordinary deck of playing cards. Find the probability that the event $A_1 \cap A_2 \cap A_3$ occurs, where A_1 is the event that the first card is a red ace, A_2 is the event that the second card is a 10 or a jack, and A_3 is the event that the third card is greater than 3 but less than 7.

SOLUTION
First we define the events

A_1:   the first card is a red ace,

A_2:   the second card is a 10 or jack,

A_3:   the third card is greater than 3 but less than 7.

Now

$$P(A_1) = \tfrac{2}{52}.$$
$$P(A_2|A_1) = \tfrac{8}{51},$$
$$P(A_3|A_1 \cap A_2) = \tfrac{12}{50},$$

and hence by Theorem 2.15,

$$P(A_1 \cap A_2 \cap A_3) = P(A_1)P(A_2|A_1)P(A_3|A_1 \cap A_2)$$
$$= (\tfrac{2}{52})(\tfrac{8}{51})(\tfrac{12}{50})$$
$$= \tfrac{8}{5525}.$$

EXAMPLE 2.37   A coin is biased so that a head is twice as likely to occur as a tail. If the coin is tossed 3 times, what is the probability of getting 2 tails and 1 head?

SOLUTION

The sample space for the experiment consists of the 8 elements

$$S = \{HHH, HHT, HTH, THH, HTT, THT, TTH, TTT\}.$$

However, with an unbalanced coin it is no longer possible to assign equal probabilities to each sample point. To find the probabilities, first consider the sample space $S_1 = \{H, T\}$, which represents the outcomes when the coin is tossed once. Assigning probabilities of w and $2w$ for getting a tail and a head, respectively, we have $3w = 1$ or $w = 1/3$. Hence $P(H) = 2/3$ and $P(T) = 1/3$. Now let A be the event of getting 2 tails and 1 head in the 3 tosses of the coin. Then

$$A = \{TTH, THT, HTT\},$$

and since the outcomes on each of the 3 tosses are independent, it follows from Theorem 2.15 that

$$P(THH) = P(T \cap T \cap H) = P(T)P(T)P(H)$$

$$= (\tfrac{1}{3})(\tfrac{1}{3})(\tfrac{2}{3}) = \tfrac{2}{27}.$$

Similarly,

$$P(THT) = P(HTT) = \tfrac{2}{27}$$

and hence

$$P(A) = \tfrac{2}{27} + \tfrac{2}{27} + \tfrac{2}{27} = \tfrac{2}{9}.$$

Exercises

1. If R is the event that a convict committed armed robbery and D is the event that the convict pushed dope, state in words what probabilities are expressed by
   (a) $P(R|D)$;
   (b) $P(D'|R)$;
   (c) $P(R'|D')$.

2. A class in advanced physics is comprised of 10 juniors, 30 seniors, and 10 graduate students. The final grades showed that 3 of the juniors, 10 of the seniors, and 5 of the graduate students received an A for the course. If a student is chosen at random from this class and is found to have earned an A, what is the probability that he or she is a senior?

3. A random sample of 200 adults are classified below according to sex and the level of education attained.

Education	Male	Female
Elementary	38	45
Secondary	28	50
College	22	17

If a person is picked at random from this group, find the probability that
   (a) the person is a male, given that the person has a secondary education;
   (b) the person does not have a college degree, given that the person is a female.

4. In an experiment to study the dependence of hypertension on smoking habits, the following data were collected on 180 individuals:

	Non-smokers	Moderate Smokers	Heavy Smokers
Hypertension	21	36	30
No hypertension	48	26	19

If one of these individuals is selected at random, find the probability that the person is

(a) experiencing hypertension, given that the person is a heavy smoker;

(b) a nonsmoker, given that the person is experiencing no hypertension.

5. In the senior year of a high school graduating class of 100 students, 42 studied mathematics, 68 studied psychology, 54 studied history, 22 studied both mathematics and history, 25 studied both mathematics and psychology, 7 studied history but neither mathematics nor psychology, 10 studied all three subjects, and 8 did not take any of the three. If a student is selected at random, find the probability that

(a) a person enrolled in psychology takes all three subjects;

(b) a person not taking psychology is taking both history and mathematics.

6. A pair of dice is thrown. If it is known that one die shows a 4, what is the probability that

(a) the other die shows a 5?

(b) the total of both dice is greater than 7?

7. A card is drawn from an ordinary deck and we are told that it is red. What is the probability that the card is greater than 2 but less than 9?

8. The probability that an automobile being filled with gasoline will also need an oil change is 0.25; the probability that it needs a new oil filter is 0.40; and the probability that both the oil and filter need changing is 0.14.

(a) If the oil had to be changed, what is the probability that a new oil filter is needed?

(b) If a new oil filter is needed, what is the probability that the oil has to be changed?

9. The probability that a married man watches a certain television show is 0.4 and the probability that a married woman watches the show is 0.5. The probability that a man watches the show, given that his wife does, is 0.7. Find the probability that

(a) a married couple watches the show;

(b) a wife watches the show given that her husband does;

(c) at least 1 person of a married couple will watch the show.

10. For married couples living in a certain city suburb the probability that the husband will vote on a bond referendum is 0.21, the probability that his wife will vote in the referendum is 0.28, and the probability that both the husband and wife will vote is 0.15. What is the probability that

(a) at least one member of a married couple will vote?

(b) a wife will vote, given that her husband will vote?

(c) a husband will vote, given that his wife does not vote?

11. The probability that a vehicle entering the Luray Caverns has Canadian license plates is 0.12; the probability that it is a camper is 0.28; and the probability that it is a camper with Canadian license plates is 0.09. What is the probability that

(a) a camper entering the Luray Caverns has Canadian license plates?

(b) a vehicle with Canadian license plates entering the Luray Caverns is a camper?

(c) a vehicle entering the Luray Caverns does not have Canadian plates or is not a camper?

12. The probability that the lady of the house is home when the Avon representative calls is 0.6. Given that the lady of the house is home, the probability that she makes a purchase is 0.4. Find the probability that the lady of the house is home and makes a purchase when the Avon representative calls.

13. The probability that a doctor correctly diagnoses a particular illness is 0.7. Given that the doctor makes an incorrect diagnosis, the probability that the patient enters a law suit is 0.9. What is the probability that the doctor makes an incorrect diagnosis and the patient sues?

14. One bag contains 4 white balls and 3 black balls, and a second bag contains 3 white balls and 5 black balls. One ball is drawn at random from the second bag and is placed unseen in the first bag. What is the probability that a ball now drawn from the first bag is white?

15. A real estate agent has 8 master keys to open several new homes. Only 1 master key will open any given house. If 40% of these homes are usually left unlocked, what is the probability that the real estate agent can get into a specific home if the agent selects 3 master keys at random before leaving the office?

16. Two cards are drawn in succession from a deck without replacement. What is the probability that
    (a) both cards are red?
    (b) both cards are greater than 3 but less than 8?

17. A town has 2 fire engines operating independently. The probability that a specific engine is available when needed is 0.96.
    (a) What is the probability that neither is available when needed?
    (b) What is the probability that a fire engine is available when needed?

18. The probability that Tom will be alive in 20 years is 0.7, and the probability that Nancy will be alive in 20 years is 0.9. If we assume independence for both, what is the probability that neither will be alive in 20 years?

19. One overnight case contains 2 bottles of aspirin and 3 bottles of thyroid tablets. A second tote bag contains 3 bottles of aspirin, 2 bottles of thyroid, and 1 bottle of laxative tablets. If 1 bottle of tablets is taken at random from each piece of luggage, find the probability that
    (a) both bottles contain thyroid tablets;
    (b) neither bottle contains thyroid tablets;
    (c) the 2 bottles contain different tablets.

20. The probability that a person visiting his dentist will have an X-ray is 0.6; the probability that a person who has an X-ray will also have a cavity filled is 0.3; and the probability that a person who has had an X-ray and a cavity filled will also have a tooth extracted is 0.1. What is the probability that a person visiting his dentist will have an X-ray, a cavity filled, and a tooth extracted?

21. Find the probability of randomly selecting 4 good quarts of milk in succession from a cooler containing 20 quarts of which 5 have spoiled, by using
    (a) the first formula of Theorem 2.15 on page 38.
    (b) the formulas of Theorems 2.8 and 2.9 on pages 23 and 27, respectively.

2.8 Bayes' Rule

Let us now return to the illustration of Section 2.6, where an individual is being selected at random from the adults of a small town to tour the country and publicize the advantages of establishing new industries in the town. Suppose that we are now given the additional information that 36 of those employed and 12 of those unemployed are members of the Rotary Club. We wish to find the probability of the event A that the individual selected is a member of the Rotary Club. Referring to Figure 2.9, we can write A as the union of the two mutually exclusive events $E \cap A$ and $E' \cap A$. Hence

$$A = (E \cap A) \cup (E' \cap A),$$

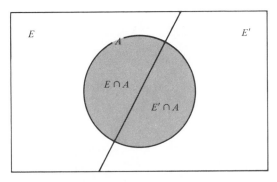

FIGURE 2.9 Venn diagram for the events A, E, and E'.

and by Corollary 1 of Theorem 2.10, and then Theorem 2.13, we can write

$$P(A) = P[(E \cap A) \cup (E' \cap A)]$$
$$= P(E \cap A) + P(E' \cap A)$$
$$= P(E)P(A|E) + P(E')P(A|E').$$

The data of Section 2.6, together with the additional data given above for the set A, enable us to compute

$$P(E) = \tfrac{600}{900} = \tfrac{2}{3}, \qquad P(A|E) = \tfrac{36}{600} = \tfrac{3}{50},$$

and

$$P(E') = \tfrac{1}{3}, \qquad P(A|E') = \tfrac{12}{300} = \tfrac{1}{25}.$$

If we display these probabilities by means of the tree diagram of Figure 2.10, in which the first branch yields the probability $P(E)P(A|E)$ and the second branch yields the probability $P(E')P(A|E')$, it follows that

$$P(A) = (\tfrac{2}{3})(\tfrac{3}{50}) + (\tfrac{1}{3})(\tfrac{1}{25})$$
$$= \tfrac{4}{75}.$$

A generalization of the foregoing illustration to the case where the sample space is partitioned into k subsets is covered by the following theorem, sometimes called the **theorem of total probability** or the **rule of elimination.**

THEOREM 2.16   *If the events $B_1, B_2, \ldots, B_k$ constitute a partition of the sample space S such that $P(B_i) \neq 0$ for $i = 1, 2, \ldots, k$, then for any event A of S,*

$$P(A) = \sum_{i=1}^{k} P(B_i \cap A) = \sum_{i=1}^{k} P(B_i)(A|B_i).$$

PROOF. Consider the Venn diagram of Figure 2.11. The event A is seen to be the union of the mutually exclusive events $B_1 \cap A$, $B_2 \cap A$, $\ldots$, $B_k \cap A$; that is,

$$A = (B_1 \cap A) \cup (B_2 \cap A) \cup \cdots \cup (B_k \cap A).$$

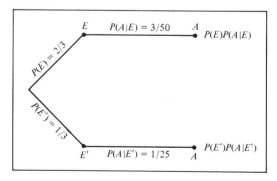

FIGURE 2.10   Tree diagram for the data on page 33.

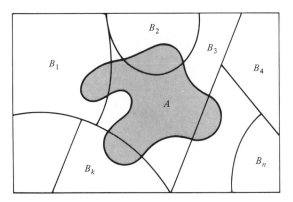

FIGURE 2.11 Partitioning the sample space S.

Using Corollary 2 of Theorem 2.10 and then Theorem 2.13, we have

$$P(A) = P[(B_1 \cap A) \cup (B_2 \cap A) \cup \cdots \cup (B_k \cap A)]$$
$$= P(B_1 \cap A) + P(B_2 \cap A) + \cdots + P(B_k \cap A)$$
$$= \sum_{i=1}^{k} P(B_i \cap A) = \sum_{i=1}^{k} P(B_i)P(A|B_i).$$

EXAMPLE 2.38   In a certain assembly plant, three machines, B_1, B_2, and B_3, make 30%, 45%, and 25%, respectively, of the products. It is known from past experience that 2%, 3%, and 2% of the products made by each machine, respectively, are defective. Now, suppose that a finished product is randomly selected. What is the probability that it is defective?

SOLUTION

Consider the following events:

A: the product is defective,

B_1:   the product is made by machine B_1,

B_2:   the product is made by machine B_2,

B_3:   the product is made by machine B_3.

Applying the rule of elimination, we can write

$$P(A) = P(B_1)P(A|B_1) + P(B_2)P(A|B_2) + P(B_3)P(A|B_3).$$

Referring to the tree diagram of Figure 2.12, we find that the three branches give the probabilities

$$P(B_1)P(A|B_1) = (0.3)(0.02) = 0.006,$$
$$P(B_2)P(A|B_2) = (0.45)(0.03) = 0.0135,$$
$$P(B_3)P(A|B_3) = (0.25)(0.02) = 0.005,$$

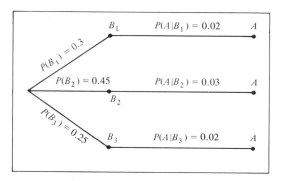

FIGURE 2.12   Tree diagram for Example 2.38.

and hence

$$P(A) = 0.006 + 0.0135 + 0.005 = 0.0245.$$

Instead of asking for $P(A)$, by the rule of elimination, suppose that we now consider the problem of finding the conditional probability $P(B_i|A)$ in Example 2.38. In other words, suppose that a product was randomly selected and it is defective. What is the probability that this product was made by machine B_i? Questions of this type can be answered by using the following theorem, called **Bayes' rule:**

THEOREM 2.17

(Bayes' Rule)   *If the events B_1, B_2, . . . , B_k constitute a partition of the sample space S, where $P(B_i) \neq 0$ for $i = 1, 2, \ldots , k$, then for any event A in S such that $P(A) \neq 0$,*

$$P(B_r|A) = \frac{P(B_r \cap A)}{\displaystyle\sum_{i=1}^{k} P(B_i \cap A)} = \frac{P(B_r)P(A|B_r)}{\displaystyle\sum_{i=1}^{k} P(B_i)P(A|B_i)}$$

for $r = 1, 2, \ldots , k$. ■

PROOF. By the definition of conditional probability,

$$P(B_r|A) = \frac{P(B_r \cap A)}{P(A)},$$

and then using Theorem 2.16 in the denominator, we have

$$P(B_r|A) = \frac{P(B_r \cap A)}{\displaystyle\sum_{i=1}^{k} P(B_i \cap A)}.$$

Applying Theorem 2.13 to both numerator and denominator, we obtain the alternative form,

$$P(B_r|A) = \frac{P(B_r)P(A|B_r)}{\displaystyle\sum_{i=1}^{k} P(B_i)P(A|B_i)},$$

which completes the proof.

EXAMPLE 2.39   With reference to Example 2.38, if a product were chosen randomly and found to be defective, what is the probability that it was made by machine B_3?

SOLUTION
Using Bayes' rule to write

$$P(B_3|A) = \frac{P(B_3)P(A|B_3)}{P(B_1)P(A|B_1) + P(B_2)P(A|B_2) + P(B_3)P(A|B_3)},$$

and then substituting the probabilities calculated in Example 2.38, we have

$$P(B_3|A) = \frac{0.005}{0.006 + 0.0135 + 0.005} = \frac{0.005}{0.0245} = \frac{10}{49}.$$

In view of the fact that a defective product was selected, this result suggests that it probably was not made by machine B_3.

Exercises

1. In a certain region of the country it is known from past experience that the probability of selecting an adult over 40 years of age with cancer is 0.02. If the probability of a doctor correctly diagnosing a person with cancer as having the disease is 0.78 and the probability of incorrectly diagnosing a person without cancer as having the disease is 0.06, what is the probability that a person is diagnosed as having cancer?

2. Police plan to enforce speed limits by using radar traps at 4 different locations within the city limits. The radar traps at each of the locations L_1, L_2, L_3, and L_4 are operated 40%, 30%, 20%, and 30% of the time, and if a person who is speeding on his way to work has probabilities of 0.2, 0.1, 0.5, and 0.2, respectively, of passing through these locations, what is the probability that he will receive a speeding ticket?

3. Referring to Exercise 1, what is the probability that a person diagnosed as having cancer actually has the disease?

4. If in Exercise 2 the person received a speeding ticket on his way to work, what is the probability that he passed through the radar trap located at L_2?

5. Suppose that the four inspectors at a film factory are supposed to stamp the expiration date on each package of film at the end of the assembly line. John, who stamps 20% of the packages, fails to stamp the expiration date once in every 200 packages; Tom, who stamps 60% of the packages, fails to stamp the expiration date once in every 100 packages; Jeff, who stamps 15% of the packages, fails to stamp the expiration date once in every 90 packages; and Pat, who stamps 5% of the packages, fails to stamp the expiration date once in every 200 packages. If a customer complains that her package of film does not show the expiration date, what is the probability that it was inspected by John?

6. A regional telephone company operates three identical relay stations at different locations. During a one-year

period, the number of malfunctions reported by each station and the causes are shown below.

	Stations		
	A	B	C
Problems with electricity supplied	2	1	1
Computer malfunction	4	3	2
Malfunctioning electrical equipment	5	4	2
Caused by other human errors	7	7	5

Suppose that a malfunction was reported and it was found to be caused by other human errors. What is the probability that it came from station C?

Review Exercises

1. A truth serum given to a suspect is known to be 90% reliable when the person is guilty and 99% reliable when the person is innocent. In other words, 10% of the guilty are judged innocent by the serum and 1% of the innocent are judged guilty. If the suspect was selected from a group of suspects of which only 5% have ever committed a crime, and the serum indicates that he is guilty, what is the probability that he is innocent?

2. An allergist claims that 50% of the patients she tests are allergic to some type of weed. What is the probability that
   (a) exactly 3 of her next 4 patients are allergic to weeds?
   (b) none of her next 4 patients are allergic to weeds?

3. By comparing appropriate regions of Venn diagrams, verify that
   (a) $(A \cap B) \cup (A \cap B') = A$;
   (b) $A' \cup (B' \cup C) = (A \cap B') \cup (A' \cup C)$.

4. The probabilities that a service station will pump gas into 0, 1, 2, 3, 4, or 5 or more cars during a certain 30-minute period are 0.03, 0.18, 0.24, 0.28, 0.10, and 0.17. Find the probability that in this 30-minute period
   (a) more than 2 cars receive gas;
   (b) at most 4 cars receive gas;
   (c) 4 or more cars receive gas.

5. How many bridge hands are possible containing 4 spades, 6 diamonds, 1 club, and 2 hearts?

6. If the probability is 0.1 that a person will make a mistake on his or her state income tax return, find the probability that

(a) four totally unrelated persons each make a mistake;
(b) Mr. Jones and Ms. Clark both make a mistake, and Mr. Roberts and Ms. Williams do not make a mistake.

7. A large industrial firm uses 3 local motels to provide overnight accommodations for its clients. From past experience it is known that 20% of the clients are assigned rooms at the Ramada Inn, 50% at the Sheraton, and 30% at the Lakeview Motor Lodge. If the plumbing is faulty in 5% of the rooms at the Ramada Inn, in 4% of the rooms at the Sheraton, and in 8% of the rooms at the Lakeview Motor Lodge, what is the probability that
   (a) a client will be assigned a room with faulty plumbing?
   (b) a person with a room having faulty plumbing was assigned accommodations at the Lakeview Motor Lodge?

8. From a group of 4 men and 5 women, how many committees of size 3 are possible
   (a) with no restrictions?
   (b) with 1 man and 2 women?
   (c) with 2 men and 1 woman if a certain man must be on the committee?

9. The probability that a patient recovers from a delicate heart operation is 0.8. What is the probability that
   (a) exactly 2 of the next 3 patients who have this operation survive?
   (b) all of the next 3 patients who have this operation survive?

10. In a certain federal prison it is known that 2/3 of the inmates are under 25 years of age. It is also known that 3/5 of the inmates are male and that 5/8 of the inmates are female or 25 years of age or older. What is the probability that a prisoner selected at random from this prison is female and at least 25 years old?

11. From 4 red, 5 green, and 6 yellow apples, how many selections of 9 apples are possible if 3 of each color are to be selected?

12. From a box containing 6 black balls and 4 green balls, 3 balls are drawn in succession, each ball being replaced in the box before the next draw is made. What is the probability that
(a) all 3 are the same color?
(b) each color is represented?

13. A shipment of 12 television sets contains 3 defective sets. In how many ways can a hotel purchase 5 of these sets and receive at least 2 of the defective sets?

3

Random Variables and Probability Distributions

3.1 Concept of a Random Variable

The field of statistics is concerned with making inferences about populations and population characteristics. Experiments are conducted with results that are subject to chance. The testing of a number of electronic components is an example of a **statistical experiment**, a term that is used to describe any process by which several chance observations are generated. It is often very important to allocate a numerical description to the outcome. For example, the sample space giving a detailed description of each possible outcome when three electronic components are tested may be written

$$S = \{NNN, NND, NDN, DNN, NDD, DND, DDN, DDD\},$$

where N denotes "nondefective" and D denotes "defective." One is naturally concerned with the number of defectives that occur. Thus each point in the sample space will be *assigned a numerical value* of 0, 1, 2, or 3. These values are, of course, random quantities *determined by the outcome of the experiment*. They may be viewed as values assumed by the *random variable, X,* the number of defective items when three electronic components are tested.

DEFINITION 3.1   *A* **random variable** *is a function that associates a real number with each element in the sample space.* ∎

We shall use a capital letter, say X, to denote a random variable and its corresponding small letter, x in this case, for one of its values. In the electronic component testing illustration above, we notice that the random variable X assumes the value 2 for all elements in the subset

$$E = \{DDN, DND, NDD\}$$

of the sample space S. That is, each possible value of X represents an event that is a subset of the sample space for the given experiment.

EXAMPLE 3.1   Two balls are drawn in succession without replacement from an urn containing 4 red balls and 3 black balls. The possible outcomes and the values y of the random variable Y, where Y is the number of red balls, are

Sample Space	y
RR	2
RB	1
BR	1
BB	0

EXAMPLE 3.2   A stockroom clerk returns three safety helmets at random to three steel mill employees, who had previously checked them. If Smith, Jones, and Brown, in that order, receive one of the three hats, list the sample points for the possible orders of returning the helmets and find the value m of the random variable M that represents the number of correct matches.

SOLUTION
If S, J, and B stand for Smith's, Jones', and Brown's helmets, respectively, then the possible arrangements in which the helmets may be returned and the number of correct matches are

Sample Space	m
SJB	3
SBJ	1
JSB	1
JBS	0
BSJ	0
BJS	1

In each of the two preceding examples the sample space contains a finite number of elements. On the other hand, when a die is thrown until a 5 occurs, we obtain a sample space with an unending sequence of elements,

$$S = \{F,\ NF,\ NNF,\ NNNF,\ \ldots\},$$

where F and N represent, respectively, the occurrence and nonoccurrence of a 5. But even in this experiment, the number of elements can be equated to the number of whole numbers so that there is a first element, a second element, a third element, and so on, and in this sense can be counted.

DEFINITION 3.2   *If a sample space contains a finite number of possibilities or an unending sequence with as many elements as there are whole numbers, it is called a* **discrete sample space**. ■

The outcomes of some statistical experiments may be either finite nor countable. Such is the case, for example, when one conducts an investigation measuring the distances that a certain make of automobile will travel over a prescribed test course on 5 liters of gasoline. Assuming distance to be a variable measured to any degree of accuracy, then clearly we have an infinite number of possible distances in the sample space that cannot be equated to the number of whole numbers. Also, if one were to record the length of time for a chemical reaction to take place, once again the possible time intervals making up our sample space are infinite in number and uncountable. We see now that all sample spaces need not be discrete.

DEFINITION 3.3   *If a sample space contains an infinite number of possibilities equal to the number of points on a line segment, it is called a* **continuous sample space**. ■

A random variable is called a **discrete random variable** if its set of possible outcomes is countable. Since the possible values of Y in Example 3.1 are 0, 1, and 2, and the possible values of M in Example 3.2 are 0, 1, and 3, it follows that Y and M are discrete random variables. When a random variable can take on values on a continuous scale, it is called a **continuous random variable**. Often the possible values of a continuous random variable are precisely the same values that are contained in the continuous sample space. Such is the case when the random variable represents the measured distance that a certain make of automobile will travel over a test course on 5 liters of gasoline.

In most practical problems, continuous random variables represent *measured* data, such as all possible heights, weights, temperatures, distance, or life periods, whereas discrete random variables represent *count* data, such as the number of defectives in a sample of k items or the number of highway fatalities per year in a given state. Note that the random variables Y and M of Examples 3.1 and 3.2 both represent count data, Y the number of red balls and M the number of correct hat matches.

3.2 Discrete Probability Distributions _____

A discrete random variable assumes each of its values with a certain probability. In the case of tossing a coin three times, the variable X, representing the number of heads, assumes the value 2 with probability 3/8, since 3 of the 8 equally likely sample points result in two heads and one tail. If one assumes equal weights for the simple events in Example 3.2, the probability that no employee gets back his right helmet, that is, the probability that M assumes the value zero, is 1/3. The possible values m of M and their probabilities are given by

m	0	1	3
$P(M = m)$	$\frac{1}{3}$	$\frac{1}{2}$	$\frac{1}{6}$

Note that the values of m exhaust all possible cases and hence the probabilities add to 1.

Frequently, it is convenient to represent all the probabilities of a random variable X by a formula. Such a formula would necessarily be a function of the numerical values x that we shall denote by $f(x)$, $g(x)$, $r(x)$, and so forth. Therefore, we write $f(x) = P(X = x)$; that is, $f(3) = P(X = 3)$. The set of ordered pairs $(x, f(x))$ is called the **probability function** or **probability distribution** of the discrete random variable X.

DEFINITION 3.4   *The set of ordered pairs $(x, f(x))$ is a* **probability function**, **probability mass function**, *or* **probability distribution** *of the discrete random variable X if, for each possible outcome x,*

1. $f(x) \geq 0$.

2. $\sum\limits_{x} f(x) = 1$.

3. $P(X = x) = f(x)$.   ∎

EXAMPLE 3.3   A shipment of 8 similar microcomputers to a retail outlet contains 3 that are defective. If a school makes a random purchase of 2 of these computers, find the probability distribution for the number of defectives.

SOLUTION
Let X be a random variable whose values x are the possible numbers of defective computers purchased by the school. Then x can be any of the numbers 0, 1, and 2. Now,

$$f(0) = P(X = 0) = \frac{\binom{3}{0}\binom{5}{2}}{\binom{8}{2}} = \frac{10}{28},$$

$$f(1) = P(X = 1) = \frac{\binom{3}{1}\binom{5}{1}}{\binom{8}{2}} = \frac{15}{28},$$

$$f(2) = P(X = 0) = \frac{\binom{3}{2}\binom{5}{0}}{\binom{8}{2}} = \frac{3}{28}.$$

Thus the probability distribution of X is

x	0	1	2
$f(x)$	$\frac{10}{28}$	$\frac{15}{28}$	$\frac{3}{28}$

EXAMPLE 3.4   If 50% of the automobiles sold by an agency for a certain foreign car are equipped with diesel engines, find a formula for the probability distribution of the number of diesel models among the next 4 cars sold by this agency.

SOLUTION
Since the probability of selling a diesel model or a gasoline model is 0.5, the $2^4 = 16$ points in the sample space are equally likely to occur. Therefore, the denominator for all probabilities, and also for our function, will be 16. To obtain the number of ways of selling 3 diesel models, we need to consider the number of ways of partitioning 4 outcomes into two cells with 3 diesel models assigned to one cell and a gasoline model assigned to the other. This can be done in $\binom{4}{3} = 4$ ways. In general, the event of selling x diesel models and $4 - x$ gasoline models can occur in $\binom{4}{x}$ ways, where x can be 0, 1, 2, 3, or 4. Thus the probability distribution $f(x) = P(X = x)$ is

$$f(x) = \frac{\binom{4}{x}}{16} \qquad \text{for } x = 0, 1, 2, 3, 4.$$

There are many problems in which we wish to compute the probability that the observed value of a random variable X will be less than or equal to some real number x. Writing $F(x) = P(X \leq x)$ for every real number x, we define $F(x)$ to be the **cumulative distribution** of the random variable X.

DEFINITION 3.5   *The **cumulative distribution** $F(x)$ of a discrete random variable X with probability distribution $f(x)$ is given by*

$$F(x) = P(X \leq x) = \sum_{t \leq x} f(t) \qquad \text{for } -\infty < x < \infty.$$ ∎

For the random variable M, the number of correct matches in Example 3.2, we have

$$F(2.4) = P(M \le 2.4) = f(0) + f(1) = (\tfrac{1}{3}) + (\tfrac{1}{2}) = \tfrac{5}{6}.$$

The cumulative distribution of M is given by

$$F(m) = \begin{cases} 0 & \text{for } m < 0 \\ \tfrac{1}{3} & \text{for } 0 \le m < 1 \\ \tfrac{5}{6} & \text{for } 1 \le m < 3 \\ 1 & \text{for } m \ge 3. \end{cases}$$

One should pay particular notice to the fact that the cumulative distribution is defined not only for the values assumed by the given random variable but for all real numbers.

EXAMPLE 3.5   Find the cumulative distribution of the random variable X in Example 3.4. Using $F(x)$, verify that $f(2) = 3/8$.

SOLUTION
Direct calculations of the probability distribution of Example 3.4 give $f(0) = 1/16$, $f(1) = 1/4$, $f(2) = 3/8$, $f(3) = 1/4$, and $f(4) = 1/16$. Therefore,

$$F(0) = f(0) = \tfrac{1}{16}$$

$$F(1) = f(0) + f(1) = \tfrac{5}{16}$$

$$F(2) = f(0) + f(1) + f(2) = \tfrac{11}{16}$$

$$F(3) = f(0) + f(1) + f(2) + f(3) = \tfrac{15}{16}$$

$$F(4) = f(0) + f(1) + f(2) + f(3) + f(4) = 1.$$

Hence

$$F(x) = \begin{cases} 0 & \text{for } x < 0 \\ \tfrac{1}{16} & \text{for } 0 \le x < 1 \\ \tfrac{5}{16} & \text{for } 1 \le x < 2 \\ \tfrac{11}{16} & \text{for } 2 \le x < 3 \\ \tfrac{15}{16} & \text{for } 3 \le x < 4 \\ 1 & \text{for } x \ge 4. \end{cases}$$

Now

$$f(2) = F(2) - F(1) = \tfrac{11}{16} - \tfrac{5}{16} = \tfrac{3}{8}.$$

It is often helpful to look at a probability distribution in graphic form. One might plot the points $(x, f(x))$ of Example 3.4 to obtain Figure 3.1. By joining the points to the x axis either with a dashed or solid line, we obtain what is commonly called a **bar chart**. Figure 3.1 makes it very easy to see what values of X are most likely to occur, and it also indicates a perfectly symmetric situation in this case.

Instead of plotting the points $(x, f(x))$, we more frequently construct rectangles, as in Figure 3.2. Here the rectangles are constructed so that their bases of equal

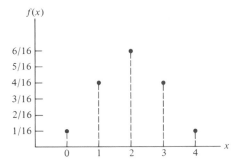

FIGURE 3.1   Bar chart.

width are centered at each value x and their heights are equal to the corresponding probabilities given by $f(x)$. The bases are constructed so as to leave no space between the rectangles. Figure 3.2 is called a **probability histogram**.

Since each base in Figure 3.2 has unit width, the $P(X = x)$ is equal to the area of the rectangle centered at x. Even if the bases were not of unit width, we could adjust the heights of the rectangles to give areas that would still equal the probabilities of X assuming any of its values x. This concept of using areas to represent probabilities is necessary for our consideration of the probability distribution of a continuous random variable.

The graph of the cumulative distribution of Example 3.2, which appears as a step function in Figure 3.3, is obtained by plotting the points $(x, F(x))$.

Certain probability distributions are applicable to more than one physical situation. The probability distribution of Example 3.4, for example, also applies to the random variable Y, where Y is the number of heads when a coin is tossed 4 times, or to the random variable W, where W is the number of red cards that occur when 4 cards are drawn at random from a deck in succession with each card replaced and the deck shuffled before the next drawing. Special discrete distributions that can be applied to many different experimental situations will be considered in Chapter 5.

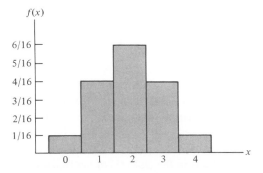

FIGURE 3.2   Probability histogram.

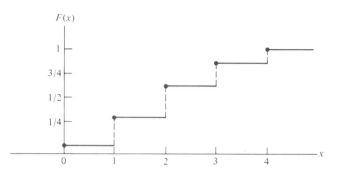

FIGURE 3.3   Discrete cumulative distribution.

3.3 Continuous Probability Distributions

A continuous random variable has a probability of zero of assuming *exactly* any of its values. Consequently, its probability distribution cannot be given in tabular form. At first this may seem startling, but it becomes more plausible when we consider a particular example. Let us discuss a random variable whose values are the heights of all people over 21 years of age. Between any two values, say 163.5 and 164.5 centimeters, or even 163.99 and 164.01 centimeters, there are an infinite number of heights, one of which is 164 centimeters. The probability of selecting a person at random who is exactly 164 centimeters tall and not one of the infinitely large set of heights so close to 164 centimeters that you cannot humanly measure the difference is remote, and thus we assign a probability of zero to the event. This is not the case, however, if we talk about the probability of selecting a person who is at least 163 centimeters but not more than 165 centimeters tall. Now we are dealing with an interval rather than a point value of our random variable.

We shall concern ourselves with computing probabilities for various intervals of continuous random variables such as $P(a < X < b)$, $P(W > c)$, and so forth. Note that when X is continuous

$$P(a < X \leq b) = P(a < X < b) + P(X = b)$$

$$= P(a < X < b).$$

That is, it does not matter whether we include an endpoint of the interval or not. This is not true, though, when X is discrete.

Although the probability distribution of a continuous random variable cannot be presented in tabular form, it can have a formula. Such a formula would necessarily be a function of the numerical values of the continuous variable X and as such will be represented by the functional notation $f(x)$. In dealing with continuous variables, $f(x)$ is usually called the **probability density function**, or simply the **density function** of X. Since X is defined over a continuous sample space, it is possible for $f(x)$ to have a finite number of discontinuities. However, most density functions that have practical applications in the analysis of statistical data are continuous and their graphs may take any of several forms, some of which are shown in Figure 3.4.

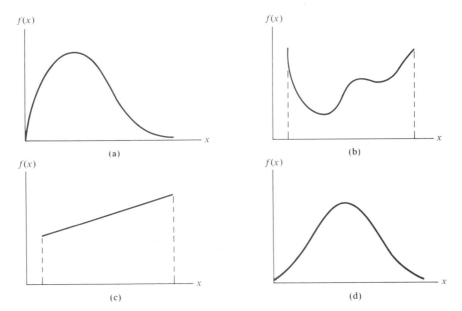

FIGURE 3.4   Typical density functions.

Because areas will be used to represent probabilities and probabilities are positive numerical values, the density function must lie entirely above the x axis.

A probability density function is constructed so that the area under its curve bounded by the x axis is equal to 1 when computed over the range of X for which $f(x)$ is defined. Should this range of X be a finite interval, it is always possible to extend the interval to include the entire set of real numbers by defining $f(x)$ to be zero at all points in the extended portions of the interval. In Figure 3.5, the probability that X assumes a value between a and b is equal to the shaded area under the

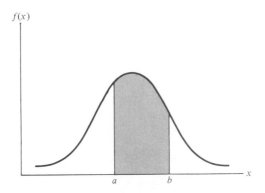

FIGURE 3.5 $P(a < X < b)$.

density function between the ordinates at $x = a$ and $x = b$, and from integral calculus is given by

$$P(a < X < b) = \int_a^b f(x)\, dx.$$

DEFINITION 3.6   *The function $f(x)$ is a **probability density function** for the continuous random variable X, defined over the set of real numbers R, if*

1. $f(x) \geq 0$, *for all $x \in R$.*

2. $\displaystyle\int_{-\infty}^{\infty} f(x)\, dx = 1.$

3. $P(a < X < b) = \displaystyle\int_a^b f(x)\, dx.$ ■

EXAMPLE 3.6   Suppose that the error in the reaction temperature, in °C, for a controlled laboratory experiment is a continuous random variable X having the probability density function

$$f(x) = \begin{cases} \dfrac{x^2}{3}, & -1 < x < 2 \\ 0, & \text{elsewhere.} \end{cases}$$

(a) Verify condition 2 of Definition 3.6.
(b) Find $P(0 < X \leq 1)$.

SOLUTION

(a) $\displaystyle\int_{-\infty}^{\infty} f(x)\, dx = \int_{-1}^{2} \frac{x^2}{3}\, dx = \frac{x^3}{9} \Big|_{-1}^{2} = \frac{8}{9} + \frac{1}{9} = 1.$

(b) $P(0 < X \leq 1) = \displaystyle\int_0^1 \frac{x^2}{3}\, dx = \frac{x^3}{9} \Big|_0^1 = \frac{1}{9}.$

DEFINITION 3.7   *The **cumulative distribution** $F(x)$ of a continuous random variable X with density function $f(x)$ is given by*

$$F(x) = P(X \leq x) = \int_{-\infty}^{x} f(t)\, dt \quad \text{for} \ -\infty < x < \infty.$$ ■

As an immediate consequence of Definition 3.7 one can write the two results

$$P(a < X < b) = F(b) - F(a)$$

and

$$f(x) = \frac{dF(x)}{dx}$$

if the derivative exists.

EXAMPLE 3.7 For the density function of Example 3.6 find $F(x)$ and use it to evaluate $P(0 < X \leq 1)$.

SOLUTION
For $-1 < x < 2$,

$$F(x) = \int_{-\infty}^{x} f(t) \; dt = \int_{-1}^{x} \frac{t^2}{3} \; dt = \frac{t^3}{9} \bigg|_{-1}^{x} = \frac{x^3 + 1}{9}.$$

Therefore,

$$F(x) = \begin{cases} 0, & x \leq -1 \\ \dfrac{x^3 + 1}{9}, & -1 \leq x < 2 \\ 1, & x \geq 2. \end{cases}$$

The cumulative distribution $F(x)$ is expressed graphically in Figure 3.6. Now,

$$P(0 < X \leq 1) = F(1) - F(0) = \tfrac{2}{9} - \tfrac{1}{9} = \tfrac{1}{9},$$

which agrees with the result obtained by using the density function in Example 3.6.

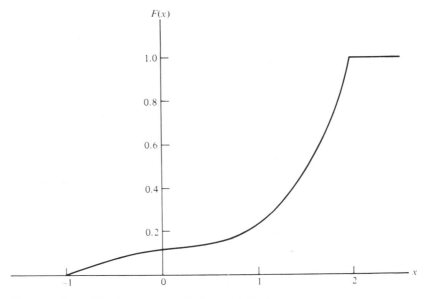

FIGURE 3.6 Continuous cumulative distribution.

Exercises

1. Classify the following random variables as discrete or continuous.

   X: the number of automobile accidents per year in Virginia.
   Y: the length of time to play 18 holes of golf.
   M: the amount of milk produced yearly by a particular cow.
   N: the number of eggs laid each month by a hen.
   P: the number of building permits issued each month in a certain city.
   Q: the weight of grain produced per acre.

2. An overseas shipment of 5 foreign automobiles contains 2 that have slight paint blemishes. If an agency receives 3 of these automobiles at random, list the elements of the sample space S using the letters B and N for "blemished" and "nonblemished," respectively; then to each sample point assign a value x of the random variable X representing the number of automobiles purchased by the agency with paint blemishes.

3. Let W be a random variable giving the number of heads minus the number of tails in three tosses of a coin. List the elements of the sample space S for the three tosses of the coin and to each sample point assign a value w of W.

4. A coin is flipped until 3 heads in succession occur. List only those elements of the sample space that require 6 or less tosses. Is this a discrete sample space? Explain.

5. Determine the value c so that each of the following functions can serve as a probability distribution of the discrete random variable X:
   (a) $f(x) = c(x^2 + 4)$ for x = 0, 1, 2, 3;

   (b) $f(x) = c\binom{2}{x}\binom{3}{3-x}$ for x = 0, 1, 2.

6. The shelf life, in days, for bottles of a certain prescribed medicine is a random variable having the density function

   $$f(x) = \begin{cases} \dfrac{20,000}{(x + 100)^3}, & x > 0 \\ 0, & \text{elsewhere.} \end{cases}$$

   Find the probability that a bottle of this medicine will have a shelf life of

   (a) at least 200 days;
   (b) anywhere from 80 to 120 days.

7. The total number of hours, measured in units of 100 hours, that a family runs a vacuum cleaner over a period of one year is a continuous random variable X that has the density function

   $$f(x) = \begin{cases} x, & 0 < x < 1 \\ 2 - x, & 1 \le x < 2 \\ 0, & \text{elsewhere.} \end{cases}$$

   Find the probability that over a period of one year, a family runs their vacuum cleaner
   (a) less than 120 hours;
   (b) between 50 and 100 hours.

8. Find the probability distribution of the random variable W in Exercise 3, assuming that the coin is biased so that a head is twice as likely to occur as a tail.

9. The proportion of people who respond to a certain mail-order solicitation is a continuous random variable X that has the density function

   $$f(x) = \begin{cases} \dfrac{2(x + 2)}{5}, & 0 < x < 1 \\ 0, & \text{elsewhere.} \end{cases}$$

   (a) Show that $P(0 < X < 1) = 1$.
   (b) Find the probability that more than 1/4 but fewer than 1/2 of the people contacted will respond to this type of solicitation.

10. Find a formula for the probability distribution of the random variable X representing the outcome when a single die is rolled once.

11. A shipment of 7 television sets contains 2 defective sets. A hotel makes a random purchase of 3 of the sets. If x is the number of defective sets purchased by the hotel, find the probability distribution of X. Express the results graphically as a probability histogram.

12. An investment firm offers its customers municipal bonds that mature after different numbers of years. Given that the cumulative distribution of T, the number of years to maturity for a randomly selected bond, is

$$F(t) = \begin{cases} 0, & t < 1 \\ \frac{1}{4}, & 1 \le t < 3 \\ \frac{1}{2}, & 3 \le t < 5 \\ \frac{3}{4}, & 5 \le t < 7 \\ 1, & t \ge 7, \end{cases}$$

find

(a) $P(T = 5)$;

(b) $P(T > 3)$;

(c) $P(1.4 < T < 6)$.

13. The probability distribution of X, the number of imperfections per 10 meters of a synthetic fabric in continuous rolls of uniform width, is given by

x	0	1	2	3	4
$f(x)$	0.41	0.37	0.16	0.05	0.01

Construct the cumulative distribution of X.

14. The waiting time, in hours, between successive speeders spotted by a radar unit is a continuous random variable with cumulative distribution

$$F(x) = \begin{cases} 0, & x \le 0 \\ 1 - e^{-8x}, & x > 0. \end{cases}$$

Find the probability of waiting less than 12 minutes between successive speeders

(a) using the cumulative distribution of X;

(b) using the probability density function of X.

15. Find the cumulative distribution of the random variable X representing the number of defectives in Exercise 11. Using $F(x)$, find

(a) $P(X = 1)$;

(b) $P(0 < X \le 2)$.

16. Construct a graph of the cumulative distribution of Exercise 15.

17. A continuous random variable X that can assume values between $x = 1$ and $x = 3$ has a density function given by $f(x) = 1/2$.

(a) Show that the area under the curve is equal to 1.

(b) Find $P(2 < X < 2.5)$.

(c) Find $P(X \le 1.6)$.

18. A continuous random variable X that can assume values between $x = 2$ and $x = 5$ has a density function given by $f(x) = 2(1 + x)/27$. Find

(a) $P(X < 4)$;

(b) $P(3 < X < 4)$.

19. For the density function of Exercise 17, find $F(x)$. Use it to evaluate $P(2 < X < 2.5)$.

20. For the density function of Exercise 18, find $F(x)$, and use it to evaluate $P(3 \le X < 4)$.

21. Consider the density function

$$f(x) = \begin{cases} k\sqrt{x}, & 0 < x < 1 \\ 0, & \text{elsewhere}. \end{cases}$$

(a) Evaluate k.

(b) Find $F(x)$ and use it to evaluate

$$P(0.3 < X < 0.6).$$

22. Three cards are drawn in succession from a deck without replacement. Find the probability distribution for the number of spades.

23. Find the cumulative distribution of the random variable W in Exercise 8. Using $F(w)$, find

(a) $P(W > 0)$;

(b) $P(-1 \le W < 3)$.

24. Find the probability distribution for the number of jazz records when 4 records are selected at random from a collection consisting of 5 jazz records, 2 classical records, and 3 polka records. Express your results by means of a formula.

25. From a box containing 4 dimes and 2 nickels, 3 coins are selected at random without replacement. Find the probability distribution for the total T of the 3 coins. Express the probability distribution graphically as a probability histogram.

26. From a box containing 4 black balls and 2 green balls, 3 balls are drawn in succession, each ball being replaced in the box before the next draw is made. Find the probability distribution for the number of green balls.

3.4 Empirical Distributions

In the previous sections the reader has been exposed to concepts dealing with discrete and continuous distributions. Methods of computing probabilities depend on knowledge of the probability mass function or probability density function. The probability function for the discrete case and the density function for the continuous case are ways of characterizing the distribution of probability for a population or system.

Obviously, the user of statistical methods cannot generate sufficient information or experimental data to characterize the distribution totally. But sets of data are often used to learn about certain properties of the distribution. Scientists and engineers are accustomed to dealing with data sets. The importance of characterizing or *summarizing* the nature of collections of data should be obvious. Often a summary of a collection of data via a graphical display can provide insight regarding the system from which the data were taken.

In later chapters, sampling from distributions and the display of data in order to enhance **statistical inference** about scientific systems will be explored in detail. In this section we merely introduce some simple but often effective displays that complement the study of statistical distributions.

Usually, in an experiment involving a continuous random variable the density function $f(x)$ is unknown and its form is assumed. For the choice of $f(x)$ to be reasonably valid, good judgment based on all available information is needed in its selection. Statistical data, generated in large masses, can be very useful in studying the behavior of the distribution if presented in a combined tabular and graphic display called a **stem and leaf plot**.

To illustrate the construction of a stem and leaf plot, consider the data of Table 3.1, which represent the lives of 40 similar car batteries recorded to the nearest tenth of a year. The batteries were guaranteed to last 3 years. First, split each observation into two parts consisting of a stem and a leaf such that the stem represents the digit preceding the decimal and the leaf corresponds to the decimal part of the number. In other words, for the number 3.7 the digit 3 is designated the stem and the digit 7 is the leaf. The four stems 1, 2, 3, and 4 for our data are listed consecutively on the left side of a vertical line in Table 3.2; the leaves are recorded on the right side of the line opposite the appropriate stem value. Thus the leaf 6 of the number 1.6 is recorded opposite the stem 1; the leaf 5 of the number 2.5 is recorded opposite the stem 2; and so forth. The number of leaves recorded opposite each stem is summarized under the frequency column.

The stem and leaf plot of Table 3.2 contains only four stems and consequently does not provide an adequate picture of the distribution. To remedy this problem, we need to increase the number of stems in our plot. One simple way to accomplish this is to write each stem value twice on the left side of the vertical line and then record the leaves 0, 1, 2, 3, and 4 opposite the appropriate stem value where it

TABLE 3.1   Car Battery Lives

2.2	4.1	3.5	4.5	3.2	3.7	3.0	2.6
3.4	1.6	3.1	3.3	3.8	3.1	4.7	3.7
2.5	4.3	3.4	3.6	2.9	3.3	3.9	3.1
3.3	3.1	3.7	4.4	3.2	4.1	1.9	3.4
4.7	3.8	3.2	2.6	3.9	3.0	4.2	3.5

TABLE 3.2 Stem and Leaf Plot of Battery Lives

Stems	Leaves	Frequency
1	69	2
2	25696	5
3	4318514723628297130097145	25
4	71354172	8

appears for the first time; and the leaves 5, 6, 7, 8, and 9 opposite this same stem value where it appears for the second time. This modified double-stem and leaf plot is illustrated in Table 3.3, where the stems corresponding to leaves 0 through 4 have been coded by the symbol $*$ and the stem corresponding to leaves 5 through 9 by the symbol $\cdot$.

A further increase in the number of stems may be achieved by writing each stem value five times on the left side of a vertical line, where we might now code the stems a for leaves 0 and 1, b for leaves 2 and 3, c for leaves 4 and 5, d for leaves 6 and 7, and e for leaves 8 and 9. For the data of Table 3.1 we would then use the stems $1d$, $1e$, $2a$, $2b$, $2c$, $2d$, and $2e$ to construct a five-stem and leaf plot.

In any given problem, we must decide on the appropriate stem values. This decision is made somewhat arbitrarily, although we are guided by the size of our sample. Usually, we choose between 5 and 20 stems. The smaller the number of data available, the smaller is our choice for the number of stems. For example, if the data consist of numbers from 1 to 21 representing the number of people in a cafeteria line on 40 randomly selected workdays and we choose a double-stem and leaf plot, the stems would be $0*$, $0\cdot$, $1*$, $1\cdot$, and $2*$ so that the smallest observation 1 has stem $0*$ and leaf 1, the number 18 has stem $1\cdot$ and leaf 8, and the largest observation 21 has stem $2*$ and leaf 1. On the other hand, if the data consist of numbers from \$8800 to \$9600 representing the best possible deals on 100 new automobiles from a certain dealership and we choose a single-stem and leaf plot, the stems would be 88, 89, 90, . . . , and 96 and the leaves would now each contain two digits. A car that sold for \$9385 would have a stem value of 93 and the two-digit leaf 85. Multiple-digit leaves belonging to the same stem are usually separated by commas in the stem and leaf plot. Decimal points in the data are

TABLE 3.3 Double-Stem and Leaf Plot of Battery Lives

Stems	Leaves	Frequency
1$\cdot$	69	2
2$*$	2	1
2$\cdot$	5696	4
3$*$	431142322130014	15
3$\cdot$	8576897975	10
4$*$	13412	5
4$\cdot$	757	3

generally ignored when all the digits to the right of the decimal represent the leaf. Such was the case in Tables 3.2 and 3.3. However, if the data consist of numbers ranging from 21.8 to 74.9, we might choose the digits 2, 3, 4, 5, 6, and 7 as our stems so that a number such as 48.3 would have a stem value of 4 and a leaf of 8.3.

A **frequency distribution** in which the data are grouped into different classes or intervals can easily be constructed by counting the leaves belonging to each stem and noting that each stem defines a class interval. In Table 3.3 the stem 1 with 2 leaves defines the interval 1.0–1.9 containing 2 observations; the stem 2 with 5 leaves defines the interval 2.0–2.9 containing 5 observations; the stem 3 with 25 leaves defines the interval 3.0–3.9 with 25 observations; and the stem 4 with 8 leaves defines the interval 4.0–4.9 containing 8 observations. For the double-stem and leaf plot of Table 3.1 the stems define the seven class intervals 1.5–1.9, 2.0–2.4, 2.5–2.9, 3.0–3.4, 3.5–3.9, 4.0–4.4, and 4.5–4.9 with frequencies 2, 1, 4, 15, 10, 5, and 3, respectively. Dividing each class frequency by the total number of observations, we obtain the proportion of the set of observations in each of the classes. A table listing relative frequencies is called a **relative frequency distribution**. The relative frequency distribution for the data of Table 3.1, showing the midpoints of each class interval, is given in Table 3.4.

The information provided by a relative frequency distribution in tabular form is easier to grasp if presented graphically. Using the midpoints of each interval and the corresponding relative frequencies, we construct a **relative frequency histogram** (Figure 3.7) in exactly the same manner that we constructed the probability histogram of Section 3.2.

In Section 3.2 we suggested that the heights of the rectangles be adjusted so that the areas would represent probabilities. Once this is done, the vertical axis may be omitted. If we wish to estimate the probability distribution $f(x)$ of a continuous random variable X by a smooth curve as in Figure 3.8, it is important that the rectangles of the relative frequency histogram be adjusted so that the total area is equal to 1.

The probability that a battery lasts between 3.45 and 4.45 years when selected at random from the infinite line of production of such batteries is given by the shaded area under the curve. Our estimated probability based on the recorded lives of the 40

TABLE 3.4 Relative Frequency Distribution of Battery Lives

Class Interval	Class Midpoint	Frequency f	Relative Frequency
1.5–1.9	1.7	2	0.050
2.0–2.4	2.2	1	0.025
2.5–2.9	2.7	4	0.100
3.0–3.4	3.2	15	0.375
3.5–3.9	3.7	10	0.250
4.0–4.4	4.2	5	0.125
4.5–4.9	4.7	3	0.075

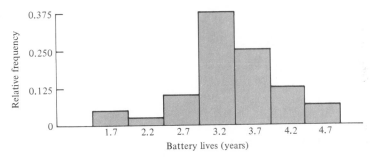

FIGURE 3.7   Relative frequency histogram.

batteries would be the sum of the areas contained in the rectangles between 3.45 and 4.45.

Although we have drawn an estimate of the shape of $f(x)$ in Figure 3.8, we still have no knowledge of its formula or equation and therefore cannot find the area that has been shaded. To help understand the method of estimating the formula for $f(x)$, let us recall some elementary analytic geometry. Parabolas, hyperbolas, circles, ellipses, and so forth, all have well-known forms of equations, and in each case we would recognize their graphs. Thinking in reverse, if we had only their graphs but recognized their form, then it is not difficult to estimate the unknown constants or parameters and arrive at the exact equation. For example, if the curve appeared to have the form of a parabola, then we know that it has an equation of the form $f(x) = ax^2 + bx + c$, where a, b, and c are parameters that can be determined by various estimation procedures.

Many continuous distributions can be represented graphically by the characteristic bell-shaped curve of Figure 3.8. The equation of the probability density function $f(x)$ in this case is as well known as that of a parabola or circle and depends only on the determination of two parameters. Once these parameters are estimated from the data, we can write the estimated equation, and then, using appropriate tables, find any probabilities we choose.

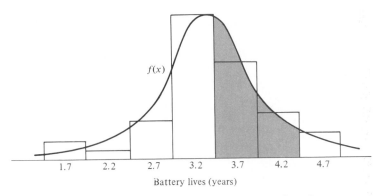

FIGURE 3.8   Estimating the probability density function.

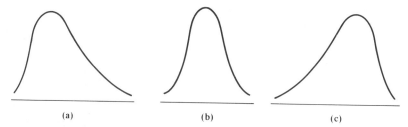

(a)   (b)   (c)

FIGURE 3.9   Skewness of data.

A distribution is said to be **symmetric** if it can be folded along a vertical axis so that the two sides coincide. A distribution that lacks symmetry with respect to a vertical axis is said to be **skewed**. The distribution illustrated in Figure 3.9(a) is said to be skewed to the right, since it has a long right tail and a much shorter left tail. In Figure 3.9(b) we see that the distribution is symmetric, while in Figure 3.9(c) it is skewed to the left.

By rotating a stem and leaf plot counterclockwise through an angle of 90°, we observe that the resulting columns of leaves form a picture that is similar to a histogram. Consequently, if our primary purpose in looking at the data is to determine the general shape or form of the density function, it will seldom be necessary to construct a relative frequency histogram. In Chapter 6 we shall consider most of the important density functions that are used in engineering and scientific investigations.

The cumulative distribution of X, where X represents the life of the car battery, can be estimated geometrically using the data of Table 3.4. To construct such a graph, we first arrange our data as in Table 3.5, a **relative cumulative frequency distribution**, and then plot the relative cumulative frequency less than each class boundary against the corresponding class boundary as in Figure 3.10. We estimate $F(x)$ by drawing a smooth curve through the points.

TABLE 3.5   Relative Cumulative Frequency Distribution of Battery Lives

Class Boundaries	Relative Cumulative Frequency
Less than 1.45	0.000
Less than 1.95	0.050
Less than 2.45	0.075
Less than 2.95	0.175
Less than 3.45	0.550
Less than 3.95	0.800
Less than 4.45	0.925
Less than 4.95	1.000

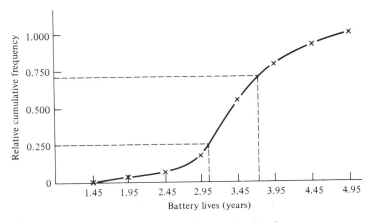

FIGURE 3.10 Continuous cumulative distribution.

Percentile, decile, and quartile points may be read quickly from the cumulative distribution. In Figure 3.10 the dashed lines indicate that the twenty-fifth percentile or first quartile and the seventh decile are approximately 3.05 and 3.70 years, respectively. This means that 25% or one-fourth of all the batteries of this type are expected to last less than 3.05 years, while 70% of such batteries can be expected to last less than 3.70 years.

Exercises

1. The following scores represent the final examination grade for an elementary statistics course:

23	60	79	32	57	74	52	70	82	36
80	77	81	95	41	65	92	85	55	76
52	10	64	75	78	25	80	98	81	67
41	71	83	54	64	72	88	62	74	43
60	78	89	76	84	48	84	90	15	79
34	67	17	82	69	74	63	80	85	61

(a) Construct a stem and leaf plot for the examination grades in which the stems are 1, 2, 3, . . . , 9.

(b) Set up a relative frequency distribution.

(c) Construct a relative frequency histogram, draw an estimate of the graph of $f(x)$, and discuss the skewness of the distribution.

(d) Construct a relative cumulative frequency distribution.

(e) Draw an estimate of the graph of $F(x)$.

(f) Estimate the first quartile and the seventh decile.

2. The following data represent the length of life in years, measured to the nearest tenth, of 30 similar fuel pumps:

2.0	3.0	0.3	3.3	1.3	0.4
0.2	6.0	5.5	6.5	0.2	2.3
1.5	4.0	5.9	1.8	4.7	0.7
4.5	0.3	1.5	0.5	2.5	5.0
1.0	6.0	5.6	6.0	1.2	0.2

(a) Construct a stem and leaf plot for the fuel pump lives using the digit to the left of the decimal point as the stem for each observation.

(b) Set up a relative frequency distribution.

(c) Construct a relative cumulative frequency distribution.

(d) Draw an estimate of the graph of $F(x)$.

(e) Estimate the value below which two-thirds of the values fall.

3. The following data represent the length of life, in seconds, of 50 fruit flies subject to a new spray in a controlled laboratory experiment:

17	20	10	9	23	13	12	19	18	24
12	14	6	9	13	6	7	10	13	7
16	18	8	13	3	32	9	7	10	11
13	7	18	7	10	4	27	19	16	8
7	10	5	14	15	10	9	6	7	15

(a) Construct a double-stem and leaf plot for the fruit fly lives using the stems 0∗, 0•, 1∗, 1•, 2∗, 2•, and 3∗ such that stems coded by the symbols ∗ and • are associated, respectively, with leaves 0 through 4 and 5 through 9.

(b) Set up a relative frequency distribution.

(c) Construct a relative frequency histogram, draw an estimate of the graph of $f(x)$, and discuss the skewness of the distribution.

(d) Construct a relative cumulative frequency distribution.

(e) Draw an estimate of the graph of $F(x)$.

(f) Estimate the seventy-fifth percentile.

4. Construct a stem and leaf plot for the data of Table 3.1 by writing each stem five times and then coding the stems as described on page 63.

5. The nicotine contents, in milligrams, for 40 cigarettes of a certain brand were recorded as follows:

1.09	1.92	2.31	1.79	2.28
1.74	1.47	1.97	0.85	1.24
1.58	2.03	1.70	2.17	2.55
2.11	1.86	1.90	1.68	1.51
1.64	0.72	1.69	1.85	1.82
1.79	2.46	1.88	2.08	1.67
1.37	1.93	1.40	1.64	2.09
1.75	1.63	2.37	1.75	1.69

(a) Construct a stem and leaf plot for the data in which the stems are the digits to the left of the decimal point, each repeated five times such that the double-digit leaves 00 through 19 are associated with stems coded by the letter a; leaves 20 through 39 are associated with stems coded by the letter b; and so forth. Thus a number such as 1.29 has a stem value of $1b$ and a leaf equal to 29.

(b) Set up a relative frequency distribution.

3.5 Joint Probability Distributions

Our study of random variables and their probability distributions in the preceding sections was restricted to one-dimensional sample spaces, in that we recorded outcomes of an experiment as values assumed by a single random variable. There will be situations, however, where we may find it desirable to record the simultaneous outcomes of several random variables. For example, we might measure the amount of precipitate P and volume V of gas released from a controlled chemical experiment giving rise to a two-dimensional sample space consisting of the outcomes (p, v), or one might be interested in the hardness H and tensile strength T of cold-drawn copper resulting in the outcomes (h, t). In a study to determine the likelihood of success in college, based on high school data, one might use a three-dimensional sample space and record for each individual his or her aptitude test score, high school rank in class, and grade-point average at the end of the freshman year in college.

If X and Y are two discrete random variables, the probability distribution for their simultaneous occurrence can be represented by a function with values $f(x, y)$ for any pair of values (x, y) within the range of the random variables X and Y. It is customary to refer to this function as the **joint probability distribution** of X and Y. Hence, in the discrete case,

$$f(x, y) = P(X = x, Y = y);$$

that is, the values $f(x, y)$ give the probability that outcomes x and y occur at the same time. For example, if a television set is to be serviced and X represents the age to the nearest year of the set and Y represents the number of defective tubes in the set, then $f(5, 3)$ is the probability that the television set is 5 years old and needs 3 new tubes.

DEFINITION 3.8   *The function $f(x, y)$ is a **joint probability distribution** or **probability mass func-tion** of the discrete random variables X and Y if*

1. $f(x, y) \geq 0$ *for all* (x, y).

2. $\sum_x \sum_y f(x, y) = 1$.

3. $P(X = x, Y = y) = f(x, y)$.

For any region A in the xy plane, $P[(X, Y) \in A] = \sum\sum_A f(x, y)$. ■

EXAMPLE 3.8   Two refills for a ballpoint pen are selected at random from a box that contains 3 blue refills, 2 red refills, and 3 green refills. If X is the number of blue refills and Y is the number of red refills selected, find (a) the joint probability function $f(x, y)$, and (b) $P[(X, Y) \in A]$, where A is the region $\{(x, y) | x + y \leq 1\}$.

SOLUTION

(a) The possible pairs of values (x, y) are $(0, 0)$, $(0, 1)$, $(1, 0)$, $(1, 1)$, $(0, 2)$, and $(2, 0)$. Now, $f(0, 1)$, for example, represents the probability that a red and a green refill are selected. The total number of equally likely ways of selecting any 2 refills from the 8 is $\binom{8}{2} = 28$. The number of ways of selecting 1 red from 2 red refills and 1 green from 3 green refills is $\binom{2}{1}\binom{3}{1} = 6$. Hence $f(0, 1), = 6/28 = 3/14$. Similar calculations yield the probabilities for the other cases, which are presented in Table 3.6. Note that the probabilities sum to 1. In Chapter 4 it will become clear that the joint probability distribution of Table 3.6 can be represented by the formula

$$f(x, y) = \frac{\binom{3}{x}\binom{2}{y}\binom{3}{2 - x - y}}{\binom{8}{2}},$$

for $x = 0, 1, 2$; $y = 0, 1, 2$; $0 \leq x + y \leq 2$.

(b) $P[(X, Y) \in A] = P(X + Y \leq 1)$
$= f(0, 0) + f(0, 1) + f(1, 0)$
$= \frac{3}{28} + \frac{3}{14} + \frac{9}{28}$
$= \frac{9}{14}$.

TABLE 3.6 Joint Probability Distribution for Example 3.8

$f(x, y)$	x			Row Totals
	0	1	2	
0	$\frac{3}{28}$	$\frac{9}{28}$	$\frac{3}{28}$	$\frac{15}{28}$
y 1	$\frac{3}{14}$	$\frac{3}{14}$		$\frac{3}{7}$
2	$\frac{1}{28}$			$\frac{1}{28}$
Column Totals	$\frac{5}{14}$	$\frac{15}{28}$	$\frac{3}{28}$	1

When X and Y are continuous random variables, the **joint density function** $f(x, y)$ is a surface lying above the xy plane, and $P[(X, Y) \in A]$, where A is any region in the xy plane, is equal to the volume of the right cylinder bounded by the base A and the surface.

DEFINITION 3.9   *The function $f(x, y)$ is a* **joint density function** *of the continuous random variables X and Y if*

1. $f(x, y) \geq 0$ *for all* (x, y).

2. $\displaystyle\int_{-\infty}^{\infty} \int_{-\infty}^{\infty} f(x, y)\, dx\, dy = 1$.

3. $\displaystyle P[(X, Y) \in A] = \iint\limits_{A} f(x, y)\, dx\, dy$.

for any region A in the xy plane. ■

EXAMPLE 3.9   A candy company distributes boxes of chocolates with a mixture of creams, toffees, and nuts coated in both light and dark chocolate. For a randomly selected box, let X and Y, respectively, be the proportions of the light and dark chocolates that are creams and suppose that the joint density function is given by

$$f(x, y) = \begin{cases} \frac{2}{5}(2x + 3y), & 0 \leq x \leq 1, 0 \leq y \leq 1 \\ 0, & \text{elsewhere.} \end{cases}$$

(a) Verify condition 2 of Definition 3.9.
(b) Find $P[(X, Y) \in A]$, where A is the region $\{(x, y) \mid 0 < x < \frac{1}{2}, \frac{1}{4} < y < \frac{1}{2}\}$.

SOLUTION

(a) $\displaystyle\int_{-\infty}^{\infty} \int_{-\infty}^{\infty} f(x, y)\, dx\, dy = \int_{0}^{1} \int_{0}^{1} \frac{2}{5}(2x + 3y)\, dx\, dy$

$\displaystyle = \int_{0}^{1} \frac{2x^2}{5} + \frac{6xy}{5} \bigg|_{x=0}^{x=1} dy$

$$= \int_0^1 \left(\frac{2}{5} + \frac{6y}{5} \right) dy = \frac{2y}{5} + \frac{3y^2}{5} \Big|_0^1$$

$$= \frac{2}{5} + \frac{3}{5} = 1.$$

(b) $P[(X, Y) \in A] = P(0 < X < \frac{1}{2}, \frac{1}{4} < Y < \frac{1}{2})$

$$= \int_{1/4}^{1/2} \int_0^{1/2} \frac{2}{5}(2x + 3y) \, dx \, dy$$

$$= \int_{1/4}^{1/2} \frac{2x^2}{5} + \frac{6xy}{5} \Big|_{x=0}^{x=1/2} dy$$

$$= \int_{1/4}^{1/2} \left(\frac{1}{10} + \frac{3y}{5} \right) dy = \frac{y}{10} + \frac{3y^2}{10} \Big|_{1/4}^{1/2}$$

$$= \frac{1}{10} \left[\left(\frac{1}{2} + \frac{3}{4} \right) - \left(\frac{1}{4} + \frac{3}{16} \right) \right] = \frac{13}{160}.$$

Given the joint probability distribution $f(x, y)$ of the discrete random variables X and Y, the probability distribution $g(x)$ of X alone is obtained by summing $f(x, y)$ over the values of Y. Similarly, the probability distribution $h(y)$ of Y alone is obtained by summing $f(x, y)$ over the values of X. We define $g(x)$ and $h(y)$ to be the **marginal distributions** of X and Y, respectively. When X and Y are continuous random variables, summations are replaced by integrals. We can now make the following general definition.

DEFINITION 3.10   *The **marginal distributions** of X alone and of Y alone are given by*

$$g(x) = \sum_y f(x, y) \qquad and \qquad h(y) = \sum_x f(x, y)$$

for the discrete case and by

$$g(x) = \int_{-\infty}^{\infty} f(x, y) \, dy \qquad and \qquad h(y) = \int_{-\infty}^{\infty} f(x, y) \, dx$$

for the continuous case.   ■

The term *marginal* is used here because, in the discrete case, the values of $g(x)$ and $h(y)$ are just the marginal totals of the respective columns and rows when the values of $f(x, y)$ are displayed in a rectangular table.

EXAMPLE 3.10   Show that the column and row totals of Table 3.6 give the marginal distribution of X alone and of Y alone.

SOLUTION

For the random variable X, we see that

$$P(X = 0) = g(0) = \sum_{y=0}^{2} f(0, y) = f(0, 0) + f(0, 1) + f(0, 2)$$

$$= \tfrac{3}{28} + \tfrac{3}{14} + \tfrac{1}{28} = \tfrac{5}{14},$$

$$P(X = 1) = g(1) = \sum_{y=0}^{2} f(1, y)$$

$$= f(1, 0) + f(1, 1) + f(1, 2)$$

$$= \tfrac{9}{28} + \tfrac{3}{14} + 0 = \tfrac{15}{28},$$

and

$$P(X = 2) = g(2) = \sum_{y=0}^{2} f(2, y) = f(2, 0) + f(2, 1) + f(2, 2)$$

$$= \tfrac{3}{28} + 0 + 0 = \tfrac{3}{28},$$

which are just the column totals of Table 3.6. In a similar manner we could show that the values of $h(y)$ are given by the row totals. In tabular form, these marginal distributions may be written as follows:

x	0	1	2
$g(x)$	$\tfrac{5}{14}$	$\tfrac{15}{28}$	$\tfrac{3}{28}$

y	0	1	2
$h(y)$	$\tfrac{15}{28}$	$\tfrac{3}{7}$	$\tfrac{1}{28}$

EXAMPLE 3.11   Find $g(x)$ and $h(y)$ for the joint density function of Example 3.9.

SOLUTION

By definition,

$$g(x) = \int_{-\infty}^{\infty} f(x, y) \, dy = \int_{0}^{1} \frac{2}{5}(2x + 3y) \, dy$$

$$= \frac{4xy}{5} + \frac{6y^2}{10} \Big|_{y=0}^{y=1} = \frac{4x + 3}{5}$$

for $0 \leq x \leq 1$, and $g(x) = 0$ elsewhere. Similarly,

$$h(y) = \int_{-\infty}^{\infty} f(x, y) \, dx = \int_{0}^{1} \frac{2}{5}(2x + 3y) \, dx$$

$$= \frac{2(1 + 3y)}{5}$$

for $0 \leq y \leq 1$, and $h(y) = 0$ elsewhere.

The fact that the marginal distributions $g(x)$ and $h(y)$ are indeed the probability distributions of the individual variables X and Y alone can easily be verified by showing that the conditions of Definition 3.4 or Definition 3.6 are satisfied. For example, in the continuous case

$$\int_{-\infty}^{\infty} g(x)\,dx = \int_{-\infty}^{\infty} \int_{-\infty}^{\infty} f(x,\,y)\,dy\,dx = 1$$

and

$$P(a < X < b) = P(a < X < b,\ -\infty < Y < \infty)$$

$$= \int_{a}^{b} \int_{-\infty}^{\infty} f(x,\,y)\,dy\,dx$$

$$= \int_{a}^{b} g(x)\,dx.$$

In Section 3.1 we stated that the value x of the random variable X represents an event that is a subset of the sample space. If we use the definition of conditional probability as given in Chapter 2,

$$P(B \mid A) = \frac{P(A \cap B)}{P(A)}, \qquad P(A) > 0,$$

where A and B are now the events defined by $X = x$ and $Y = y$, respectively, then

$$P(Y = y \mid X = x) = \frac{P(X = x,\ Y = y)}{P(X = x)}$$

$$= \frac{f(x,\,y)}{g(x)}, \qquad g(x) > 0,$$

when X and Y are discrete random variables.

It is not difficult to show that the function $f(x,\,y)/g(x)$, which is strictly a function of y with x fixed, satisfies all the conditions of a probability distribution. This is also true when $f(x,\,y)$ and $g(x)$ are the joint density and marginal distribution of continuous random variables. Expressing such a probability distribution by the symbol $f(y \mid x)$, we have the following definition.

DEFINITION 3.11   *Let X and Y be two random variables, discrete or continuous. The* **conditional distribution** *of the random variable Y, given that X = x, is given by*

$$f(y \mid x) = \frac{f(x,\,y)}{g(x)}, \qquad g(x) > 0.$$

Similarly, the **conditional distribution** *of the random variable X, given that Y = y, is given by*

$$f(x \mid y) = \frac{f(x,\,y)}{h(y)}, \qquad h(y) > 0.$$ ■

If one wished to find the probability that the discrete random variable X falls between a and b when it is known that the discrete variable $Y = y$, we evaluate

$$P(a < X < b | Y = y) = \sum_x f(x|y),$$

where the summation extends over all values of X between a and b. When X and Y are continuous, we evaluate

$$P(a < X < b | Y = y) = \int_a^b f(x|y) \, dx.$$

EXAMPLE 3.12   Referring to Example 3.8, find the conditional distribution of X, given that $Y = 1$, and use it to determine $P(X = 0 | Y = 1)$.

SOLUTION
We need to find $f(x|y)$, where $y = 1$. First we find that

$$h(1) = \sum_{x=0}^{2} f(x, 1) = \tfrac{3}{14} + \tfrac{3}{14} + 0 = \tfrac{3}{7}.$$

Now

$$f(x|1) = \frac{f(x, 1)}{h(1)} = \frac{7}{3} f(x, 1), \qquad x = 0, 1, 2.$$

Therefore,

$$f(0|1) = \tfrac{7}{3} f(0, 1) = (\tfrac{7}{3})(\tfrac{3}{14}) = \tfrac{1}{2}$$
$$f(1|1) = \tfrac{7}{3} f(1, 1) = (\tfrac{7}{3})(\tfrac{3}{14}) = \tfrac{1}{2}$$
$$f(2|1) = \tfrac{7}{3} f(2, 1) = (\tfrac{7}{3})(0) = 0$$

and the conditional distribution of X, given that $Y = 1$, is

x	0	1	2
$f(x\|1)$	$\tfrac{1}{2}$	$\tfrac{1}{2}$	0

Finally,

$$P(X = 0 | Y = 1) = f(0|1) = \tfrac{1}{2}.$$

Therefore, if it is known that 1 of the 2 pen refills selected is red, we have a probability equal to 1/2 that the other refill is not blue.

EXAMPLE 3.13   The joint density for the random variables (X, Y), where X is the unit temperature change and Y is the proportion of spectrum shift that a certain atomic particle produces is given by

$$f(x, y) = \begin{cases} 10xy^2, & 0 < x < y < 1 \\ 0, & \text{elsewhere.} \end{cases}$$

(a) Find the marginal densities $g(x)$, $h(y)$, and the conditional density $f(y|x)$.
(b) Find the probability that the spectrum shifts more than half of the total observations given the temperature was increased to 0.25 unit.

SOLUTION

(a) By definition,

$$g(x) = \int_{-\infty}^{\infty} f(x, y)\, dy = \int_{x}^{1} 10xy^2\, dy = \frac{10}{3} xy^3 \Big|_{y=x}^{y=1} = \frac{10}{3} x(1 - x^3), \qquad 0 < x < 1$$

$$h(y) = \int_{-\infty}^{\infty} f(x, y)\, dx = \int_{0}^{y} 10xy^2\, dx = 5x^2y^2 \Big|_{x=0}^{x=y} = 5y^4, \qquad\qquad 0 < y < 1.$$

Now

$$f(y\,|\,x) = \frac{f(x, y)}{g(x)} = \frac{10xy^2}{\frac{10}{3} x(1 - x^3)} = \frac{3y^2}{(1 - x^3)}, \qquad 0 < x < y < 1.$$

(b) Therefore,

$$P\left(Y > \frac{1}{2}\,\Big|\, X = 0.25\right) = \int_{1/2}^{1} f(y\,|\,x = 0.25)\, dy = \int_{1/2}^{1} \frac{3y^2}{(1 - 0.25^3)}\, dy = \frac{8}{9}.$$

EXAMPLE 3.14   Given the joint density function

$$f(x, y) = \begin{cases} \dfrac{x(1 + 3y^2)}{4}, & 0 < x < 2,\ 0 < y < 1 \\ 0, & \text{elsewhere,} \end{cases}$$

find $g(x)$, $h(y)$, $f(x\,|\,y)$, and evaluate $P(\frac{1}{4} < X < \frac{1}{2}\,|\,Y = \frac{1}{3})$.

SOLUTION

By definition,

$$g(x) = \int_{-\infty}^{\infty} f(x, y)\, dy = \int_{0}^{1} \frac{x(1 + 3y^2)}{4}\, dy$$

$$= \frac{xy}{4} + \frac{xy^3}{4} \Big|_{y=0}^{y=1} = \frac{x}{2}, \qquad 0 < x < 2,$$

and

$$h(y) = \int_{-\infty}^{\infty} f(x, y)\, dx = \int_{0}^{2} \frac{x(1 + 3y^2)}{4}\, dx$$

$$= \frac{x^2}{8} + \frac{3x^2y^2}{8} \Big|_{x=0}^{x=2} = \frac{1 + 3y^2}{2}, \qquad 0 < y < 1.$$

Therefore,

$$f(x\,|\,y) = \frac{f(x, y)}{h(y)} = \frac{x(1 + 3y^2)/4}{(1 + 3y^2)/2} = \frac{x}{2}, \qquad 0 < x < 2,$$

and

$$P\left(\frac{1}{4} < X < \frac{1}{2}\,\Big|\, Y = \frac{1}{3}\right) = \int_{1/4}^{1/2} \frac{x}{2}\, dx = \frac{3}{64}.$$

Statistical Independence

If $f(x|y)$ does not depend on y, as was the case in Example 3.14, then $f(x|y) = g(x)$ and $f(x, y) = g(x)h(y)$. The proof follows by substituting

$$f(x, y) = f(x|y)h(y)$$

into the marginal distribution of X. That is,

$$g(x) = \int_{-\infty}^{\infty} f(x, y) \, dy = \int_{-\infty}^{\infty} f(x|y)h(y) \, dy.$$

If $f(x|y)$ does not depend on y, we may write

$$g(x) = f(x|y) \int_{-\infty}^{\infty} h(y) \, dy.$$

Now

$$\int_{-\infty}^{\infty} h(y) \, dy = 1,$$

since $h(y)$ is the probability density function of Y. Therefore,

$$g(x) = f(x|y)$$

and then

$$f(x, y) = g(x)h(y).$$

It should make sense to the reader that if $f(x|y)$ does not depend on y, then of course the outcome of the random variable Y has no impact on the outcome of the random variable X. In other words, we say that X and Y are independent random variables. We now offer the following formal definition of statistical independence.

DEFINITION 3.12   *Let X and Y be two random variables, discrete or continuous, with joint probability distribution f(x, y) and marginal distributions g(x) and h(y), respectively. The random variables X and Y are said to be* **statistically independent** *if and only if*

$$f(x, y) = g(x)h(y)$$

for all (x, y) within their range. ■

The continuous random variables of Example 3.14 are statistically independent, since the product of the two marginal distributions gives the joint density function. This is obviously not the case, however, for the continuous variables of Example 3.13. Checking for statistical independence of discrete random variables requires a more thorough investigation, since it is possible to have the product of the marginal distributions equal to the joint probability distribution for some but not all combinations of (x, y). If you can find any point (x, y) for which $f(x, y)$ is defined such that $f(x, y) \neq g(x)h(y)$, the discrete variables X and Y are not statistically independent.

EXAMPLE 3.15   Show that the random variables of Example 3.8 are not statistically independent.

SOLUTION

Let us consider the point $(0, 1)$. From Table 3.6 we find the three probabilities $f(0, 1)$, $g(0)$, and $h(1)$ to be

$$f(0, 1) = \tfrac{3}{14}$$

$$g(0) = \sum_{y=0}^{2} f(0, y) = \tfrac{3}{28} + \tfrac{3}{14} + \tfrac{1}{28} = \tfrac{5}{14}$$

$$h(1) = \sum_{x=0}^{2} f(x, 1) = \tfrac{3}{14} + \tfrac{3}{14} + 0 = \tfrac{3}{7}.$$

Clearly,

$$f(0, 1) \neq g(0)h(1),$$

and therefore X and Y are not statistically independent.

All the preceding definitions concerning two random variables can be generalized to the case of n random variables. Let $f(x_1, x_2, \ldots, x_n)$ be the joint probability function of the random variables $X_1, X_2, \ldots, X_n$. The marginal distribution of X_1, for example, is given by

$$g(x_1) = \sum_{x_2} \cdots \sum_{x_n} f(x_1, x_2, \ldots, x_n)$$

for the discrete case and by

$$g(x_1) = \int_{-\infty}^{\infty} \cdots \int_{-\infty}^{\infty} f(x_1, x_2, \ldots, x_n) \, dx_2 \, dx_3 \cdots dx_n$$

for the continuous case. We can now obtain **joint marginal distributions** such as $\phi(x_1, x_2)$, where

$$\phi(x_1, x_2) =
\begin{cases}
\displaystyle\sum_{x_3} \cdots \sum_{x_n} f(x_1, x_2, \ldots, x_n) & \text{(discrete case)} \\[2ex]
\displaystyle\int_{-\infty}^{\infty} \cdots \int_{-\infty}^{\infty} f(x_1, x_2, \ldots, x_n) \, dx_3 \, dx_4 \cdots dx_n & \text{(continuous case)}.
\end{cases}$$

One could consider numerous conditional distributions. For example, the **joint conditional distribution** of X_1, X_2, and X_3, given that $X_4 = x_4$, $X_5 = x_5$, $\ldots$, $X_n = x_n$, is written

$$f(x_1, x_2, x_3 \mid x_4, x_5, \ldots, x_n) = \frac{f(x_1, x_2, \ldots, x_n)}{g(x_4, x_5, \ldots, x_n)},$$

where $g(x_4, x_5, \ldots, x_n)$ is the joint marginal distribution of the random variables $X_4, X_5, \ldots, X_n$.

A generalization of Definition 3.12 leads to the following definition for the mutually statistical independence of the variables $X_1, X_2, \ldots, X_n$.

DEFINITION 3.13   *Let $X_1, X_2, \ldots, X_n$ be n random variables, discrete or continuous, with joint probability distribution $f(x_1, x_2, \ldots, x_n)$ and marginal distributions $f_1(x_1)$, $f_2(x_2), \ldots, f_n(x_n)$, respectively. The random variables $X_1, X_2, \ldots, X_n$ are said to be mutually* **statistically independent** *if and only if*

$$f(x_1, x_2, \ldots, x_n) = f_1(x_1)f_2(x_2)\cdots f_r(x_n)$$

for all $(x_1, x_2, \ldots, x_n)$ within their range. ■

EXAMPLE 3.16   Suppose that the shelf life, in years, of a certain perishable food product packaged in cardboard containers is a random variable whose probability density function is given by

$$f(x) = \begin{cases} e^{-x}, & x > 0 \\ 0, & \text{elsewhere.} \end{cases}$$

Let X_1, X_2, and X_3 represent the shelf lives for three of these containers selected independently and find $P(X_1 < 2, 1 < X_2 < 3, X_3 > 2)$.

SOLUTION
Since the containers were selected independently, we can assume that the random variables X_1, X_2, and X_3 are statistically independent, having the joint probability density

$$f(x_1, x_2, x_3) = f(x_1)f(x_2)f(x_3)$$

$$= e^{-x_1}e^{-x_2}e^{-x_3}$$

$$= e^{-x_1-x_2-x_3}$$

for $x_1 > 0$, $x_2 > 0$, $x_3 > 0$, and $f(x_1, x_2, x_3) = 0$ elsewhere. Hence

$$P(X_1 < 2, 1 < X_2 < 3, X_3 > 2) = \int_2^\infty \int_1^3 \int_0^2 e^{-x_1-x_2-x_3} \, dx_1 \, dx_2 \, dx_3$$

$$= (1 - e^{-2})(e^{-1} - e^{-3})e^{-2}$$

$$= 0.0376.$$

Exercises

1. Determine the value of c so that the following functions represent joint probability distributions of the random variables X and Y:
   (a) $f(x, y) = cxy$, for $x = 1, 2, 3$; $y = 1, 2, 3$.
   (b) $f(x, y) = c|x - y|$, for $x = -2, 0, 2$; $y = -2, 3$.

2. If the joint probability distribution of X and Y is given by

$$f(x, y) = \frac{(x + y)}{30}, \quad \text{for } x = 0, 1, 2, 3; y = 0, 1, 2,$$

find
   (a) $P(X \le 2, Y = 1)$;   (b) $P(X > 2, Y \le 1)$;
   (c) $P(X > Y)$;   (d) $P(X + Y = 4)$.

3. From a sack of fruit containing 3 oranges, 2 apples, and 3 bananas a random sample of 4 pieces of fruit is selected. If X is the number of oranges and Y is the number of apples in the sample, find
   (a) the joint probability distribution of X and Y;
   (b) $P[(X, Y) \in A]$, where A is the region given by $\{(x, y) | x + y \le 2\}$.

4. A privately owned liquor store operates both a drive-up facility and a walk-in facility. On a randomly selected day, let X and Y, respectively, be the proportions of the time that the drive-up and walk-in facilities are in use and suppose that the joint density function of these random variables is given by

$$f(x, y) = \begin{cases} \frac{2}{3}(x + 2y), & 0 \le x \le 1, 0 \le y \le 1 \\ 0, & \text{elsewhere.} \end{cases}$$

(a) Find the marginal density of X.
(b) Find the marginal density of Y.
(c) Find the probability that the drive-in facility is busy less than one-half of the time.

5. A candy company distributes boxes of chocolates with a mixture of creams, toffees, and cordials. Suppose that the weight of each box is 1 kilogram, but the individual weights of the creams, toffees, and cordials vary from box to box. For a randomly selected box, let X and Y represent the weights of the creams and the toffees, respectively, and suppose that the joint density function of these variables is given by

$$f(x, y) = \begin{cases} 24xy, & 0 \le x \le 1, 0 \le y \le 1, x + y \le 1 \\ 0, & \text{elsewhere.} \end{cases}$$

(a) Find the probability that in a given box the cordials account for more than 1/2 of the weight.
(b) Find the marginal density for the weight of the creams.
(c) Find the probability that the weight of the toffees in a box is less than 1/8 of a kilogram if it is known that creams constitute 3/4 of the weight.

6. Let X and Y denote the lengths of life, in years, of two components in an electronic system. If the joint density function of these variables is given by

$$f(x, y) = \begin{cases} e^{-(x+y)}, & x > 0, y > 0 \\ 0, & \text{elsewhere,} \end{cases}$$

find $P(0 < X < 1 \mid Y = 2)$.

7. Let X denote the reaction time, in seconds, to a certain stimulant and Y denote the temperature (°F) at which a certain reaction starts to take place. Suppose that two random variables X and Y have the joint density given by

$$f(x, y) = \begin{cases} 4xy, & 0 < x, y < 1 \\ 0, & \text{elsewhere.} \end{cases}$$

Find
(a) $P(0 \le X \le 1/2$ and $1/4 \le Y \le 1/2)$;
(b) $P(X < Y)$.

8. Each rear tire on an experimental airplane is supposed to be filled to a pressure of 40 psi. Let X denote the actual air pressure for the right tire and Y denote the actual air pressure for the left tire. Suppose that X and Y are random variables with the joint density given by

$$f(x, y) = \begin{cases} k(x^2 + y^2), & 30 \le x, y < 50 \\ 0, & \text{elsewhere.} \end{cases}$$

(a) Find k.
(b) Find $P(30 \le X \le 40$ and $40 \le Y < 50)$.
(c) Find the probability that both tires are underfilled.

9. Let X denote the diameter of an armored electric cable and Y denote the diameter of the ceramic mold that makes the cable. Both X and Y are scaled so that they range between 0 and 1. Suppose that X and Y have the joint density given by

$$f(x, y) = \begin{cases} \dfrac{1}{y}, & 0 < x < y < 1 \\ 0, & \text{elsewhere.} \end{cases}$$

Find $P(X + Y > 1/2)$.

10. Referring to Exercise 2, find
(a) the marginal distribution of X;
(b) the marginal distribution of Y.

11. The amount of kerosene, in thousands of liters, in a tank at the beginning of any day is a random amount Y from which a random amount X is sold during that day. Suppose that the tank is not resupplied during the day so that $x \le y$, and assume that the joint density function of these variables is given by

$$f(x, y) = \begin{cases} 2, & 0 < x < y, 0 < y < 1 \\ 0, & \text{elsewhere.} \end{cases}$$

(a) Determine if X and Y are independent.
(b) Find $P(1/4 < X < 1/2 \mid Y = 3/4)$.

12. Referring to Exercise 3, find
(a) $f(y \mid 2)$ for all values of y;
(b) $P(y = 0 \mid X = 2)$.

13. Let X denote the number of times a certain numerical control machine will malfunction: 1, 2, or 3 times on any given day. Let Y denote the number of times a technician is called on an emergency call. Their joint probability distribution is given as:

		x	
$f(x, y)$	1	2	3
1	0.05	0.05	0.1
y 2	0.05	0.1	0.35
3	0	0.2	0.1

(a) Evaluate the marginal distribution of X.
(b) Evaluate the marginal distribution of Y.
(c) Find $P(Y = 3 | X = 2)$.

14. Suppose that X and Y have the following joint probability distribution:

		x
$f(x, y)$	2	4
1	0.10	0.15
y 3	0.20	0.30
5	0.10	0.15

(a) Find the marginal distribution of X.
(b) Find the marginal distribution of Y.

15. Consider an experiment that consists of 2 rolls of a balanced die. If X is the number of 4's and Y is the number of 5's obtained in the 2 rolls of the die, find
(a) the joint probability distribution of X and Y;
(b) $P[(X, Y) \in A]$, where A is the region given by $\{(x, y) | 2x + y < 3\}$.

16. Let X denote the number of heads and Y the number of heads minus the number of tails when 3 coins are tossed. Find the joint probability distribution of X and Y.

17. Three cards are drawn without replacement from the 12 face cards (jacks, queens, and kings) of an ordinary deck of 52 playing cards. Let X be the number of kings selected and Y the number of jacks. Find
(a) the joint probability distribution of X and Y;
(b) $P[(X, Y) \in A]$, where A is the region given by $\{(x, y) | x + y \geq 2\}$.

18. A coin is tossed twice. Let Z denote the number of heads on the first toss and W the total number of heads on the 2 tosses. If the coin is unbalanced and a head has a 40% chance of occurring, find

(a) the joint probability distribution of W and Z;
(b) the marginal distribution of W;
(c) the marginal distribution of Z;
(d) the probability that at least 1 head occurs.

19. Given the joint density function

$$f(x, y) = \begin{cases} \dfrac{6 - x - y}{8}, & 0 < x < 2, \, 2 < y < 4 \\ 0, & \text{elsewhere,} \end{cases}$$

find $P(1 < Y < 3 | X = 2)$.

20. Determine whether the two random variables of Exercise 13 are dependent or independent.

21. Determine whether the two random variables of Exercise 14 are dependent or independent.

22. The joint density function of the random variables X and Y is given by

$$f(x, y) = \begin{cases} 6x, & 0 < x < 1, \, 0 < y < 1 - x \\ 0, & \text{elsewhere.} \end{cases}$$

(a) Show that X and Y are not independent.
(b) Find $P(X > 0.3 | Y = 0.5)$.

23. Let X, Y, and Z have the joint probability density function

$$f(x, y, z) = \begin{cases} kxy^2z, & 0 < x < 1, \, 0 < y < 1, \\ & \quad 0 < z < 2 \\ 0, & \text{elsewhere.} \end{cases}$$

(a) Find k.
(b) Find $P(X < 1/4, \, Y > 1/2, \, 1 < Z < 2)$.

24. Determine whether the two random variables of Exercise 7 are dependent or independent.

25. Determine whether the two random variables of Exercise 8 are dependent or independent.

26. The joint probability density function of the random variables X, Y, and Z is given by

$$f(x, y, z) = \begin{cases} \dfrac{4xyz^2}{9}, & 0 < x < 1, \, 0 < y < 1, \\ & \quad 0 < z < 3 \\ 0, & \text{elsewhere.} \end{cases}$$

Find
(a) the joint marginal density function of Y and Z;
(b) the marginal density of Y;
(c) $P(1/4 < X < 1/2, \, Y > 1/3, \, 1 < Z < 2)$;
(d) $P(0 < X < 1/2 | Y = 1/4, \, Z = 2)$.

Review Exercises

1. A tobacco company produces blends of tobacco with each blend containing various proportions of Turkish, domestic, and other tobaccos. The proportion of Turkish and domestic in a blend are random variables with joint density function (X = Turkish and Y = domestic)

$$f(x, y) = \begin{cases} 24xy, & 0 \le x \le 1, \, 0 \le y \le 1, \\ & x + y \le 1 \\ 0, & \text{elsewhere.} \end{cases}$$

(a) Find the probability that in a given box the Turkish tobacco accounts for over half the blend.
(b) Find the marginal density function for the proportion of the domestic tobacco.
(c) Find the probability that the proportion of Turkish tobacco is less than 1/8 if it is known that the blend contains 3/4 domestic tobacco.

2. An insurance company offers its policyholders a number of different premium payment options. For a randomly selected policyholder, let X be the number of months between successive payments. The cumulative distribution function of X is

$$F(x) = \begin{cases} 0 & \text{if } x < 1 \\ 0.4 & \text{if } 1 \le x < 3 \\ 0.6 & \text{if } 3 \le x < 5 \\ 0.8 & \text{if } 5 \le x < 7 \\ 1.0 & \text{if } x \ge 7. \end{cases}$$

(a) What is the probability mass function of X?
(b) Compute $P(4 < X \le 7)$.

3. Two electronic components of a missile system work in harmony for the success of the total system. Let X and Y denote the life in hours of the two systems. The joint density of X and Y is given by

$$f(x, y) = \begin{cases} ye^{-y(1+x)}, & x \ge 0, \, y \ge 0 \\ 0, & \text{elsewhere.} \end{cases}$$

(a) Give the marginal density functions for both random variables.
(b) What is the probability that both components will exceed 2 hours?

4. A service facility operates with two service lines. On a randomly selected day, let X be the proportion of time that the first line is in use while Y is the proportion of time that the second line is in use. Suppose that the joint probability density function for (X, Y) is given by

$$f(x, y) = \begin{cases} \frac{3}{2}(x^2 + y^2), & 0 \le x \le 1, \, 0 \le y \le 1 \\ 0, & \text{elsewhere.} \end{cases}$$

(a) Compute the probability that neither line is busy more than half the time.
(b) Find the probability that the first line is busy more than 75% of the time.

5. Let the number of phone calls received by a switchboard during a 5-minute interval be a random variable X with probability function

$$f(x) = \frac{e^2 2^x}{x!} \quad \text{for } x = 0, 1, 2, \ldots.$$

(a) Determine the probability that X equals 0, 1, 2, 3, 4, 5, and 6.
(b) Graph the probability mass function for these values of x.
(c) Determine the cumulative distribution function for these values of X.

6. Consider the random variables X and Y with joint density function given by

$$f(x, y) = \begin{cases} x + y, & 0 \le x, \, y \le 1 \\ 0, & \text{elsewhere.} \end{cases}$$

(a) Give the marginal distributions of X and Y.
(b) Find $P(X > 0.5, Y > 0.5)$.

7. An industrial process manufactures items that can be classified as either defective or not defective. The probability that an item is defective is 0.1. An experiment is conducted in which 5 items are drawn randomly from the process. Let the random variable X be the number of defectives in this sample of 5. What is the probability mass function of X?

8. Consider the following joint probability density function of the random variables X and Y:

$$f(x, y) = \begin{cases} \dfrac{3x - y}{9}, & 1 < x < 3, \, 1 < y < 2 \\ 0, & \text{elsewhere.} \end{cases}$$

(a) Find the marginal distributions of X and Y.

(b) Are X and Y independent?

(c) Find $P(X > 2)$.

9. The life span in hours of an electrical component is a random variable with cumulative distribution function

$$F(x) = \begin{cases} 1 - e^{-x/50}, & x > 0 \\ 0, & \text{elsewhere.} \end{cases}$$

(a) Determine the probability density function.

(b) Determine the probability that the life span of such a component will exceed 70 hours.

10. Pairs of pants are being produced by a particular outlet facility. The pants are ''checked'' by a group of 10 workers. The workers inspect pairs of pants taken randomly from the production line. Each inspector is assigned a number from 1 through 10. A buyer selects a pair of pants for purchase. Let the random variable X be the inspector number.

(a) Give a reasonable probability mass function for X.

(b) Plot the cumulative distribution function for X.

Mathematical Expectation

4.1 Mean of a Random Variable

If two coins are tossed 16 times and X is the number of heads that occur per toss, then the values of X can be 0, 1, and 2. Suppose that the experiment yields no heads, one head, and two heads a total of 4, 7, and 5 times, respectively. The average number of heads per toss of the two coins is then

$$\frac{(0)(4) + (1)(7) + (2)(5)}{16} = 1.06.$$

This is an average value and is not necessarily a possible outcome for the experiment. For instance, a salesman's average monthly income is not likely to be equal to any of his monthly paychecks.

Let us now restructure our computation for the average number of heads so as to have the following equivalent form:

$$(0)(\tfrac{4}{16}) + (1)(\tfrac{7}{16}) + (2)(\tfrac{5}{16}) = 1.06.$$

The numbers 4/16, 7/16, and 5/16 are the fractions of the total tosses resulting in 0, 1, and 2 heads, respectively. These fractions are also the relative frequencies for the different values of X in our experiment. In effect, then, we can calculate the mean or average of a set of data by knowing the distinct values that occur and their relative frequencies, without any knowledge of the total number of observations in our set of data. Therefore, if 4/16 or 1/4 of the tosses result in no heads, 7/16 of the tosses result in one head, and 5/16 of the tosses result in two heads, the mean number of heads per toss would be 1.06 no matter whether the total number of tosses was 16, 1000, or even 10,000.

Let us now use this method of relative frequencies to calculate the average number of heads per toss of two coins that we might expect in the long run. We shall refer to this average value as the **mean of the random variable X** or the **mean of the probability distribution of X** and write it as μ_X or simply as μ when it is clear to which random variable we refer. It is also common among statisticians to refer to this mean as the **mathematical expectation** or the **expected value** of the random variable X and denote it as $E(X)$.

Assuming that fair coins were tossed, we find that the sample space for our experiment is given by

$$S = \{HH, HT, TH, TT\}.$$

Since the 4 sample points are all equally likely, it follows that

$$P(X = 0) = P(TT) = \tfrac{1}{4},$$

$$P(X = 1) = P(TH) + P(HT) = \tfrac{1}{2},$$

and

$$P(X = 2) = P(HH) = \tfrac{1}{4},$$

where a typical element, say *TH*, indicates that the first toss resulted in a tail followed by a head on the second toss. Now, these probabilities are just the relative frequencies for the given events in the long run. Therefore,

$$\mu = E(X) = (0)(\tfrac{1}{4}) + (1)(\tfrac{1}{2}) + (2)(\tfrac{1}{4})$$

$$= 1.$$

This means that a person who tosses 2 coins over and over again will, on the average, get 1 head per toss.

The method described above for calculating the expected number of heads per toss of 2 coins suggests that the mean or expected value of any discrete random variable may be obtained by multiplying each of the values $x_1, x_2, \ldots, x_n$ of the random variable X by its corresponding probability $f(x_1), f(x_2), \ldots, f(x_n)$ and summing the products. This is true, however, only if the random variable is discrete. In the case of continuous random variables, the definition of an expected value is essentially the same with summations replaced by integrations.

DEFINITION 4.1   *Let X be a random variable with probability distribution $f(x)$. The **mean** or **expected value** of X is*

$$\mu = E(X) = \sum_x xf(x)$$

if X is discrete, and

$$\mu = E(X) = \int_{-\infty}^{\infty} xf(x)\ dx$$

if X is continuous. ∎

EXAMPLE 4.1   A lot containing seven components is sampled by a quality inspector; the lot contains 4 good components and 3 defective components. A sample of 3 is taken by the inspector. Find the expected value of the number of good components in this sample.

SOLUTION
Let X represent the number of good components in the sample. The probability distribution of X is given by

$$f(x) = \frac{\binom{4}{x}\binom{3}{3-x}}{\binom{7}{3}}, \qquad x = 0,\ 1,\ 2,\ 3.$$

A few simple calculations yield $f(0) = 1/35$, $f(1) = 12/35$, $f(2) = 18/35$, and $f(3) = 4/35$. Therefore,

$$\mu = E(X) = (0)(\tfrac{1}{35}) + (1)(\tfrac{12}{35}) + (2)(\tfrac{18}{35}) + (3)(\tfrac{4}{35})$$

$$= \tfrac{12}{7} = 1.7.$$

Thus, if a sample of size 3 is selected at random over and over again from a lot of 4 good components and 3 defective components, it would contain, on the average, 1.7 good components.

EXAMPLE 4.2   In a gambling game a man is paid $5 if he gets all heads or all tails when three coins are tossed and he pays out $3 if either one or two heads show. What is his expected gain?

SOLUTION
The sample space for the possible outcomes when three coins are tossed simultaneously, or equivalently if 1 coin is tossed three times, is

$$S = \{HHH,\ HHT,\ HTH,\ THH,\ HTT,\ THT,\ TTH,\ TTT\}.$$

One can argue that each of these possibilities is equally likely and occurs with probability equal to 1/8. An alternative approach would be to apply the multiplicative rule of probability for independent events to each element of S. For example,

$$P(HHT) = P(H)P(H)P(T)$$

$$= (\tfrac{1}{2})(\tfrac{1}{2})(\tfrac{1}{2}) = \tfrac{1}{8}.$$

The random variable of interest is Y, the amount the gambler can win; and the possible values of Y are $5 if event $E_1 = \{HHH,\ TTT\}$ occurs and $-$3 if event

$E_2 = \{HHT, HTH, THH, HTT, THT, TTH\}$ occurs. Since E_1 and E_2 occur with probabilities 1/4 and 3/4, respectively, it follows that

$$\mu = E(Y) = (5)(\tfrac{1}{4}) + (-3)(\tfrac{3}{4}) = -1.$$

In this game the gambler will on the average, lose $1 per toss of the three coins. A game is considered "fair" if the gambler will, on the average, come out even. Therefore, an expected gain of zero defines a fair game.

Examples 4.1 and 4.2 are designed to allow the reader to gain some insight into what we mean by the expected value of a random variable. In both cases the random variables were discrete. We follow with an example of a continuous random variable in which an engineer is interested in the *mean life* of a certain type of electronic device. This is an illustration of a *time to failure* problem that occurs often in practice. The expected value of the life of the device is an important parameter in its evaluation.

EXAMPLE 4.3   Let X be the random variable that denotes the life in hours of a certain electronic device. The probability density function is given by

$$f(x) = \begin{cases} \dfrac{20{,}000}{x^3}, & x > 100 \\[2mm] 0, & \text{elsewhere.} \end{cases}$$

Find the expected life of this type of device.

SOLUTION
Using Definition 4.1, we have

$$\mu = E(X) = \int_{100}^{\infty} x \frac{20{,}000}{x^3}\, dx$$

$$= \int_{100}^{\infty} \frac{20{,}000}{x^2}\, dx$$

$$= 200.$$

Therefore, we can expect this type of device to last, *on the average*, 200 hours.

Now let us consider a new random variable $g(X)$, which depends on X; that is, each value of $g(X)$ is determined by knowing the values of X. For instance, $g(X)$ might be X^2 or $3X - 1$, so that whenever X assumes the value 2, $g(X)$ assumes the value $g(2)$. In particular, if X is a discrete random variable with probability distribution $f(x)$, $x = -1, 0, 1, 2$, and $g(X) = X^2$, then

$$P[g(X) = 0] = P(X = 0) = f(0)$$

$$P[g(X) = 1] = P(X = -1) + P(X = 1) = f(-1) + f(1)$$

$$P[g(X) = 4] = P(X = 2) = f(2),$$

so that the probability distribution of $g(X)$ may be written

$g(x)$	0	1	4
$P[g(X) = g(x)]$	$f(0)$	$f(-1) + f(1)$	$f(2)$

By the definition of an expected value of a random variable, we obtain

$$\mu_{g(X)} = E[g(x)]$$
$$= 0f(0) + 1[f(-1) + f(1)] + 4f(2)$$
$$= (-1)^2 f(-1) + (0)^2 f(0) + (1)^2 f(1) + (2)^2 f(2)$$
$$= \sum_x g(x)f(x).$$

This result is generalized in Theorem 4.1 for both discrete and continuous random variables.

THEOREM 4.1   *Let X be a random variable with probability distribution f(x). The mean or expected value of the random variable g(X) is*

$$\mu_{g(X)} = E[g(X)] = \sum g(x)f(x)$$

if X is discrete, and

$$\mu_{g(X)} = E[g(X)] = \int_{-\infty}^{\infty} g(x)f(x)\ dx$$

if X is continuous. ∎

EXAMPLE 4.4   Suppose that the number of cars, X, that pass through a car wash between 4:00 P.M. and 5:00 P.M. on any sunny Friday has the following probability distribution:

x	4	5	6	7	8	9
$P(X = x)$	$\frac{1}{12}$	$\frac{1}{12}$	$\frac{1}{4}$	$\frac{1}{4}$	$\frac{1}{6}$	$\frac{1}{6}$

Let $g(X) = 2X - 1$ represent the amount of money in dollars, paid to the attendant by the manager. Find the attendant's expected earnings for this particular time period.

SOLUTION
By Theorem 4.1, the attendant can expect to receive

$$E[g(X)] = E(2X - 1) = \sum_{x=4}^{9} (2x - 1)f(x)$$

$$= (7)(\tfrac{1}{12}) + (9)(\tfrac{1}{12}) + (11)(\tfrac{1}{4}) + (13)(\tfrac{1}{4})$$
$$+ (15)(\tfrac{1}{6}) + (17)(\tfrac{1}{6})$$
$$= \$12.67.$$

EXAMPLE 4.5   Let X be a random variable with density function

$$f(x) = \begin{cases} \dfrac{x^2}{3}, & -1 < x < 2 \\ 0, & \text{elsewhere.} \end{cases}$$

Find the expected value of $g(X) = 4X + 3$.

SOLUTION
By Theorem 4.1, we have

$$E(4X + 3) = \int_{-1}^{2} \frac{(4x + 3)x^2}{3} \, dx$$

$$= \frac{1}{3} \int_{-1}^{2} (4x^3 + 3x^2) \, dx$$

$$= 8.$$

We shall now extend our concept of mathematical expectation to the case of two random variables X and Y with joint probability distribution $f(x, y)$.

DEFINITION 4.2   *Let X and Y be random variables with joint probability distribution $f(x, y)$. The mean or expected value of the random variable $g(X, Y)$ is*

$$\mu_{g(X, Y)} = E[g(X, Y)] = \sum_x \sum_y g(x, y) f(x, y)$$

if X and Y are discrete, and

$$\mu_{g(X, Y)} = E[g(X, Y)] = \int_{-\infty}^{\infty} \int_{-\infty}^{\infty} g(x, y) f(x, y) \, dx \, dy$$

if X and Y are continuous.   ■

Generalization of Definition 4.2 for the calculation of mathematical expectations of functions of several random variables is straightforward.

EXAMPLE 4.6   Let X and Y be the random variables with joint probability distribution given by Table 3.6 on page 70. Find the expected value of $g(X, Y) = XY$.

SOLUTION
By Definition 4.2, we write

$$E(XY) = \sum_{x=0}^{2} \sum_{y=0}^{2} xy f(x, y)$$

$$= (0)(0)f(0, 0) + (0)(1)f(0, 1) + (0)(2)f(0, 2)$$

$$+ (1)(0)f(1, 0) + (1)(1)f(1, 1)$$

$$+ (2)(0)f(2, 0)$$

$$= f(1, 1) = \tfrac{3}{14}.$$

EXAMPLE 4.7 Find $E\left(\dfrac{Y}{X}\right)$ for the density function

$$f(x, y) = \begin{cases} \dfrac{x(1 + 3y^2)}{4}, & 0 < x < 2,\ 0 < y < 1 \\ 0, & \text{elsewhere.} \end{cases}$$

SOLUTION
We have

$$E\left(\frac{Y}{X}\right) = \int_0^1 \int_0^2 \frac{y(1 + 3y^2)}{4}\, dx\, dy$$

$$= \int_0^1 \frac{y + y^3}{2}\, dy$$

$$= \frac{5}{8}.$$

Note that if $g(X, Y) = X$ in Definition 4.2, we have

$$E(X) = \begin{cases} \displaystyle\sum_x \sum_y xf(x, y) = \sum_x xg(x) & \text{(discrete case)} \\ \displaystyle\int_{-\infty}^{\infty} \int_{-\infty}^{\infty} xf(x, y)\, dx\, dy = \int_{-\infty}^{\infty} xg(x)\, dx & \text{(continuous case),} \end{cases}$$

where $g(x)$ is the marginal distribution of X. Therefore, in calculating $E(X)$ over a two-dimensional space, one may use either the joint probability distribution of X and Y or the marginal distribution of X.

Similarly, we define

$$E(Y) = \begin{cases} \displaystyle\sum_x \sum_y yf(x, y) = \sum_y yh(y) & \text{(discrete case)} \\ \displaystyle\int_{-\infty}^{\infty} \int_{-\infty}^{\infty} yf(x, y)\, dx\, dy = \int_{-\infty}^{\infty} yh(y)\, dy & \text{(continuous case),} \end{cases}$$

where $h(y)$ is the marginal distribution of the random variable Y.

Exercises

1. Assume that two variables (X, Y) are uniformly distributed on a circle with radius a. Then the joint probability density function is

$$f(x, y) = \begin{cases} \dfrac{1}{\pi a^2}, & x^2 + y^2 \le a^2 \\ 0, & \text{otherwise} \end{cases}$$

Find the expected value of X, μ_x.

2. The probability distribution of the discrete random variable X is

$$f(x) = \binom{3}{x}\left(\frac{1}{4}\right)^x\left(\frac{3}{4}\right)^{3-x}, \qquad x = 0, 1, 2, 3.$$

Find the mean of X.

3. Find the mean of the random variable T representing the total of the three coins in Exercise 25 on page 61.

4. A coin is biased so that a head is three times as likely to occur as a tail. Find the expected number of tails when this coin is tossed twice.

5. The probability distribution of X, the number of imperfections per 10 meters of a synthetic fabric in continuous rolls of uniform width, was given in Exercise 13 on page 61 as

x	0	1	2	3	4
$f(x)$	0.41	0.37	0.16	0.05	0.01

Find the average number of imperfections per 10 meters of this fabric.

6. An attendant at a car wash is paid according to the number of cars that pass through. Suppose the probabilities are 1/12, 1/12, 1/4, 1/4, 1/6, and 1/6, respectively, that the attendant receives $7, $9, $11, $13, $15, or $17 between 4:00 P.M. and 5:00 P.M. on any sunny Friday. Find the attendant's expected earnings for this particular period.

7. By investing in a particular stock, a person can make a profit in one year of $4000 with probability 0.3 or take a loss of $1000 with probability 0.7. What is this person's expected gain?

8. Suppose that an antique jewelry dealer is interested in purchasing a gold necklace for which the probabilities are 0.22, 0.36, 0.28, and 0.14, respectively, that she will be able to sell it for a profit of $250, sell it for a profit of $150, break even, or sell it for a loss of $150. What is her expected profit?

9. In a gambling game a woman is paid $3 if she draws a jack or a queen and $5 if she draws a king or an ace from an ordinary deck of 52 playing cards. If she draws any other card, she loses. How much should she pay to play if the game is fair?

10. Two tire-quality experts examine stacks of tires and give quality ratings to each tire on a 3-point scale. Let X denote the grade given by expert A and Y denote the grade given by B. The following table gives the joint distribution for X and Y.

			Y	
		1	2	3
X	1	0.1	0.05	0.02
	2	0.1	0.35	0.05
	3	0.03	0.1	0.2

Find μ_x and μ_y.

11. A private pilot wishes to insure his airplane for $50,000. The insurance company estimates that a total loss may occur with probability 0.002, a 50% loss with probability 0.01, and a 25% loss with probability 0.1. Ignoring all other partial losses, what premium should the insurance company charge each year to realize an average profit of $500?

12. If a dealer's profit, in units of $1000, on a new automobile can be looked upon as a random variable X having the density function

$$f(x) = \begin{cases} 2(1-x), & 0 < x < 1 \\ 0, & \text{elsewhere,} \end{cases}$$

find the average profit per automobile.

13. The density function of coded measurements of pitch diameter of threads of a fitting is given by

$$f(x) = \begin{cases} \dfrac{4}{\pi(1+x^2)}, & 0 < x < 1 \\ 0, & \text{elsewhere.} \end{cases}$$

Find the expected value of X.

14. What proportion of the people can be expected to respond to a certain mail-order solicitation if the proportion X has the density function

$$f(x) = \begin{cases} \dfrac{2(x+2)}{5}, & 0 < x < 1 \\ 0, & \text{elsewhere?} \end{cases}$$

15. The density function of the continuous random variable X, the total number of hours, in units of 100 hours, that a family runs a vacuum cleaner over a period of one year, was given in Exercise 7 on page 60 as

$$f(x) = \begin{cases} x, & 0 < x < 1 \\ 2-x, & 1 \le x < 2 \\ 0, & \text{elsewhere.} \end{cases}$$

Find the average number of hours per year that families run their vacuum cleaners.

16. Suppose that you are inspecting a lot of 1000 light bulbs, among which 20 are defectives. Choose two light bulbs randomly from the lot without replacement. Let

$$X_1 = \begin{cases} 1, & \text{if the first light bulb is defective} \\ 0, & \text{otherwise,} \end{cases}$$

$$X_2 = \begin{cases} 1, & \text{if the second light bulb is defective} \\ 0, & \text{otherwise.} \end{cases}$$

Find the probability that either light bulb chosen is defective. [HINT: Compute $P(X_1 + X_2 = 1)$.]

17. Let X be a random variable with the following probability distribution:

x	-3	6	9
$f(x)$	$\frac{1}{6}$	$\frac{1}{2}$	$\frac{1}{3}$

Find $\mu_{g(X)}$, where $g(X) = (2X + 1)^2$.

18. Find the expected value of the random variable $g(X) = X^2$, where X has the probability distribution of Exercise 2.

19. A large industrial firm purchases several new typewriters at the end of each year, the exact number depending on the frequency of repairs in the previous year. Suppose that the number of typewriters, X, that are purchased each year has the following probability distribution:

x	0	1	2	3
$f(x)$	$\frac{1}{10}$	$\frac{3}{10}$	$\frac{2}{5}$	$\frac{1}{5}$

If the cost of the desired model will remain fixed at $1200 throughout this year and a discount of $50X^2$ dollars is credited toward any purchase, how much can this firm expect to spend on new typewriters at the end of this year?

20. A continuous random variable X has the density function

$$f(x) = \begin{cases} e^{-x}, & x > 0 \\ 0 & \text{elsewhere.} \end{cases}$$

Find the expected value of $g(X) = e^{2X/3}$.

21. What is the dealer's average profit per automobile if the profit on each automobile is given by $g(X) = X^2$, where X is a random variable having the density function of Exercise 12?

22. The hospital period, in days, for patients following treatment for a certain type of kidney disorder is a random variable $Y = X + 4$, where X has the density function

$$f(x) = \begin{cases} \dfrac{32}{(x + 4)^3}, & x > 0 \\ 0, & \text{elsewhere.} \end{cases}$$

Find the average number of days that a person is hospitalized following treatment for this disorder.

23. Suppose that X and Y have the following joint probability function:

$f(x, y)$		x	
		2	4
	1	0.10	0.15
y	3	0.20	0.30
	5	0.10	0.15

(a) Find the expected value of $g(X, Y) = XY^2$.
(b) Find μ_X and μ_Y.

24. Referring to the random variables whose joint probability distribution is given in Exercise 3 on page 78,
(a) find $E(X^2Y - 2XY)$;
(b) find $\mu_X - \mu_Y$.

25. Referring to the random variables whose joint probability distribution is given in Exercise 17 on page 80, find the mean for the total number of jacks and kings when 3 cards are drawn without replacement from the 12 face cards of an ordinary deck of 52 playing cards.

26. Let X and Y be random variables with joint density function

$$f(x, y) = \begin{cases} 4xy, & 0 < x < 1, 0 < y < 1 \\ 0, & \text{elsewhere.} \end{cases}$$

Find the expected value of $Z = \sqrt{X^2 + Y^2}$.

4.2 Variance and Covariance

The mean or expected value of a random variable X is of special importance in statistics because it describes where the probability distribution is centered. By itself, however, the mean does not give adequate distribution of the shape of the distribution. We need to characterize the variability in the distribution. In Figure 4.1 we have the histograms of two discrete probability distributions with the same mean

$\mu = 2$ that differ considerably in the variability or dispersion of their observations about the mean.

The most important measure of variability of a random variable X is obtained by letting $g(X) = (X - \mu)^2$ in Theorem 4.1. Because of its importance in statistics, it is referred to as the **variance of the random variable** X or the **variance of the probability distribution of** X and is denoted by $\text{Var}(X)$ or the symbol σ_X^2, or simply by σ^2 when it is clear to which random variable we refer.

DEFINITION 4.3   *Let X be a random variable with probability distribution $f(x)$ and mean μ. The* **variance** *of X is*

$$\sigma^2 = E[(X - \mu)^2] = \sum_x (x - \mu)^2 f(x)$$

if X is discrete, and

$$\sigma^2 = E[(X - \mu)^2] = \int_{-\infty}^{\infty} (x - \mu)^2 f(x) \, dx$$

if X is continuous. The positive square root of the variance, σ, is called the **standard deviation** *of X.* ■

The quantity $x - \mu$ in Definition 4.3 is called the **deviation of an observation from its mean**. Since these deviations are being squared and then averaged, σ^2 will be much smaller for a set of x values that are close to μ than it would be for a set of values that vary considerably from μ.

EXAMPLE 4.8   Let the random variable X represent the number of automobiles that are used for official business purposes on any given workday. The probability distribution for company A [Figure 4.1(a)] is given by

x	1	2	3
$f(x)$	0.3	0.4	0.3

and for company B [Figure 4.1(b)] by

x	0	1	2	3	4
$f(x)$	0.2	0.1	0.3	0.3	0.1

Show that the variance of the probability distribution for company B is greater than that of company A.

SOLUTION
For company A, we find that

$$\mu = E(X) = (1)(0.3) + (2)(0.4) + (3)(0.3) = 2.0$$

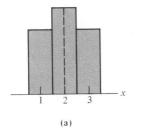 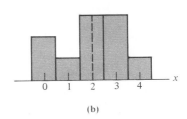

(a)   (b)

FIGURE 4.1 Distributions with equal means and different dispersions.

and then

$$\sigma^2 = \sum_{x=1}^{3} (x - 2)^2 f(x)$$

$$= (1 - 2)^2(0.3) + (2 - 2)^2(0.4) + (3 - 2)^2(0.3)$$

$$= 0.6.$$

For company B, we have

$$\mu = E(X) = (0)(0.2) + (1)(0.1) + (2)(0.3) + (3)(0.3) + (4)(0.1)$$

$$= 2.0$$

and then

$$\sigma^2 = \sum_{x=0}^{4} (x - 2)^2 f(x)$$

$$= (0 - 2)^2(0.2) + (1 - 2)^2(0.1) + (2 - 2)^2(0.3) + (3 - 2)^2(0.3)$$

$$+ (4 - 2)^2(0.1)$$

$$= 1.6.$$

Clearly, the variance of the number of automobiles that are used for official business purposes is greater for company B than for company A.

An alternative and preferred formula for finding σ^2, which often simplifies the calculations, is given in the following theorem.

THEOREM 4.2   *The variance of a random variable X is given by*

$$\sigma^2 = E(X^2) - \mu^2.$$ ■

PROOF. For the discrete case we can write

$$\sigma^2 = \sum_{x} (x - \mu)^2 f(x) = \sum_{x} (x^2 - 2\mu x + \mu^2) f(x)$$

$$= \sum_{x} x^2 f(x) - 2\mu \sum_{x} x f(x) + \mu^2 \sum_{x} f(x).$$

Since $\mu = \sum_x xf(x)$ by definition, and $\sum_x f(x) = 1$ for any discrete probability distribution, it follows that

$$\sigma^2 = \sum_x x^2 f(x) - \mu^2$$
$$= E(X^2) - \mu^2$$

For the continuous case the proof is step by step the same, with summations replaced by integrations.

EXAMPLE 4.9   Let the random variable X represent the number of defective parts for a machine when 3 parts are sampled from a production line and tested. The following is the probability distribution of X.

x	0	1	2	3
$f(x)$	0.51	0.38	0.10	0.01

Using Theorem 4.2, calculate σ^2.

SOLUTION
First, we compute

$$\mu = (0)(0.51) + (1)(0.38) + (2)(0.10) + (3)(0.01)$$
$$= 0.61.$$

Now

$$E(X^2) = (0)(0.51) + (1)(0.38) + (4)(0.10) + (9)(0.01)$$
$$= 0.87.$$

Therefore,

$$\sigma^2 = 0.87 - (0.61)^2 = 0.4979.$$

EXAMPLE 4.10   The weekly demand for Pepsi, in thousands of liters, from a local chain of efficiency stores, is a continuous random variable X having the probability density

$$f(x) = \begin{cases} 2(x - 1), & 1 < x < 2 \\ 0, & \text{elsewhere.} \end{cases}$$

Find the mean and variance of X.

SOLUTION

$$\mu = E(X) = 2 \int_1^2 x(x - 1) \, dx = \tfrac{5}{3}$$

and

$$E(X^2) = 2 \int_1^2 x^2(x - 1) \, dx = \tfrac{17}{6}.$$

Therefore,

$$\sigma^2 = \tfrac{17}{6} - (\tfrac{5}{3})^2 = \tfrac{1}{18}.$$

At this point the variance or standard deviation only has meaning when we compare two or more distributions that have the same units of measurement. Therefore, we could compare the variances of the distributions of contents, measured in liters, for two companies bottling orange juice, and the larger value would indicate the company whose product is more variable or less uniform. It would not be meaningful to compare the variance of a distribution of heights to the variance of a distribution of aptitude scores. In Section 4.4 we show how the standard deviation can be used to describe a single distribution of observations.

We shall now extend our concept of the variance of a random variable X to also include random variables related to X. For the random variable $g(X)$, the variance will be denoted by $\sigma^2_{g(X)}$ and is calculated by means of the following theorem.

THEOREM 4.3   *Let X be a random variable with probability distribution $f(x)$. The variance of the random variable $g(X)$ is*

$$\sigma^2_{g(X)} = E\{[g(X) - \mu_{g(X)}]^2\} = \sum_x [g(x) - \mu_{g(x)}]^2 f(x)$$

if X is discrete, and

$$\sigma^2_{g(X)} = E\{[g(X) - \mu_{g(X)}]^2\} = \int_{-\infty}^{\infty} [g(x) - \mu_{g(x)}]^2 f(x)\ dx$$

if X is continuous.   ■

PROOF. Since $g(X)$ is itself a random variable with mean $\mu_{g(X)}$ as defined in Theorem 4.1, it follows from Definition 4.3 that

$$\sigma^2_{g(X)} = E\{[g(X) - \mu_{g(X)}]^2\}.$$

Now, applying Theorem 4.1 again to the random variable $[g(X) - \mu_{g(X)}]^2$, the proof is complete.

EXAMPLE 4.11   Calculate the variance of $g(X) = 2X + 3$, where X is a random variable with probability distribution

x	0	1	2	3
$f(x)$	$\frac{1}{4}$	$\frac{1}{8}$	$\frac{1}{2}$	$\frac{1}{8}$

SOLUTION
First let us find the mean of the random variable $2X + 3$. According to Theorem 4.1,

$$\mu_{2X+3} = E(2X + 3) = \sum_{x=0}^{3} (2x + 3)f(x) = 6.$$

Now, using Theorem 4.3, we have

$$\sigma^2_{2X+3} = E\{[(2X + 3) - \mu_{2X+3}]^2\}$$

$$= E\{[2X + 3 - 6]^2\}$$

$$= E(4X^2 - 12X + 9)$$

$$= \sum_{x=0}^{3} (4x^2 - 12x + 9)f(x)$$

$$= 4.$$

EXAMPLE 4.12   Let X be a random variable having the density function given in Example 4.5 on page 88. Find the variance of the random variable $g(X) = 4X + 3$.

SOLUTION
In Example 4.5 we found $\mu_{4x+3} = 8$. Now using Theorem 4.3,

$$\sigma^2_{4X+3} = E\{[(4X + 3) - 8]^2\}$$

$$= E[(4X - 5)^2]$$

$$= \int_{-1}^{2} (4x - 5)^2 \frac{x^2}{3} \, dx$$

$$= \frac{1}{3}\int_{-1}^{2} (16x^4 - 40x^3 + 25x^2) \, dx$$

$$= \frac{51}{5}.$$

If $g(X, Y) = (X - \mu_X)(Y - \mu_Y)$, where $\mu_X = E(X)$ and $\mu_Y = E(Y)$, Definition 4.2 yields an expected value called the **covariance** of X and Y, which we denote by σ_{XY} or $\text{cov}(X, Y)$.

DEFINITION 4.4   *Let X and Y be random variables with joint probability distribution $f(x, y)$. The* **covariance** *of X and Y is*

$$\sigma_{XY} = E[(X - \mu_X)(Y - \mu_Y)] = \sum_x \sum_y (x - \mu_X)(y - \mu_Y)f(x, y)$$

if X and Y are discrete, and

$$\sigma_{XY} = E[(X - \mu_X)(Y - \mu_Y)] = \int_{-\infty}^{\infty} \int_{-\infty}^{\infty} (x - \mu_X)(y - \mu_Y)f(x, y) \, dx \, dy$$

if X and Y are continuous.   ∎

The covariance between two random variables is a measurement of the nature of the association between the two. If large values of X often result in large values of Y or small values of X result in small values of Y, positive $X - \mu_X$ will often result in positive $Y - \mu_Y$ and negative $X - \mu_X$ will often result in negative $Y - \mu_Y$. Thus the product $(X - \mu_X)(Y - \mu_Y)$ will tend to be positive. On the other hand, if large X values often result in small Y values, the product $(X - \mu_X)(Y - \mu_Y)$ will tend to be negative. Thus the *sign* of the covariance indicates whether the relationship between two dependent random variables is positive or negative. When X and Y are statistically independent, it can be shown that the covariance is zero (see Theorem 4.10, Corollary 1). The converse, however, is not generally true. Two variables may have zero covariance and still not be statistically independent.

The alternative and preferred formula for σ_{XY} is given in the following theorem.

THEOREM 4.4   *The covariance of two random variables X and Y with means μ_X and μ_Y, respectively, is given by*

$$\sigma_{XY} = E(XY) - \mu_X\mu_Y.$$ ∎

PROOF. For the discrete case we can write

$$\sigma_{XY} = \sum_x \sum_y (x - \mu_X)(y - \mu_Y)f(x, y)$$

$$= \sum_x \sum_y (xy - \mu_X y - \mu_Y x + \mu_X\mu_Y)f(x, y)$$

$$= \sum_x \sum_y xyf(x, y) - \mu_X \sum_x \sum_y yf(x, y)$$

$$- \mu_Y \sum_x \sum_y xf(x, y) + \mu_X\mu_Y \sum_x \sum_y f(x, y).$$

Since $\mu_X = \sum_x \sum_y xf(x, y)$ and $\mu_Y = \sum_x \sum_y yf(x, y)$ by definition, and in addition

$\sum_x \sum_y f(x, y) = 1$ for any joint discrete distribution, it follows that

$$\sigma_{XY} = E(XY) - \mu_X\mu_Y - \mu_Y\mu_X + \mu_X\mu_Y$$

$$= E(XY) - \mu_X\mu_Y.$$

For the continuous case the proof is identical with summations replaced by integrals.

EXAMPLE 4.13   The number of blue refills X and the number of red refills Y, when 2 refills for a ballpoint pen are selected at random from a certain box, was described in Example 3.8 on page 69 by the following joint probability distribution:

		x		
$f(x, y)$	0	1	2	$h(y)$
0	$\frac{3}{28}$	$\frac{9}{28}$	$\frac{3}{28}$	$\frac{15}{28}$
y 1	$\frac{3}{14}$	$\frac{3}{14}$		$\frac{3}{7}$
2	$\frac{1}{28}$			$\frac{1}{28}$
$g(x)$	$\frac{5}{14}$	$\frac{15}{28}$	$\frac{3}{28}$	1

Find the covariance of X and Y.

SOLUTION

From Example 4.6, we see that $E(XY) = 3/14$. Now

$$\mu_X = E(X) = \sum_{x=0}^{2} \sum_{y=0}^{2} xf(x, y) = \sum_{x=0}^{2} xg(x)$$

$$= (0)(\tfrac{5}{14}) + (1)(\tfrac{15}{28}) + (2)(\tfrac{3}{28})$$

$$= \tfrac{3}{4}$$

and

$$\mu_Y = E(Y) = \sum_{x=0}^{2} \sum_{y=0}^{2} yf(x, y) = \sum_{y=0}^{2} yh(y)$$

$$= (0)(\tfrac{15}{28}) + (1)(\tfrac{3}{7}) + (2)(\tfrac{1}{28})$$

$$= \tfrac{1}{2}.$$

Therefore,

$$\sigma_{XY} = E(XY) - \mu_X \mu_Y$$

$$= \tfrac{3}{14} - (\tfrac{3}{4})(\tfrac{1}{2})$$

$$= -\tfrac{9}{56}.$$

EXAMPLE 4.14   The fraction X of male runners and the fraction Y of female runners who complete marathon races is described by the joint density function

$$f(x, y) = \begin{cases} 8xy, & 0 \le x \le 1,\ 0 \le y \le x \\ 0, & \text{elsewhere.} \end{cases}$$

Find the covariance of X and Y.

SOLUTION

We first must compute the marginal density functions. They are given by

$$g(x, y) = \begin{cases} 4x^3, & 0 \le x \le 1 \\ 0, & \text{elsewhere} \end{cases}$$

and

$$h(y) = \begin{cases} 4y(1 - y^2), & 0 \le y \le 1 \\ 0, & \text{elsewhere.} \end{cases}$$

From the marginal density functions given above, we compute

$$\mu_X = E(X) = \int_0^1 4x^4\, dx = \tfrac{4}{5}$$

$$\mu_Y = E(Y) = \int_0^1 4y^2(1 - y^2)\, dy = \tfrac{8}{15}.$$

From the joint density functions given, we have

$$E(XY) = \int_0^1 \int_y^1 8x^2y^2 \, dx \, dy = \tfrac{4}{9}.$$

Then

$$\sigma_{XY} = E(XY) - \mu_X\mu_Y$$

$$= \tfrac{4}{9} - (\tfrac{4}{5})(\tfrac{8}{15})$$

$$= \tfrac{4}{225}.$$

Although the covariance between two random variables does give information regarding the nature of the relationship, the magnitude of σ_{XY} *does not indicate anything regarding the strength of the relationship*, since σ_{XY} is not scale free. Its magnitude will depend on the units measured for both X and Y. There is a scale-free version of the covariance called the **correlation coefficient** that is used widely in statistics. We shall defer discussion of the correlation coefficient until Chapter 11, where we deal with correlation in conjunction with linear regression.

Exercises

1. Use Definition 4.3 on page 92 to find the variance of the random variable X of Exercise 7 on page 90.

2. Let X be a random variable with the following probability distribution:

x	-2	3	5
$f(x)$	0.3	0.2	0.5

   Find the standard deviation of X.

3. The random variable X, representing the number of errors per 100 lines of software code, has the following probability distribution:

x	2	3	4	5	6
$f(x)$	0.01	0.25	0.4	0.3	0.04

   Using Theorem 4.2, find the variance of X.

4. Suppose that the probabilities are 0.4, 0.3, 0.2, and 0.1, respectively, that 0, 1, 2, or 3 power failures will hit a certain subdivision in any given year. Find the mean and variance of the random variable X representing the number of power failures hitting this subdivision.

5. The dealer's profit, in units of $1000, on a new automobile is a random variable X having the density function given in Exercise 12 on page 90. Find the variance of X.

6. The proportion of people who respond to a certain mail-order solicitation is a random variable X having the density function given in Exercise 14 on page 90. Find the variance of X.

7. The total number of hours, in units of 100 hours, that a family runs a vacuum cleaner over a period of one year is a random variable X having the density function given in Exercise 15 on page 90. Find the variance of X.

8. Referring to Exercise 14 on page 90, find $\sigma^2_{g(X)}$ for the function $g(X) = 3X^2 + 4$.

9. Find the standard deviation of the random variable $g(X) = (2X + 1)^2$ in Exercise 17 on page 91.

10. Using the results of Exercise 21 on page 91, find the variance of $g(X) = X^2$, where X is a random variable having the density function given in Exercise 12 on page 90.

11. The length of time, in minutes, for an airplane to wait for clearance to take off at a certain airport is a ran-

dom variable $Y = 3X - 2$, where X has the density function

$$f(x) = \begin{cases} \frac{1}{4}e^{-x/4}, & x > 0 \\ 0, & \text{elsewhere.} \end{cases}$$

Find the mean and variance of the random variable Y.

12. Find the covariance of the random variables X and Y of Exercise 3 on page 78.

13. Find the covariance of the random variables X and Y of Exercise 13 on page 79.

14. Find the covariance of the random variables X and Y of Exercise 8 on page 79.

15. Referring to the random variables whose joint density function is given in Exercise 4 on page 78, find the covariance of X and Y.

4.3 Means and Variances of Linear Combinations of Random Variables

We shall now develop some useful properties that will simplify the calculations of means and variances of random variables that appear in later chapters. These properties will permit us to deal with expectations in terms of other parameters that are either known or are easily computed. All the results that we present here are valid for both discrete and continuous random variables. Proofs are given only for the continuous case. We begin with a theorem and two corollaries that should be intuitively very reasonable to the reader.

THEOREM 4.5   *If a and b are constant, then*

$$E(aX + b) = aE(X) + b.$$

PROOF. By the definition of an expected value,

$$E(aX + b) = \int_{-\infty}^{\infty} (ax + b)f(x)\, dx$$

$$= a \int_{-\infty}^{\infty} xf(x)\, dx + b \int_{-\infty}^{\infty} f(x)\, dx.$$

The first integral on the right is $E(X)$ and the second integral equals 1. Therefore, we have

$$E(aX + b) = aE(X) + b.$$

COROLLARY 1   *Setting $a = 0$, we see that $E(b) = b$.*

COROLLARY 2   *Setting $b = 0$, we see that $E(aX) = aE(X)$.*

EXAMPLE 4.15   Applying Theorem 4.5 to the discrete random variable $g(X) = 2X - 1$, rework Example 4.4.

SOLUTION
According to Theorem 4.5, we can write

$$E(2X - 1) = 2E(X) - 1.$$

Now

$$\mu = E(X) = \sum_{x=4}^{9} xf(x)$$

$$= (4)(\tfrac{1}{12}) + (5)(\tfrac{1}{12}) + (6)(\tfrac{1}{4}) + (7)(\tfrac{1}{4}) + (8)(\tfrac{1}{6}) + (9)(\tfrac{1}{6})$$

$$= \tfrac{41}{6}$$

Therefore,

$$\mu_{2X-1} = (2)(\tfrac{41}{6}) - 1 = \$12.67,$$

as before.

EXAMPLE 4.16   Applying Theorem 4.5 to the continuous random variable $g(X) = 4X + 3$, rework Example 4.5.

SOLUTION
In Example 4.5 we may use Theorem 4.5 to write

$$E(4X + 3) = 4E(X) + 3.$$

Now

$$E(X) = \int_{-1}^{2} x\left(\frac{x^2}{3}\right) dx = \int_{-1}^{2} \frac{x^3}{3} \, dx = \frac{5}{4}.$$

Therefore,

$$E(4X + 3) = (4)\left(\frac{5}{4}\right) + 3 = 8,$$

as before.

THEOREM 4.6   *The expected value of the sum or difference of two or more functions of a random variable X is the sum or difference of the expected values of the functions. That is,*

$$E[g(X) \pm h(X)] = E[g(X)] \pm E[h(X)]. \qquad ∎$$

PROOF. By definition,

$$E[g(X) \pm h(X)] = \int_{-\infty}^{\infty} [g(x) \pm h(x)]f(x) \, dx$$

$$= \int_{-\infty}^{\infty} g(x)f(x) \, dx \pm \int_{-\infty}^{\infty} h(x)f(x) \, dx$$

$$= E[g(X)] \pm E[h(X)].$$

EXAMPLE 4.17   Let X be a random variable with probability distribution as follows:

x	0	1	2	3
$f(x)$	$\tfrac{1}{3}$	$\tfrac{1}{2}$	0	$\tfrac{1}{6}$

Find the expected value of $Y = (X - 1)^2$.

SOLUTION

Applying Theorem 4.6 to the function $Y = (X - 1)^2$, we can write

$$E[(X - 1)^2] = E(X^2 - 2X + 1) = E(X^2) - 2E(X) + E(1).$$

From Corollary 1 of Theorem 4.5, $E(1) = 1$, and by direct computation

$$E(X) = (0)(\tfrac{1}{3}) + (1)(\tfrac{1}{2}) + (2)(0) + (3)(\tfrac{1}{6}) = 1$$

and

$$E(X^2) = (0)(\tfrac{1}{3}) + (1)(\tfrac{1}{2}) + (4)(0) + (9)(\tfrac{1}{6}) = 2.$$

Hence

$$E[(X - 1)^2] = 2 - (2)(1) + 1 = 1.$$

EXAMPLE 4.18 The weekly demand for a certain drink, in thousands of liters, from a local chain of efficiency stores is a continuous random variable $g(X) = X^2 + X - 2$, where X has the density function

$$f(x) = \begin{cases} 2(x - 1), & 1 < x < 2 \\ 0, & \text{elsewhere.} \end{cases}$$

Find the expected value of the weekly demand of the drink.

SOLUTION

By Theorem 4.6, we write

$$E(X^2 + X - 2) = E(X^2) + E(X) - E(2).$$

From Corollary 1 of Theorem 4.5, $E(2) = 2$, and by direct integration

$$E(X) = \int_1^2 2x(x - 1) \, dx = 2 \int_1^2 (x^2 - x) \, dx = \tfrac{5}{3}$$

and

$$E(X^2) = \int_1^2 2x^2(x - 1) \, dx = 2 \int_1^2 (x^3 - x^2) \, dx = \tfrac{17}{6}.$$

Now

$$E(X^2 + X - 2) = \tfrac{17}{6} + \tfrac{5}{3} - 2 = \tfrac{5}{2},$$

so that the average weekly demand for the drink from this chain of efficiency stores is 2500 liters.

Suppose that we have two random variables X and Y with joint probability distribution $f(x, y)$. Two additional properties that will be very useful in succeeding chapters involve the expected values of the sum, difference, and product of these two random variables. First, however, let us prove a theorem on the expected value of the sum or difference of functions of the given variables. This, of course, is merely an extension of Theorem 4.6.

THEOREM 4.7

The expected value of the sum or difference of two or more functions of the random variables X and Y is the sum or difference of the expected values of the functions. That is,

$$E[g(X, Y) \pm h(X, Y)] = E[g(X, Y)] \pm E[h(X, Y)].$$ ∎

PROOF. By Definition 4.2,

$$E[g(X, Y) \pm h(X, Y)] = \int_{-\infty}^{\infty} \int_{-\infty}^{\infty} [g(x, y) \pm h(x, y)] f(x, y) \, dx \, dy$$

$$= \int_{-\infty}^{\infty} \int_{-\infty}^{\infty} g(x, y) f(x, y) \, dx \, dy$$

$$\pm \int_{-\infty}^{\infty} \int_{-\infty}^{\infty} h(x, y) f(x, y) \, dx \, dy$$

$$= E[g(X, Y)] \pm E[h(X, Y)].$$

COROLLARY 1

Setting $g(X, Y) = g(X)$ and $h(X, Y) = h(Y)$, we see that

$$E[g(X) \pm h(Y)] = E[g(X)] \pm E[h(Y)].$$ ∎

COROLLARY 2

Setting $g(X, Y) = X$ and $h(X, Y) = Y$, we see that

$$E(X \pm Y) = E(X) \pm E(Y).$$ ∎

If X represents the daily production of some item from machine A and Y the daily production of the same kind of item from machine B, then $X + Y$ represents the total number of items produced daily from both machines. The second corollary of Theorem 4.7 states that the average daily production for both machines is equal to the sum of the average daily production of each machine.

THEOREM 4.8

Let X and Y be two independent random variables. Then

$$E(XY) = E(X)E(Y).$$ ∎

PROOF. By Definition 4.2

$$E(XY) = \int_{-\infty}^{\infty} \int_{-\infty}^{\infty} xy f(x, y) \, dx \, dy.$$

Since X and Y are independent, we may write

$$f(x, y) = g(x)h(y),$$

where $g(x)$ and $h(y)$ are the marginal distributions of X and Y, respectively. Hence

$$E(XY) = \int_{-\infty}^{\infty} \int_{-\infty}^{\infty} xyg(x)h(y) \, dx \, dy$$

$$= \int_{-\infty}^{\infty} xg(x) \, dx \int_{-\infty}^{\infty} yh(y) \, dy$$

$$= E(X)E(Y).$$

Theorem 4.8 can be illustrated for discrete variables by tossing a green die and a red die. Let the random variable X represent the outcome on the green die and the random variable Y represent the outcome on the red die. Then XY represents the product of the numbers that occur on the pair of dice. In the long run, the average of the products of the numbers is equal to the product of the average number that occurs on the green die and the average number that occurs on the red die.

EXAMPLE 4.19   In producing gallium–arsenide microchips, it is known that the ratio between gallium and arsenide is independent of producing a high percentage of workable wafers, which are the main building blocks of microchips. Let X denote the ratio of gallium to arsenide and Y denote the percentage of workable microwafers retrieved during a 1-hour period. X and Y are independent random variables with the joint density being known as

$$f(x, y) = \begin{cases} \dfrac{x(1 + 3y^2)}{4}, & 0 < x < 2,\ 0 < y < 1 \\ 0, & \text{elsewhere.} \end{cases}$$

Illustrate that $E(XY) = E(X)E(Y)$, as Theorem 4.8 suggests.

SOLUTION
By definition,

$$E(XY) = \int_0^1 \int_0^2 xyf(x, y) \, dx \, dy = \int_0^1 \int_0^2 \frac{x^2y(1 + 3y^2)}{4} \, dx \, dy$$

$$= \int_0^1 \frac{x^3y(1 + 3y^2)}{12} \Big|_{x=0}^{x=2} \, dy = \int_0^1 \frac{2y(1 + 3y^2)}{3} \, dy = \frac{5}{6}$$

$$E(X) = \int_0^1 \int_0^2 xf(x, y) \, dx \, dy = \int_0^1 \int_0^2 \frac{x^2(1 + 3y^2)}{4} \, dx \, dy$$

$$= \int_0^1 \frac{x^3(1 + 3y^2)}{12} \Big|_{x=0}^{x=2} \, dy = \int_0^1 \frac{2(1 + 3y^2)}{3} \, dy = \frac{4}{3}$$

$$E(Y) = \int_0^1 \int_0^2 yf(x, y) \, dx \, dy = \int_0^1 \int_0^2 \frac{xy(1 + 3y^2)}{4} \, dx \, dy$$

$$= \int_0^1 \frac{x^2y(1 + 3y^2)}{8} \Big|_{x=0}^{x=2} \, dy = \int_0^1 \frac{y(1 + 3y^2)}{2} \, dy = \frac{5}{8}.$$

Hence $E(X)E(Y) = (\frac{4}{3})(\frac{5}{8}) = \frac{5}{6} = E(XY)$.

We conclude this section by proving two theorems that are useful in calculating variances or standard deviations.

THEOREM 4.9   *If a and b are constants, then*

$$\sigma^2_{aX+b} = a^2 \sigma^2_X = a^2 \sigma^2.$$ ■

PROOF.   By definition

$$\sigma^2_{aX+b} = E\{[(aX + b) - \mu_{aX+b}]^2\}.$$

Now

$$\mu_{aX+b} = E(aX + b) = a\mu + b$$

by Theorem 4.5. Therefore,

$$\sigma^2_{aX+b} = E[(aX + b - a\mu - b)^2]$$
$$= a^2 E[(X - \mu)^2]$$
$$= a^2 \sigma^2$$

COROLLARY 1   *Setting a = 1, we see that*

$$\sigma^2_{X+b} = \sigma^2_X = \sigma^2.$$ ■

COROLLARY 2   *Setting b = 0, we see that*

$$\sigma^2_{aX} = a^2 \sigma^2_X = a^2 \sigma^2.$$ ■

Corollary 1 states that the variance is unchanged if a constant is added to or subtracted from a random variable. The addition or subtraction of a constant simply shifts the values of X to the right or to the left but does not change their variability. However, if a random variable is multiplied or divided by a constant, then Corollary 2 states that the variance is multiplied or divided by the square of the constant.

THEOREM 4.10   *If X and Y are random variables with joint probability distribution f(x, y), then*

$$\sigma^2_{aX+bY} = a^2 \sigma^2_X + b^2 \sigma^2_Y + 2ab\sigma_{XY}.$$ ■

PROOF.   By definition

$$\sigma^2_{aX+bY} = E\{[(aX + bY) - \mu_{aX+bY}]^2\}.$$

Now

$$\mu_{aX+bY} = E(aX + bY) = aE(X) + bE(Y) = a\mu_X + b\mu_Y,$$

by using Corollary 2 of Theorem 4.7 followed by Corollary 2 of Theorem 4.5. Therefore,

$$\sigma^2_{aX+bY} = E\{[(aX + bY) - (a\mu_X + b\mu_Y)]^2\}$$

$$= E\{[a(X - \mu_X) + b(Y - \mu_Y)]^2\}$$

$$= a^2E[(X - \mu_X)^2] + b^2E[(Y - \mu_Y)^2] + 2abE[(X - \mu_X)(Y - \mu_Y)]$$

$$= a^2\sigma^2_X + b^2\sigma^2_Y + 2ab\sigma_{XY}.$$

COROLLARY 1   *If X and Y are independent random variables, then*

$$\sigma^2_{aX+bY} = a^2\sigma^2_X + b^2\sigma^2_Y. \qquad \blacksquare$$

The result given in Corollary 1 is obtained from Theorem 4.10 by proving the covariance of the independent variables X and Y to be zero. Hence, from Theorem 4.4,

$$\sigma_{XY} = E(XY) - \mu_X\mu_Y$$

$$= 0,$$

since $E(XY) = E(X)E(Y)$ for independent variables.

COROLLARY 2   *If X and Y are independent random variables, then*

$$\sigma^2_{aX-bY} = a^2\sigma^2_X + b^2\sigma^2_Y. \qquad \blacksquare$$

Corollary 2 follows by replacing b by $-b$ in Corollary 1. Generalizing to a linear combination of n independent random variables, we write

COROLLARY 3   *If $X_1, X_2, \ldots, X_n$ are independent random variables, then*

$$\sigma^2_{a_1X_1+a_2X_2+\cdots+a_nX_n} = a_1^2\sigma^2_{X_1} + a_2^2\sigma^2_{X_2} + \cdots + a_n^2\sigma^2_{X_n}. \qquad \blacksquare$$

EXAMPLE 4.20   If X and Y are random variables with variances $\sigma^2_X = 2$, $\sigma^2_Y = 4$, and covariance $\sigma_{XY} = -2$, find the variance of the random variable $Z = 3X - 4Y + 8$.

SOLUTION

$$\sigma^2_Z = \sigma^2_{3X-4Y+8}$$

$$= \sigma^2_{3X-4Y} \qquad \text{(by Theorem 4.9)}$$

$$= 9\sigma^2_X + 16\sigma^2_Y - 24\sigma_{XY} \qquad \text{(by Theorem 4.10)}$$

$$= (9)(2) + (16)(4) - (24)(-2)$$

$$= 130.$$

EXAMPLE 4.21   Let X and Y denote the amount of two different types of impurities in a batch of a certain chemical product. Suppose that X and Y are independent random variables with variances $\sigma^2_x = 2$ and $\sigma^2_y = 3$. Find the variance of the random variable $Z = 3X - 2Y + 5$.

SOLUTION

$$\sigma_Z^2 = \sigma_{3x-2y+5}^2$$
$$= \sigma_{3x-2y}^2 \qquad \text{(by Theorem 4.9, Corollary 1)}$$
$$= 9\sigma_x^2 + 4\sigma_y^2 \qquad \text{(by Theorem 4.10, Corollary 2)}$$
$$= (9)(2) + (4)(3)$$
$$= 30.$$

4.4 Chebyshev's Theorem

In Section 4.2 we stated that the variance of a random variable tells us something about the variability of the observations about the mean. If a random variable has a small variance or standard deviation, we would expect most of the values to be grouped around the mean. Therefore, the probability that a random variable assumes a value within a certain interval about the mean is greater than for a similar random variable with a larger standard deviation. If we think of probability in terms of area, we would expect a continuous distribution with a small standard deviation to have most of its area close to μ, as in Figure 4.2(a). However, a large value of σ indicates a greater variability, and therefore we should expect the area to be more spread out, as in Figure 4.2(b).

We can argue the same way for a discrete distribution. The area in the probability histogram in Figure 4.3(b) is spread out much more than that of Figure 4.3(a) indicating a more variable distribution of measurements or outcomes.

The Russian mathematician P. L. Chebyshev (1821–1894) discovered that the fraction of the area between any two values symmetric about the mean is related to the standard deviation. Since the area under a probability distribution curve or in a probability histogram adds to 1, the area between any two numbers is the probability of the random variable assuming a value between these numbers.

The following theorem, due to Chebyshev, gives a conservative estimate of the probability that a random variable assumes a value within k standard deviations of its mean for any real number k. We shall give the proof only for the continuous case, leaving the discrete case as an exercise.

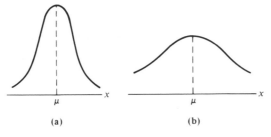

(a)   (b)

FIGURE 4.2 Variability of continuous observations about the mean.

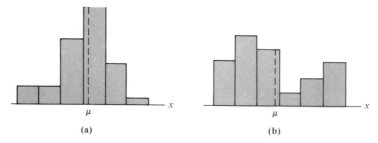

FIGURE 4.3   Variability of discrete observations about the mean.

THEOREM 4.11   **(Chebyshev's Theorem)**   *The probability that any random variable X will assume a value within k standard deviations of the mean is* **at least** $1 - 1/k^2$. *That is,*

$$P(\mu - k\sigma < X < \mu + k\sigma) \geq 1 - \frac{1}{k^2}.$$ ■

PROOF.   By our previous definition of the variance of X we can write

$$\sigma^2 = E[(X - \mu)^2]$$

$$= \int_{-\infty}^{\infty} (x - \mu)^2 f(x)\ dx$$

$$= \int_{-\infty}^{\mu - k\sigma} (x - \mu)^2 f(x)\ dx + \int_{\mu - k\sigma}^{\mu + k\sigma} (x - \mu)^2 f(x)\ dx + \int_{\mu + k\sigma}^{\infty} (x - \mu)^2 f(x)\ dx$$

$$\geq \int_{-\infty}^{\mu - k\sigma} (x - \mu)^2 f(x)\ dx + \int_{\mu + k\sigma}^{\infty} (x - \mu)^2 f(x)\ dx,$$

since the second of the three integrals is nonnegative. Now, since $|x - \mu| \geq k\sigma$ wherever $x \geq \mu + k\sigma$ or $x \leq \mu - k\sigma$, we have $(x - \mu)^2 \geq k^2\sigma^2$ in both remaining integrals. It follows that

$$\sigma^2 \geq \int_{-\infty}^{\mu - k\sigma} k^2\sigma^2 f(x)\ dx + \int_{\mu + k\sigma}^{\infty} k^2\sigma^2 f(x)\ dx$$

and that

$$\int_{-\infty}^{\mu - k\sigma} f(x)\ dx + \int_{\mu + k\sigma}^{\infty} f(x)\ dx \leq \frac{1}{k^2}.$$

Hence

$$P(\mu - k\sigma < X < \mu + k\sigma) = \int_{\mu - k\sigma}^{\mu + k\sigma} f(x)\ dx \geq 1 - \frac{1}{k^2}$$

and the theorem is established.

For $k = 2$ the theorem states that the random variable X has a probability of at least $1 - 1/2^2 = 3/4$ of falling within two standard deviations of the mean. That is, three-fourths or more of the observations of any distribution lie in the interval $\mu \pm 2\sigma$. Similarly, the theorem says that at least eight-ninths of the observations of any distribution fall in the interval $\mu \pm 3\sigma$.

EXAMPLE 4.22   A random variable X has a mean $\mu = 8$, a variance $\sigma^2 = 9$, and an unknown probability distribution. Find (a) $P(-4 < X < 20)$, and (b) $P(|X - 8| \geq 6)$.

SOLUTION
(a) $P(-4 < X < 20) = P[8 - (4)(3) < X < 8 + (4)(3)]$

$$\geq \tfrac{15}{16}.$$

(b) $P(|X - 8| \geq 6) = 1 - P(|X - 8| < 6)$

$$= 1 - P(-6 < X - 8 < 6)$$

$$= 1 - P[8 - (2)(3) < X < 8 + (2)(3)]$$

$$\leq \tfrac{1}{4}.$$

Chebyshev's theorem holds for any distribution of observations and, for this reason, the results are usually weak. The value given by the theorem is a lower bound only. That is, we know that the probability of a random variable falling within two standard deviations of the mean can be *no less* than 3/4, but we never know how much more it might actually be. Only when the probability distribution is know can we determine exact probabilities. For this reason we call the theorem a *distribution-free* result. When specific distributions are assumed as in future chapters, the results will be less conservative. The use of Chebyshev's theorem is relegated to situations in which the form of the distribution is unknown.

Exercises

1. Referring to Exercise 3 on page 99, find the mean and variance of the discrete random variable $Z = 3X - 2$, where X represents the number of errors per 100 lines of code.

2. Using Theorems 4.5 and 4.9, find the mean and variance of the random variable $Z = 5X + 3$, where X has the probability distribution of Exercise 4 on page 99.

3. Suppose that a grocery store purchases 5 cartons of skim milk at the wholesale price of $1.20 per carton and retails the milk at $1.65 per carton. After the expiration date, the unsold milk is removed from the shelf and the grocer receives a credit from the distributor equal to three-fourths of the wholesale price. If the probability distribution of the random variable X, the number of cartons that are sold from this lot, is given by

x	0	1	2	3	4	5
$f(x)$	$\frac{1}{15}$	$\frac{2}{15}$	$\frac{2}{15}$	$\frac{3}{15}$	$\frac{4}{15}$	$\frac{3}{15}$

find the expected profit.

4. Repeat Exercise 11 on page 99, by applying Theorems 4.5 and 4.9.

5. Let X be a random variable with the following probability distribution:

x	-3	6	9
$f(x)$	$\frac{1}{6}$	$\frac{1}{2}$	$\frac{1}{3}$

Find $E(X)$ and $E(X^2)$ and then, using these values, evaluate $E[(2X + 1)^2]$.

6. The total time, measured in units of 100 hours, that a teenager runs her stereo set over a period of one year is a continuous random variable X that has the density function

$$f(x) = \begin{cases} x, & 0 < x < 1 \\ 2 - x, & 1 \le x < 2 \\ 0, & \text{elsewhere.} \end{cases}$$

Use Theorem 4.6 to evaluate the mean of the random variable $Y = 60X^2 + 39X$, where Y is equal to the number of kilowatt hours expended annually.

7. If a random variable X is defined such that $E[(X - 1)^2] = 10$, $E[(X - 2)^2] = 6$, find μ and σ^2.

8. Suppose that X and Y are independent random variables having the joint probability distribution

$f(x, y)$		x	
		2	4
	1	0.10	0.15
y	3	0.20	0.30
	5	0.10	0.15

Find
(a) $E(2X - 3Y)$;
(b) $E(XY)$.

9. Use Theorem 4.7 to evaluate $E(2XY^2 - X^2Y)$ for the joint probability distribution given in Table 3.6.

10. Seventy new jobs are opening up at an automobile manufacturing plant, but 1000 applicants show up for the 70 positions. To select the best 70 from among the applicants, the company gives a test that covers mechanical skill, manual dexterity, and mathematical ability. The mean grade on this test turns out to be 60, and the scores have a standard deviation 6. Can a person who has an 84 score count on getting one of the jobs? [HINT: Use Chebyshev's theorem.]

11. An electrical firm manufactures a 100-watt light bulb, which, according to specifications written on the package, has a mean life of 900 hours with a standard deviation of 50 hours. At most, what percentage of the bulbs fail to last even 700 hours?

12. A local company manufactures telephone wire. The average length of the wire is 52 inches with a standard deviation of 6.5 inches. At most, what percentage of

the telephone wire from this company exceeds 71.5 inches?

13. Suppose that you roll a 10-sided die (0, 1, 2, . . . , 9) 500 times. Using Chebyshev's theorem, compute the probability that the sample mean, $\overline{X}$, is between 4 and 5.

14. If X and Y are independent random variables with variances $\sigma_X^2 = 5$ and $\sigma_Y^2 = 3$, find the variance of the random variable $Z = -2X + 4Y - 3$.

15. Repeat Exercise 14 if X and Y are not independent and $\sigma_{XY} = 1$.

16. A random variable X has a mean $\mu = 12$, a variance $\sigma^2 = 9$, and an unknown probability distribution. Using Chebyshev's theorem, find
(a) $P(6 < X < 18)$;
(b) $P(3 < X < 21)$.

17. A random variable X has a mean $\mu = 10$ and a variance $\sigma^2 = 4$. Using Chebyshev's theorem, find
(a) $P(|X - 10| \ge 3)$;
(b) $P(|X - 10| < 3)$;
(c) $P(5 < X < 15)$;
(d) the value of the constant c such that

$$P(|X - 10| \ge c) \le 0.04.$$

18. Compute the $P(\mu - 2\sigma < X < \mu + 2\sigma)$, where X has the density function

$$f(x) = \begin{cases} 6x(1 - x), & 0 < x < 1 \\ 0, & \text{elsewhere} \end{cases}$$

and compare with the result given in Chebyshev's theorem.

19. Let X represent the number that occurs when a red die is tossed and Y the number that occurs when a green die is tossed. Find
(a) $E(X + Y)$;
(b) $E(X - Y)$;
(c) $E(XY)$.

20. Suppose that X and Y are independent random variables with probability densities

$$g(x) = \begin{cases} 8/x^3, & x > 2 \\ 0, & \text{elsewhere} \end{cases}$$

and

$$h(y) = \begin{cases} 2y, & 0 < y < 1 \\ 0, & \text{elsewhere.} \end{cases}$$

Find the expected value of $Z = XY$.

21. If the joint density function of X and Y is given by

$$f(x, y) = \begin{cases} \frac{2}{7}(x + 2y), & 0 < x < 1, \, 1 < y < 2 \\ 0, & \text{elsewhere}, \end{cases}$$

find the expected value of $g(X, Y) = (X/Y^3) + X^2Y$.

22. Let X represent the number that occurs when a green die is tossed and Y the number that occurs when a red die is tossed. Find the variance of the random variable
(a) $2X - Y$;
(b) $X + 3Y - 5$.

Review Exercises _____

1. Prove Chebyshev's theorem when X is a discrete random variable.

2. Find the covariance of the random variables X and Y having the joint probability density function.

$$f(x, y) = \begin{cases} x + y, & 0 < x < 1, \, 0 < y < 1 \\ 0, & \text{elsewhere}. \end{cases}$$

3. Referring to the random variables whose joint probability density function is given in Exercise 11 on page 79, find the average amount of kerosene left in the tank at the end of the day.

4. Assume the length X in minutes of a particular type of telephone conversation is a random variable with probability density function

$$f(X) = \tfrac{1}{5}e^{-x/5} \qquad X > 0$$

$$= 0 \qquad \text{elsewhere}.$$

Determine
(a) The mean length $[E(X)]$ of this type of telephone conversation.
(b) Find the variance and standard deviation of X.
(c) Find $E(X + 5)^2$.

5. Referring to the random variables whose joint density function is given in Exercise 5 on page 79, find the covariance between the weight of the creams and the weight of the toffees in these boxes of chocolates.

6. Referring to the random variables whose joint density function is given in Exercise 4 on page 79,
(a) find μ_X and μ_Y;
(b) find $E[(X + Y)/2]$.

7. Suppose it is known that the life X of a particular compressor in hours has the density function

$$f(X) = \frac{1}{900}e^{-X/900} \qquad X > 0$$

$$= 0 \qquad \text{elsewhere}.$$

(a) Find the mean life of the compressor.
(b) Find $E(X^2)$.
(c) Find the variance and standard deviation of the random variable X.

8. Referring to the random variables whose joint probability density function is given in Exercise 5 on page 79, find the expected weight for the sum of the creams and toffees if one purchased a box of these chocolates.

9. Show that $\text{Cov}(aX, bY) = ab \, \text{Cov}(X, Y)$.

5

Some Discrete Probability Distributions

5.1 Introduction

No matter whether a discrete probability distribution is represented graphically by a histogram, in tabular form, or by means of a formula, the behavior of a random variable is described. Often, the observations generated by different statistical experiments have the same general type of behavior. Consequently, discrete random variables associated with these experiments can be described by essentially the same probability distribution and therefore can be represented by a single formula. In fact, one needs only a handful of important probability distributions to describe many of the discrete random variables encountered in practice.

5.2 Discrete Uniform Distribution

The simplest of all discrete probability distributions is one in which the random variable assumes each of its values with an equal probability. Such a probability distribution is called the **discrete uniform distribution**.

**DISCRETE
UNIFORM
DISTRIBUTION**

If the random variable X assumes the values $x_1, x_2, \ldots, x_k$, with equal probabilities, then the discrete uniform distribution is given by

$$f(x; k) = \frac{1}{k}, \qquad x = x_1, x_2, \ldots, x_k. \qquad ■$$

We have used the notation $f(x; k)$ instead of $f(x)$ to indicate that the uniform distribution depends on the *parameter k*.

EXAMPLE 5.1   When a light bulb is selected at random from a box that contains a 40-watt bulb, a 60-watt bulb, a 75-watt bulb, and a 100-watt bulb, each element of the sample space $S = \{40, 60, 75, 100\}$ occurs with probability 1/4. Therefore, we have a uniform distribution, with

$$f(x; 4) = \tfrac{1}{4}, \qquad x = 40, 60, 75, 100.$$

EXAMPLE 5.2   When a die is tossed, each element of the sample space $S = \{1, 2, 3, 4, 5, 6\}$ occurs with probability 1/6. Therefore, we have a uniform distribution, with

$$f(x; 6) = \tfrac{1}{6}, \qquad x = 1, 2, 3, 4, 5, 6.$$

The graphic representation of the uniform distribution by means of a histogram always turns out to be a set of rectangles with equal heights. The histogram for Example 5.2 is shown in Figure 5.1.

THEOREM 5.1   *The mean and variance of the discrete uniform distribution $f(x; k)$ are*

$$\mu = \frac{\sum_{i=1}^{k} x_i}{k} \qquad and \qquad \sigma^2 = \frac{\sum_{i=1}^{k} (x_i - \mu)^2}{k}. \qquad ■$$

PROOF.   By definition

$$\mu = E(X) = \sum_{i=1}^{k} x_i f(x_i; k) = \sum_{i=1}^{k} \frac{x_i}{k}$$

$$= \frac{\sum_{i=1}^{k} x_i}{k}.$$

Also, by definition,

$$\sigma^2 = E[(X - \mu)^2] = \sum_{i=1}^{k} (x_i - \mu)^2 f(x_i; k)$$

$$= \sum_{i=1}^{k} \frac{(x_i - \mu)^2}{k} = \frac{\sum_{i=1}^{k} (x_i - \mu)^2}{k}.$$

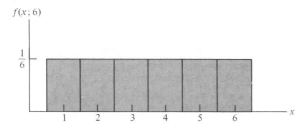

FIGURE 5.1   Histogram for the tossing of a die.

EXAMPLE 5.3   Referring to Example 5.2, we find that

$$\mu = \frac{1 + 2 + 3 + 4 + 5 + 6}{6} = 3.5$$

and

$$\sigma^2 = \frac{(1 - 3.5)^2 + (2 - 3.5)^2 + \cdots + (6 - 3.5)^2}{6} = \frac{35}{12}.$$

5.3 Binomial and Multinomial Distributions _____

An experiment often consists of repeated trials, each with two possible outcomes that may be labeled **success** and **failure**. The most obvious application deals with the testing of items as they come off an assembly line, where each test or trial may indicate a defective or a nondefective item. We may choose to define either outcome as a success. One may consider cards drawn in succession from an ordinary deck and each trial is labeled a success or a failure, depending on whether the card is a heart or not a heart. If each card is replaced and the deck shuffled before the next drawing, the two experiments just described have similar properties, in that the repeated trials are independent and the probability of a success remains constant from trial to trial. The process is referred to as a **Bernoulli process**. Each trial is called a **Bernoulli trial**. Observe in the card-drawing example that the probabilities of a success for the repeated trials change if the cards are not replaced. This is, the probability of selecting a heart on the first draw is 1/4, but on the second draw it is a conditional probability having a value of 13/51 or 12/51, depending on whether or not a heart occurred on the first draw. This, then, would no longer be considered a set of Bernoulli trials.

The Bernoulli Process

Strictly speaking, the Bernoulli process must possess the following properties:

1. The experiment consists of n repeated trials.
2. Each trial results in an outcome that may be classified as a success or a failure.
3. The probability of success, denoted by p, remains constant from trial to trial.
4. The repeated trials are independent.

Consider the set of Bernoulli trials where three items are selected at random from a manufacturing process, inspected, and classified defective or nondefective. A defective item is designated a success. The number of successes is a random variable X assuming integral values from zero through 3. The eight possible outcomes and the corresponding values of X are

Outcome	x
NNN	0
NDN	1
NND	1
DNN	1
NDD	2
DND	2
DDN	2
DDD	3

Since the items are selected independently from a process that we shall assume produces 25% defectives,

$$P(NDN) = P(N)P(D)P(N) = (\tfrac{3}{4})(\tfrac{1}{4})(\tfrac{3}{4}) = \tfrac{9}{64}.$$

Similar calculations yield the probabilities for the other possible outcomes. The probability distribution of X is therefore given by

x	0	1	2	3
$f(x)$	$\tfrac{27}{64}$	$\tfrac{27}{64}$	$\tfrac{9}{64}$	$\tfrac{1}{64}$

The number X of successes in n Bernoulli trials is called a **binomial random variable**. The probability distribution of this discrete random variable is called the **binomial distribution** and its values will be denoted by $b(x; n, p)$, since they depend on the number of trials and the probability of a success on a given trial. Thus, for the probability distribution of X, the number of defectives,

$$P(X = 2) = f(2) = b(2; 3, \tfrac{1}{4}) = \tfrac{9}{64}.$$

Let us now generalize the above illustration to yield a formula for $b(x; n, p)$. That is, we wish to find a formula that gives the probability of x successes in n trials for a binomial experiment. First, consider the probability of x successes and $n - x$ failures in a specified order. Since the trials are independent, we can multiply all the probabilities corresponding to the different outcomes. Each success occurs with probability p and each failure with probability $q = 1 - p$. Therefore, the probability for the specified order is $p^x q^{n-x}$. We must now determine the total number of sample points in the experiment that have x successes and $n - x$ failures. This number is equal to the number of partitions of n outcomes into two groups with x in one group and $n - x$ in the other and is given by $\binom{n}{x}$. Because these partitions are mutually exclusive, we add the probabilities of all the different partitions to obtain the general formula, or simply multiply $p^x q^{n-x}$ by $\binom{n}{x}$.

BINOMIAL DISTRIBUTION

A Bernoulli trial can result in a success with probability p and a failure with probability $q = 1 - p$. Then the probability distribution of the binomial random variable X, the number of successes in n independent trials, is

$$b(x; n, p) = \binom{n}{x} p^x q^{n-x}, \qquad x = 0, 1, 2, \ldots, n. \qquad \blacksquare$$

Note that when $n = 3$ and $p = 1/4$, the probability distribution of X, the number of defectives, may be written as

$$b\left(x; 3, \frac{1}{4}\right) = \binom{3}{x}\left(\frac{1}{4}\right)^x\left(\frac{3}{4}\right)^{3-x}, \qquad x = 0, 2, 2, 3,$$

rather than in the tabular form above.

EXAMPLE 5.4 The probability that a certain kind of component will survive a given shock test is 3/4. Find the probability that exactly 2 of the next 4 components tested survive.

SOLUTION
Assuming that the tests are independent and $p = 3/4$ for each of the 4 tests, we obtain

$$b\left(2; 4, \frac{3}{4}\right) = \binom{4}{2}\left(\frac{3}{4}\right)^2\left(\frac{1}{4}\right)^2$$

$$= \frac{4!}{2!\,2!} \cdot \frac{3^2}{4^4}$$

$$= \frac{27}{128}.$$

The binomial distribution derives its name from the fact that the $n + 1$ terms in the binomial expansion of $(q + p)^n$ correspond to the various values of $b(x; n, p)$ for $x = 0, 1, 2, \ldots, n$. That is,

$$(q + p)^n = \binom{n}{0}q^n + \binom{n}{1}pq^{n-1} + \binom{n}{2}p^2q^{n-2} + \cdots + \binom{n}{n}p^n$$

$$= b(0; n, p) + b(1; n, p) + b(2; n, p) + \cdots + b(n; n, p).$$

Since $p + q = 1$, we see that $\sum_{x=0}^{n} b(x; n, p) = 1$, a condition that must hold for any probability distribution.

Frequently, we are interested in problems where it is necessary to find $P(X < r)$ or $P(a \le X \le b)$. Fortunately, binomial sums $B(r; n, p) = \sum_{x=0}^{r} b(x; n, p)$ are available and are given in Table A.1 of the Appendix for $n = 1, 2, \ldots, 20$, and selected values of p from 0.1 to 0.9. We illustrate the use of Table A.1 with the following example.

EXAMPLE 5.5   The probability that a patient recovers from a rare blood disease is 0.4. If 15 people are known to have contracted this disease, what is the probability that (a) at least 10 survive, (b) from 3 to 8 survive, and (c) exactly 5 survive?

SOLUTION
Let X be the number of people that survive.

(a) $P(X \geq 10) = 1 - P(X < 10)$

$$= 1 - \sum_{x=0}^{9} b(x; 15, 0.4)$$

$$= 1 - 0.9662$$

$$= 0.0338.$$

(b) $P(3 \leq X \leq 8) = \sum_{x=3}^{8} b(x; 15, 0.4)$

$$= \sum_{x=0}^{8} b(x; 15, 0.4) - \sum_{x=0}^{2} b(x; 15, 0.4)$$

$$= 0.9050 - 0.0271$$

$$= 0.8779.$$

(c) $P(X = 5) = b(5; 15, 0.4)$

$$= \sum_{x=0}^{5} b(x; 15, 0.4) - \sum_{x=0}^{4} b(x; 15, 0.4)$$

$$= 0.4032 - 0.2173$$

$$= 0.1859.$$

Areas of Application

From Examples 5.4 and 5.5, it should be clear that the binomial distribution finds applications in many scientific fields. An industrial engineer is keenly interested in the "proportion defective" in an industrial process. Often, quality control measures and sampling schemes for processes are based on the binomial distribution. The binomial applies in any industrial situation where an outcome of a process is dichotomous and the results of the process are independent, with the probability of a success being constant from trial to trial. The binomial distribution is also used extensively in medical and military applications. In both cases, a success or failure result is important. For example, "cure" and "no cure" is important in pharmaceutical work, while "hit" or "miss" is often the interpretation of the result of firing a guided missile.

Since the probability distribution of any binomial random variable depends only on the values assumed by the parameters n, p, and q, it would seem reasonable to assume that the mean and variance of a binomial random variable also depend on the values assumed by these parameters. Indeed, this is true, and in Theorem 5.2 we derive general formulas as functions of n, p, and q that can be used to compute the mean and variance of any binomial random variable.

THEOREM 5.2   *The mean and variance of the binomial distribution $b(x; n, p)$ are*

$$\mu = np \quad and \quad \sigma^2 = npq.$$ ∎

PROOF. Let the outcome on the jth trial be represented by a Bernoulli random variable I_j, which assumes the values 0 and 1 with probabilities q and p, respectively.

Therefore, in a binomial experiment the number of successes can be written as the sum of the n independent indicator variables. Hence

$$X = I_1 + I_2 + \cdots + I_n.$$

The mean of any I_j is $E(I_j) = (0 \cdot q) + (1 \cdot p) = p$. Therefore, using Corollary 2 of Theorem 4.7, the mean of the binomial distribution is

$$\mu = E(X) = E(I_1) + E(I_2) + \cdots + E(I_n)$$

$$= \underbrace{p + p + \cdots + p}_{n \text{ terms}}$$

$$= np.$$

The variance of any I_j is given by

$$\sigma_{I_j}^2 = E[(I_j - p)^2] = E(I_j^2) - p^2$$

$$= (0)^2 q + (1)^2 p - p^2$$

$$= p(1 - p) = pq.$$

By extending Corollary 2 of Theorem 4.10 to the case of n independent variables, the variance of the binomial distribution is

$$\sigma_X^2 = \sigma_{I_1}^2 + \sigma_{I_2}^2 + \cdots + \sigma_{I_n}^2$$

$$= \underbrace{pq + pq + \cdots + pq}_{n \text{ terms}}$$

$$= npq.$$

EXAMPLE 5.6   Find the mean and variance of the binomial random variable of Example 5.5, and then use Chebyshev's theorem to interpret the interval $\mu \pm 2\sigma$.

SOLUTION
Since Example 5.5 was a binomial experiment with $n = 15$ and $p = 0.4$, by Theorem 5.2, we have

$$\mu = (15)(0.4) = 6 \quad and \quad \sigma^2 = (15)(0.4)(0.6) = 3.6.$$

Taking the square root of 3.6, we find that $\sigma = 1.897$. Hence the required interval is $6 \pm (2)(1.897)$, or from 2.206 to 9.794. Chebyshev's theorem states that the number of recoveries among 15 patients subjected to the given disease has a probability of at least 3/4 of falling between 2.206 and 9.794, or, because the data are discrete, between 3 and 9 inclusive.

As the reader should realize by now, in many applications there are more than two possible outcomes. To borrow an example from the field of genetics, the color of guinea pigs produced as offspring may be red, black, or white. Often the "defective" or "not defective" dichotomy in engineering situations is truly an oversimplification. Indeed, there are often more than two categories that characterize items or parts coming off an assembly line.

Multinomial Experiments

The binomial experiment becomes a **multinomial experiment** if we let each trial have more than 2 possible outcomes. Hence the classification of a manufactured product as being light, heavy, or acceptable and the recording of accidents at a certain intersection according to the day of the week constitute multinomial experiments. The drawing of a card from a deck *with replacement* is also a multinomial experiment if the 4 suits are the outcomes of interest.

In general, if a given trial can result in any one of k possible outcomes E_1, $E_2, \ldots, E_k$ with probabilities $p_1, p_2, \ldots, p_k$, then the **multinomial distribution** will give the probability that E_1 occurs x_1 times; E_2 occurs x_2 times; $\ldots$; and E_k occurs x_k times in n independent trials, where

$$x_1 + x_2 + \cdots + x_k = n.$$

We shall denote this joint probability distribution by $f(x_1, x_2, \ldots, x_k; p_1, p_2, \ldots, p_k, n)$. Clearly, $p_1 + p_2 + \cdots + p_k = 1$, since the result of each trial must be one of the k possible outcomes.

To derive the general formula, we proceed as in the binomial case. Since the trials are independent, any specified order yielding x_1 outcomes for E_1, x_2 for $E_2, \ldots, x_k$ for E_k will occur with probability $p_1^{x_1} p_2^{x_2} \cdots p_k^{x_2}$. The total number of orders yielding similar outcomes for the n trials is equal to the number of partitions of n items into k groups with x_1 in the first group; x_2 in the second group; $\ldots$; and x_k in the kth group. This can be done in

$$\binom{n}{x_1, x_2, \ldots, x_k} = \frac{n!}{x_1!\, x_2! \cdots x_k!}$$

ways. Since all the partitions are mutually exclusive and occur with equal probability, we obtain the multinomial distribution by multiplying the probability for a specified order by the total number of partitions.

MULTINOMIAL DISTRIBUTION

If a given trial can result in the k outcomes $E_1, E_2, \ldots, E_k$ with probabilities $p_1, p_2, \ldots, p_k$, then the probability distribution of the random variables $X_1, X_2, \ldots, X_k$, representing the number of occurrences for $E_1, E_2, \ldots, E_k$ in n independent trials is

$$f(x_1, x_2, \ldots, x_k; p_1, p_2, \ldots, p_k, n) = \binom{n}{x_1, x_2, \ldots x_k} p_1^{x_1} p_2^{x_2} \cdots p_k^{x_k}$$

with

$$\sum_{i=1}^{k} x_i = n \quad and \quad \sum_{i=1}^{k} p_i = 1.$$ ∎

The multinomial distribution derives its name from the fact that the terms of the multinomial expansion of $(p_1 + p_2 + \cdots + p_k)^n$ correspond to all the possible values of $f(x_1, x_2, \ldots, x_k; p_1, p_2, \ldots, p_k, n)$.

EXAMPLE 5.7   If a pair of dice are tossed 6 times, what is the probability of obtaining a total of 7 or 11 twice, a matching pair once, and any other combination 3 times?

SOLUTION
We list the following possible events,

E_1:   a total of 7 or 11 occurs,

E_2:   a matching pair occurs,

E_3:   neither a pair nor a total of 7 or 11 occurs.

The corresponding probabilities for a given trial are $p_1 = 2/9$, $p_2 = 1/6$, and $p_3 = 11/18$. These values remain constant for all 6 trials. Using the multinomial distribution with $x_1 = 2$, $x_2 = 1$, and $x_3 = 3$, we find that the required probability is

$$f\left(2, 1, 3; \frac{2}{9}, \frac{1}{6}, \frac{11}{18}, 6\right) = \binom{6}{2, 1, 3}\left(\frac{2}{9}\right)^2\left(\frac{1}{6}\right)^1\left(\frac{11}{18}\right)^3$$

$$= \frac{6!}{2!\ 1!\ 3!} \cdot \frac{2^2}{9^2} \cdot \frac{1}{6} \cdot \frac{11^3}{18^3}$$

$$= 0.1127.$$

Exercises

1. An employee is selected from a staff of 10 to supervise a certain project by selecting a tag at random from a box containing 10 tags numbered from 1 to 10. Find the formula for the probability distribution of X representing the number on the tag that is drawn. What is the probability that the number drawn is less than 4?

2. Twelve people are given two identical speakers to listen for differences, if any. Suppose that these people answered by guessing only. Find the probability that three people claim to have heard a difference between the two speakers.

3. Find the mean and variance of the random variable X of Exercise 1.

4. In a certain city district the need for money to buy drugs is given as the reason for 75% of all thefts. Find the probability that among the next 5 theft cases reported in this district,
   (a) exactly 2 resulted from the need for money to buy drugs;
   (b) at most 3 resulted from the need for money to buy drugs.

5. A fruit grower claims that 2/3 of his peach crop has been contaminated by the medfly infestation. Find the probability that among 4 peaches inspected by this grower
   (a) all 4 have been contaminated by the medfly;
   (b) anywhere from 1 to 3 have been contaminated.

6. According to a survey by the Administrative Management Society, 1/3 of U.S. companies give employees four weeks of vacation after they have been with the company for 15 years. Find the probability that among 6 companies surveyed at random, the number that give employees 4 weeks of vacation after 15 years of employment is
   (a) anywhere from 2 to 5;
   (b) fewer than 3.

7. One prominent physician claims that 70% of those with lung cancer are chain smokers. If his assertion is correct:
   (a) find the probability that of 10 such patients recently admitted to a hospital, fewer than half are chain smokers.
   (b) find the probability that of 20 such patients recently admitted to a hospital, fewer than half are chain smokers.

8. According to a study published by a group of University of Massachusetts sociologists, approximately 60% of the Valium users in the state of Massachusetts first took Valium for psychological problems. Find the probability that among the next 8 users interviewed from this state.
   (a) exactly 3 began taking Valium for psychological problems;
   (b) at least 5 began taking Valium for problems that were not psychological.

9. In testing a certain kind of truck tire over a rugged terrain, it is found that 25% of the trucks fail to complete the test run without a blowout. Of the next 15 trucks tested, find the probability that
   (a) from 3 to 6 have blowouts;
   (b) fewer than 4 have blowouts;
   (c) more than 5 have blowouts.

10. A nationwide survey of seniors by the University of Michigan reveals that almost 70% disapprove of daily pot smoking according to a report in *Parade*, September 14, 1980. If 12 seniors are selected at random and asked their opinion, find the probability that the number who disapprove of smoking pot daily is
    (a) anywhere from 7 to 9;
    (b) at most 5;
    (c) not less than 8.

11. The probability that a patient recovers from a delicate heart operation is 0.9. What is the probability that exactly 5 of the next 7 patients having this operation survive?

12. A traffic control engineer reports that 75% of the vehicles passing through a checkpoint are from within the state. What is the probability that fewer than 4 of the next 9 vehicles are from out of the state?

13. A study conducted at George Washington University and the National Institutes of Health examined national attitudes about tranquilizers. The study revealed that approximately 70% believe "tranquilizers don't really cure anything, they just cover up the real trouble." According to this study what is the probability that at least 3 of the next 5 people selected at random will be of this opinion?

14. A survey of the residents in a U.S. city showed that 20% preferred a white telephone over any other color available. What is the probability that more than half of the next 20 telephones installed in this city will be white?

15. It is known that 40% of mice inoculated with a serum are protected from a certain disease. If 5 mice are inoculated, find the probability that
    (a) none contracts the disease;
    (b) fewer than 2 contract the disease;
    (c) more than 3 contract the disease.

16. Suppose that airplane engines operate independently and fail with probability equal to 0.4. Assuming that a plane makes a safe flight if at least one-half of its engines run, determine whether a 4-engine plane or a 2-engine plane has the higher probability for a successful flight.

17. If X represents the number of people in Exercise 13 who believe that tranquilizers do not cure but only cover up the real problem, find the mean and variance of X when 5 people are selected at random and then use Chebyshev's theorem to interpret the interval $\mu \pm 2\sigma$.

18. (a) In Exercise 9 how many of the 15 trucks would you expect to have blowouts?
    (b) According to Chebyshev's theorem, there is a probability of at least 3/4 that the number of trucks among the next 15 that have blowouts will fall in what interval?

19. As a student drives to school, he encounters a traffic signal. This traffic signal stays green for 35 seconds, yellow for 5 seconds, and red for 60 seconds. Assume that the student goes to school each weekday between 8:00 and 8:30. Let X_1 be the number of times he encounters a green light, X_2 be the number of times he encounters a yellow light, and X_3 be the number of

times he encounters a red light. Find the joint distribution of X_1, X_2, and X_3.

20. A card is drawn from a well-shuffled deck of 52 playing cards, the result recorded, and the card replaced. If the experiment is repeated 5 times, what is the probability of obtaining 2 spades and 1 heart?

21. The surface of a circular dart board has a small center circle called the bull's-eye and 20 pie-shaped regions numbered from 1 to 20. Each of the pie-shaped regions is further divided into three parts such that a person throwing a dart that lands on a specified number scores the value of the number, double the number, or triple the number, depending on which of the three parts the dart falls. If a person hits the bull's-eye with probability 0.01, hits a double with probability 0.10, hits a triple with probability 0.05, and misses the dart board with probability 0.02, what is the probability that 7 throws will result in no bull's-eyes, no triples, a double twice, and a complete miss once?

22. According to the theory of genetics, a certain cross of guinea pigs will result in red, black, and white offspring in the ratio 8:4:4. Find the probability that among 8 offspring 5 will be red, 2 black, and 1 white.

23. The probabilities are 0.4, 0.2, 0.3, and 0.1, respectively, that a delegate to a certain convention arrived by air, bus, automobile, or train. What is the probability that among 9 delegates randomly selected at this convention, 3 arrived by air, 3 arrived by bus, 1 arrived by automobile, and 2 arrived by train?

24. A safety engineer claims that only 40% of all workers wear safety helmets when they eat lunch at the workplace. Assuming that his claim is right, find the probability that 4 of 6 workers randomly chosen will be wearing their helmets while having lunch at the workplace.

25. Suppose that for a very large shipment of integrated-circuit chips, the probability of failure for any one chip is 0.10. Assuming that the assumptions underlying the binomial distributions are met, find the probability that at most 3 chips fail in a random sample of 20.

26. Assuming that 6 in 10 automobile accidents are due mainly to a speed violation, find the probability that among 8 automobile accidents 6 will be due mainly to a speed violation
    (a) by using the formula for the binomial distribution;
    (b) by using the binomial table.

27. If the probability that a fluorescent light has a useful life of at least 800 hours is 0.9, find the probabilities that among 20 such lights
    (a) exactly 18 will have a useful life of at least 800 hours;
    (b) at least 15 will have a useful life of at least 800 hours;
    (c) at least 2 will *not* have a useful life of at least 800 hours.

28. A manufacturer knows that on the average 20% of the electric toasters which he makes will require repairs within 1 year after they are sold. When 20 toasters are randomly selected, find appropriate numbers x and y such that
    (a) the probability that at least x of them will require repairs is less than 0.5;
    (b) the probability that at least y of them will *not* require repairs is greater than 0.8.

5.4 Hypergeometric Distribution

The simplest way to view the distinction between the binomial distribution of Section 5.3 and the hypergeometric distribution lies in the way the sampling is done. The types of applications of the hypergeometric are very similar to those of the binomial distribution. We are interested in computing probabilities for the number of observations that fall into a particular category. But in the case of the binomial, independence among trials is required. As a result, if the binomial is applied to, say, sampling from a lot of items (deck of cards, batch of production items), the sampling must be done **with replacement** of each item after it is observed. On the other hand, the hypergeometric distribution does not require independence and is based on the sampling done **without replacement**.

Applications for the hypergeometric distribution are found in many areas, with heavy uses in acceptance sampling, electronic testing, and quality assurance. Obviously, in many of these fields testing is done at the expense of the item being tested. The item is destroyed and hence cannot be replaced in the sample. Thus sampling without replacement is necessary. In the following paragraphs a simple example with playing cards is used for illustration.

If we wish to find the probability of observing 3 red cards in 5 draws from an ordinary deck of 52 playing cards, the binomial distribution of Section 5.3 does not apply unless each card is replaced and the deck reshuffled before the next drawing is made. To solve the problem of sampling without replacement, let us restate the problem. If 5 cards are drawn at random, we are interested in the probability of selecting 3 red cards from the 26 available and 2 black cards from the 26 black cards available in the deck. There are $\binom{26}{3}$ ways of selecting 3 red cards, and for each of these ways we can choose 2 black cards in $\binom{26}{2}$ ways. Therefore, the total number of ways to select 3 red and 2 black cards in 5 draws is the product $\binom{26}{3}\binom{26}{2}$. The total number of ways to select any 5 cards from the 52 that are available is $\binom{52}{5}$. Hence the probability of selecting 5 cards without replacement of which 3 are red and 2 are black is given by

$$\frac{\binom{26}{3}\binom{26}{2}}{\binom{52}{5}} = \frac{(26!/3!\ 23!)(26!/2!\ 24!)}{(52!/5!\ 47!)} = 0.3251.$$

In general, we are interested in the probability of selecting x successes from the k items labeled success and $n - x$ failures from the $N - k$ items labeled failures when a random sample of size n is selected from N items. This is known as a **hypergeometric experiment**.

A hypergeometric experiment is one that possesses the following two properties:

1. A random sample of size n is selected without replacement from N items.
2. k of the N items may be classified as successes and $N - k$ are classified as failures.

The number X of successes in a hypergeometric experiment is called a **hypergeometric random variable**. Accordingly, the probability distribution of the hypergeometric variable is called the **hypergeometric distribution** and its values will be denoted by $h(x; N, n, k)$, since they depend on the number of successes k in the set N from which we select n items.

EXAMPLE 5.8   A committee of size 5 is to be selected at random from 3 chemists and 5 physicists. Find the probability distribution for the number of chemists on the committee.

SOLUTION

Let the random variable X be the number of chemists on the committee. The two properties of a hypergeometric experiment are satisfied. Hence

$$P(X = 0) = h(0; 8, 5, 3) = \frac{\binom{3}{0}\binom{5}{5}}{\binom{8}{5}} = \frac{1}{56}$$

$$P(X = 1) = h(1; 8, 5, 3) = \frac{\binom{3}{1}\binom{5}{4}}{\binom{8}{5}} = \frac{15}{56}$$

$$P(X = 2) = h(2; 8, 5, 3) = \frac{\binom{3}{2}\binom{5}{3}}{\binom{8}{5}} = \frac{30}{56}$$

$$P(X = 3) = h(3; 8, 5, 3) = \frac{\binom{3}{3}\binom{5}{2}}{\binom{8}{5}} = \frac{10}{56}.$$

In tabular form the hypergeometric distribution of X is as follows:

x	0	1	2	3
$h(x; 8, 5, 3)$	$\frac{1}{56}$	$\frac{15}{56}$	$\frac{30}{56}$	$\frac{10}{56}$

It is not difficult to see that the probability distribution can be given by the formula

$$h(x; 8, 5, 3) = \frac{\binom{3}{x}\binom{5}{5-x}}{\binom{8}{5}}, \qquad x = 0, 1, 2, 3.$$

Let us now generalize Example 5.8 to find a formula for $h(x; N, n, k)$. The total number of samples of size n chosen from N items is $\binom{N}{n}$. These samples are assumed to be equally likely. There are $\binom{k}{x}$ ways of selecting x successes from the k that are available and for each of these ways we can choose the $n - x$ failures in $\binom{N-k}{n-x}$ ways. Thus the total number of favorable samples among the $\binom{N}{n}$ possible samples is given by $\binom{k}{x}\binom{N-k}{n-x}$. Hence we have the following definition.

HYPERGEOMETRIC
DISTRIBUTION

The probability distribution of the hypergeometric random variable X, the number of successes in a random sample of size n selected from N items of which k are labeled **success** *and N − k labeled* **failure**, *is*

$$h(x; N, n, k) = \frac{\binom{k}{x}\binom{N-k}{n-x}}{\binom{N}{n}}, \qquad x = 0, 1, 2, \ldots, n.$$ ■

EXAMPLE 5.9   Lots of 40 components each are called acceptable if they contain no more than 3 defectives. The procedure for sampling the lot is to select 5 components at random and to reject the lot if a defective is found. What is the probability that exactly 1 defective will be found in the sample if there are 3 defectives in the entire lot?

SOLUTION
Using the hypergeometric distribution with $n = 5$, $N = 40$, $k = 3$, and $x = 1$, we find the probability of obtaining one defective to be

$$h(1; 40, 5, 3) = \frac{\binom{3}{1}\binom{37}{4}}{\binom{40}{5}} = 0.3011.$$

THEOREM 5.3   *The mean and variance of the hypergeometric distribution h(x; N, n, k) are*

$$\mu = \frac{nk}{N}$$

and

$$\sigma^2 = \frac{N-n}{N-1} \cdot n \cdot \frac{k}{N}\left(1 - \frac{k}{N}\right).$$ ■

We shall proceed to prove the result for the mean only.

PROOF.   To find the mean of the hypergeometric distribution, we write

$$E(X) = \sum_{x=0}^{n} x\frac{\binom{k}{x}\binom{N-k}{n-x}}{\binom{N}{n}}$$

$$= k\sum_{x=1}^{n} \frac{(k-1)!}{(x-1)!(k-x)!} \cdot \frac{\binom{N-k}{n-x}}{\binom{N}{n}}$$

$$= k \sum_{x=1}^{n} \frac{\binom{k-1}{x-1}\binom{N-k}{n-x}}{\binom{N}{n}}.$$

Letting $y = x - 1$, we find that this becomes

$$E(X) = k \sum_{y=0}^{n-1} \frac{\binom{k-1}{y}\binom{N-k}{n-1-y}}{\binom{N}{n}}.$$

Writing

$$\binom{N-k}{n-1-y} = \binom{(N-1)-(k-1)}{n-1-y}$$

and

$$\binom{N}{n} = \frac{N!}{n!(N-n)!} = \frac{N}{n}\binom{N-1}{n-1},$$

we obtain

$$E(X) = \frac{nk}{N} \sum_{y=0}^{n-1} \frac{\binom{k-1}{y}\binom{(N-1)-(k-1)}{n-1-y}}{\binom{N-1}{n-1}}$$

$$= \frac{nk}{N},$$

since the summation represents the total of all probabilities in a hypergeometric experiment when $N - 1$ items are selected at random from $N - 1$, of which $k - 1$ are labeled success.

EXAMPLE 5.10   Find the mean and variance of the random variable of Example 5.9, and then use Chebyshev's theorem to interpret the interval $\mu \pm 2\sigma$.

SOLUTION
Since Example 5.9 was a hypergeometric experiment with $n = 40$, $n = 5$, and $k = 3$, then by Theorem 5.3 we have

$$\mu = \frac{(5)(3)}{40} = \frac{3}{8} = 0.375$$

and

$$\sigma^2 = \left(\frac{40-5}{39}\right)(5)\left(\frac{3}{40}\right)\left(1 - \frac{3}{40}\right)$$

$$= 0.3113.$$

Taking the square root of 0.3113, we find that $\sigma = 0.558$. Hence the required interval is $0.375 \pm (2)(0.558)$, or from -0.741 to 1.491. Chebyshev's theorem states that the number of defectives obtained when 5 components are selected at random from a lot of 40 components of which 3 are defective has a probability of at least 3/4 of falling between -0.741 and 1.491. That is, at least three-fourths of the time, the 5 components include less than 2 defectives.

Relationship to the Binomial Distribution

In this chapter we discuss several important discrete distributions that have wide applicability. Many of these distributions relate nicely to each other. The beginning student should gain a clear understanding of these relationships. There is an interesting relationship between the hypergeometric and the binomial distribution. As one might expect, if n is small compared to N, the nature of the N items changes very little in each draw. Thus the quantity k/n plays the role of the binomial parameter p. As a result, the binomial distribution may be viewed as a large population edition of the hypergeometric distributions. The mean and variance then come from the formulas

$$\mu = np = \frac{nk}{N}$$

$$\sigma^2 = npq = n \cdot \frac{k}{N}\left(1 - \frac{k}{N}\right).$$

Comparing these formulas with those of Theorem 5.3, we see that the mean is the same while the variance differs by a correction factor of $(N - n)/(N - 1)$. This is negligible when n is small relative to N.

EXAMPLE 5.11   A manufacturer of automobile tires reports that among a shipment of 5000 sent to a local distributor, 1000 are slightly blemished. If one purchases 10 of these tires at random from the distributor, what is the probability that exactly 3 will be blemished?

SOLUTION
Since $N = 5000$ is large relative to the sample size $n = 10$, we shall approximate the desired probability by using the binomial distribution. The probability of obtaining a blemished tire is 0.2. Therefore, the probability of obtaining exactly 3 blemished tires is

$$h(3; 5000, 10, 1000) \simeq b(3; 10, 0.2)$$

$$= \sum_{x=0}^{3} b(x; 10, 0.2) - \sum_{x=0}^{2} b(x; 10, 0.2)$$

$$= 0.8791 - 0.6778$$

$$= 0.2013.$$

The hypergeometric distribution can be extended to treat the case where the N items can be partitioned into k cells $A_1, A_2, \ldots, A_k$ with a_1 elements in the first cell, a_2 elements in the second cell, $\ldots$, a_k elements in the kth cell. We are now interested in the probability that a random sample of size n yields x_1 elements from A_1, x_2 elements from A_2, $\ldots$, and x_k elements from A_k. Let us represent this probability by

$$f(x_1, x_2, \ldots, x_k; a_1, a_2, \ldots, a_k, N, n).$$

To obtain a general formula, we note that the total number of samples that can be chosen of size n from N items is still $\binom{N}{n}$. There are $\binom{a_1}{x_1}$ ways of selecting x_1 items from the items in A_1, and for each of these we can choose x_2 items from the items in A_2 in $\binom{a_2}{x_2}$ ways. Therefore, we can select x_1 items from A_1 and x_2 items from A_2 in $\binom{a_1}{x_1}\binom{a_2}{x_2}$ ways. Continuing in this way, we can select all n items consisting of x_1 from A_1, x_2 from $A_2, \ldots$, and x_k from A_k in $\binom{a_1}{x_1}\binom{a_2}{x_2}\cdots\binom{a_k}{x_k}$ ways. The required probability distribution is now defined as follows.

MULTIVARIATE HYPERGEOMETRIC DISTRIBUTION

If N items can be partitioned into the k cells $A_1, A_2, \ldots, A_k$ with $a_1, a_2, \ldots, a_k$ elements, respectively, then the probability distribution of the random variables $X_1, X_2, \ldots, X_k$, representing the number of elements selected from $A_1, A_2, \ldots, A_k$ in a random sample of size n, is

$$f(x_1, x_2, \ldots, x_k; a_1, a_2, \ldots, a_k, N, n) = \frac{\binom{a_1}{x_1}\binom{a_2}{x_2}\cdots\binom{a_k}{x_k}}{\binom{N}{n}}$$

with $\displaystyle\sum_{i=1}^{k} x_i = n$ and $\displaystyle\sum_{i=1}^{k} a_i = N$. ■

EXAMPLE 5.12 A group of 10 individuals is being used in a biological case study. The group contains 3 people with blood type O, 4 with blood type A, and 3 with blood type B. What is the probability that a random sample of 5 will contain 1 person with blood type O, 2 people with blood type A, and 2 people with blood type B?

SOLUTION
Using the extension of the hypergeometric distribution with $x_1 = 1$, $x_2 = 2$, $x_3 = 2$, $a_1 = 3$, $a_2 = 4$, $a_3 = 3$, $N = 10$, and $n = 5$, we find that the desired probability is

$$f(1, 2, 2; 3, 4, 3, 10, 5) = \frac{\binom{3}{1}\binom{4}{2}\binom{3}{2}}{\binom{10}{5}} = \frac{3}{14}.$$

Exercises

1. If 7 cards are dealt from an ordinary deck of 52 playing cards, what is the probability that
   (a) exactly 2 of them will be face cards?
   (b) at least 1 of them will be a queen?

2. To avoid detection at customs, a traveler has placed 6 narcotic tablets in a bottle containing 9 vitamin pills that are similar in appearance. If the customs official selects 3 of the tablets at random for analysis, what is the probability that the traveler will be arrested for illegal possession of narcotics?

3. A homeowner plants 6 bulbs selected at random from a box containing 5 tulip bulbs and 4 daffodil bulbs. What is the probability that he planted 2 daffodil bulbs and 4 tulip bulbs?

4. From a lot of 10 missiles, 4 are selected at random and fired. If the lot contains 3 defective missiles that will not fire, what is the probability that
   (a) all 4 will fire?
   (b) at most 2 will not fire?

5. A random committee of size 3 is selected from 4 doctors and 2 nurses. Write a formula for the probability distribution of the random variable X representing the number of doctors on the committee. Find $P(2 \leq X \leq 3)$.

6. What is the probability that a waitress will refuse to serve alcoholic beverages to only 2 minors if she randomly checks the IDs of 5 students from among 9 students of which 4 are not of legal age?

7. A company is interested in evaluating its current inspection procedure on shipments of 50 identical items. The procedure is to take a sample of 5 and pass the shipment if no more than 2 are found to be defective. What proportion of 20% defective shipments will be accepted?

8. A manufacturing company uses an acceptance scheme on production items before they are shipped. The plan is a two-stage one. Boxes of 25 are readied for shipment and a sample of 3 are tested for defectives. If any defectives are found, the entire box is sent back for 100% screening. If no defectives are found, the box is shipped.
   (a) What is the probability that a box containing 3 defectives will be shipped?
   (b) What is the probability that a box containing only 1 defective will be sent back for screening?

9. Suppose that the manufacturing company in Exercise 8 decided to change its acceptance scheme. Under the new scheme an inspector takes one at random, inspects it, and then replaces it in the box; a second inspector does likewise. Finally, a third inspector goes through the same procedure. The box is not shipped if any of the three find a defective. Answer Exercise 8 under this new plan.

10. In Exercise 4 how many defective missiles might we expect to be included among the 4 that are selected? Use Chebyshev's theorem to describe the variability of the number of defective missiles included when 4 are selected from several lots each of size 10 containing 3 defective missiles.

11. If a person is dealt 13 cards from an ordinary deck of 52 playing cards several times, how many hearts per hand can he expect? Between what two values would you expect the number of hearts to fall at least 75% of the time?

12. It is estimated that 4000 of the 10,000 voting residents of a town are against a new sales tax. If 15 eligible voters are selected at random and asked their opinion, what is the probability that at most 7 favor the new tax?

13. An annexation suit is being considered against a county subdivision of 1200 residences by a neighboring city. If the occupants of half the residences object to being annexed, what is the probability that in a random sample of 10 at least 3 favor the annexation suit?

14. Among 150 IRS employees in a large city, only 30 are women. If 10 of the applicants are chosen at random to provide free tax assistance for the residents of this city, use the binomial approximation to the hypergeometric to find the probability that at least 3 women are selected.

15. A nationwide survey of 17,000 seniors by the University of Michigan reveals that almost 70% disapprove of daily pot smoking according to a report in *Parade*, September 14, 1980. If 18 of these seniors are selected at random and asked their opinions, what is the probability that more than 9 but less than 14 disapprove of smoking pot?

16. Find the probability of being dealt a bridge hand of 13 cards containing 5 spades, 2 hearts, 3 diamonds, and 3 clubs.

17. A foreign student club lists as its members 2 Canadians, 3 Japanese, 5 Italians, and 2 Germans. If a committee of 4 is selected at random, find the probability that
(a) all nationalities are represented;
(b) all nationalities except the Italians are represented.

18. An urn contains 3 green balls, 2 blue balls, and 4 red balls. In a random sample of 5 balls, find the probability that both blue balls and at least 1 red ball are selected.

5.5   Negative Binomial and Geometric Distributions ———

Let us consider an experiment in which the properties are the same as those listed for a binomial experiment, with the exception that the trials will be repeated until a *fixed* number of successes occur. Therefore, instead of finding the probability of x successes in n trials, where n is fixed, we are now interested in the probability that the kth success occurs on the xth trial. Experiments of this kind are called **negative binomial experiments**.

As an illustration, consider the use of a drug that is known to be effective in 60% of the cases in which it is used. The use of the drug will be considered a success if it is effective in bringing some degree of relief to the patient. We are interested in finding the probability that the fifth patient to experience relief is the seventh patient to receive the drug during a given week. Designating a success by S and a failure by F, a possible order of achieving the desired result is $SFSSSFS$, which occurs with probability $(0.6)(0.4)(0.6)(0.6)(0.6)(0.4)(0.6) = (0.6)^5(0.4)^2$. We could list all possible orders by rearranging the F's and S's except for the last outcome, which must be the fifth success. The total number of possible orders is equal to the number of partitions of the first six trials into two groups with 2 failures assigned to the one group and the 4 successes assigned to the other group. This can be done in $\binom{6}{4} = 15$ mutually exclusive ways. Hence, if X represents the outcome on which the fifth success occurs, then

$$P(X = 7) = \binom{6}{4}(0.6)^5(0.4)^2 = 0.1866.$$

The number X of trials to produce k successes in a negative binomial experiment is called a **negative binomial random variable** and its probability distribution is called the **negative binomial distribution**. Since its probabilities depend on the number of successes desired and the probability of a success on a given trial, we shall denote them by the symbol $b^*(x; k, p)$. To obtain the general formula for $b^*(x; k, p)$, consider the probability of a success on the xth trial preceded by $k - 1$ successes and $x - k$ failures in some specified order. Since the trials are independent, we can multiply all the probabilities corresponding to each desired outcome. Each success occurs with probability p and each failure with probability $q = 1 - p$. Therefore, the probability for the specified order, ending in success, is $p^{k-1}q^{x-k}p = p^kq^{x-k}$. The total number of sample points in the experiment ending in a success, after the occurrence of $k - 1$ successes and $x - k$ failures in any order, is equal to the number of partitions of $x - 1$ trials into two groups with $k - 1$ successes corresponding to one group and $x - k$ failures corresponding to the other group. This number is given by the term $\binom{x - 1}{k - 1}$, each mutually exclusive and occurring with equal probability $p^k q^{x-k}$. We obtain the general formula by multiplying $p^k q^{x-k}$ by $\binom{x - 1}{k - 1}$.

NEGATIVE BINOMIAL DISTRIBUTION

If repeated independent trials can result in a success with probability p and a failure with probability q = 1 − p, then the probability distribution of the random variable X, the number of the trial on which the kth success occurs, is given by

$$b^*(x; k, p) = \binom{x - 1}{k - 1} p^k q^{x-k}, \qquad x = k, \, k + 1, \, k + 2, \, \ldots\ldots \quad ■$$

EXAMPLE 5.13   Find the probability that a person tossing three coins will get either all heads or all tails for the second time on the fifth toss.

SOLUTION
Using the negative binomial distribution with $x = 5$, $k = 2$, and $p = 1/4$, we have

$$b^*\left(5; 2, \frac{1}{4}\right) = \binom{4}{1}\left(\frac{1}{4}\right)^2\left(\frac{3}{4}\right)^3$$

$$= \frac{4!}{1! \, 3!} \cdot \frac{3^3}{4^5}$$

$$= \frac{27}{256}.$$

The negative binomial distribution derives its name from the fact that each term in the expansion of $p^k(1 - q)^{-k}$ corresponds to the values of $b^*(x; k, p)$ for $x = k$, $k + 1$, $k + 2$, $\ldots\ldots$.

If we consider the special case of the negative binomial distribution where $k = 1$, we have a probability distribution for the number of trials required for a single success. An example would be the tossing of a coin until a head occurs. We might be interested in the probability that the first head occurs on the fourth toss. The negative binomial distribution reduces to the form $b^*(x; 1, p) = pq^{x-1}$, $x = 1, 2, 3, \ldots\ldots$. Since the successive terms constitute a geometric progression, it is customary to refer to this special case as the **geometric distribution** and denote its values by $g(x; p)$.

GEOMETRIC DISTRIBUTION

If repeated independent trials can result in a success with probability p and a failure with probability q = 1 − p, then the probability distribution of the random variable X, the number of the trial on which the first success occurs, is given by

$$g(x; p) = pq^{x-1}, \qquad x = 1, 2, 3, \ldots\ldots \quad ■$$

EXAMPLE 5.14   In a certain manufacturing process it is known that, on the average, 1 in every 100 items is defective. What is the probability that the fifth item inspected is the first defective item found?

SOLUTION
Using the geometric distribution with $x = 5$ and $p = 0.01$, we have

$$g(5; 0.01) = (0.01)(0.99)^4$$

$$= 0.0096.$$

EXAMPLE 5.15   At *busy time* a telephone exchange is very near capacity, so people cannot find a line to use. It may be of interest to know the number of attempts necessary in order to gain a connection. Suppose that we let $p = 0.05$ be the probability of a connection during busy time. We are interested in knowing the probability that 5 attempts are necessary for a successful call.

SOLUTION
Using the geometric distribution with $x = 5$ and $p = 0.05$ yields

$$P(X = x) = g(5; 0.05) = (0.05)(0.95)^4$$

$$= 0.041.$$

Quite often in applications dealing with the geometric distribution, the mean and variance are important. For example, in Example 5.15 the *expected* number of calls necessary to make a connection is quite important. The following gives without proof the mean and variance of the geometric distribution.

THEOREM 5.4   *The mean and variance of a random variable following the geometric distribution are given by*

$$\mu = \frac{1}{p}$$

$$\sigma^2 = \frac{1 - p}{p^2}.$$ ∎

5.6 Poisson Distribution and the Poisson Process ——————

Experiments yielding numerical values of a random variable X, the number of outcomes occurring during a given time interval or in a specified region, are often called **Poisson experiments**. The given time interval may be of any length, such as a minute, a day, a week, a month, or even a year. Hence a Poisson experiment might generate observations for the random variable X representing the number of telephone calls per hour received by an office, the number of days school is closed due to snow during the winter, or the number of postponed games due to rain during a baseball season. The specified region could be a line segment, an area, a volume, or perhaps a piece of material. In this case X might represent the number of field mice per acre, the number of bacteria in a given culture, or the number of typing errors per page.

A Poisson experiment is derived from the **Poisson process** and possesses the following properties:

Properties of Poisson Process

1. The number of outcomes occurring in one time interval or specified region is independent of the number that occurs in any other disjoint time interval or region of space. In this way we say that the Poisson process has no memory.

2. The probability that a single outcome will occur during a very short time interval or in a small region is proportional to the length of the time interval or the size of the region and does not depend on the number of outcomes occurring outside this time interval or region.

3. The probability that more than one outcome will occur in such a short time interval or fall in such a small region is negligible.

The number X of outcomes occurring in a Poisson experiment is called a **Poisson random variable** and its probability distribution is called the **Poisson distribution**. The mean number of outcomes is computed from $\mu = \lambda t$, where t is the specific "time" or "region" of interest. Since its probabilities depend on λ, the rate of occurrence of outcomes, we shall denote them by the symbol $P(x; \lambda t)$. The derivation of the formula for $p(x; \lambda t)$, based on properties of a Poisson process listed above, is beyond the scope of this book. The following is used for computing Poisson probabilities.

POISSON DISTRIBUTION

The probability distribution of the Poisson random variable X, representing the number of outcomes occurring in a given time interval or specified region denoted by t, is given by

$$p(x; \lambda t) = \frac{e^{-\lambda t}(\lambda t)^x}{x!}, \qquad x = 0, 1, 2, \ldots ,$$

where λ is the average number of outcomes per unit time or region and $e = 2.71828 \ldots .$ ■

Table A.2 contains Poisson probability sum $P(r; \lambda t) = \displaystyle\sum_{x=0}^{r} p(x; \lambda t)$ for a few selected values of λt ranging from 0.1 to 18. We illustrate the use of this table with the following two examples.

EXAMPLE 5.16   The average number of radioactive particles passing through a counter during 1 millisecond in a laboratory experiment is 4. What is the probability that 6 particles enter the counter in a given millisecond?

SOLUTION

Using the Poisson distribution with $x = 6$ and $\lambda t = 4$, we find from Table A.2 that

$$p(6; 4) = \frac{e^{-4}4^6}{6!} = \sum_{x=0}^{6} p(x; 4) - \sum_{x=0}^{5} p(x; 4) = 0.8893 - 0.7851$$

$$= 0.1042.$$

EXAMPLE 5.17   The average number of oil tankers arriving each day at a certain port city is known to be 10. The facilities at the port can handle at most 15 tankers per day. What is the probability that on a given day tankers will have to be sent away?

SOLUTION

Let X be the number of tankers arriving each day. Then, using Table A.2, we have

$$P(X > 15) = 1 - P(X \le 15)$$

$$= 1 - \sum_{x=0}^{15} p(x; 10)$$

$$= 1 - 0.9513$$

$$= 0.0487.$$

Like the binomial distribution, the Poisson distribution enjoys application in quality control, quality assurance, and acceptance sampling. In addition, certain important continuous distributions used in reliability theory and queuing theory depend on the Poisson process. Some of these distributions are discussed and developed in Chapter 6.

THEOREM 5.5   *The mean and variance of the Poisson distribution $p(x; \lambda t)$ both have the value λt.*
∎

PROOF.   To verify that the mean is indeed λt, let $\mu = \lambda t$. We can write

$$E(X) = \sum_{x=0}^{\infty} x \cdot \frac{e^{-\mu}\mu^x}{x!} = \sum_{x=1}^{\infty} x \cdot \frac{e^{-\mu}\mu^x}{x!}$$

$$= \mu \sum_{x=1}^{\infty} \frac{e^{-\mu}\mu^{x-1}}{(x-1)!}.$$

Now, let $y = x - 1$ to give

$$E(X) = \mu \sum_{y=0}^{\infty} \frac{e^{-\mu}\mu^y}{y!} = \mu,$$

since

$$\sum_{y=0}^{\infty} \frac{e^{-\mu}\mu^y}{y!} = \sum_{y=0}^{\infty} p(y; \mu) = 1.$$

The variance of the Poisson distribution is obtained by first finding

$$E[X(X-1)] = \sum_{x=0}^{\infty} x(x-1)\frac{e^{-\mu}\mu^x}{x!} = \sum_{x=2}^{\infty} x(x-1)\frac{e^{-\mu}\mu^x}{x!}$$

$$= \mu^2 \sum_{x=2}^{\infty} \frac{e^{-\mu}\mu^{x-2}}{(x-2)!}.$$

Setting $y = x - 2$, we have

$$E[X(X-1)] = \mu^2 \sum_{y=0}^{\infty} \frac{e^{-\mu}\mu^y}{y!} = \mu^2.$$

Hence

$$\sigma^2 = E[X(X-1)] + \mu - \mu^2$$

$$= \mu^2 + \mu - \mu^2$$

$$= \mu$$

$$= \lambda t.$$

In Example 5.16, where $\lambda t = 4$, we also have $\sigma^2 = 4$ and hence $\sigma = 2$. Using Chebyshev's theorem, we can state that our random variable has a probability of at least 3/4 of falling in the interval $\mu \pm 2\sigma = 4 \pm (2)(2)$, or from 0 to 8. Therefore, we conclude that at least three-fourths of the time the number of radioactive particles entering the counter will be anywhere from 0 to 8 during a given millisecond.

The Poisson Distribution As a Limiting Form of the Binomial

It should be evident from the three principles of the Poisson process that the Poisson distribution relates to the binomial distribution. Although the Poisson usually finds applications in space and time problems as illustrated by Examples 5.16 and 5.17, it can be viewed as a limiting form of the binomial distribution. In the case of the binomial, if n is quite large and p is small, the conditions begin to simulate the *continuous space or time region* implications of the Poisson process. The independence among Bernoulli trials in the binomial case is consistent with property 2 of the Poisson process. Allowing the parameter p to be close to zero relates to property 3. Indeed, we shall now derive the Poisson distribution as a limiting form of the binomial distribution when $n \to \infty$, $p \to 0$, and np remains constant. Hence, if n is large and p is close to 0, the Poisson distribution can be used, with $\mu = np$, to approximate binomial probabilities. If p is close to 1, we can still use the Poisson distribution to approximate binomial probabilities by interchanging what we have defined to be a success and a failure, thereby changing p to a value close to 0.

THEOREM 5.6   *Let X be a binomial random variable with probability distribution b(x; n, p). When* $n \to \infty$, $p \to 0$, *and* $\mu = np$ *remains constant,*

$$b(x; n, p) \to p(x; \mu). \qquad \blacksquare$$

PROOF. The binomial distribution can be written

$$b(x; n, p) = \binom{n}{x} p^x q^{n-x}$$

$$= \frac{n!}{x!(n-x)!} p^x (1-p)^{n-x}$$

$$= \frac{n(n-1)\cdots(n-x+1)}{x!} p^x (1-p)^{n-x}.$$

Substituting $p = \mu/n$, we have

$$b(x; n, p) = \frac{n(n-1)\cdots(n-x+1)}{x!} \left(\frac{\mu}{n}\right)^x \left(1 - \frac{\mu}{n}\right)^{n-x}$$

$$= 1\left(1 - \frac{1}{n}\right)\cdots\left(1 - \frac{x-1}{n}\right) \frac{\mu^x}{x!} \left(1 - \frac{\mu}{n}\right)^n \left(1 - \frac{\mu}{n}\right)^{-x}.$$

As $n \to \infty$ while x and μ remain constant,

$$\lim_{n \to \infty} 1\left(1 - \frac{1}{n}\right)\cdots\left(1 - \frac{x-1}{n}\right) = 1,$$

$$\lim_{n \to \infty} \left(1 - \frac{\mu}{n}\right)^{-x} = 1,$$

and from the definition of the number e,

$$\lim_{n \to \infty} \left(1 - \frac{\mu}{n}\right)^n = \lim_{n \to \infty} \left\{ \left[1 + \frac{1}{(-n)/\mu}\right]^{-n/\mu} \right\}^{-\mu} = e^{-\mu}.$$

Hence, under the given limiting conditions,

$$b(x; n, p) \to \frac{e^{-\mu}\mu^x}{x!}, \qquad x = 0, 1, 2, \ldots.$$

EXAMPLE 5.18   In a manufacturing process in which glass items are being produced, defects or bubbles occur, occasionally rendering the piece undesirable for marketing. It is known that on the average 1 in every 1000 of these items produced has one or more bubbles. What is the probability that a random sample of 8000 will yield fewer than 7 items possessing bubbles?

SOLUTION
This is essentially a binomial experiment with $n = 8000$ and $p = 0.001$. Since p is very close to zero and n is quite large, we shall approximate with the Poisson

distribution using $\mu = (8000)(0.001) = 8$. Hence, if X represents the number of bubbles, we have

$$P(X < 7) = \sum_{x=0}^{6} b(x; 8000, 0.001)$$

$$\simeq \sum_{x=0}^{6} p(x; 8)$$

$$= 0.3134.$$

Exercises

1. The probability that a person living in a certain city owns a dog is estimated to be 0.3. Find the probability that the tenth person randomly interviewed in this city is the fifth one to own a dog.

2. A scientist inoculates several mice, one at a time, with a disease germ until he finds 2 that have contracted the disease. If the probability of contracting the disease is 1/6, what is the probability that 8 mice are required?

3. In an inventory study it was determined that, on the average, demands for a particular item at a warehouse were made 5 times per day. What is the probability that on a given day this item is requested
(a) more than 5 times?
(b) not at all?

4. Find the probability that a person flipping a coin gets
(a) the third head on the seventh flip;
(b) the first head on the fourth flip.

5. Three people toss a coin and the odd man pays for the coffee. If the coins all turn up the same, they are tossed again. Find the probability that fewer than 4 tosses are needed.

6. According to a study published by a group of University of Massachusetts sociologists, about two-thirds of the 20 million persons in this country who take Valium are women. Assuming this figure to be a valid estimate, find the probability that on a given day the fifth prescription written by a doctor for Valium is
(a) the first prescribing Valium for a woman;
(b) the third prescribing Valium for a woman.

7. The probability that a student pilot passes the written test for his private pilot's license is 0.7. Find the probability that a person passes the test

(a) on the third try;
(b) before the fourth try.

8. On the average a certain intersection results in 3 traffic accidents per month. What is the probability that in any given month at this intersection
(a) exactly 5 accidents will occur?
(b) less than 3 accidents will occur?
(c) at least 2 accidents will occur?

9. A secretary makes 2 errors per page on the average. What is the probability that on the next page he or she makes
(a) 4 or more errors?
(b) no errors?

10. A certain area of the eastern United States is, on the average, hit by 6 hurricanes a year. Find the probability that in a given year this area will be hit by
(a) fewer than 4 hurricanes;
(b) anywhere from 6 to 8 hurricanes.

11. Suppose the probability is 0.8 that any given person will believe a tale about the transgressions of a famous actress. What is the probability that
(a) the sixth person to hear this tale is the fourth one to believe it?
(b) the third person to hear this tale is the first one to believe it?

12. The average number of field mice per acre in a 5-acre wheat field is estimated to be 12. Find the probability that fewer than 7 field mice are found
(a) on a given acre;
(b) on 2 of the next 3 acres inspected.

13. A restaurant prepares a tossed salad containing on the average 5 vegetables. Find the probability that the salad contains more than 5 vegetables

(a) on a given day;

(b) on 3 of the next 4 days;

(c) for the first time in April on April 5.

14. The probability that a person dies from a certain respiratory infection is 0.002. Find the probability that fewer than 5 of the next 2000 so infected will die.

15. Suppose that on the average 1 person in 1000 makes a numerical error in preparing his or her income tax return. If 10,000 forms are selected at random and examined, find the probability that 6, 7, or 8 of the forms will be in error.

16. The probability that a student fails the screening test for scoliosis (curvature of the spine) at a local high school is known to be 0.004. Of the next 1875 students who are screened for scoliosis, find the probability that

(a) fewer than 5 fail the test;

(b) 8, 9, or 10 fail the test.

17. (a) Find the mean and variance in Exercise 14 of the random variable X representing the number of persons among 2000 that die from the respiratory infection.

(b) According to Chebyshev's theorem, there is a probability of at least 3/4 that the number of persons to die among 2000 persons infected will fall within what interval?

18. (a) Find the mean and variance in Exercise 15 of the random variable X representing the number of persons among 10,000 who make an error in preparing their income tax returns.

(b) According to Chebyshev's theorem, there is a probability of at least 8/9 that the number of persons who make errors in preparing their income tax returns among 10,000 returns will be within what interval?

Review Exercises

1. In a manufacturing process 15 units are randomly selected each day from the production line to check the percent defective of the process. From historical information it is known that the probability of a defective unit is 0.05. Any time that two or more defectives are found in the sample of 15, the process is stopped. This procedure is used to provide a signal in case the probability of a defective has increased.

(a) What is the probability that on any given day the production process will be stopped? (Assume 5% defective.)

(b) Suppose that the probability of a defective has increased to 0.07. What is the probability that on any given day the production process will not be stopped?

2. An automatic welding machine is being considered for production. It will be considered successful if it is successful on 99% of its welds. Otherwise, it will not be considered efficient. A test is conducted on a prototype and it is to perform 100 welds. The machine will be accepted for manufacture if it misses no more than 3 welds.

(a) What is the probability that a good machine will be rejected?

(b) What is the probability that an inefficient machine with 95% welding success will be accepted?

3. A car rental agency at a local airport has available 5 Fords, 7 Chevrolets, 4 Dodges, 3 Hondas, and 4 Toyotas. If the agency randomly selects 9 of these cars to chauffeur delegates from the airport to the downtown convention center, find the probability that 2 Fords, 3 Chevrolets, 1 Dodge, 1 Honda, and 2 Toyotas are used.

4. Service calls come to a maintenance center according to a Poisson process and on the average 2.7 calls come per minute. Find the probability that

(a) no more than 4 calls come in any minute;

(b) fewer than 2 calls come in any minute;

(c) more than 10 calls come in a 5-minute period.

5. An electronics firm claims that the proportion of defective units of a certain process is 5%. A buyer has a standard procedure of inspecting 15 units selected randomly from a large lot. On a particular occasion, the buyer found 5 items defective.

(a) What is the probability of this occurrence, given that the claim of 5% defective is correct?

(b) What would be your reaction if you were the buyer?

6. An electronic switching device occasionally malfunctions and may need to be replaced. It is known that the device is satisfactory if it makes, on the average, no more than 0.20 error per hour. A particular 5-hour

period is chosen as a "test" on the device. If no more than 1 error occurs, the device is considered satisfactory.

(a) What is the probability that a satisfactory device will be considered unsatisfactory on the basis of the test? Assume that a Poisson process exists.

(b) What is the probability that a device will be accepted as satisfactory when, in fact, the mean number of errors is 0.25? Again, assume that a Poisson process exists.

7. A company generally purchases large lots of a certain type of electronic device. A method is used that rejects a lot if two or more defective units are found in a random sample of 100 units.

(a) What is the probability of rejecting a lot that is 1% defective?

(b) What is the probability of accepting a lot that is 5% defective?

8. A local drugstore owner knows that on average 100 people stop by his store per hour.

(a) Find the probability that in a given 3-minute period nobody enters the store.

(b) Find the probability that in a given 3-minute period more than 5 people enter the store.

9. (a) Suppose that you throw 4 dice. Find the probability that you get at least one 1.

(b) Suppose that you throw 2 dice 24 times. Find the probability that you get at least one (1, 1); that is, you roll *snake-eyes*.

[NOTE: The probability of part (a) is greater than the probability of part (b).]

10. Suppose that 500 lottery tickets are sold. Among them, 200 tickets *at least* pay off the cost of the ticket. Now suppose that you buy 5 tickets. Find the probability that you win back at least the cost of 3 tickets.

Some Continuous Probability Distributions

6.1 Normal Distribution

The most important continuous probability distribution in the entire field of statistics is the **normal distribution**. Its graph, called the **normal curve**, is the bell-shaped curve of Figure 6.1, which approximately describes many phenomena that occur in nature, industry, and research. Physical measurements in areas such as meteorological experiments, rainfall studies, and measurements on manufactured parts are often more than adequately explained with a normal distribution. In addition, errors in scientific measurements are extremely well approximated by a normal distribution. In 1733, Abraham DeMoivre developed the mathematical equation of the normal curve. It provided a basis on which much of the theory of inductive statistics is founded. The normal distribution is often referred to as the **Gaussian distribution**, in honor of Karl Friedrich Gauss (1777–1855), who also derived its equation from a study of errors in repeated measurements of the same quantity.

A continuous random variable X having the bell-shaped distribution of Figure 6.1 is called a **normal random variable**. The mathematical equation for the probability

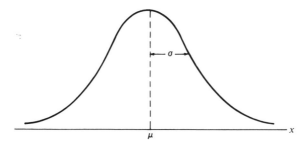

FIGURE 6.1   The normal curve.

distribution of the normal variable depends upon the two parameters μ and σ, its mean and standard deviation. Hence we denote the values of the density of X by $n(x; \mu, \sigma)$.

**NORMAL
DISTRIBUTION**

The density function of the normal random variable X, with mean μ and variance σ^2, is

$$n(x; \mu, \sigma) = \frac{1}{\sqrt{2\pi}\,\sigma}\, e^{-(1/2)[(x-\mu)/\sigma]^2}, \qquad -\infty < x < \infty,$$

where $\pi = 3.14159\ldots$ and $e = 2.71828\ldots$.   ■

Once μ and σ are specified, the normal curve is completely determined. For example, if $\mu = 50$ and $\sigma = 5$, then the ordinates $n(x; 50, 5)$ can easily be computed for various values of x and the curve drawn. In Figure 6.2 we have sketched two normal curves having the same standard deviation but different means. The two curves are identical in form but are centered at different positions along the horizontal axis.

In Figure 6.3 we have sketched two normal curves with the same mean but different standard deviations. This time we see that the two curves are centered at exactly the same position on the horizontal axis, but the curve with the larger standard deviation is lower and spreads out farther. Remember that the area under a probability curve must be equal to 1 and therefore the more variable the set of observations the lower and wider the corresponding curve will be.

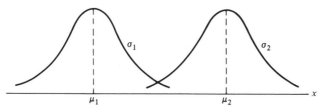

FIGURE 6.2   Normal curves with $\mu_1 < \mu_2$ and $\sigma_1 = \sigma_2$.

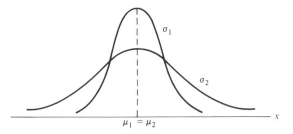

FIGURE 6.3   Normal curves with $\mu_1 = \mu_2$ and $\sigma_1 < \sigma_2$.

Figure 6.4 shows the results of sketching two normal curves having different means and different standard deviations. Clearly, they are centered at different positions on the horizontal axis and their shapes reflect the two different values of σ.

From an inspection of Figures 6.1 through 6.4 and by examination of the first and second derivatives of $n(x; \mu, \sigma)$, we list the following properties of the normal curve:

1. The mode, which is the point on the horizontal axis where the curve is a maximum, occurs at $x = \mu$.
2. The curve is symmetric about a vertical axis through the mean μ.
3. The curve has its points of inflection at $x = \mu \pm \sigma$, is concave downward if $\mu - \sigma < X < \mu + \sigma$, and is concave upward otherwise.
4. The normal curve approaches the horizontal axis asymptotically as we proceed in either direction away from the mean.
5. The total area under the curve and above the horizontal axis is equal to 1.

We shall now show that the parameters μ and σ^2 are indeed the mean and the variance of the normal distribution. To evaluate the mean, we write

$$E(X) = \frac{1}{\sqrt{2\pi}\,\sigma} \int_{-\infty}^{\infty} x e^{-(1/2)[(x-\mu)/\sigma]^2}\,dx.$$

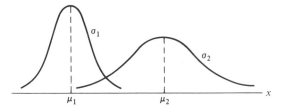

FIGURE 6.4   Normal curves with $\mu_1 < \mu_2$ and $\sigma_1 < \sigma_2$.

Setting $z = (x - \mu)/\sigma$ and $dx = \sigma\, dz$, we obtain

$$E(X) = \frac{1}{\sqrt{2\pi}} \int_{-\infty}^{\infty} (\mu + \sigma z) e^{-z^2/2}\, dz$$

$$= \mu \frac{1}{\sqrt{2\pi}} \int_{-\infty}^{\infty} e^{-z^2/2}\, dz + \frac{\sigma}{\sqrt{2\pi}} \int_{-\infty}^{\infty} z e^{-z^2/2}\, dz.$$

The first term on the right is μ times the area under a normal curve with mean zero and variance 1, and hence equal to μ. By straightforward integration, the second term is equal to 0. Hence

$$E(X) = \mu.$$

The variance of the normal distribution is given by

$$E[(X - \mu)^2] = \frac{1}{\sqrt{2\pi}\,\sigma} \int_{-\infty}^{\infty} (x - \mu)^2 e^{-(1/2)[(x-\mu)/\sigma]^2}\, dx.$$

Again setting $z = (x - \mu)/\sigma$ and $dx = \sigma\, dz$, we obtain

$$E[(X - \mu)^2] = \frac{\sigma^2}{\sqrt{2\pi}} \int_{-\infty}^{\infty} z^2 e^{-z^2/2}\, dz.$$

Integrating by parts with $u = z$ and $dv = z e^{-z^2/2}\, dz$ so that $du = dz$ and $v = -e^{-z^2/2}$, we find that

$$E[(X - \mu)^2] = \frac{\sigma^2}{\sqrt{2\pi}} \left(-z e^{-z^2/2} \Big|_{-\infty}^{\infty} + \int_{-\infty}^{\infty} e^{-z^2/2}\, dz \right)$$

$$= \sigma^2(0 + 1)$$

$$= \sigma^2.$$

Many random variables have probability distributions that can be described adequately by the normal curve once μ and σ^2 are specified. In this chapter we shall assume that these two parameters are known, perhaps from previous investigations. Later we shall make statistical inferences when μ and σ^2 are unknown and have been estimated from the available experimental data.

We pointed out earlier the role that the normal distribution plays as a reasonable approximation of scientific variables in real-life experiments. There are other applications of the normal distribution that the reader will appreciate as he or she moves on in the book. The normal distribution finds enormous application as a *limiting distribution*. Under certain conditions the normal distribution provides a good continuous approximation to the binomial and hypergeometric distributions. The case of the approximation to the binomial is covered in Section 6.4. In Chapter 8 the reader will learn about **sampling distributions**. It turns out that the limiting distribution of sample averages is normal. This provides a broad base for statistical inference that proves very valuable to the data analyst interested in estimation and hypothesis testing. The important areas of analysis of variance (Chapters 13, 14, and 15) and quality control (Chapter 17) have their theory based on assumptions that make use of the normal distribution.

6.2 Areas Under the Normal Curve

The curve of any continuous probability distribution or density function is constructed so that the area under the curve bounded by the two ordinates $x = x_1$ and $x = x_2$ equals the probability that the random variable X assumes a value between $x = x_1$ and $x = x_2$. Thus, for the normal curve in Figure 6.5,

$$P(x_1 < X < x_2) = \int_{x_1}^{x_1} n(x; \mu, \sigma) \, dx$$

$$= \frac{1}{\sqrt{2\pi}\,\sigma} \int_{x_1}^{x_2} e^{-(1/2)[(x-\mu)/\sigma]^2} \, dx$$

is represented by the area of the shaded region.

In Figures 6.2, 6.3, and 6.4 we saw how the normal curve is dependent on the mean and the standard deviation of the distribution under investigation. The area under the curve between any two ordinates must then also depend on the values μ and σ. This is evident in Figure 6.6, where we have shaded regions corre-

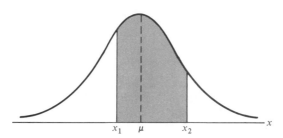

FIGURE 6.5 $P(x_1 < X < x_2)$ = area of the shaded region.

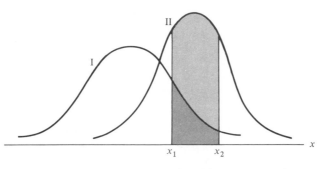

FIGURE 6.6 $P(x_1 < X < x_2)$ for different normal curves.

sponding to $P(x_1 < X < x_2)$ for two curves with different means and variances. The $P(x_1 < X < x_2)$, where X is the random variable describing distribution I, is indicated by the darker shaded area. If X is the random variable describing distribution II, then $P(x_1 < X < x_2)$ is given by the entire shaded region. Obviously, the two shaded regions are different in size; therefore, the probability associated with each distribution will be different for the two given values of X.

The difficulty encountered in solving integrals of normal density functions necessitates the tabulation of normal curve areas for quick reference. However, it would be a hopeless task to attempt to set up separate tables for every conceivable value of μ and σ. Fortunately, we are able to transform all the observations of any normal random variable X to a new set of observations of a normal random variable Z with mean zero and variance 1. This can be done by means of the transformation

$$Z = \frac{X - \mu}{\sigma}.$$

Whenever X assumes a value x, the corresponding value of Z is given by $z = (x - \mu)/\sigma$. Therefore, if X falls between the values $x = x_1$ and $x = x_2$, the random variable Z will fall between the corresponding values $z_1 = (x_1 - \mu)/\sigma$ and $z_2 = (x_2 - \mu)/\sigma$. Consequently, we may write

$$P(x_1 < X < x_2) = \frac{1}{\sqrt{2\pi}\,\sigma} \int_{x_1}^{x_2} e^{-(1/2)[(x-\mu)/\sigma]^2}\, dx$$

$$= \frac{1}{\sqrt{2\pi}} \int_{z_1}^{z_2} e^{-z^2/2}\, dz$$

$$= \int_{z_1}^{z_2} n(z; 0, 1)\, dz = P(z_1 < Z < z_2),$$

where Z is seen to be a normal random variable with mean zero and variance 1.

DEFINITION 6.1   *The distribution of a normal random variable with mean zero and variance 1 is called a* **standard normal distribution**.   ■

The original and transformed distributions are illustrated in Figure 6.7. Since all the values of X falling between x_1 and x_2 have corresponding z values between z_1

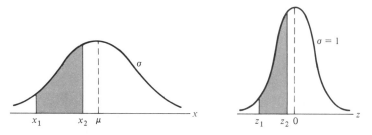

FIGURE 6.7   The original and transformed normal distributions.

and z_2, the area under the X-curve between the ordinates $x = x_1$ and $x = x_2$ in Figure 6.7 equals the area under the Z-curve between the transformed ordinates $z = z_1$ and $z = z_2$.

We have now reduced the required number of tables of normal-curve areas to one, that of the standard normal distribution. Table A.3 gives the area under the standard normal curve corresponding to $P(Z < z)$ for values of z ranging from -3.49 to 3.49. To illustrate the use of this table, let us find the probability that Z is less than 1.74. First, we locate a value of z equal to 1.7 in the left column, then move across the row to the column under 0.04, where we read 0.9591. Therefore, $P(Z < 1.74) = 0.9591$. To find a z-value corresponding to a given probability, the process is reversed. For example, the z-value leaving an area of 0.2148 under the curve to the left of z is seen to be -0.79.

EXAMPLE 6.1   Given a standard normal distribution, find the area under the curve that lies (a) to the right of $z = 1.84$, and (b) between $z = -1.97$ and $z = 0.86$.

SOLUTION
(a) The area in Figure 6.8(a) to the right of $z = 1.84$ is equal to 1 minus the area in Table A.3 to the left of $z = 1.84$, namely, $1 - 0.9671 = 0.0329$.
(b) The area in Figure 6.8(b) between $z = -1.97$ and $z = 0.86$ is equal to the area to the left of $z = 0.86$ minus the area to the left of $z = -1.97$. From Table A.3 we find the desired area to be $0.8051 - 0.0244 = 0.7807$.

EXAMPLE 6.2   Given a standard normal distribution, find the value of k such that (a) $P(Z > k) = 0.3015$, and (b) $P(k < Z < -0.18) = 0.4197$.

SOLUTION
(a) In Figure 6.9(a) we see that the k-value leaving an area of 0.3015 to the right must then leave an area of 0.6985 to the left. From Table A.3 it follows that $k = 0.52$.
(b) From Table A.3 we note that the total area to the left of -0.18 is equal to 0.4286. In Figure 6.9(b) we see that the area between k and -0.18 is 0.4197 so that the area to the left of k must be $0.4286 - 0.4197 = 0.0089$. Hence, from Table A.3, we have $k = -2.37$.

EXAMPLE 6.3   Given a normal distribution with $\mu = 50$ and $\sigma = 10$, find the probability that X assumes a value between 45 and 62.

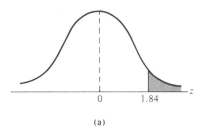

(a)

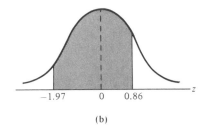

(b)

FIGURE 6.8   Areas for Example 6.1.

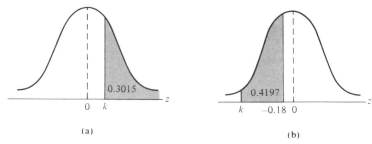

(a)   (b)

FIGURE 6.9   Areas for Example 6.2.

SOLUTION
The z-values corresponding to $x_1 = 45$ and $x_2 = 62$ are

$$z_1 = \frac{45 - 50}{10} = -0.5$$

and

$$z_2 = \frac{62 - 50}{10} = 1.2.$$

Therefore,

$$P(45 < X < 62) = P(-0.5 < Z < 1.2).$$

The $P(-0.5 < Z < 1.2)$ is given by the area of the shaded region in Figure 6.10. This area may be found by subtracting the area to the left of the ordinate $z = -0.5$ from the entire area to the left of $z = 1.2$. Using Table A.3, we have

$$P(45 < X < 62) = P(-0.5 < Z < 1.2)$$

$$= P(Z < 1.2) - P(Z < -0.5)$$

$$= 0.8849 - 0.3085$$

$$= 0.5764.$$

EXAMPLE 6.4   Given a normal distribution with $\mu = 300$ and $\sigma = 50$, find the probability that X assumes a value greater than 362.

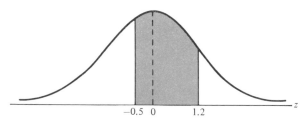

FIGURE 6.10   Area for Example 6.3.

SOLUTION

The normal probability distribution showing the desired area is given in Figure 6.11. To find the $P(X > 362)$, we need to evaluate the area under the normal curve to the right of $x = 362$. This can be done by transforming $x = 362$ to the corresponding z-value, obtaining the area to the left of z from Table A.3, and then subtracting this area from 1. We find that

$$z = \frac{362 - 300}{50} = 1.24.$$

Hence

$$P(X > 362) = P(Z > 1.24)$$
$$= 1 - P(Z < 1.24)$$
$$= 1 - 0.8925$$
$$= 0.1075.$$

According to Chebyshev's theorem, the probability that a random variable assumes a value within 2 standard deviations of the mean is at least 3/4. If the random variable has a normal distribution, the z-values corresponding to $x_1 = \mu - 2\sigma$ and $x_2 = \mu + 2\sigma$ are easily computed to be

$$z_1 = \frac{(\mu - 2\sigma) - \mu}{\sigma} = -2$$

and

$$z_2 = \frac{(\mu + 2\sigma) - \mu}{\sigma} = 2.$$

Hence

$$P(\mu - 2\sigma < X < \mu + 2\sigma) = P(-2 < Z < 2)$$
$$= P(Z < 2) - P(Z < -2)$$
$$= 0.9772 - 0.0228$$
$$= 0.9544,$$

which is a much stronger statement than that given by Chebyshev's theorem.

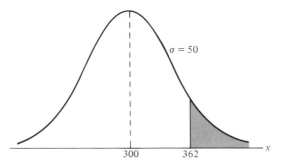

FIGURE 6.11 Area for Example 6.4.

Occasionally, we are required to find the value of z corresponding to a specified probability that falls between values listed in Table A.3 (see Example 6.5). For convenience, we shall always choose the z-value corresponding to the tabular probability that comes closest to the specified probability. However, if the given probability falls midway between two tabular probabilities, we shall choose for z the value falling midway between the corresponding values of z. For instance, to find the z-value corresponding to a probability of 0.7975, which falls between 0.7967 and 0.7995 in Table A.3, we choose $z = 0.83$, since 0.7975 is closer to 0.7967. On the other hand, for a probability of 0.7981, which falls midway between 0.7967 and 0.7995, we take $z = 0.835$.

The preceding two examples were solved by going first from a value of x to a z-value and then computing the desired area. In Example 6.5 we reverse the process and begin with a known area or probability, find the z-value, and then determine x by rearranging the formula

$$z = \frac{x - \mu}{\sigma} \quad \text{to give} \quad x = \sigma z + \mu.$$

EXAMPLE 6.5   Given a normal distribution with $\mu = 40$ and $\sigma = 6$, find the value of x that has (a) 45% of the area to the left, and (b) 14% of the area to the right.

SOLUTION

(a) An area of 0.45 to the left of the desired x-value is shaded in Figure 6.12(a). We require a z-value that leaves an area of 0.45 to the left. From Table A.3 we find $P(Z < -0.13) = 0.45$ so that the desired z-value is -0.13. Hence

$$x = (6)(-0.13) + 40$$

$$= 39.22.$$

(b) In Figure 6.12(b) we shade an area equal to 0.14 to the right of the desired x-value. This time we require a z-value that leaves 0.14 of the area to the right and hence an area of 0.86 to the left. Again, from Table A.3, we find $P(Z > 1.08) = 0.86$ so that the desired z-value is 1.08 and

$$x = (6)(1.08) + 40$$

$$= 46.48.$$

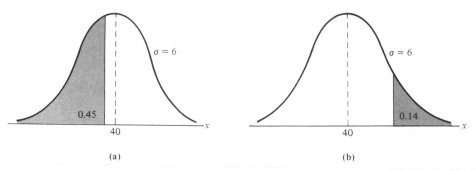

FIGURE 6.12   Areas for Example 6.5.

6.3 Applications of the Normal Distribution _____

Some of the many problems to which the normal distribution is applicable are treated in the following examples. The use of the normal curve to approximate binomial probabilities will be considered in Section 6.4.

EXAMPLE 6.6   A certain type of storage battery lasts on the average 3.0 years, with a standard deviation of 0.5 year. Assuming that the battery lives are normally distributed, find the probability that a given battery will last less than 2.3 years.

SOLUTION

First construct a diagram such as Figure 6.13, showing the given distribution of battery lives and the desired area. To find the $P(X < 2.3)$, we need to evaluate the area under the normal curve to the left of 2.3. This is accomplished by finding the area to the left of the corresponding x-value. Hence we find that

$$z = \frac{2.3 - 3}{0.5} = -1.4,$$

and then using Table A.3 we have

$$P(X < 2.3) = P(Z < -1.4)$$

$$= 0.0808.$$

EXAMPLE 6.7   An electrical firm manufactures light bulbs that have a length of life that is normally distributed with mean equal to 800 hours and a standard deviation of 40 hours. Find the probability that a bulb burns between 778 and 834 hours.

SOLUTION

The distribution of light bulbs is illustrated in Figure 6.14. The z-values corresponding to $x_1 = 778$ and $x_2 = 834$ are

$$z_1 = \frac{778 - 800}{40} = -0.55$$

and

$$z_2 = \frac{834 - 800}{40} = 0.85.$$

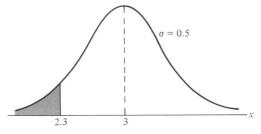

FIGURE 6.13   Area for Example 6.6.

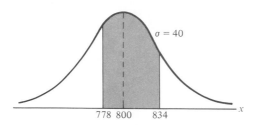

FIGURE 6.14   Area for Example 6.7.

Hence

$$P(778 < X < 834) = P(-0.55 < Z < 0.85)$$

$$= P(Z < 0.85) - P(Z < -0.55)$$

$$= 0.8023 - 0.2912$$

$$= 0.5111.$$

EXAMPLE 6.8   In an industrial process the diameter of a ball bearing is an important component part. The buyer sets specifications on the diameter to be 3.0 ± 0.01 cm. The implication is that no part falling outside these specifications will be accepted. It is known that in the process the diameter of a ball bearing has a normal distribution with mean 3.0 and standard deviation $\sigma = 0.005$. On the average, how many manufactured ball bearings will be scrapped?

SOLUTION

The distribution of diameters is illustrated in Figure 6.15. The values corresponding to the specification limits are $x_1 = 2.99$ and $x_2 = 3.01$. The corresponding z-values are

$$z_1 = \frac{2.99 - 3.0}{0.005} = -2.0$$

$$z_2 = \frac{3.01 - 3.00}{0.005} = +2.0.$$

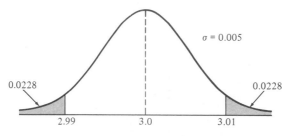

FIGURE 6.15   Area for Example 6.8.

Hence

$$P(2.99 < X < 3.01) = P(-2.0 < Z < 2.0).$$

From Table A.3, $P(Z < -2.0) = 0.0228$. Due to symmetry of the normal distribution, we find that

$$P(-2.0 < Z < 2.0) = 1 - 2(0.0228)$$
$$= 0.9544.$$

As a result, it is anticipated that on the average, 4.56% of manufactured ball bearings will be scrapped.

EXAMPLE 6.9   Gauges are used to reject all components in which a certain dimension is not within the specification $1.50 \pm d$. It is known that this measurement is normally distributed with mean 1.50 and standard deviation 0.2. Determine the value d such that the specifications "cover" 95% of the measurements.

SOLUTION
From Table A.3 we know that

$$P(-1.96 < Z < 1.96) = 0.95.$$

Therefore,

$$1.96 = \frac{(1.50 + d) - 1.50}{0.2},$$

from which we obtain

$$d = (0.2)(1.96) = 0.392.$$

An illustration of the specifications is given in Figure 6.16.

EXAMPLE 6.10   A certain machine makes electrical resistors having a mean resistance of 40 ohms and a standard deviation of 2 ohms. Assuming that the resistance follows a normal distribution and can be measured to any degree of accuracy, what percentage of resistors will have a resistance that exceeds 43 ohms?

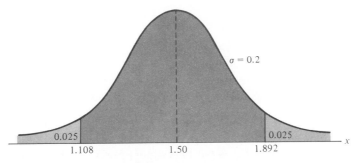

FIGURE 6.16 Specifications for Example 6.9.

SOLUTION

A percentage is found by multiplying the relative frequency by 100%. Since the relative frequency for an interval is equal to the probability of falling in the interval, we must find the area to the right of $x = 43$ in Figure 6.17. This can be done by transforming $x = 43$ to the corresponding z-value, obtaining the area to the left of z from Table A.3, and then subtracting this area from 1. We find

$$z = \frac{43 - 40}{2} = 1.5.$$

Hence

$$P(X > 43) = P(Z > 1.5)$$
$$= 1 - P(Z < 1.5)$$
$$= 1 - 0.9332$$
$$= 0.0668.$$

Therefore, 6.68% of the resistors will have a resistance exceeding 43 ohms.

EXAMPLE 6.11   Find the percentage of resistances exceeding 43 ohms in Example 6.10 if the resistance is measured to the nearest ohm.

SOLUTION

This problem differs from Example 6.10 in that we now assign a measurement of 43 ohms to all resistors whose resistances are greater than 42.5 and less than 43.5. We are actually approximating a discrete distribution by means of a continuous normal distribution. The required area is the region shaded to the right of 43.5 in Figure 6.18. We now find that

$$z = \frac{43.5 - 40}{2} = 1.75.$$

Hence

$$P(X > 43.5) = P(Z > 1.75)$$
$$= 1 - P(Z < 1.75)$$
$$= 1 - 0.9599$$
$$= 0.0401.$$

Therefore, 4.01% of the resistances exceed 43 ohms when measured to the nearest ohm. The difference 6.68% − 4.01% = 2.67% between this answer and that of Example 6.10 represents all those resistors having a resistance greater than 43 and less than 43.5 that are now being recorded as 43 ohms.

EXAMPLE 6.12   On an examination the average grade was 74 and the standard deviation was 7. If 12% of the class are given A's, and the grades are curved to follow a normal distribution, what is the lowest possible A and the highest possible B?

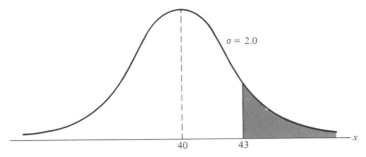

FIGURE 6.17   Area for Example 6.10.

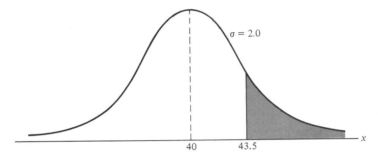

FIGURE 6.18   Area for Example 6.11.

SOLUTION
In this example we begin with a known area of probability, find the z-value, and then determine x from the formula $x = \sigma z + \mu$. An area of 0.12, corresponding to the fraction of students receiving A's, is shaded in Figure 6.19. We require a z-value that leaves 0.12 of the area to the right and hence an area of 0.88 to the left. From Table A.3, $P(Z < 1.175) = 0.88$, so that the desired z-value is 1.175. Hence

$$x = (7)(1.175) + 74$$

$$= 82.225.$$

Therefore, the lowest A is 83 and the highest B is 82.

EXAMPLE 6.13   Refer to Example 6.12 and find the sixth decile.

SOLUTION
The sixth decile, written D_6, is the x-value that leaves 60% of the area to the left as shown in Figure 6.20. From Table A.3 we find $P(Z < 0.25) = 0.6$, so that the desired z-value is 0.25. Now

$$x = (7)(0.25) + 74$$

$$= 75.75.$$

Hence $D_6 = 75.75$. That is, 60% of the grades are 75 or less.

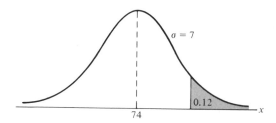

FIGURE 6.19   Area for Example 6.12.

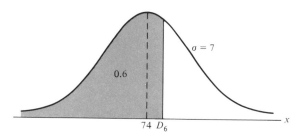

FIGURE 6.20   Area for Example 6.13.

Exercises

1. Given a standard normal distribution, find the area under the curve which lies
   (a) to the left of $z = 1.43$;
   (b) to the right of $z = -0.89$;
   (c) between $z = -2.16$ and $z = -0.65$;
   (d) to the left of $z = -1.39$;
   (e) to the right of $z = 1.96$;
   (f) between $z = -0.48$ and $z = 1.74$.

2. Find the value of z if the area under a standard normal curve
   (a) to the right of z is 0.3622;
   (b) to the left of z is 0.1131;
   (c) between 0 and z, with $z > 0$, is 0.4838;
   (d) between $-z$ and z, with $z > 0$, is 0.9500.

3. Given a standard normal distribution, find the value of k such that
   (a) $P(Z < k) = 0.0427$;
   (b) $P(Z > k) = 0.2946$;
   (c) $P(-0.93 < Z < k) = 0.7235$.

4. Given a normal distribution with $\mu = 30$ and $\sigma = 6$, find
   (a) the normal-curve area to the right of $x = 17$;
   (b) the normal-curve area to the left of $x = 22$;

   (c) the normal-curve area between $x = 32$ and $x = 41$;
   (d) the value of x that has 80% of the normal-curve area to the left;
   (e) the two values of x that contain the middle 75% of the normal-curve area.

5. Given the normally distributed variable X with mean 18 and standard deviation 2.5, find
   (a) $P(X < 15)$;
   (b) the value of k such that $P(X < k) = 0.2236$;
   (c) the value of k such that $P(X > k) = 0.1814$;
   (d) $P(17 < X < 21)$.

6. According to Chebyshev's theorem, the probability that any random variable assumes a value within 3 standard deviations of the mean is at least 8/9. If it is known that the probability distribution of a random variable X is normal with mean μ and variance σ^2, what is the exact value of $P(\mu - 3\sigma < X < \mu + 3\sigma)$?

7. A UCLA research scientist reports that mice will live an average of 40 months when their diets are sharply restricted and then enriched with vitamins and proteins. Assuming that the lifetimes of such mice are

normally distributed with a standard deviation of 6.3 months, find the probability that a given mouse will live

(a) more than 32 months;

(b) less than 28 months;

(c) between 37 and 49 months.

8. The loaves of rye bread distributed to local stores by a certain bakery have an average length of 30 centimeters and a standard deviation of 2 centimeters. Assuming that the lengths are normally distributed, what percentage of the loaves are

(a) longer than 31.7 centimeters?

(b) between 29.3 and 33.5 centimeters in length?

(c) shorter than 25.5 centimeters?

9. A soft-drink machine is regulated so that it discharges an average of 200 milliliters per cup. If the amount of drink is normally distributed with a standard deviation equal to 15 milliliters,

(a) what fraction of the cups will contain more than 224 milliliters?

(b) what is the probability that a cup contains between 191 and 209 milliliters?

(c) how many cups will probably overflow if 230-milliliter cups are used for the next 1000 drinks?

(d) below what value do we get the smallest 25% of the drinks?

10. The finished inside diameter of a piston ring is normally distributed with a mean of 10 centimeters and a standard deviation of 0.03 centimeter.

(a) What proportion of rings will have inside diameters exceeding 10.075 centimeters?

(b) What is the probability that a piston ring will have an inside diameter between 9.97 and 10.03 centimeters?

(c) Below what value of inside diameter will 15% of the piston rings fall?

11. A lawyer commutes daily from his suburban home to his midtown office. On the average trip one way takes 24 minutes, with a standard deviation of 3.8 minutes. Assume the distribution of trip times to be normally distributed.

(a) What is the probability that a trip will take at least 1/2 hour?

(b) If the office opens at 9:00 A.M. and he leaves his house at 8:45 A.M. daily, what percentage of the time is he late for work?

(c) If he leaves the house at 8:35 A.M. and coffee is served at the office from 8:50 A.M. until 9:00

A.M., what is the probability that he misses coffee?

(d) Find the length of time above which we find the slowest 15% of the trips.

(e) Find the probability that 2 of the next 3 trips will take at least 1/2 hour.

12. If a set of grades on a statistics examination is approximately normally distributed with a mean of 74 and a standard deviation of 7.9, find

(a) the lowest passing grade if the lowest 10% of the students are given F's;

(b) the highest B if the top 5% of the students are given A's;

(c) the lowest B if the top 10% of the students are given A's and the next 25% are given B's.

13. The average life of a certain type of small motor is 10 years with a standard deviation of 2 years. The manufacturer replaces free all motors that fail while under guarantee. If he is willing to replace only 3% of the motors that fail, how long a guarantee should he offer? Assume that the lives of the motors follow a normal distribution.

14. The heights of 1000 students are normally distributed with a mean of 174.5 centimeters and a standard deviation of 6.9 centimeters. Assuming that the heights are recorded to the nearest half-centimeter, how many of these students would you expect to have heights

(a) less than 160.0 centimeters?

(b) between 171.5 and 182.0 centimeters inclusive?

(c) equal to 175.0 centimeters?

(d) greater than or equal to 188.0 centimeters?

15. A company pays its employees an average wage of $9.25 an hour with a standard deviation of 60 cents. If the wages are approximately normally distributed and paid to the nearest cent,

(a) what percentage of the workers receive wages between $8.75 and $9.69 an hour inclusive?

(b) the highest 5% of the employee hourly wages is greater than what amount?

16. The weights of a large number of miniature poodles are approximately normally distributed with a mean of 8 kilograms and a standard deviation of 0.9 kilogram. If measurements are recorded to the nearest tenth of a kilogram, find the fraction of these poodles with weights

(a) over 9.5 kilograms;

(b) at most 8.6 kilograms;

(c) between 7.3 and 9.1 kilograms inclusive.

17. The tensile strength of a certain metal component is normally distributed with a mean 10,000 kilograms per square centimeter and a standard deviation of 100 kilograms per square centimeter. Measurements are recorded to the nearest 50 kilograms per square centimeter.
    (a) What proportion of these components exceed 10,150 kilograms per square centimeter in tensile strength?
    (b) If specifications require that all components have tensile strength between 9800 and 10,200 kilograms per square centimeter inclusive, what proportion of pieces would we expect to scrap?

18. If a set of observations are normally distributed, what percent of these differ from the mean by
    (a) more than 1.3σ?
    (b) less than 0.52σ?

19. The IQs of 600 applicants to a certain college are approximately normally distributed with a mean of 115 and a standard deviation of 12. If the college requires an IQ of at least 95, how many of these students will be rejected on this basis regardless of their other qualifications?

20. The average rainfall, recorded to the nearest hundredth of a centimeter, in Roanoke, Virginia, for the month of March is 9.22 centimeters. Assuming a normal distribution with a standard deviation of 2.83 centimeters, find the probability that next March Roanoke receives
    (a) less than 1.84 centimeters of rain;
    (b) more than 5 centimeters but not over 7 centimeters of rain;
    (c) more than 13.8 centimeters of rain.

21. In a mathematics examination the average grade was 82 and the standard deviation was 5. All students with grades from 88 to 94 received a grade of B. If the grades are approximately normally distributed and 8 students received a B grade, how many students took the examination?

6.4 Normal Approximation to the Binomial

Probabilities associated with binomial experiments are readily obtainable from the formula $b(x; n, p)$ of the binomial distribution or from Table A.1 when n is small. In addition, binomial probabilities are readily available in many software computer packages. However, it is instructive to learn the relationship between the binomial and normal distribution. In Section 5.6 we illustrated how the Poisson distribution can be used to approximate binomial probabilities when n is quite large and p is very close to 0 or 1. Both the binomial and Poisson distributions are discrete. The first application of a continuous probability distribution to approximate probabilities over a discrete sample space was demonstrated in Example 6.11, where the normal curve was used. The normal distribution is often a good approximation to a discrete distribution when the latter takes on a symmetric bell shape. From a theoretical point of view, some distributions converge to the normal as their parameters approach certain limits. The normal distribution is a convenient approximating distribution because the cumulative distribution function is so easily tabled. The binomial distribution is nicely approximated by the normal in practical problems when one works with the cumulative distribution function. We now state a theorem that allows us to use areas under the normal curve to approximate binomial properties when n is sufficiently large.

THEOREM 6.1   *If X is a binomial random variable with mean $\mu = np$ and variance $\sigma^2 = npq$, then the limiting form of the distribution of*

$$Z = \frac{X - np}{\sqrt{npq}},$$

as $n \to \infty$, is the standard normal distribution $n(z; 0, 1)$. ■

It turns out that the normal distribution with $\mu = np$ and $\sigma^2 = np(1 - p)$ not only provides a very accurate approximation to the binomial distribution when n is large and p is not extremely close to 0 or 1, but also provides a fairly good approximation even when n is small and p is reasonably close to 1/2.

To illustrate the normal approximation to the binomial distribution, we first draw the histogram for $b(x; 15, 0.4)$ and then superimpose the particular normal curve having the same mean and variance as the binomial variable X. Hence we draw a normal curve with

$$\mu = np = (15)(0.4) = 6$$

and

$$\sigma^2 = npq = (15)(0.4)(0.6) = 3.6.$$

The histogram of $b(x; 15, 0.4)$ and the corresponding superimposed normal curve, which is completely determined by its mean and variance, are illustrated in Figure 6.21.

The exact probability that the binomial random variable X assumes a given value x is equal to the area whose base is centered at x. For example, the exact probability that X assumes the value 4 is equal to the area of the rectangle with base centered at $x = 4$. Using Table A.1, we find this area to be

$$P(X = 4) = b(4; 15, 0.4) = 0.1268,$$

which is approximately equal to the area of the shaded region under the normal curve between the two ordinates $x_1 = 3.5$ and $x_2 = 4.5$ in Figure 6.22. Converting to z-values, we have

$$z_1 = \frac{3.5 - 6}{1.897} = -1.32$$

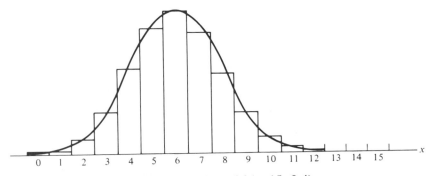

FIGURE 6.21   Normal approximation of $b(x; 15, 0.4)$.

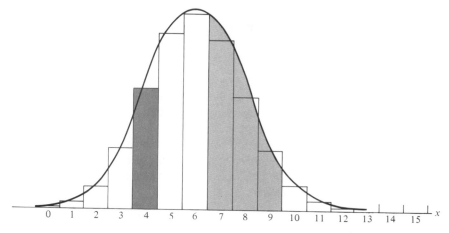

FIGURE 6.22   Normal approximation of $b(4; 15, 0.4)$ and $\sum\limits_{x=7}^{9} b(x; 15, 0.4)$.

and

$$z_2 = \frac{4.5 - 6}{1.897} = -0.79.$$

If X is a binomial random variable and Z a standard normal variable, then

$$P(X = 4) = b(4; 15, 0.4)$$

$$\approx P(-1.32 < Z < -0.79)$$

$$= P(Z < -0.79) - P(Z < -1.32)$$

$$= 0.2148 - 0.0934$$

$$= 0.1214.$$

This agrees very closely with the exact value of 0.1268.

The normal approximation is most useful in calculating binomial sums for large values of n. Referring to Figure 6.22, we might be interested in the probability that X assumes a value from 7 to 9 inclusive. The exact probability is given by

$$P(7 \leq X \leq 9) = \sum_{x=7}^{9} b(x; 15, 0.4)$$

$$= \sum_{x=0}^{9} b(x; 15, 0.4) - \sum_{x=0}^{6} b(x; 15, 0.4)$$

$$= 0.9662 - 0.6098$$

$$= 0.3564,$$

which is equal to the sum of the areas of the rectangles with bases centered at $x = 7$, 8, and 9. For the normal approximation we find the area of the shaded region under the curve between the ordinates $x_1 = 6.5$ and $x_2 = 9.5$ in Figure 6.22. The corresponding z-values are

$$z_1 = \frac{6.5 - 6}{1.897} = 0.26$$

and

$$z_2 = \frac{9.5 - 6}{1.897} = 1.85.$$

Now,

$$P(7 \le X \le 9) \simeq P(0.26 < Z < 1.85)$$
$$= P(Z < 1.85) - P(Z < 0.26)$$
$$= 0.9678 - 0.6026$$
$$= 0.3652.$$

Once again, the normal-curve approximation provides a value that agrees very closely with the exact value of 0.3564. The degree of accuracy, which depends on how well the curve fits the histogram, will increase as n increases. This is particularly true when p is not very close to $1/2$ and the histogram is no longer symmetric. Figures 6.23 and 6.24 show the histograms for $b(x; 6, 0.2)$ and $b(x; 15, 0.2)$, respectively. It is evident that a normal curve would fit the histogram when $n = 15$ considerably better than when $n = 6$.

In summary, we use the normal approximation to evaluate binomial probabilities whenever p is not close to 0 or 1. The approximation is excellent when n is large and fairly good for small values of n if p is reasonably close to $1/2$. One possible guide to determining when the normal approximation may be used is provided by calculating np and nq. If both np and nq are greater than or equal to 5, the approximation will be good.

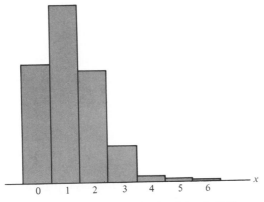

FIGURE 6.23 Histogram for $b(x; 6, 0.2)$.

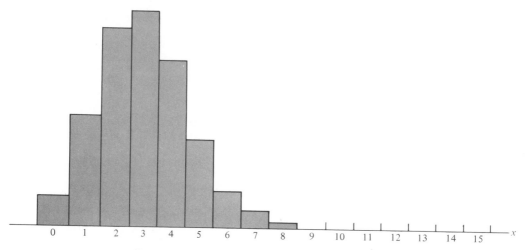

FIGURE 6.24   Histogram for $b(x; 15, 0.2)$.

As we indicated earlier, the quality of the approximation is quite good for large n. If p is close to 1/2, a moderate or small sample size will be sufficient for a reasonable approximation. We offer Table 6.1 as an indication of the quality of the approximation. Both the normal approximation and the true binomial cumulative probabilities are given. Notice that at $p = 0.05$ and $p = 0.10$, the approximation is fairly crude for $n = 10$. However, even for $n = 10$, note the improvement for $p = 0.50$. On the other hand, when p is fixed at $p = 0.05$, note the improvement of the approximation as we go from $n = 20$ to $n = 100$.

EXAMPLE 6.14   The probability that a patient recovers from a rare blood disease is 0.4. If 100 people are known to have contracted this disease, what is the probability that less than 30 survive?

SOLUTION
Let the binomial variable X represent the number of patients that survive. Since $n = 100$, we should obtain fairly accurate results using the normal-curve approximation with

$$\mu = np = (100)(0.4) = 40$$

and

$$\sigma = \sqrt{npq} = \sqrt{(100)(0.4)(0.6)} = 4.899.$$

To obtain the desired probability, we have to find the area to the left of $x = 29.5$. The z-value corresponding to 29.5 is

$$z = \frac{29.5 - 40}{4.899} = -2.14,$$

and the probability of fewer than 30 of the 100 patients surviving is given by the shaded region in Figure 6.25. Hence

$$P(X < 30) \simeq P(Z < -2.14) = 0.0162.$$

TABLE 6.1 Normal Approximation and True Cumulative Binomial Probabilities

r	p = 0.05, n = 10 Binomial	Normal	p = 0.10, n = 10 Binomial	Normal	p = 0.50, n = 10 Binomial	Normal
0	0.5987	0.5000	0.3487	0.2981	0.0010	0.0022
1	0.9139	0.9265	0.7361	0.7019	0.0107	0.0136
2	0.9885	0.9981	0.9298	0.9429	0.0547	0.0571
3	0.9990	1.0000	0.9872	0.9959	0.1719	0.1711
4	1.0000	1.0000	0.9984	0.9999	0.3770	0.3745
5			1.0000	1.0000	0.6230	0.6255
6					0.8281	0.8289
7					0.9453	0.9429
8					0.9893	0.9864
9					0.9990	0.9978
10					1.0000	0.9997

r	p = 0.05, n = 20 Binomial	Normal	p = 0.05, n = 50 Binomial	Normal	p = 0.05, n = 100 Binomial	Normal
0	0.3585	0.3015	0.0769	0.0968	0.0059	0.0197
1	0.7358	0.6985	0.2794	0.2578	0.0371	0.0537
2	0.9245	0.9382	0.5405	0.5000	0.1183	0.1251
3	0.9841	0.9948	0.7604	0.7422	0.2578	0.2451
4	0.9974	0.9998	0.8964	0.9032	0.4360	0.4090
5	0.9997	1.0000	0.9622	0.9744	0.6160	0.5910
6	1.0000	1.0000	0.9882	0.9953	0.7660	0.7549
7			0.9968	0.9994	0.8720	0.8749
8			0.9992	0.9999	0.9369	0.9463
9			0.9998	1.0000	0.9718	0.9803
10			1.0000	1.0000	0.9885	0.9941

EXAMPLE 6.15   A multiple-choice quiz has 200 questions each with 4 possible answers of which only 1 is the correct answer. What is the probability that sheer guesswork yields from 25 to 30 correct answers for 80 of the 200 problems about which the student has no knowledge?

SOLUTION
The probability of a correct answer for each of the 80 questions is $p = 1/4$. If X represents the number of correct answers due to guesswork, then

$$P(25 \leq X \leq 30) = \sum_{x=25}^{30} b(x; 80, \tfrac{1}{4}).$$

Using the normal curve approximation with

$$\mu = np = (80)(\tfrac{1}{4}) = 20$$

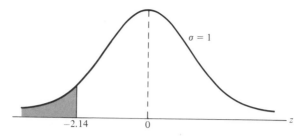

FIGURE 6.25   Area for Example 6.14.

and

$$\sigma = \sqrt{npq} = \sqrt{(180)(\tfrac{1}{4})(\tfrac{3}{4})} = 3.873,$$

we need the area between $x_1 = 24.5$ and $x_2 = 30.5$. The corresponding z-values are

$$z_1 = \frac{24.5 - 20}{3.873} = 1.16$$

and

$$z_2 = \frac{30.5 - 20}{3.873} = 2.71.$$

The probability of correctly guessing from 25 to 30 questions is given by the shaded region in Figure 6.26. From Table A.3 we find that

$$P(25 \leq X \leq 30) = \sum_{x=25}^{30} b(x; 80, \tfrac{1}{4})$$

$$\approx P(1.16 < Z < 2.71)$$

$$= P(X < 2.71) - P(Z < 1.16)$$

$$= 0.9966 - 0.8770$$

$$= 0.1196.$$

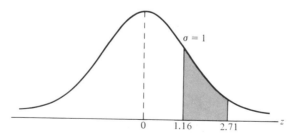

FIGURE 6.26   Area for Example 6.15.

Exercises

1. Evaluate $P(1 \leq X \leq 4)$ for a binomial variable with $n = 15$ and $p = 0.2$ by using
  (a) Table A.1 in the Appendix;
  (b) the normal-curve approximation.

2. A coin is tossed 400 times. Use the normal-curve approximation to find the probability of obtaining
  (a) between 185 and 210 heads inclusive;
  (b) exactly 205 heads;
  (c) less than 176 or more than 227 heads.

3. A process for manufacturing an electronic component is 0.1% defective. That is, on the average 1 component in 1000 is not usable. A quality control plan is to select 100 items from the process, and if none are defective, the process continues. Use the normal approximation to the binomial to find
  (a) the probability that the process continues for the sampling plan described;
  (b) the probability that the process continues even if the process has gone bad (i.e., if the frequency of defective components has shifted to 0.5% defective).

4. A process yields 10% defective items. If 100 items are randomly selected from the process, what is the probability that the number of defectives
  (a) exceeds 13?
  (b) is less than 8?

5. The probability that a patient recovers from a delicate heart operation is 0.9. Of the next 100 patients having this operation, what is the probability that
  (a) between 84 and 95 inclusive survive?
  (b) fewer than 86 survive?

6. Researchers at George Washington University and the National Institutes of Health claim that approximately 75% of the people believe "tranquilizers work very well to make a person more calm and relaxed." Of the next 80 people interviewed, what is the probability that
  (a) at least 50 are of this opinion?
  (b) at most 56 are of this opinion?

7. If 20% of the residents in a U.S. city prefer a white telephone over any other color available, what is the probability that among the next 1000 telephones installed in this city
  (a) between 170 and 185 inclusive will be white?
  (b) at least 210 but not more than 225 will be white?

8. A drug manufacturer claims that a certain drug cures a blood disease on the average 80% of the time. To check the claim, government testers used the drug on a sample of 100 individuals and decided to accept the claim if 75 or more are cured.
  (a) What is the probability that the claim will be rejected when the cure probability is, in fact, 0.8?
  (b) What is the probability that the claim will be accepted by the government when the cure probability is as low as 0.7?

9. One-sixth of the male freshmen entering a large state school are out-of-state students. If the students are assigned at random to the dormitories, 180 to a building, what is the probability that in a given dormitory at least one-fifth of the students are from out of state?

10. A pharmaceutical company knows that approximately 5% of its birth-control pills have an ingredient that is below the minimum strength, thus rendering the pill ineffective. What is the probability that fewer than 10 in a sample of 200 pills will be ineffective?

11. Statistics released by the National Highway Traffic Safety Administration and the National Safety Council show that on an average weekend night, 1 out of every 10 drivers on the road is drunk. If 400 drivers are randomly checked next Saturday night, what is the probability that the number of drunk drivers will be
  (a) less than 32?
  (b) more than 49?
  (c) at least 35 but less than 47?

12. A pair of dice is rolled 180 times. What is the probability that a total of 7 occurs
  (a) at least 25 times?
  (b) between 33 and 41 times inclusive?
  (c) exactly 30 times?

6.5 Gamma and Exponential Distributions _____

Although the normal distribution can be used to solve many problems in engineering and science, there are still numerous situations that require different types of density functions. Two such density functions, the **gamma** and **exponential distributions**, are discussed in this section. Additional densities are presented in Exercises 9 and 15 on pages 175 and 176 and in Sections 6.7, 8.6, 8.7, and 8.8.

It turns out that the exponential is a special case of the gamma distribution. Both find a large number of applications. The exponential and gamma distributions play an important role in both queuing theory and reliability problems. Time between arrivals at service facilities, and time to failure of component parts and electrical systems, often are nicely modeled by the exponential distribution. The relationship between the gamma and the exponential allows the gamma to be involved in similar types of problems. More details and illustrations will be supplied in Section 6.6.

The gamma distribution derives its name from the well-known **gamma function,** studied in many areas of mathematics. Before we proceed to the gamma distribution, let us review this function and some of its important properties.

DEFINITION 6.2   *The* **gamma function** *is defined by*

$$\Gamma(\alpha) = \int_0^\infty x^{\alpha-1} e^{-x} \, dx$$

for $\alpha > 0$.   ∎

Integrating by parts with $u = x^{\alpha-1}$ and $dv = e^{-x} \, dx$, we obtain

$$\Gamma(\alpha) = -e^{-x} x^{\alpha-1} \Big|_0^\infty + \int_0^\infty e^{-x} (\alpha - 1) x^{\alpha-2} \, dx$$

$$= (\alpha - 1) \int_0^\infty x^{\alpha-2} e^{-x} \, dx$$

for $\alpha > 1$, which yields the recursion formula

$$\Gamma(\alpha) = (\alpha - 1)\Gamma(\alpha - 1).$$

Repeated application of the recursion formula gives

$$\Gamma(\alpha) = (\alpha - 1)(\alpha - 2)\Gamma(\alpha - 2)$$

$$= (\alpha - 1)(\alpha - 2)(\alpha - 3)\Gamma(\alpha - 3),$$

and so forth. Note that when $\alpha = n$, where n is a positive integer,

$$\Gamma(n) = (n - 1)(n - 2), \ldots, \Gamma(1).$$

However, by Definition 6.2,

$$\Gamma(1) = \int_0^\infty e^{-x} \, dx = 1$$

and hence

$$\Gamma(n) = (n - 1)!.$$

One important property of the gamma function, left for the reader to verify (see Exercise 3 on page 175), is that $\Gamma(1/2) = \sqrt{\pi}$.

We shall now include the gamma function in our definition of the gamma distribution.

GAMMA DISTRIBUTION

The continuous random variable X has a **gamma distribution,** *with parameters α and β, if its density function is given by*

$$f(x) = \begin{cases} \dfrac{1}{\beta^\alpha \Gamma(\alpha)} x^{\alpha-1} e^{-x/\beta}, & x > 0 \\ 0, & elsewhere, \end{cases}$$

where $\alpha > 0$ and $\beta > 0$. ∎

Graphs of several gamma distributions are shown in Figure 6.27 for certain specified values of the parameters α and β. The special gamma distribution for which $\alpha = 1$ is called the **exponential distribution**.

EXPONENTIAL DISTRIBUTION

The continuous random variable X has an **exponential distribution**, *with parameter β, if its density function is given by*

$$f(x) = \begin{cases} \dfrac{1}{\beta} e^{-x/\beta}, & x > 0 \\ 0, & elsewhere, \end{cases}$$

where $\beta > 0$. ∎

The following theorem and corollary give the mean and variance of the gamma and exponential distributions.

THEOREM 6.2

The mean and variance of the gamma distribution are

$$\mu = \alpha\beta \quad and \quad \sigma^2 = \alpha\beta^2.$$ ∎

PROOF. To find the mean of the gamma distribution, we write

$$\mu = E(X) = \frac{1}{\beta^\alpha \Gamma(\alpha)} \int_0^\infty x^\alpha e^{-x/\beta} \, dx.$$

Now, let $y = x/\beta$, to give

$$\mu = \frac{\beta}{\Gamma(\alpha)} \int_0^\infty y^\alpha e^{-y} \, dy$$

$$= \frac{\beta \Gamma(\alpha + 1)}{\Gamma(\alpha)} = \alpha\beta.$$

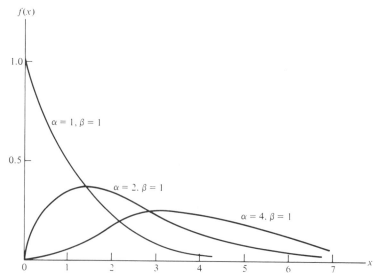

FIGURE 6.27   Gamma distributions.

To find the variance of the gamma distribution, we proceed as above to obtain

$$E(X^2) = \frac{\beta^2 \Gamma(\alpha + 2)}{\Gamma(\alpha)} = (\alpha + 1)\alpha\beta^2,$$

and then

$$\sigma^2 = E(X^2) - \mu^2 = (\alpha + 1)\alpha\beta^2 - \alpha^2\beta^2$$
$$= \alpha\beta^2.$$

COROLLARY 1   *The mean and variance of the exponential distribution are*

$$\mu = \beta \qquad and \qquad \sigma^2 = \beta^2.$$ ■

Relationship to the Poisson Process

We shall pursue applications of the exponential distribution, then return to the gamma distribution. The most important applications of the exponential distribution are situations where the Poisson process applies (see Chapter 5). The reader should recall that the Poisson process allows for the use of the discrete distribution called the Poisson distribution. Recall that the Poisson distribution is used to compute the probability of specific numbers of "events" during a particular *period of time or space*. In many applications, the time period or span of space is the random variable. For example, an industrial engineer may be interested in modeling the time T between arrivals at a congested intersection during rush hour in a large city. An arrival represents the Poisson event.

The relationship between the exponential distribution (often called the negative exponential) and the Poisson process is quite simple. In Chapter 5 the Poisson distribution was developed as a single-parameter distribution with parameter λ, where λ may be interpreted as the mean number of events *per unit "time."* Consider now the *random variable* described by the time required for the first event to occur. Using the Poisson distribution, we find that the probability of no events occurring in the span up to time t is given by

$$p(0; \lambda t) = \frac{e^{-\lambda t}(\lambda t)^0}{0!} = e^{-\lambda t}.$$

We can now make use of the above and let X be the time to the first Poisson event. The probability that the length of time until the first event will exceed x is the same as the probability that no Poisson events will occur in x. The latter, of course, is given by $e^{-\lambda x}$. As a result,

$$P(X \geq x) = e^{-\lambda x}.$$

Thus the cumulative distribution function for X is given by

$$P(0 \leq X \leq x) = 1 - e^{-\lambda x}.$$

Now, in order that we recognize the presence of the exponential distribution, we may differentiate the cumulative distribution function above to obtain the density function

$$f(x) = \lambda e^{-\lambda x},$$

which is the density function of the exponential distribution with $\lambda = 1/\beta$.

6.6 Applications of the Exponential and Gamma Distributions

In the foregoing we provided the foundation for the application of the exponential distribution in "time to arrival" or time to Poisson event problems. We will display illustrations here and then proceed to discuss the role of the gamma distribution in these modeling applications. Notice that the mean of the exponential distribution is the parameter β, the reciprocal of the parameter in the Poisson distribution. The reader should recall that it is often said that the Poisson distribution has no memory, implying that occurrences in successive time periods are independent. The important parameter β is the mean time between events. In reliability theory where equipment failure often conforms to this Poisson process, β is called *mean time between failures*. Many equipment breakdowns do follow the Poisson process, and thus the exponential distribution does apply. Other applications include survival times in biomedical experiments and computer response time.

In the following example we show a simple application of the exponential distribution to a problem in reliability. The binomial distribution also plays a role in the solution.

EXAMPLE 6.16   Suppose that a system contains a certain type of component whose time in years to failure is given by T. The random variable T is modeled nicely by the exponential distribution with mean time to failure $\beta = 5$. If 5 of these components are installed in different systems, what is the probability that at least 2 are still functioning at the end of 8 years?

SOLUTION

The probability that a given component is still functioning after 8 years is given by

$$P(T > 8) = \frac{1}{5} \int_{8}^{\infty} e^{-t/5} \, dt$$

$$= e^{-8/5}$$

$$\simeq 0.2.$$

Let X represent the number of components functioning after 8 years. Then using the binomial distribution

$$P(X \geq 2) = \sum_{x=2}^{5} b(x; 5, 0.2)$$

$$= 1 - \sum_{x=0}^{1} b(x; b, 0.2)$$

$$= 1 - 0.7373$$

$$= 0.2627.$$

Exercises 1, 7, 8, and Review Exercise 2 provide additional examples of the use of the exponential distribution in waiting time and reliability problems.

The importance of the gamma distribution lies in the fact that it defines a family of which other distributions are special cases. But the gamma itself has important applications in waiting time and reliability theory. Whereas the exponential distribution describes the time until the occurrence of a Poisson event (or the time between Poisson events) the time (or space) occurring until a *specified number of Poisson events occur* is a random variable whose density function is described by that of the gamma distribution. This specific number of events is the parameter α in the gamma density function. Thus it becomes easy to understand that when $\alpha = 1$, the special case of the exponential distribution occurs. The gamma density can be developed from its relationship to the Poisson process in much the same manner as we developed the exponential density. The details are left to the reader. The following is a numerical example of the use of the gamma distribution in a waiting time application.

EXAMPLE 6.17   Suppose that telephone calls arriving at a particular switchboard follow a Poisson process with an average of 5 calls coming per minute. What is the probability that up to a minute will elapse until 2 calls have come in to the switchboard?

SOLUTION

The Poisson process applies with **time until 2 Poisson events** following a gamma distribution with $\beta = 1/5$ and $\alpha = 2$. Let the random variable X be the time in minutes that transpires before 2 calls come. The required probability is given by

$$P(X \le x) = \int_0^x \frac{1}{\beta^2} x e^{-x/\beta} \, dx$$

$$P(X \le 1) = 25 \int_0^1 x e^{-5x} \, dx$$

$$= [1 - e^{-5(1)}(1 + 5)] = 0.96.$$

6.7 Chi-Squared Distribution

Another very important special case of the gamma distribution is obtained by letting $\alpha = \nu/2$ and $\beta = 2$, where ν is a positive integer. The result is called the **chi-squared distribution**. The distribution has a single parameter, ν, called the **degrees of freedom**.

CHI-SQUARED DISTRIBUTION

The continuous random variable X has a chi-squared distribution, with ν degrees of freedom, if its density function is given by

$$f(x) = \begin{cases} \dfrac{1}{2^{\nu/2}\Gamma(\nu/2)} x^{\nu/2-1} e^{-x/2}, & x > 0 \\ 0, & elsewhere, \end{cases}$$

where ν is a positive integer. ■

The chi-squared distribution plays a vital role in statistical inference. Material in Chapter 8 will be used not only to relate the chi-squared distribution to the normal distribution, but also to build a foundation for its use in hypothesis testing and estimation in Chapters 9 and 10. Since the chi-squared distribution is a special case of the gamma, its mean and variance are easily obtained. As a result, we state an additional corollary to Theorem 6.2.

COROLLARY 2

The mean and variance of the chi-squared distribution are

$$\mu = \nu \quad and \quad \sigma^2 = 2\nu.$$ ■

6.8 Weibull Distribution

Modern technology has enabled us to design many complicated systems whose operation, or perhaps safety, depend on the reliability of the various components making up the systems. For example, a fuse may burn out, a steel column may buckle, or a heat-sensing device may fail. Identical components subjected to identical environmental conditions will fail at different and unpredictable times. We have seen the role that the gamma and exponential distributions play in these types of problems. Another distribution that has been used extensively in recent years to deal with such problems is the **Weibull distribution** introduced by the Swedish physicist Waloddi Weibull in 1939.

WEIBULL DISTRIBUTION

*The continuous random variable X has a **Weibull distribution**, with parameters α and β, if its density function is given by*

$$f(x) = \begin{cases} \alpha\beta x^{\beta-1} e^{-\alpha x^\beta} & x > 0 \\ 0, & elsewhere, \end{cases}$$

where α > 0 and β > 0.   ■

The graphs of the Weibull distribution for $\alpha = 1$ and various values of the parameter β are illustrated in Figure 6.28. We see that the curves change in shape considerably for different values of the parameters, particularly the parameter β. If we let $\beta = 1$, the Weibull distribution reduces to the exponential distribution. For values of $\beta > 1$, the curves become somewhat bell-shaped and resemble the normal curves, but display some skewness.

The mean and variance of the Weibull distribution are stated in the following theorem. The reader is asked to provide the proof in Exercise 13 on page 175.

THEOREM 6.3

The mean and variance of the Weibull distribution are

$$\mu = \alpha^{-1/\beta} \Gamma\left(1 + \frac{1}{\beta}\right)$$

$$\sigma^2 = \alpha^{-2/\beta}\left\{\Gamma\left(1 + \frac{2}{\beta}\right) - \left[\Gamma\left(1 + \frac{1}{\beta}\right)\right]^2\right\}.$$ ■

Like the gamma and exponential distribution, the Weibull distribution is also applied to reliability and life testing problems such as the **time to failure** or **life length** of a component, measured from some specified time until it fails. Let us represent this time to failure by the continuous random variable T with probability density function $f(t)$, where $f(t)$ is the Weibull distribution.

To apply the Weibull distribution to reliability theory, let us first define the reliability of a component or product as the *probability that it will function properly for at least a specified time under specified experimental conditions.* Therefore, if $R(t)$ is defined to be the reliability of the given component at time t, we may write

$$R(t) = P(T > 1)$$

$$= \int_t^\infty f(t) \, dt$$

$$= 1 - F(t),$$

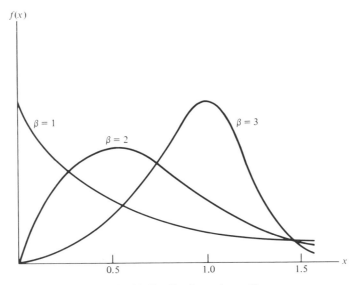

FIGURE 6.28 Weibull distributions ($\alpha = 1$).

where $F(t)$ is the cumulative distribution of T. The conditional probability that a component will fail in the interval from $T = t$ to $T = t + \Delta t$, given that it survived to time t, is given by

$$\frac{F(t + \Delta t) - F(t)}{R(t)}.$$

Dividing this ratio by Δt and taking the limit as $\Delta t \to 0$, we get the **failure rate**, denoted by $Z(t)$. Hence

$$Z(t) = \lim_{\Delta t \to 0} \frac{F(t + \Delta t) - F(t)}{\Delta t} \frac{1}{R(t)}$$

$$= \frac{F'(t)}{R(t)} = \frac{f(t)}{R(t)} = \frac{f(t)}{1 - F(t)},$$

which expresses the failure rate in terms of the distribution of the time to failure.

From the fact that $R(t) = 1 - F(t)$ and then $R'(t) = -F'(t)$, we can write the differential equation

$$Z(t) = \frac{-R'(t)}{R(t)} = \frac{-d[\ln R(t)]}{dt}$$

and then solving,

$$\ln R(t) = -\int Z(t)\, dt + \ln c$$

or

$$R(t) = ce^{-\int Z(t)\, dt},$$

where c satisfies the initial assumption that $R(0) = 1$ or $F(0) = 1 - R(0) = 0$. Thus we see that a knowledge of either the density function $f(t)$ or the failure rate $Z(t)$ uniquely determines the other.

EXAMPLE 6.18   Show that the failure-rate function is given by

$$Z(t) = \alpha\beta t^{\beta-1}, \qquad t > 0,$$

if and only if the time to failure distribution is the Weibull distribution

$$f(t) = \alpha\beta t^{\beta-1}e^{-\alpha t^{\beta}}, \qquad t > 0.$$

SOLUTION
Assume that $Z(t) = \alpha\beta t^{\beta-1}$, $t > 0$. Then we can write

$$f(t) = Z(t)R(t),$$

where

$$R(t) = ce^{-\int Z(t)\, dt} = ce^{-\int \alpha\beta t^{\beta-1}\, dt} = ce^{-\alpha t^{\beta}}.$$

From the condition that $R(0) = 1$, we find that $c = 1$. Hence

$$R(t) = e^{-\alpha t^{\beta}}$$

and

$$f(t) = \alpha\beta t^{\beta-1}e^{-\alpha t^{\beta}}, \qquad t > 0.$$

Now, if we assume that

$$f(t) = \alpha\beta t^{\beta-1}e^{-\alpha t^{\beta}}, \qquad t > 0,$$

then $Z(t)$ is determined by writing

$$Z(t) = \frac{f(t)}{R(t)},$$

where

$$R(t) = 1 - F(t) = 1 - \int_{0}^{t} \alpha\beta x^{\beta-1}e^{-\alpha x^{\beta}}\, dx$$

$$= 1 + \int_{0}^{t} de^{-\alpha x^{\beta}}$$

$$= e^{-\alpha t^{\beta}}.$$

Then

$$Z(t) = \frac{\alpha\beta t^{\beta-1}e^{-\alpha t^{\beta}}}{e^{-\alpha t^{\beta}}} = \alpha\beta t^{\beta-1}, \qquad t > 0.$$

In Example 6.18 the failure rate is seen to decrease with time if $\beta < 1$, increase with time if $\beta > 1$, and is constant if $\beta = 1.0$. It is, then, easy to see that the simpler form, namely the exponential distribution, applies for constant failure rate (via the Poisson process).

Exercises

1. If a random variable X has the gamma distribution with $\alpha = 2$ and $\beta = 1$, find $P(1.8 < X < 2.4)$.

2. In a certain city, the daily consumption of water (in millions of liters) follows approximately a gamma distribution with $\alpha = 2$ and $\beta = 3$. If the daily capacity of this city is 9 million liters of water, what is the probability that on any given day the water supply is inadequate?

3. Use the gamma function with $y = \sqrt{2x}$ to show that $\Gamma(1/2) = \sqrt{\pi}$.

4. Suppose that the time, in hours, taken to repair a heat pump is a random variable X having a gamma distribution with parameters $\alpha = 2$ and $\beta = 1/2$. What is the probability that the next service call will require
(a) at most 1 hour to repair the heat pump?
(b) at least 2 hours to repair the heat pump?

5. (a) Find the mean and variance of the daily water consumption in Exercise 2.
(b) According to Chebyshev's theorem, there is a probability of at least 3/4 that the water consumption on any given day will fall within what interval?

6. In a certain city, the daily consumption of electric power, in millions of kilowatt-hours, is a random variable X having a gamma distribution with mean $\mu = 6$ and variance $\sigma^2 = 12$.
(a) Find the values of α and β.
(b) Find the probability that on any given day the daily power consumption will exceed 12 million kilowatt-hours.

7. The length of time for one individual to be served at a cafeteria is a random variable having an exponential distribution with a mean of 4 minutes. What is the probability that a person is served in less than 3 minutes on at least 4 of the next 6 days?

8. The life in years of a certain type of electrical switch has an exponential distribution with an average life of $\beta = 2$. If 100 of these switches are installed in different systems, what is the probability that at most 30 fail during the first year?

9. A random variable X has the **continuous uniform distribution** if its density function is given by

$$f(x) = \begin{cases} \dfrac{1}{\beta - \alpha}, & \alpha < x < \beta \\ 0, & \text{elsewhere.} \end{cases}$$

For a continuous uniform distribution with $\alpha = 2$ and $\beta = 7$, find
(a) $P(X \geq 4)$;
(b) $P(3 < X < 5.5)$.

10. The daily amount of coffee, in liters, dispensed by a machine located in an airport lobby is a random variable X having a continuous uniform distribution (see Exercise 9) with $\alpha = 7$ and $\beta = 10$. Find the probability that on a given day the amount of coffee dispensed by this machine will be
(a) at most 8.8 liters;
(b) more than 7.4 liters but less than 9.5 liters;
(c) at least 8.5 liters.

11. Given a continuous uniform distribution, show that
(a) $\mu = \dfrac{\alpha + \beta}{2}$;
(b) $\sigma^2 = \dfrac{(\beta - \alpha)^2}{12}$.

12. Suppose that the service life, in years, of a hearing aid battery is a random variable having a Weibull distribution with $\alpha = 1/2$ and $\beta = 2$.
(a) How long can such a battery be expected to last?
(b) What is the probability that such a battery will still be operating after 2 years?

13. Derive the mean and variance of the Weibull distribution.

14. The lives of a certain automobile seal have the Weibull distribution with failure rate $Z(t) = 1/\sqrt{t}$. Find the probability that such a seal is still in use after 4 years.

15. The continuous random variable X has the **beta distribution** with parameters α and β if its density function is given by

$f(x) =$

$$\begin{cases} \dfrac{\Gamma(\alpha + \beta)}{\Gamma(\alpha)\Gamma(\beta)} x^{\alpha-1}(1 - x)^{\beta-1}, & 0 < x < 1 \\ 0, & \text{elsewhere,} \end{cases}$$

where $\alpha > 0$ and $\beta > 0$. If the proportion of a brand of television sets requiring service during the first year of operation is a random variable having a beta distribution with $\alpha = 3$ and $\beta = 2$, what is the probability that at least 80% of the new models sold this year of this brand will require service during their first year of operation?

16. In a biomedical research activity it was determined that the survival time in weeks of an animal when subjected to a certain exposure to gamma radiation has a gamma distribution with $\alpha = 5$ and $\beta = 10$.
    (a) What is the mean survival time of a randomly selected animal of the type used in the experiment?
    (b) What is the standard deviation of survival time?
    (c) What is the probability that an animal survives more than 30 weeks?

17. The lifetime in weeks of a certain type of transistor is known to follow a gamma distribution with mean 10 weeks and standard deviation $\sqrt{50}$ weeks.
    (a) What is the probability that the transistor will last at most 50 weeks?
    (b) What is the probability that the transistor will not survive the first 10 weeks?

18. Computer response time is an important application of the gamma and exponential distribution. Suppose that a study of a certain computer system reveals that the response time in seconds has an exponential distribution with a mean of 3 seconds.
    (a) What is the probability that response time exceeds 5 seconds?
    (b) What is the probability that response time exceeds 10 seconds?

Review Exercises

1. According to a study published by a group of sociologists at the University of Massachusetts, approximately 49% of the Valium users in the state of Massachusetts are white-collar workers. What is the probability that between 482 and 510, inclusive, of the next 100 randomly selected Valium users from this state will be white-collar workers?

2. The exponential distribution is frequently applied to the waiting times between successes in a Poisson process. If the number of calls received per hour by a telephone answering service is a Poisson random variable with parameter $\lambda = 6$, we know that the time, in hours, between successive calls has an exponential distribution with parameter $\beta = 1/6$. What is the probability of waiting more than 15 minutes between any two successive calls?

3. When α is a positive integer n, the gamma distribution is also known as the **Erlang distribution.** Setting $\alpha = n$ in the gamma distribution on page 167, the Erlang distribution is

$$f(x) = \begin{cases} \dfrac{x^{n-1}e^{-x/\beta}}{\beta^n(n - 1)!}, & x > 0 \\ 0, & \text{elsewhere.} \end{cases}$$

It can be shown that if the times between successive events are independent, each having an exponential distribution with parameter β, then the total elapsed waiting time X until all n events occur has the Erlang distribution. Referring to Exercise 2, what is the probability that the next 3 calls will be received within the next 30 minutes?

4. A manufacturer of a certain type of large machine wishes to buy rivets from one of two manufacturers. It is important that the breaking strength of each rivet exceed 10,000 psi. Two manufacturers (A and B) offer this type of rivet. Both have rivets whose breaking strength is normally distributed. The mean breaking strengths for manufacturers A and B are 14,000 psi and 13,000 psi, respectively. The standard deviations are 2000 psi and 1000 psi, respectively. Which manufacturer will produce, on the average, the fewest number of defective rivets?

5. According to the 1981 May/June issue of *Consumers' Digest*, census figures show that in 1978 almost 53% of all households in the United States were composed of only one or two persons. Assuming that this percentage is still valid today, what is the probability that between 490 and 515 inclusive of the next 1000 randomly selected households in America will consist of either one or two persons?

6. The life of a certain type of device has an advertised failure rate of 0.01 per hour. The failure rate is con-

stant and the exponential distribution applies.

(a) What is the mean time to failure?

(b) What is the probability that 200 hours will pass before a failure is observed?

7. In a chemical processing plant it is important that the yield of a certain type of batch product stay above 80%. If it stays below 80% for an extended period of time, the company loses money. Occasional defective manufactured batches are of little concern. But if several batches per day are defective, the plant shuts down and adjustments are made. It is known that the yield is normally distributed with standard deviation 4%.

(a) What is the probability of a "false alarm" (yield below 80%) when the mean yield is 85%?

(b) What is the probability that a manufactured batch will have a yield that exceeds 80% when in fact the mean yield is 79%?

8. Consider an electrical component failure rate of once every 5 hours. It is important to consider the time that it takes for 2 components to fail.

(a) Assuming that the gamma distribution applies, what is the mean time that it takes for failure of 2 components?

(b) What is the probability that 12 hours will elapse before 2 components fail?

9. The elongation of a steel bar under a particular load has been established to be normally distributed with a mean of 0.05 inch and $\sigma = 0.01$ inch. Find the probability that the elongation is

(a) above 0.1 inch;

(b) below 0.04 inch;

(c) between 0.025 and 0.065 inch.

10. A controlled satellite is known to have an error (distance from target) that is normally distributed with mean zero and standard deviation 4 feet. The manufacturer of the satellite defines a "success" as a firing in which the satellite comes within 10 feet of the target. Compute the probability that the satellite fails.

11. A technician plans to test a certain type of resin developed in the laboratory to determine the nature of the time it takes before bonding takes place. It is known that the mean time to bonding is 3 hours and the standard deviation is 0.5 hour. It will be considered an undesirable product if the bonding time either is less than 1 hour or more than 4 hours. Comment on the utility of the resin. How often would its performance be considered undesirable? Assume that time to bonding is normally distributed.

7

Functions of Random Variables

7.1 Introduction

This chapter contains a broad spectrum of material. Chapters 5 and 6 dealt with specific types of distributions, both discrete and continuous. These are distributions that find use in many subject matter applications, including reliability, quality control, and acceptance sampling. In the present chapter we begin with a more general topic, that of distributions of functions of random variables. General techniques are given and illustrated by examples. This is followed by a related concept, *moment-generating functions*, which can be helpful in learning about distributions of linear functions of random variables.

7.2 Transformations of Variables

Frequently, in statistics, one encounters the need to derive the probability distribution of a function of one or more random variables. For example, suppose that X is a discrete random variable with probability distribution $f(x)$ and suppose further that

$Y = u(X)$ defines a one-to-one transformation between the values of X and Y. We wish to find the probability distribution of Y. It is important to note that the one-to-one transformation implies that each value x is related to one, and only one, value $y = u(x)$ and that each value y is related to one, and only one, value $x = w(y)$, where $w(y)$ is obtained by solving $y = u(x)$ for x terms of y.

From our discussion of discrete probability distributions in Chapter 3 it is clear that the random variable Y assumes the value y when X assumes the value $w(y)$. Consequently, the probability distribution of Y is given by

$$g(y) = P(Y = y) = P[X = w(y)] = f[w(y)].$$

THEOREM 7.1   *Suppose that X is a **discrete** random variable with probability distribution $f(x)$. Let $Y = u(X)$ define a one-to-one transformation between the values of X and Y so that the equation $y = u(x)$ can be uniquely solved for x in terms of y, say $x = w(y)$. Then the probability distribution of Y is*

$$g(y) = f[w(y)].$$ ∎

EXAMPLE 7.1   Let X be a geometric random variable with probability distribution $f(x) = \frac{3}{4}(\frac{1}{4})^{x-1}$, $x = 1, 2, 3, \ldots$. Find the probability distribution of the random variable $Y = X^2$.

SOLUTION
Since the values of X are all positive, the transformation defines a one-to-one correspondence between the x and y values, $y = x^2$ and $x = \sqrt{y}$. Hence

$$g(y) = \begin{cases} f(\sqrt{y}) = \frac{3}{4}(\frac{1}{4})^{\sqrt{y}-1}, & y = 1, 4, 9, \ldots \\ 0, & \text{elsewhere.} \end{cases}$$

Consider a problem where X_1 and X_2 are two discrete random variables with joint probability distribution $f(x_1, x_2)$ and we wish to find the joint probability distribution $g(y_1, y_2)$ of the two new random variables $Y_1 = u_1(X_1, X_2)$ and $Y_2 = u_2(X_1, X_2)$, which define a one-to-one transformation between the set of points (x_1, x_2) and (y_1, y_2). Solving the equations $y_1 = u_1(x_1, x_2)$ and $y_2 = u_2(x_1, x_2)$ simultaneously, we obtain the unique inverse solution $x_1 = w_1(y_1, y_2)$ and $x_2 = w_2(y_1, y_2)$. Hence the random variables Y_1 and Y_2 assume the values y_1 and y_2, respectively, when X_1 assumes the value $w_1(y_1, y_2)$ and X_2 assumes the value $w_2(y_1, y_2)$. The joint probability distribution of Y_1 and Y_2 is then

$$g(y_1, y_2) = P(Y_1 = y_1, Y_2 = y_2)$$
$$= P[X_1 = w_1(y_1, y_2), X_2 = w_2(y_1, y_2)]$$
$$= f[w_1(y_1, y_2), w_2(y_1, y_2)].$$

THEOREM 7.2   *Suppose that X_1 and X_2 are **discrete** random variables with joint probability distribution $f(x_1, x_2)$. Let $Y_1 = u_1(X_1, X_2)$ and $Y_2 = u_2(X_1, X_2)$ define a one-to-one transformation between the points (x_1, x_2) and (y_1, y_2) so that the equations $y_1 = u_1(x_1, x_2)$ and $y_2 = u_2(x_1, x_2)$ may be uniquely solved for x_1 and x_2 in terms of y_1 and y_2, say $x_1 = w_1(y_1, y_2)$ and $x_2 = w_2(y_1, y_2)$. Then the joint probability distribution of Y_1 and Y_2 is*

$$g(y_1, y_2) = f[w_1(y_1, y_2), w_2(y_1, y_2)].$$ ∎

Theorem 7.2 is extremely useful in finding the distribution of some random variable $Y_1 = u_1(X_1, X_2)$, where X_1 and X_2 are discrete random variables with joint probability distribution $f(x_1, x_2)$. We simply define a second function, say $Y_2 = u_2(X_1, X_2)$, maintaining a one-to-one correspondence between the points (x_1, x_2) and (y_1, y_2), and obtain the joint probability distribution $g(y_1, y_2)$. The distribution of Y_1 is just the marginal distribution of $g(y_1, y_2)$, found by summing over the y_2 values. Denoting the distribution of Y_1 by $h(y_1)$, we can then write

$$h(y_1) = \sum_{y_2} g(y_1, y_2).$$

EXAMPLE 7.2   Let X_1 and X_2 be two independent random variables having Poisson distributions with parameters μ_1 and μ_2, respectively. Find the distribution of the random variable $Y_1 = X_1 + X_2$.

SOLUTION
Since X_1 and X_2 are independent, we can write

$$f(x_1, x_2) = f(x_1)f(x_2)$$

$$= \frac{e^{-\mu_1}\mu_1^{x_1}}{x_1!} \frac{e^{-\mu_2}\mu_2^{x_2}}{x_2!}$$

$$= \frac{e^{-(\mu_1 + \mu_2)}\mu_1^{x_1}\mu_2^{x_2}}{x_1!\, x_2!},$$

where $x_1 = 0, 1, 2, \ldots$ and $x_2 = 0, 1, 2, \ldots$. Let us now define a second random variable, say $Y_2 = X_2$. The inverse functions are given by $x_1 = y_1 - y_2$ and $x_2 = y_2$. Using Theorem 7.2, we find the joint probability distribution of Y_1 and Y_2 to be

$$g(y_1, y_2) = \frac{e^{-(\mu_1 + \mu_2)}\mu_1^{y_1 - y_2}\mu_2^{y_2}}{(y_1 - y_2)!\, y_2!},$$

where $y_1 = 0, 1, 2, \ldots$ and $y_2 = 0, 1, 2, \ldots, y_1$. Note that since $x_1 > 0$, the transformation $x_1 = y_1 - x_2$ implies that x_2 and hence y_2 must always be less than or equal to y_1. Consequently, the marginal probability distribution of Y_1 is

$$h(y_1) = \sum_{y_2=0}^{y_1} g(y_1, y_2)$$

$$= e^{-(\mu_1 + \mu_2)} \sum_{y_2=0}^{y_1} \frac{\mu_1^{y_1 - y_2}\mu_2^{y_2}}{(y_1 - y_2)!\, y_2!}$$

$$= \frac{e^{-(\mu_1 + \mu_2)}}{y_1!} \sum_{y_2=0}^{y_1} \frac{y_1!}{y_2!\, (y_1 - y_2)!}\mu_1^{y_1 - y_2}\mu_2^{y_2}$$

$$= \frac{e^{-(\mu_1 + \mu_2)}}{y_1!} \sum_{y_2=0}^{y_1} \binom{y_1}{y_2}\mu_1^{y_1 - y_2}\mu_2^{y_2}.$$

Recognizing this sum as the binomial expansion of $(\mu_1 + \mu_2)^{y_1}$, we obtain

$$h(y_1) = \frac{e^{-(\mu_1 + \mu_2)}(\mu_1 + \mu_2)^{y_1}}{y_1!}, \qquad y_1 = 0, 1, 2, \ldots,$$

from which we conclude that the sum of the two independent random variables having Poisson distributions, with parameters μ_1 and μ_2, has a Poisson distribution with parameter $\mu_1 + \mu_2$.

To find the probability distribution of the random variable $Y = u(X)$ when X is a continuous random variable and the transformation is one to one, we shall need Theorem 7.3.

THEOREM 7.3   *Suppose that X is a **continuous** random variable with probability distribution $f(x)$. Let $Y = u(X)$ define a one-to-one correspondence between the values of X and Y so that the equation $y = u(x)$ can be uniquely solved for x in terms of y, say $x = w(y)$. Then the probability distribution of Y is*

$$g(y) = f[w(y)]|J|,$$

*where $J = w'(y)$ and is called the **Jacobian** of the transformation.*   ■

PROOF.   Suppose that $y = u(x)$ is an increasing function as in Figure 7.1. Then we see that whenever Y falls between a and b, the random variable X must fall between $w(a)$ and $w(b)$. Hence

$$P(a < Y < b) = P[w(a) < X < w(b)]$$

$$= \int_{w(a)}^{w(b)} f(x) \ dx.$$

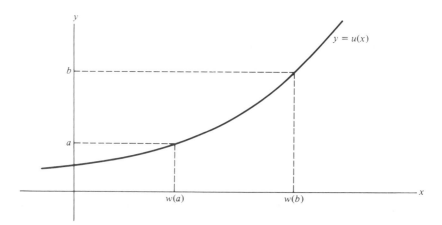

FIGURE 7.1   Increasing function.

Changing the variable of integration from x to y by the relation $x = w(y)$, we obtain $dx = w'(y)\ dy$, and hence

$$P(a < Y < b) = \int_a^b f[w(y)]w'(y)\ dy.$$

Since the integral gives the desired probability for every $a < b$ within the permissible set of y values, then the probability distribution of Y is

$$g(y) = f[w(y)]w'(y) = f[w(y)]J.$$

If we recognize $J = w'(y)$ as the reciprocal of the slope of the tangent line to the curve of the increasing function $y = u(x)$, it is then obvious that $J = |J|$. Hence

$$g(y = f[w(y)]|J|.$$

Suppose that $y = u(x)$ is a decreasing function as in Figure 7.2. Then we write

$$P(a < Y < b) = P[w(b) < X < w(a)]$$

$$= \int_{w(b)}^{w(a)} f(x)\ dx.$$

Again changing the variable of integration to y, we obtain

$$P(a < Y < b) = \int_b^a f[w(y)]w'(y)\ dy$$

$$= -\int_a^b f[w(y)]w'(y)\ dy,$$

from which we conclude that

$$g(y) = -f[w(y)]w'(y) = -f[w(y)]J.$$

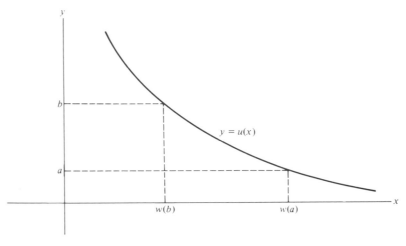

FIGURE 7.2   Decreasing function.

In this case the slope of the curve is negative and $J = -|J|$. Hence

$$g(y) = f[w(y)]|J|,$$

as before.

EXAMPLE 7.3   Let X be a continuous random variable with probability distribution

$$f(x) = \begin{cases} \dfrac{x}{12}, & 1 < x < 5 \\ 0, & \text{elsewhere.} \end{cases}$$

Find the probability distribution of the random variable $Y = 2X - 3$.

SOLUTION
The inverse solution of $y = 2x - 3$ yields $x = (y + 3)/2$, from which we obtain $J = w'(y) = dx/dy = 1/2$. Therefore, using Theorem 7.3, we find the density function of Y to be

$$g(y) = \begin{cases} \dfrac{(y + 3)/2}{12}\left(\dfrac{1}{2}\right) = \dfrac{y + 3}{48}, & -1 < y < 7 \\ 0, & \text{elsewhere.} \end{cases}$$

To find the joint probability distribution of the random variables $Y_1 = u_1(X_1, X_2)$ and $Y_2 = u_2(X_1, X_2)$ when X_1 and X_2 are continuous and the transformation is one to one, we need an additional theorem, analogous to Theorem 7.2, which we state without proof.

THEOREM 7.4   *Suppose that X_1 and X_2 are* **continuous** *random variables with joint probability distribution $f(x_1, x_2)$. Let $Y_1 = u_1(X_1, X_2)$ and $Y_2 = u_2(X_1, X_2)$ define a one-to-one transformation between the points (x_1, x_2) and (y_1, y_2) so that the equations $y_1 = u_1(x_1, x_2)$ and $y_2 = u_2(x_1, x_2)$ may be uniquely solved for x_1 and x_2 in terms of y_1 and y_2, say $x_1 = w_1(y_1, y_2)$ and $x_2 = w_2(y_1, y_2)$. Then the joint probability distribution of Y_1 and Y_2 is*

$$g(y_1, y_2) = f[w_1(y_1, y_2), w_2(y_1, y_2)]|J|,$$

where the Jacobian is the 2×2 determinant

$$J = \begin{vmatrix} \partial x_1/\partial y_1 & \partial x_1/\partial y_2 \\ \partial x_2/\partial y_1 & \partial x_2/\partial y_2 \end{vmatrix}$$

and $\partial x_1/\partial y_1$ is simply the derivative of $x_1 = w_1(y_1, y_2)$ with respect to y_1 with y_2 held constant, referred to in calculus as the partial derivative of x_1 with respect to y_1. The other partial derivatives are defined in a similar manner. ■

EXAMPLE 7.4   Let X_1 and X_2 be two continuous random variables with joint probability distribution

$$f(x_1, x_2) = \begin{cases} 4x_1x_2, & 0 < x_1 < 1, 0 < x_2 < 1 \\ 0, & \text{elsewhere.} \end{cases}$$

Find the joint probability distribution of $Y_1 = X_1^2$ and $Y_2 = X_1X_2$.

SOLUTION
The inverse solutions of $y_1 = x_1^2$ and $y_2 = x_1 x_2$ are $x_1 = \sqrt{y_1}$ and $x_2 = y_2/\sqrt{y_1}$, from which we obtain

$$J = \begin{vmatrix} 1/(2\sqrt{y_1}) & 0 \\ -y_2/2y_1^{3/2} & 1/\sqrt{y_1} \end{vmatrix} = \frac{1}{2y_1}.$$

To determine the set B of points in the $y_1 y_2$-plane into which the set A of points in the $x_1 x_2$-plane is mapped, we write

$$x_1 = \sqrt{y_1} \quad \text{and} \quad x_2 = y_2/\sqrt{y_1}$$

and then setting $x_1 = 0$, $x_2 = 0$, $x_1 = 1$, and $x_2 = 1$, the boundaries of set A are transformed to $y_1 = 0$, $y_2 = 0$, $y_1 = 1$, and $y_2 = \sqrt{y_1}$ or $y_2^2 = y_1$. The two regions are illustrated in Figure 7.3. Clearly, the transformation is one to one, mapping the set $A = \{(x_1, x_2) \mid 0 < x_1 < 1, 0 < x_2 < 1\}$ into the set $B = \{(y_1, y_2) \mid y_2^2 < y_1 < 1, 0 < y_2 < 1\}$. From Theorem 7.4 the joint probability distribution of Y_1 and Y_2 is

$$g(y_1, y_2) = 4(\sqrt{y_1}) \frac{y_2}{\sqrt{y_1}} \frac{1}{2y_1}$$

$$= \begin{cases} \dfrac{2y_2}{y_1}, & y_2^2 < y_1 < 1, 0 < y_2 < 1 \\ 0, & \text{elsewhere.} \end{cases}$$

Problems frequently arise when we wish to find the probability distribution of the random variable $Y = u(X)$ when X is a continuous random variable and the transformation is not one to one. That is, to each value x there corresponds exactly one value y, but to each y-value there corresponds more than one x-value. For example, suppose that $f(x)$ is positive over the interval $-1 < x < 2$ and zero elsewhere. Consider the transformation $y = x^2$. In this case $x = \pm\sqrt{y}$ for $0 < y < 1$ and

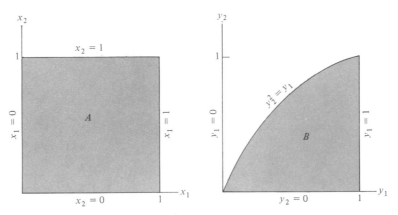

FIGURE 7.3 Mapping set A into set B.

$x = \sqrt{y}$ for $1 < y < 4$. For the interval $1 < y < 4$, the probability distribution of Y is found as before, using Theorem 7.3. That is,

$$g(y) = f[w(y)]|J| = \frac{f(\sqrt{y})}{2\sqrt{y}}, \qquad 1 < y < 4.$$

However, when $0 < y < 1$, we may partition the interval $-1 < x < 1$ to obtain the two inverse functions

$$x = -\sqrt{y}, \qquad -1 < x < 0$$

and

$$x = \sqrt{y}, \qquad 0 < x < 1.$$

Then to every y-value there corresponds a single x-value for each partition. From Figure 7.4 we see that

$$P(a < Y < b) = P(-\sqrt{b} < X < -\sqrt{a}) + P(\sqrt{a} < X < \sqrt{b})$$

$$= \int_{-\sqrt{b}}^{-\sqrt{a}} f(x)\ dx + \int_{\sqrt{a}}^{\sqrt{b}} f(x)\ dx.$$

Changing the variable of integration from x to y, we obtain

$$P(a < Y < b) = \int_{b}^{a} f(-\sqrt{y})J_1\ dy + \int_{a}^{b} f(\sqrt{y})J_2\ dy$$

$$= -\int_{a}^{b} f(-\sqrt{y})J_1\ dy + \int_{a}^{b} f(\sqrt{y})J_2\ dy,$$

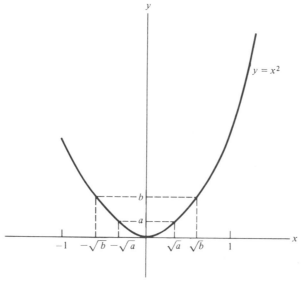

FIGURE 7.4   Decreasing and increasing function.

where

$$J_1 = \frac{d(-\sqrt{y})}{dy} = \frac{-1}{2\sqrt{y}} = -|J_1|$$

and

$$J_2 = \frac{d(\sqrt{y})}{dy} = \frac{1}{2\sqrt{y}} = |J_2|.$$

Hence we can write

$$P(a < Y < b) = \int_a^b [f(-\sqrt{y})|J_1| + f(\sqrt{y})|J_2|] \, dy,$$

and then

$$g(y) = f(-\sqrt{y})|J_1| + f(\sqrt{y})|J_2|$$
$$= \frac{[f(-\sqrt{y}) + f(\sqrt{y})]}{2\sqrt{y}}, \qquad 0 < y < 1.$$

The probability distribution of Y for $0 < y < 4$ may now be written

$$g(y) = \begin{cases} \dfrac{[f(-\sqrt{y}) + f(\sqrt{y})]}{2\sqrt{y}}, & 0 < y < 1 \\[2mm] \dfrac{f(\sqrt{y})}{2\sqrt{y}}, & 1 < y < 4 \\[2mm] 0, & \text{elsewhere.} \end{cases}$$

This procedure for finding $g(y)$ when $0 < y < 1$ is generalized in Theorem 7.5 for k inverse functions. For transformations not one to one of functions of several variables, the reader is referred to *Introduction to Mathematical Statistics* by Hogg and Craig (see the Bibliography).

THEOREM 7.5   *Suppose that X is a* **continuous** *random variable with probability distribution f(x). Let Y = u(X) define a transformation between the value of X and Y that is not one to one. If the interval over which X is defined can be partitioned into k mutually disjoint sets such that each of the inverse functions $x_1 = w_1(y)$, $x_2 = w_2(y)$, . . . , $x_k = w_k(y)$ of y = u(x) defines a one-to-one correspondence, then the probability distribution of Y is*

$$g(y) = \sum_{i=1}^{k} f[w_i(y)]|J_i|,$$

where $J_i = w_i'(y)$, i = 1, 2, . . . , k.   ■

EXAMPLE 7.5   Show that $Y = (X - \mu)^2/\sigma^2$ has a chi-squared distribution with 1 degree of freedom when X has a normal distribution with mean μ and variance σ^2.

SOLUTION

Let $Z = (X - \mu)/\sigma$, where the random variable Z has the standard normal distribution

$$f(z) = \frac{1}{\sqrt{2\pi}} e^{-z^2/2}, \qquad -\infty < z < \infty.$$

We shall now find the distribution of the random variable $Y = Z^2$. The inverse solutions of $y = z^2$ are $z = \pm\sqrt{y}$. If we designate $z_1 = -\sqrt{y}$ and $z_2 = \sqrt{y}$, then $J_1 = -1/2\sqrt{y}$ and $J_2 = 1/2\sqrt{y}$. Hence, by Theorem 7.5, we have

$$g(y) = \frac{1}{\sqrt{2\pi}} e^{-y/2} \left| \frac{-1}{2\sqrt{y}} \right| + \frac{1}{\sqrt{2\pi}} e^{-y/2} \left| \frac{1}{2\sqrt{y}} \right|$$

$$= \frac{1}{2^{1/2}\sqrt{\pi}} y^{1/2-1} e^{-y/2}, \qquad y > 0.$$

Since $g(y)$ is a density function, it follows that

$$1 = \frac{1}{2^{1/2}\sqrt{\pi}} \int_0^\infty y^{1/2-1} e^{-y/2} \, dy$$

$$= \frac{\Gamma(1/2)}{\sqrt{\pi}} \int_0^\infty \frac{1}{2^{1/2}\Gamma(1/2)} y^{1/2-1} e^{-y/2} \, dy$$

$$= \frac{\Gamma(1/2)}{\sqrt{\pi}},$$

the integral being the area under a gamma probability curve with parameters $\alpha = 1/2$ and $\beta = 2$. Therefore, $\sqrt{\pi} = \Gamma(1/2)$ and the probability distribution of Y is given by

$$g(y) = \begin{cases} \dfrac{1}{2^{1/2}\Gamma(1/2)} y^{1/2-1} e^{-y/2}, & y > 0 \\ 0, & \text{elsewhere,} \end{cases}$$

which is seen to be a chi-squared distribution with 1 degree of freedom.

7.3 Moments and Moment-Generating Functions _____

In this section we concentrate on applications of moment-generating functions. The obvious purpose of the moment-generating function is in determining moments of distributions. However, the most important contribution is to establish distributions of functions of random variables.

If $g(X) = X^r$ for $r = 0, 1, 2, 3, \ldots$, Theorem 4.1 yields an expected value called the rth **moment about the origin** of the random variable X, which we denote by μ_r'.

DEFINITION 7.1   *The* r*th* **moment about the origin** *of the random variable X is given by*

$$\mu'_r = E(X^r) = \begin{cases} \displaystyle\sum_x x^r f(x) & \text{if } X \text{ is discrete} \\ \displaystyle\int_{-\infty}^{\infty} x^r f(x)\, dx & \text{if } X \text{ is continuous.} \end{cases}$$ ■

Since the first and second moments about the origin are given by $\mu'_1 = E(X)$ and $\mu'_2 = E(X^2)$, we can write the mean and variance of a random variable as

$$\mu = \mu'_1 \qquad \text{and} \qquad \sigma^2 = \mu'_2 - \mu^2.$$

Although the moments of a random variable can be determined directly from Definition 7.1, an alternative procedure exists. This procedure requires us to utilize a **moment-generating function**.

DEFINITION 7.2   *The* **moment-generating function** *of the random variable X is given by* $E(e^{tx})$ *and is denoted by* $M_X(t)$. *Hence*

$$M_X(t) = E(e^{tx}) = \begin{cases} \displaystyle\sum_x e^{tx} f(x) & \text{if } X \text{ is discrete} \\ \displaystyle\int_{-\infty}^{\infty} e^{tx} f(x)\, dx & \text{if } X \text{ is continuous} \end{cases}$$ ■

Moment-generating functions will exist only if the sum or integral of Definition 7.2 converges. If a moment-generating function of a random variable X does exist, it can be used to generate all the moments of that variable. The method is described in Theorem 7.6.

THEOREM 7.6   *Let X be a random variable with moment-generating function* $M_X(t)$. *Then*

$$\left. \frac{d^r M_X(t)}{dt^r} \right|_{t=0} = \mu'_r.$$ ■

PROOF. Assuming that we can differentiate inside summation and integral signs, we obtain

$$\frac{d^r M_X(t)}{dt^r} = \begin{cases} \displaystyle\sum_x x^r e^{tx} f(x) & \text{if } X \text{ is discrete} \\ \displaystyle\int_{-\infty}^{\infty} x^r e^{tx} f(x)\, dx & \text{if } X \text{ is continuous.} \end{cases}$$

Setting $t = 0$, we see that both cases reduce to $E(X^r) = \mu'_r$.

EXAMPLE 7.6   Find the moment-generating function of the binomial random variable X and then use it to verify that $\mu = np$ and $\sigma^2 = npq$.

SOLUTION
From Definition 7.2 we have

$$M_X(t) = \sum_{x=0}^{n} e^{tx} \binom{n}{x} p^x q^{n-x}$$

$$= \sum_{x=0}^{n} \binom{n}{x} (pe^t)^x q^{n-x}.$$

Recognizing this last sum as the binomial expansion of $(pe^t + q)^n$, we obtain

$$M_X(t) = (pe^t + q)^n.$$

Now

$$\frac{dM_X(t)}{dt} = n(pe^t + q)^{n-1} pe^t$$

and

$$\frac{d^2 M_X(t)}{dt^2} = np[e^t(n - 1)(pe^t + q)^{n-2} pe^t + (pe^t + q)^{n-1} e^t].$$

Setting $t = 0$, we get

$$\mu_1' = np \qquad \text{and} \qquad \mu_2' = np[(n - 1)p + 1].$$

Therefore,

$$\mu = \mu_1' = np$$

and

$$\sigma^2 = \mu_2' - \mu^2 = np(1 - p) = npq,$$

which agrees with the results obtained in Chapter 5.

EXAMPLE 7.7   Show that the moment-generating function of the random variable X having a normal probability distribution with mean μ and variance σ^2 is given by $M_X(t) = e^{\mu t + \sigma^2 t^2/2}$.

SOLUTION
From Definition 7.2 the moment-generating function of the normal random variable X is

$$M_X(t) = \int_{-\infty}^{\infty} e^{tx} \frac{1}{\sqrt{2\pi}\, \sigma} e^{-(1/2)[(x-\mu)/\sigma]^2} \, dx$$

$$= \int_{-\infty}^{\infty} \frac{1}{\sqrt{2\pi}\, \sigma} e^{-[x^2 - 2(\mu + t\sigma^2)x + \mu^2]/2\sigma^2} \, dx.$$

Completing the square in the exponent, we can write

$$x^2 - 2(\mu + t\sigma^2)x + \mu^2 = [x - (\mu + t\sigma^2)]^2 - 2\mu t\sigma^2 - t^2\sigma^4$$

and then

$$M_X(t) = \int_{-\infty}^{\infty} \frac{1}{\sqrt{2\pi}\,\sigma} e^{-\{[x-(\mu+t\sigma^2)]^2 - 2\mu t\sigma^2 - t^2\sigma^4\}/2\sigma^2}\, dx$$

$$= e^{\mu t + \sigma^2 t^2/2} \int_{-\infty}^{\infty} \frac{1}{\sqrt{2\pi}\,\sigma} e^{-(1/2)\{[x-(\mu+t\sigma^2)]/\sigma\}^2}\, dx.$$

Let $w = [x - (\mu + t\sigma^2)]/\sigma$; then $dx = \sigma\, dw$ and

$$M_X(t) = e^{\mu t + \sigma^2 t^2/2} \int_{-\infty}^{\infty} \frac{1}{\sqrt{2\pi}} e^{-w^2/2}\, dw$$

$$= e^{\mu t + \sigma^2 t^2/2},$$

since the last integral represents the area under a standard normal density curve and hence equals 1.

EXAMPLE 7.8 Show that the moment-generating function of the random variable X having a chi-squared distribution with v degrees of freedom is $M_X(t) = (1 - 2t)^{-v/2}$.

SOLUTION
The chi-squared distribution was obtained as a special case of the gamma distribution by setting $\alpha = v/2$ and $\beta = 2$. Substituting for $f(x)$ in Definition 7.2, we obtain

$$M_X(t) = \int_0^{\infty} e^{tx} \frac{1}{2^{v/2}\Gamma(v/2)} x^{v/2-1} e^{-x/2}\, dx$$

$$= \frac{1}{2^{v/2}\Gamma(v/2)} \int_0^{\infty} x^{v/2-1} e^{-x(1-2t)/2}\, dx.$$

Writing $y = x(1 - 2t)/2$ and $dx = [2/(1 - 2t)]\, dy$, we get

$$M_X(t) = \frac{1}{2^{v/2}\Gamma(v/2)} \int_0^{\infty} \left(\frac{2y}{1-2t}\right)^{v/2-1} e^{-y} \frac{2}{1-2t}\, dy$$

$$= \frac{1}{\Gamma(v/2)(1-2t)^{v/2}} \int_0^{\infty} y^{v/2-1} e^{-y}\, dy$$

$$= (1 - 2t)^{-v/2},$$

since the last integral equals $\Gamma(v/2)$.

Although the method of transforming variables provides an effective way of finding the distribution of a function of several variables, there is an alternative and often preferred procedure when the function in question is a linear combination of independent random variables. This procedure utilizes the properties of moment-generating functions discussed in the following four theorems. In keeping with the mathematical scope of this book, we state Theorem 7.7 without proof.

THEOREM 7.7
(UNIQUENESS
THEOREM)
Let X and Y be two random variables with moment-generating functions $M_X(t)$ and $M_Y(t)$, respectively. If $M_X(t) = M_Y(t)$ for all values of t, then X and Y have the same probability distribution. ■

THEOREM 7.8 $M_{X+a}(t) = e^{at}M_X(t).$ ∎

PROOF

$$M_{X+a}(t) = E[e^{t(X+a)}]$$

$$= e^{at}E(e^{tX}) = e^{at}M_X(t).$$

THEOREM 7.9 $M_{aX}(t) = M_X(at).$

PROOF

$$M_{aX}(t) = E[e^{t(aX)}] = E[e^{(at)X}]$$

$$= M_X(at).$$

THEOREM 7.10   *If X_1, X_2, . . . , X_n are independent random variables with moment-generating functions $M_{X_1}(t)$, $M_{X_2}(t)$, . . . , $M_{X_n}(t)$, respectively, and $Y = X_1 + X_2 + \cdots + X_n$, then*

$$M_Y(t) = M_{X_1}(t)M_{X_2}(t) \cdots M_{X_n}(t).$$ ∎

PROOF.   For the continuous case

$$M_Y(t) = E(e^{tY}) = E[e^{t(X_1+X_2+\cdots+X_n)}]$$

$$= \int_{-\infty}^{\infty} \cdots \int_{-\infty}^{\infty} e^{t(X_1+X_2+\cdots+X_n)}f(x_1, x_2, \ldots, x_n) \, dx_1 \, dx_2 \cdots dx_n.$$

Since the variables are independent, we have

$$f(x_1, x_2, \ldots, x_n) = f_1(x_1)f_2(x_2) \cdots f_n(x_n)$$

and then

$$M_Y(t) = \int_{-\infty}^{\infty} e^{tx_1}f_1(x_1) \, dx_1 \int_{-\infty}^{\infty} e^{tx_2}f_2(x_2) \, dx_2 \cdots \int_{-\infty}^{\infty} e^{tx_n}f_n(x_n) \, dx_n$$

$$= M_{X_1}(t)M_{X_2}(t) \cdots M_{X_n}(t).$$

The proof for the discrete case is obtained in a similar manner by replacing integrations with summations.

Theorems 7.7 through 7.10 are vital in the understanding of moment-generating functions. An example will follow to illustrate. There are many situations in which we need to know the distribution of the sum of random variables. We may use Theorems 7.7 and 7.10 and the result of Exercise 19 following this section to find the distribution of a sum of two independent Poisson random variables with moment-generating functions given by

$$M_{X_1}(t) = e^{\mu_1(e^t-1)} \quad \text{and} \quad M_{X_2}(t) = e^{\mu_2(e^t-1)},$$

respectively. According to Theorem 7.10, the moment-generating function of the random variable $Y_1 = X_1 + X_2$ is

$$M_{Y_1}(t) = M_{X_1}(t)M_{X_2}(t)$$

$$= e^{\mu_1(e^t-1)}e^{\mu_2(e^t-1)}$$

$$= e^{(\mu_1+\mu_2)(e^t-1)},$$

which we immediately identify as the moment-generating function of a random variable having a Poisson distribution with the parameter $\mu_1 + \mu_2$. Hence, according to Theorem 7.7, we again conclude that the sum of two independent random variables having Poisson distributions, with parameters μ_1 and μ_2, has a Poisson distribution with parameter $\mu_1 + \mu_2$.

Linear Combinations of Random Variables

In applied statistics one frequently needs to know the probability distribution of a linear combination of independent normal random variables. Let us obtain the distribution of the random variable $Y = a_1X_1 + a_2X_2$ when X_1 is a normal variable with mean μ_1 and variance σ_1^2 and X_2 is also a normal variable but independent of X_1, with mean μ_2 and variance σ_2^2. First, by Theorem 7.10, we find

$$M_Y(t) = M_{a_1X_1}(t)M_{a_2X_2}(t),$$

and then, using Theorem 7.9,

$$M_Y(t) = M_{X_1}(a_1t)M_{X_2}(a_2t).$$

Substituting a_1t for t in a moment-generating function of the normal distribution derived in Example 7.7 and then a_2t for t, we have

$$M_Y(t) = \exp(a_1\mu_1t + a_1^2\sigma_1^2t^2/2 + a_2\mu_2t + a_2^2\sigma_2^2t^2/2)$$

$$= \exp[(a_1\mu_1 + a_2\mu_2)t + (a_1^2\sigma_1^2 + a_2^2\sigma_2^2)t^2/2],$$

which we recognize as the moment-generating function of a distribution that is normal with $a_1\mu_1 + a_2\mu_2$ and variance $a_1^2\sigma_1^2 + a_2^2\sigma_2^2$.

Generalizing to the case of n independent normal variables, we state the following result.

THEOREM 7.11   *If $X_1, X_2, \ldots, X_n$ are independent random variables having normal distributions with means $\mu_1, \mu_2, \ldots, \mu_n$ and variances $\sigma_1^2, \sigma_2^2, \ldots, \sigma_n^2$, respectively, then the random variable*

$$Y = a_1X_1 + a_2X_2 + \cdots + a_nX_n$$

has a normal distribution with mean

$$\mu_Y = a_1\mu_1 + a_2\mu_2 + \cdots + a_n\mu_n$$

and variance

$$\sigma_Y^2 = a_1^2\sigma_1^2 + a_2^2\sigma_2^2 + \cdots + a_n^2\sigma_n^2.$$ ■

It is now evident that the Poisson distribution and the normal distribution possess a reproductive property in that the sum of independent random variables having either of these distributions is a random variable that also has the same type of distribution. This reproductive property is also possessed by the chi-squared distribution.

THEOREM 7.12

If $X_1, X_2, \ldots, X_n$ are mutually independent random variables that have, respectively, chi-squared distributions with $v_1, v_2, \ldots, v_n$ degrees of freedom, then the random variable

$$Y = X_1 + X_2 + \cdots + X_n$$

has a chi-squared distribution with $v = v_1 + v_2 + \cdots + v_n$ degrees of freedom.   ∎

PROOF. By Theorem 7.10,

$$M_Y(t) = M_{X_1}(t)M_{X_2}(t) \cdots M_{X_n}(t).$$

From Example 7.8,

$$M_{X_i}(t) = (1 - 2t)^{-v_i/2}, \qquad i = 1, 2, \ldots, n.$$

Therefore,

$$M_Y(t) = (1 - 2t)^{-v_1/2}(1 - 2t)^{-v_2/2} \cdots (1 - 2t)^{-v_n/2}$$

$$= (1 - 2t)^{-(v_1 + v_2 + \cdots + v_n)/2},$$

which we recognize as the moment-generating function of a chi-squared distribution with $v = v_1 + v_2 + \cdots + v_n$ degrees of freedom.

COROLLARY

If $X_1, X_2, \ldots, X_n$ are independent random variables having identical normal distributions with mean μ and variance σ^2, then the random variable

$$Y = \sum_{i=1}^{n} \left(\frac{X_i - \mu}{\sigma} \right)^2$$

has a chi-squared distribution with $v = n$ degrees of freedom.   ∎

The above is an immediate consequence of Example 7.5, which states that each of the n independent random variables $[(X_i - \mu)/\sigma]^2$, $i = 1, 2, \ldots, n$, has a chi-squared distribution with 1 degree of freedom. This corollary is extremely important. It establishes a relationship between the very important chi-squared distribution and the normal distribution. It also should provide the reader a clear idea of what we mean by the parameter that we call *degrees of freedom*. As we move into future chapters, the notion of degrees of freedom plays an increasingly important role. We see from the corollary that if $Z_1, Z_2, \ldots, Z_n$ are independent standard normal random variables, then $\sum_{i=1}^{n} Z_i^2$ has a chi-squared distribution and the *single parameter*, v, the degrees of freedom, is n, the *number* of standard normal variates.

Exercises

1. Let X be a random variable with probability distribution

$$f(x) = \begin{cases} \frac{1}{3}, & x = 1, 2, 3 \\ 0, & \text{elsewhere.} \end{cases}$$

Find the probability distribution of the random variable $Y = 2X - 1$.

2. Let X be a binomial random variable with probability distribution

$$f(x) = \begin{cases} \binom{3}{x}\left(\frac{2}{5}\right)^x \left(\frac{3}{5}\right)^{3-x}, & x = 0, 1, 2, 3 \\ 0, & \text{elsewhere.} \end{cases}$$

Find the probability distribution of the random variable $Y = X^2$.

3. Let X_1 and X_2 be discrete random variables with the multinomial distribution

$$f(x_1, x_2)$$
$$= \binom{2}{x_1, x_2, 2 - x_1 - x_2}\left(\frac{1}{4}\right)^{x_1}\left(\frac{1}{3}\right)^{x_2}\left(\frac{5}{12}\right)^{2-x_1-x_2}$$

for $x_1 = 0, 1, 2; x_2 = 0, 1, 2; x_1 + x_2 \le 2$; and zero elsewhere. Find the joint probability distribution of $Y_1 = X_1 + X_2$ and $Y_2 = X_1 - X_2$.

4. Let X_1 and X_2 be discrete random variables with joint probability distribution

$$f(x_1, x_2) = \begin{cases} \dfrac{x_1 x_2}{18}, & x_1 = 1, 2; x_2 = 1, 2, 3 \\ 0, & \text{elsewhere.} \end{cases}$$

Find the probability distribution of the random variable $Y = X_1 X_2$.

5. Let X have the probability distribution

$$f(x) = \begin{cases} 1, & 0 < x < 1 \\ 0, & \text{elsewhere.} \end{cases}$$

Show that the random variable $Y = -2 \ln X$ has a chi-squared distribution with 2 degrees of freedom.

6. Given the random variable X with probability distribution

$$f(x) = \begin{cases} 2x, & 0 < x < 1 \\ 0, & \text{elsewhere,} \end{cases}$$

find the probability distribution of Y, where $Y = 8X^3$.

7. The speed of a molecule in a uniform gas at equilibrium is a random variable V whose probability distribution is given by

$$f(v) = \begin{cases} kv^2 e^{-bv^2}, & v > 0 \\ 0, & \text{elsewhere,} \end{cases}$$

where k is an appropriate constant and b depends on the absolute temperature and mass of the molecule. Find the probability distribution of the kinetic energy of the molecule W, where $W = mV^2/2$.

8. In Exercise 21 on page 91 a dealer's profit, in units of $1000, on a new automobile is given by $Y = X^2$, where X is a random variable having the density function

$$f(x) = \begin{cases} 2(1 - x), & 0 < x < 1 \\ 0, & \text{elsewhere.} \end{cases}$$

(a) Find the probability density function of the random variable Y.

(b) Using the density function of Y, find the probability that the profit will be less than $250 on the next new automobile sold by this dealership.

9. In Exercise 22 on page 91 the hospital period, in days, for patients following treatment for a certain type of kidney disorder is a random variable $Y = X + 4$, where X has the density function

$$f(x) = \begin{cases} \dfrac{32}{(x + 4)^3}, & x > 0 \\ 0, & \text{elsewhere.} \end{cases}$$

(a) Find the probability density function of the random variable Y.

(b) Using the density function of Y, find the probability that the hospital period for a patient following this treatment will exceed 8 days.

10. In Exercise 5 on page 79 the random variables X and Y, representing the weights of creams and toffees in 1-kilogram boxes of chocolates containing a mixture of creams, toffees, and cordials, have the joint density function

$$f(x, y) = \begin{cases} 24xy, & 0 \le x \le 1, 0 \le y \le 1, \\ & x + y \le 1 \\ 0, & \text{elsewhere.} \end{cases}$$

(a) Find the probability density function of the random variable $Z = X + Y$.

(b) Using the density function of Z, find the probability that in a given box the sum of the creams and toffees accounts for at least 1/2 but less than 3/4 of the total weight.

11. In Exercise 11 on page 79 the amount of kerosene, in thousands of liters, in a tank of the beginning of any day is a random amount Y from which a random amount X is sold during that day. Assume that the joint density function of these variables is given by

$$f(x, y) = \begin{cases} 2, & 0 < x < y, \, 0 < y < 1 \\ 0, & \text{elsewhere.} \end{cases}$$

Find the probability density function for the amount of kerosene left in the tank at the end of the day.

12. Let X_1 and X_2 be independent random variables each having the probability distribution

$$f(x) = \begin{cases} e^{-x}, & x > 0 \\ 0, & \text{elsewhere.} \end{cases}$$

Show that the random variables Y_1 and Y_2 are independent when $Y_1 = X_1 + X_2$ and $Y_2 = X_1/(X_1 + X_2)$.

13. A current of I amperes flowing through a resistance of R ohms varies according to the probability distribution

$$f(i) = \begin{cases} 6i(1 - i), & 0 < i < 1 \\ 0, & \text{elsewhere.} \end{cases}$$

If the resistance varies independently of the current according to the probability distribution

$$g(r) = \begin{cases} 2r, & 0 < r < 1 \\ 0, & \text{elsewhere,} \end{cases}$$

find the probability distribution for the power $W = I^2 R$ watts.

14. Let X be a random variable with probability distribution

$$f(x) = \begin{cases} \dfrac{1 + x}{2}, & -1 < x < 1 \\ 0, & \text{elsewhere.} \end{cases}$$

Find the probability distribution of the random variable $Y = X^2$.

15. Let X have the probability distribution

$$f(x) = \begin{cases} \dfrac{2(x + 1)}{9}, & -1 < x < 2 \\ 0, & \text{elsewhere.} \end{cases}$$

Find the probability distribution of the random variable $Y = X^2$.

16. Show that the rth moment about the origin of the gamma distribution is given by

$$\mu_r' = \frac{\beta^r \Gamma(\alpha + r)}{\Gamma(\alpha)}.$$

HINT: Substitute $y = x/\beta$ in the integral defining μ_r' and then use the gamma function to evaluate the integral.

17. A random variable X has the discrete uniform distribution

$$f(x; k) = \begin{cases} \dfrac{1}{k}, & x = 1, 2, 3, \ldots, k \\ 0, & \text{elsewhere.} \end{cases}$$

Show that the moment-generating function of X is given by

$$M_X(t) = \frac{e^t(1 - e^{kt})}{k(1 - e^t)}.$$

18. A random variable X has the geometric distribution $g(x; p) = pq^{x-1}$ for $x = 1, 2, 3, \ldots$. Show that the moment-generating function of X is given by

$$M_X(t) = \frac{pe^t}{1 - qe^t}$$

and then use $M_X(t)$ to find the mean and variance of the geometric distribution.

19. A random variable X has the Poisson distribution $p(x; \mu) = e^{-\mu} \mu^x/x!$ for $x = 0, 1, 2, \ldots$. Show that the moment-generating function of X is given by

$$M_X(t) = e^{\mu(e^t - 1)}.$$

Using $M_X(t)$, find the mean and variance of the Poisson distribution.

20. The moment-generating function of a certain Poisson random variable X is given by

$$M_X(t) = e^{\mu(e^t - 1)}.$$

Find $P(\mu - 2\sigma < X < \mu + 2\sigma)$.

21. Using the moment-generating function of Example 7.8, show that the mean and variance of the chi-squared distribution with v degrees of freedom are, respectively, v and $2v$.

22. By expanding e^{tx} in a Maclaurin series and integrating term by term, show that

$$M_X(t) = \int_{-\infty}^{\infty} e^{tx} f(x) \, dx$$

$$= 1 + \mu t + \mu_2' \frac{t^2}{2!} + \cdots + \mu_r' \frac{t^r}{r!} + \cdots.$$

8

Random Sampling, Data Description, and Some Fundamental Sampling Distributions

8.1 Random Sampling

The outcome of a statistical experiment may be recorded either as a numerical value or as a descriptive representation. When a pair of dice are tossed and the total is the outcome of interest, we record a numerical value. However, if the students in a certain school are given blood tests and the type of blood is of interest, then a descriptive representation might be the most useful. A person's blood can be classified in 8 ways. It must be AB, A, B, or O, with a plus or minus sign, depending on the presence or absence of the Rh antigen.

The statistician is primarily concerned with the analysis of numerical data. For the classification of blood types, it may be convenient to use numbers from 1 to 8 to represent the blood types and then record the appropriate number for each student. In any particular study, the number of possible observations may be small, large but finite, or infinite. For example, in the classification of blood types we can only have as many observations as there are students in the school. The project, therefore, results in a finite number of observations. On the other hand, if we could toss a pair of dice indefinitely and record the totals that occur, we would obtain an infinite set of values, each value representing the result of a single toss of a pair of dice.

In this chapter we focus on sampling from distributions or populations and study such important quantities as the *sample mean* and *sample variance*, which are of vital importance in future chapters. In addition, we attempt to give the reader an introduction to the role that the sample mean and variance will play in later chapters in statistical inference. Introductions are given to *data displays*, through which one may make graphical assessments of what can be learned from experimental data.

The use of the modern high-speed computer has allowed the scientist or engineer to greatly enhance his or her use of formal statistical inference with graphical techniques. Much of the time formal inference appears quite dry and perhaps even abstract to the practitioner or the manager who wishes to let statistical analysis be a guide to decision making.

We begin this section by discussing the notions of *populations* and *samples*. Both were mentioned in a broad fashion in the introduction given in Chapter 1. The totality of observations with which we are concerned, whether their number be finite or infinite, constitutes what we call a **population**. There was a time when the word *population* referred to observations obtained from statistical studies about people. Today, the statistician uses the term to refer to observations relevant to anything of interest, whether it be groups of people, animals, or all possible outcomes from some complicated biological or engineering system.

DEFINITION 8.1   *A **population** consists of the totality of the observations with which we are concerned.*   ■

The number of observations in the population is defined to be the **size** of the population. If there are 600 students in the school that we classified according to blood type, we say that we have a population of size 600. The numbers on the cards in a deck, the heights of residents in a certain city, and the lengths of fish in a particular lake are examples of populations with finite size. In each case the total number of observations is a finite number. The die-tossing experiment generates a population whose size is infinite. Similarly, the observations obtained by measuring the atmospheric pressure every day from the past on into the future, or all measurements on the depth of a lake from any conceivable position, are examples of populations whose sizes are infinite. Some finite populations are so large that in theory we assume them to be infinite. This is true if you consider the population of lifetimes of a certain type of storage battery being manufactured for mass distributions throughout the country.

Each observation in a population is a value of a random variable X having some probability distribution $f(x)$. If one is inspecting items coming off an assembly line for defects, then each observation in the population might be a value 0 or 1 of the binomial random variable X with probability distribution

$$b(x; 1, p) = p^x q^{1-x}, \qquad x = 0, 1,$$

where 0 indicates a nondefective item and 1 indicates a defective item. Of course, it is assumed that p, the probability of any item being defective, remains constant from trial to trial. In the blood-type experiment the random variable X represents the type of blood by assuming a value from 1 to 8. Each student is given one of the values of the discrete random variable. The lives of the storage batteries are values assumed by a continuous random variable having perhaps a normal distribution. When we refer hereafter to a ''binomial population,'' a ''normal population,'' or, in general, the ''population $f(x)$,'' we shall mean a population whose observations are values of a random variable having a binomial distribution, a normal distribution, or the probability distribution $f(x)$. Hence the mean and variance of a random variable or probability distribution are also referred to as the mean and variance of the corresponding population.

In the field of statistical inference the statistician is interested in arriving at conclusions concerning a population when it is impossible or impractical to observe the entire set of observations that make up the population. For example, in attempting to determine the average length of life of a certain brand of light bulb, it would be impossible to test all such bulbs if we are to have any left to sell. Exorbitant costs can also be a prohibitive factor in studying the entire population. Therefore, we must depend on a subset of observations from the population to help us make inferences concerning that same population. This brings us to consider the notion of sampling.

DEFINITION 8.2   *A **sample** is a subset of a population.*   ■

If our inferences from the sample to the population are to be valid, we must obtain samples that are representative of the population. All too often we are tempted to choose a sample by selecting the most convenient members of the population. Such a procedure may lead to erroneous inferences concerning the population. Any sampling procedure that produces inferences that consistently overestimate or consistently underestimate some characteristic of the population is said to be **biased**. To eliminate any possibility of bias in the sampling procedure, it is desirable to choose a **random sample** in the sense that the observations are made independently and at random.

In selecting a random sample of size n from a population $f(x)$, let us define the random variable X_i, $i = 1, 2, \ldots, n$, to represent the ith measurement or sample value that we observe. The random variables $X_1, X_2, \ldots, X_n$ will then constitute a random sample from the population $f(x)$ with numerical values $x_1, x_2, \ldots, x_n$ if the measurements are obtained by repeating the experiment n independent times under essentially the same conditions. Because of the identical conditions under which the elements of the sample are selected, it is reasonable to assume that the n random variables $X_1, X_2, \ldots, X_n$ are independent and that each has the same probability distribution $f(x)$. That is, the probability distributions of $X_1, X_2, \ldots, X_n$ are, respectively, $f(x_1), f(x_2), \ldots, f(x_n)$ and their joint probability distribution is

$$f(x_1, x_2, \ldots, x_n) = f(x_1)f(x_2) \cdots f(x_n).$$

The concept of a random sample is defined formally in the following definition.

DEFINITION 8.3   *Let $X_1, X_2, \ldots, X_n$ be n independent random variables each having the same probability distribution $f(x)$. We then define $X_1, X_2, \ldots, X_n$ to be a* **random sample** *of size n from the population $f(x)$ and write its joint probability distribution as*

$$f(x_1, x_2, \ldots, x_n) = f(x_1)f(x_2) \cdots f(x_n). \qquad \blacksquare$$

If one makes a random selection of $n = 8$ storage batteries from a manufacturing process, which has maintained the same specifications, and records the length of life for each battery with the first measurement x_1 being a value of X_1, the second measurement x_2 a value of X_2, and so forth, then $x_1, x_2, \ldots, x_8$ are the values of the random sample $X_1, X_2, \ldots, X_8$. If we assume the population of battery lives to be normal, the possible values of any X_i, $i = 1, 2, \ldots, 8$, will be precisely the same as those in the original population, and hence X_i has the same identical normal distribution as X.

8.2 Some Important Statistics

Our main purpose in selecting random samples is to elicit information about the unknown population parameters. Suppose, for example, that we wish to arrive at a conclusion concerning the proportion of coffee-drinking people in the United States who prefer a certain brand of coffee. It would be impossible to question every coffee-drinking American in order to compute the value of the parameter p representing the population proportion. Instead, a large random sample is selected and the proportion $\hat{p}$ of people in this sample favoring the brand of coffee in question is calculated. The value $\hat{p}$ is now used to make an inference concerning the true proportion p.

Now, $\hat{p}$ is a function of the observed values in the random sample; since many random samples are possible from the same population, we would expect $\hat{p}$ to vary somewhat from sample to sample. That is, $\hat{p}$ is a value of a random variable that we represent by $\hat{P}$. Such a random variable is called a **statistic**.

DEFINITION 8.4   *Any function of the random variables constituting a random sample is called a* **statistic**. $\qquad \blacksquare$

Central Tendency in the Sample

In Chapter 4 we introduced the two parameters μ and σ^2, which measure the center of location and the variability of a probability distribution. These are constant population parameters and are in no way affected or influenced by the observations of a random sample. We shall, however, define some important statistics that describe corresponding measures of a random sample. The most commonly used statistics for measuring the center of a set of data, arranged in order of magnitude, are the **mean, median,** and **mode**. The most important of these and the one we shall consider first is the mean.

DEFINITION 8.5   *If $X_1, X_2, \ldots, X_n$ represent a random sample of size n, then the **sample mean** is defined by the statistic*

$$\overline{X} = \frac{\sum\limits_{i=1}^{n} X_i}{n}.$$ ■

Note that the statistic $\overline{X}$ assumes the value $\overline{x} = \sum\limits_{i=1}^{n} x_i/n$ when X_1 assumes the value x_1, X_2 assumes the value x_2, and so forth. In practice the value of a statistic is usually given the same name as the statistic. For instance, the term *sample mean* is applied to both the statistic $\overline{X}$ and its computed value $\overline{x}$.

There was an earlier reference made to the sample mean in Chapter 1. In that context we were discussing computed values based on specific data sets.

EXAMPLE 8.1   A food inspector examined a random sample of 7 cans of a certain brand of tuna to determine the percent of foreign impurities. The following data were recorded: 1.8, 2.1, 1.7, 1.6, 0.9, 2.7, and 1.8. Compute the sample mean.

SOLUTION
The observed value $\overline{x}$ of the statistic $\overline{X}$ is

$$\overline{x} = \frac{1.8 + 2.1 + 1.7 + 1.6 + 0.9 + 2.7 + 1.8}{7} = 1.8\%.$$

The second most useful statistic for measuring the center of a set of data is the median. We shall designate the median by the symbol $\tilde{X}$.

DEFINITION 8.6   *If $X_1, X_2, \ldots, X_n$ represent a random sample of size n, arranged in increasing order of magnitude, then the **sample median** is defined by the statistic*

$$\tilde{X} = \begin{cases} X_{(n+1)/2} & \textit{if n is odd} \\ \dfrac{X_{n/2} + X_{(n/2)+1}}{2} & \textit{if n is even.} \end{cases}$$ ■

EXAMPLE 8.2   The number of foreign ships arriving at an east coast port on 7 randomly selected days were 8, 3, 9, 5, 6, 8, and 5. Find the sample median.

SOLUTION
Arranging the observations in increasing order of magnitude, we get

$$3 \quad 5 \quad 5 \quad 6 \quad 8 \quad 8 \quad 9$$

and hence $\tilde{x} = 6$.

EXAMPLE 8.3   The nicotine contents for a random sample of 6 cigarettes of a certain brand are found to be 2.3, 2.7, 2.5, 2.9, 3.1, and 1.9 milligrams. Find the median.

SOLUTION
If we arrange these nicotine contents in an increasing order of magnitude, we get

$$1.9 \quad 2.3 \quad 2.5 \quad 2.7 \quad 2.9 \quad 3.1$$

and the median is then the mean of 2.5 and 2.7. Therefore,

$$\tilde{x} = \frac{2.5 + 2.7}{2} = 2.6 \text{ milligrams.}$$

The third and final statistic for measuring the center of a random sample that we shall discuss is the mode, designated by the statistic M.

DEFINITION 8.7   *If $X_1, X_2, \ldots, X_n$, not necessarily all different, represent a random sample of size n, then the* **mode** *M is that value of the sample that occurs most often or with the greatest frequency. The mode may not exist, and when it does it is not necessarily unique.* ∎

The mode does not always exist. This is certainly true when all observations occur with the same frequency. For some sets of data there may be several values occurring with the greatest frequency, in which case we have more than one mode.

EXAMPLE 8.4   If the donations of a random sample of residents of Fairway Forest toward the Virginia Lung Association are recorded as 9, 10, 5, 9, 9, 7, 8, 6, 10, and 11 dollars, then the mode is $m = \$9$, the value that occurs with the greatest frequency.

EXAMPLE 8.5   The number of movies attended last month by a random sample of 12 high school students were recorded as follows: 2, 0, 3, 1, 2, 4, 2, 5, 4, 0, 1, and 4. In this case, there are two modes, 2 and 4, since both 2 and 4 occur with the greatest frequency. The distribution is said to be **bimodal**.

EXAMPLE 8.6   No mode exists for the nicotine contents of Example 8.3, since each measurement occurs only once.

In summary, let us consider the relative merits of the mean, median, and mode. The mean is the most commonly used measure of central location in statistics. It employs all available information. The distributions of means obtained in repeated sampling from a population are well known, and consequently the methods used in statistical inference for estimating μ are based on the sample mean. This will become apparent in Section 8.3. The only real disadvantage to the mean is that it may be affected adversely by extreme values. In Example 8.4 the mean contribution to the Virginia Lung Association was $8.40, which is fairly close to the mode or median, both of which are $9. However, if one of the contributions had been much larger, say $90 instead of $11, then the mean contribution is $16.30, a value considerably higher than the majority of gifts.

The median has the advantage of being easy to compute if the number of observations is relatively small. It is not influenced by extreme values and consequently in Example 8.4 gives a better *center of the data,* namely $9, if the highest contribution is $90 rather than $11. In dealing with samples selected from populations, the sample means usually will not vary as much from sample to sample as will the medians. Therefore, if we are attempting to estimate the center of a population based on a sample value, the mean is more stable than the median. Hence a sample mean is likely to be closer than the sample median to the population mean.

The mode is the least used measure of the three. For small sets of data its value is almost useless if, in fact, it exists at all. Only in the case of a large mass of data does it have a significant meaning. Its two main advantages are that (1) it requires no calculation, and (2) it can be used for qualitative as well as quantitative data. Thus, if jogging is the preferred form of exercise expressed by most people, we say that jogging is the **modal choice**.

Variability in the Sample

The three measures of central location defined above do not by themselves give an adequate description of our data. We need to know how the observations spread out from the average. The reader is referred to Chapter 1 for more discussion. It is quite possible to have two sets of observations with the same mean or median that differ considerably in the variability of their measurements about the average.

Consider the following measurements, in liters, for two samples of orange juice bottled by companies A and B:

Sample A	0.97	1.00	0.94	1.03	1.11
Sample B	1.06	1.01	0.88	0.91	1.14

Both samples have the same mean, 1.00 liter. It is quite obvious that company A bottles orange juice with a more uniform content than company B. We say that the variability or the dispersion of the observations from the average is less for sample A than for sample B. Therefore, in buying orange juice, we would feel more confident that the bottle we select will be closer to the advertised average if we buy from company A.

The most important statistics for measuring the variability of a random sample are the **range** and the **variance**. The simpler of these to compute is the range.

DEFINITION 8.8   *The **range** of a random sample $X_1, X_2, \ldots, X_n$ is defined by the statistic $X_{(n)} - X_{(1)}$, where $X_{(n)}$ and $X_{(1)}$ are, respectively, the largest and smallest observations in the sample.* ∎

EXAMPLE 8.7   The IQs of a random sample of five members of a sorority are 108, 112, 127, 118, and 113. Find the range.

SOLUTION

The range of the five IQs is $127 - 108 = 19$.

In the case of the companies bottling orange juice, the range for company A is 0.17 liter compared to a range of 0.26 liter for company B, indicating a greater spread in the values for company B.

The range can be a poor measure of variability, particularly if the size of the sample or population is large. It considers only the extreme values and tells us nothing about the distribution of values in between. Consider, for example, the following two sets of data, both with a range of 12:

					Observed Data				
Set A	3	4	5	6	8	9	10	12	15
Set B	3	7	7	7	8	8	8	9	15

In the first set the mean and median are both 8, but the numbers vary over the entire interval from 3 to 15. In the second set the mean and median are also both 8, but most of the values are closer to the center of the data. Although the range fails to measure this variability between the upper and lower observations, it does have some useful applications. In industry the range might be predetermined by specifying in advance that a particular measurement on items coming off an assembly line must fall within a certain interval. As long as all the measurements fall within the specified interval, the process is said to be in control. The use of the sample range for *control charts* will be discussed at length in Chapter 17.

To overcome the disadvantage of the range, we shall consider a measure of variability, namely, the **sample variance**, that considers the position of each observation relative to the sample mean.

DEFINITION 8.9   *If $X_1, X_2, \ldots, X_n$ represent a random sample of size n, then the* **sample variance** *is defined by the statistic*

$$S^2 = \frac{\sum_{i=1}^{n} (X_i - \overline{X})^2}{n - 1}.$$ ■

The computed value of S^2 for a given sample is denoted by s^2. Note that S^2 is essentially defined to be the average of the squares of the deviations of the observations from their mean. The reason for using $n - 1$ as a divisor rather than the more obvious choice n will become apparent in Chapter 9.

EXAMPLE 8.8   A comparison of coffee prices at 4 randomly selected grocery stores in San Diego showed increases from the previous month of 12, 15, 17, and 20 cents for a 200-gram jar. Find the variance of this random sample of price increases.

SOLUTION

Calculating the sample mean, we get

$$\bar{x} = \frac{12 + 15 + 17 + 20}{4} = 16 \text{ cents.}$$

Therefore,

$$s^2 = \frac{\sum_{i=1}^{4} (x_i - 16)^2}{3}$$

$$= \frac{(12 - 16)^2 + (15 - 16)^2 + (17 - 16)^2 + (20 - 16)^2}{3}$$

$$= \frac{(-4)^2 + (-1)^2 + (1)^2 + (4)^2}{3}$$

$$= \frac{34}{3}.$$

If $\bar{x}$ is a decimal number that has been rounded off, we accumulate a large error using the sample variance formula of Definition 8.9. To avoid this, let us derive a very useful formula, as given in the following theorem.

THEOREM 8.1   *If S^2 is the variance of a random sample of size n, we may write*

$$S^2 = \frac{n \sum_{i=1}^{n} X_i^2 - \left(\sum_{i=1}^{n} X_i \right)^2}{n(n-1)}.$$ ■

PROOF. By definition,

$$S^2 = \frac{\sum_{i=1}^{n} (X_i - \bar{X})^2}{n-1}$$

$$= \frac{\sum_{i=1}^{n} (X_i^2 - 2\bar{X}X_i + \bar{X}^2)}{n-1}$$

$$= \frac{\sum_{i=1}^{n} X_i^2 - 2\bar{X} \sum_{i=1}^{n} X_i + n\bar{X}^2}{n-1}$$

Replacing $\bar{X}$ by $\sum_{i=1}^{n} X_i/n$ and multiplying numerator and denominator by n, we obtain the more useful computational formula

$$S^2 = \frac{n \sum\limits_{i=1}^{n} X_i^2 - \left(\sum\limits_{i=1}^{n} X_i\right)^2}{n(n-1)}.$$

DEFINITION 8.10   *The* **sample standard deviation***, denoted by S, is the positive square root of the sample variance.* ∎

EXAMPLE 8.9   Find the variance of the data 3, 4, 5, 6, 6, and 7, representing the number of trout caught by a random sample of 6 fishermen on June 19, 1984, at Lake Muskoka.

SOLUTION

We find that $\sum\limits_{i=1}^{6} x_i^2 = 171$, $\sum\limits_{i=1}^{6} x_i = 31$, $n = 6$. Hence

$$s^2 = \frac{(6)(171) - (31)^2}{(6)(5)} = \frac{13}{6}.$$

Exercises

1. Defined suitable populations from which the following samples are selected:
   (a) Persons in 200 homes are called by telephone in the city of Richmond and asked to name the candidate that they favor for election to the school board.
   (b) A coin is tossed 100 times and 34 tails are recorded.
   (c) Two hundred pairs of a new type of tennis shoe were tested on the professional tour and, on the average, lasted 4 months.
   (d) On five different occasions it took a lawyer 21, 26, 24, 22, and 21 minutes to drive from her suburban home to her midtown office.

2. The number of tickets issued for traffic violations by 8 state troopers during the Memorial Day weekend are 5, 4, 7, 7, 6, 3, 8, and 6.
   (a) If these values represent the number of tickets issued by a random sample of 8 state troopers from Montgomery County in Virginia, define a suitable population.
   (b) If the values represent the number of tickets issued by a random sample of 8 state troopers from South Carolina, define a suitable population.

3. The numbers of incorrect answers on a true-false competency test for a random sample of 15 students

were recorded as follows: 2, 1, 3, 0, 1, 3, 6, 0, 3, 3, 5, 2, 1, 4, and 2. Find
   (a) the mean;
   (b) the median;
   (c) the mode.

4. The lengths of time, in minutes, that 10 patients waited in a doctor's office before receiving treatment were recorded as follows: 5, 11, 9, 5, 10, 15, 6, 10, 5, and 10. Treating the data as a random sample, find
   (a) the mean;
   (b) the median;
   (c) the mode.

5. The reaction times for a random sample of 9 subjects to a stimulant were recorded as 2.5, 3.6, 3.1, 4.3, 2.9, 2.3, 2.6, 4.1, and 3.4 seconds. Calculate
   (a) the mean;
   (b) the median.

6. According to ecology writer Jacqueline Killeen, phosphates contained in household detergents pass right through our sewer systems, causing lakes to turn into swamps that eventually dry up into deserts. The following data show the amount of phosphates per load of laundry, in grams, for a random sample of various types of detergents used according to the prescribed directions:

Laundry Detergent	Phosphates per Load (grams)
A & P Blue Sail	48
Dash	47
Concentrated All	42
Cold Water All	42
Breeze	41
Oxydol	34
Ajax	31
Sears	30
Fab	29
Cold Power	29
Bold	29
Rinso	26

For the given phosphate data, find
(a) the mean;
(b) the median;
(c) the mode.

7. A random sample of employees from a local manufacturing plant pledged the following donations, in dollars, to the United Fund: 10, 40, 25, 5, 20, 10, 25, 50, 30, 10, 5, 15, 25, 50, 10, 30, 5, 25, 45, and 15. Calculate
(a) the mean;
(b) the mode.

8. Find the mean, median, and mode for the sample whose observations, 15, 7, 8, 95, 19, 12, 8, 22, and 14, represent the number of sick days claimed on 9 federal income tax returns. Which value appears to be the best measure of the center of our data? Give reasons for your preference.

9. With reference to the lengths of time that 10 patients waited in a doctor's office before receiving treatment in Exercise 4, find
(a) the range;
(b) the standard deviation.

10. With reference to the sample of reaction times for the 9 subjects receiving the stimulant in Exercise 5, calculate
(a) the range;
(b) the variance using the formula of Definition 8.9.

11. With reference to the random sample of incorrect answers on a true-false competency test for the 15 students in Exercise 3, calculate the variance using the formula
(a) of Definition 8.9;
(b) of Theorem 8.1.

12. The tar contents of 8 brands of cigarettes selected at random from the latest list released by the Federal Trade Commission are as follows: 7.3, 8.6, 10.4, 16.1, 12.2, 15.1, 14.5, and 9.3 milligrams. Calculate
(a) the mean;
(b) the variance.

13. The grade-point averages of 20 college seniors selected at random from a graduating class are as follows:

3.2	1.9	2.7	2.4
2.8	2.9	3.8	3.0
2.5	3.3	1.8	2.5
3.7	2.8	2.0	3.2
2.3	2.1	2.5	1.9

Calculate the standard deviation.

14. (a) Show that the sample variance is unchanged if a constant c is added to or subtracted from each value in the sample.
(b) Show that the sample variance becomes c^2 times its original value if each observation in the sample is multiplied by c.

15. Verify that the variance of the sample 4, 9, 3, 6, 4, and 7 is 5.1, and using this fact along with the results of Exercise 14, find
(a) the variance of the sample 12, 27, 9, 18, 12, and 21;
(b) the variance of the sample 9, 14, 8, 11, 9, and 12.

8.3 Data Displays and Graphical Methods _____

In Section 3.4 we introduced the reader to empirical distributions. The motivation was to use creative displays to extract information about properties of a set of data. For example, the stem and leaf plots provide the viewer a look at symmetry and other properties of the data. In this chapter we deal with samples which, of course, are collections of experimental data from which we draw conclusions about populations. Often the appearance of the sample will provide information about the distribution from which the data were taken. For example, in Chapter 1 we illustrated the general nature of pairs of samples with point plots that displayed a relative comparison between central tendency and variability among two samples.

In chapters that follow, we often make the assumption that the distribution is normal. Graphical information regarding the validity of this assumption can be retrieved from displays like the stem and leaf plots and frequency histograms. In addition, we will introduce the notion of *normal probability plots* and *quantile plots* in this section. These plots are used in studies that have varying degrees of complexity, with the main objective of the plots being to provide a diagnostic check on the assumption that the data came from a normal distribution.

We can characterize statistical analysis as the process of drawing conclusions about systems in the presence of system variability. An engineer's attempt to learn about a chemical process is often clouded by *process variability*. A study involving the number of defective items in a production process is often made more difficult by variability in the method of manufacture of the items. In what has preceded, we have learned about samples and statistics that express center of location and variability in the sample. These statistics provide single measures, whereas a graphical display adds additional information in terms of a picture.

Box and Whisker Plots

Another display that is helpful in reflecting properties of a sample is the **box and whisker plot**. This plot encloses the *interquartile range* of the data in a box that has the median displayed within. The interquartile range has as its extremes the 75th percentile (upper quartile) and the 25th percentile (lower quartile). In addition to the box, "whiskers" extend, showing *extreme observations* in the sample. For reasonably large samples the display shows center of location, variability, and the degree of asymmetry.

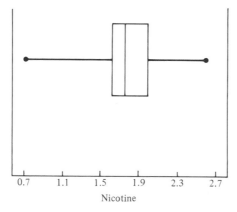

FIGURE 8.1 Box and whisker plot for nicotine data of Exercise 5 following Section 2.4.

EXAMPLE 8.10   Consider the data in Exercise 5 following Section 3.4. Nicotine content was measured in a random sample of 40 cigarettes. Figure 8.1 shows the box and whisker plot of the data.

EXAMPLE 8.11   Consider the following data, consisting of 30 samples measuring the thickness of paint can ears (see the work by Hogg and Ledolter in the Bibliography). Figure 8.2 depicts a box and whisker plot for this asymmetric set of data.

Sample	Measurements				
1	29	36	39	34	34
2	29	29	28	32	31
3	34	34	39	38	37
4	35	37	33	38	41
5	30	29	31	38	29
6	34	31	37	39	36
7	30	35	33	40	36
8	28	28	31	34	30
9	32	36	38	38	35
10	35	30	37	35	31
11	35	30	35	38	35
12	38	34	35	35	31
13	34	35	33	30	34
14	40	35	34	33	35
15	34	35	38	35	30
16	35	30	35	29	37
17	40	31	38	35	31
18	35	36	30	33	32
19	35	34	35	30	36
20	35	35	31	38	36
21	32	36	36	32	36
22	36	37	32	34	34
23	29	34	33	37	35
24	36	36	35	37	37
25	36	30	35	33	31
26	35	30	29	38	35
27	35	36	30	34	36
28	35	30	36	29	35
29	38	36	35	31	31
30	30	34	40	28	30

Notice that the left block is considerably larger than the block on the right. The median is 35. The lower quartile is 31, while the upper quartile is 36. Notice also that the extreme observation on the right is farther away from the box than the extreme observation on the left.

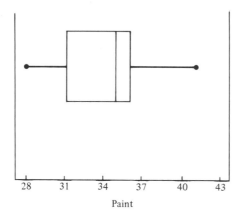

28   31   34   37   40   43

Paint

FIGURE 8.2   Box and whisker plot for thickness of paint can ears.

There are additional ways that box and whisker plots and other graphical displays can aid the analyst. Multiple samples can be compared graphically. Plots of data can suggest relationships between variables. Graphs can aid in the detection of anomalies or outlying observations in samples.

Another type of plot that can be particularly useful in characterizing the nature of a data set is the *quantile plot*. As in the case of the box and whisker plot, one can use the basic ideas in the quantile plot to *compare samples of data*, where the goal of the analyst is to draw distinctions. Further illustrations of this type of usage will be given in future chapters where the formal statistical inference associated with comparing samples is discussed. At that point, case studies will be demonstrated in which the reader is exposed to both the formal inference and the diagnostic graphics for the same data set.

Quantile Plot

The purpose of the quantile plot is to depict, in sample form, the cumulative distribution function discussed in Chapter 3.

DEFINITION 8.11   A **quantile** *of a sample,* $q(f)$, *is a value for which a specified fraction,* f, *of the data values is less than or equal to* $q(f)$.

Obviously, a quantile represents an estimate of a characteristic of a population, or rather, the theoretical distribution. The sample median is $q(0.5)$. The 75th percentile (upper quartile) is $q(0.75)$ and the lower quartile is $q(0.25)$.

A **quantile plot** simply *plots the data values on the vertical axis against an empirical assessment of the fraction of observations exceeded by the data value.* For theoretical purposes this fraction is computed as

$$f_i = \frac{i - \frac{3}{8}}{n + \frac{1}{4}},$$

where i is the order of the observations when they are ranked from low to high. In other words, if we denote the ranked observations as

$$y_{(1)} \leq y_{(2)} \leq y_{(3)} \cdots \leq y_{(n-1)} \leq y_{(n)},$$

then the quantile plot depicts a plot of $y_{(i)}$ against f_i. In Figure 8.3 the quantile plot is given for the paint can ear data discussed previously.

Unlike the box and whisker plot the quantile plot actually shows all observations. All quantiles, including the median and the upper and lower quantile, can be approximated visually. For example, we readily observe a median of 35 and an upper quartile of about 36. Indications of relatively large clusters around specific values are indicated by slopes near zero, while sparse data in certain areas produce steeper slopes. Figure 8.3 depicts sparsity of data from the values 28 through 30 but relatively high density at 36 through 38. In Chapters 9 and 10 we pursue quantile plotting further by illustrating useful ways of comparing distinct samples.

Detection of Deviations from Normality

It should be somewhat evident to the reader that detection of whether or not a data set came from a normal distribution can be an important tool for the data analyst. As we indicated earlier in this section, we often make the assumption that all or subsets of observations in a data set are realizations of independent identically distributed normal random variables. Once again, the diagnostic plot can often nicely augment (for display purposes) a formal *goodness-of-fit test* on the data. Goodness-of-fit tests are discussed in Chapter 10. For the reader of a scientific paper or report, diagnostic information is much clearer, less dry, and perhaps not boring. In later chapters (Chapters 9 through 13) we will focus again on methods of detecting deviations from normality as an augmentation of formal statistical inference. The following subsection provides a discussion and illustration of a diagnostic plot called the *normal quantile–quantile plot*.

Normal Quantile–Quantile Plot

The normal quantile–quantile plot takes advantage of what is known about the quantiles of the normal distribution. The methodology involves a plot of the empirical quantiles recently discussed against the corresponding quantile of the normal distribution. Now, the expression for a quantile of an $N(\mu, \sigma)$ random variable is very complicated. However, a good approximation is given by

$$q_{\mu,\sigma}(f) = \mu + \sigma\{4.91[f^{0.14} - (1 - f)^{0.14}]\}.$$

The expression in brackets (the multiple of σ) is the approximation for the corresponding quantile for the $N(0, 1)$ random variable, that is,

$$q_{0,1}(f) = 4.91[f^{0.14} - (1 - f)^{0.14}].$$

DEFINITION 8.12   *The **normal quantile–quantile plot** is a plot of $y_{(i)}$ (ordered observations) against* $q_{0,1}(f_i)$, *where* $f_i = \dfrac{i - \frac{3}{8}}{n + \frac{1}{4}}$.

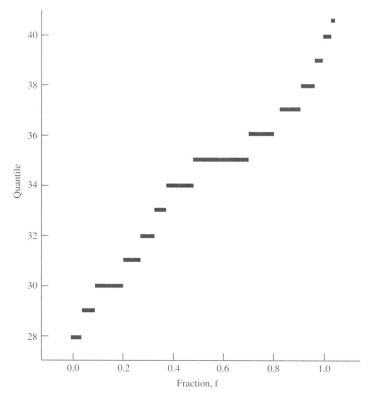

FIGURE 8.3   Quantile plot for paint data on page 209.

A nearly straight-line relationship suggests that the data came from a normal distribution. The intercept on the vertical axis is an estimate of the population mean and the slope is an estimate of the standard deviation σ. Figure 8.4 shows a normal quantile–quantile plot for the paint can data.

Normal Probability Plotting

Notice how the deviation from normality becomes clear from the appearance of the plot. The asymmetry exhibited in the data results in changes in the slope.

The ideas of probability plotting are manifested in plots other than the normal quantile–quantile plot discussed here. For example, much attention is given to the so-called **normal probability plot**, in which the vertical axis contains f plotted on special paper and the scale used results in a straight line when plotted against the ordered data values. In addition, an alternative plot makes use of the expected values of the ranked observations for the normal distribution and plots the ranked observations against their expected value, under the assumption of data from $N(\mu, \sigma)$. Once again, the straight line is the graphical yardstick used. We continue to suggest that the foundation in graphical analytical methods developed in this section aid in the illustration of formal methods of distinguishing between distinct samples of data.

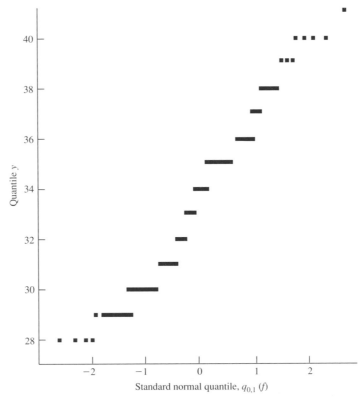

FIGURE 8.4 Normal quantile-quantile plot for paint can data.

EXAMPLE 8.12 Consider the data in Exercise 23 on page 332 in Chapter 10. In a study "*Nutrient Retention and Macro Invertebrate Community Response to Sewage Stress in a Stream Ecosystem*" conducted in the Department of Zoology at the Virginia Polytechnic Institute and State University, data were collected on density measurements (number of organisms per square meter) at two different collecting stations. Details are given in Chapter 10 regarding analytical methods of comparing samples to determine if both are from the same $N(\mu, \sigma)$ distribution. The data are as follows:

Number of Organisms per Square Meter			
Station 1		Station 2	
5030	4980	2800	2810
13,700	11,910	4670	1330
10,730	8130	6890	3320
11,400	26,850	7720	1230
860	17,660	7030	2130
2200	22,800	7330	2190
4250	1130		
15,040	1690		

Construct a normal quantile–quantile plot and draw conclusions regarding whether or not it is reasonable to assume that the two samples are from the same $N(\mu, \sigma)$ distribution.

SOLUTION
Figure 8.5 shows the normal quantile–quantile plot for the density measurements. The plot shows an appearance that is far from a single straight line. In fact, the data from station 1 reflect a few values in the lower tail of the distribution and several in the upper tail. The "clustering" of observations would make it seem unlikely that the two samples came from a common $N(\mu, \sigma)$ distribution.

Although we have concentrated our development and illustration on probability plotting for the normal distribution, one could focus on any distribution. One would merely need to compute quantities analytically for the theoretical distribution in question.

8.4 Sampling Distributions

The field of statistical inference is basically concerned with generalizations and predictions. For example, we might claim, based on the opinions of several people interviewed on the street, that in a forthcoming election 60% of the eligible voters in the city of Detroit favor a certain candidate. In this case we are dealing with a random sample of opinions from a very large finite population. As a second illustration we might state that the average cost to build a residence in Charleston, South Carolina, is between $90,000 and $95,000, based on the estimates of 3 contractors selected at random from the 30 now building in this city. The population being sampled here is again finite but very small. Finally, let us consider a soft-drink dispensing machine in which the average amount of drink dispensed is being held to 240 milliliters. A company official computes the mean of 40 drinks to obtain $\bar{x} = 236$ milliliters, and on the basis of this value decides that the machine is still dispensing drinks with an average content of $\mu = 240$ milliliters. The 40 drinks represent a sample from the infinite population of possible drinks that will be dispensed by this machine.

In each of the examples above we have computed a statistic from a sample selected from the population, and from these statistics we made various statements concerning the values of population parameters that may or may not be true. The company official made the decision that the soft-drink machine dispenses drinks with an average content of 240 milliliters, even though the sample mean was 236 milliliters, because he knows from sampling theory that such a sample value is likely to occur. In fact, if he ran similar tests, say every hour, he would expect the values of $\bar{x}$ to fluctuate above and below $\mu = 240$ milliliters. Only when the value of $\bar{x}$ is substantially different from 240 milliliters will the company official initiate action to adjust the machine.

Since a statistic is a random variable that depends only on the observed sample, it must have a probability distribution.

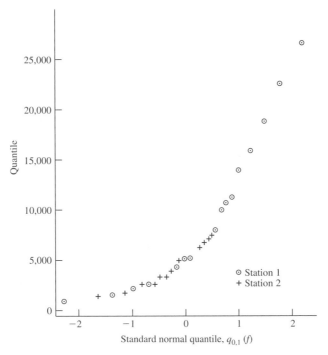

FIGURE 8.5 Normal quantile-quantile plot for density data of Example 8.12.

DEFINITION 8.13   *The probability distribution of a statistic is called a* **sampling distribution**. ∎

The probability distribution of $\overline{X}$ is called the **sampling distribution of the mean**.

The sampling distribution of a statistic will depend on the size of the population, the size of the samples, and the method of choosing the samples. For the remainder of this chapter we study several of the more important sampling distributions of frequently used statistics. Applications of these sampling distributions to problems of statistical inference are considered throughout most of the remaining chapters.

One should view the sampling distributions of $\overline{X}$ and S^2 as the mechanism from which we are eventually to make inferences on the parameters μ and σ^2. The sampling distribution of $\overline{X}$ with sample size n is the distribution that results when an experiment is conducted over and over (always with sample size n) and the many values of $\overline{X}$ result. This sampling distribution, then, describes the variability of sample averages around the population mean μ. In the case of the soft-drink machine, knowledge of the sampling distribution of $\overline{X}$ arms the analyst with the knowledge of a "typical" discrepancy between an observed $\bar{x}$ value and μ. The same principle applies in the case of the distribution of S^2. The sampling distribution produces information about the variability of s^2 values around σ^2 in repeated experiments.

8.5 Sampling Distributions of Means

The first important sampling distribution to be considered is that of the mean $\overline{X}$. Suppose that a random sample of n observations is taken from a normal population with mean μ and variance σ^2. Each observation X_i, $i = 1, 2, \ldots, n$, of the random sample will then have the same normal distribution as the population being sampled. Hence, by the reproductive property of the normal distribution established in Theorem 7.11, we conclude that

$$\overline{X} = \frac{X_1 + X_2 + \cdots + X_n}{n}$$

has a normal distribution with mean

$$\mu_{\overline{X}} = \frac{\mu + \mu + \cdots + \mu}{n} = \mu$$

and variance

$$\sigma_{\overline{X}}^2 = \frac{\sigma^2 + \sigma^2 + \cdots + \sigma^2}{n^2} = \frac{\sigma^2}{n}.$$

If we are sampling from a population with unknown distribution, either finite or infinite, the sampling distribution of $\overline{X}$ will still be approximately normal with mean μ and variance σ^2/n provided that the sample size is large. This amazing result is an immediate consequence of the following theorem, called the **central limit theorem**. The proof is outlined in Exercise 11 on page 223.

THEOREM 8.2 (CENTRAL LIMIT THEOREM)

If $\overline{X}$ is the mean of a random sample of size n taken from a population with mean μ and finite variance σ^2, then the limiting form of the distribution of

$$Z = \frac{\overline{X} - \mu}{\sigma/\sqrt{n}},$$

as $n \rightarrow \infty$, is the standard normal distribution $n(z; 0, 1)$. ∎

The normal approximation for $\overline{X}$ will generally be good if $n \geq 30$ regardless of the shape of the population. If $m < 30$, the approximation is good only if the population is not too different from a normal distribution and, as stated above, if the population is known to be normal, the sampling distribution of $\overline{X}$ will follow a normal distribution exactly, no matter how small the size of the samples.

EXAMPLE 8.13

An electrical firm manufactures light bulbs that have a length of life that is approximately normally distributed, with mean equal to 800 hours and a standard deviation of 40 hours. Find the probability that a random sample of 16 bulbs will have an average life of less than 775 hours.

SOLUTION

The sampling distribution of $\overline{X}$ will be approximately normal, with $\mu_{\overline{x}} = 800$ and $\sigma_{\overline{x}} = 40/\sqrt{16} = 10$. The desired probability is given by the area of the shaded region in Figure 8.6.

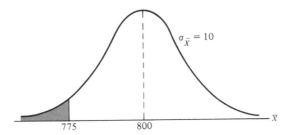

FIGURE 8.6 Area for Example 8.13.

Corresponding to $\bar{x} = 775$, we find that

$$z = \frac{775 - 800}{10} = -2.5,$$

and therefore

$$P(\bar{X} < 775) = P(Z < -2.5)$$
$$= 0.0062.$$

Inferences on the Population Mean

One very important application of the central limit theorem is the determination of reasonable values of the population mean μ. Topics such as hypothesis testing, estimation, quality control, and others make use of the central limit theorem. The following example illustrates the use of the central limit theorem in that regard, although the formal application to the foregoing topics is relegated to future chapters.

EXAMPLE 8.14   An important manufacturing process produces cylindrical component parts for the automotive industry. It is important that the process produce parts having a mean of 5 millimeters. The engineer involved conjectures that the population mean is 5.0 millimeters. An experiment is conducted in which 100 parts produced by the process are selected randomly and the diameter measured on each. It is known that the population standard deviation $\sigma = 0.1$. The experiment gives a sample average diameter $\bar{x} = 5.027$ millimeters. Does this sample information appear to support or refute the engineer's conjecture?

SOLUTION
This example reflects the kind of problem often posed and solved with hypothesis-testing machinery introduced in future chapters. We will not use the formality associated with hypothesis testing here but we will illustrate the principles and logic used.

Whether or not the data support or refute the conjecture depends on the probability that data similar to that obtained in this experiment ($\bar{x} = 5.027$) can readily occur when in fact $\mu = 5.0$ (Figure 8.7). In other words, how likely is it that one can obtain $\bar{x} \geq 5.027$ with $n = 100$ if the population mean $\mu = 5.0$? If this prob-

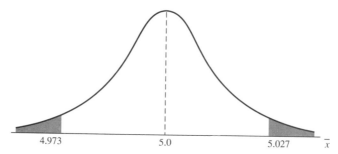

FIGURE 8.7   Area for Example 8.14.

ability suggests that $\bar{x} = 5.027$ is not unreasonable, the conjecture is not refuted. If the probability is quite low, one can certainly argue that the data do not support the conjecture that $\mu = 5.0$. The probability that we choose to compute is given by

$$\Pr[|(\overline{X} - 5)| \geq 0.027].$$

In other words, if the mean μ is 5, what is the chance that $\overline{X}$ will deviate by as much as 0.027 millimeter?

$$P[|(\overline{X} - 5)| \geq 0.027] = P[(\overline{X} - 5) \geq 0.027] + P[(\overline{X} - \mu) \leq -0.027]$$

$$= 2P\left(\frac{\overline{X} - 5}{0.1/\sqrt{100}}\right) \geq 2.7.$$

Here we are simply standardizing $\overline{X}$ according to the central limit theorem. *If the conjecture $\mu = 5.0$ is true,* $\dfrac{\overline{X} - 5.0}{0.1/\sqrt{100}}$ *is N(0, 1). Thus*

$$2P\left[\frac{\overline{X} - 5.0}{0.1/\sqrt{100}} \geq 2.7\right] = 2P[Z \geq 2.7] = 2(0.0035) = 0.007$$

Thus one would experience by chance an $\bar{x}$ that is 0.027 millimeter from the mean in only 7 in 1000 experiments. As a result, this experiment with $\bar{x} = 5.027$ certainly does not give supporting evidence to the conjecture that $\mu = 5.0$.

Sampling Distribution of the Difference Between Two Averages

The illustration in Example 8.14 deals with notions of statistical inference on a single mean μ. The engineer was interested in supporting a conjecture regarding a single population mean. A far more important application involves two populations. A scientist or engineer is interested in a comparative experiment in which two manufacturing methods, 1 and 2, are to be compared. The basis for that comparison is $\mu_1 - \mu_2$, the difference in the population means.

Suppose that we have two populations, the first with mean μ_1 and variance σ_1^2, and the second with mean μ_2 and variance σ_2^2. Let the statistic $\overline{X}_1$ represent the mean of a random sample of size n_1 selected from the first population, and the statistic $\overline{X}_2$ represent the mean of a random sample selected from the second population, independent of the sample from the first population. What can we say about the sampling distribution of the difference $\overline{X}_1 - \overline{X}_2$ for repeated samples of size n_1 and n_2? According to Theorem 8.2, the variables $\overline{X}_1$ and $\overline{X}_2$ are both approximately normally distributed with means μ_1 and μ_2 and variances σ_1^2/n_1 and σ_2^2/n_2, respectively. This approximation improves as n_1 and n_2 increase. By choosing independent samples from the two populations the variables $\overline{X}_1$ and $\overline{X}_2$ will be independent, and then using Theorem 7.11, with $a_1 = 1$ and $a_2 = -1$, we can conclude that $\overline{X}_1 - \overline{X}_2$ is approximately normally distributed with mean

$$\mu_{\overline{X}_1-\overline{X}_2} = \mu_{\overline{X}_1} - \mu_{\overline{X}_2} = \mu_1 - \mu_2$$

and variance

$$\sigma_{\overline{X}_1-\overline{X}_2}^2 = \sigma_{\overline{X}_1}^2 + \sigma_{\overline{X}_2}^2 = \frac{\sigma_1^1}{n_1} + \frac{\sigma_2^2}{n_2}.$$

THEOREM 8.3

If independent samples of size n_1 and n_2 are drawn at random from two populations, discrete or continuous, with means μ_1 and μ_2 and variances σ_1^2 and σ_2^2, respectively, then the sampling distribution of the differences of means, $\overline{X}_1 - \overline{X}_2$, is approximately normally distributed with mean and variance given by

$$\mu_{\overline{X}_1-\overline{X}_2} = \mu_1 - \mu_2 \qquad and \qquad \sigma_{\overline{X}_1-\overline{X}_2}^2 = \frac{\sigma_1^2}{n_1} + \frac{\sigma_2^2}{n_2}.$$

Hence

$$Z = \frac{(\overline{X}_1 - \overline{X}_2) - (\mu_1 - \mu_2)}{\sqrt{(\sigma_1^2/n_1) + (\sigma_2^2/n_2)}}$$

is approximately a standard normal variable.   ■

If both n_1 and n_2 are greater than or equal to 30, the normal approximation for the distribution of $\overline{X}_1 - \overline{X}_2$ is very good regardless of the shapes of the two populations. However, even when n_1 and n_2 are less than 30, the normal approximation is reasonably good except when the populations are decidedly nonnormal. Of course if both populations are normal, then $\overline{X}_1 - X_2$ has a normal distribution no matter what the sizes are of n_1 and n_2.

EXAMPLE 8.15   Two independent experiments are being run in which two different types of paints are compared. Eighteen specimens are painted using type A and the drying time in hours is recorded on each. The same is done with type B. The population standard deviations are both known to be 1.0.

Assuming that the mean drying time is equal for the two types of paint, find $\Pr(\overline{X}_A - \overline{X}_B > 1.0)$, where $\overline{X}_A$ and $\overline{X}_B$ are average drying times for samples of size $n_A = n_B = 18$.

SOLUTION

From the sampling distribution of $\overline{X}_A - \overline{X}_B$, we know that the distribution is normal with mean

$$\mu_{\overline{X}_A - \overline{X}_B} = \mu_A - \mu_B = 0$$

and variance

$$\sigma^2_{\overline{X}_A - \overline{X}_B} = \frac{\sigma^2_A}{n_A} + \frac{\sigma^2_B}{n_B}$$

$$= \tfrac{1}{18} + \tfrac{1}{18} = \tfrac{1}{9}.$$

The desired probability is given by the shaded region in Figure 8.8. Corresponding to the value $\overline{X}_A - \overline{X}_B = 1.0$, we have

$$Z = \frac{1 - (\mu_A - \mu_B)}{\sqrt{1/9}}$$

$$= \frac{1 - 0}{\sqrt{1/9}} = 3.0;$$

so

$$Pr(z > 3.0) = 1 - Pr(z < 3.0)$$

$$= 1 - 0.9987$$

$$= 0.0013.$$

What might we learn from the result above? The machinery in the calculation is based on the presumption that $\mu_A = \mu_B$. Suppose, however, that the experiment is actually conducted for the purposes of drawing an inference regarding the equality of μ_A and μ_B, the two population mean drying times. If the two averages differ by as much as 1 hour (or more), this clearly is evidence that would lead one to conclude that the population mean drying time is not equal for the two types of paint. On the other hand, suppose that the difference in the two sample averages is as small as, say, 15 minutes. If $\mu_A = \mu_B$,

$$Pr[|(\overline{X}_A - \overline{X}_B)| > 0.25 \text{ hour}] = Pr\left(\frac{\overline{X}_A - \overline{X}_B - 0}{\sqrt{1/9}}\right) > \frac{3}{4}$$

$$= Pr(Z > \tfrac{3}{4})$$

$$= 1 - Pr(Z < 0.75)$$

$$= 1 - 0.7734$$

$$= 0.2266.$$

Since this probability is not excessively low, one would conclude that a difference in sample means of 15 minutes can happen by chance (i.e., it happens frequently even though $\mu_A = \mu_B$). As a result, that type of difference in average drying time certainly is *not a clear signal* that $\mu_A \neq \mu_B$.

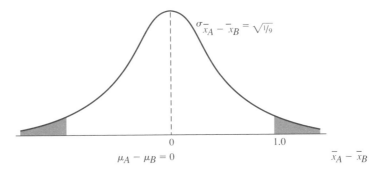

FIGURE 8.8   Area for Example 8.15.

As we indicated earlier, more formalism regarding this and other types of statistical inference (e.g., hypothesis testing) will be supplied in detail in future chapters. The central limit theorem and sampling distributions discussed in the next three sections will also play a vital role.

EXAMPLE 8.16   The television picture tubes of manufacturer A have a mean lifetime of 6.5 years and a standard deviation of 0.9 year, while those of manufacturer B have a mean lifetime of 6.0 years and a standard deviation of 0.8 year. What is the probability that a random sample of 36 tubes from manufacturer A will have a mean lifetime that is at least 1 year more than the mean lifetime of a sample of 49 tubes from manufacturer B?

SOLUTION
We are given the following information:

Population 1	Population 2
$\mu_1 = 6.5$	$\mu_2 = 6.0$
$\sigma_1 = 0.9$	$\sigma_2 = 0.8$
$n_1 = 36$	$n_2 = 49$

If we use Theorem 8.3, the sampling distribution of $\bar{X}_1 - \bar{X}_2$ will be approximately normal and will have a mean and standard deviation given by

$$\mu_{\bar{X}_1 - \bar{X}_2} = 6.5 - 6.0 = 0.5$$

and

$$\sigma_{\bar{X}_1 - \bar{X}_2} = \sqrt{\frac{0.81}{36} + \frac{0.64}{49}} = 0.189.$$

The probability that the mean of 36 tubes from manufacturer A will be at least 1 year longer than the mean of 49 tubes from manufacturer B is given by the area of

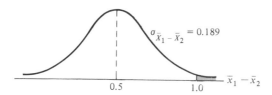

FIGURE 8.9   Area for Example 8.16.

the shaded region in Figure 8.9. Corresponding to the value $\bar{x}_1 - \bar{x}_2 = 1.0$, we find that

$$z = \frac{1.0 - 0.5}{0.189} = 2.65,$$

and hence

$$P(\bar{X}_1 - \bar{X}_2 \geq 1.0) = P(Z > 2.65)$$
$$= 1 - P(Z < 2.65)$$
$$= 1 - 0.9960$$
$$= 0.0040.$$

Exercises

1. If all possible samples of size 16 are drawn from a normal population with mean equal to 50 and standard deviation equal to 5, what is the probability that a sample mean $\bar{X}$ will fall in the interval from $\mu_{\bar{X}} - 1.9\sigma_{\bar{X}}$ to $\mu_{\bar{X}} - 0.4\sigma_{\bar{X}}$? Assume that the sample means can be measured to any degree of accuracy.

2. Given the discrete uniform population

$$f(x) = \begin{cases} \frac{1}{3}, & x = 2, 4, 6 \\ 0, & \text{elsewhere,} \end{cases}$$

find the probability that a random sample of size 54, selected with replacement, will yield a sample mean greater than 4.1 but less than 4.4. Assume the means to be measured to the nearest tenth.

3. A certain type of thread is manufactured with a mean tensile strength of 78.3 kilograms and a standard deviation of 5.6 kilograms. How is the variance of the sample mean changed when the sample size is
   (a) increased from 64 to 196?
   (b) decreased from 784 to 49?

4. If the standard deviation of the mean for the sampling distribution of random samples of size 36 from a large or infinite population is 2, how large must the size of the sample become if the standard deviation is to be reduced to 1.2?

5. A soft-drink machine is being regulated so that the amount of drink dispensed averages 240 milliliters with a standard deviation of 15 milliliters. Periodically, the machine is checked by taking a sample of 40 drinks and computing the average content. If the mean of the 40 drinks is a value within the interval $\mu_{\bar{X}} \pm 2\sigma_{\bar{X}}$, the machine is thought to be operating satisfactorily; otherwise, adjustments are made. In Section 6.7, the company official found the mean of 40 drinks to be $\bar{x} = 236$ milliliters and concluded that the machine needed no adjustment. Was this a reasonable decision?

6. The heights of 1000 students are approximately normally distributed with a mean of 174.5 centimeters and a standard deviation of 6.9 centimeters. If 200 random samples of size 25 are drawn from this population and the means recorded to the nearest tenth of a centimeter, determine
   (a) the mean and standard deviation of the sampling distribution of $\bar{X}$;

(b) the number of sample means that fall between 172.5 and 175.8 centimeters inclusive;

(c) the number of sample means falling below 172.0 centimeters.

7. The random variable X, representing the number of cherries in a cherry puff, has the following probability distribution:

x	4	5	6	7
$P(X = x)$	0.2	0.4	0.3	0.1

(a) Find the mean μ and the variance σ^2 of X.

(b) Find the mean $\mu_{\bar{X}}$ and the variance $\sigma_{\bar{X}}^2$ of the mean $\bar{X}$ for random samples of 36 cherry puffs.

(c) Find the probability that the average number of cherries in 36 cherry puffs will be less than 5.5.

8. If a certain machine makes electrical resistors having a mean resistance of 40 ohms and a standard deviation of 2 ohms, what is the probability that a random sample of 36 of these resistors will have a combined resistance of more than 1458 ohms?

9. The average life of a manufacturer's blender is 5 years, with a standard deviation of 1 year. Assuming that the lives of these blenders follow approximately a normal distribution, find

(a) the probability that the mean life of a random sample of 9 such blenders falls between 4.4 and 5.2 years;

(b) The value of $\bar{x}$ to the right of which 15% of the means computed from random samples of size 9 would fall.

10. The amount of time that a bank teller spends on a customer is a random variable with a mean $\mu = 3.2$ minutes and a standard variable $\sigma = 1.6$ minutes. If a random sample of 64 customers is observed, find the probability that their mean time at the teller's counter is

(a) at most 2.7 minutes;

(b) more than 3.5 minutes;

(c) at least 3.2 minutes but less than 3.4 minutes.

11. Central Limit Theorem

(a) Using Theorems 7.8, 7.9, and 7.10, show that

$$M_{(\bar{X}-\mu)/(\sigma/\sqrt{n})}(t) = e^{-\mu\sqrt{nt}/\sigma}\left[M_x\left(\frac{t}{\sigma\sqrt{n}}\right)\right]^n,$$

where $\bar{X}$ is the mean of a random sample of size n from a population $f(x)$ with mean μ and variance σ^2, and hence

$$\ln M_{(\bar{X}-\mu)/\sqrt{n}}(t) = \frac{-\mu\sqrt{n}\,t}{6} + n \ln M_X\left(\frac{t}{\sigma\sqrt{n}}\right).$$

(b) Use the result of Exercise 22 on page 196 to expand $M_X(t/\sigma\sqrt{n})$ as an infinite series in powers of t. We can then write $M_X(t/\sigma\sqrt{n}) = 1 + v$, where v is an infinite series.

(c) Assuming n sufficiently large, expand $\ln(1 + v)$ in a Maclaurin series and then show that

$$\lim_{n\to\infty} \ln M_{(\bar{X}-\mu)/(\sigma/\sqrt{n})}(t) = \frac{t^2}{2}$$

and hence

$$\lim_{n\to\infty} M_{(\bar{X}-\mu)/(\sigma/\sqrt{n})}(t) = e^{t^2/2}.$$

12. A random sample of size 25 is taken from a normal population having a mean of 80 and a standard deviation of 5. A second random sample of size 36 is taken from a different normal population having a mean of 75 and a standard deviation of 3. Find the probability that the sample mean computed from the 25 measurements will exceed the sample mean computed from the 36 measurements by at least 3.4 but less than 5.9. Assume the means to be measured to the nearest tenth.

13. The distribution of heights of a certain breed of terrier dogs has a mean height of 72 centimeters and a standard deviation of 10 centimeters, whereas the distribution of heights of a certain breed of poodles has a mean height of 28 centimeters with a standard deviation of 5 centimeters. Assuming that the sample means can be measured to any degree of accuracy, find the probability that the sample mean for a random sample of heights of 64 terriers exceeds the sample mean for a random sample of heights of 100 poodles by at most 44.2 centimeters.

14. The mean score for freshmen on an aptitude test, at a certain college, is 540, with a standard deviation of 50. What is the probability that two groups of students selected at random, consisting of 32 and 50 students, respectively, will differ in their mean scores by

(a) more than 20 points?

(b) an amount between 5 and 10 points?

Assume the means to be measured to any degree of accuracy.

15. Construct a quantile plot of these data. The lifetimes in hours of fifty 40-watt 110-volt internally frosted incandescent lamps taken from forced life tests:

919 1196 785 1126 936 918 1156 920 948
1067 1092 1162 1170 929 950 905 972 1035
1045 855 1195 1195 1340 1122 938 970 1237
956 1102 1157 978 832 1009 1157 1151 1009
765 958 902 1022 1333 811 1217 1085 896
958 1311 1037 702 923

16. Construct a normal quantile–quantile plot of these data. Diameters of 36 rivet heads in 1/100 of an inch:

6.72 6.77 6.82 6.70 6.78 6.70 6.62 6.75 6.66
6.66 6.64 6.76 6.73 6.80 6.72 6.76 6.76 6.68
6.66 6.62 6.72 6.76 6.70 6.78 6.76 6.67 6.70
6.72 6.74 6.81 6.79 6.78 6.66 6.76 6.76 6.72

8.6 Sampling Distribution of S^2

In the preceding section we learned about the sampling distribution of $\overline{X}$. The central limit theorem allowed us to make use of the fact that

$$\frac{\overline{X} - \mu}{\sigma/\sqrt{n}}$$

tends toward $N(0, 1)$ as the sample size grows large. Examples 8.15 and 8.16 illustrated application of the central limit theorem, which allowed conclusions to be drawn about the population mean or the difference in two population means. *Sampling distributions of important statistics* allow us to learn information about parameters. Usually, the parameters are the counterpart to the statistic in question. If an engineer is interested in the population mean resistance of a certain type of resistor, the sampling distribution of $\overline{X}$ will be exploited once the sample information is gathered. On the other hand, if the variability in resistance is to be studied, clearly the sampling distribution of S^2 will be used in learning about the parametric counterpart, the population variance σ^2.

If a random sample of size n is drawn from a normal population with mean μ and variance σ^2, and the sample variance σ^2 is computed, we obtain a value of the statistic S^2. We shall proceed to consider the distribution of the statistic $(n - 1)S^2/\sigma^2$.

By the addition and subtraction of the sample mean $\overline{X}$, it is easy to see that

$$\sum_{i=1}^{n} (X_i - \mu)^2 = \sum_{i=1}^{n} [(X_i - \overline{X}) + (\overline{X} - \mu)]^2$$

$$= \sum_{i=1}^{n} (X_i - \overline{X})^2 + \sum_{i=1}^{n} (\overline{X} - \mu)^2 + 2(\overline{X} - \mu) \sum_{i=1}^{n} (X_i - \overline{X})$$

$$= \sum_{i=1}^{n} (X_i - \overline{X})^2 + n(\overline{X} - \mu)^2.$$

Dividing each term of the equality by σ^2 and substituting $(n - 1)S^2$ for $\sum_{i=1}^{n} (X_i - \overline{X})^2$, we obtain

$$\frac{\sum_{i=1}^{n} (X_i - \mu)^2}{\sigma^2} = \frac{(n-1)S^2}{\sigma^2} + \frac{(\overline{X} - \mu)^2}{\sigma^2/n}.$$

Now, according to the corollary of Theorem 7.12 we know that $\sum_{i=1}^{n} (X_i - \mu)^2/\sigma^2$ is a chi-squared random variable with n degrees of freedom. The second term on the right of the equality is the square of a standard normal variable, since $\overline{X}$ is a normal random variable with mean $\mu_{\overline{X}} = \mu$ and variance $\sigma_{\overline{X}}^2 = \sigma^2/n$. Therefore, we may conclude from Example 8.1 that $(\overline{X} - \mu)^2/(\sigma^2/n)$ is a chi-squared random variable with 1 degree of freedom. Using advanced techniques beyond the scope of this book, one can also show that the two chi-squared variables $\sum_{i=1}^{n} (X_i - \mu)^2/\sigma^2$ and $(\overline{X} - \mu)^2/(\sigma^2/n)$ are independent. Owing to the reproductive property of independent chi-squared random variables, established in Theorem 7.12, it would seem reasonable to presume that $(n-1)S^2/\sigma^2$ is also a chi-squared random variable with $v = n - 1$ degrees of freedom. We state this result in the following theorem.

THEOREM 8.4
If S^2 is the variance of a random sample of size n taken from a normal population having the variance σ^2, then the statistic

$$X^2 = \frac{(n-1)S^2}{\sigma^2} = \sum_{i=1}^{n} (X_i - \overline{X})^2/\sigma^2$$

has a chi-squared distribution with $v = n - 1$ degrees of freedom. ■

The values of the random variable X^2 are calculated from each sample by the formula

$$\chi^2 = \frac{(n-1)s^2}{\sigma^2}.$$

The probability that a random sample produces a χ^2-value greater than some specified value is equal to the area under the curve to the right of this value. It is customary to let χ_α^2 represent the χ^2-value above which we find an area of α. This is illustrated by the shaded region in Figure 8.10.

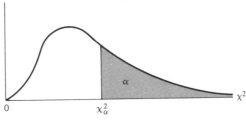

FIGURE 8.10 Tabulated values of the chi-squared distribution.

Table A.5 gives values of χ_α^2 for various values of α and v. The areas, α, are the column headings; the degrees of freedom, v, are given in the left column; and the table entries are the χ^2-values. Hence the χ^2-values with 7 degrees of freedom, leaving an area of 0.05 to the right, is $\chi_{0.05}^2 = 14.067$. Owing to lack of symmetry, we must also use the tables to find $\chi_{0.95}^2 = 2.167$ for $v = 7$.

Exactly 95% of a chi-squared distribution lies between $\chi_{0.975}^2$ and $\chi_{0.025}^2$. A χ^2-value falling to the right of $\chi_{0.025}^2$ is not likely to occur unless our assumed value of σ^2 is too small. Similarly, a χ^2-value falling to the left of $\chi_{0.975}^2$ is unlikely unless our assumed value of σ^2 is too large. In other words, it is possible to have a χ^2-value to the left of $\chi_{0.975}^2$ or to the right of $\chi_{0.025}^2$ when σ^2 is correct, but if this should occur, it is more probable that the assumed value of σ^2 is in error.

EXAMPLE 8.17   A manufacturer of car batteries guarantees that his batteries will last, on the average, 3 years with a standard deviation of 1 year. If five of these batteries have lifetimes of 1.9, 2.4, 3.0, 3.5, and 4.2 years, is the manufacturer still convinced that his batteries have a standard deviation of 1 year? Assume that the battery lifetime follows a normal distribution.

SOLUTION
We first find the sample variance:

$$s^2 = \frac{(5)(48.26) - (15)^2}{(5)(4)} = 0.815.$$

Then

$$\chi^2 = \frac{(4)(0.815)}{1} = 3.26$$

is a value from a chi-squared distribution with 4 degrees of freedom. Since 95% of the χ^2-values with 4 degrees of freedom fall between 0.484 and 11.143, the computed value with $\sigma^2 = 1$ is reasonable, and therefore the manufacturer has no reason to suspect that the standard deviation is other than 1 year.

Degrees of Freedom As a Measure of Sample Information

The reader may gain some insight by considering Theorem 8.4 and the corollary following Theorem 7.12 in Section 7.3. We know that with the conditions of Theorem 7.12, namely a random sample is taken from a normal distribution, that the random variable

$$\sum_{i=1}^{n} \frac{(X_i - \mu)^2}{\sigma^2}$$

has a χ^2-distribution with *n degrees of freedom*. Now note Theorem 8.4, which indicates that with the same conditions of Theorem 7.12, the random variable

$$\frac{(n-1)S^2}{\sigma^2} = \sum_{i=1}^{n} \frac{(X_i - \bar{X})^2}{\sigma^2}$$

has a χ^2-distribution with $n - 1$ *degrees of freedom*. The reader may recall that the term *degrees of freedom*, used in this identical context, was discussed in Chapter 1.

As we indicated earlier, the proof of Theorem 8.4 will not be given. However, the reader can view Theorem 8.4 as indicating that when μ is not known and one considers the distribution of

$$\sum_{i=1}^{n} \frac{(X_i - \overline{X})^2}{\sigma^2},$$

there is 1 less degree of freedom, or a degree of freedom is lost in the estimation of μ (i.e., when μ is replaced by $\overline{x}$). In other words, there are n degrees of freedom or independent *pieces of information* in the random sample from the normal distribution. When the data (the values in the sample) are used to compute the mean, there is 1 less degree of freedom in the information used to estimate σ^2.

8.7   *t*-Distribution

In Section 8.6 we discussed the utility of the central limit theorem. Its applications revolve around inferences on a population mean or the difference between two population means. Use of the central limit theorem and the normal distribution is certainly helpful in this context. However, it was assumed that the population standard deviation *is known*. This assumption may not be unreasonable in situations where the engineer is quite familiar with the system or process. However, in many experimental scenarios knowledge of σ is certainly no more reasonable than knowledge of the population mean μ. Often, in fact, an estimate of σ must be supplied by the same sample information that produced the sample average $\overline{x}$. As a result, a natural statistic to consider to deal with inferences on μ is

$$T = \frac{\overline{X} - \mu}{S/\sqrt{n}}$$

since S is the sample analog to σ. If the sample size is small, the values of S^2 fluctuate considerably from sample to sample (see Exercise 7 on page 235) and the distribution of T deviates appreciably from that of a standard normal distribution.

If the sample size is large enough, say $n \geq 30$, the distribution of T does not differ considerably from the standard normal. However, for $n < 30$, it is useful to deal with the exact distribution of T. In developing the sampling distribution of T, we shall assume that our random sample was selected from a normal population. We can then write

$$T = \frac{(\overline{X} - \mu)/(\sigma/\sqrt{n})}{\sqrt{S^2/\sigma^2}} = \frac{Z}{\sqrt{V/(n-1)}},$$

where

$$Z = \frac{\overline{X} - \mu}{\sigma/\sqrt{n}}$$

has the standard normal distribution, and

$$V = \frac{(n-1)S^2}{\sigma^2}$$

has a chi-squared distribution with $v = n - 1$ degrees of freedom. In sampling from normal populations, one can show that $\overline{X}$ and S^2 are independent, and consequently so are Z and V. The following theorem gives the density function of the T random variable.

THEOREM 8.5   *Let Z be a standard normal random variable and V a chi-squared random variable with v degrees of freedom. If Z and V are independent, then the distribution of the random variable T, where*

$$T = \frac{Z}{\sqrt{V/v}},$$

is given by

$$h(t) = \frac{\Gamma[(v+1)/2]}{\Gamma(v/2)\sqrt{\pi v}} \left(1 + \frac{t^2}{v}\right)^{-(v+1)/2}, \qquad -\infty < t < \infty.$$

*This is known as the **t-distribution** with v degrees of freedom.*   ■

COROLLARY   Let $X_1, X_2, \ldots, X_n$ be independent random variables that are all normal with mean μ and standard deviation σ. Let $\overline{X} = \sum\limits_{i=1}^{n} X_i/n$ and $s^2 = \sum\limits_{i=1}^{n} (X_i - \overline{X})^2/(n-1)$. Then the random variable $T = \dfrac{\overline{X} - \mu}{S/\sqrt{n}}$ has a t-distribution with $v = n - 1$ degrees of freedom.   ■

The probability distribution of T was first published in 1908 in a paper by W. S. Gosset. At the time, Gosset was employed by an Irish brewery that disallowed publication of research by members of its staff. To circumvent this restriction, he published his work secretly under the name "Student." Consequently, the distribution of T is usually called the **Student t-distribution**, or simply the **t-distribution**. In deriving the equation of this distribution, Gosset assumed that the samples were selected from a normal population. Although this would seem to be a very restrictive assumption, it can be shown that nonnormal populations possessing bell-shaped distributions will still provide values of T that approximate the t-distribution very closely.

The distribution of T is similar to the distribution of Z in that they both are symmetric about a mean of zero. Both distributions are bell-shaped, but the t-distribution is more variable, owing to the fact that the T-values depend on the fluctuations of two quantities, $\overline{X}$ and S^2, whereas the Z-values depend only on the changes of $\overline{X}$ from sample to sample. The distribution of T differs from that of Z in that the variance of T depends on the sample size n and is always greater than 1. Only when the sample size $n \rightarrow \infty$ will the two distributions become the same. In Figure 8.11

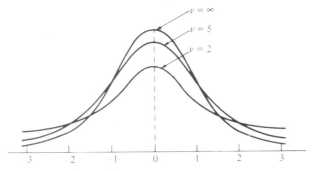

FIGURE 8.11 The *t*-distribution curves for $v = 2$, 5, and ∞.

we show the relationship between a standard normal distribution ($v = \infty$) and *t*-distributions with 2 and 5 degrees of freedom.

It is customary to let t_α represent the *t*-value above which we find an area equal to α. Hence the *t*-value with 10 degrees of freedom leaving an area of 0.025 to the right is $t = 2.228$. Since the *t*-distribution is symmetric about a mean of zero, we have $t_{1-\alpha} = -t_\alpha$; that is, the *t*-value leaving an area of $1 - \alpha$ to the right and therefore an area of α to the left is equal to the negative *t*-value that leaves an area of α in the right tail of the distribution (see Figure 8.12). That is, $t_{0.95} = -t_{0.05}$, $t_{0.99} = -t_{0.01}$, and so forth.

EXAMPLE 8.18   The *t*-value with $v = 14$ degrees of freedom that leaves an area of 0.025 to the left, and therefore an area of 0.975 to the right, is

$$t_{0.975} = -t_{0.025} = -2.145.$$

EXAMPLE 8.19   Find $P(-t_{0.025} < T < t_{0.05})$.

SOLUTION
Since $t_{0.05}$ leaves an area of 0.05 to the right, and $-t_{0.025}$ leaves an area of 0.025 to the left, we find a total area of

$$1 - 0.05 - 0.025 = 0.925$$

between $-t_{0.025}$ and $t_{0.05}$. Hence

$$P(-t_{0.025} < T < t_{0.05}) = 0.925.$$

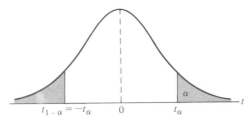

FIGURE 8.12 Symmetry property of the *t*-distribution.

EXAMPLE 8.20   Find k such that $P(k < T < -1.761) = 0.045$, for a random sample of size 15 selected from a normal distribution.

SOLUTION

From Table A.4 we note that 1.761 corresponds to $t_{0.05}$ when $v = 14$. Therefore, $-t_{0.05} = -1.761$. Since k in the original probability statement is to the left of $-t_{0.05} = -1.761$, let $k = -t_\alpha$. Then, from Figure 8.13, we have

$$0.045 = 0.05 - \alpha$$

or

$$\alpha = 0.005.$$

Hence, from Table A.4 with $v = 14$,

$$k = -t_{0.005} = -2.977$$

and

$$P(-2.977 < T < -1.761) = 0.045.$$

Exactly 95% of the values of a t-distribution with $v = n - 1$ degrees of freedom lie between $-t_{0.025}$ and $t_{0.025}$. Of course, there are other t-values that contain 95% of the distribution, such as $-t_{0.02}$ and $t_{0.03}$, but these values do not appear in Table A.4, and furthermore the shortest possible interval is obtained by choosing t-values that leave exactly the same area in the two tails of our distribution. A t-value that falls below $-t_{0.025}$ or above $t_{0.025}$ would tend to make us believe that either a very rare event has taken place or perhaps our assumption about μ is in error. Should this happen, we shall make the latter decision and claim that our assumed value of μ is in error. In fact, a t-value falling below $-t_{0.01}$ or above $t_{0.01}$ would provide even stronger evidence that our assumed value of μ is quite unlikely. General procedures for testing claims concerning the value of the parameter μ will be treated in Chapter 10. A preliminary look into the foundation of these procedures is illustrated in the following example.

EXAMPLE 8.21   A chemical engineer claims that the yield of a certain batch process is 500 grams per milliliter of raw material. To check this claim he samples 25 batches each month. If the computed t-value falls between $-t_{0.05}$ and $t_{0.05}$, he is satisfied with his claim.

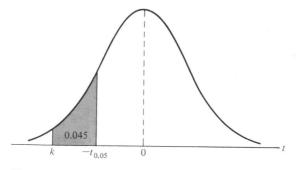

FIGURE 8.13   The t-values for Example 8.20.

What conclusion should he draw from a sample that has a mean $\bar{x} = 518$ grams per milliliter and a sample standard deviation $s = 40$ grams? Assume the distribution of yields to be approximately normal.

SOLUTION
From Table A.4 we find that $t_{0.05} = 1.711$ for 24 degrees of freedom. Therefore, the manufacturer is satisfied with his claim if a sample of 25 batches yields a t-value between -1.711 and 1.711. If $\mu = 500$, then

$$t = \frac{518 - 500}{40/\sqrt{25}} = 2.25,$$

a value well above 1.711. The probability of obtaining a t-value, with $v = 24$, equal to or greater than 2.25 is approximately 0.02. If $\mu > 500$, the value of t computed from the sample would be more reasonable. Hence the manufacturer is likely to conclude that the process produces a better product than he thought.

What Is the T-Distribution Used For?

The t-distribution is used extensively in problems that deal with inference about the population mean (as illustrated in Example 8.21) or in problems that involve comparative samples (i.e., in cases where one is trying to determine if means from two samples are significantly different). The reader should note that use of the t-distribution for the statistic

$$T = \frac{\bar{X} - \mu}{S/\sqrt{n}}$$

requires that $X_1, X_2, \ldots, X_n$ be normal. The use of the t-distribution and the sample size consideration do not relate to the central limit theorem. The use of the standard normal distribution rather than T for $n \geq 30$ merely implies that S is a sufficiently good estimator of σ in this case. In chapters that follow the t-distribution finds extensive usage.

8.8 F-Distribution

We have motivated the t-distribution in part on the basis of application to problems in which there is comparative sampling (i.e., a comparison between two sample means). Some of our examples in future chapters will provide the formalism. A chemical engineer collects data on two catalysts. A biologist collects data on two growth media. A chemist gathers data on two methods of coating material to inhibit corrosion. While it is of interest to let sample information shed light on two population means, it is often the case that a comparison, in some sense, of variability is equally important, if not more so. The F-distribution finds enormous application in comparing sample variances. The applications of the F-distribution are found in problems involving two or more samples.

The statistic F is defined to be the ratio of two independent chi-squared random variables, each divided by its number of degrees of freedom. Hence we can write

$$F = \frac{U/v_1}{V/v_2},$$

where U and V are independent random variables having chi-squared distributions with v_1 and v_2 degrees of freedom, respectively. We shall now state the sampling distribution of F.

THEOREM 8.6

Let U and V be two independent random variables having chi-squared distributions with v_1 and v_2 degrees of freedom, respectively. Then the distribution of the random variable

$$F = \frac{U/v_1}{V/v_2}$$

is given by

$$h(f) = \begin{cases} \dfrac{\Gamma[(v_1 + v_2)/2](v_1/v_2)^{v_1/2}}{\Gamma(v_1/2)\Gamma(v_2/2)} \dfrac{f^{v_1/2-1}}{(1 + v_1 f/v_2)^{(v_1+v_2)/2}} & 0 < f < \infty \\ 0, & elsewhere. \end{cases}$$

*This is known as the **F-distribution** with v_1 and v_2 degrees of freedom.* ■

The curve of the F-distribution depends not only on the two parameters v_1 and v_2 but also on the order in which we state them. Once these two values are given, we can identify the curve. Typical F curves are shown in Figure 8.14.

Let f_α be the f-value above which we find an area equal to α. This is illustrated by the shaded region in Figure 8.15. Table A.6 gives values of f_α only for $\alpha = 0.05$ and $\alpha = 0.01$ for various combinations of the degrees of freedom v_1 and v_2. Hence the f-value with 6 and 10 degrees of freedom, leaving an area of 0.05 to the right, is $f_{0.05} = 3.22$. By means of the following theorem, Table A.6 can also be used to find values of $f_{0.95}$ and $f_{0.99}$. The proof is left for the reader.

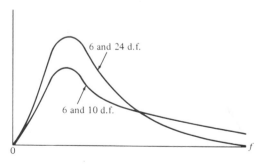

6 and 24 d.f.

6 and 10 d.f.

f

0

FIGURE 8.14 Typical F-distributions.

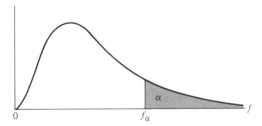

FIGURE 8.15 Tabulated values of the *F*-distribution.

THEOREM 8.7   *Writing $f_\alpha(v_1, v_2)$ for f_α with v_1 and v_2 degrees of freedom, we obtain*

$$f_{1-\alpha}(v_1, v_2) = \frac{1}{f_\alpha(v_2, v_1)}. \qquad \blacksquare$$

Thus the *f*-value with 6 and 10 degrees of freedom, leaving an area of 0.95 to the right, is

$$f_{0.95}(6, 10) = \frac{1}{f_{0.05}(10, 6)} = \frac{1}{4.06} = 0.246.$$

Suppose that random samples of size n_1 and n_2 are selected from two normal populations with variances σ_1^2 and σ_2^2, respectively. From Theorem 8.4, we know that

$$X_1^2 = \frac{(n_1 - 1)S_1^2}{\sigma_1^2}$$

and

$$X_2^2 = \frac{(n_2 - 1)S_2^2}{\sigma_2^2}$$

are random variables having chi-squared distributions with $v_1 = n_1 - 1$ and $v_2 = n_2 - 1$ degrees of freedom. Furthermore, since the samples are selected at random, we are dealing with independent random variables, and then using Theorem 8.6 with $X_1^2 = U$ and $X_2^2 = V$, we obtain the following result.

THEOREM 8.8   *If S_1^2 and S_2^2 are the variances of independent random samples of size n_1 and n_2 taken from normal populations with variances σ_1^2 and σ_2^2, respectively, then*

$$F = \frac{S_1^2/\sigma_1^2}{S_2^2/\sigma_2^2} = \frac{\sigma_2^2 S_1^2}{\sigma_1^2 S_2^2}$$

has an F-distribution with $v_1 = n_1 - 1$ and $v_2 = n_2 - 1$ degrees of freedom. $\blacksquare$

What Is the *F*-Distribution Used For?

We answered this question in part at the beginning of this section. The *F*-distribution is used in two-sample situations to draw inferences about the population variances. This involves the application of the result of Theorem 8.8. However, the *F*-distribution is applied to many other types of problems in which the sample variances are involved. In fact, the *F*-distribution is called the *variance ratio distribution*. As an illustration, consider Example 8.15. Two paints, *A* and *B*, were compared with regard to mean drying time. The normal distribution applies nicely (assuming that σ_A and σ_B are known). However, suppose that there are three types of paints to compare, say *A*, *B*, and *C*. We wish to determine if the population means are equivalent. Suppose, in fact, that important summary information from the experiment is as follows:

Paint	Sample Mean	Sample Variance	Sample Size
A	$\bar{X}_A = 4.5$	$s_A^2 = 0.2$	10
B	$\bar{X}_B = 5.5$	$s_B^2 = 0.14$	10
C	$\bar{X}_C = 6.5$	$s_C^2 = 0.11$	10

The problem centers around whether or not the sample averages $(\bar{x}_A, \bar{x}_B, \bar{x}_C)$ are far enough apart. The implication of "far enough apart" is very important. It would seem reasonable that if the variability between sample averages is larger than what one would expect by chance, the data do not support the conclusion that $\mu_A = \mu_B = \mu_C$. Whether or not these sample averages could have occurred by chance depends on the *variability within samples*, as quantified by s_A^2, s_B^2, and s_C^2. The notion of the important components of variability is best seen through some simple graphics. Consider the plot of raw data from samples *A*, *B*, and *C*, shown in Figure 8.16. These data could easily have generated the above summary information.

It appears evident that the data came from distributions with different population means, although there is some overlap between the samples. An analysis that involves all of the data would attempt to determine if the variability between the sample averages *and* the variability within the samples could have occurred jointly *if in fact the populations have a common mean*. Notice that the key to this analysis centers around the two following sources of variability.

1. Variability within samples (between observations in distinct samples).
2. Variability between samples (between sample averages).

Clearly, if the variability in (1) is considerably larger than that in (2) there will be considerable overlap in the sample data and a signal that the data could all have come from a common distribution. An example is found in the data set containing three samples, shown in Figure 8.17. On the other hand, it is very unlikely that data

FIGURE 8.16 Data from three distinct samples.

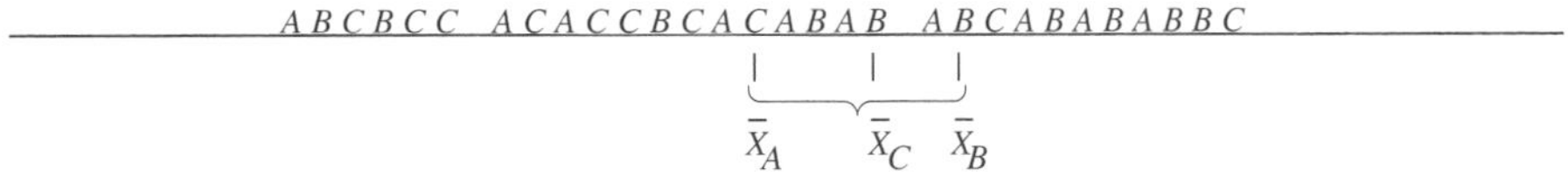

$$A\,B\,C\,B\,C\,C \quad A\,C\,A\,C\,C\,B\,C\,A\,C\,A\,B\,A\,B \quad A\,B\,C\,A\,B\,A\,B\,A\,B\,B\,C$$

FIGURE 8.17   Data that easily could have come from the same population.

from a distribution with a common mean could have variability between sample averages that is considerably larger than the variability within samples.

The sources of variability in (1) and (2) above generate important ratios of *sample variances* and ratios are used in conjunction with the *F*-distribution. The general procedure involved is called **analysis of variance**. It is interesting that in the paint example described here, we are dealing with inferences on three population means, but two sources of variability are used. We will not supply details here but in Chapters 13, 14, and 15 we make extensive use of analysis of variance, and, of course, the *F*-distribution plays an important role.

Exercises

1. For a chi-squared distribution find
   (a) $\chi^2_{0.025}$ when $v = 15$;
   (b) $\chi^2_{0.01}$ when $v = 7$;
   (c) $\chi^2_{0.05}$ when $v = 24$.

2. For a chi-squared distribution find the following:
   (a) $\chi^2_{0.005}$ when $v = 5$;
   (b) $\chi^2_{0.05}$ when $v = 19$;
   (c) $\chi^2_{0.01}$ when $v = 12$.

3. For a chi-squared distribution find χ^2_α such that
   (a) $P(X^2 > \chi^2_\alpha) = 0.99$ when $v = 4$;
   (b) $P(X^2 > \chi^2_\alpha) = 0.025$ when $v = 19$;
   (c) $P(37.652 < X^2 < \chi^2_\alpha) = 0.045$ when $v = 25$.

4. For a chi-squared distribution find χ^2_α such that
   (a) $P(X^2 > \chi^2_\alpha) = 0.01$ when $v = 21$;
   (b) $P(X^2 < \chi^2_\alpha) = 0.95$ when $v = 6$;
   (c) $P(\chi^2_\alpha < X^2 < 23.209) = 0.015$ when $v = 10$.

5. Find the probability that a random sample of 25 observations, from a normal population with variance $\sigma^2 = 6$, will have a variance s^2
   (a) greater than 9.1;
   (b) between 3.462 and 10.745.
   Assume the sample variances to be continuous measurements.

6. The scores on a placement test given to college freshmen for the past five years are approximately normally distributed with a mean $\mu = 74$ and a variance $\sigma^2 = 8$. Would you still consider $\sigma^2 = 8$ to be a valid value of the variance if a random sample of 20 stu-

dents who take this placement test this year obtain a value of $s^2 = 20$?

7. Show that the variance of S^2 for random samples of size n from a normal population decreases as n becomes large. HINT: First find the variance of $(n - 1)S^2/\sigma^2$.

8. (a) Find $t_{0.025}$ when $v = 14$.
   (b) Find $-t_{0.10}$ when $v = 10$.
   (c) Find $t_{0.995}$ when $v = 7$.

9. (a) Find $P(T < 2.365)$ when $v = 7$.
   (b) Find $P(T > 1.318)$ when $v = 24$.
   (c) Find $P(-1.356 < T < 2.179)$ when $v = 12$.
   (d) Find $P(T > -2.567)$ when $v = 17$.

10. (a) Find $P(-t_{0.005} < T < t_{0.01})$.
    (b) Find $P(T > -t_{0.025})$.

11. Given a random sample of size 24 from a normal distribution, find k such that
    (a) $P(-2.069 < T < k) = 0.965$;
    (b) $P(k < T < 2.807) = 0.095$;
    (c) $P(-k < T < k) = 0.90$.

12. A manufacturing firm claims that the batteries used in their electronic games will last an average of 30 hours. To maintain this average, 16 batteries are tested each month. If the computed *t*-value falls between $-t_{0.025}$ and $t_{0.025}$, the firm is satisfied with its claim. What conclusion should the firm draw from a sample that has a mean $\bar{x} = 27.5$ hours and a standard

deviation $s = 5$ hours? Assume the distribution of battery lives to be approximately normal.

13. A normal population with unknown variance has a mean of 20. Is one likely to obtain a random sample of size 9 from this population with a mean of 24 and a standard deviation of 4.1? If not, what conclusion would you draw?

14. A cigarette manufacturer claims that his cigarettes have an average nicotine content of 1.83 milligrams. If a random sample of 8 cigarettes of this type have nicotine contents of 2.0, 1.7, 2.1, 1.9, 2.2, 2.1, 2.0, and 1.6 milligrams, would you agree with the manufacturer's claim?

15. For an F-distribution find:
   (a) $f_{0.05}$ with $v_1 = 7$ and $v_2 = 15$;
   (b) $f_{0.05}$ with $v_1 = 15$ and $v_2 = 7$;
   (c) $f_{0.01}$ with $v_1 = 24$ and $v_2 = 19$;
   (d) $f_{0.95}$ with $v_1 = 19$ and $v_2 = 24$;
   (e) $f_{0.99}$ with $v_1 = 28$ and $v_2 = 12$.

16. Pull-strength tests on 10 soldered leads for a semiconductor device yield the following results in pounds

force required to rupture the bond:

19.8   12.7   13.2   16.9   10.6   18.8   11.1   14.3
17.0   12.5

Another set of 8 leads was tested after encapsulation to determine whether the pull strength has been increased by encapsulation of the device, with the following results:

24.9   22.8   23.6   22.1   20.4   21.6   21.8   22.5

Comment on the evidence available concerning equality of the two population variances.

17. Consider the following measurements of the heat-producing capacity of the coal produced by two mines (in millions of calories per ton):

| Mine 1: | 8260 | 8130 | 8350 | 8070 | 8340 | |
| Mine 2: | 7950 | 7890 | 7900 | 8140 | 7920 | 7840 |

Can it be concluded that the two population variances are equal?

Review Exercises

1. Consider the data displayed in Exercise 3 following Section 3.4. Construct a box and whisker plot and comment on the nature of the sample. Compute the sample mean and sample standard deviation.

2. If $X_1, X_2, \ldots, X_n$ are independent random variables having identical exponential distributions with parameter θ, show that the density function of the random variable $Y = X_1 + X_2 + \cdots + X_n$ is that of a gamma distribution with parameters $\alpha = n$ and $\beta = \theta$.

3. In testing for carbon monoxide in a certain brand of cigarette, the data, in milligrams per cigarette, were coded by subtracting 12 from each observation. Use the results of Exercise 14 on p. 207 to find the standard deviation for the carbon monoxide contents of a random sample of 15 cigarettes of this brand if the coded measurements are 3.8, -0.9, 5.4, 4.5, 5.2, 5.6, 2.7, -0.1, -0.3, -1.7, 5.7, 3.3, 4.4, -0.5, and 1.9.

4. If S_1^2 and S_2^2 represent the variances of independent random samples of size $n_1 = 8$ and $n_2 = 12$, taken

from normal populations with equal variances, find the $P(S_1^2/S_2^2 < 4.89)$.

5. A random sample of 5 bank presidents indicated annual salaries of $63,000, $48,000, $52,000, $35,000, and $41,000. Use the results of Exercise 14 on p. 000 to find the variance of this set of data by first dividing each salary by 1000 and then subtracting 50.

6. If the number of hurricanes that hit a certain area of the eastern United States per year is a random variable having a Poisson distribution with $\mu = 6$, find the probability that this area will be hit by
   (a) exactly 15 hurricanes in 2 years;
   (b) at most 9 hurricanes in 2 years.

7. A taxi company tested a random sample of 10 steel-belted radial tires of a certain brand and recorded the following tread wear: 48,000, 53,000, 45,000, 61,000, 59,000, 56,000, 63,000, 49,000, 53,000, and 54,000 kilometers. Use the results of Exercise 14 on p. 207 to find the standard deviation of this set of

data by first dividing each observation by 1000 and then subtracting 55.

8. If the number of minutes it takes a pharmacist to fill a prescription is a random variable having an exponential distribution with parameter $\theta = 5$, use the results of Exercise 2 on p. 176 to find the probability that the pharmacist will take
(a) less than 8 minutes to fill 2 prescriptions;
(b) at least 12 minutes to fill 3 prescriptions.

9. Consider the data of Exercise 2 following Section 3.4. Construct a box and whisker plot. Comment. Compute the sample mean and sample standard deviation.

10. If S_1^2 and S_2^2 represent the variances of independent random samples of size $n_1 = 25$ and $n_2 = 31$, taken from normal populations with variances $\sigma_1^2 = 10$ and $\sigma_2^2 = 15$, respectively, find the $P(S_1^2/S_2^2 > 1.26)$.

9

One- and Two-Sample Estimation Problems

9.1 Introduction

In previous chapters we have emphasized sampling properties of the sample mean and sample variance. We have also emphasized displays of data in various forms. The purpose of these presentations is to build a foundation that allows statisticians to draw conclusions about the population parameters from experimental data. For example, the central limit theorem provides information about the distribution of the sample mean $\overline{X}$. The distribution involves the population mean μ. Thus any conclusions drawn concerning μ from an observed sample average must depend on knowledge of this sampling distribution. Similar comments could apply for S^2 and σ^2. Clearly, any conclusions we draw about the variance of a normal distribution would likely involve the sampling distribution of S^2.

In this chapter we begin by formally outlining the purpose of statistical inference. We follow this by discussing the problem of **estimation of population parameters**. We confine our formal developments of specific estimation procedures to problems involving one and two samples.

9.2 Statistical Inference

In Chapter 1 we discussed the general philosophy of formal statistical inference. The theory of **statistical inference** consists of those methods by which one makes inferences or generalizations about a population. The trend of today is to distinguish between the **classical method** of estimating a population parameter, whereby inferences are based strictly on information obtained from a random sample selected from the population, and the **Bayesian method**, which utilizes prior subjective knowledge about the probability distribution of the unknown parameters in conjunction with the information provided by the sample data. Throughout most of this chapter we shall use classical methods to estimate unknown population parameters such as the mean, proportion, and the variance by computing statistics from random samples and applying the theory of sampling distributions, much of which was covered in Chapter 8. For completeness, a brief discussion of the Bayesian approach to statistical decision theory is presented in Sections 9.13 and 9.14.

Statistical inference may be divided into two major areas: **estimation** and **tests of hypotheses.** We treat these two areas separately, dealing with theory and applications of estimation in this chapter and hypothesis testing in Chapter 10. To distinguish clearly between the two areas, consider the following examples. A candidate for public office may wish to estimate the true proportion of voters favoring him by obtaining the opinions from a random sample of 100 eligible voters. The fraction of voters in the sample favoring the candidate could be used as an estimate of the true proportion in the population of voters. A knowledge of the sampling distribution of a proportion enables one to establish the degree of accuracy of our estimate. This problem falls in the area of estimation.

Now consider the case in which one is interested in finding out whether brand A floor wax is more scuff-resistant than brand B floor wax. He or she might hypothesize that brand A is better than brand B and, after proper testing, accept or reject this hypothesis. In this example we do not attempt to estimate a parameter, but instead we try to arrive at a correct decision about a prestated hypothesis. Once again we are dependent on sampling theory and the use of data to provide us with some measure of accuracy for our decision.

9.3 Classical Methods of Estimation

A **point estimate** of some population parameter θ is a single value $\hat{\theta}$ of a statistic $\hat{\Theta}$. For example, the value $\bar{x}$ of the statistic $\bar{X}$, computed from a sample of size n, is a point estimate of the population parameter μ. Similarly, $\hat{p} = x/n$ is a point estimate of the true proportion p for a binomial experiment.

An estimator is not expected to estimate the population parameter without error. We do not expect $\bar{X}$ to estimate μ exactly, but we certainly hope that it is not far off. For a particular sample it is possible to obtain a closer estimate of μ by using the sample median $\tilde{X}$ as an estimator. Consider, for instance, a sample consisting of the values 2, 5, and 11 from a population whose mean is 4 but supposedly unknown.

We would estimate μ to be $\bar{x} = 6$, using the sample mean as our estimate, or $\tilde{x} = 5$, using the sample median as our estimate. In this case the estimator $\bar{X}$ produces an estimate closer to the true parameter than that of the estimator $\tilde{X}$. On the other hand, if our random sample contains the values 2, 6, and 7, then $\bar{x} = 5$ and $\tilde{x} = 6$, so that $\bar{X}$ is now the better estimator. Not knowing the true value of μ, we must decide in advance whether to use $\bar{X}$ or $\tilde{X}$ as our estimator.

Unbiased Estimator

What are the desirable properties of a ''good'' decision function that would influence us to choose one estimator rather than another? Let $\hat{\Theta}$ be an estimator whose value $\hat{\theta}$ is a point estimate of some unknown population parameter θ. Certainly, we would like the sampling distribution of $\hat{\Theta}$ to have a mean equal to the parameter estimated. An estimator possessing this property is said to be **unbiased**.

DEFINITION 9.1   *A statistic $\hat{\Theta}$ is said to be an* **unbiased** *estimator of the parameter θ if*

$$\mu_{\hat{\Theta}} = E(\hat{\Theta}) = \theta.$$

■

EXAMPLE 9.1   Show that S^2 is an unbiased estimator of the parameter σ^2.

SOLUTION
Let us write

$$\sum_{i=1}^{n} (X_i - \bar{X})^2 = \sum_{i=1}^{n} [(X_i - \mu) - (\bar{X} - \mu)]^2$$

$$= \sum_{i=1}^{n} (X_i - \mu)^2 - 2(\bar{X} - \mu) \sum_{i=1}^{n} (X_i - \mu) + n(\bar{X} - \mu)^2$$

$$= \sum_{i=1}^{n} (X_i - \mu)^2 - n(\bar{X} - \mu)^2.$$

Now

$$E(S^2) = E\left[\frac{\sum_{i=1}^{n} (X_i - \bar{X})^2}{n - 1} \right]$$

$$= \frac{1}{n - 1} \left[\sum_{i=1}^{n} E(X_i - \mu)^2 - nE(\bar{X} - \mu)^2 \right]$$

$$= \frac{1}{n - 1} \left(\sum_{i=1}^{n} \sigma_{X_i}^2 - n\sigma_{\bar{X}}^2 \right).$$

However,

$$\sigma_{X_i}^2 = \sigma^2 \quad \text{for } i = 1, 2, \ldots, n$$

and

$$\sigma_{\bar{X}}^2 = \frac{\sigma^2}{n}.$$

Therefore,

$$E(S^2) = \frac{1}{n-1}\left(n\sigma^2 - n\frac{\sigma^2}{n}\right) = \sigma^2.$$

Although S^2 is an unbiased estimator of σ^2, S, on the other hand, is a biased estimator of σ with the bias becoming insignificant for large samples.

Variance of a Point Estimator

If $\hat{\Theta}_1$ and $\hat{\Theta}_2$ are two unbiased estimators of the same population parameter θ, we would choose the estimator whose sampling distribution has the smaller variance. Hence, if $\sigma_{\hat{\Theta}_1}^2 < \sigma_{\hat{\Theta}_2}^2$, we say that $\hat{\Theta}_1$ is a **more efficient estimator** of θ than $\hat{\Theta}_2$.

DEFINITION 9.2   *If we consider all possible unbiased estimators of some pararameter θ, the one with the smallest variance is called the* **most efficient estimator** *of θ.*   ■

In Figure 9.1 we illustrate the sampling distributions of 3 different estimators $\hat{\Theta}_1$, $\hat{\Theta}_2$, and $\hat{\Theta}_3$, all estimating θ. It is clear that only $\hat{\Theta}_1$ and $\hat{\Theta}_2$ are unbiased, since their distributions are centered at θ. The estimator $\hat{\Theta}_1$ has a smaller variance than $\hat{\Theta}_2$ and is therefore more efficient. Hence our choice for an estimator of θ, among the three considered, would be $\hat{\Theta}_1$.

For normal populations one can show that both $\bar{X}$ and $\tilde{X}$ are unbiased estimators of the population mean μ, but the variance of $\bar{X}$ is smaller than the variance of $\tilde{X}$. Thus both estimates $\bar{x}$ and $\tilde{x}$ will, on the average, equal the population mean μ, but $\bar{x}$ is likely to be closer to μ for a given sample, and thus $\bar{X}$ is more efficient than $\tilde{X}$.

Even the most efficient unbiased estimator is unlikely to estimate the population parameter exactly. It is true that our accuracy increases with large samples, but there is still no reason why we should expect a **point** estimate from a given sample to be exactly equal to the population parameter it is supposed to estimate. There are many situations in which it is preferable to determine an interval within which we would expect to find the value of the parameter. Such an interval is called an **interval estimate**.

Interval Estimation

An interval estimate of a population parameter θ is an interval of the form $\hat{\theta}_L < \theta < \hat{\theta}_U$, where $\hat{\theta}_L$ and $\hat{\theta}_U$ depend on the value of the statistic $\hat{\Theta}$ for a particular sample and also on the sampling distribution of $\hat{\Theta}$. Thus a random sample of SAT verbal scores for students of the entering freshman class might produce an interval from 530 to 550 within which we expect to find the true average of all SAT verbal scores for the freshman class. The values of the end points, 530 and 550, will depend on the computed sample mean $\bar{x}$ and the sampling distribution of $\bar{X}$. As the sample size

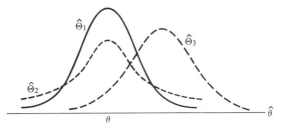

FIGURE 9.1   Sampling distributions of different estimators of θ.

increases, we know that $\sigma_{\bar{X}}^2 = \sigma^2/n$ decreases, and consequently our estimate is likely to be closer to the parameter μ, resulting in a shorter interval. Thus the interval estimate indicates, by its length, the accuracy of the point estimate. An engineer will gain some insight into the population proportion defective by taking a sample and computing the *sample proportion defectives*. But an interval estimate might be more informative.

Since different samples will generally yield different values of $\hat{\Theta}$ and, therefore, different values of $\hat{\theta}_L$ and $\hat{\theta}_U$, these end points of the interval are values of corresponding random variables $\hat{\Theta}_L$ and $\hat{\Theta}_U$. From the sampling distribution of $\hat{\Theta}$ we shall be able to determine $\hat{\theta}_L$ and $\hat{\theta}_U$ such that the $P(\hat{\Theta}_L < \theta < \hat{\Theta}_U)$ is equal to any positive fractional value we care to specify. If, for instance, we find $\hat{\theta}_L$ and $\hat{\theta}_U$ such that

$$P(\hat{\Theta}_L < \theta < \hat{\Theta}_U) = 1 - \alpha,$$

For $0 < \alpha < 1$, then we have a probability of $1 - \alpha$ of selecting a random sample that will produce an interval containing θ. The interval $\hat{\theta}_L < \theta < \hat{\theta}_U$, computed from the selected sample, is then called a $(1 - \alpha)100\%$ **confidence interval**, the fraction $1 - \alpha$ is called the **confidence coefficient** or the **degree of confidence**, and the endpoints, $\hat{\theta}_L$ and $\hat{\theta}_U$, are called the lower and upper **confidence limits**. Thus, when $\alpha = 0.05$, we have a 95% confidence interval, and when $\alpha = 0.01$ we obtain a wider 99% confidence interval. The wider the confidence interval is, the more confident we can be that the given interval contains the unknown parameter. Of course, it is better to be 95% confident that the average life of a certain television transistor is between 6 and 7 years than to be 99% confident that it is between 3 and 10 years. Ideally, we prefer a short interval with a high degree of confidence. Sometimes, restrictions on the size of our sample prevent us from achieving short intervals without sacrificing some of our degree of confidence.

In the sections that follow we pursue the notions of point and interval estimation, with each section representing a different special case. The reader should notice that while point and interval estimation represent different approaches to gain information regarding a parameter, they are related in the sense that confidence interval estimators are based on point estimators. In the following section, for example, we should see that the estimator $\bar{X}$ is a very reasonable point estimator of μ. As a result, the important confidence interval estimator of μ depends on knowledge of the sampling distribution of $\bar{X}$.

9.4 Single Sample: Estimating the Mean

The sampling distribution of $\overline{X}$ is centered at μ and in most applications the variance is smaller than that of any other estimators of μ. Thus the sample mean $\overline{x}$ will be used as a point estimate for the population mean μ. Recall that $\sigma_{\overline{X}}^2 = \sigma^2/n$, so that a large sample will yield a value of $\overline{X}$ that comes from a sampling distribution with a small variance. Hence $\overline{x}$ is likely to be a very accurate estimate of μ when n is large.

Let us now consider the interval estimate of μ. If our sample is selected from a normal population or, failing this, if n is sufficiently large, we can establish a confidence interval for μ by considering the sampling distribution of $\overline{X}$. According to the central limit theorem, we can expect the sampling distribution of $\overline{X}$ to be approximately normally distributed with mean $\mu_{\overline{X}} = \mu$ and standard deviation $\sigma_{\overline{X}} = \sigma/\sqrt{n}$. Writing $z_{\alpha/2}$ for the z-value above which we find an area of $\alpha/2$, we can see from Figure 9.2 that

$$P(-z_{\alpha/2} < Z < z_{\alpha/2}) = 1 - \alpha,$$

where

$$Z = \frac{\overline{X} - \mu}{\sigma/\sqrt{n}}.$$

Hence

$$P\left(-z_{\alpha/2} < \frac{\overline{X} - \mu}{\sigma/\sqrt{n}} < z_{\alpha/2}\right) = 1 - \alpha.$$

Multiplying each term in the inequality by $\sigma/\sqrt{n}$, and then subtracting $\overline{X}$ from each term and multiplying by -1 (reversing the sense of the inequalities), we obtain

$$P\left(\overline{X} - z_{\alpha/2}\frac{\sigma}{\sqrt{n}} < \mu < \overline{X} + z_{\alpha/2}\frac{\sigma}{\sqrt{n}}\right) = 1 - \alpha.$$

A random sample of size n is selected from a population whose variance σ^2 is known and the mean $\overline{x}$ is computed to give the following $(1 - \alpha)100\%$ confidence interval.

CONFIDENCE INTERVAL OF μ; σ KNOWN

If $\overline{x}$ is the mean of a random sample of size n from a population with known variance σ^2, a $(1 - \alpha)100\%$ confidence interval for μ is given by

$$\overline{x} - z_{\alpha/2}\frac{\sigma}{\sqrt{n}} < \mu < \overline{x} + z_{\alpha/2}\frac{\sigma}{\sqrt{n}},$$

where $z_{\alpha/2}$ is the z-value leaving an area of $\alpha/2$ to the right. ∎

For small samples selected from nonnormal populations, we cannot expect our degree of confidence to be accurate. However, for samples of size $n \geq 30$, regardless of the shape of most populations, sampling theory guarantees good results.

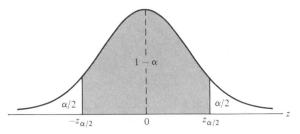

FIGURE 9.2 $P(-z_{\alpha/2} < Z < z_{\alpha/2}) = 1 - \alpha$.

Clearly, the values of the random variables $\hat{\Theta}_L$ and $\hat{\Theta}_U$, defined in Section 9.3, are the confidence limits

$$\hat{\theta}_L = \bar{x} - z_{\alpha/2}\frac{\sigma}{\sqrt{n}} \qquad \text{and} \qquad \hat{\theta}_U = \bar{x} + z_{\alpha/2}\frac{\sigma}{\sqrt{n}}.$$

Different samples will yield different values of $\bar{x}$ and therefore produce different interval estimates of the parameter μ as shown in Figure 9.3. The circular dots at the center of each interval indicate the position of the point estimate $\bar{x}$ for each random sample. Most of the intervals are seen to contain μ, but not in every case. Note that all of these intervals are of the same width, since their widths depend only on the choice of $z_{\alpha/2}$ once $\bar{x}$ is determined. The larger the value we choose for $z_{\alpha/2}$, the wider we make all the intervals and the more confident we can be that the particular sample selected will produce an interval that contains the unknown parameter μ.

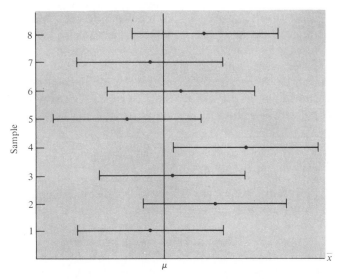

FIGURE 9.3   Interval estimates of μ for different samples.

EXAMPLE 9.2   The average zinc concentration recovered from a sample of zinc measurements in 36 different locations was found to be 2.6 grams per milliliter. Find the 95% and 99% confidence intervals for the mean zinc concentration in the river. Assume that the population standard deviation is 0.3.

SOLUTION
The point estimate of μ is $\bar{x} = 2.6$. The z-value, leaving an area of 0.025 to the right and therefore an area of 0.975 to the left, is $z_{0.025} = 1.96$ (Table A.3). Hence the 95% confidence interval is

$$2.6 - (1.96)\left(\frac{0.3}{\sqrt{36}}\right) < \mu < 26 + (1.96)\left(\frac{0.3}{\sqrt{36}}\right),$$

which reduces to

$$2.50 < \mu < 2.70.$$

To find a 99% confidence interval, we find the z-value leaving an area of 0.005 to the right and 0.995 to the left. Therefore, using Table A.3 again, $z_{0.005} = 2.575$, and the 99% confidence interval is

$$2.6 - (2.575)\left(\frac{0.3}{\sqrt{36}}\right) < \mu < 2.6 + (2.575)\left(\frac{0.3}{\sqrt{36}}\right),$$

or simply

$$2.47 < \mu < 2.73.$$

We now see that a longer interval is required to estimate μ with a higher degree of accuracy.

The $(1 - \alpha)100\%$ confidence interval provides an estimate of the accuracy of our point estimate. If μ is actually the center value of the interval, then $\bar{x}$ estimates μ without error. Most of the time, however, $\bar{x}$ will not be exactly equal to μ and the point estimate is in error. The size of this error will be the absolute value of the difference between μ and $\bar{x}$, and we can be $(1 - \alpha)100\%$ confident that this difference will not exceed $z_{\alpha/2}\sigma/\sqrt{n}$. We can readily see this if we draw a diagram of a hypothetical confidence interval as in Figure 9.4.

THEOREM 9.1   *If $\bar{x}$ is used as an estimate of μ, we can then be $(1 - \alpha)100\%$ confident that the error will not exceed $z_{\alpha/2}\sigma/\sqrt{n}$.*   ∎

In Example 9.2 we are 95% confident that the sample mean $\bar{x} = 2.6$ differs from the true mean μ by an amount less than 0.1 and 99% confident that the difference is less than 0.13.

Frequently, we wish to know how large a sample is necessary to ensure that the error in estimating μ will be less than a specified amount e. By Theorem 9.1 this means that we must choose n such that $z_{\alpha/2}\sigma/\sqrt{n} = e$. Solving this equation gives the following formula for n.

FIGURE 9.4 Error in estimating μ by $\bar{x}$.

THEOREM 9.2   *If $\bar{x}$ is used as an estimate of μ, we can be $(1 - \alpha)100\%$ confident that the error will not exceed a specified amount e when the sample size is*

$$n = \left(\frac{z_{\alpha/2}\sigma}{e}\right)^2.$$ ■

When solving for the sample size, n, all fractional values are rounded up to the next whole number. By adhering to this principle, we can be sure that our degree of confidence never falls below $(1 - \alpha)100\%$.

Strictly speaking, the formula in Theorem 9.2 is applicable only if we know the variance of the population from which we are to select our sample. Lacking this information, we could take a preliminary sample of size $n \geq 30$ to provide an estimate of σ. Then, using s as an approximation for σ in Theorem 9.2 we could determine approximately how many observations are needed to provide the desired degree of accuracy.

EXAMPLE 9.3   How large a sample is required in Example 9.2 if we want to be 95% confident that our estimate of μ is off by less than 0.05?

SOLUTION
The population standard deviation is $\sigma = 0.3$. Then, by Theorem 9.2,

$$n = \left[\frac{(1.96)(0.3)}{0.05}\right]^2 = 138.3.$$

Therefore, we can be 95% confident that a random sample of size 139 will provide an estimate $\bar{x}$ differing from μ by an amount less than 0.05.

Frequently, we are attempting to estimate the mean of a population when the variance is unknown. The reader should recall that in Chapter 8 we learned that if we have a random sample from a *normal distribution*, then the random variable

$$T = \frac{\bar{X} - \mu}{S/\sqrt{n}}$$

has a Student's t-distribution with $n - 1$ degrees of freedom. Here S is the sample standard deviation. In this situation with σ unknown, T can be used to construct a confidence interval on μ. The procedure is the same as that with known σ except that σ is replaced by S and the standard normal distribution is replaced by the t-distribution. Referring to Figure 9.5, we can assert that

$$P(-t_{\alpha/2} < T < t_{\alpha/2}) = 1 - \alpha,$$

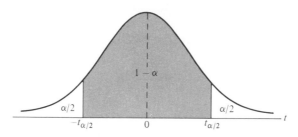

FIGURE 9.5 $P(-t_{\alpha/2} < T < t_{\alpha/2}) = 1 - \alpha.$

where $t_{\alpha/2}$ is the *t*-value with $n - 1$ degrees of freedom, above which we find an area of $\alpha/2$. Because of symmetry, an equal area of $\alpha/2$ will fall to the left of $-t_{\alpha/2}$. Substituting for T, we write

$$P\left(-t_{\alpha/2} < \frac{\overline{X} - \mu}{S/\sqrt{n}} < t_{\alpha/2}\right) = 1 - \alpha.$$

Multiplying each term in the inequality by $S/\sqrt{n}$, and then subtracting $\overline{X}$ from each term and multiplying by -1, we obtain

$$P\left(\overline{X} - t_{\alpha/2}\frac{S}{\sqrt{n}} < \mu < \overline{X} + t_{\alpha/2}\frac{S}{\sqrt{n}}\right) = 1 - \alpha.$$

For our particular random sample of size n, the mean $\bar{x}$ and standard deviation s are computed and the following $(1 - \alpha)100\%$ confidence interval for μ is obtained.

CONFIDENCE INTERVAL FOR μ; σ UNKNOWN

If $\bar{x}$ and s are the mean and standard deviation of a random sample from a normal population with unknown variance σ^2, a $(1 - \alpha)100\%$ confidence interval for μ is given by

$$\bar{x} - t_{\alpha/2}\frac{s}{\sqrt{n}} < \mu < \bar{x} + t_{\alpha/2}\frac{s}{\sqrt{n}},$$

where $t_{\alpha/2}$ is the t-value with $v = n - 1$ degrees of freedom, leaving an area of $\alpha/2$ to the right. ∎

We have made a distinction between the cases of σ known and σ unknown in computing the confidence interval estimates. We should emphasize that for the σ known case we exploited the central limit theorem, whereas for σ unknown we made use of the sampling distribution of the random variable T. However, the use of the *t*-distribution is based on the premise that the sampling is from a normal distribution. As long as the distribution is approximately bell-shaped, confidence intervals can be computed when σ^2 is unknown by using the *t*-distribution and one may expect very good results.

Quite often statisticians recommend that even when normality cannot be assumed, σ is unknown, and $n \geq 30$, s can replace σ and the confidence interval

$$\bar{x} \pm z_{\alpha/2}s/\sqrt{n}$$

may be used. This is often referred to as a *large-sample confidence interval*. The justification lies only in the presumption that with a sample as large as 30, s will be very close to the true σ and thus the central limit theorem prevails. It should be emphasized that this is only an approximation and the quality of the approach becomes better as the sample size grows larger.

EXAMPLE 9.4   The contents of 7 similar containers of sulfuric acid are 9.8, 10.2, 10.4, 9.8, 10.0, 10.2, and 9.6 liters. Find a 95% confidence interval for the mean of all such containers, assuming an approximate normal distribution.

SOLUTION
The sample mean and standard deviation for the given data are

$$\bar{x} = 10.0 \quad \text{and} \quad s = 0.283.$$

Using Table A.4, we find $t_{0.025} = 2.447$ for $v = 6$ degrees of freedom. Hence the 95% confidence interval for μ is

$$10.0 - (2.477)\left(\frac{0.283}{\sqrt{7}}\right) < \mu < 10.0 + (2.447)\left(\frac{0.283}{\sqrt{7}}\right),$$

which reduces to

$$9.74 < \mu < 10.26.$$

9.5 Standard Error of a Point Estimate

We have made a rather sharp distinction between the goals of point estimates and confidence interval estimates. The former supplies a single number extracted from a set of experimental data, and the latter provides an interval *given the experimental data* that is reasonable for the parameter; that is, $100(1 - \alpha)\%$ of such computed intervals "cover" the parameter.

These two approaches to estimation are related to each other. The "common thread" is the sampling distribution of the point estimator. Consider, for example, the estimator $\bar{X}$ of μ with σ known. We indicated earlier that a measure of quality of an unbiased estimator is its variance. The variance of $\bar{X}$ is given by

$$\sigma_{\bar{X}}^2 = \frac{\sigma^2}{n}.$$

Thus the standard deviation of $\bar{X}$ or *standard error of $\bar{X}$* is $\sigma/\sqrt{n}$. Simply put, the standard error of an estimator is its standard deviation. For the case of $\bar{X}$, the computed confidence limit

$$\bar{x} \pm z_{\alpha/2}\frac{\sigma}{\sqrt{n}}$$

is written as

$$\bar{x} \pm z_{\alpha/2}\text{s.e.}(\bar{x}),$$

where s.e. is the standard error. The important point to make is that the width of the confidence interval on μ is dependent on the quality of the point estimator through its standard error. In the case where σ is unknown and sampling is from a normal distribution, s replaces σ and the *estimated standard error* $s/\sqrt{n}$ is involved. Thus the confidence limits on μ are given by

$$\bar{x} \pm t_{\alpha/2}\frac{s}{\sqrt{n}} = \bar{x} \pm t_{\alpha/2}\hat{s}.e.(\bar{x}).$$

Again, the confidence interval is *no better* (in terms of width) *than the quality of the point estimate*, in this case through its estimated standard error. Computer packages often refer to estimated standard errors merely as "standard errors."

As we move to more complex confidence intervals, there is a prevailing concept that widths of confidence intervals become shorter as the quality of the corresponding point estimate becomes better, although it is not always quite as simple as what we have illustrated here. It can be argued that a confidence interval is merely an augmentation of the point estimate to take into account precision of the point estimate.

9.6 Tolerance Limits

Confidence interval estimation can be very useful in an attempt to pinpoint the value of a parameter, say a population mean. However, the interpretation of a confidence interval on a mean is often confusing for the engineer or scientist. It is often the case that the analyst is attracted to the *error bounds* given by a confidence interval for a mean when, in fact, the problem requires something quite different.

In Section 9.3 we defined a confidence interval for a parameter θ that depends on the sampling distribution of $\hat{\Theta}$. That is, we computed confidence limits so that a specified fraction of the intervals computed from all possible samples of the same size contain the population parameter θ. For example, when all possible samples of the same size n are selected from a normal distribution, 95% of the intervals determined by the confidence limits $\bar{x} \pm 1.96\sigma/\sqrt{n}$ will contain the parameter μ. This fact was demonstrated in Figure 9.3. Hence we are 95% confident that the interval $\bar{x} \pm 1.96\sigma/\sqrt{n}$, computed from a particular sample, will contain the parameter μ.

Bounds on Future Single Observations

Quite often the scientist or engineer is less interested in estimating parameters and more concerned about gaining a notion about where *individual observations* or measurements might fall. If an engineer is manufacturing component parts and he or she has specifications on a certain dimension for the manufacture of the part, there is little concern about the mean of the dimension. Rather, one must attempt to determine bounds which in some probabilistic sense "cover" the individual values in the population (i.e., the measured values of the dimension).

One method of establishing a bound on single values in the population is to determine a confidence interval on a *fixed proportion* of the measurements. This is best motivated by visualizing a situation in which we are doing random sampling from a normal distribution with known mean μ and variance σ^2. Clearly, a bound that covers the middle 95% of the population of observations is given by

$$\mu \pm 1.96\sigma.$$

This is called a *tolerance interval,* and indeed the coverage of 95% of measured observations is **exact**. However, in practice μ and σ are seldom known; thus the user must apply

$$\bar{x} \pm ks,$$

and now, of course, the interval is a random variable and hence the *coverage* of a proportion of the population enjoyed by the interval is not exact. As a result a $100(1 - \gamma)\%$ confidence interval is applied to the statement since $\bar{x} \pm ks$ cannot be expected to cover any specified proportion all the time. As a result we have the following definition.

TOLERANCE LIMITS   *For a normal distribution of measurements with unknown mean μ and unknown standard deviation σ,* **tolerance limits** *are given by $\bar{x} \pm ks$, where k is determined so that one can assert with $100(1 - \gamma)\%$ confidence that the given limits contain* **at least** *the proportion $1 - \alpha$ of the measurements.*   ■

Table A.7 gives values of k for $1 - \alpha = 0.90, 0.95, 0.99$; $\gamma = 0.05, 0.01$; and for selected values of n from 2 to 1000.

EXAMPLE 9.5   A machine is producing metal pieces that are cylindrical in shape. A sample of these pieces is taken and the diameters are found to be $1.01, 0.97, 1.03, 1.04, 0.99, 0.98, 0.99, 1.01$, and 1.03 centimeters. Find the 99% tolerance limits that will contain 95% of the metal pieces produced by this machine, assuming an approximate normal distribution.

SOLUTION
The sample mean and standard deviation for the given data are

$$\bar{x} = 1.0056 \qquad \text{and} \qquad s = 0.0245.$$

From Table A.7 for $n = 9$, $1 - \gamma = 0.99$, and $1 - \alpha = 0.95$, we find $k = 4.662$ for two-sided limits. Hence the 99% tolerance limits are given by

$$1.0056 \pm (4.622)(0.0245).$$

That is, we are 99% confident that the tolerance interval from 0.894 to 1.117 will contain 95% of the metal pieces produced by this machine. It is interesting to note that the corresponding 99% confidence interval for μ (see Exercise 13 on page 253) has a lower limit of 0.978 and an upper limit of 1.033, verifying our earlier statement that a tolerance interval must necessarily be longer than a confidence interval with the same degree of confidence.

Distinction Between Confidence Intervals and Tolerance Intervals

It is important to reemphasize the difference between confidence limits and tolerance limits and to explain where each is used. The confidence interval on the mean is not useful unless the data analyst is *interested in the population mean*. The population mean needs to be estimated and the confidence interval produces appropriate bounds. The tolerance interval is much more attentive to *where individual observations fall*. Where are the majority of values going to be? Consider Exercise 15 on page 253. There is an engineering process that produces these shearing pins. Process capability is quite important to the engineer, and variability in Rockwell hardness may well be troublesome. Indeed, the buyer will probably set a lower specification limit below which he will not accept any pins. Here, of course, tolerance limits are considerably more relevant and what the process produces on the average must take a backseat. Suppose that a process is producing cylindrical pieces for a automobile (as in Exercise 13, page 253) and the target is 1.00 centimeter in diameter. In addition, suppose that specification limits are set at 1.00 ± 0.05. After a sample is taken from the process, the computed 99% tolerance limits ($\gamma = 0.01$) containing, say, 99% of individual diameters is given by the interval [0.96, 1.02]. The tolerance limits are tighter than the process requirements. The engineer may then feel quite confident that it would be quite rare to experience a part that falls outside requirements set by the consumer.

The reader should understand that the tolerance interval has features that are not unlike those of the confidence interval. In fact, the $100(1 - \alpha)\%$ tolerance interval on, say, the proportion 0.95 can be viewed as a confidence interval *on the middle* 95% of the corresponding normal distribution. One-sided tolerance limits are also relevant. For example, in the Rockwell hardness illustration one would want the tolerance limit to reflect the following: "We are, say, 99% confident that at least 99% of Rockwell hardness values will exceed the computed value"; the form of the interval is then

$$\bar{x} - ks.$$

Here the one-sided tolerance is a one-sided confidence limit on the lower 1% point of the normal distribution.

Exercises

1. Let us define $S'^2 = \sum_{i=1}^{n} (X_i - \bar{X})^2/n$. Show that $E(S'^2) = [(n - 1)/n]\sigma^2$, and hence S'^2 is a biased estimator for σ^2.

2. If X is a binomial random variable, show that

   (a) $\hat{P} = \dfrac{X}{n}$ is an unbiased estimator of p;

   (b) $P' = \dfrac{X + \sqrt{n}/2}{n + \sqrt{n}}$ is a biased estimator of p.

3. Show that the estimator P' of Exercise 2(b) becomes unbiased as $n \to \infty$.

4. An electrical firm manufactures light bulbs that have a length of life that is approximately normally distributed with a standard deviation of 40 hours. If a sam-

ple of 30 bulbs has an average life of 780 hours, find a 96% confidence interval for the population mean of all bulbs produced by this firm.

5. A soft-drink machine is regulated so that the amount of drink dispensed is approximately normally distributed with a standard deviation equal to 0.15 deciliter. Find a 95% confidence interval for the mean of all drinks dispensed by this machine if a random sample of 36 drinks has an average content of 2.25 deciliters.

6. The heights of a random sample of 50 college students showed a mean of 174.5 centimeters and a standard deviation of 6.9 centimeters.
   (a) Construct a 98% confidence interval for the mean height of all college students.
   (b) What can we assert with 98% confidence about the possible size of our error if we estimate the mean height of all college students to be 174.5 centimeters?

7. A random sample of 100 automobile owners shows that, in the state of Virginia, an automobile is driven on the average 23,500 kilometers per year with a standard deviation of 3900 kilometers.
   (a) Construct a 99% confidence interval for the average number of kilometers an automobile is driven annually in Virginia.
   (b) What can we assert with 99% confidence about the possible size of our error if we estimate the average number of kilometers driven by car owners in Virginia to be 23,500 kilometers per year?

8. How large a sample is needed in Exercise 4 if we wish to be 96% confident that our sample mean will be within 10 hours of the true mean?

9. How large a sample is needed in Exercise 5 if we wish to be 95% confident that our sample mean will be within 0.09 deciliter of the true mean?

10. An efficiency expert wishes to determine the average time that it takes to drill three holes in a certain metal clamp. How large a sample will he need to be 95% confident that his sample mean will be within 15 seconds of the true mean? Assume that it is known from previous studies that $\sigma = 40$ seconds.

11. A UCLA researcher claims that the life span of mice can be extended by as much as 25% when the calories in their food are reduced by approximately 40% from the time they are weaned. The restricted diets are enriched to normal levels by vitamins and protein. Assuming that it is known from previous studies that $\sigma = 5.8$ months, how many mice should be included

in our sample if we wish to be 99% confident that the mean life span of the sample will be within 2 months of the population mean for all mice subjected to this reduced diet?

12. Regular consumption of presweetened cereals contributes to tooth decay, heart disease, and other degenerative diseases according to studies conducted by Dr. W. H. Bowen of the National Institutes of Health and Dr. J. Yudben, Professor of Nutrition and Dietetics at the University of London. In a random sample of 20 similar single servings of Alpha-Bits the average sugar content was 11.3 grams with a standard deviation of 2.45 grams. Assuming that the sugar contents are normally distributed, construct a 95% confidence interval for the mean sugar content for single servings of Alpha-Bits.

13. A machine is producing metal pieces that are cylindrical in shape. A sample of pieces is taken and the diameters are 1.01, 0.97, 1.03, 1.04, 0.99, 0.98, 0.99, 1.01, and 1.03 centimeters. Find a 99% confidence interval for the mean diameter of pieces from this machine, assuming an approximate normal distribution.

14. A random sample of 8 cigarettes of a certain brand has an average nicotine content of 2.6 milligrams and a standard deviation of 0.9 milligram. Construct a 99% confidence interval for the true average nicotine content of this particular brand of cigarettes, assuming the distribution of nicotine contents to be approximately normal.

15. A random sample of 12 shearing pins are taken in a study of the Rockwell hardness of the head on the pin. Measurements on the Rockwell hardness were made for each of the 12, yielding an average value of 48.50 with a sample standard deviation of 1.5. Assuming the measurements to be normally distributed, construct a 90% confidence interval for the mean Rockwell hardness.

16. A random sample of 12 graduates of a certain secretarial school typed an average of 79.3 words per minute with a standard deviation of 7.8 words per minute. Assuming a normal distribution for the number of words typed per minute, find a 95% confidence interval for the average number of words typed by all graduates of this school.

17. A random sample of 25 cigarettes of a certain brand has an average nicotine content of 1.3 milligrams and a standard deviation of 0.17 milligram. Find the 95%

tolerance limits that will contain 90% of the nicotine contents for this brand of cigarettes, assuming the measurements to be normally distributed.

18. The following measurements were recorded for the drying time, in hours, of a certain brand of latex paint:

3.4	2.5	4.8	2.9	3.6
2.8	3.3	5.6	3.7	2.8
4.4	4.0	5.2	3.0	4.8

Assuming that the measurements represent a random sample from a normal population, find the 99% tolerance limits that will contain 95% of the drying times.

19. Referring to Exercise 7, construct a 99% tolerance interval containing 99% of the miles traveled by automobiles annually in Virginia. Assume the distribution of measurements to be approximately normal.

20. Referring to Exercise 15, construct a 95% interval containing 90% of the measurements.

9.7   Two Samples: Estimating the Difference Between Two Means

If we have two populations with means μ_1 and μ_2 and variances σ_1^2 and σ_2^2, respectively, a point estimator of the difference between μ_1 and μ_2 is given by the statistic $\bar{X}_1 - \bar{X}_2$. Therefore, to obtain a point estimate of $\mu_1 - \mu_2$, we shall select two independent random samples, one from each population, of size n_1 and n_2, and compute the difference $\bar{x}_1 - \bar{x}_2$, of the sample means. Clearly, we must consider the sampling distributions of $\bar{X}_1 - \bar{X}_2$.

According to Theorem 8.3, we can expect the sampling distribution of $\bar{X}_1 - \bar{X}_2$ to be approximately normally distributed with mean $\mu_{\bar{X}_1 - \bar{X}_2} = \mu_1 - \mu_2$ and standard deviation $\sigma_{\bar{X}_1 - \bar{X}_2} = \sqrt{(\sigma_1^2/n_1) + (\sigma_2^2/n_2)}$. Therefore, we can assert with a probability of $1 - \alpha$ that the standard normal variable

$$Z = \frac{(\bar{X}_1 - \bar{X}_2) - (\mu_1 - \mu_2)}{\sqrt{(\sigma_1^2/n_1) + (\sigma_2^2/n_2)}}$$

will fall between $-z_{\alpha/2}$ and $z_{\alpha/2}$. Referring once again to Figure 9.2, we write

$$P(-z_{\alpha/2} < Z < z_{\alpha/2}) = 1 - \alpha.$$

Substituting for Z, we state equivalently that

$$P\left[-z_{\alpha/2} < \frac{(\bar{X}_1 - \bar{X}_2) - (\mu_1 - \mu_2)}{\sqrt{(\sigma_1^2/n_1) + (\sigma_2^2/n_2)}} < z_{\alpha/2}\right] = 1 - \alpha,$$

which leads to the following $(1 - \alpha)100\%$ confidence interval for $\mu_1 - \mu_2$.

CONFIDENCE INTERVAL FOR $\mu_1 - \mu_2$; σ_1^2 **AND** σ_2^2 **KNOWN**	*If $\bar{x}_1$ and $\bar{x}_2$ are the means of independent random samples of size n_1 and n_2 from populations with known variances σ_1^2 and σ_2^2, respectively, a $(1 - \alpha)100\%$ confidence interval for $\mu_1 - \mu_2$ is given by*

$$(\bar{x}_1 - \bar{x}_2) - z_{\alpha/2}\sqrt{\frac{\sigma_1^2}{n_1} + \frac{\sigma_2^2}{n_2}} < \mu_1 - \mu_2 < (\bar{x}_1 - \bar{x}_2) + z_{\alpha/2}\sqrt{\frac{\sigma_1^2}{n_1} + \frac{\sigma_2^2}{n_2}},$$

where $z_{\alpha/2}$ is the z-value leaving an area of $\alpha/2$ to the right.   ■

The degree of confidence is exact when samples are selected from normal populations. For nonnormal populations the central limit theorem allows for a good approximation for reasonable size samples.

EXAMPLE 9.6   An experiment was conducted in which two types of engines, A and B, were compared. Gas mileage in miles per gallon was measured. Fifty experiments were conducted using engine type A and 75 experiments were done for engine type B. The gasoline used and other conditions were held constant. The average gas mileage for engine A was 36 miles per gallon and the average for machine B was 42 miles per gallon. Find a 96% confidence interval on $\mu_B - \mu_A$, where μ_B and μ_A are population mean gas mileage for machines B and A, respectively. Assume that the population standard deviations are 6 and 8 for machines A and B, respectively.

SOLUTION
The point estimate of $\mu_B - \mu_A$ is $\bar{x}_B - \bar{x}_A = 42 - 36 = 6$. Using $\alpha = 0.04$, we find $z_{0.02} = 2.05$ from Table A.3. Hence with substitution in the formula above, the 96% confidence interval is

$$6 - 2.05\sqrt{\frac{64}{75} + \frac{36}{50}} < \mu_1 - \mu_2 < 6 + 2.05\sqrt{\frac{64}{75} + \frac{36}{50}}$$

or simply

$$3.43 < \mu_B - \mu_A < 8.57.$$

This procedure for estimating the difference between two means is applicable if σ_1^2 and σ_2^2 are known. If the variances are not known and the two distributions involved are approximately normal, the t-distribution becomes involved as in the case of a single sample. If one is not willing to assume normality, large samples (say greater than 30) will allow the use of s_1 and s_2 in place of σ_1 and σ_2, respectively, with the rationale that $s_1 \simeq \sigma_1$ and $s_2 \simeq \sigma_2$. Again, of course, the confidence interval is an approximate one.

Variances Unknown

Consider the case in which σ_1^2 and σ_2^2 are unknown. If $\sigma_1^2 = \sigma_2^2 = \sigma^2$, we obtain a standard normal variable in the form

$$Z = \frac{(\bar{X}_1 - \bar{X}_2) - (\mu_1 - \mu_2)}{\sqrt{\sigma^2[(1/n_1) + (1/n_2)]}}.$$

According to Theorem 8.4, the two random variables $(n_1 - 1)S_1^2/\sigma^2$ and $(n_2 - 1)S_2^2/\sigma^2$ have chi-squared distributions with $n_1 - 1$ and $n_2 - 1$ degrees of freedom, respectively. Furthermore, they are independent chi-squared variables, since the random samples were selected independently. Consequently, their sum

$$V = \frac{(n_1 - 1)S_1^2}{\sigma^2} + \frac{(n_2 - 1)S_2^2}{\sigma^2} = \frac{(n_1 - 1)S_1^2 + (n_2 - 1)S_2^2}{\sigma^2}$$

has a chi-squared distribution with $v_1 = n_1 + n_2 - 2$ degrees of freedom.

Since the preceding expressions for Z and V can be shown to be independent, it follows from Theorem 8.5 that the statistic

$$T = \frac{(\bar{X}_1 - \bar{X}_2) - (\mu_1 - \mu_2)}{\sqrt{\sigma^2[(1/n_1) + (1/n_2)]}} \bigg/ \sqrt{\frac{(n_1 - 1)S_1^2 + (n_2 - 1)S_2^2}{\sigma^2(n_1 + n_2 - 2)}}$$

has the t-distribution with $v = n_1 + n_2 - 2$ degrees of freedom.

A point estimate of the unknown common variance σ^2 can be obtained by pooling the sample variances. Denoting the pooled estimator by S_p^2, we write

$$S_p^2 = \frac{(n_1 - 1)S_1^2 + (n_2 - 1)S_2^2}{n_1 + n_2 - 2}.$$

Substituting S_p^2 in the T statistic, we obtain the less cumbersome form

$$T = \frac{(\bar{X}_1 - \bar{X}_2) - (\mu_1 - \mu_2)}{S_p\sqrt{(1/n_1) + (1/n_2)}}.$$

Using the statistic T, we have

$$P(-t_{\alpha/2} < T < t_{\alpha/2}) = 1 - \alpha,$$

where $t_{\alpha/2}$ is the t-value with $n_1 + n_2 - 2$ degrees of freedom, above which we find an area of $\alpha/2$. Substituting for T in the inequality, we write

$$P\left[-t_{\alpha/2} < \frac{(\bar{X}_1 - \bar{X}_2) - (\mu_1 - \mu_2)}{S_p\sqrt{(1/n_1) + (1/n_2)}} < t_{\alpha/2}\right] = 1 - \alpha.$$

After performing the usual mathematical manipulations, the difference of the sample means $\bar{x}_1 - \bar{x}_2$ and the pooled variance

$$s_p^2 = \frac{(n_1 - 1)s_1^2 + (n_2 - 1)s_2^2}{n_1 + n_2 - 2}$$

are computed and then the following $(1 - \alpha)100\%$ confidence interval for $\mu_1 - \mu_2$ is obtained.

The value for s_p^2 above is easily seen to be a weighted average of the two sample variances s_1^2 and s_2^2, where the weights are the degrees of freedom.

CONFIDENCE INTERVAL FOR $\mu_1 - \mu_2$; $\sigma_1^2 = \sigma_2^2$ BUT UNKNOWN

If $\bar{x}_1$ and $\bar{x}_2$ are the means of independent random samples of size n_1 and n_2, respectively, from approximate normal populations with unknown but equal variances, a $(1 - \alpha)100\%$ confidence interval for $\mu_1 - \mu_2$ is given by

$$(\bar{x}_1 - \bar{x}_2) - t_{\alpha/2}s_p\sqrt{\frac{1}{n_1} + \frac{1}{n_2}} < \mu_1 - \mu_2 < (\bar{x}_1 - \bar{x}_2) + t_{\alpha/2}s_p\sqrt{\frac{1}{n_1} + \frac{1}{n_2}},$$

where s_p is the pooled estimate of the population standard deviation and $t_{\alpha/2}$ is the t-value with $v = n_1 + n_2 - 2$ degrees of freedom, leaving an area of $\alpha/2$ to the right. ∎

For the case of confidence interval estimation on the difference between two means, we need to consider the experimental conditions in the data-taking process. It is assumed that we have two independent random samples from distributions with means μ_1 and μ_2, respectively. It is important that experimental conditions emulate this "ideal" described by the assumptions as closely as possible. Quite often the experimenter should plan the strategy of the experiment accordingly. For almost any study of this type, there is a so-called *experimental unit*, which is that part of the experiment that produces experimental error and is responsible for the population variance we refer to as σ^2. In a drug study the experimental unit is the patient or subject. In an agricultural experiment, it may be a plot of ground. In a chemical experiment, it may be a quantity of raw materials. It is important that differences between these units have minimal impact on the results. The experimenter will have a degree of insurance that experimental units will not bias results if the conditions that define the two populations are *randomly assigned* to the experimental units. We shall again focus on randomization in future chapters that deal in hypothesis testing.

EXAMPLE 9.7   In the article "*Macroinvertebrate Community Structure as an Indicator of Acid Mine Pollution*" published in the *Journal of Environmental Pollution* (Vol. 6, 1974), we are given a report on an investigation undertaken in Cane Creek, Alabama, to determine the relationship between selected physiochemical parameters and different measures of macroinvertebrate community structure. One facet of the investigation was an evaluation of the effectiveness of a numerical species diversity index to indicate aquatic degradation due to acid mine drainage. Conceptually, a high index of macroinvertebrate species diversity should indicate an unstressed aquatic system, while a low diversity index should indicate a stressed aquatic system.

Two independent sampling stations were chosen for this study, one located downstream from the acid mine discharge point and the other located upstream. For 12 monthly samples collected at the downstream station the species diversity index had a mean value $\bar{x}_1 = 3.11$ and a standard deviation $s_1 = 0.771$, while 10 monthly samples collected at the upstream station had a mean index value $\bar{x}_2 = 2.04$ and a standard deviation $s_2 = 0.448$. Find a 90% confidence interval for the difference between the population means for the two locations, assuming that the populations are approximately normally distributed with equal variances.

SOLUTION

Let μ_1 and μ_2 represent the population means, respectively, for the species diversity index at the downstream and upstream stations. We wish to find a 90% confidence interval for $\mu_1 - \mu_2$. Our point estimate of $\mu_1 - \mu_2$ is $\bar{x}_1 - \bar{x}_2 = 3.11 - 2.04 = 1.07$. The pooled estimate, s_p^2, of the common variance, σ^2, is

$$s_p^2 = \frac{(n_1 - 1)s_1^2 + (n_2 - 1)s_2^2}{n_1 + n_2 - 2}$$

$$= \frac{(11)(0.771^2) + (9)(0.448^2)}{12 + 10 - 2} = 0.417.$$

Taking the square root, we obtain $s_p = 0.646$. Using $\alpha = 0.1$, we find in Table A.4 that $t_{0.05} = 1.725$ for $v = n_1 + n_2 - 2 = 20$ degrees of freedom. Therefore, the 90% confidence interval for $\mu_1 - \mu_2$ is

$$1.07 - (1.725)(0.646)\sqrt{\tfrac{1}{12} + \tfrac{1}{10}} < \mu_1 - \mu_2 < 1.07 + (1.725)(0.646)\sqrt{\tfrac{1}{12} + \tfrac{1}{10}},$$

which simplifies to

$$0.593 < \mu_1 - \mu_2 < 1.547.$$

Interpretation of the Confidence Interval

In the case of a single parameter, the confidence interval simply produces error bounds on the parameter. Values contained in the interval should be viewed as reasonable values given the experimental data. In the case of a difference between two means, the interpretation can be extended to one of comparing the two means. For example, if we have high confidence that a difference $\mu_1 - \mu_2$ is positive, we would certainly infer that $\mu_1 > \mu_2$ with little risk of being in error. For example, in Example 9.7, we are 90% confident that the interval from 0.593 to 1.547 contains the difference of the population means for values of the species diversity index at the two stations. The fact that both confidence limits are positive indicates that, on the average, the index for the station located downstream from the discharge point is greater than the index for the station located upstream.

Equal Sample Sizes

The procedure for constructing confidence intervals for $\mu_1 - \mu_2$ with $\sigma_1 = \sigma_2 = \sigma$ unknown requires the assumption that the populations are normal. Slight departures from either the equal variance or normality assumption do not seriously alter the degree of confidence for our interval. (A procedure will be presented in Chapter 10 for testing the equality of two unknown population variances based on the information provided by the sample variances.) If the population variances are considerably different, we still obtain reasonable results when the populations are normal, provided that $n_1 = n_2$. Therefore, in a planned experiment, one should make every effort to equalize the size of the samples.

Unequal Variances

Let us now consider the problem of finding an interval estimate of $\mu_1 - \mu_2$ when the unknown population variances are not likely to be equal. The statistic most often used in this case is

$$T' = \frac{(\overline{X}_1 - \overline{X}_2) - (\mu_1 - \mu_2)}{\sqrt{(S_1^2/n_1) + (S_2^2/n_2)}},$$

which has approximately a t-distribution with v degrees of freedom, where

$$v = \frac{(s_1^2/n_1 + s_2^2/n_2)^2}{[(s_1^2/n_1)^2/(n_1 - 1)] + [(s_2^2/n_2)^2/(n_2 - 1)]}.$$

Since v is seldom an integer, we round it off to the nearest whole number.

Using the statistic T', we write

$$P(-t_{\alpha/2} < T' < t_{\alpha/2}) \simeq 1 - \alpha,$$

where $t_{\alpha/2}$ is the value of the t-distribution with v degrees of freedom, above which we find an area of $\alpha/2$. Substituting for T' in the inequality, and following the exact steps as before, we state the final result.

CONFIDENCE INTERVAL FOR $\mu_1 - \mu_2$; $\sigma_1^2 \neq \sigma_2^2$ AND UNKNOWN

If $\bar{x}_1$ and s_1^2, and $\bar{x}_2$ and s_2^2, are the means and variances of small independent samples of size n_1 and n_2, respectively, from approximate normal distributions with unknown and unequal variances, an approximate $(1 - \alpha)100\%$ confidence interval for $\mu_1 - \mu_2$ is given by

$$(\bar{x}_1 - \bar{x}_2) - t_{\alpha/2}\sqrt{\frac{s_1^2}{n_1} + \frac{s_2^2}{n_2}} < \mu_1 - \mu_2 < (\bar{x}_1 - \bar{x}_2) + t_{\alpha/2}\sqrt{\frac{s_1^2}{n_1} + \frac{s_2^2}{n_2}},$$

where $t_{\alpha/2}$ is the t-value with

$$v = \frac{(s_1^2/n_1 + s_2^2/n_2)^2}{[(s_1^2/n_1)^2/(n_1 - 1)] + [(s_2^2/n_2)^2/(n_2 - 1)]}$$

degrees of freedom, leaving an area $\alpha/2$ to the right. ■

Before we illustrate the confidence interval above with an example, we should point out that all the confidence intervals on $\mu_1 - \mu_2$ are of the same general form as those on a single mean; namely, they can be written

$$\text{point estimate} \pm t_{\alpha/2} \; \hat{\text{s}}.\text{e.} \; (\text{point estimate})$$

or

$$\text{point estimate} \pm z_{\alpha/2} \; \text{s.e.} \; (\text{point estimate}).$$

For example, in the case where $\sigma_1 = \sigma_2 = \sigma$, the estimated standard error of $\bar{x}_1 - \bar{x}_2$ is $s_p\sqrt{1/n_1 + 1/n_2}$. For the case where $\sigma_1^2 \neq \sigma_2^2$,

$$\hat{\text{s}}.\text{e.}(\bar{x}_1 - \bar{x}_2) = \sqrt{\frac{s_1^2}{n_1} + \frac{s_2^2}{n_2}}.$$

EXAMPLE 9.8   A study on the "*Nutrient Retention and Macroinvertebrate Community Response to Sewage Stress in a Stream Ecosystem*" was conducted by the Department of Zoology at the Virginia Polytechnic Institute and State University in 1980 to estimate the difference in the amount of the chemical orthophosphorus measured at two different stations on the James River. Orthophosphorus is measured in milligrams per liter. Fifteen samples were collected from station 1 and 12 samples were obtained from station 2. The 15 samples from station 1 had an average orthophosphorus content of 3.84 milligrams per liter and a standard deviation of 3.07 milligrams per liter, while the 12 samples from station 2 had an average content of 1.49 milligrams per liter and a standard deviation of 0.80 milligram per liter. Find a 95% confidence interval for the difference in the true average orthophosphorus contents at these two stations, assuming that the observations came from normal populations with different variances.

SOLUTION

For station 1 we have $\bar{x}_1 = 3.84$, $s_1 = 3.07$, and $n_1 = 15$. For station 2, $\bar{x}_2 = 1.49$, $s_2 = 0.80$, and $n_2 = 12$. We wish to find a 95% confidence interval for $\mu_1 - \mu_2$. Since the population variances are assumed to be unequal, we can only find an approximate 95% confidence interval based on the t-distribution with v degrees of freedom, where

$$v = \frac{(3.07^2/15 + 0.80^2/12)^2}{[(3.07^2/15)^2/14] + [(0.80^2/12)^2/11]}$$

$$= 16.3 \simeq 16.$$

Our point estimate of $\mu_1 - \mu_2$ is $\bar{x}_1 - \bar{x}_2 = 3.84 - 1.49 = 2.35$. Using $\alpha = 0.05$, we find in Table A.4 that $t_{0.025} = 2.120$ for $v = 16$ degrees of freedom. Therefore, the 95% confidence interval for $\mu_1 - \mu_2$ is

$$2.35 - 2.120\sqrt{\frac{3.07^2}{15} + \frac{0.80^2}{12}} < \mu_1 - \mu_2 < 2.35 + 2.120\sqrt{\frac{3.07^2}{15} + \frac{0.80^2}{12}},$$

which simplifies to

$$0.60 < \mu_1 - \mu_2 < 4.10.$$

Hence we are 95% confident that the interval from 0.60 to 4.10 milligrams per liter contains the difference of the true average orthophosphorus contents for these two locations.

9.8 Paired Observations

At this point we shall consider estimation procedures for the difference of two means when the samples are not independent and the variances of the two populations are not necessarily equal. The situation considered here deals with a very special experimental situation, namely that of *paired observations*. Unlike the situation described earlier, the conditions of the two populations are not assigned randomly to experimental units. Rather, each homogeneous experimental unit receives both population conditions; as a result, each experimental unit has a pair of observations, one for each population. For example, if we run a test on a new diet using 15 individuals, the weight before and after going on the diet form the information for our two samples. These two populations are "before" and "after" and the experimental unit is the individual. Obviously, the observations in a pair have something in common. To determine if the diet is effective, we consider the differences $d_1, d_2, \ldots, d_n$ in the paired observations. These differences are the values of a random sample $D_1, D_2, \ldots, D_n$ from a population of differences that we shall assume to be normally distributed with mean $\mu_D = \mu_1 - \mu_2$ and variance σ_D^2. We estimate σ_D^2 by s_d^2, the variance of the differences that constitute our sample. The point estimator of μ_D is given by $\bar{D}$.

Pairing observations in an experiment is a strategy that can be employed in many fields of application. The reader will be exposed to this concept in material related to hypothesis testing in Chapter 10 and experimental design issues in Chapters 13 and 15. By selecting experimental units that are relatively homogeneous (within the units) and allowing each unit to experience both population conditions, the effective "experimental error variance" (in this case σ_D^2) is reduced. The reader may visualize that the ith pair consists of the measurement

$$D_i = X_{1i} - X_{2i}.$$

Since the two observations are taken on the sample experimental unit, they are not independent and, in fact,

$$\text{Var } D_i = \text{Var}(X_{1i} - X_{2i})$$
$$= \sigma_1^2 + \sigma_2^2 - 2 \text{ Cov}(X_{1i}, X_{2i}).$$

Now, intuitively, it is expected that σ_D^2 should be reduced because of the similarity in nature of the "errors" of the two observations within an experimental unit, and this comes through in the expression above. One certainly expects that if the unit is homogeneous, the covariance is positive. As a result, the gain in quality of the confidence interval over that of not pairing will be greatest when there is homogeneity within units and large differences as one goes from unit to unit. One should keep in mind that the performance of the confidence interval will depend on the standard error of $\overline{D}$, which is, of course, $\sigma_D/\sqrt{n}$, where n is the number of pairs. As we indicated earlier, the intent of pairing is to reduce σ_D.

In comparing the confidence interval against that of the unpaired situation, it is apparent that there is a "trade-off" involved. Although pairing should indeed reduce variance and hence reduce the standard error of the point estimate, the degrees of freedom are reduced by reducing the problem to a one-sample problem. As a result, the $t_{\alpha/2}$ point attached to the standard error is adjusted accordingly. Thus pairing may be counterproductive. This would certainly be the case if one experiences only a modest reduction in variance (through σ_D^2) by pairing.

Another illustration of pairing might involve the choice of n pairs of subjects with each pair having a similar characteristic, such as IQ, age, breed, and so on; then for each pair one member is selected at random to yield a value of X_1, leaving the other member to provide the value of X_2. In this case X_1 and X_2 might represent the grades obtained by two individuals of equal IQ when one of the individuals is assigned at random to a class using the conventional lecture approach while the other individual is assigned to a class using programmed materials.

A $(1 - \alpha)100\%$ confidence interval for μ_D can be established by writing

$$P(-t_{\alpha/2} < T < t_{\alpha/2}) = 1 - \alpha,$$

where

$$T = \frac{\overline{D} - \mu_D}{S_d/\sqrt{n}}$$

and $t_{\alpha/2}$, as before, is a value of the t-distribution with $n - 1$ degrees of freedom.

It is now a routine procedure to replace T, by its definition, in the inequality above and carry out the mathematical steps that lead to the following $(1 - \alpha)100\%$ confidence interval for $\mu_1 - \mu_2 = \mu_D$.

CONFIDENCE INTERVAL FOR $\mu_D = \mu_1 - \mu_2$ FOR PAIRED OBSERVATIONS

If $\bar{d}$ and s_d are the mean and standard deviation of the normally distributed differences of n random pairs of measurements, a $(1 - \alpha)100\%$ confidence interval for $\mu_D = \mu_1 - \mu_2$ is

$$\bar{d} - t_{\alpha/2}\frac{s_d}{\sqrt{n}} < \mu_D < \bar{d} + t_{\alpha/2}\frac{s_d}{\sqrt{n}},$$

where $t_{\alpha/2}$ is the t-value with $v = n - 1$ degrees of freedom, leaving an area of $\alpha/2$ to the right. ∎

EXAMPLE 9.9   In the article *"Essential Elements in Fresh and Canned Tomatoes,"* published in the *Journal of Food Science* (Vol. 46, 1981), the contents of essential elements were determined in fresh and canned tomatoes by atomic absorption spectrophotometry. The copper contents in fresh tomatoes compared to the copper contents for the same tomatoes after being canned were recorded as follows:

Pair	Fresh Tomatoes	Canned Tomatoes	d_i
1	0.066	0.085	0.019
2	0.079	0.088	0.009
3	0.069	0.091	0.022
4	0.076	0.096	0.020
5	0.071	0.093	0.022
6	0.087	0.095	0.008
7	0.071	0.079	0.008
8	0.073	0.078	0.005
9	0.067	0.065	−0.002
10	0.062	0.068	0.006

Find a 98% confidence interval for the true difference in the mean copper contents of fresh and canned tomatoes assuming the distribution of differences to be normal.

SOLUTION
We wish to find a 98% confidence interval for $\mu_1 - \mu_2$, where μ_1 and μ_2 represent the true average copper contents for canned tomatoes and fresh tomatoes, respectively. Since the observations are paired, $\mu_1 - \mu_2 = \mu_D$. The point estimate of μ_D is given by $\bar{d} = 0.0117$. The standard deviation, s_d, of the sample differences is

$$s_d = \sqrt{\frac{n\sum_{i=1}^{n} d_i^2 - \left(\sum_{i=1}^{n} d_i\right)^2}{n(n-1)}} = \sqrt{\frac{(10)(0.002003) - (0.117)^2}{(10)(9)}}$$

$$= 0.0084.$$

Using $\alpha = 0.02$, we find in Table A.4 that $t_{0.01} = 2.81$ for $v = n - 1 = 9$ degrees of freedom. Therefore, the 98% confidence interval is

$$0.0117 - (2.821)\left(\frac{0.0084}{\sqrt{10}}\right) < \mu_D < 0.0117 + (2.821)\left(\frac{0.0084}{\sqrt{10}}\right)$$

or simply

$$0.0042 < \mu_D < 0.0192,$$

from which we conclude that there are significantly higher amounts of copper in canned tomatoes than in fresh tomatoes.

Exercises

1. A random sample size $n_1 = 25$ taken from a normal population with a standard deviation $\sigma_1 = 5$ has a mean $\bar{x}_1 = 80$. A second random sample size $n_2 = 36$, taken from a different normal population with a standard deviation $\sigma_2 = 3$, has a mean $\bar{x}_2 = 75$. Find a 94% confidence interval for $\mu_1 - \mu_2$.

2. Two kinds of thread are being compared for strength. Fifty pieces of each type of threaead are tested under similar conditions. Brand A had an average tensile strength of 78.3 kilograms with a standard deviation of 5.6 kilograms, while brand B had an average tensile strength of 87.2 kilograms with a standard deviation of 6.3 kilograms. Construct a 95% confidence interval for the difference of the population means.

3. A study was made to determine if a certain metal treatment has any effect on the amount of metal removed in a pickling operation. A random sample of 100 pieces was immersed in a bath for 24 hours without the treatment, yielding an average of 12.2 millimeters of metal removed and a sample standard deviation of 1.1 millimeters. A second sample of 200 pieces was exposed to the treatment followed by the 24-hour immersion in the bath, resulting in an average removal of 9.1 millimeters of metal with a sample standard deviation of 0.9 millimeter. Compute a 98% confidence interval estimate for the difference between the population means. Does the treatment appear to reduce the mean amount of metal removed?

4. In a batch chemical process, two catalysts are being compared for their effect on the output of the process reaction. A sample of 12 batches was prepared using catalyst 1 and a sample of 10 batches was obtained using catalyst 2. The 12 batches for which catalyst 1 was used gave an average yield of 85 with a sample standard deviation of 4, while the average for the second sample gave an average of 81 and a sample standard deviation of 5. Find a 90% confidence interval for the difference between the population means, assuming that the populations are approximately normally distributed with equal variances.

5. Students may choose between a 3-semester-hour course in physics without labs and a 4-semester-hour course with labs. The final written examination is the same for each section. If 12 students in the section with labs made an average examination grade of 84 with a standard deviation of 4, and 18 students in the section without labs made an average grade of 77 with a standard deviation of 6, find a 99% confidence interval for the difference between the average grades for the two courses. Assume the populations to be approximately normally distributed with equal variances.

6. In a study conducted at the Virginia Polytechnic Institute and State University in 1983 on the development of ectomycorrhizal, a symbiotic relationship between the roots of trees and a fungus in which minerals are transferred from the fungus to the trees and sugars from the trees to the fungus, 20 northern red oak seedlings with the fungus *Pisolithus tinctorus* were grown in a greenhouse. All seedlings were planted in the same type of soil and received the same amount of sunshine and water. Half received no nitrogen at planting time to serve as a control and the other half received 368 ppm of nitrogen in the form $NaNO_3$. The stem weights, recorded in grams, at the end of 140 days were recorded as follows:

No Nitrogen	Nitrogen
0.32	0.26
0.53	0.43
0.28	0.47
0.37	0.49
0.47	0.52
0.43	0.75
0.36	0.79
0.42	0.86
0.38	0.62
0.43	0.46

Construct a 95% confidence interval for the difference in the mean stem weights between seedlings that receive no nitrogen and those that receive 368 ppm of nitrogen. Assume the populations to be normally distributed with equal variances.

7. The following data, recorded in days, represent the length of time to recovery for patients randomly treated with one of two medications to clear up severe bladder infections:

Medication 1	Medication 2
$n_1 = 14$	$n_2 = 16$
$\bar{x}_1 = 17$	$\bar{x}_2 = 19$
$s_1^2 = 1.5$	$s_2^2 = 1.8$

Find a 99% confidence interval for the difference $\mu_2 - \mu_1$ in the mean recovery time for the two medications, assuming normal populations with equal variances.

8. An experiment reported in *Popular Science,* in 1981, compared fuel economies for two types of similarly equipped diesel mini-trucks. Let us suppose that 12 Volkswagen and 10 Toyota trucks are used in 90-kilometer per hour steady-speed tests. If the 12 Volkswagen trucks average 16 kilometers per liter with a standard deviation of 1.0 kilometer per liter and the 10 Toyota trucks average 11 kilometers per liter with a standard deviation of 0.8 kilometer per liter, construct a 90% confidence interval for the difference between the average kilometers per liter of these two mini-trucks. Assume that the distances per

liter for each truck model are approximately normally distributed with equal variances.

9. A taxi company is trying to decide whether to purchase brand A or brand B tires for its fleet of taxis. To estimate the difference in the two brands, an experiment is conducted using 12 of each brand. The tires are run until they wear out. The results are

Brand A: $\bar{x}_1 = 36,300$ kilometers,
$s_1 = 5000$ kilometers.

Brand B: $\bar{x}_2 = 38,100$ kilometers,
$s_2 = 6100$ kilometers

Compute a 95% confidence interval for $\mu_1 - \mu_2$, assuming the populations to be approximately normally distributed.

10. The following data represent the running times of films produced by two motion-picture companies.

Company	Time (minutes)						
I	103	94	110	87	98		
II	97	82	123	92	175	88	118

Compute a 90% confidence interval for the difference between the average running times of films produced by the two companies. Assume that the running-time differences are approximately normally distributed.

11. The government awarded grants to the agricultural departments of 9 universities to test the yield capabilities of two new varieties of wheat. Each variety was planted on plots of equal area at each university and the yields, in kilograms per plot, recorded as follows:

Variety	University								
	1	2	3	4	5	6	7	8	9
1	38	23	35	41	44	29	37	31	38
2	45	25	31	38	50	33	36	40	43

Find a 95% confidence interval for the mean difference between the yields of the two varieties, assuming the differences of yields to be approximately normally distributed. Explain why pairing is necessary in this problem.

12. Referring to Exercise 9, find a 99% confidence interval for $\mu_1 - \mu_2$ if a tire from each company is assigned at random to the rear wheels of 8 taxis and the following distance, in kilometers, recorded:

Taxi	Brand A	Brand B
1	34,400	36,700
2	45,500	46,800
3	36,700	37,700
4	32,000	31,100
5	48,400	47,800
6	32,800	36,400
7	38,100	38,900
8	30,100	31,500

Assume that the differences of the distances are approximately normally distributed.

9.9 Single Sample: Estimating a Proportion

A point estimator of the proportion p in a binomial experiment is given by the statistic $\hat{P} = X/n$, where X represents the number of successes in n trials. Therefore, the sample proportion $\hat{p} = x/n$ will be used as the point estimate of the parameter p.

If the unknown proportion p is not expected to be too close to zero or 1, we can establish a confidence interval for p by considering the sampling distribution of $\hat{P}$. Designating a failure in each binomial trial by the value 0 and a success by the value 1, the number of successes, x, can be interpreted as the sum of n values consisting only of zeros and ones, and $\hat{p}$ is just the sample mean of these n values. Hence, by the central limit theorem, for n sufficiently large, $\hat{P}$ is approximately normally distributed with mean

$$\mu_{\hat{P}} = E(\hat{P}) = E\left[\frac{X}{n}\right] = \frac{np}{n} = p$$

and variance

$$\sigma_{\hat{P}}^2 = \sigma_{X/n}^2 = \frac{\sigma_X^2}{n^2} = \frac{npq}{n^2} = \frac{pq}{n}.$$

Therefore, we can assert that

$$P(-z_{\alpha/2} < Z < z_{\alpha/2}) = 1 - \alpha,$$

where

$$Z = \frac{\hat{P} - p}{\sqrt{pq/n}}$$

and $z_{\alpha/2}$ is the value of the standard normal curve above which we find an area of $\alpha/2$. Substituting for Z, we write

$$P\left(-z_{\alpha/2} < \frac{\hat{P} - p}{\sqrt{pq/n}} < z_{\alpha/2}\right) = 1 - \alpha.$$

Multiplying each term of the inequality by $\sqrt{pq/n}$, and then subtracting $\hat{P}$ and multiplying by -1, we obtain

$$P\left(\hat{P} - z_{\alpha/2}\sqrt{\frac{pq}{n}} < p < \hat{P} + z_{\alpha/2}\sqrt{\frac{pq}{n}}\right) = 1 - \alpha.$$

It is difficult to manipulate the inequalities so as to obtain a random interval whose end points are independent of p, the unknown parameter. When n is large, very little error is introduced by substituting the point estimate $\hat{p} = x/n$ for the p under the radical sign. Then we can write

$$P\left(\hat{P} - z_{\alpha/2}\sqrt{\frac{\hat{p}\hat{q}}{n}} < p < \hat{P} + z_{\alpha/2}\sqrt{\frac{\hat{p}\hat{q}}{n}}\right) \simeq 1 - \alpha.$$

For our particular random sample of size n, the sample proportion $\hat{p} = x/n$ is computed, and the following approximate $(1 - \alpha)100\%$ confidence interval for p is obtained.

LARGE-SAMPLE CONFIDENCE INTERVAL FOR p

If $\hat{p}$ is the proportion of successes in a random sample of size n, and $\hat{q} = 1 - \hat{p}$, an approximate $(1 - \alpha)100\%$ confidence interval for the binomial parameter p is given by

$$\hat{p} - z_{\alpha/2}\sqrt{\frac{\hat{p}\hat{q}}{n}} < p < \hat{p} + z_{\alpha/2}\sqrt{\frac{\hat{p}\hat{q}}{n}},$$

where $z_{\alpha/2}$ is the z-value leaving an area of $\alpha/2$ to the right. ∎

When n is small and the unknown proportion p is believed to be close to 0 or to 1, the confidence-interval procedure established here is unreliable and, therefore, should not be used. To be on the safe side, one should require both $n\hat{p}$ or $n\hat{q}$ to be greater than or equal to 5. The method for finding a confidence interval for the binomial parameter p is also applicable when the binomial distribution is being used to approximate the hypergeometric distribution, that is, when n is small relative to N, as illustrated in Example 9.10.

EXAMPLE 9.10   In a random sample of $n = 500$ families owning television sets in the city of Hamilton, Canada, it was found that $x = 340$ subscribed to HBO. Find a 95% confidence interval for the actual proportion of families in this city who subscribe to HBO.

SOLUTION
The point estimate of p is $\hat{p} = 340/500 = 0.68$. Using Table A.3, we find $z_{0.025} = 1.96$. Therefore, the 95% confidence interval for p is

$$0.68 - 196\sqrt{\frac{(0.68)(0.32)}{500}} < p < 0.68 + 1.96\sqrt{\frac{(0.68)(0.32)}{500}},$$

which simplifies to

$$0.64 < p < 0.72.$$

If p is the center value of a $(1 - \alpha)100\%$ confidence interval, then $\hat{p}$ estimates p without error. Most of the time, however, $\hat{p}$ will not be exactly equal to p and the point estimate is in error. The size of this error will be the positive difference that separates p and $\hat{p}$, and we can be $(1 - \alpha)100\%$ confident that this difference will not exceed $z_{\alpha/2}\sqrt{\hat{p}\hat{q}/n}$. We can readily see this if we draw a diagram of a typical confidence interval as in Figure 9.6.

THEOREM 9.3   *If $\hat{p}$ is used as an estimate of p, we can be $(1 - \alpha)100\%$ confident that the error will not exceed $z_{\alpha/2}\sqrt{\hat{p}\hat{q}/n}$.* ∎

In Example 9.10 we are 95% confident that the sample proportion $\hat{p} = 0.68$ differs from the true proportion p by an amount not exceeding 0.04.

Let us now determine how large a sample is necessary to ensure that the error in estimating p will be less than a specified amount e. By Theorem 7.3, this means we must choose n such that $z_{\alpha/2}\sqrt{\hat{p}\hat{q}/n} = e$.

THEOREM 9.4   *If $\hat{p}$ is used as an estimate of p, we can be $(1 - \alpha)100\%$ confident that the error will be less than a specified amount e when the sample size is approximately*

$$n = \frac{z_{\alpha/2}^2\hat{p}\hat{q}}{e^2}.$$ ∎

Theorem 9.4 is somewhat misleading in that we must use $\hat{p}$ to determine the sample size n, but $\hat{p}$ is computed from the sample. If a crude estimate of p can be made without taking a sample, we could use this value for $\hat{p}$ and then determine n. Lacking such an estimate, we could take a preliminary sample of size $n \geq 30$ to provide an estimate of p. Then, using Theorem 9.4 we could determine approximately how many observations are needed to provide the desired degree of accuracy. Once again, all fractional values of n are rounded up to the next whole number.

EXAMPLE 9.11   How large a sample is required in Example 9.10 if we want to be 95% confident that our estimate of p is within 0.02?

SOLUTION

Let us treat the 500 families as a preliminary sample providing an estimate $\hat{p} = 0.68$. Then, by Theorem 9.4,

$$n = \frac{(1.96)^2(0.68)(0.32)}{(0.02)^2} = 2090.$$

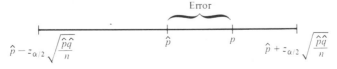

FIGURE 9.6   Error in estimating p by $\hat{p}$.

Therefore, if we base our estimate of p on a random sample of size 2090, we can be 95% confident that our sample proportion will not differ from the true proportion by more than 0.02.

Occasionally, it will be impractical to obtain an estimate of p to be used in determining the sample size for a specified degree of confidence. If this happens, an upper bound for n is established by noting that $\hat{p}\hat{q} = \hat{p}(1 - \hat{p})$, which must be at most equal to 1/4, since $\hat{p}$ must lie between 0 and 1. This fact may be verified by completing the square. Hence

$$\hat{p}(1 - \hat{p}) = -(\hat{p}^2 - \hat{p}) = \tfrac{1}{4} - (\hat{p}^2 - \hat{p} + \tfrac{1}{4})$$
$$= \tfrac{1}{4} - (\hat{p} - \tfrac{1}{2})^2,$$

which is always less than 1/4 except when $\hat{p} = 1/2$ and then $\hat{p}\hat{q} = 1/4$. Therefore, if we substitute $\hat{p} = 1/2$ into the formula for n in Theorem 9.4, when, in fact, p actually differs from 1/2, then n will turn out to be larger than necessary for the specified degree of confidence and as a result our degree of confidence will increase.

THEOREM 9.5   *If $\hat{p}$ is used as an estimate of p, we can be **at least** $(1 - \alpha)100\%$ confident that the error will not exceed a specified amount e when the sample size is*

$$n = \frac{z_{\alpha/2}^2}{4e^2}.$$

EXAMPLE 9.12   How large a sample is required in Example 9.10 if we want to be at least 95% confident that our estimate of p is within 0.02?

SOLUTION
Unlike Example 9.11, we shall now assume that no preliminary sample has been taken to provide an estimate of p. Consequently, we can be at least 95% confident that our sample proportion will not differ from the true proportion by more than 0.02 if we choose a sample of size

$$n = \frac{(1.96)^2}{(4)(0.02)^2} = 2401.$$

Comparing the results of Examples 9.11 and 9.12, we see that information concerning p, provided by a preliminary sample, or perhaps from past experience, enables us to choose a smaller sample while maintaining our required degree of accuracy.

9.10 Two Samples: Estimating the Difference Between Two Proportions

Consider the problem in which we wish to estimate the difference between two binomial parameters p_1 and p_2. For example, we might let p_1 be the proportion of smokers with lung cancer and p_2 the proportion of nonsmokers with lung cancer.

Our problem, then, is to estimate the difference between these two proportions. First, we select independent random samples of size n_1 and n_2 from the two binomial populations with means $n_1 p_1$ and $n_2 p_2$ and variances $n_1 p_1 q_1$ and $n_2 p_2 q_2$, respectively, then determine the numbers x_1 and x_2 of people in each sample with lung cancer, and form the proportions $\hat{p}_1 = x_1/n_1$ and $\hat{p}_2 = x_2/n_2$. A point estimator of the difference between the two proportions, $p_1 - p_2$, is given by the statistic $\hat{P}_1 - \hat{P}_2$. Therefore, the difference of the sample proportions, $\hat{p}_1 - \hat{p}_2$, will be used as the point estimate of $p_1 - p_2$.

A confidence interval for $p_1 - p_2$ can be established by considering the sampling distribution of $\hat{P}_1 - \hat{P}_2$. From Section 9.9 we know that $\hat{P}_1$ and $\hat{P}_2$ are each approximately normally distributed, with means p_1 and p_2 and variances $p_1 q_1/n_1$ and $p_2 q_2/n_2$, respectively. By choosing independent samples from the two populations, the variables $\hat{P}_1$ and $\hat{P}_2$ will be independent, and then by the reproductive property of the normal distribution established in Theorem 7.11, we conclude that $\hat{P}_1 - \hat{P}_2$ is approximately normally distributed with mean

$$\mu_{\hat{P}_1 - \hat{P}_2} = p_1 - p_2$$

and variance

$$\sigma^2_{\hat{P}_1 - \hat{P}_2} = \frac{p_1 q_1}{n_1} + \frac{p_2 q_2}{n_2}.$$

Therefore, we can assert that

$$P(-z_{\alpha/2} < Z < z_{\alpha/2}) = 1 - \alpha,$$

where

$$Z = \frac{(\hat{P}_1 - \hat{P}_2) - (p_1 - p_2)}{\sqrt{(p_1 q_1/n_1) + (p_2 q_2/n_2)}}$$

and $z_{\alpha/2}$ is a value of the standard normal curve above which we find an area of $\alpha/2$. Substituting for Z, we write

$$P\left[-z_{\alpha/2} < \frac{(\hat{P}_1 - \hat{P}_2) - (p_1 - p_2)}{\sqrt{(p_1 q_1/n_1) + (p_2 q_2/n_2)}} < z_{\alpha/2}\right] = 1 - \alpha.$$

After performing the usual mathematical manipulations, we replace p_1, p_2, q_1, and q_2 under the radical sign by their estimates $\hat{p}_1 = x_1/n_1$, $\hat{p}_2 = x_2/n_2$, $\hat{q}_1 = 1 - \hat{p}_1$, and $\hat{q}_2 = 1 - \hat{p}_2$, provided that $n_1 \hat{p}_1$, $n_1 \hat{q}_1$, $n_2 \hat{p}_2$, and $n_2 \hat{q}_2$ are all greater than or equal to 5, and the following approximate $(1 - \alpha)100\%$ confidence interval for $p_1 - p_2$ is obtained.

LARGE-SAMPLE CONFIDENCE INTERVAL FOR $p_1 - p_2$

If $\hat{p}_1$ and $\hat{p}_2$ are the proportion of successes in random samples of size n_1 and n_2, respectively, $\hat{q}_1 = 1 - \hat{p}_1$ and $\hat{q}_2 = 1 - \hat{p}_2$, an approximate $(1 - \alpha)100\%$ confidence interval for the difference of two binomial parameters $p_1 - p_2$, is given by

$$(\hat{p}_1 - \hat{p}_2) - z_{\alpha/2}\sqrt{\frac{\hat{p}_1 \hat{q}_1}{n_1} + \frac{\hat{p}_2 \hat{q}_2}{n_2}} < p_1 - p_2 < (\hat{p}_1 - \hat{p}_2) + z_{\alpha/2}\sqrt{\frac{\hat{p}_1 \hat{q}_1}{n_1} + \frac{\hat{p}_2 \hat{q}_2}{n_2}},$$

where $z_{\alpha/2}$ is the z-value leaving an area of $\alpha/2$ to the right. ∎

EXAMPLE 9.13   A certain change in a manufacturing procedure for component parts is being considered. Samples are taken using both the existing and the new procedure in order to
determine if the new procedure results in an improvement. If 75 of 1500 items from
the existing procedure were found to be defective and 80 of 2000 items from the
new procedure were found to be defective, find a 90% confidence interval for the
true difference in the fraction of defectives between the existing and the new
process.

SOLUTION
Let p_1 and p_2 be the true proportions of defectives for the existing and new procedures, respectively. Hence $\hat{p}_1 = 75/1500 = 0.05$ and $\hat{p}_2 = 80/2000 = 0.04$, and
the point estimate of $p_1 - p_2 + \hat{p}_1 - \hat{p}_2 = 0.05 - 0.04 = 0.01$. Using Table A.3,
we find $z_{0.05} = 1.645$. Therefore, substituting into this formula we obtain the 90%
confidence interval

$$0.01 - 1.645\sqrt{\frac{(0.05)(0.95)}{1500} + \frac{(0.04)(0.96)}{2000}} < p_1 - p_2$$

$$< 0.01 + 1.645\sqrt{\frac{(0.05)(0.95)}{1500} + \frac{(0.04)(0.96)}{2000}},$$

which simplifies to

$$-0.0017 < p_1 - p_2 < 0.0217.$$

Since the interval contains the value 0, there is no reason to believe that the new
procedure produced a significant decrease in the proportion of defectives over the
existing method.

Exercises

1. (a) A random sample of 200 voters is selected and
114 are found to support an annexation suit. Find
the 96% confidence interval for the fraction of the
voting population favoring the suit.
   (b) What can we assert with 96% confidence about
the possible size of our error if we estimate the
fraction of voters favoring the annexation suit to
be 0.57?

2. (a) A random sample of 500 cigarette smokers is selected and 86 are found to have a preference for
brand X. Find the 90% confidence interval for the
fraction of the population of cigarette smokers
who prefer brand X.
   (b) What can we assert with 90% confidence about
the possible size of our error if we estimate the
fraction of cigarette smokers who prefer brand X
to be 0.172?

3. In a random sample of 1000 homes in a certain city, it
is found that 228 are heated by oil. Find the 99%
confidence interval for the proportion of homes in this
city that are heated by oil.

4. Compute a 98% confidence interval for the proportion of defective items in a process when it is found
that a sample of size 100 yields 8 defectives.

5. A new rocket-launching system is being considered
for deployment of small, short-range rockets. The
existing system has $p = 0.8$ as the probability of a
successful launch. A sample of 40 experimental
launches is made with the new system and 34 are
successful.
   (a) Construct a 95% confidence interval for p.
   (b) Would you conclude that the new system is better?

6. A geneticist is interested in the proportion of African males that have a certain minor blood disorder. In a random sample of 100 African males, 24 are found to be afflicted.

   (a) Compute a 99% confidence interval for the proportion of African males that have this blood disorder.

   (b) What can we assert with 99% confidence about the possible size of our error if we estimate the proportion of African males with this blood disorder to be 0.24?

7. (a) According to a report in the *Roanoke Times & World-News*, August 20, 1981, approximately 2/3 of the 1600 adults polled by telephone said they think the space shuttle program is a good investment for the country. Find a 95% confidence interval for the proportion of American adults who think the space shuttle program is a good investment for the country.

   (b) What can we assert with 95% confidence about the possible size of our error if we estimate the proportion of American adults who think the space shuttle program is a good investment to be 2/3?

8. In the newspaper article referred to in Exercise 7, 32% of the 1600 adults polled said the U.S. space program should emphasize scientific exploration. How large a sample of adults is needed in the poll if one wishes to be 95% confident that the estimated percentage will be within 2% of the true percentage?

9. How large a sample is needed in Exercise 1 if we wish to be 96% confident that our sample proportion will be within 0.02 of the true fraction of the voting population?

10. How large a sample is needed in Exercise 3 if we wish to be 99% confident that our sample proportion will be within 0.05 of the true proportion of homes in this city that are heated by oil?

11. How large a sample is needed in Exercise 4 if we wish to be 98% confident that our sample proportion will be within 0.05 of the true proportion defective?

12. A study is to be made to estimate the percentage of citizens in a town who favor having their water fluoridated. How large a sample is needed if one wishes to be at least 95% confident that our estimate is within 1% of the true percentage?

13. According to Dr. Memory Elvin-Lewis, head of the microbiology department at Washington University

School of Dental Medicine in St. Louis, a couple of cups of either green or oolong tea each day will provide sufficient fluoride to protect your teeth from decay. People who do not like tea and who live in unfluoridated areas should ask their local governments to consider having their water fluoridated. How large a sample is needed to estimate the percentage of citizens in a certain town who favor having their water fluoridated if one wishes to be at least 99% confident that the estimate is within 1% of the true percentage?

14. A study is to be made to estimate the proportion of residents in a certain city and its suburbs who favor the construction of a nuclear power plant. How large a sample is needed if one wishes to be at least 95% confident that the estimate is within 0.04 of the true proportion of residents in this city and its suburbs that favor the construction of the nuclear power plant?

15. A certain geneticist is interested in the proportion of males and females in the population that have a certain minor blood disorder. In a random sample of 1000 males, 250 are found to be afflicted, whereas 275 of 1000 females tested appear to have the disorder. Compute a 95% confidence interval for the difference between the proportion of males and females that have the blood disorder.

16. A cigarette-manufacturing firm claims that its brand A line of cigarettes outsells its brand B line by 8%. If it is found that 42 of 200 smokers prefer brand A and 18 of 150 smokers prefer brand B, compute a 94% confidence interval for the difference between the proportions of sales of the 2 brands and decide if the 8% difference is a valid claim.

17. A clinical trial is conducted to determine if a certain type of inoculation has an effect on the incidence of a certain disease. A sample of 1000 rats was kept in a controlled environment for a period of 1 year and 500 of the rats were given the inoculation. Of the group not given the drug, there were 120 incidences of the disease, while 98 of the inoculated group contracted it. If we call p_1 the probability of incidence of the disease in uninoculated rats and p_2 the probability of incidence after receiving the drug, compute a 90% confidence interval for $p_1 - p_2$.

18. In a study, *"Germination and Emergence of Broccoli,"* conducted by the Department of Horticulture at Virginia Polytechnic Institute and State University, a researcher found that at 5°C, 10 seeds out of 20

germinated, while at 15°C, 15 out of 20 seeds germinated. Compute a 95% confidence interval for the difference between the proportion of germination at the two different temperatures and decide if there is a significant difference.

19. In 1988 a survey of 1000 students concluded that 274 students chose a professional baseball team, A, as his or her favorite team. In 1991, the same survey was conducted involving 760 students. It concluded that 240 of them also chose team A as their favorite. Compute a 95% confidence interval for the difference between the proportion of students favoring team A between the two surveys. Is there a significant difference?

20. A well-known sock company wants to determine if their medium-size socks satisfy the same proportion of boys and girls who wear their socks. In a survey of 1000 randomly selected teenagers, 462 were boys and 538 were girls. Among the 282 boys who wear the company's medium-size socks, 267 say they feel comfortable with its size, while 290 girls out of 313 who wear the company's socks say they were satisfied. The rest say that the socks are either too small or too big. Compute a 95% confidence interval for the difference between the proportion of boys and girls who were satisfied with the company's medium-size socks.

9.11   Single Sample: Estimating the Variance

If a sample size of n is drawn from a normal population with variance σ^2, and the sample variance s^2 is computed, we obtain a value of the statistic S^2. This computed sample variance will be used as a point estimate of σ^2. Hence the statistic S^2 is called an estimator of σ^2.

An interval estimate of σ^2 can be established by using the statistic

$$X^2 = \frac{(n-1)S^2}{\sigma^2}.$$

According to Theorem 8.4, the statistic X^2 has a chi-squared distribution with $n-1$ degrees of freedom when samples are chosen from a normal population. We may write (see Figure 9.7)

$$P(\chi_{1-\alpha/2}^2 < X^2 < \chi_{\alpha/2}^2) = 1 - \alpha,$$

where $\chi_{1-\alpha/2}^2$ and $\chi_{\alpha/2}^2$ are values of the chi-squared distribution with $n-1$ degrees of freedom, leaving areas of $1 - \alpha/2$ and $\alpha/2$, respectively, to the right. Substituting for X^2, we write

$$P\left[\chi_{1-\alpha/2}^2 < \frac{(n-1)S^2}{\sigma^2} < \chi_{\alpha/2}^2\right] = 1 - \alpha.$$

Dividing each term in the inequality by $(n-1)S^2$, and then inverting each term (thereby changing the sense of the inequalities), we obtain

$$P\left[\frac{(n-1)S^2}{\chi_{\alpha/2}^2} < \sigma^2 < \frac{(n-1)S^2}{\chi_{1-\alpha/2}^2}\right] = 1 - \alpha.$$

For our particular random sample of size n, the sample variance s^2 is computed, and the following $(1 - \alpha)100\%$ confidence interval for σ^2 is obtained.

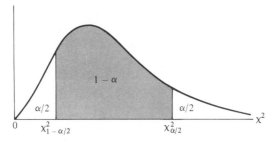

FIGURE 9.7 $P(\chi^2_{1-\alpha/2} < X^2 < \chi^2_{\alpha/2}) = 1 - \alpha.$

CONFIDENCE INTERVAL FOR σ^2

If s^2 is the variance of a random sample of size n from a normal population, a $(1 - \alpha)100\%$ confidence interval for σ^2 is given by

$$\frac{(n - 1)s^2}{\chi^2_{\alpha/2}} < \sigma^2 < \frac{(n - 1)s^2}{\chi^2_{1-\alpha/2}},$$

where $\chi^2_{\alpha/2}$ and $\chi^2_{1-\alpha/2}$ are χ^2-values with $v = n - 1$ degrees of freedom, leaving areas of $\alpha/2$ and $1 - \alpha/2$, respectively, to the right. ∎

A $(1 - \alpha)100\%$ confidence interval for σ is obtained by taking the square root of each endpoint of the interval for σ^2.

EXAMPLE 9.14   The following are the weights, in decagrams, of 10 packages of grass seed distributed by a certain company: 46.4, 46.1, 45.8, 47.0, 46.1, 45.9, 45.8, 46.9, 45.2, and 46.0. Find a 95% confidence interval for the variance of all such packages of grass seed distributed by this company, assuming a normal population.

SOLUTION
First we find

$$s^2 = \frac{n \sum\limits_{i=1}^{n} x_i^2 - \left(\sum\limits_{i=1}^{n} x_i\right)^2}{n(n - 1)} = \frac{(10)(21{,}273.12) - (461.2)^2}{(10)(9)} = 0.286.$$

To obtain a 95% confidence interval, we choose $\alpha = 0.05$. Then, using Table A.5 with $v = 9$ degrees of freedom we find $\chi^2_{0.025} = 19.023$ and $\chi^2_{0.975} = 2.700$. Therefore, the 95% confidence interval for σ^2 is given by

$$\frac{(9)(0.286)}{19.023} < \sigma^2 < \frac{(9)(0.286)}{2.700},$$

or simply

$$0.135 < \sigma^2 < 0.953.$$

9.12 Two Samples: Estimating the Ratio of Two Variances

A point estimate of the ratio of two population variances σ_1^2/σ_2^2 is given by the ratio s_1^2/s_2^2 of the sample variances. Hence the statistic S_1^2/S_2^2 is called an estimator of σ_1^2/σ_2^2.

If σ_1^2 and σ_2^2 are the variances of normal populations, we can establish an interval estimate of σ_1^2/σ_2^2 by using the statistic

$$F = \frac{\sigma_2^2 S_1^2}{\sigma_1^2 S_2^2}.$$

According to Theorem 8.8, the random variable F has an F-distribution with $v_1 = n_1 - 1$ and $v_2 = n_2 - 1$ degrees of freedom. Therefore, we may write (see Figure 9.8)

$$P[f_{1-\alpha/2}(v_1, v_2) < F < f_{\alpha/2}(v_1, v_2)] = 1 - \alpha,$$

where $f_{1-\alpha/2}(v_1, v_2)$ and $f_{\alpha/2}(v_1, v_2)$ are the values of the F-distribution with v_1 and v_2 degrees of freedom, leaving areas of $1 - \alpha/2$ and $\alpha/2$, respectively, to the right. Substituting for F, we write

$$P\left[f_{1-\alpha/2}(v_1, v_2) < \frac{\sigma_2^2 S_1^2}{\sigma_1^2 S_2^2} < f_{\alpha/2}(v_1, v_2)\right] = 1 - \alpha.$$

Multiplying each term in the inequality by S_2^2/S_1^2, and then inverting each term (again changing the sense of the inequalities), we obtain

$$P\left[\frac{S_1^2}{S_2^2} \frac{1}{f_{\alpha/2}(v_1, v_2)} < \frac{\sigma_1^2}{\sigma_2^2} < \frac{S_1^2}{S_2^2} \frac{1}{f_{1-\alpha/2}(v_1, v_2)}\right] = 1 - \alpha.$$

The results of Theorem 8.7 enable us to replace the quantity $f_{1-\alpha/2}(v_1, v_2)$ by $1/f_{\alpha/2}(v_2, v_1)$. Therefore,

$$P\left[\frac{S_1^2}{S_2^2} \frac{1}{f_{\alpha/2}(v_1, v_2)} < \frac{\sigma_1^2}{\sigma_2^2} < \frac{S_1^2}{S_2^2} f_{\alpha/2}(v_2, v_1)\right] = 1 - \alpha.$$

For any two independent random samples of size n_1 and n_2 selected from two normal populations, the ratio of the sample variances s_1^2/s_2^2, is computed and the following $(1 - \alpha)100\%$ confidence interval for σ_1^2/σ_2^2 is obtained.

CONFIDENCE INTERVAL FOR σ_1^2/σ_2^2

If s_1^2 and s_2^2 are the variances of independent samples of size n_1 and n_2, respectively, from normal populations, then a $(1 - \alpha)100\%$ confidence interval for σ_1^2/σ_2^2 is

$$\frac{s_1^2}{s_2^2} \frac{1}{f_{\alpha/2}(v_1, v_2)} < \frac{\sigma_1^2}{\sigma_2^2} < \frac{s_1^2}{s_2^2} f_{\alpha/2}(v_2, v_1),$$

where $f_{\alpha/2}(v_1, v_2)$ is an f-value with $v_1 = n_1 - 1$ and $v_2 = n_2 - 1$ degrees of freedom leaving an area of $\alpha/2$ to the right, and $f_{\alpha/2}(v_2, v_1)$ is a similar f-value with $v_2 = n_2 - 1$ and $v_1 = n_1 - 1$ degrees of freedom. ■

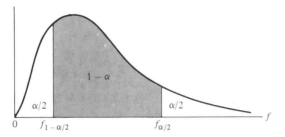

FIGURE 9.8 $P[f_{1-\alpha/2}(v_1, v_2) < F < f_{\alpha/2}(v_1, v_2)] = 1 - \alpha.$

As in Section 9.11, a $(1 - \alpha)100\%$ confidence interval for σ_1/σ_2 is obtained by taking the square root of each endpoint of the interval for σ_1^2/σ_2^2.

EXAMPLE 9.15   A confidence interval for the difference in the mean orthophosphorus contents, measured in milligrams per liter, at two stations on the James River was constructed in Example 9.8 on page 259 by assuming the normal population variance to be unequal. Justify this assumption by constructing a 98% confidence interval for σ_1^2/σ_2^2 and for σ_1/σ_2, where σ_1^2 and σ_2^2 are the variances of the populations of orthophosphorus contents at station 1 and station 2, respectively.

SOLUTION
From Example 9.8, we have $n_1 = 15$, $n_2 = 12$, $s_1 = 3.07$, and $s_2 = 0.80$. For a 98% confidence interval, $\alpha = 0.02$. Interpolating in Table A.6, we find $f_{0.01}(14, 11) \simeq 4.30$ and $f_{0.01}(11, 14) \simeq 3.87$. Therefore, the 98% confidence interval for σ_1^2/σ_2^2 is

$$\frac{3.07^2}{0.80^2}\left(\frac{1}{4.30}\right) < \frac{\sigma_1^2}{\sigma_2^2} < \frac{3.07^2}{0.80^2}(3.87),$$

which simplifies to

$$3.425 < \frac{\sigma_1^2}{\sigma_2^2} < 56.991.$$

Taking square roots of the confidence limits, we find that a 98% confidence interval or σ_1/σ_2 is

$$1.851 < \frac{\sigma_1}{\sigma_2} < 7.549.$$

Since this interval does not allow for the possibility of σ_1/σ_2 being equal to 1, we were correct in assuming that $\sigma_1 \neq \sigma_2$ or $\sigma_1^2 \neq \sigma_2^2$ in Example 9.8.

Up to this point all confidence intervals presented were of the form

point estimate $\pm K$ s.e.(point estimate),

where K is a constant (either t or normal percent point). This is the case when the parameter is a mean, difference between means, proportion, or difference between proportions. However, it does not extend to variances and ratios of variances.

Exercises

1. A manufacturer of car batteries claims that his batteries will last, on average, 3 years with a variance of 1 year. If 5 of these batteries have lifetimes of 1.9, 2.4, 3.0, 3.5, and 4.2 years, construct a 95% confidence interval for σ^2 and decide if the manufacturer's claim that $\sigma^2 = 1$ is valid. Assume the population of battery lives to be approximately normally distributed.

2. A random sample of 20 students obtained a mean of $\bar{x} = 72$ and a variance of $s^2 = 16$ on a college placement test in mathematics. Assuming the scores to be normally distributed, construct a 98% confidence interval for σ^2.

3. Construct a 95% confidence interval for σ in Exercise 12 on page 253.

4. Construct a 99% confidence interval for σ^2 in Exercise 13 on page 253.

5. Construct a 99% confidence interval for σ in Exercise 14 on page 253.

6. Construct a 90% confidence interval for σ^2 in Exercise 15 on page 253.

7. Construct a 98% confidence interval for σ_1/σ_2 in Exercise 8 on page 264, where σ_1 and σ_2 are, respectively, the standard deviations for the distances obtained per liter of fuel by the Volkswagen and Toyota mini-trucks.

8. Construct a 90% confidence interval for σ_1^2/σ_2^2 in Exercise 9 on page 264. Were we justified in assuming that $\sigma_1^2 = \sigma_2^2$ when we constructed our confidence interval for $\mu_1 - \mu_2$?

9. Construct a 90% confidence interval for σ_1^2/σ_2^2 in Exercise 10 on page 264. Should we have assumed $\sigma_1^2 = \sigma_2^2$ in constructing our confidence interval for $\mu_{II} - \mu_{I}$?

9.13   Bayesian Methods of Estimation

The classical methods of estimation that we have studied so far are based solely on information provided by the random sample. These methods essentially interpret probabilities as relative frequencies. For example, in arriving at a 95% confidence interval for μ, we interpret the statement $P(-1.96 < Z < 1.96) = 0.95$ to mean that 95% of the time in repeated experiments Z will fall between -1.96 and 1.96. Probabilities of this type that can be interpreted in the frequency sense will be referred to as **objective probabilities**. The Bayesian approach to statistical methods of estimation combines sample information with other available prior information that may appear to be pertinent.

Subjective Probability

Consider the problem of finding a point estimate of the parameter θ for the population $f(x; \theta)$. The classical approach would be to take a random sample of size n and substitute the information provided by the sample into the appropriate estimator or decision function. Thus in the case of a binomial population $b(x; n, p)$ our estimate of p, the proportion of successes, would be $\hat{p} = x/n$. Now, suppose that additional information is given about θ, namely, that it is known to vary according to some probability distribution $f(\theta)$, often called a **prior distribution**, with **prior mean** μ_0 and **prior variance** σ_0^2. That is, we are now assuming θ to be a value of a random variable Θ with probability distribution $f(\theta)$ and we wish to estimate the particular value θ for the population from which we selected our random sample. The probabilities associated with this prior distribution are called **subjective probabilities**, in that they measure a person's *degree of belief* in the location of the parameter. The person uses his or her own experience and knowledge as the basis for arriving at the subjective probabilities given by the prior distribution. Bayesian techniques use the prior distribution $f(\theta)$ along with the joint distribution of the sample $f(x_1, x_2, \ldots, x_n; \theta)$ to compute the **posterior distribution** $f(\theta | x_1, x_2, \ldots, x_n)$. The posterior distribution consists of information from both the subjective prior distribution and the objective sampling distribution and expresses our degree of belief in the location of the parameter θ after we have observed the sample.

Let us begin with $f(x_1, x_2, \ldots, x_n | \theta)$ when we need the joint probability distribution of our sample whenever we wish to indicate that the parameter is also a random variable. The joint distribution of the sample $X_1, X_2, \ldots, X_n$ and the parameter Θ is then

$$f(x_1, x_2, \ldots, x_n, \theta) = f(x_1, x_2, \ldots, x_n | \theta) f(\theta),$$

from which we readily obtain the marginal distribution

$$g(x_1, x_2, \ldots, x_n) = \begin{cases} \displaystyle\sum_\theta f(x_1, x_2, \ldots, x_n, \theta) & \text{(discrete case)} \\[2ex] \displaystyle\int_{-\infty}^{\infty} f(x_1, x_2, \ldots, x_n, \theta) \, d\theta & \text{(continuous case).} \end{cases}$$

Hence the posterior distribution may be written

$$f(\theta | x_1, x_2, \ldots, x_n) = \frac{f(x_1, x_2, \ldots, x_n, \theta)}{g(x_1, x_2, \ldots, x_n)}.$$

DEFINITION 9.3   *The mean of the posterior distribution $f(\theta | x_1, x_2, \ldots, x_n)$, denoted by θ^*, is called the* **Bayes estimate** *of θ.* ∎

EXAMPLE 9.16   Let us assume that the prior distribution for the proportion p of defectives produced by a machine is

p	0.1	0.2
$f(p)$	0.6	0.4

Find the Bayes estimate for the proportion of defectives being produced by this machine if a random sample of size 2 yields 1 defective.

SOLUTION
Let X be the number of defectives in our sample. Then the probability distribution for our sample is

$$f(x | p) = b(x; n, p) = \binom{2}{x} p^x q^{2-x}, \qquad x = 0, 1, 2.$$

If $p = 0.1$, the probability that the random sample of size 2 yields 1 defective is found to be

$$f(1 \mid 0.1) = b(1; 2, 0.1) = \binom{2}{1}(0.1)(0.9) = 0.18.$$

Similarly, when $p = 0.2$, we find that

$$f(1 \mid 0.2) = b(1; 2, 0.2) = \binom{2}{1}(0.2)(0.8) = 0.32.$$

From the fact that $f(x, p) = f(x \mid p)f(p)$, we can set up the table

p	0.1	0.2
$f(1, p)$	0.108	0.128

from which we get

$$g(1) = \sum_p f(1, p) = 0.236.$$

We obtain the posterior distribution for the proportion of defectives, p, when $x = 1$, from the formula $f(p \mid x) = f(x, p)/g(x)$. Hence we have

p	0.1	0.2
$f(p \mid x = 1)$	0.458	0.542

from which we get the Bayes estimate of p, denoted by p^*, to be

$$p^* = (0.1)(0.458) + (0.2)(0.542) = 0.1542.$$

Note that the Bayes estimate is much smaller than the 1/2 value given by the classical estimate $\hat{p} = x/n$.

EXAMPLE 9.17   Repeat Example 9.16 using the uniform prior distribution $f(p) = 1, 0 < p < 1$.

SOLUTION
As before, we have

$$f(x \mid p) = \binom{2}{x}p^x q^{2-x}, \qquad x = 0, 1, 2.$$

From the fact that $f(x, p) = f(x \mid p)f(p)$, we can write

$$f(1, p) = \binom{2}{1}pq$$

$$= 2p(1 - p), \qquad 0 < p < 1$$

and then

$$g(1) = \int_0^1 2p(1 - p)\, dp = \tfrac{1}{3}.$$

The posterior distribution for the proportion of defectives, p, when $x = 1$, is then

$$f(p \mid x = 1) = \frac{2p(1 - p)}{\frac{1}{3}}$$

$$= 6p(1 - p), \qquad 0 < p < 1,$$

from which we get the Bayes estimate of p to be

$$p^* = 6 \int_0^1 p^2(1 - p) \ dp = \tfrac{1}{2}.$$

In this case we see that the Bayes estimate p^* and the classical estimate $\hat{p}$ are equivalent.

A $(1 - \alpha)100\%$ **Bayesian interval** for the parameter θ can be constructed by finding an interval centered at the posterior mean that contains $(1 - \alpha)100\%$ of the posterior probability.

DEFINITION 9.4   *The interval $a < \theta < b$ will be called a $(1 - \alpha)100\%$ **Bayes interval** for θ if*

$$\int_{\theta^*}^b f(\theta \mid x_1, x_2, \ldots, x_n) \ d\theta = \int_a^{\theta^*} f(\theta \mid x_1, x_2, \ldots, x_n) \ d\theta = \frac{1 - \alpha}{2}. \quad \blacksquare$$

Bayesian methods of estimation concerning the mean μ of a normal population are based on the following theorem.

THEOREM 9.6   *If $\bar{x}$ is the mean of a random sample of size n from a normal population with known variance σ^2, and the prior distribution of the population mean is a normal distribution with mean μ_0 and variance σ_0^2, then the posterior distribution of the population mean is also a normal distribution with mean μ^* and the standard deviation σ^*, where*

$$\mu^* = \frac{n\bar{x}\sigma_0^2 + \mu_0\sigma^2}{n\sigma_0^2 + \sigma^2} \qquad and \qquad \sigma^* = \sqrt{\frac{\sigma_0^2\sigma^2}{n\sigma_0^2 + \sigma^2}}. \quad \blacksquare$$

PROOF. Multiplying the density of our sample

$$f(x_1, x_2, \ldots, x_n \mid \mu) = \frac{1}{(2\pi)^{n/2} \cdot \sigma^n} \exp\left[-\left(\frac{1}{2}\right) \sum_{i=1}^n \left(\frac{x_i - \mu}{\sigma}\right)^2\right],$$

for $-\infty < x_i < \infty$ and $i = 1, 2, \ldots, n$ by our prior

$$f(\mu) = \frac{1}{\sqrt{2\pi}\,\sigma_0} e^{-(1/2)[(\mu - \mu_0)/\sigma_0]^2}, \qquad -\infty < \mu < \infty,$$

we obtain the joint density of the random sample and the mean of the population from which the sample is selected. That is,

$$f(x_1, x_2, \ldots, x_n, \mu) = \frac{1}{(2\pi)^{(n+1)/2} \cdot \sigma^n \sigma_0}$$

$$\times \exp\left\{-\left(\frac{1}{2}\right)\left[\sum_{i=1}^{n}\left(\frac{x_i - \mu}{\sigma}\right)^2 + \left(\frac{\mu - \mu_0}{\sigma_0}\right)^2\right]\right\}.$$

In Section 8.4 we established the identity

$$\sum_{i=1}^{n}(x_i - \mu)^2 = \sum_{i=1}^{n}(x_i - \bar{x})^2 + n(\bar{x} - \mu)^2,$$

which enables us to write

$$f(x_1, x_2, \ldots, x_n, \mu) = \frac{1}{(2\pi)^{(n+1)/2} \cdot \sigma^n \sigma_0} \exp\left[-\left(\frac{1}{2}\right)\sum_{i=1}^{n}\left(\frac{x_i - \bar{x}}{\sigma}\right)^2\right]$$

$$\times e^{-(1/2)\{n[(\bar{x}-\mu)/\sigma]^2 + [(\mu-\mu_0)/\sigma_0]^2\}}.$$

Completing the square in the second exponent, we can write the joint density of the random sample and the population mean in the form

$$f(x_1, x_2, \ldots, x_n, \mu) = Ke^{-(1/2)[(\mu-\mu^*)/\sigma^*]^2},$$

where

$$\mu^* = \frac{n\bar{x}\sigma_0^2 + \mu_0\sigma^2}{n\sigma_0^2 + \sigma^2},$$

$$\sigma^* = \sqrt{\frac{\sigma_0^2\sigma^2}{n\sigma_0^2 + \sigma^2}},$$

and K is a function of the sample values and the known parameters. The marginal distribution of the sample is then

$$g(x_1, x_2, \ldots, x_n) = K\sqrt{2\pi}\,\sigma^* \int_{-\infty}^{\infty} \frac{1}{\sqrt{2\pi}\,\sigma^*} e^{-(1/2)[(\mu-\mu^*)/\sigma^*]^2}\,d\mu$$

$$= K\sqrt{2\pi}\,\sigma^*,$$

and the posterior distribution is

$$f(\mu \,|\, x_1, x_2, \ldots, x_n) = \frac{f(x_1, x_2, \ldots, x_n, \mu)}{g(x_1, x_2, \ldots, x_n)}$$

$$= \frac{1}{\sqrt{2\pi}\,\sigma^*} e^{-(1/2)[(\mu-\mu^*)/\sigma^*]^2}, \qquad -\infty < \mu < \infty,$$

which is easily identified as a normal distribution with mean μ^* and standard deviation σ^*, where μ^* and σ^* are defined above.

The central limit theorem allows us to use Theorem 9.6 also when we select random samples of size $n \geq 30$ from nonnormal populations, and when the prior distribution of the mean is approximately normal.

To compute μ^* and σ^* by the formulas of Theorem 9.6 we have assumed that σ^2 is known. Since this is generally not the case, we shall replace σ^2 by the sample variance s^2 whenever $n \geq 30$. The posterior mean μ^* is the Bayes estimate of the population mean μ, and a $(1 - \alpha)100\%$ **Bayesian interval** for μ can be constructed by computing the interval

$$\mu^* - z_{\alpha/2}\sigma^* < \mu < \mu^* + z_{\alpha/2}\sigma^*,$$

which is centered at the posterior mean and contains $(1 - \alpha)100\%$ of the posterior probability.

EXAMPLE 9.18   An electrical firm manufactures light bulbs that have a length of life that is approximately normally distributed with a standard deviation of 100 hours. Prior experience leads us to believe that μ is a value of a normal random variable with a mean $\mu_0 = 800$ hours and a standard deviation $\sigma_0 = 10$ hours. If a random sample of 25 bulbs has an average life of 780 hours, find a 95% Bayesian interval for μ.

SOLUTION
According to Theorem 9.6, the posterior distribution of the mean is also a normal distribution with mean

$$\mu^* = \frac{(25)(780)(10)^2 + (800)(100)^2}{(25)(10)^2 + (100)^2} = 796$$

and standard deviation

$$\sigma^* = \sqrt{\frac{(10)^2(100)^2}{(25)(10)^2 + (100)^2}} = \sqrt{80}.$$

The 95% Bayesian interval for μ is then given by

$$796 - 1.96\sqrt{80} < \mu < 796 + 1.96\sqrt{80}$$

or

$$778.5 < \mu < 813.5.$$

By ignoring the prior information about μ in Example 9.18, we could proceed as in Section 9.4 and construct the classical 95% confidence interval

$$780 - (1.96)\left(\frac{100}{\sqrt{25}}\right) < \mu < 780 + (1.96)\left(\frac{100}{\sqrt{25}}\right)$$

or

$$740.8 < \mu < 819.2,$$

which is seen to be wider than the corresponding Bayesian interval.

9.14 Decision Theory

In our discussion of the classical approach to point estimation, we adopted the criterion that selects the decision function that is most efficient. That is, we choose from all possible unbiased estimators the one with the smallest variance as our "best" estimator. In *decision theory* we also take into account the rewards for making correct decisions and the penalties for making incorrect decisions. This leads to a new criterion that chooses the decision function $\hat{\Theta}$ that penalizes us the least when the action taken is incorrect. It is convenient now to introduce a **loss function** whose values depend on the true value of the parameter θ and the action $\hat{\theta}$. This is usually written in functional notation as $L(\hat{\Theta}; \theta)$. In many decision-making problems it is desirable to use a loss function of the form

$$L(\hat{\Theta}; \theta) = |\hat{\Theta} - \theta|$$

or perhaps

$$L(\hat{\Theta}; \theta) = (\hat{\Theta} - \theta)^2$$

in arriving at a choice between two or more decision functions.

Since θ is unknown, it must be assumed that it can equal any of several possible values. The set of all possible values that θ can assume is called the **parameter space**. For each possible value of θ in the parameter space, the loss function will vary from sample to sample. We define the **risk function** for the decision function $\hat{\Theta}$ to be the expected value of the loss function when the value of the parameter is θ and denote this function by $R(\hat{\Theta}; \theta)$. Hence we have

$$R(\hat{\Theta}; \theta) = E[L(\hat{\Theta}; \theta)].$$

One method of arriving at a choice between $\hat{\Theta}_1$ and $\hat{\Theta}_2$ as an estimator for θ would be to apply the **minimax criterion**. Essentially, we determine the maximum value of $R(\hat{\Theta}_1; \theta)$ and the maximum value of $R(\hat{\Theta}_2; \theta)$ in the parameter space and then choose the decision function that provided the minimum of these two maximum risks.

EXAMPLE 9.19   According to the minimax criterion, is $\overline{X}$ or $\tilde{X}$ a better estimator of the mean μ of a normal population with known variance σ^2, based on a random sample of size n when the loss function is of the form $L(\hat{\Theta}; \theta) = (\hat{\Theta} - \theta)^2$?

SOLUTION
The loss function corresponding to $\overline{X}$ is given by

$$L(\overline{X}; \mu) = (\overline{X} - \mu)^2.$$

Hence the risk function is

$$R(\overline{X}; \mu) = E[(\overline{X} - \mu)^2] = \frac{\sigma^2}{n}$$

for every μ in the parameter space. Similarly, one can show that the risk function corresponding to $\tilde{X}$ is given by

$$R(\bar{X}; \mu) = E[(\bar{X} - \mu)^2] \simeq \frac{\pi\sigma^2}{2n}$$

for every μ in the parameter space. In view of the fact that $\sigma^2/n < \pi\sigma^2/2n$, the minimax criterion selects $\bar{X}$ rather than $\tilde{X}$ as the better estimator for μ.

In some practical situations we may have additional information concerning the unknown parameter θ. For example, suppose that we wish to estimate the binomial parameter p, the proportion of defectives produced by a machine during a certain day when we know that p varies from day to day. If we can write down the prior distribution $f(p)$, then it is possible to determine the expected value of the risk function for each decision function. The expected risk corresponding to the estimator $\hat{P} = X/n$, often referred to as the **Bayes risk**, is written $B(\hat{P}) = E[R(\hat{P}; P)]$, where we are now treating the true proportion of defectives as a random variable. In general, when the unknown parameter is treated as a random variable with a prior distribution given by $f(\theta)$, the Bayes risk in estimating θ by means of the estimator $\hat{\Theta}$ is given by

$$B(\hat{\Theta}) = E[R(\hat{\Theta}; \Theta)] = \begin{cases} \sum_i R(\hat{\Theta}; \theta_i)f(\theta_1) & \text{(discrete case)} \\ \int_{-\infty}^{\infty} R(\hat{\Theta}; \theta)f(\theta)\, d\theta & \text{(continuous case)}. \end{cases}$$

The decision function $\hat{\Theta}$ that minimizes $B(\hat{\Theta})$ is called the **Bayes estimator** of θ. We shall make no attempt in this book to derive a Bayes estimator, but instead we shall employ the Bayes risk to establish a criterion for choosing between two estimators.

BAYES' CRITERION *Suppose that the parameter θ is a value of the random variable Θ and $f(\theta)$ is the value of its probability distribution at θ. If $\hat{\Theta}_1$ and $\hat{\Theta}_2$ are two estimators of θ and $B(\hat{\Theta}_1) < B(\hat{\Theta}_2)$, then $\hat{\Theta}_1$ is selected as the better estimator for θ.* ∎

The foregoing discussion on decision theory might be better understood if one considers the following two examples.

EXAMPLE 9.20   Suppose that a friend has three similar coins except for the fact that the first one has two heads, the second one has two tails, and the third one is honest. We wish to estimate which coin our friend is flipping on the basis of two flips of the coin. Let θ be the number of heads on the coin. Consider two decision functions $\hat{\Theta}_1$ and $\hat{\Theta}_2$, where $\hat{\Theta}_1$ is the estimator that assigns to θ the number of heads that occur when the coin is flipped twice and $\hat{\Theta}_2$ is the estimator that assigns the value of 1 to θ no matter what the experiment yields. If the loss function is of the form $L(\hat{\Theta}; \theta) = (\hat{\Theta} - \theta)^2$, which estimator is better according to the minimax procedure?

SOLUTION
For the estimator $\hat{\Theta}_1$, the loss function assumes the values $L(\hat{\theta}_1; \theta) = (\hat{\theta}_1 - \theta)^2$, where $\hat{\theta}_1$ may be 0, 1, or 2, depending on the true value of θ. Clearly, if $\theta = 0$ or 2, both flips will yield all tails or all heads and our decision will be a correct one.

Hence $L(0; 0) = 0$ and $L(2; 2) = 0$, from which one may easily conclude that $R(\hat{\Theta}_1; 0) = 0$ and $R(\hat{\Theta}_1; 2) = 0$. However, when $\theta = 1$ we could obtain 0, 1, or 2 heads in the two flips with probabilities 1/4, 1/2, and 1/4, respectively. In this case we have $L(0; 1) = 1$, $L(1; 1) = 0$, and $L(2; 1) = 1$, from which we find that

$$R(\hat{\Theta}_1; 1) = 1 \times \tfrac{1}{4} + 0 \times \tfrac{1}{2} + 1 \times \tfrac{1}{4} = \tfrac{1}{2}.$$

For the estimator $\hat{\Theta}_2$, the loss function assumes values given by $L(\hat{\theta}_2; \theta) = (\hat{\theta}_2 - \theta)^2 = (1 - \theta)^2$. Hence $L(1; 0) = 1$, $L(1; 1) = 0$, and $L(1; 2) = 1$, and the corresponding risks are $R(\hat{\Theta}_2; 0) = 1$, $R(\hat{\Theta}_2; 1) = 0$, and $R(\hat{\Theta}_2; 2) = 1$. Since the maximum risk is 1/2 for the estimator $\hat{\Theta}_1$ compared to a maximum risk of 1 for $\hat{\Theta}_2$, the minimax criterion selects $\hat{\Theta}_1$ as the better of the two estimators.

EXAMPLE 9.21   Referring to Example 9.20, let us suppose that our friend flips the honest coin 80% of the time and the other two coins each about 10% of the time. Does the Bayes criterion select $\hat{\Theta}_1$ or $\hat{\Theta}_2$ as the better estimator?

SOLUTION

The parameter $\hat{\Theta}$ may now be treated as a random variable with the following probability distribution:

θ	0	1	2
$f(\theta)$	0.1	0.8	0.1

For the estimator Θ_1, the Bayes risk is

$$B(\hat{\Theta}_1) = R(\hat{\Theta}_1; 0)f(0) + R(\hat{\Theta}_1; 1)f(1) + R(\hat{\Theta}_1; 2)f(2)$$
$$= (0)(0.1) + \tfrac{1}{2}(0.8) + (0)(0.1) = 0.4.$$

Similarly, for the estimator $\hat{\Theta}_2$, we have

$$B(\hat{\Theta}_2) = R(\hat{\Theta}_2; 0)f(0) + R(\hat{\Theta}_2; 1)f(1) + R(\hat{\Theta}_2; 2)f(2)$$
$$= (1)(0.1) + (0)(0.8) + (1)(0.1) = 0.2.$$

Since $B(\hat{\Theta}_2) < B(\hat{\Theta}_1)$, the Bayes criterion selects $\hat{\Theta}_2$ as the better estimator for the parameter θ.

Exercises

1. Estimate the proportion of defectives being produced by the machine in Example 9.16 if the random sample of size 2 yields 2 defectives.

2. Let us assume that the prior distribution for the proportion p of drinks from a vending machine that overflow is

p	0.05	0.10	0.15
$f(p)$	0.3	0.5	0.2

If 2 of the next 9 drinks from this machine overflow, find
(a) the posterior distribution for the proportion p;
(b) the Bayes estimate of p.

3. Repeat Exercise 2 when 1 of the next 4 drinks overflows and the uniform prior distribution is

$$f(p) = 10, \qquad 0.05 < p < 0.15.$$

4. The developer of a new condominium complex claims that 3 out of 5 buyers will prefer a two-

bedroom unit, while his banker claims that it would be more correct to say that 7 out of 10 buyers will prefer a two-bedroom unit. In previous predictions of this type the banker has been twice as reliable as the developer. If 12 of the next 15 condominiums sold in this complex are two-bedroom units, find
   (a) the posterior probabilities associated with the claims of the developer and banker;
   (b) a point estimate of the proportion of buyers who prefer a two-bedroom unit.

5. The burn time for the first stage of a rocket is a normal random variable with a standard deviation of 0.8 minute. Assume a normal prior distribution for μ with a mean of 8 minutes and a standard deviation of 0.2 minute. If 10 of these rockets are fired and the first stage has an average burn time of 9 minutes, find a 95% Bayesian interval for μ.

6. The daily profit from a cigarette vending machine placed in a restaurant is a value of a normal random variable with unknown mean μ and variance σ^2. Of course, the mean will vary somewhat from restaurant to restaurant, and the distributor feels that these average daily profits can best be described by a normal distribution with mean $\mu_0 = \$8.00$ and standard deviation $\sigma_0 = \$0.40$. If one of these cigarette machines, placed in a certain restaurant, showed an average daily profit of $\bar{x} = \$6.75$ during the first 30 days with a standard deviation of $s = \$1.20$, find
   (a) a Bayes estimate of the true average daily profit for this restaurant;
   (b) a 96% Bayesian interval of μ for this restaurant;
   (c) the probability that the average daily profit from the machine in this restaurant is between $6.59 and $7.12.

7. The mathematics department of a large university is designing a placement test to be given to the incoming freshman classes. Members of the department feel that the average grade for this test will vary from one freshman class to another. This variation of the average class grade is expressed subjectively by a normal distribution with mean $\mu_0 = 72$ and variance $\sigma_0^2 = 5.76$.
   (a) What prior probability does the department assign to the actual average grade being somewhere between 71.8 and 73.4 for next year's freshman class?
   (b) If the test is tried on a random sample of 100 freshman students from the next incoming freshman class resulting in an average grade of 70 with

a variance of 64, construct a 95% Bayesian interval for μ.
   (c) What posterior probability should the department assign to the event of part (a)?

8. Suppose that in Example 9.18 the electrical firm does not have enough prior information regarding the population mean length of life to be able to assume a normal distribution for μ. The firm believes, however, that μ is surely between 770 and 830 hours and it is felt that a more realistic Bayesian approach would be to assume the prior distribution

$$f(\mu) = \tfrac{1}{60}, \qquad 770 < \mu < 830.$$

If a random sample of 25 bulbs gives an average life of 780 hours, follow the steps of the proof for Theorem 9.6 to find the posterior distribution

$$f(\mu \mid x_1, x_2, \ldots , x_{25}).$$

9. Suppose that the time to failure T of a certain hinge is an exponential random variable with probability density

$$f(t) = \theta e^{-\theta t}, \qquad t > 0.$$

From prior experience we are led to believe that θ is a value of an exponential random variable with probability density

$$f(\theta) = 2e^{-2\theta}, \qquad \theta > 0.$$

If we have a sample of n observations on T, show that the posterior distribution of Θ is a gamma distribution with parameters

$$\alpha = n + 1 \quad \text{and} \quad \beta = 1 \bigg/ \left(\sum_{i=1}^{n} t_i + 2 \right).$$

10. We wish to estimate the binomial parameter p by the decision function $\widehat{P}$, the proportion of successes in a binomial experiment consisting of n trials. Find $R(\widehat{P}; p)$ when the loss function is of the form $L(\widehat{P}; p) = (\widehat{P} - p)^2$.

11. Suppose that an urn contains three balls, of which θ are red and the remainder black, where θ can vary from 0 to 3. We wish to estimate θ by selecting two balls in succession without replacement. Let $\widehat{\Theta}_1$ be the decision function that assigns to θ the value 0 if neither ball is red, the value 1 if the first ball only is red, the value 2 if the second ball only is red, and the value 3 if both balls are red. Using a loss function of the form $L(\widehat{\Theta}_1; \theta) = |\widehat{\Theta}_1 - \theta|$, find $R(\widehat{\Theta}_1; \theta)$.

12. In Exercise 11 consider the estimator $\hat{\Theta}_2 = X(X + 1)/2$, where X is the number of red balls in our sample. Find $R(\hat{\Theta}_2; \theta)$.

13. Use the minimax criterion to determine whether the estimator $\hat{\Theta}_1$ of Exercise 11 or the estimator $\hat{\Theta}_2$ of Exercise 12 is the better estimator.

14. Use the Bayes criterion to determine whether the estimator $\hat{\Theta}_1$ of Exercise 11 or the estimator $\hat{\Theta}_2$ of Exercise 12 is the better estimator, given the following additional information:

θ	0	1	2	3
$f(\theta)$	0.1	0.5	0.1	0.3

9.15 Maximum Likelihood Estimation

Often the estimators of parameters have been those that appeal to intuition. The estimator $\overline{X}$ certainly seems reasonable as an estimator of a population mean μ. The virtue of s^2 as an estimator of σ^2 was underscored through the discussion of unbiasedness in Section 9.3. The estimator for a binomial parameter p is merely a sample proportion, which of course is an *average* and appeals to common sense. But there are many situations in which it is not at all obvious what the proper estimator should be. As a result, there is much to be learned by the student in statistics concerning different philosophies that produce difference methods of estimation. In this section we deal with the **method of maximum likelihood**.

Maximum likelihood estimation represents one of the most important approaches to estimation in all of statistical inference. We will not give a thorough development of the method. Rather, we will attempt to communicate the philosophy of maximum likelihood and illustrate with examples that relate to other estimation problems discussed in this chapter.

The Likelihood Function

As the name implies, the method of maximum likelihood is that for which the *likelihood function* is maximized. The likelihood function is best illustrated through the use of an example with a discrete distribution and a single parameter. Let $X_1, X_2, \ldots, X_n$ denote independent random variables taken from a discrete probability distribution represented by $f(X, \theta)$, where θ is a single parameter of the distribution. Now

$$L(X_1, X_2, \ldots, X_n; \theta) = f(X_1, X_2, \ldots, X_n; \theta)$$
$$= f(X_1, \theta) \cdot f(X_2, \theta) \cdot \ldots \cdot f(X_n, \theta)$$

is the *joint distribution of the random variables*. This is often referred to as the **likelihood function**. Let $x_1, x_2, \ldots, x_n$ denote observed values in a sample. In the case of a discrete random variable the interpretation is very clear. The quantity $L(x_1, x_2, \ldots, x_n; \theta)$, *the likelihood of the sample*, is the following joint probability:

$$P(X_1 = x_1; X_2 = x_2; \ldots; X_n = x_n).$$

In other words, it is the *probability of obtaining the sample values* $x_1, x_2, \ldots, x_n$. For the discrete case the maximum likelihood estimator is one that results in a maximum value for this joint probability, or *maximizes* the likelihood of the sample.

Consider a fictitious example in which three items from an assembly line are inspected. The items are ruled either defective or nondefective and thus the Bernoulli process applies. Testing the three items results in two nondefective items followed by a defective item. It is of interest to estimate p, the proportion nondefective in the process. The likelihood of the sample for this illustration is given by

$$p \cdot p \cdot q = p^2 q$$
$$= p^2 - p^3,$$

where $q = 1 - p$. Maximum likelihood estimation would give an estimate of p for which the likelihood is maximized. It is clear that if we differentiate the likelihood with respect to p, set the derivative to zero, and solve, we obtain the value

$$\hat{p} = \tfrac{2}{3}.$$

Now, of course, in this situation $\hat{p} = 2/3$ is the sample proportion defective and is thus a reasonable estimator of the probability of a defective. The reader should attempt to understand that the philosophy of maximum likelihood estimation evolves from the notion that the reasonable estimator of a parameter based on sample information *is that parameter value that produces the largest probability of obtaining the sample*. This is, indeed, the interpretation for the discrete case, since the likelihood is the probability of jointly observing the values in the sample.

Now, while the interpretation of the likelihood function as a joint probability is confined to the discrete case, the notion of maximum likelihood extends to the estimation of parameters of a continuous distribution. We now present a formal definition of maximum likelihood estimation.

DEFINITION 9.5 *Given independent observations $x_1, x_2, \ldots, x_n$ from a probability density function (continuous case) or probability mass function (discrete case) $f(x, \theta)$, the maximum likelihood estimator, $\hat{\theta}$, is that which maximizes the likelihood function*

$$L(x_1, x_2, \ldots, x_n, \theta) = f(x_1, \theta) \cdot f(x_2, \theta) \cdot \ldots \cdot f(x_n, \theta). \quad \blacksquare$$

Quite often it is convenient to work with the natural log of the likelihood function in finding the maximum of the likelihood function. Consider the following example dealing with the parameter μ of a Poisson distribution.

EXAMPLE 9.22 Consider a Poisson distribution with probability mass function

$$f(x, \mu) = \frac{e^{-\mu}\mu^x}{x!}, \qquad x = 0, 1, 2, \ldots.$$

Suppose that a random sample $x_1, x_2, \ldots, x_n$ is taken from the distribution. What is the maximum likelihood estimate of μ?

SOLUTION

The likelihood function is given by

$$L(x_1, x_2, \ldots, x_n, \mu) = \prod_{i=1}^{n} f(x_i, \mu)$$

$$= \frac{e^{-n\mu}\mu^{\Sigma x_1}}{\prod_{i=1}^{n} x_i!}.$$

Now consider

$$\ln L(x_1, x_2, \ldots, x_n, \mu) = -n\mu + \sum_{i=1}^{n} x_i \ln \mu - \ln \prod_{i=1}^{n} x_i!$$

$$\frac{\partial \ln L(x_1, x_2, \ldots, x_n, \mu)}{\partial \mu} = -n + \sum_{i=1}^{n} \frac{x_i}{\mu}.$$

Solving for $\hat{\mu}$, the maximum likelihood estimator involves setting the derivative to zero and solving for the parameter. Thus

$$\hat{\mu} = \sum_{i=1}^{n} \frac{x_i}{n}$$

$$= \bar{x}.$$

Since μ is the mean of the Poisson distribution (Chapter 5), the sample average would certainly seem like a reasonable estimator.

The following example shows the use of the method of maximum likelihood for finding estimates of two parameters. We simply find the values of the parameters that maximize (jointly) the likelihood function.

EXAMPLE 9.23   Consider a random sample $x_1, x_2, \ldots, x_n$ from a normal distribution $N(\mu, \sigma)$. Find the maximum likelihood estimators for μ and σ^2.

SOLUTION

The likelihood function for the normal distribution is given by

$$L(x_1, x_2, \ldots, x_n, \mu, \sigma^2) = \frac{1}{(2\pi)^{n/2}(\sigma^2)^{n/2}} e^{\left[-1/2 \sum_{i=1}^{n} ((x_1 - \mu)\sigma)^2\right]}$$

Taking logs gives us

$$\ln L(x_1, x_2, \ldots, x_n, \mu, \sigma^2) = -\frac{n}{2} \ln 2\pi - \frac{n}{2} \ln \sigma^2 - \frac{1}{2} \sum_{i=1}^{n} \left(\frac{x_i - \mu}{\sigma}\right)^2$$

$$\frac{\partial \ln L}{\partial \mu} = \sum_{i=1}^{n} \frac{x_i - \mu}{\sigma}.$$

Setting the derivative to zero, one obtains

$$\sum_{i=1}^{n} x_i - n\mu = 0.$$

Thus the maximum likelihood estimator is given by

$$\hat{\mu} = \sum_{i=1}^{n} \frac{x_i}{n}$$

$$= \bar{x}.$$

This result is pleasing since $\bar{x}$ has played such an important role in this chapter as a point estimate of μ.

For the case of the estimator of σ^2, straightforward calculus and algebraic manipulation gives

$$\hat{\sigma}^2 = \sum_{i=1}^{n} \frac{(x_i - \bar{x})^2}{n}.$$

It is interesting to note the distinction between the maximum likelihood estimator of σ^2 and the unbiased estimator S^2 developed earlier in this chapter. The numerator is identical, of course, and the denominator is the "degrees of freedom" $n - 1$ for the unbiased estimator and n for the maximum likelihood estimator. Maximum likelihood estimators do not necessarily enjoy the property of unbiasedness. However, the maximum likelihood estimators do have very important asymptotic properties. This is discussed briefly in the section that follows.

Additional Comments Concerning Maximum Likelihood Estimation

A thorough discussion of the properties of maximum likelihood estimation is beyond the scope of this book and is usually a major topic in a course in the theory of statistical inference. The method of maximum likelihood allows the analyst to make use of knowledge of the distribution in determining an appropriate estimator. *The method of maximum likelihood cannot be applied without knowledge of the underlying distribution.* We learned in Example 9.23 that the maximum likelihood estimator is not necessarily unbiased. The maximum likelihood estimator is unbiased *in the limit*; that is, the amount of bias approaches zero as the sample size grows large. Earlier in this chapter the notion of efficiency was discussed, efficiency being linked to the variance property of an estimator. Maximum likelihood estimators possess desirable variance properties *in the limit*. The reader should consult Lehman or Bickel and Doksum for details.

Exercises

1. Consider the exponential distribution with density function given by

$$f(x) = \begin{cases} \dfrac{1}{\beta} e^{-x/\beta}, & x \geq 0 \\ 0, & \text{elsewhere.} \end{cases}$$

Suppose that a sample of independent observations $x_1, x_2, \ldots, x_n$ is taken from this distribution.

(a) What is the likelihood function?

(b) What function of the observations is the maximum likelihood estimator of β?

2. Suppose that there are n trials $x_1, x_2, \ldots, x_n$ from a Bernoulli process with parameter p, the probability of a success. That is, the probability of r successes is given by $\binom{n}{r} p^r q^{n-r}$. Work out the maximum likelihood estimator for the parameter p.

Review Exercises

1. Consider two estimators of σ^2 in sample $x_1, x_2, \ldots, x_n$, which is drawn from a normal distribution with mean μ and variance σ^2. The estimators are

$$s^2 = \sum_{i=1}^{n} \frac{(x_i - \bar{x})^2}{n-1} \quad \text{unbiased estimator}$$

$$\sigma^2 = \sum_{i=1}^{n} \frac{(x_i - \bar{x})^2}{n} \quad \begin{array}{l} \text{maximum likelihood} \\ \text{estimator.} \end{array}$$

Discuss the variance properties of these two estimators.

2. It is claimed that a new diet will reduce a person's weight by 4.5 kilograms on the average in a period of 2 weeks. The weights of 7 women who followed this diet were recorded before and after a 2-week period.

	Woman						
	1	2	3	4	5	6	7
Weight before	58.5	60.3	61.7	69.0	64.0	62.6	56.7
Weight after	60.0	54.9	58.1	62.1	58.5	59.9	54.4

Test a manufacturer's claim by computing a 95% confidence interval for the mean difference in the weight. Assume the differences of weights to be approximately normally distributed.

3. In a study to estimate the proportion of residents in a certain city and its suburbs who favor the construction

of a nuclear power plant, it is found that 168 of 400 urban residents favor the construction while only 145 of 500 suburban residents are in favor. Find a 95% confidence interval for the difference between the proportion of urban and suburban residents who favor construction of the nuclear plant.

4. A health spa claims that a new exercise program will reduce a person's waist size by 2 centimeters on the average over a 5-day period. The waist sizes of 6 men who participated in this exercise program are recorded before and after the 5-day period in the following table:

	Man					
	1	2	3	4	5	6
Waist size before	90.4	95.5	98.7	115.9	104.0	85.6
Waist size after	91.7	93.9	97.4	112.8	101.3	84.0

By computing a 95% confidence interval for the mean reduction in waist size, determine whether the health spa's claim is valid. Assume the distribution of differences of waist sizes before and after the program to be approximately normal.

5. The study "*Evaluation of Prescribed Burning in Relation to Available Deer Browse*" was undertaken at the Virginia Polytechnic Institute and State University in 1964 to determine if fire can be used as a

viable management tool to increase the amount of forage available to deer during the critical months in late winter and early spring. Calcium is a required element for plants and animals. The amount taken up and stored in the plant is closely correlated to the amount present in the soil. It was hypothesized that a fire may change the calcium levels present in the soil and thus affect the amount available to the deer. A large tract of land in the Fishburn Forest was selected for a prescribed burn. Soil samples were taken from 12 plots of equal area just prior to the burn on May 20, 1964, and analyzed for calcium. On July 16, 1964, postburn calcium levels were analyzed from the same plots. These values, in kilograms per plot, are presented in the following table:

Plot	Calcium Level (kg/plot)	
	Preburn	Postburn
1	50	9
2	50	18
3	82	45
4	64	18
5	82	18
6	73	9
7	77	32
8	54	9
9	23	18
10	45	9
11	36	9
12	54	9

Construct a 95% confidence interval for the mean difference in the calcium level that is present in the soil prior to and after the prescribed burn. Assume the distribution of differences of calcium levels to be approximately normal.

6. In the article *"A Study of a Modified Membrane Filter Technique for the Enumeration of Stressed Fecal Coliforms in Urban Runoff"* (1977), the Department of Civil Engineering at the Virginia Polytechnic Institute and State University compared a modified (M-5 hr) assay technique for recovering fecal coliforms in stormwater runoff from an urban area to a most probable number (MPN) technique. A total of 12 runoff samples were collected between May 25 and June 28, 1977, and analyzed by the two techniques. Fecal coliform counts per 100 milliliters are recorded in the following table:

Sample	MPN Count	M-5 hr Count
1	2300	2010
2	1200	930
3	450	400
4	210	436
5	270	4100
6	450	2090
7	154	219
8	179	169
9	192	194
10	230	174
11	340	274
12	194	183

Construct a 90% confidence interval for the difference in the mean fecal coliform counts between the M-5 hr and the MPN techniques. Assume that the count differences are approximately normally distributed.

7. An anthropologist is interested in the proportion of individuals in two Indian tribes with double occipital hair whorls. Suppose that independent samples are taken from each of the two tribes, and it is found that 24 of 100 Indians from tribe A and 36 of 120 Indians from tribe B possess this characteristic. Construct a 95% confidence interval for the difference $p_B - p_A$ between the proportions of these two tribes with occipital hair whorls.

8. An experiment was conducted to determine whether surface finish has an effect on the endurance limit of steel. An existing theory says that polishing increases the average endurance limit (reverse bending). An experiment was performed on 0.4% carbon steel using both unpolished and polished smooth-turned specimens. The finish on the smooth-turned polished specimens was obtained by polishing with No. 0 and No. 00 emery cloth. The data are as follows:

Endurance Limit (psi) for:	
Polished 0.4% Carbon	Unpolished 0.4% Carbon
85,500	82,600
91,900	82,400
89,400	81,700
84,000	79,500
89,900	79,400
78,700	69,800
87,500	79,900
83,100	83,400

From a practical point of view, polishing should not have any effect on the standard deviation of the endurance limit, which is known from the performance of numerous endurance limit experiments to be 4000 psi. Find a 95% confidence interval for the difference between the population means for the two methods, assuming that the populations are approximately normally distributed.

9. A manufacturer of electric irons produces these items in two plants. Both plants have the same suppliers of small parts. A saving can be made by purchasing thermostats for plant B from a local supplier. A single lot was purchased from the local supplier and it was desired to test whether or not these new thermostats were as accurate as the old. The thermostats were to be tested on the irons on the 550°F setting, and the actual temperatures were to be read to the nearest 0.1°F with a thermocouple. The data are as follows:

New supplier (°F):

530.3	559.3	549.4	544.0	551.7	566.3	549.9
556.9	536.7	558.8	538.8	543.3	559.1	555.0
538.6	551.1	565.4	554.9	550.0	554.9	554.7
536.1	569.1					

Old supplier (°F):

559.7	534.7	554.8	545.0	544.6	538.0	550.7
563.1	551.1	553.8	538.8	564.6	554.5	553.0
538.4	548.3	552.9	535.1	555.0	544.8	558.4
548.7	560.3					

Find a 95% confidence interval for σ_1^2/σ_2^2 and for σ_1/σ_2, where σ_1^2 and σ_2^2 are the population variances of the thermostat readings from the new and old suppliers, respectively.

10. It is argued that the resistance of wire A is greater than the resistance of wire B. An experiment on the wires shows the following results (in ohms):

Wire A	Wire B
0.140	0.135
0.138	0.140
0.143	0.136
0.142	0.142
0.144	0.138
0.137	0.140

Assuming equal variances, what conclusions do you draw? Justify your answer.

10

One- and Two-Sample Tests of Hypotheses

10.1 Statistical Hypotheses: General Concepts

Often, the problem confronting the scientist or engineer is not so much the estimation of a population parameter as discussed in Chapter 9, but rather the formation of a data-based decision procedure that can produce a conclusion about some scientific system. For example, a medical researcher may decide on the basis of experimental evidence whether or not coffee drinking increases the risk of cancer in humans; an engineer might have to decide on the basis of sample data whether there is a difference between the accuracy of two kinds of gauges; or a sociologist might wish to collect appropriate data to enable him or her to decide whether a person's blood type and eye color are independent variables. In each of these cases the scientist or engineer *postulates* or *conjectures* something about a system. In addition, each must involve the use of experimental data and decision making that is based on the data. Formally, in each case, the conjecture can be put in the form of a statistical hypothesis. Procedures that lead to the acceptance or rejection of statistical hypotheses such as these comprise a major area of statistical inference. First, let us define precisely what we mean by a **statistical hypothesis**.

DEFINITION 10.1   *A* **statistical hypothesis** *is an assertion or conjecture concerning one or more populations.*   ■

The truth or falsity of a statistical hypothesis is never known with absolute certainty unless we examine the entire population. This, of course, would be impractical in most situations. Instead, we take a random sample from the population of interest and use the data contained in this sample to provide evidence that either supports or does not support the hypothesis. Evidence from the sample that is inconsistent with the stated hypothesis leads to a rejection of the hypothesis, whereas evidence supporting the hypothesis leads to its acceptance.

It should be made clear to the reader that the design of a decision procedure must be done with the notion in mind of the *probability of a wrong conclusion*. For example, suppose that the conjecture (the hypothesis) postulated by the engineer is that the fraction defective p in a certain process is 0.10. The experiment is to observe a random sample of the product in question. Suppose that 100 items are tested and 12 items are found defective. It is reasonable to conclude that this evidence does not refute the condition $p = 0.10$, and thus it may lead to an acceptance of the hypothesis. However, it also does not refute $p = 0.12$ or perhaps even $p = 0.15$. As a result, the reader must be accustomed to understanding that **the acceptance of a hypothesis merely implies that the data do not give sufficient evidence to refute it. On the other hand, rejection implies that the sample evidence refutes it.** Put another way, **rejection means that there is a small probability of obtaining the sample information observed when, in fact, the hypothesis is true.** For example, in our proportion-defective hypothesis, a sample of 100 revealing 20 defective items is certainly evidence of rejection. Why? If, indeed, $p = 0.10$, the probability of obtaining 20 or more defectives is approximately 0.0035. With the resulting small risk of a *wrong conclusion*, it would seem safe to **reject the hypothesis** that $p = 0.10$. In other words, rejection of a hypothesis tends to all but "rule out" the hypothesis. On the other hand, it is very important to emphasize that the acceptance or, rather, failure to reject does not rule out other possibilities. As a result, the *firm conclusion is established by the data analyst when a hypothesis is rejected.*

The formal statement of a hypothesis is often influenced by the structure of the probability of a wrong conclusion. If the scientist is interested in *strongly supporting* a contention, he or she hopes to arrive at the contention in the form of rejection of a hypothesis. If the medical researcher wishes to show strong evidence in favor of the contention that coffee drinking increases the risk of cancer, the hypothesis tested should be of the form "there is no increase in cancer risk produced by drinking coffee." As a result, the contention is reached via a rejection. Similarly, to support the claim that one kind of gauge is more accurate than another, the engineer tests the hypothesis that there is no difference in the accuracy of the two kinds of gauges.

The Null and Alternative Hypotheses

The structure of hypothesis testing will be formulated with the use of the term **null hypothesis**. This refers to any hypothesis we wish to test and is denoted by H_0. The rejection of H_0 leads to the acceptance of an **alternative hypothesis**, denoted by H_1.

A null hypothesis concerning a population parameter will always be stated so as to specify an exact value of the parameter, whereas the alternative hypothesis allows for the possibility of several values. Hence, if H_0 is the null hypothesis $p = 0.5$ for a binomial population, the alternative hypothesis H_1 would be one of the following: $p > 0.5$, $p < 0.5$, or $p \neq 0.5$.

10.2 Testing a Statistical Hypothesis

To illustrate the concepts used in testing a statistical hypothesis about a population, consider the following example. A certain type of cold vaccine is known to be only 25% effective after a period of 2 years. To determine if a new and somewhat more expensive vaccine is superior in providing protection against the same virus for a longer period of time, suppose that 20 people are chosen at random and inoculated. In an actual study of this type the participants receiving the new vaccine might number several thousand. The number 20 is being used here only to demonstrate the basic steps in carrying out a statistical test. If more than 8 of those receiving the new vaccine surpass the 2-year period without contracting the virus, the new vaccine will be considered superior to the one presently in use. The requirement that the number exceed 8 is somewhat arbitrary but appears reasonable in that it represents a modest gain over the 5 people that could be expected to receive protection if the 20 people had been inoculated with the vaccine already in use. We are essentially testing the null hypothesis that the new vaccine is equally effective after a period of 2 years as the one now commonly used. The alternative hypothesis is that the new vaccine is in fact superior. This is equivalent to testing the hypothesis that the binomial parameter for the probability of a success on a given trial is $p = 1/4$ against the alternative that $p > 1/4$. This is usually written as follows:

$$H_0: \quad p = \tfrac{1}{4},$$

$$H_1: \quad p > \tfrac{1}{4}.$$

The **test statistic** on which we base our decision is X, the number of individuals in our test group who receive protection from the new vaccine for a period of at least 2 years. The possible values of X, from 0 to 20, are divided into two groups: those numbers less than or equal to 8 and those greater than 8. All possible scores greater than 8 constitute the **critical region**, and all possible scores less than or equal to 8 determine the **acceptance region**. The last number that we observe in passing from the acceptance region into the critical region is called the **critical value**. In our illustration the critical value is the number 8. Therefore, if $x > 8$, we reject H_0 in favor of the alternative hypothesis H_1. If $x \leq 8$, we accept H_0. This decision criterion is illustrated in Figure 10.1.

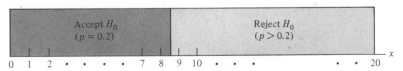

FIGURE 10.1   Decision criterion for testing $p = 0.2$ versus $p > 0.2$.

The decision procedure just described could lead to either of two wrong conclusions. For instance, the new vaccine may be no better than the one now in use and, for this particular randomly selected group of individuals, more than 8 surpass the 2-year period without contracting the virus. We would be committing an error by rejecting H_0 in favor of H_1 when, in fact, H_0 is true. Such an error is called a **type I error**.

DEFINITION 10.2   *Rejection of the null hypothesis when it is true is called a* **type I error**.   ■

A second kind of error is committed if 8 or fewer of the group surpass the 2-year period successfully and we conclude that the new vaccine is no better when it actually is better. In this case we would accept H_0 when it is false. This is called a **type II error**.

DEFINITION 10.3   *Acceptance of the null hypothesis when it is false is called a* **type II error**.   ■

In testing any statistical hypothesis, there are four possible situations that determine whether our decision is correct or in error. These four situations are summarized in Table 10.1.

The probability of committing a type I error, also called the **level of significance**, is denoted by the Greek letter α. In our illustration, a type I error will occur when more than 8 individuals surpass the 2-year period without contracting the virus using a new vaccine that is actually equivalent to the one in use. Hence, If X is the number of individuals who remain free of the virus for at least 2 years,

$$\alpha = P(\text{type I error})$$

$$= P(X > 8 \text{ when } p = \tfrac{1}{4})$$

$$= \sum_{x=9}^{20} b(x; 20, \tfrac{1}{4})$$

$$= 1 - \sum_{x=0}^{8} b(x; 20, \tfrac{1}{4})$$

$$= 1 - 0.9591$$

$$= 0.0409.$$

We say that the null hypothesis, $p = 1/4$, is being tested at the $\alpha = 0.0409$ level of significance. Sometimes the level of significance is called the **size** of the critical region. A critical region of size 0.0409 is very small and therefore it is unlikely that a type I error will be committed. Consequently, it would be most unusual for more than 8 individuals to remain immune to a virus for a 2-year period using a new vaccine that is essentially equivalent to the one now on the market.

The probability of committing a type II error, denoted by β, is impossible to compute unless we have a specific alternative hypothesis. If we test the null hypothesis that $p = 1/4$ against the alternative hypothesis that $p = 1/2$, then we are able to compute the probability of accepting H_0 when it is false. We simply find the prob-

TABLE 10.1   Possible Situations in Testing a Statistical
    Hypothesis

	H_0 *Is True*	H_0 *Is False*
Accept H_0	Correct decision	Type II error
Reject H_0	Type I error	Correct decision

ability of obtaining 8 or fewer in the group that surpass the 2-year period when $p = 1/2$. In this case

$$\beta = P(\text{type II error})$$

$$= P(X \leq 8 \text{ when } p = \tfrac{1}{2})$$

$$= \sum_{x=0}^{8} b(x; 20, \tfrac{1}{2})$$

$$= 0.2517.$$

This is a rather high probability, indicating a test procedure in which it is quite likely that we shall reject the new vaccine when, in fact, it is superior to that now in use. Ideally, we like to use a test procedure for which both the type I and type II errors are small.

It is possible that the director of the testing program is willing to make a type II error if the more expensive vaccine is not significantly superior. In fact the only time he wishes to guard against the type II error is when the true value of p is at least 0.7. If $p = 0.07$, this test procedure gives

$$\beta = P(\text{type II error})$$

$$= P(X \leq 8 \text{ when } p = 0.7)$$

$$= \sum_{x=0}^{8} b(x; 20, 0.7)$$

$$= 0.0051.$$

With such a small probability of committing a type II error, it is extremely unlikely that the new vaccine would be rejected when it is 70% effective after a period of 2 years. As the alternative hypothesis approaches unity, the value of β diminishes to zero.

Let us assume that the director of the testing program is unwilling to commit a type II error when the alternative hypothesis $p = 1/2$ is true even though we have found the probability of such an error to be $\beta = 0.2517$. A reduction in β is always possible by increasing the size of the critical region. For example, consider what happens to the values of α and β when we change our critical value to 7 so that all scores greater than 7 fall in the critical region and those less than or equal to 7 fall in the acceptance region. Now, in testing $p = 1/4$ against the alternative hypothesis that $p = 1/2$, we find that

$$\alpha = \sum_{x=8}^{20} b(x; 20, \tfrac{1}{4})$$

$$= 1 - \sum_{x=0}^{7} b(x; 20, \tfrac{1}{4})$$

$$= 1 - 0.8982$$

$$= 0.1018$$

and

$$\beta = \sum_{x=0}^{7} b(x; 20, \tfrac{1}{2})$$

$$= 0.1316.$$

By adopting a new decision procedure, we have reduced the probability of committing a type II error at the expense of increasing the probability of committing a type I error. For a fixed sample size, a decrease in the probability of one error will usually result in an increase in the probability of the other error. Fortunately, the probability of committing both types of error can be reduced by increasing the sample size. Consider the same problem using a random sample of 100 individuals. If more than 36 of the group surpass the 2-year period, we reject the null hypothesis that $p = 1/4$ and accept the alternative hypothesis that $p > 1/4$. The critical value is now 36. All possible scores above 36 constitute the critical region and all possible scores less than or equal to 36 fall in the acceptance region.

To determine the probability of committing a type I error, we shall use the normal-curve approximation with

$$\mu = np = (100)(\tfrac{1}{4}) = 25$$

and

$$\sigma = \sqrt{npq} = \sqrt{(100)(\tfrac{1}{4})(\tfrac{3}{4})} = 4.33.$$

Referring to Figure 10.2, we need the area under the normal curve to the right of $x = 36.5$. The corresponding z-value is

$$z = \frac{36.5 - 25}{4.33} = 2.66.$$

From Table A.3 we find that

$$\alpha = p(\text{type I error})$$

$$= P(X > 36 \text{ when } p = \tfrac{1}{4})$$

$$\approx P(Z > 2.66)$$

$$= 1 - P(Z < 2.66)$$

$$= 1 - 0.9961$$

$$= 0.0039.$$

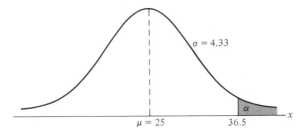

FIGURE 10.2 Probability of a type I error.

If H_0 is false and the true value of H_1 is $p = 1/2$, we can determine the probability of a type II error using the normal-curve approximation with

$$\mu = np = (100)(\tfrac{1}{2}) = 50$$

and

$$\sigma = \sqrt{npq} = \sqrt{(100)(\tfrac{1}{2})(\tfrac{1}{2})} = 5.$$

The probability of falling in the acceptance region when H_1 is true is given by the area of the shaded region to the left of $x = 36.5$ in Figure 10.3. The z-value corresponding to $x = 36.5$ is

$$z = \frac{36.5 - 50}{5} = -2.7.$$

Therefore,

$$\begin{aligned}
\beta &= P(\text{type II error}) \\
&= P(X \le 36 \text{ when } p = \tfrac{1}{2}) \\
&\approx P(Z < -2.7) \\
&= 0.0035.
\end{aligned}$$

Obviously, the type I and type II errors will rarely occur if the experiment consists of 100 individuals.

The illustration above underscores the strategy of the scientist in hypothesis testing. After the null and alternative hypotheses are stated, it is important to consider

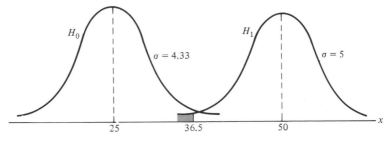

FIGURE 10.3 Probability of a type II error.

the sensitivity of the test procedure. By this we mean that there should be a determination, for a fixed α, of a reasonable value for the probability of wrongly accepting H_0 (i.e., the value of β) when the true situation represents some *important deviation from H_0*. The value of the sample size can usually be determined for which there is a reasonable balance between α and the value of β computed in this fashion. The vaccine problem is an illustration.

The concepts discussed here for a discrete population can equally well be applied to continuous populations. Consider the null hypothesis that the average weight of male students in a certain college is 68 kilograms against the alternative hypothesis that it is unequal to 68. That is, we wish to test

$$H_0: \quad \mu = 68,$$

$$H_1: \quad \mu \neq 68.$$

The alternative hypothesis allows for the possibility that $\mu < 68$ or $\mu > 68$.

A sample mean that falls close to the hypothesized value of 68 would be considered evidence in favor of H_0. On the other hand, a sample mean that is considerably less than or more than 68 would be evidence inconsistent with H_0 and therefore favoring H_1. The sample mean is the test statistic in this case. A critical region for the test statistic might arbitrarily be chosen to be the two intervals $\bar{x} < 67$ and $\bar{x} > 69$. The acceptance region will then be the interval $67 \leq \bar{x} \leq 69$. This decision criterion is illustrated in Figure 10.4.

Let us now use the decision criterion of Figure 10.4 to calculate the probabilities of committing type I and type II errors when testing the null hypothesis that $\mu = 68$ kilograms against the alternative that $\mu \neq 68$ kilograms for the continuous population of students' weights.

Assume the standard deviation of the population of weights to be $\sigma = 3.6$. For large samples we may substitute s for σ if no other estimate of σ is available. Our decision statistic, based on a random sample of size $n = 36$, will be $\bar{X}$, the most efficient estimator of μ. From the central limit theorem, we know that the sampling distribution of $\bar{X}$ is approximately normal with standard deviation $\sigma_{\bar{X}} = \sigma/\sqrt{n} = 3.6/6 = 0.6$.

The probability of committing a type I error, or the level of significance of our test, is equal to the sum of the areas that have been shaded in each tail of the distribution in Figure 10.5. Therefore,

$$\alpha = P(\bar{X} < 67 \text{ when } \mu = 68) + P(\bar{X} > 69 \text{ when } \mu = 68).$$

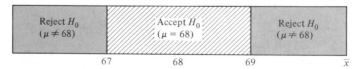

FIGURE 10.4   Probability of a type II error.

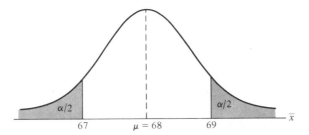

FIGURE 10.5   Critical region for testing $\mu = 68$ versus $\mu \neq 68$.

The z-values corresponding to $\bar{x}_1 = 67$ and $\bar{x}_2 = 69$ when H_0 is true are

$$z_1 = \frac{67 - 68}{0.6} = -1.67$$

and

$$z_2 = \frac{69 - 68}{0.6} = 1.67.$$

Therefore,

$$\alpha = P(Z < -1.67) + P(Z > 1.67)$$
$$= 2P(Z < -1.67)$$
$$= 0.0950.$$

Thus 9.5% of all samples of size 36 would lead us to reject $\mu = 68$ kilograms when it is true. To reduce α, we have a choice of increasing the sample size or widening the acceptance region. Suppose that we increase the sample size to $n = 64$. Then $\sigma_{\bar{X}} = 3.6/8 = 0.45$. Now

$$z_1 = \frac{67 - 68}{0.45} = -2.22$$

and

$$z_2 = \frac{69 - 68}{0.45} = 2.22.$$

Hence

$$\alpha = P(Z < -2.22) + P(Z > 2.22)$$
$$= 2P(Z < -2.22)$$
$$= 0.0264.$$

The reduction in α is not sufficient by itself to guarantee a good testing procedure. We must evaluate β for various alternative hypotheses that we feel should be accepted if true. Therefore, if it is important to reject H_0 when the true mean is some value $\mu \geq 70$ or $\mu \leq 66$, then the probability of committing a type II error should be computed and examined for the alternatives $\mu = 66$ and $\mu = 70$. Because of symmetry, it is only necessary to consider the probability of accepting the null hypothesis that $\mu = 68$ when the alternative $\mu = 70$ is true. A type II error will result when the sample mean $\bar{x}$ falls between 67 and 69 when H_1 is true. Therefore, referring to Figure 10.6, we find that

$$\beta = P(67 \leq \bar{X} \leq 69 \text{ when } \mu = 70).$$

The z-values corresponding to $\bar{x}_1 = 67$ and $\bar{x}_2 = 69$ when H_1 is true are

$$z_1 = \frac{67 - 70}{0.45} = -6.67$$

and

$$z_2 = \frac{69 - 70}{0.45} = -2.22.$$

Therefore,

$$\beta = P(-6.67 < Z < -2.22)$$
$$= P(Z < -2.22) - P(Z < -6.67)$$
$$= 0.0132 - 0.0000$$
$$= 0.0132.$$

If the true value of μ is the alternative $\mu = 66$, the value of β will again be 0.0132. For all possible values of $\mu < 66$ or $\mu > 70$, the value of β will be even smaller when $n = 64$, and consequently there would be little chance of accepting H_0 when it is false.

The probability of committing a type II error increases rapidly when the true value of μ approaches, but is not equal to, the hypothesized value. Of course, this is usually the situation where we do not mind making a type II error. For example, if the alternative hypothesis $\mu = 68.5$ is true, we do not mind committing a type II error by concluding that the true answer is $\mu = 68$. The probability of making such an error will be high when $n = 64$. Referring to Figure 10.7, we have

$$\beta = P(67 \leq \bar{X} \leq 69 \text{ when } \mu = 68.5).$$

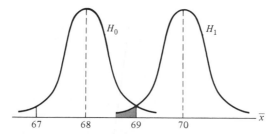

FIGURE 10.6   Type II error for testing $\mu = 68$ versus $\mu = 70$.

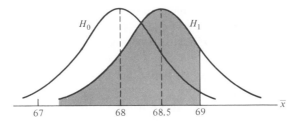

FIGURE 10.7   Type II error for testing $\mu = 68$ versus $\mu = 68.5$.

The z-values corresponding to $\bar{x}_1 = 67$ and $\bar{x}_2 = 69$ when $\mu = 68.5$ are

$$z_1 = \frac{67 - 68.5}{0.45} = -3.33 \quad \text{and} \quad z_2 = \frac{69 - 68.5}{0.45} = 1.11.$$

Therefore,

$$\begin{aligned}
\beta &= P(-3.33 < Z < 1.11) \\
&= P(Z < 1.11) - P(Z < -3.33) \\
&= 0.8665 - 0.0004 \\
&= 0.8661.
\end{aligned}$$

The preceding examples illustrate the following important properties:

1. The type I error and type II error are related. A decrease in the probability of one generally results in an increase in the probability of the other.
2. The size of the critical region, and therefore the probability of committing a type I error, can always be reduced by adjusting the critical value(s).
3. An increase in the sample size n will reduce α and β simultaneously.
4. If the null hypothesis is false, β is a maximum when the true value of a parameter approaches the hypothesized value. The greater the distance between the true value and the hypothesized value, the smaller β will be.

One very important concept that relates to error probabilities is the notion of the power of a test.

DEFINITION 10.4   *The **power** of a test is the probability of rejecting H_0 given that a specific alternative is true.*   ■

The power of a test can be computed as $1 - \beta$. Often different types of tests are compared by contrasting power properties. Consider the previous illustration in which we were testing H_0: $\mu = 68$ and H_1: $\mu \neq 68$. As before, suppose we are interested in assessing the sensitivity of the test. The test is governed by the rule that we accept if $67 \leq \bar{x} \leq 69$. We seek the capability of the test for properly rejecting H_0 when indeed $\mu = 68.5$. We have seen that the probability of a type II error is given by $\beta = 0.8661$. Thus the **power** of the test is $1 - 0.8661 = 0.1339$. In a sense, the power is a more succinct measure of how sensitive the test is for "detecting differences'' between a mean of 68 and 68.5. In this case, if μ is truly 68.5, the test as described will *properly reject H_0 only 13.39% of the time*. As a result, the test would not be a good one if it is important that the analyst have a reasonable chance of truly distinguishing between a mean of 68.0 (specified by H_0) and a mean of 68.5. From the foregoing, it is clear that to produce a desirable power (say, greater than 0.8), one must either increase α or increase the sample size.

In what has preceded in this chapter, much of the text on hypothesis testing revolves around foundations and definitions. In the sections that follow we get more specific and put hypotheses in categories as well as discuss tests of hypotheses on various parameters of interest. We begin by drawing the distinction between a one-sided and a two-sided hypothesis.

10.3 One- and Two-Tailed Tests

A test of any statistical hypothesis, where the alternative is **one-sided**, such as

$$H_0: \quad \theta = \theta_0,$$
$$H_1: \quad \theta > \theta_0,$$

or perhaps

$$H_0: \quad \theta = \theta_0,$$
$$H_1: \quad \theta < \theta_0,$$

is called a **one-tailed test**.

In Section 10.2, we made reference to the **test statistic** for a hypothesis. Generally, the critical region for the alternative hypothesis $\theta > \theta_0$ lies in the right tail of the distribution of the test statistic, while the critical region for the alternative hypothesis $\theta < \theta_0$ lies entirely in the left tail. In a sense, the inequality symbol points in the direction where the critical region lies. A one-tailed test was used in the vaccine experiment of Section 10.2 to test the hypothesis $p = 1/4$ against the one-sided alternative $p > 1/4$ for the binomial distribution. The one-tailed critical region is usually obvious. For an understanding the reader should visualize the behavior of the test statistic and notice the obvious *signal* that would produce evidence supporting the alternative hypothesis.

A test of any statistical hypothesis where the alternative is **two-sided**, such as

$$H_0: \quad \theta = \theta_0,$$
$$H_1: \quad \theta \neq \theta_0,$$

is called a **two-tailed test**, since the critical region is split into two parts, often having equal probabilities placed in each tail of the distribution of the test statistic. The alternative hypothesis $\theta \neq \theta_0$ states that either $\theta < \theta_0$ or $\theta > \theta_0$. A two-tailed test was used to test the null hypothesis that $\mu = 68$ kilograms against the two-sided alternative $\mu \neq 68$ kilograms for the continuous population of student weights in Section 10.2.

The null hypothesis, H_0, will always be stated using the equality sign so as to specify a single value. In this way the probability of committing a type I error can be controlled. Whether one sets up a one-tailed or a two-tailed test will depend on the conclusion to be drawn if H_0 is rejected. The location of the critical region can be determined only after H_1 has been stated. For example, in testing a new drug, one sets up the hypothesis that it is no better than similar drugs now on the market and tests this against the alternative hypothesis that the new drug is superior. Such an alternative hypothesis will result in a one-tailed test with the critical region in the right tail. However, if we wish to compare a new teaching technique with the conventional classroom procedure, the alternative hypothesis should allow for the new approach to be either inferior or superior to the conventional procedure. Hence the test is two-tailed with the critical region divided equally so as to fall in the extreme left and right tails of the distribution of our statistic.

Certain guidelines are desirable in determining which hypothesis should be stated as H_0 and which should be stated as H_1. First, read the problem carefully and determine the claim that you want to test. Should the claim suggest a simple direction such as *more than*, *less than*, *superior to*, *inferior to*, and so on, then H_1 will be stated using the inequality symbol ($<$ or $>$) corresponding to the suggested direction. If, for example, in testing a new drug we wish to show strong evidence that *more than* 30% of the people will be helped, we immediately write $H_1: p > 0.3$ and then the null hypothesis is written $H_0: p = 0.3$. Should the claim suggest a compound direction (equality as well as direction) such as *at least*, *equal to or greater*, *at most*, *no more than*, and so on, then this entire compound direction ($\leq$ or $\geq$) is expressed as H_0, but using only the equality sign, and H_1 is given by the opposite direction. Finally, if no direction whatsoever is suggested by the claim, then H_1 is stated using the *not equal* symbol ($\neq$).

EXAMPLE 10.1   The manufacturer of a certain brand of cigarettes claims that the average nicotine content does not exceed 2.5 milligrams. State the null and alternative hypotheses to be used in testing this claim and determine where the critical region is located.

SOLUTION
The manufacturer's claim should be rejected only if μ is greater than 2.5 milligrams and should be accepted if μ is less than or equal to 2.5 milligrams. Since the null hypothesis always specifies a single value of the parameter, we test

$$H_0: \quad \mu = 2.5,$$
$$H_1: \quad \mu > 2.5.$$

Although we have stated the null hypothesis with an equal sign, it is understood to include any value not specified by the alternative hypothesis. Consequently, the acceptance of H_0 does not imply that μ is exactly equal to 2.5 milligrams but rather that we do not have sufficient evidence favoring H_1. Since we have a one-tailed test, the greater than symbol indicates that the critical region lies entirely in the right tail of the distribution of our test statistic $\overline{X}$.

EXAMPLE 10.2   A real estate agent claims that 60% of all private residences being built today are 3-bedroom homes. To test this claim, a large sample of new residences is inspected;

the proportion of these homes with 3 bedrooms is recorded and used as our test statistic. State the null and alternative hypotheses to be used in this test and determine the location of the critical region.

SOLUTION
If the test statistic is substantially higher or lower than $p = 0.6$, we would reject the agent's claim. Hence we should make the test

$$H_0: \quad p = 0.6,$$

$$H_1: \quad p \neq 0.6.$$

The alternative hypothesis implies a two-tailed test with the critical region divided equally in both tails of the distribution of $\hat{P}$, our test statistic.

10.4 The Use of P-Values in Decision Making

In testing hypotheses in which the test statistic is discrete, the critical region may be chosen arbitrarily and its size determined. If α is too large, it can be reduced by making an adjustment in the critical value. It may be necessary to increase the sample size to offset the decrease that occurs automatically in the power of the test.

Over a number of generations of statistical analysis, it had become customary to choose an α of 0.05 or 0.01 and select the critical region accordingly. Then, of course, strict rejection or nonjection of H_0 would depend on that critical region. For example, if the test is two-tailed and α is set at the 0.05 level of significance and the test statistic involves, say, the standard normal distribution, then a z-value is observed from the data and the critical region is

$$z > 1.96; \qquad z < -1.96,$$

where the value 1.96 is found as $z_{0.025}$ in Table A.3. A value of z in the critical region prompts the statement: "The value of the test statistic is significant." We can translate that into the user's language. For example, if the hypothesis is given by

$$H_0: \quad \mu = 10,$$

$$H_1: \quad \mu \neq 10,$$

one might say: "The mean differs significantly from the value 10."

This preselection of a significance level α has its roots in the philosophy that the maximum risk of making a type I error should be controlled. However, this approach does not account for values of test statistics that are "close" to the critical region. Suppose, for example, in the illustration with $H_0: \mu = 10$; $H_1: \mu \neq 10$, a value of $z = 1.87$ is observed; strictly speaking, with $\alpha = 0.05$ the value is not significant. But the risk of committing a type I error if one rejects H_0 in this case could hardly be considered severe. In fact, in a **two-tailed** scenario one can quantify this risk as

$$P = 2P(z > 1.87 \text{ when } \mu = 10)$$

$$= 2(0.0307)$$

$$= 0.0614.$$

As a result, 0.0614 is the probability of obtaining a value of z as large or larger (in magnitude) than 1.87 when in fact $\mu = 10$. Although this evidence against H_0 is not as strong as that which would result from a rejection at an $\alpha = 0.05$ level, it is important information to the user. Indeed, continued use of $\alpha = 0.05$ or 0.01 is only a result of what standards have been passed through the generations. The **P-value** approach has been adopted extensively by users in applied statistics. The approach is designed to give the user an alternative (in terms of a probability) to a mere "reject" or "do not reject" conclusion. The P-value computation also gives the user important information when the z-value falls *well into the ordinary critical region*. For example, if z is 2.73, it is informative for the user to observe that

$$P = 2(0.0032)$$

$$= 0.0064$$

and thus the z-value is significant at a level considerably less than 0.05. It is important to know that under the condition of H_0, a value of $z = 2.73$ is an extremely rare event. Namely, a value at least that large in magnitude would only occur 64 times in 10,000 experiments.

One very simple way of explaining a P-value graphically is to consider two distinct samples prematurely. Suppose that two materials are considered for coating a particular type of metal in order to inhibit corrosion. Specimens are obtained and one collection is coated with material 1 and one collection coated with material 2. The sample sizes are $n_1 = n_2 = 10$ for each sample and corrosion was measured in percent of surface area affected. The hypothesis is that the samples came from common distributions with mean $\mu = 10$. Let us assume that the population variance is 1.0. Then we are testing

$$H_0: \quad \mu_1 = \mu_2 = 10.$$

Let Figure 10.8 represent a point plot of the data; the data are placed on the distribution stated by the null hypothesis. Now it seems clear that the data do refute the null hypothesis. But how can this be summarized in one number? The P-value can be viewed as simply the probability of obtaining this data set *given* that the samples come from the distribution depicted. Clearly, this probability is quite small, say 0.00000001! Thus the small P-value H_0, and the conclusion is that the population means are significantly different.

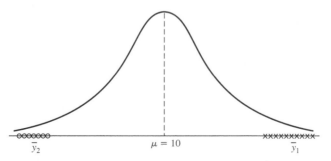

FIGURE 10.8   Data that are likely generated from populations with two different means.

The *P*-value approach as an aid in decision making is quite natural because nearly all computer packages that provide hypothesis-testing computation print out *P*-values along with values of the appropriate test statistic. The following is a formal definition of a *P*-value.

DEFINITION 10.5   *A P-value is the lowest level (of significance) at which the observed value of the test statistic is significant.* ◼

It might be appropriate at this point to summarize the procedures for hypothesis testing. This may serve as a foundation on which special cases are based in succeeding sections. For this summary, assume that the hypothesis is H_0: $\theta = \theta_0$.

1. State the null hypothesis H_0 that $\theta = \theta_0$.
2. Choose an appropriate alternative hypothesis H_1 from one of the alternatives $\theta < \theta_0$, $\theta > \theta_0$, or $\theta \neq \theta_0$.
3. Choose a significance level of size α.
4. Select the appropriate test statistic and establish the critical region. (If the decision is to be based on a *P*-value, it is not necessary to state the critical region.)
5. Compute the value of the test statistic from the sample data.
6. Decision: Reject H_0 if the test statistic has a value in the critical region (or if the computed *P*-value is less than or equal to the desired significance level α); otherwise, do not reject H_0.

The reader should realize that the conclusions drawn by the analyst may be affected by computed *P*-values. In other words, one may not have a preselected α level in mind and thus draw conclusions based on the information provided by the *P*-value. As indicated earlier, this is the approach often taken in real-life situations.

Exercises

1. Suppose that an allergist wishes to test the hypothesis that at least 30% of the public is allergic to some cheese products. Explain how the allergist could commit
   (a) a type I error;
   (b) a type II error.

2. A sociologist is concerned about the effectiveness of a training course designed to get more drivers to use seat belts in automobiles.
   (a) What hypothesis is she testing if she commits a type I error by erroneously concluding that the training course is ineffective?
   (b) What hypothesis is she testing if she commits a type II error by erroneously concluding that the training course is effective?

3. A large manufacturing firm is being charged with discrimination in its hiring practices.

   (a) What hypothesis is being tested if a jury commits a type I error by finding the firm guilty?
   (b) What hypothesis is being tested if a jury commits a type II error by finding the firm guilty?

4. The proportion of adults living in a small town who are college graduates is estimated to be $p = 0.3$. To test this hypothesis, a random sample of 15 adults is selected. If the number of college graduates in our sample is anywhere from 2 to 7, we shall accept the null hypothesis that $p = 0.3$; otherwise, we shall conclude that $p \neq 0.3$.
   (a) Evaluate α assuming that $p = 0.3$.
   (b) Evaluate β for the alternative $p = 0.2$ and $p = 0.4$.
   (c) Is this a good test procedure?

5. Repeat Exercise 4 when 200 adults are selected and the acceptance region is defined to be $48 \leq x \leq 72$,

where x is the number of college graduates in our sample.

6. The proportion of families buying milk from company A in a certain city is believed to be $p = 0.6$. If a random sample of 10 families shows that 3 or less buy milk from company A, we shall reject the hypothesis that $p = 0.6$ in favor of the alternative $p < 0.6$.
   (a) Find the probability of committing a type I error if the true proportion is $p = 0.6$.
   (b) Find the probability of committing a type II error for the alternatives $p = 0.3$, $p = 0.4$, and $p = 0.5$.

7. Repeat Exercise 6 when 50 families are selected and the critical region is defined to be $x \leq 24$, where x is the number of families in our sample that buy milk from company A.

8. A dry cleaning establishment claims that a new spot remover will remove more than 70% of the spots to which it is applied. To check this claim, the spot remover will be used on 12 spots chosen at random. If fewer than 11 of the spots are removed, we shall accept the null hypothesis that $p = 0.7$; otherwise, we conclude that $p > 0.7$.
   (a) Evaluate α, assuming that $p = 0.7$.
   (b) Evaluate β for the alternative $p = 0.9$.

9. Repeat Exercise 8 when 100 spots are treated and the critical region is defined to be $x > 82$, where x is the number of spots removed.

10. In the publication *Relief from Arthritis* by Thorsons Publishers, Ltd. (1979), John E. Croft claims that over 40% of the sufferers from osteoarthritis received measurable relief from an ingredient produced by a particular species of mussel found off the coast of New Zealand. To test this claim, the mussel extract is to be given to a group of 7 osteoarthritic patients. If 3 or more of the patients receive relief, we shall accept the null hypothesis that $p = 0.4$; otherwise, we conclude that $p < 0.4$.
    (a) Evaluate α, assuming that $p = 0.4$.
    (b) Evaluate β for the alternative $p = 0.3$.

11. Repeat Exercise 10 when 70 patients are given the mussel extract and the critical region is defined to be $x < 24$, where x is the number of osteoarthritic patients who receive relief.

12. A random sample of 400 voters in a certain city are asked if they favor an additional 4% gasoline sales tax to provide badly needed revenues for street repairs. If

more than 220 but fewer than 260 favor the sales tax, we shall conclude that 60% of the voters are for it.
   (a) Find the probability of committing a type I error if 60% of the voters favor the increased tax.
   (b) What is the probability of committing a type II error using this test procedure if actually only 48% of the voters are in favor of the additional gasoline tax?

13. Suppose, in Exercise 12, we conclude that 60% of the voters favor the gasoline sales tax if more than 214 but fewer than 266 voters in our sample favor it. Show that this new acceptance region results in a smaller value for α at the expense of increasing β.

14. A manufacturer has developed a new fishing line, which he claims has a mean breaking strength of 15 kilograms with a standard deviation of 0.5 kilogram. To test the hypothesis that $\mu = 15$ kilograms against the alternative that $\mu < 15$ kilograms, a random sample of 50 lines will be tested. The critical region is defined to be $\bar{x} < 14.9$.
   (a) Find the probability of committing a type I error when H_0 is true.
   (b) Evaluate β for the alternatives $\mu = 14.8$ and $\mu = 14.9$ kilograms.

15. A soft-drink machine at the Longhorn Steak House is regulated so that the amount of drink dispensed is approximately normally distributed with a mean of 200 milliliters and a standard deviation of 15 milliliters. The machine is checked periodically by taking a sample of 9 drinks and computing the average content. If $\bar{x}$ falls in the interval $191 < \bar{x} < 209$, the machine is thought to be operating satisfactorily; otherwise, we conclude that $\mu \neq 200$ milliliters.
   (a) Find the probability of committing a type I error when $\mu = 200$ milliliters.
   (b) Find the probability of committing a type II error when $\mu = 215$ milliliters.

16. Repeat Exercise 15 for samples of size $n = 25$. Use the same critical region.

17. A new cure has been developed for a certain type of cement that results in a compressive strength of 5000 kilograms per square centimeter and a standard deviation of 120. To test the hypothesis that $\mu = 5000$ against the alternative that $\mu < 5000$, a random sample of 50 pieces of cement is tested. The critical region is defined to be $\bar{x} < 4970$.
   (a) Find the probability of committing a type I error when H_0 is true.

(b) Evaluate β for the alternatives $\mu = 4970$ and $\mu = 4960$.

18. If we plot the probabilities of accepting H_0 corresponding to various alternatives for μ (including the value specified by H_0) and connect all the points by a smooth curve, we obtain the **operating characteristic curve** of the test criterion, or simply the **OC curve**. Note that the probability of accepting H_0 when

it is true is simply $1 - \alpha$. Operating characteristic curves are widely used in industrial applications to provide a visual display of the merits of the test criterion. With reference to Exercise 15, find the probabilities of accepting H_0 for the following 9 values of μ and plot the OC curve: 184, 188, 192, 196, 200, 204, 208, 212, and 216.

10.5 Single Sample: Tests Concerning a Single Mean (Variance Known)

In this section we consider formally tests of hypotheses on a single population mean. Many of the illustrations from previous sections involved tests on the mean, so the reader should already have insight into some of the details that are outlined here. We should first describe the assumptions on which the experiment is based. The model for the underlying situation centers around an experiment with $X_1, X_2, \ldots, X_n$ representing a random sample from a distribution with mean μ and variance $\sigma^2 > 0$. Consider first the hypothesis

$$H_0: \quad \mu = \mu_0,$$

$$H_1: \quad \mu \neq \mu_0.$$

The appropriate test statistic should be based on the random variable $\overline{X}$. In Chapter 8, the central limit theorem was introduced which essentially stated that despite the distribution of X, the random variable $\overline{X}$ has approximately a normal distribution with mean μ and variance σ^2/n. So, $\mu_{\overline{x}} = \mu$ and $\sigma_{\overline{X}}^2 = \sigma^2/n$. We can then determine a critical region based on the computed sample average, $\overline{x}$. It should be clear to the reader by now that there will be a two-tailed critical region for the test.

It is convenient to standardize $\overline{X}$ and formally involve the **standard normal** random variable Z, where

$$Z = \frac{\overline{X} - \mu}{\sigma/\sqrt{n}}.$$

We know that *under H_0*, that is, if $\mu = \mu_0$, then $(\overline{X} - \mu_0)/\sigma/\sqrt{n}$ has an $N(0, 1)$ distribution, and hence the expression

$$P\left(-z_{\alpha/2} < \frac{\overline{X} - \mu_0}{\sigma/\sqrt{n}} < z_{\alpha/2}\right) = 1 - \alpha$$

can be used to write an appropriate acceptance region. The reader should keep in mind that, formally, the critical region is designed to control α, the probability of type I error. It should be obvious that a *two-tailed signal* of evidence is needed to support H_1. Thus, given a computed value $\overline{x}$, the formal test involves rejecting H_0 if the computed *test statistic* $z = (\overline{x} - \mu_0)/\sigma/\sqrt{n} > z_{\alpha/2}$ or $z < -z_{\alpha/2}$. If $-z_{\alpha/2} < z <$

$z_{\alpha/2}$, do not reject H_0. Rejection of H_0, of course, implies acceptance of the alternative hypothesis $\mu \neq \mu_0$. With this definition of the critical region it should be clear that there will be probability α of rejecting H_0 (falling into the critical region) when, indeed, $\mu = \mu_0$.

Although it is easier to understand the critical region written in terms of z, we write the same critical region in terms of the computed average $\bar{x}$. The following can be written as an identical decision procedure:

$$\text{reject } H_0 \text{ if } \bar{x} > b \text{ or } \bar{x} < a,$$

where

$$a = \mu_0 - z_{\alpha/2}\frac{\sigma}{\sqrt{n}}$$

$$b = \mu_0 + z_{\alpha/2}\frac{\sigma}{\sqrt{n}}.$$

Hence, for an α level of significance, the critical values of the random variable z and $\bar{x}$ are both depicted in Figure 10.9.

Tests of one-sided hypotheses on the mean involve the same statistic described in the two-sided case. The difference, of course, is that the critical region is only in one tail of the standard normal distribution. As a result, for example, suppose that we seek to test

$$H_0: \quad \mu = \mu_0,$$

$$H_1: \quad \mu > \mu_0.$$

The signal that favors H_1 comes from *large values* of z. Thus rejection of H_0 results when the computed $z > z_\alpha$. Obviously, if the alternative is $H_1: \mu < \mu_0$, the critical region is entirely in the lower tail and thus rejection results from $z < -z_\alpha$.

The following two examples illustrate tests on means for the case in which σ is known.

EXAMPLE 10.3   A random sample of 100 recorded deaths in the United States during the past year showed an average life span of 71.8 years. Assuming a population standard deviation of 8.9 years, does this seem to indicate that the average life span today is greater than 70 years? Use a 0.05 level of significance.

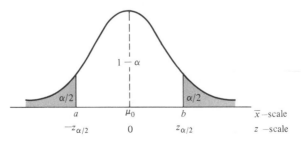

FIGURE 10.9 Critical region for the alternative hypothesis $\mu \neq \mu_0$.

SOLUTION

1. H_0: $\mu = 70$ years.
2. H_1: $\mu > 70$ years.
3. $\alpha = 0.05$.
4. Critical region: $z > 1.645$, where

$$z = \frac{\bar{x} - \mu_0}{\sigma/\sqrt{n}}.$$

5. Computations: $\bar{x} = 71.8$ years, $\sigma = 8.9$ years, and

$$z = \frac{71.8 - 70}{8.9/\sqrt{100}} = 2.02.$$

6. Decision: Reject H_0 and conclude that the average life span today is greater than 70 years.

In Example 10.3 the P-value corresponding to $z = 2.02$ is given by the area of the shaded region in Figure 10.10.
Using Table A.3, we have

$$P = P(Z > 2.02) = 0.0217.$$

As a result, the evidence in favor of H_1 is even stronger than that suggested by a 0.05 level of significance.

EXAMPLE 10.4   A manufacturer of sports equipment has developed a new synthetic fishing line that he claims has a mean breaking strength of 8 kilograms with a standard deviation of 0.5 kilogram. Test the hypothesis that $\mu = 8$ kilograms against the alternative that $\mu \neq 8$ kilograms if a random sample of 50 lines is tested and found to have a mean breaking strength of 7.8 kilograms. Use a 0.01 level of significance.

SOLUTION

1. H_0: $\mu = 8$ kilograms.
2. H_1: $\mu \neq 8$ kilograms.
3. $\alpha = 0.01$.
4. Critical region: $z < -2.575$ and $z > 2.575$, where

$$z = \frac{\bar{x} - \mu_0}{\sigma/\sqrt{n}}.$$

5. Computations: $\bar{x} = 7.8$ kilograms, $n = 50$, and hence

$$z = \frac{7.8 - 8}{0.5/\sqrt{50}} = -2.83.$$

6. Decision: Reject H_0 and conclude that the average breaking strength is not equal to 8 but is, in fact, less than 8 kilograms.

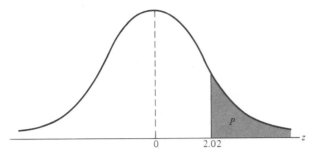

FIGURE 10.10   *P*-value for Example 8.3.

Since the test in Example 10.4 is two-tailed, the desired *P*-value is twice the area of the shaded region in Figure 10.11 to the left of $z = -2.83$. Therefore, using Table A.3, we have

$$P = P(|Z| > 2.83) = 2P(Z < -2.83) = 0.0046,$$

which allows us to reject the null hypothesis that $\mu = 8$ kilograms at a level of significance smaller than 0.01.

10.6 Relationship to Confidence Interval Estimation

The reader should realize by now that the hypothesis-testing approach to statistical inference in this chapter is very closely related to the confidence interval approach in Chapter 9. Confidence interval estimation involves computation of bounds for which it is "reasonable" that the parameter in question is inside the bounds. For the case of a single population mean μ with σ^2 known, the structure of both hypothesis testing and confidence interval estimation is based on the random variable

$$Z = \frac{\overline{X} - \mu}{\sigma/\sqrt{n}}.$$

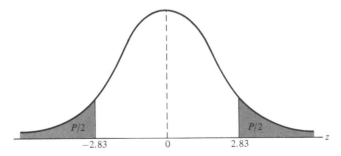

FIGURE 10.11   *P*-value for Example 8.4.

It turns out that the testing of H_0: $\mu = \mu_0$ against H_1: $\mu \neq \mu_0$ at a significance level α is *equivalent to computing a* $100(1 - \alpha)\%$ *confidence interval on* μ *and rejecting H_0 if μ_0 is not inside the confidence interval*. If μ_0 is inside the confidence interval, the hypothesis is not rejected. The equivalence is very intuitive and quite simple to illustrate. Recall that with an observed value $\bar{x}$ failure to reject H_0 at significance level α implies that

$$-z_{\alpha/2} \leq \frac{\bar{x} - \mu_0}{\sigma/\sqrt{n}} \leq z_{\alpha/2},$$

which is equivalent to

$$\bar{x} - z_{\alpha/2}\frac{\sigma}{\sqrt{n}} \leq \mu_0 \leq \bar{x} + z_{\alpha/2}\frac{\sigma}{\sqrt{n}}.$$

The confidence interval equivalence to hypothesis testing extends to differences between two means, variances, ratios of variances, and so on. As a result, the student of statistics should not consider confidence interval estimation and hypothesis testing as separate forms of statistical inference. For example, consider Example 9.2. The 95% confidence interval on the mean is given by the bounds [2.50, 2.70]. Thus with the same sample information, a two-sided hypothesis on μ involving any hypothesized value between 2.50 and 2.70 will not be rejected. As we turn to different areas of hypothesis testing, the equivalence to the confidence interval estimation will continue to be exploited.

10.7 Single Sample: Tests on a Single Mean (Variance Unknown)

One would certainly suspect that tests on a population mean μ with σ^2 unknown, like confidence interval estimation, should involve the use of Student's t-distribution. Strictly speaking, the application of Student's t for both confidence intervals and hypothesis testing is developed under the following assumptions. The random variables $X_1, X_2, \ldots, X_n$ represent a random sample from a normal distribution with unknown μ and σ^2. Then the random variable $\sqrt{n}(\bar{X} - \mu)/S$ has a Student's t-distribution with $n - 1$ degrees of freedom. The structure of the test is identical to that for the case of σ known with the exception that the value σ in the test statistic is replaced by the computed estimate S and the standard normal distribution is replaced by a t-distribution. As a result, for the two-sided hypothesis

$$H_0: \quad \mu = \mu_0,$$

$$H_1: \quad \mu \neq \mu_0$$

rejection of H_0 at significance level α results when a computed t-statistic

$$t = \frac{\bar{x} - \mu_0}{s/\sqrt{n}}$$

exceeds $t_{\alpha/2, n-1}$ or is less than $-t_{\alpha/2, n-1}$. The reader should recall from Chapters 8 and 9 that the t-distribution is symmetric around the value zero. Thus this two-tailed critical region applies in a fashion similar to that for the case of known σ. For the two-sided hypothesis at significance level α, the two-tailed critical regions apply. For H_1: $\mu > \mu_0$, rejection results when $t > t_{\alpha, n-1}$. For H_1: $\mu < \mu_0$, the critical region is given by $t < -t_{\alpha, n-1}$.

EXAMPLE 10.5   The *Edison Electric Institute* has published figures on the annual number of kilo-watt-hours expended by various home appliances. It is claimed that a vacuum cleaner expends an average of 46 kilowatt-hours per year. If a random sample of 12 homes included in a planned study indicates that vacuum cleaners expend an aver-age of 42 kilowatt-hours per year with a standard deviation of 11.9 kilowatt-hours, does this suggest at the 0.05 level of significance that vacuum cleaners expend, on the average, less than 46 kilowatt-hours annually? Assume the population of kilo-watt-hours to be normal.

SOLUTION

1. H_0: $\mu = 46$ kilowatt-hours.
2. H_1: $\mu < 46$ kilowatt-hours.
3. $\alpha = 0.05$.
4. Critical region: $t < -1.796$, where

$$t = \frac{\bar{x} - \mu_0}{s/\sqrt{n}}$$

   with $v = 11$ degrees of freedom.
5. Computations: $\bar{x} = 42$ kilowatt-hours, $s = 11.9$ kilowatt-hours, and $n = 12$. Hence

$$t = \frac{42 - 46}{11.9/\sqrt{12}} = -1.16$$

$$P = P(T < -1.16) = 0.135.$$

6. Decision: Do not reject H_0 and conclude that the average number of kilowatt-hours expended annually by home vacuum cleaners is not significantly less than 46.

Comment on the Single-Sample T-Test

The reader has probably already noticed that the equivalence of the two-tailed t-test for a single mean and the computation of a confidence interval on μ with σ replaced by s is maintained. For example, consider Example 9.4. Essentially, we can view that computation as one in which we have found all values of μ_0, the hypothesized mean volume of containers of sulfuric acid, for which the hypothesis H_0: $\mu = \mu_0$ *will not be rejected* at $\alpha = 0.05$. Again, this is consistent with the statement: "Based on the sample information, values of the population mean volume between 9.74 and 10.26 liters are not unreasonable."

Comments regarding the normality assumption are worth emphasis at this point. We have indicated that when σ is known, the central limit theorem allows for the use of a test statistic or a confidence interval which is based on Z, the standard normal random variable. Strictly speaking, of course, the central limit theorem and thus the use of the standard normal does not apply unless σ is known. Now, in Chapter 8, the development of the t-distribution is given. At that point it was stated that normality on $X_1, X_2, \ldots, X_n$ was an underlying assumption. Thus, *strictly speaking*, the Student's t-tables of percentage points for tests or confidence intervals should not be used unless it is known that the sample comes from a normal population. In practice, σ can rarely be assumed known. However, a very good estimate may be available from previous experiments. Many statistics textbooks suggest that one can safely replace σ by s in the test statistic

$$z = \frac{\bar{x} - \mu_0}{\sigma/\sqrt{n}}$$

when $n \geq 30$ and still use the Z-tables for the appropriate critical region. The implication here is that the central limit theorem is indeed being invoked and one is relying on the fact that $s \approx \sigma$. Obviously when this is done the results must be viewed as being approximate. Thus a computed P-value (from the Z-distribution) of 0.15 may be 0.12 or perhaps 0.17, or a computed confidence interval may be a 93% confidence interval rather than a 95% interval as desired. Now what about situations where $n \leq 30$? The user cannot rely on s being close to σ, and in order to take into account the inaccuracy of the estimate, the confidence interval should be wider or the critical value larger in magnitude. The t-distribution percentage points accomplish this but are correct only when the sample is from a normal distribution.

For small samples, it is often difficult to detect deviations from a normal distribution. (Goodness-of-fit tests are discussed in a later section of this chapter.) For bell-shaped distributions of the random variables $X_1, X_2, \ldots, X_n$, the use of the t-distribution for tests or confidence intervals is likely to be quite good. When in doubt, the user should resort to nonparametric procedures which are presented in Chapter 16.

Annotated Computer Printout for Single-Sample T-Test

It should be of interest for the reader to see annotated computer printout showing the result of a single-sample t-test. Suppose that an engineer is interested in testing the bias in a pH meter. Data are collected on a neutral substance (pH = 7.0). A sample of the measurements were taken with the data as follows:

7.07	7.00
7.10	6.97
7.00	7.03
7.01	7.01
6.98	7.08

It is, then, of interest to test

$$H_0: \quad \mu = 7.0,$$

$$H_i: \quad \mu \neq 7.0.$$

In this illustration we use the computer package MINITAB to illustrate the analysis of the data set above. Notice the key components of the printout shown in Figure 10.12. Of course, the mean $\bar{y} = 7.0250$, STDEV is simply the sample standard deviation $s = 0.440$, and SE MEAN is the estimated standard error of the mean and is computed as $s/\sqrt{n} = 0.0139$. The t-value is the ratio $(7.0250 - 7)/(0.0139) = 1.80$. The P-value of 0.11 suggests results that are inconclusive. There is not a strong rejection of H_0 (based on an α of 0.05 or 0.10), yet one certainly cannot truly conclude that the pH meter is unbiased. Notice that the sample size of 10 is rather small. An increase in sample size (perhaps another experiment) may sort things out. A discussion regarding appropriate sample size appears in Section 10.10.

```
          H_0:  μ = 7  vs.  H_1:  μ ≠ 7

pH-meter
  7.07  7.10  7.00  7.01  6.98  7.00  6.97  7.03  7.01  7.08
MTB > ttest mu = 7 'pH-meter'
TEST OF MU = 7.0000 VS MU N.E. 7.0000

          N     MEAN   STDEV   SE MEAN     T   P VALUE
pH-meter  10   7.0250  0.0440   0.0139   1.80    0.11
```

FIGURE 10.12 Minitab printout for one sample t-test for pH meter.

10.8 Two Samples: Tests on Two Means

The reader has already come to understand the relationship between tests and confidence intervals. The reader should rely largely on details supplied by the confidence interval material in Chapter 9. Tests concerning two means represent a set of very important analytical tools for the scientist or engineer. The experimental setting is very much like that described in Section 9.7. Two independent random samples of size n_1 and n_2, respectively, are drawn from two populations with means μ_1 and μ_2 and variances σ_1^2 and σ_2^2. We know that the random variable

$$Z = \frac{(\bar{X}_1 - \bar{X}_2) - (\mu_1 - \mu_2)}{\sqrt{\sigma_1^2/n_1 + \sigma_2^2/n_2}}$$

has a standard normal distribution. Obviously, if we can assume that $\sigma_1 = \sigma_2 = \sigma$, the statistic above reduces to

$$Z = \frac{(\bar{X}_1 - \bar{X}_2) - (\mu_1 - \mu_2)}{\sigma\sqrt{1/n_1 + 1/n_2}}.$$

The two statistics above serve as a basis for the development of the test procedures involving two means. The confidence interval equivalence and the ease in the transition from the case of tests on one mean provide simplicity.

The two-sided hypothesis on two means can be written quite generally as

$$H_0: \quad \mu_1 - \mu_2 = d_0.$$

Obviously, the alternative can be two-sided or one-sided. Again, the distribution used is the distribution of the test statistic under H_0. Values $\bar{x}_1$ and $\bar{x}_2$ are computed and for σ_1 and σ_2 known, the test statistic is given by

$$z = \frac{(\bar{x}_1 - \bar{x}_2) - d_0}{\sqrt{\sigma_1^2/n_1 + \sigma_2^2/n_2}}$$

with a two-tailed critical region in the case of a two-sided alternative. That is, reject H_0 in favor of H_1: $\mu_1 - \mu_2 \neq d_0$ if $z > z_{\alpha/2}$ or $z < -z_{\alpha/2}$. One-tailed critical regions are used in the case of the one-sided alternatives. The reader should, as before, study the test statistic and be satisfied that for, say, H_1: $\mu_1 - \mu_2 > d_0$, the signal favoring H_1 comes from large values of z. Thus the upper-tailed critical region applies.

Unknown Variances

The more prevalent situations involving tests on two means are those in which variances are unknown. If the scientist involved is willing to assume that both distributions are normal and that $\sigma_1 = \sigma_2 = \sigma$, the *pooled t-test* (often called the two-sample *t*-test) may be used. The test statistic (see Section 9.7) is given by the following test.

Two-Sample Pooled T-Test

$$t = \frac{(\bar{x}_1 - \bar{x}_2) - d_0}{s_p\sqrt{1/n_1 + 1/n_2}},$$

where

$$s_p^2 = \frac{s_1^2(n_1 - 1) + s_2^2(n_2 - 1)}{n_1 + n_2 - 2}.$$

The *t*-distribution is involved and the two-sided hypothesis is *not rejected* when

$$-t_{\alpha/2, \, n_1+n_2-2} < t < t_{\alpha/2, \, n_1+n_2-2}.$$

The reader should recall from material in Chapter 9 that the degrees of freedom for the *t*-distribution are a result of pooling of information from the two samples to estimate σ^2. One-sided alternatives suggest one-sided critical regions, as one might expect. For example, for H_1: $\mu_1 - \mu_2 > d_0$, reject H_0: $\mu_1 - \mu_2 = d_0$ when $t > t_{\alpha, \, n_1+n_2-2}$.

EXAMPLE 10.6   An experiment was performed to compare the abrasive wear of two different laminated materials. Twelve pieces of material 1 were tested, by exposing each piece to a machine measuring wear. Ten pieces of material 2 were similarly tested. In each case, the depth of wear was observed. The samples of material 1 gave an average (coded) wear of 85 units with a sample standard deviation of 4, while the samples of material 2 gave an average of 81 and a sample standard deviation of 5. Can we conclude at the 0.05 level of significance that the abrasive wear of material 1 exceeds that of material 2 by more than 2 units? Assume the populations to be approximately normal with equal variances.

SOLUTION
Let μ_1 and μ_2 represent the population means of the abrasive wear for material 1 and material 2, respectively.

1. H_0: $\mu_1 - \mu_2 = 2$.
2. H_1: $\mu_1 - \mu_2 > 2$.
3. $\alpha = 0.05$.
4. Critical region: $t > 1.725$, where

$$t = \frac{(\bar{x}_1 - \bar{x}_2) - d_0}{s_p\sqrt{1/n_1 + 1/n_2}}$$

with $v = 20$ degrees of freedom.
5. Computations:

$$\bar{x}_1 = 85, \qquad s_1 = 4, \qquad n_1 = 12,$$
$$\bar{x}_2 = 81, \qquad s_2 = 5, \qquad n_2 = 10.$$

Hence

$$s_p = \sqrt{\frac{(11)(16) + (9)(25)}{12 + 10 - 2}} = 4.478$$

$$t = \frac{(85 - 81) - 2}{4.478\sqrt{(1/12) + (1/10)}} = 1.04$$

$$P = P(T > 1.04) \approx 0.16.$$

6. Decision: Do not reject H_0. We are unable to conclude that the abrasive wear of material 1 exceeds that of material 2 by more than 2 units.

Paired Observations

When the student of statistics studies the two-sample t-test or confidence interval on the difference between means, he or she should realize that some elementary notions dealing in experimental design become relevant and must be addressed. Recall the discussion of experimental units in Chapter 9. It was suggested at that point that the condition of the two populations (often referred to as the two treatments) should be assigned randomly to the experimental units. This is done to avoid biased results due to systematic differences between experimental units. In other words, in terms of hypothesis-testing jargon, it is important that the significant difference found between means (or not found) be due to the different conditions of the populations and not due to the experimental units in the study. For example, consider Exercise 6 that follows Section 9.8. The 20 seedlings play the role of the experimental units. Ten of them are to be treated with nitrogen and 10 with no nitrogen. It may be very important that this assignment to the ''nitrogen'' and ''no nitrogen'' treatment be random to ensure that systematic differences between the seedlings do not interfere with a valid comparison between the means.

In Example 10.6, time of measurement is the most likely choice of the experimental unit. The 22 pieces of material should be measured in random order. We need to guard against the possibility that wear measurements made close together in time might tend to give similar results. Systematic (nonrandom) differences in experimental units *are not expected*. However, random assignments guard against the problem.

References to planning of experiments, randomization, choice of sample size, and so on, will continue to influence much of the development in Chapters 13, 14, and 15. Any scientist or engineer whose interest lies in analysis of real data should study this material. The pooled *t*-test is extended in Chapter 13 to cover more than two means.

Testing of two means can be accomplished when data are in the form of paired observations as discussed in Chapter 9. In this pairing structure, the conditions of the two populations (treatments) are assigned randomly within homogeneous units. Computation of the confidence interval for $\mu_1 - \mu_2$ in the situation with paired observations is based on the random variable

$$T = \frac{\overline{D} - \mu_D}{S_d/\sqrt{n}},$$

where $\overline{D}$ and S_d are random variables representing the sample mean and standard deviations of the differences of the observations in the experimental units. As in the case of the *pooled t-test*, the assumption is that the observations from each population are normal. This two-sample problem is essentially reduced to a one-sample problem by using the computed differences $d_1, d_2, \ldots, d_n$. Thus the hypothesis reduces to

$$H_0: \quad \mu_D = d_0.$$

The computed test statistic is then given by

$$t = \frac{\overline{d} - d_0}{s_d/\sqrt{n}}.$$

Critical regions are constructed using the *t*-distribution with $n - 1$ degrees of freedom.

EXAMPLE 10.7   In the paper "*Influence of Physical Restraint and Restraint-Facilitating Drugs on Blood Measurements of White-Tailed Deer and Other Selected Mammals*," Virginia Polytechnic Institute and State University (1976), J. A. Wesson examined the influence of the drug *succinylcholine* on the circulation levels of androgens in the blood. Blood samples from wild, free-ranging deer were obtained via the jugular vein immediately after an intramuscular injection of succinylcholine using darts and a capture gun. Deer were bled again approximately 30 minutes after the injection and then released. The levels of androgens at time of capture and 30 minutes later, measured in nanograms per milliliter (ng/ml), for 15 deer are as follows:

| | | Androgen (ng/ml) | |
Deer	Time of Injection	30 Minutes After Injection	d_i
1	2.76	7.02	4.26
2	5.18	3.10	−2.08
3	2.68	5.44	2.76
4	3.05	3.99	0.94
5	4.10	5.21	1.11
6	7.05	10.26	3.21
7	6.60	13.91	7.31
8	4.79	18.53	13.74
9	7.39	7.91	0.52
10	7.30	4.85	−2.45
11	11.78	11.10	−0.68
12	3.90	3.74	−0.16
13	26.00	94.03	68.03
14	67.48	94.03	26.55
15	17.04	41.70	24.66

Assuming that the populations of androgen at time of injection and 30 minutes later are normally distributed, test at the 0.05 level of significance whether the androgen concentrations are altered after 30 minutes of restraint.

SOLUTION

Let μ_1 and μ_2 be the average androgen concentration at the time of injection and 30 minutes later, respectively. We proceed as follows:

1. H_0: $\mu_1 = \mu_2$ or $\mu_D = \mu_1 - \mu_2 = 0$.
2. H_1: $\mu_1 \neq \mu_2$ or $\mu_D = \mu_1 - \mu_2 \neq 0$.
3. $\alpha = 0.05$.
4. Critical region: $t < -2.145$ and $t > 2.145$, where

$$t = \frac{\bar{d} - d_0}{s_d/\sqrt{n}}$$

with $v = 14$ degrees of freedom.

5. Computations: The sample mean and standard deviation for the d_i's are

$$\bar{d} = 9.848 \quad \text{and} \quad s_d = 18.474.$$

Therefore,

$$t = \frac{9.848 - 0}{18.474/\sqrt{15}} = 2.06.$$

6. Though the *t*-statistic is not significant at the 0.05 level, $P = P(|T| > 2.06) \simeq$ 0.06. As a result, there is some evidence that there is a difference in mean circulating levels of androgen.

In the case of paired observations, it is important that there be no interaction between the treatments and the experimental units. This was discussed in Chapter 9 in the development of confidence intervals. The no interaction assumption implies that the effect of the experimental unit is the same for each of the two treatments. In Example 10.7, we are assuming that the effect of the deer is the same for the two conditions under study, namely "at injection" and 30 minutes after injection.

Summary of Test Procedures

As we complete the formal development of tests on population means, we offer Table 10.2, which summarizes the test procedure for the cases of a single mean and two means. Notice the approximate procedure when distributions are normal and variances are unknown but not assumed to be equal. This statistic was introduced in Chapter 9.

10.9 Choice of Sample Size for Testing Means _____

In Section 10.2 we demonstrated how the analyst can exploit relationships among the sample size, the significance level α, and the power of the test to achieve a certain standard of quality. In most practical circumstances the experiment should be planned with a choice of sample size made prior to the data-taking process if possible. The sample size is usually made to achieve good power for a fixed α and fixed specific alternative. This fixed alternative may be in the form of $\mu - \mu_0$ in the case of a hypothesis involving a single mean or $\mu_1 - \mu_2$ in the case of a problem involving two means. Specific cases will provide illustrations.

Suppose that we wish to test the hypothesis

$$H_0: \quad \mu = \mu_0,$$
$$H_1: \quad \mu > \mu_0,$$

with a significance level α when the variance σ^2 is known. For a specific alternative, say $\mu = \mu_0 + \delta$, the power of our test is shown in Figure 10.13 to be

$$1 - \beta = P(\overline{X} > a \text{ when } \mu = \mu_0 + \delta).$$

Therefore,

$$\beta = P(\overline{X} < a \text{ when } \mu = \mu_0 + \delta)$$
$$= P\left[\frac{\overline{X} - (\mu_0 + \delta)}{\sigma/\sqrt{n}} < \frac{a - (\mu_0 + \delta)}{\sigma/\sqrt{n}} \text{ when } \mu = \mu_0 + \delta\right].$$

TABLE 10.2   Tests Concerning Means

H_0	Value of Test Statistic	H_1	Critical Region
$\mu = \mu_0$	$z = \dfrac{\bar{x} - \mu_0}{\sigma/\sqrt{n}}$; σ known	$\mu < \mu_0$ $\mu > \mu_0$ $\mu \neq \mu_0$	$z < -z_\alpha$ $z > z_\alpha$ $z < -z_{\alpha/2}$ and $z > z_{\alpha/2}$
$\mu = \mu_0$	$t = \dfrac{\bar{x} - \mu_0}{s/\sqrt{n}}$; $v = n - 1$, σ unknown	$\mu < \mu_0$ $\mu > \mu_0$ $\mu \neq \mu_0$	$t < -t_\alpha$ $t > t_\alpha$ $t < -t_{\alpha/2}$ and $t > t_{\alpha/2}$
$\mu_1 - \mu_2 = d_0$	$z = \dfrac{(\bar{x}_1 - \bar{x}_2) - d_0}{\sqrt{(\sigma_1^2/n_1) + (\sigma_2^2/n_2)}}$; σ_1 and σ_2 known	$\mu_1 - \mu_2 < d_0$ $\mu_1 - \mu_2 > d_0$ $\mu_1 - \mu_2 \neq d_0$	$z < -z_\alpha$ $z > z_\alpha$ $z < -z_{\alpha/2}$ and $z > z_{\alpha/2}$
$\mu_1 - \mu_2 = d_0$	$t = \dfrac{(\bar{x}_1 - \bar{x}_2) - d_0}{s_p\sqrt{(1/n_1) + (1/n_2)}}$; $v = n_1 + n_2 - 2$, $\sigma_1 = \sigma_2$ but unknown $s_p^2 = \dfrac{(n_1 - 1)s_1^2 + (n_2 - 1)s_2^2}{n_1 + n_2 - 2}$	$\mu_1 - \mu_2 < d_0$ $\mu_1 - \mu_2 > d_0$ $\mu_1 - \mu_2 \neq d_0$	$t < -t_\alpha$ $t > t_\alpha$ $t < -t_{\alpha/2}$ and $t > t_{\alpha/2}$
$\mu_1 - \mu_2 = d_0$	$t' = \dfrac{(\bar{x}_1 - \bar{x}_2) - d_0}{\sqrt{(s_1^2/n_1) + (s_2^2/n_2)}}$; $v = \dfrac{(s_1^2/n_1 + s_2^2/n_2)^2}{\dfrac{(s_1^2/n_1)^2}{n_1 - 1} + \dfrac{(s_2^2/n_2)^2}{n_2 - 1}}$; $\sigma_1 \neq \sigma_2$ and unknown	$\mu_1 - \mu_2 < d_0$ $\mu_1 - \mu_2 > d_0$ $\mu_1 - \mu_2 \neq d_0$	$t' < -t_\alpha$ $t' > t_\alpha$ $t' < -t_{\alpha/2}$ and $t' > t_{\alpha/2}$
$\mu_D = d_0$	$t = \dfrac{\bar{d} - d_0}{s_d/\sqrt{n}}$; $v - 1$, paired observations	$\mu_D < d_0$ $\mu_D > d_0$ $\mu_D \neq d_0$	$t < -t_\alpha$ $t > t_\alpha$ $t < -t_{\alpha/2}$ and $t > t_{\alpha/2}$

Under the alternative hypothesis $\mu = \mu_0 + \delta$, the statistic

$$\frac{\overline{X} - (\mu_0 + \delta)}{\sigma/\sqrt{n}}$$

is the standard normal variable Z. Therefore,

$$\beta = P\left(Z < \frac{a - \mu_0}{\sigma/\sqrt{n}} - \frac{\delta}{\sigma/\sqrt{n}}\right)$$

$$= P\left(Z < z_\alpha - \frac{\delta}{\sigma/\sqrt{n}}\right),$$

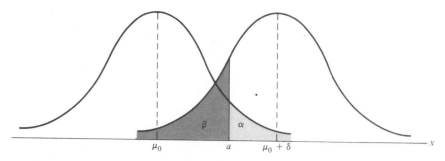

FIGURE 10.13   Testing $\mu = \mu_0$ versus $\mu = \mu_0 + \delta$.

from which we conclude that

$$-z_\beta = z_\alpha - \frac{\delta\sqrt{n}}{\sigma}$$

and hence

CHOICE OF
SAMPLE SIZE

$$n = \frac{(z_\alpha + z_\beta)^2\sigma^2}{\delta^2},$$

a result that is also true when the alternative hypothesis is $\mu < \mu_0$.

In the case of a two-tailed test we obtain the power $1 - \beta$ for a specified alternative when

$$n \simeq \frac{(z_{\alpha/2} + z_\beta)^2\sigma^2}{\delta^2}.$$

EXAMPLE 10.8   Suppose that we wish to test the hypothesis

$$H_0: \quad \mu = 68 \text{ kilograms},$$

$$H_1: \quad \mu > 68 \text{ kilograms}$$

for the weights of male students at a certain college using an $\alpha = 0.05$ level of significance when it is known that $\sigma = 5$. Find the sample size required if the power of our test is to be 0.95 when the true mean is 69 kilograms.

SOLUTION
Since $\alpha = \beta = 0.05$, we have $z_\alpha = z_\beta = 1.645$. For the alternative $\mu = 69$, we take $\delta = 1$ and then

$$n = \frac{(1.645 + 1.645)^2(25)}{1} = 270.6.$$

Therefore, it requires 271 observations if the test is to reject the null hypothesis 95% of the time when, in fact, μ is as large as 69 kilograms.

A similar procedure can be used to determine the sample size $n = n_1 = n_2$ required for a specific power of the test in which two population means are being compared. For example, suppose that we wish to test the hypothesis

$$H_0: \quad \mu_1 - \mu_2 = d_0,$$
$$H_1: \quad \mu_1 - \mu_2 \neq d_0,$$

when σ_1 and σ_2 are known. For a specific alternative, say $\mu_1 - \mu_2 = d_0 + \delta$, the power of our test is shown in Figure 10.14 to be

$$1 - \beta = P(|\overline{X}_1 - \overline{X}_2| > a \text{ when } \mu_1 - \mu_2 = d_0 + \delta).$$

Therefore,

$$\beta = P(-a < \overline{X}_1 - \overline{X}_2 < a \text{ when } \mu_1 - \mu_2 = d_0 + \delta)$$

$$= P\left[\frac{-a - (d_0 + \delta)}{\sqrt{(\sigma_1^2 + \sigma_2^2)/n}} < \frac{\overline{X}_1 - \overline{X}_2 - (d_0 + \delta)}{\sqrt{(\sigma_1^2 + \sigma_2^2)/n}} \right.$$

$$\left. < \frac{a - (d_0 + \delta)}{\sqrt{(\sigma_1^2 + \sigma_2^2)/n}} \text{ when } \mu_1 - \mu_2 = d_0 + \delta \right].$$

Under the alternative hypothesis $\mu_1 - \mu_2 = d_0 + \delta$, the statistic

$$\frac{\overline{X}_1 - \overline{X}_2 - (d_0 + \delta)}{\sqrt{(\sigma_1^2 + \sigma_2^2)/n}}$$

is the standard normal variable Z. Now, writing

$$-z_{\alpha/2} = \frac{-a - d_0}{\sqrt{(\sigma_1^2 + \sigma_2^2)/n}} \quad \text{and} \quad z_{\alpha/2} = \frac{a - d_0}{\sqrt{(\sigma_1^2 + \sigma_2^2)/n}},$$

we have

$$\beta = P\left[-z_{\alpha/2} - \frac{\delta}{\sqrt{(\sigma_1^2 + \sigma_2^2)/n}} < Z < z_{\alpha/2} - \frac{\delta}{\sqrt{(\sigma_1^2 + \sigma_2^2)/n}} \right],$$

from which we conclude that

$$-z_\beta \cong z_{\alpha/2} - \frac{\delta}{\sqrt{(\sigma_1^2 + \sigma_2^2)/n}}$$

and hence

$$n \cong \frac{(z_{\alpha/2} + z_\beta)^2(\sigma_1^2 + \sigma_2^2)}{\delta^2}.$$

For the one-tailed test, the expression for the required sample size when $n = n_1 = n_2$ is given by

CHOICE OF SAMPLE SIZE

$$n = \frac{(z_\alpha + z_\beta)^2(\sigma_1^2 + \sigma_2^2)}{\delta^2}.$$

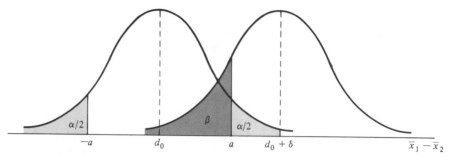

FIGURE 10.14 Testing $\mu_1 - \mu_2 = d_0$ versus $\mu_1 - \mu_2 = d_0 + \delta$.

When the population variance (or variances in the two-sample situation) is unknown, the choice of sample size is not straightforward. In testing the hypothesis $\mu = \mu_0$ when the true value is $\mu = \mu_0 + \delta$, the statistic

$$\frac{\overline{X} - (\mu_0 + \delta)}{S/\sqrt{n}}$$

does not follow the t-distribution, as one might expect, but instead follows the **noncentral t-distribution**. However, tables or charts based on the noncentral t-distribution do exist for determining the appropriate sample size if some estimate of σ is available or if δ is a multiple of σ. Table A.8 gives the sample sizes needed to control the values of α and β for various values of

$$\Delta = \frac{|\delta|}{\sigma} = \frac{|\mu - \mu_0|}{\sigma}$$

for both one- and two-tailed tests. In the case of the two-sample t-test in which the variances are unknown but assumed equal, we obtain the sample sizes $n = n_1 = n_2$ needed to control the values of α and β for various values of

$$\Delta = \frac{|\delta|}{\sigma} = \frac{|\mu_1 - \mu_2 - d_0|}{\sigma}$$

from Table A.9.

EXAMPLE 10.9   In comparing the performance of two catalysts on the effect of a reaction yield, a two-sample t-test is to be conducted with $\alpha = 0.05$. The variances in the yields are considered to be the same for the two catalysts. How large a sample for each catalyst is needed to test the hypothesis

$$H_0: \quad \mu_1 = \mu_2,$$

$$H_1: \quad \mu_1 \neq \mu_2$$

if it is essential to detect a difference of 0.8σ between the catalysts with probability 0.9?

SOLUTION
From Table A.9, with $\alpha = 0.05$ for a two-tailed test, $\beta = 0.1$, and

$$\Delta = \frac{|0.8\sigma|}{\sigma} = 0.8,$$

we find the required sample size to be $n = 34$.

It should be emphasized that in practical situations it might be difficult to force a scientist or engineer to make a commitment on information from which a value of Δ can be found. The reader should be reminded that the Δ-value quantifies the kind of difference between the means that the scientist considers important, that is, a difference considered *significant* from a scientific, not a statistical, point of view. Example 10.9 illustrates how this choice is often made, namely, by selecting a fraction of σ. Obviously, if the sample size is based on a choice of $|\delta|$ that is a small fraction of σ, the resulting sample size may be quite large compared to what the study allows.

10.10 Graphical Methods for Comparing Means _____

In Chapter 3 considerable attention was directed toward displaying data in graphical form. Stem and leaf displays and, in Chapter 8, box and whisker, quantile plots, and quantile–quantile normal plots were used to provide a "picture" to summarize a set of experimental data. Many computer software packages produce graphical displays. As we proceed to other forms of data analysis (e.g., regression analysis and analysis of variance), graphical methods become even more informative.

Graphical aids used in conjunction with hypothesis testing are not used as a replacement of the test procedure. Certainly, the value of the test statistic indicates the proper type of evidence in support of H_0 or H_1. However, a pictorial display provides a good illustration and is often a better communicator of evidence to the beneficiary of the analysis. Also, a picture will often shed some light on why a significant difference was found. A failure of an important assumption may be uncovered through a summary type of graphical display.

For the comparison of means, side-by-side box and whisker plots provide a telling display. The reader should recall that these plots display the 25th percentile, 75th percentile, and the median in a data set. In addition, the whiskers display the extremes in a data set. Consider Exercise 22 following this section. Plasma ascorbic acid levels were measured in two groups of pregnant women, smokers and nonsmokers. Figure 10.15 shows the box and whisker plots for both groups of women. Two things are very apparent. Taking into account variability, there appears to be a negligible difference in the sample means. In addition, the variability in the two groups appears to be somewhat different. Of course, the analyst must keep in mind the rather sizable differences between the sample sizes in this case.

Consider Exercise 6 following Section 9.8. Figure 10.16 shows the multiple box and whisker plot for the data on 10 seedlings, half given nitrogen and half given no nitrogen. The display reveals a smaller variability for the group containing no nitrogen. In addition, the lack of overlap of the box plots suggests a significant differ-

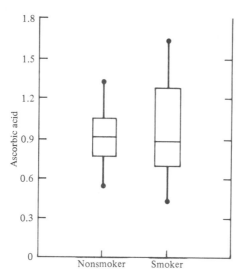

FIGURE 10.15   Multiple box and whisker plot for plasma ascorbic acid in smokers and nonsmokers.

ence between the mean stem weights between the two groups. It would appear that the presence of nitrogen increases the stem weights and perhaps increases the variability in the weights.

There are no certain rules of thumb regarding when two box and whisker plots give evidence of significant difference between the means. However, a rough guideline is that if the 25th percentile line for one sample exceeds the median line for the other sample, there is strong evidence of a difference between means.

More emphasis is placed on graphical methods in a real-life case study demonstrated later in this chapter.

Annotated Computer Printout for Two-Sample T-Test

Consider the data of Exercise 6 following Section 9.8 in which seedling data under conditions of nitrogen and no nitrogen were collected. Test

$$H_0:\quad \mu_{\text{NIT}} = \mu_{\text{NON}}$$

$$H_i:\quad \mu_{\text{NIT}} > \mu_{\text{NON}},$$

where the population means indicate mean weights. Figure 10.17 is an annotated computer printout using the SAS package. Notice that sample standard deviations and standard error are shown for both samples. The t-statistic under the assumption of "equal variance" and "unequal variance" are both given. From the box and whisker plot of Figure 10.16 it would certainly appear that the equal variance assumption is violated. A P-value of 0.0229 suggests a conclusion of unequal means. This concurs with the diagnostic information given in Figure 10.16. Incidentally, notice that t and t' are equal in this case, since $n_1 = n_2$.

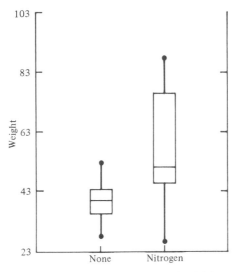

FIGURE 10.16   Multiple box and whisker plots or seedling data.

TTEST PROCEDURE

Variable: WEIGHT

MINERAL	N	Mean	Std Dev	Std Error
no nitrogen	10	0.39900000	0.07279347	0.02301932
nitrogen	10	0.56500000	0.18674106	0.05905271

Variances	T	DF	Prob > |T|
Unequal	-2.6191	11.7	0.0229
Equal	-2.6191	18.0	0.0174

For H0: Variances are equal, F' = 6.58 DF = (9,9)
Prob > F' = 0.0098

FIGURE 10.17   SAS printout for two-sample t-test.

Exercises

1. An electrical firm manufactures light bulbs that have a length of life that is approximately normally distributed with a mean of 800 hours and a standard deviation of 40 hours. Test the hypothesis that $\mu = 800$ hours against the alternative $\mu \neq 800$ hours if a random sample of 30 bulbs has an average life of 788 hours. Use a 0.04 level of significance.

2. A random sample of 36 drinks from a soft-drink machine has an average content of 21.9 deciliters, with a standard deviation of 1.42 deciliters. Test the hypoth-

esis that $\mu = 22.2$ deciliters against the alternative hypothesis, $\mu < 22.2$ at the 0.05 level of significance.

3. In a research report by Richard H. Weindruch of the UCLA Medical School, it is claimed that mice with an average life span of 32 months will live to be about 40 months old when 40% of the calories in their food are replaced by vitamins and protein. Is there any reason to believe that $\mu < 40$ if 64 mice that are placed on this diet have an average life of 38 months

with a standard deviation of 5.8 months? Use a 0.025 level of significance.

4. The average height of females in the freshman class of a certain college has been 162.5 centimeters with a standard deviation of 6.9 centimeters. Is there reason to believe that there has been a change in the average height if a random sample of 50 females in the present freshman class has an average height of 165.2 centimeters? Use a P-value in your conclusion.

5. It is claimed that an automobile is driven on the average more than 20,000 kilometers per year. To test this claim, a random sample of 100 automobile owners are asked to keep a record of the kilometers they travel. Would you agree with this claim if the random sample showed an average of 23,500 kilometers and a standard deviation of 3900 kilometers? Use a P-value in your conclusion.

6. The *Edison Electric Institute* has published figures on the annual hours of use of various home appliances. It is claimed that a trash compactor is run an average of 125 hours per year. If a random sample of 49 homes equipped with trash compactors indicates an annual average use of 126.9 hours with a standard deviation of 8.4 hours, does this suggest that trash compactors are used, on the average, more than 125 hours per year? Use a P-value in your conclusion.

7. Test the hypothesis that the average content of containers of a particular lubricant is 10 liters if the contents of a random sample of 10 containers are 10.2, 9.7, 10.1, 10.3, 10.1, 9.8, 9.9, 10.4, 10.3, and 9.8 liters. Use a 0.01 level of significance and assume that the distribution of contents is normal.

8. According to *Dietary Goals for the United States* (1977), high sodium intake may be related to ulcers, stomach cancer, and migraine headaches. The human requirement for salt is only 220 milligrams per day, which is surpassed in most single servings of ready-to-eat cereals. If a random sample of 20 similar servings of Special K has a mean sodium content of 244 milligrams of sodium and a standard deviation of 24.5 milligrams, does this suggest at the 0.05 level of significance that the average sodium content for single servings of Special K is greater than 220 milligrams? Assume the distribution of sodium contents to be normal.

9. A random sample of 8 cigarettes of a certain brand has an average nicotine content of 4.2 milligrams and

a standard deviation of 1.4 milligrams. Is this in line with the manufacturer's claim that the average nicotine content does not exceed 3.5 milligrams? Use a P-value in your conclusion and assume the distribution of nicotine contents to be normal.

10. Last year the employees of the city sanitation department donated an average of $10.00 to the volunteer rescue squad. Test the hypothesis at the 0.01 level of significance that the average contribution this year is still $10.00 if a random sample of 12 employees showed an average donation of $10.90 with a standard deviation of $1.75. Assume that the donations are approximately normally distributed.

11. Past experience indicates that the time for high school seniors to complete a standardized test is a normal random variable with a mean of 35 minutes. If a random sample of 20 high school seniors took an average of 33.1 minutes to complete this test with a standard deviation of 4.3 minutes, test the hypothesis at the 0.025 level of significance that $\mu = 35$ minutes against the alternative that $\mu < 35$ minutes.

12. A random sample of size $n_1 = 25$, taken from a normal population with a standard deviation $\sigma_1 = 5.2$, has a mean $\bar{x}_1 = 81$. A second random sample of size $n_2 = 36$, taken from a different normal population with a standard deviation $\sigma_2 = 3.4$, has a mean $\bar{x}_2 = 76$. Test the hypothesis that $\mu_1 = \mu_2$ against the alternative $\mu_1 \neq \mu_2$. Quote a P-value in your conclusion.

13. A manufacturer claims that the average tensile strength of thread A exceeds the average tensile strength of thread B by at least 12 kilograms. To test his claim, 50 pieces of each type of thread are tested under similar conditions. Type A thread had an average tensile strength of 86.7 kilograms with a standard deviation of 6.28 kilograms, while type B thread had an average tensile strength of 77.8 kilograms with a standard deviation of 5.61 kilograms. Test the manufacturer's claim using a 0.05 level of significance.

14. A study was made to estimate the difference in salaries of college professors in the private and state colleges of North Carolina. A random sample of 100 professors in private colleges showed an average 9-month salary of $32,000 with a standard deviation of $1300. A random sample of 200 professors in state colleges showed an average salary of $32,900 with a standard deviation of $1400. Test the hypothesis that the average salary for professors teaching in state col-

leges does not exceed the average salary for professors teaching in private colleges by more than $500. Use a 0.01 level of significance.

15. A study is made to see if increasing the substrate concentration has an appreciable effect on the velocity of a chemical reaction. With a substrate concentration of 1.5 moles per liter, the reaction was run 15 times with an average velocity of 7.5 micromoles per 30 minutes and a standard deviation of 1.5. With a substrate concentration of 2.0 moles per liter, 12 runs were made, yielding an average velocity of 8.8 micromoles per 30 minutes and a sample standard deviation of 1.2. Is there any reason to believe that this increase in substrate concentration causes an increase in the mean velocity by more than 0.5 micromole per 30 minutes? Use a 0.01 level of significance and assume the populations to be approximately normally distributed with equal variances.

16. A study was made to determine if the subject matter in a physics course is better understood when a lab constitutes part of the course. Students were randomly selected to participate in either a 3-semester-hour course without labs or a 4-semester-hour course with labs. In the section with labs 11 students made an average grade of 85 with a standard deviation of 4.7, and in the section without labs 17 students made an average grade of 79 with a standard deviation of 6.1. Would you say that the laboratory course increase the average grade by as much as 8 points? Use a P-value in your conclusion and assume the populations to be approximately normally distributed with equal variances.

17. To find out whether a new serum will arrest leukemia, 9 mice, which have all reached an advanced stage of the disease, are selected. Five mice receive the treatment and 4 do not. The survival times, in years, from the time the experiment commenced are as follows:

Treatment	2.1	5.3	1.4	4.6	0.9
No treatment	1.9	0.5	2.8	3.1	

At the 0.05 level of significance can the serum be said to be effective? Assume the two distributions to be normally distributed with equal variances.

18. A large automobile manufacturing company is trying to decide whether to purchase brand A or brand B tires for its new models. To help arrive at a decision, an experiment is conducted using 12 of each brand. The tires are run until they wear out. The results are

Brand A: $\bar{x}_1 = 37{,}900$ kilometers,
   $s_1 = 5100$ kilometers
Brand B: $\bar{x}_2 = 39{,}800$ kilometers,
   $s_2 = 5900$ kilometers.

Test the hypothesis at the 0.05 level of significance that there is no difference in the 2 brands of tires. Assume the populations to be approximately normally distributed.

19. In Exercise 8 on page 264, test the hypothesis that Volkswagen mini-trucks, on the average, exceed similarly equipped Toyota mini-trucks by 4 kilometers per liter. Use a 0.10 level of significance.

20. A UCLA researcher claims that the average life span of mice can be extended by as much as 8 months when the calories in their food are reduced by approximately 40% from the time they are weaned. The restricted diets are enriched to normal levels by vitamins and protein. Suppose that a random sample of 10 mice are fed a normal diet and live an average life span of 32.1 months with a standard deviation of 3.2 months, while a random sample of 15 mice are fed the restricted diet and live an average life span of 37.6 months with a standard deviation of 2.8 months. Test the hypothesis at the 0.05 level of significance that the average life span of mice on this restricted diet is increased by 8 months against the alternative that the increase is less than 8 months. Assume the distributions of life spans for the regular and restricted diets are approximately normal with equal variances.

21. The following data represent the running times of films produced by 2 motion-picture companies:

Company	Time (*minutes*)						
1	102	86	98	109	92		
2	81	165	97	134	92	87	114

Test the hypothesis that the average running time of films produced by company 2 exceeds the average running time of films produced by company 1 by 10 minutes against the one-sided alternative that the difference is more than 10 minutes. Use a 0.1 level of significance and assume the distributions of times to be approximately normal with unequal variances.

22. In the study *"Interrelationships Between Stress, Dietary Intake, and Plasma Ascorbic Acid During Pregnancy"* conducted at the Virginia Polytechnic Institute and State University in May 1983, the plasma ascorbic acid levels of pregnant women were compared for smokers versus nonsmokers. Thirty-two women in the last three months of pregnancy, free of major health disorders, and ranging in age from 15 to 32 years were selected for the study. Prior to the collection of 20 ml of blood, the participants were told to avoid breakfast, forego their vitamin supplements, and avoid foods high in ascorbic acid content. From the blood samples, the following plasma ascorbic acid values of each subject were determined in milligrams per 100 milliliters:

Plasma Ascorbic Acid Values

Nonsmokers		Smokers
0.97	1.16	0.48
0.72	0.86	0.71
1.00	0.85	0.98
0.81	0.58	0.68
0.62	0.57	1.18
1.32	0.64	1.36
1.24	0.98	0.78
0.99	1.09	1.64
0.90	0.92	
0.74	0.78	
0.88	1.24	
0.94	1.18	

Is there sufficient evidence to conclude that there is a difference between plasma ascorbic acid levels of smokers and nonsmokers? Assume that the two sets of data came from normal populations with unequal variances. Use a *P*-value.

23. A study on the *"Nutrient Retention and Macroinvertebrate Community Response to Sewage Stress in a Stream Ecosystem"* was conducted by the Department of Zoology at the Virginia Polytechnic Institute and State University in 1980 to determine if there is a significant difference in the density of organisms at two different stations located on Cedar Run, a sec-

ondary stream located in the Roanoke River drainage basin. Sewage from a sewage treatment plant and overflow from the Federal Mogul Corporation settling pond enter the stream near its headwaters. The following data give the density measurements, in number of organisms per square meter, at the two different collecting stations:

Number of Organisms per Square Meter

Station 1		Station 2	
5030	4980	2800	2810
13,700	11,910	4670	1330
10,730	8130	6890	3320
11,400	26,850	7720	1230
860	17,660	7030	2130
2200	22,800	7330	2190
4250	1130		
15,040	1690		

Can we conclude, at the 0.05 level of significance, that the average densities at the two stations are equal? Assume that the observations come from normal populations with different variances.

24. Five samples of a ferrous-type substance are to be used to determine if there is a difference between a laboratory chemical analysis and an X-ray fluorescence analysis of the iron content. Each sample was split into two subsamples and the two types of analysis were applied. Following are the coded data showing the iron content analysis:

Analysis	Sample				
	1	2	3	4	5
X-ray	2.0	2.0	2.3	2.1	2.4
Chemical	2.2	1.9	2.5	2.3	2.4

Assuming that the populations are normal, test at the 0.05 level of significance whether the two methods of analysis give, on the average, the same result.

25. A taxi company is trying to decide whether the use of radial tires instead of regular belted tires improves fuel economy. Twelve cars were equipped with radial tires and driven over a prescribed test course. Without

changing drivers, the same cars were then equipped with regular belted tires and driven once again over the test course. The gasoline consumption, in kilometers per liter, was recorded as follows:

| Car | Kilometers per Liter | |
	Radial Tires	Belted Tires
1	4.2	4.1
2	4.7	4.9
3	6.6	6.2
4	7.0	6.9
5	6.7	6.8
6	4.5	4.4
7	5.7	5.7
8	6.0	5.8
9	7.4	6.9
10	4.9	4.7
11	6.1	6.0
12	5.2	4.9

At the 0.025 level of significance, can we conclude that cars equipped with radial tires give better fuel economy than those equipped with belted tires? Assume the populations to be normally distributed.

26. In Exercise 2 on page 290, use the t-distribution to test the hypothesis that the diet reduces a person's weight by 4.5 kilograms on the average against the alternative hypothesis that the mean difference in weight is less than 4.5 kilograms. Use a P-value.

27. According to the article "*Practice and Fatigue Effects on the Programming of a Coincident Timing Response*," published in the *Journal of Human Movement Studies* in 1976, practice under fatigued conditions distorts mechanisms which govern performance. An experiment was conducted using 15 college males who were trained to make a continuous horizontal right-to-left arm movement from a microswitch to a barrier, knocking over the barrier coincident with the arrival of a clock sweephand to the 6 o'clock position. The absolute value of the difference between the time, in milliseconds, that it took to knock over the barrier and the time for the sweephand to reach the 6 o'clock position (500 msec) was recorded. Each participant performed the task five times under prefatigue and postfatigue conditions, and the sums of the absolute differences for the five performances were recorded as follows:

| Subject | Absolute Time Differences (msec) | |
	Prefatigue	Postfatigue
1	158	91
2	92	59
3	65	215
4	98	226
5	33	223
6	89	91
7	148	92
8	58	177
9	142	134
10	117	116
11	74	153
12	66	219
13	109	143
14	57	164
15	85	100

An increase in the mean absolute time differences when the task is performed under postfatigue conditions would support the claim that practice under fatigued conditions distorts mechanisms that govern performance. Assuming the populations to be normally distributed, test this claim.

28. In the study "*Comparison of Sorbic Acid in Country Ham Before and After Storage*" conducted by the Department of Human Nutrition and Foods at the Virginia Polytechnic Institute and State University in 1983, the following data on the comparison of sorbic acid residuals in parts per million in ham immediately after dipping in a sorbate solution and after 60 days of storage were recorded:

| Slice | Sorbic Acid Residuals in Ham | |
	Before Storage	After Storage
1	224	116
2	270	96
3	400	239
4	444	329
5	590	437
6	660	597
7	1400	689
8	680	576

Assuming the populations to be normally distributed, is there sufficient evidence, at the 0.05 level of significance, to say that the length of storage influences sorbic acid residual concentrations?

29. How large a sample is required in Exercise 2 if the power of our test is to be 0.90 when the true mean is 21.3? Assume that $\sigma = 1.42$.

30. If the distribution of life spans in Exercise 3 is approximately normal, how large a sample is required in order that the probability of committing a type II error be 0.1 when the true mean is 35.9 months? Assume that $\sigma = 5.8$ months.

31. How large a sample is required in Exercise 4 if the power of our test is to be 0.95 when the true average height differs from 162.5 by 3.1 centimeters?

32. How large should the samples be in Exercise 13 if the power of our test is to be 0.95 when the true difference between thread types A and B is 8 kilograms?

33. How large a sample is required in Exercise 9 if the power of our test is to be 0.8 when the true nicotine content exceeds the hypothesized value by 1.2σ?

34. On testing

$$H_0: \quad \mu = 14,$$

$$H_1: \quad \mu = 14,$$

an $\alpha = 0.05$ level t-test is being considered. What sample size is necessary in order that the probability is 0.1 of falsely accepting H_0 when the true population mean differs from 14 by 0.5? From a preliminary sample we estimate σ to be 1.25.

35. A study was conducted at the Department of Veterinary Medicine at Virginia Polytechnic Institute and State University to determine if the ''strength'' of a wound from surgical incision is affected by the temperature of the knife. Eight dogs were used in the experiment. The incision was performed in the abdomen of the animals. A ''hot'' and ''cold'' incision was made on each dog and the strength was measured. The resulting data appear below.
(a) Write an appropriate hypothesis to determine if there is a significant difference in strength between the hot and cold incisions.
(b) Test the hypothesis using a paired t-test. Use a P-value in your conclusion.

Dog	Knife	Strength
1	Hot	5,120
1	Cold	8,200
2	Hot	10,000
2	Cold	8,600
3	Hot	10,000
3	Cold	9,200
4	Hot	10,000
4	Cold	6,200
5	Hot	10,000
5	Cold	10,000
6	Hot	7,900
6	Cold	5,200
7	Hot	510
7	Cold	885
8	Hot	1,020
8	Cold	460

36. Nine subjects were used in an experiment to determine if an atmosphere involving exposure to carbon monoxide has an impact on breathing capability. The data were collected by personnel in the Health and Physical Education Department at Virginia Polytechnic Institute and State University. The data were analyzed in the Statistics Consulting Center at VPI & SU. The subjects were exposed to breathing chambers, one of which contained a high concentration of CO. Several breathing measures were made for each subject for each chamber. The subjects were exposed to the breathing chambers in random sequence. The following data give the breathing frequency in number of breaths taken per minute.

Subject	With CO	Without CO
1	30	30
2	45	40
3	26	25
4	25	23
5	34	30
6	51	49
7	46	41
8	32	35
9	30	28

Make a one-sided test of the hypothesis that mean breathing frequency is the same for the two environments. Use $\alpha = 0.05$.

10.11 One Sample: Test on a Single Proportion ——————

Tests of hypotheses concerning proportions are required in many areas. The politician is certainly interested in knowing what fraction of the voters will favor him in the next election. All manufacturing firms are concerned about the proportion of defective items when a shipment is made. The gambler depends on a knowledge of the proportion of outcomes that he considers favorable.

We shall consider the problem of testing the hypothesis that the proportion of successes in a binomial experiment equals some specified value. That is, we are testing the null hypothesis H_0 that $p = p_0$, where p is the parameter of the binomial distribution. The alternative hypothesis may be one of the usual one-sided or two-sided alternatives: $p < p_0$, $p > p_0$, or $p \neq p_0$.

The appropriate random variable on which we base our decision criterion is the binomial random variable X, although we could just as well use the statistic $\hat{p} = X/n$. Values of X that are far from the mean $\mu = np_0$ will lead to the rejection of the null hypothesis. Because X is a discrete binomial variable, it is unlikely that a critical region can be established whose size is *exactly* equal to a prespecified value of α. For this reason it is preferable, in dealing with small samples, to base our decisions on P-values. To test the hypothesis

$$H_0: \quad p = p_0,$$
$$H_1: \quad p < p_0,$$

we use the binomial distribution to compute the P-value

$$P = P(X \leq x \text{ when } p = p_0).$$

The value x is the number of successes in our sample of size n. If this P-value is less than or equal to α, our test is significant at the α level and we reject H_0 in favor of H_1. Similarly, to test the hypothesis

$$H_0: \quad p = p_0,$$
$$H_1: \quad p > p_0,$$

at the α-level of significance, we compute

$$P = P(X \geq x \text{ when } p = p_0)$$

and reject H_0 in favor of H_1 if this P-value is less than or equal to α. Finally, to test the hypothesis

$$H_0: \quad p = p_0,$$
$$H_1: \quad p \neq p_0,$$

at the α-level of significance, we compute

$$P = 2P(X \leq x \text{ when } p = p_0)$$

if $x < np_0$ or

$$P = 2P(X \geq x \text{ when } p = p_0)$$

if $x > np_0$ and reject H_0 in favor of H_1 if the computed P-value is less than or equal to α.

The steps for testing a null hypothesis about a proportion against various alternatives using the binomial probabilities of Table A.1 are as follows:

TESTING A PROPORTION: SMALL SAMPLES

1. H_0: $p = p_0$.
2. H_1: *Alternatives are $p < p_0$, $p > p_0$, or $p \neq p_0$.*
3. *Choose a level of significance equal to α.*
4. *Test statistic: Binomial variable X with $p = p_0$.*
5. *Computations: Find x, the number of successes and compute the appropriate P-value.*
6. *Decision: Draw appropriate conclusions based on the P-value.* ■

EXAMPLE 10.10   A builder claims that heat pumps are installed in 70% of all homes being constructed today in the city of Richmond. Would you agree with this claim if a random survey of new homes in this city shows that 8 out of 15 had heat pumps installed? Use a 0.10 level of significance.

SOLUTION

1. H_0: $p = 0.7$.
2. H_1: $p \neq 0.7$.
3. $\alpha = 0.10$.
4. Test statistic: Binomial variable X with $p = 0.7$ and $n = 15$.
5. Computations: $x = 8$ and $np_0 = (15)(0.7) = 10.5$. Therefore, from Table A.1, the computed P-value is

$$P = 2P(X \leq 8 \text{ when } p = 0.7)$$

$$= 2 \sum_{x=0}^{8} b(x; 15, 0.7)$$

$$= 0.2622 > 0.10.$$

6. Decision: Do not reject H_0. Conclude that there is insufficient reason to doubt the builder's claim.

In Section 5.3, we saw that binomial probabilities were obtainable from the actual binomial formula or from Table A.1 when n is small. For large n, approximation procedures are required. When the hypothesized value p_0 is very close to 0 or 1, the Poisson distribution, with parameter $\mu = np_0$, may be used. However, the normal-curve approximation, with parameters $\mu = np_0$ and $\sigma^2 = np_0q_0$, is usually preferred for large n and is very accurate as long as p_0 is not extremely close to 0 or to 1. If we use the normal approximation, the **z-value for testing $p = p_0$** is given by

$$z = \frac{x - np_0}{\sqrt{np_0q_0}},$$

which is a value of the standard normal variable Z. Hence, for a two-tailed test at the α-level of significance, the critical region is $z < -z_{\alpha/2}$ and $z > z_{\alpha/2}$. For the one-sided alternative $p < p_0$, the critical region is $z < -z_\alpha$, and for the alternative $p > p_0$, the critical region is $z > z_\alpha$.

EXAMPLE 10.11   A commonly prescribed drug on the market for relieving nervous tension is believed to be only 60% effective. Experimental results with a new drug administered to a random sample of 100 adults who were suffering from nervous tension showed that 70 received relief. Is this sufficient evidence to conclude that the new drug is superior to the one commonly prescribed? Use a 0.05 level of significance.

SOLUTION

1. H_0: $p = 0.6$.
2. H_1: $p > 0.6$.
3. $\alpha = 0.05$.
4. Critical region: $z > 1.645$.
5. Computations: $x = 70$, $n = 100$, $np_0 = (100)(0.6) = 60$, and

$$z = \frac{70 - 60}{\sqrt{(100)(0.6)(0.4)}} = 2.04$$

$$P = P(Z > 2.04) < 0.025.$$

6. Decision: Reject H_0 and conclude that the new drug is superior.

10.12   Two Samples: Tests on Two Proportions _____

Situations often arise where we wish to test the hypothesis that two proportions are equal. For example, we might try to show evidence that the proportion of doctors who are pediatricians in one state is equal to the proportion of pediatricians in another state. A person may decide to give up smoking only if he or she is convinced that the proportion of smokers with lung cancer exceeds the proportion of nonsmokers with lung cancer.

In general, we wish to test the null hypothesis that two proportions, or binomial parameters, are equal. That is, we are testing $p_1 = p_2$ against one of the alternatives $p_1 < p_2$, $p_1 > p_2$, or $p_1 \neq p_2$. Of course, this is equivalent to testing the null hypothesis that $p_1 - p_2 = 0$ against one of the alternatives $p_1 - p_2 < 0$, $p_1 - p_2 > 0$, or $p_1 - p_2 \neq 0$. The statistic on which we base our decision is the random variable $\hat{P}_1 - \hat{P}_2$. Independent samples of size n_1 and n_2 are selected at random from two binomial populations and the proportion of successes $\hat{P}_1$ and $\hat{P}_2$ for the two samples are computed.

In our construction of confidence intervals for p_1 and p_2 we noted, for n sufficiently large, that the point estimator $\hat{P}_1$ and $\hat{P}_2$ was approximately normally distributed with mean

$$\mu_{\hat{P}_1 - \hat{P}_2} = p_1 - p_2$$

and variance

$$\sigma^2_{\hat{P}_1 - \hat{P}_2} = \frac{p_1 q_1}{n_1} + \frac{p_2 q_2}{n_2}.$$

Therefore, our acceptance and critical regions can be established by using the standard normal variable

$$Z = \frac{(\hat{P}_1 - \hat{P}_2) - (p_1 - p_2)}{\sqrt{(p_1 q_1/n_1) + (p_2 q_2/n_2)}}.$$

When H_0 is true, we can substitute $p_1 = p_2 = p$ and $q_1 = q_2 = q$ (where p and q are the common values) in the preceding formula for Z to give the form

$$Z = \frac{\hat{P}_1 - \hat{P}_2}{\sqrt{pq[(1/n_1) + (1/n_2)]}}.$$

To compute a value of Z, however, we must estimate the parameters p and q that appear in the radical. Upon pooling the data from both samples, the **pooled estimate of the proportion p** is

$$\hat{p} = \frac{x_1 + x_2}{n_1 + n_2},$$

where x_1 and x_2 are the number of successes in each of the two samples. Substituting $\hat{p}$ for p and $\hat{q} = 1 - \hat{p}$ for q, the z-**value for testing $p_1 = p_2$** is determined from the formula

$$z = \frac{\hat{p}_1 - \hat{p}_2}{\sqrt{\hat{p}\hat{q}[(1/n_1) + (1/n_2)]}}.$$

The critical regions for the appropriate alternative hypotheses are set up as before using critical points of the standard normal curve. Hence, for the alternative $p_1 \neq p_2$ at the α-level of significance, the critical region is $z < -z_{\alpha/2}$ and $z > z_{\alpha/2}$. For a test where the alternative is $p_1 < p_2$, the critical region is $z < -z_\alpha$, and when the alternative is $p_1 > p_2$, the critical region is $z > z_\alpha$.

EXAMPLE 10.12   A vote is to be taken among the residents of a town and the surrounding county to determine whether a proposed chemical plant should be constructed. The construction site is within the town limits and for this reason many voters in the county feel that the proposal will pass because of the large proportion of town voters who favor the construction. To determine if there is a significant difference in the proportion of town voters and county voters favoring the proposal, a poll is taken. If 120 of 200 town voters favor the proposal and 240 of 500 county residents favor it, would you agree that the proportion of town voters favoring the proposal is higher than the proportion of county voters? Use a 0.025 level of significance.

SOLUTION

Let p_1 and p_2 be the true proportion of voters in the town and county, respectively, favoring the proposal.

1. H_0: $p_1 = p_2$.

2. H_1: $p_1 > p_2$.

3. $\alpha = 0.025$.

4. Critical region: $z > 1.96$.

5. Computations:

$$\hat{p}_1 = \frac{x_1}{n_1} = \frac{120}{200} = 0.60$$

$$\hat{p}_2 = \frac{x_2}{n_2} = \frac{240}{500} = 0.48$$

$$\hat{p} = \frac{x_1 + x_2}{n_1 + n_2} = \frac{120 + 240}{200 + 500} = 0.51.$$

Therefore,

$$z = \frac{0.60 - 0.48}{\sqrt{(0.51)(0.49)[(1/200) + (1/500)]}} = 2.9$$

$$P = P(Z > 2.9) = 0.0019.$$

6. Decision: Reject H_0 and agree that the proportion of town voters favoring the proposal is higher than the proportion of county voters.

Exercises

1. A distributor of cigarettes claims that 20% of the smokers in Miami prefer Kent cigarettes. To test this claim, 20 cigarette smokers are selected at random and asked what brand they prefer. If 6 of the 20 named Kent as their preference, what conclusion do we draw? Use a 0.05 level of significance.

2. Suppose that, in the past, 40% of all adults favored capital punishment. Do we have reason to believe that the proportion of adults favoring capital punishment today has increased if, in a random sample of 15 adults, 8 favor capital punishment? Use a 0.05 level of significance.

3. A coin is tossed 20 times, resulting in 5 heads. Is this sufficient evidence to reject the hypothesis that the coin is balanced in favor of the alternative that heads occur less than 50% of the time? Quote a *P*-value.

4. It is believed that at least 60% of the residents in a certain area favor an annexation suit by a neighboring city. What conclusion would you draw if only 110 in a sample of 200 voters favor the suit? Use a 0.04 level of significance.

5. A fuel oil company claims that one-fifth of the homes in a certain city are heated by oil. Do we have reason to doubt this claim if, in a random sample of 1000 homes in this city, it is found that 236 are heated by oil? Use a 0.01 level of significance.

6. At a certain college it is estimated that at most 25% of the students ride bicycles to class. Does this seem to be a valid estimate if, in a random sample of 90 college students, 28 are found to ride bicycles to class? Use a 0.05 level of significance.

7. A new radar device is being considered for a certain defense missile system. The system is checked by experimenting with actual aircraft in which a *kill* or a *no kill* is simulated. If in 300 trials, 250 kills occur, accept or reject, at the 0.04 level of significance, the claim that the probability of a kill with the new system does not exceed the 0.8 probability of the existing device.

8. In a controlled laboratory experiment scientists at the University of Minnesota discovered that 25% of a certain strain of rats subjected to a 20% coffee bean diet and then force-fed a powerful cancer-causing chemical later developed cancerous tumors. Would we have reason to believe that the proportion of rats developing tumors when subjected to this diet has increased if the experiment were repeated and 16 of 48 rats developed tumors? Use a 0.05 level of significance.

9. In a study to estimate the proportion of residents in a certain city and its suburbs who favor the construction of a nuclear power plant, it is found that 63 of 100 urban residents favor the construction while only 59 of 125 suburban residents are in favor. Is there a significant difference between the proportion of urban and suburban residents who favor construction of the nuclear plant? Make use of a *P*-value.

10. In a study on the fertility of married women conducted by Martin O'Connell and Carolyn C. Rogers for the Census Bureau in 1979, two groups of childless wives aged 25 to 29 were selected at random and each wife was asked if she eventually planned to have a child. One group was selected from among those wives married less than two years and the other from among those wives married five years. Suppose that 240 of 300 wives married less than two years planned to have children some day compared to 288 of the 400 wives married five years. Can we conclude that the proportion of wives married less than two years who planned to have children is significantly higher than the proportion of wives married five years? Make use of a *P*-value.

11. A cigarette manufacturing firm distributes two brands of cigarettes. If it is found that 56 of 200 smokers prefer brand A and that 29 of 150 smokers prefer brand B, can we conclude at the 0.06 level of significance that brand A outsells brand B?

12. In a winter of an epidemic flu, 2000 babies were surveyed by a well-known pharmaceutical company to determine if the company's new medicine was effective after two days. Among 120 babies who had the flu and were given the medicine, 29 were cured within two days. Among 280 babies who had the flu but were not given the medicine, 56 were cured within two days. Is there any significant indication that supports the company's claim of the effectiveness of the medicine?

10.13 One- and Two-Sample Tests Concerning Variances

In this section we are concerned with testing hypotheses concerning population variances or standard deviations. One- and two-sample tests on variances are certainly not difficult to motivate. Engineers and scientists are constantly confronted with studies in which they are required to demonstrate that measurements involving products or processes fall inside specifications set by consumers. The specifications are often met if the process variance is sufficiently small. Attention is also focused on comparative experiments between methods or processes where inherent reproducibility or variability must formally be compared. In addition, a test comparing two variances is often applied prior to conducting a *t*-test on two means. The goal is to determine if the equal variance assumption is violated.

Let us first consider the problem of testing the null hypothesis H_0 that the population variance σ^2 equals a specified value σ_0^2 against one of the usual alternatives $\sigma^2 < \sigma_0^2$, $\sigma^2 > \sigma_0^2$, or $\sigma^2 \neq \sigma_0^2$. The appropriate statistic on which we base our decision is the same chi-squared statistic of Theorem 8.4 that was used in Chapter 9 to construct a confidence interval for σ^2. Therefore, if we assume that the distribution of the population being sampled is normal, the chi-squared value for testing $\sigma^2 = \sigma_0^2$ is given by

$$\chi^2 = \frac{(n - 1)s^2}{\sigma_0^2},$$

where n is the sample size, s^2 is the sample variance, and σ_0^2 is the value of σ^2 given by the null hypothesis. If H_0 is true, χ^2 is a value of the chi-squared distribution with

$v = n - 1$ degrees of freedom. Hence, for a two-tailed test at the α-level of significance, the critical region is $\chi^2 < \chi^2_{1-\alpha/2}$ and $\chi^2 > \chi^2_{\alpha/2}$. For the one-sided alternative $\sigma^2 < \sigma^2_0$, the critical region is $\chi^2 < \chi^2_{1-\alpha}$, and for the one-sided alternative $\sigma^2 > \sigma^2_0$, the critical region is $\chi^2 > \chi^2_{\alpha}$.

EXAMPLE 10.13 A manufacturer of car batteries claims that the life of his batteries is approximately normally distributed with a standard deviation equal to 0.9 year. If a random sample of 10 of these batteries has a standard deviation of 1.2 years, do you think that $\sigma > 0.9$ year? Use a 0.05 level of significance.

SOLUTION

1. H_0: $\sigma^2 = 0.81$.
2. H_1: $\sigma^2 > 0.81$.
3. $\alpha = 0.05$.
4. Critical region: From Figure 10.18 we see that the null hypothesis is rejected when $\chi^2 > 16.919$, where

$$\chi^2 = \frac{(n-1)s^2}{\sigma^2_0}$$

   with $v = 9$ degrees of freedom.
5. Computations: $s^2 = 1.44$, $n = 10$, and

$$\chi^2 = \frac{(9)(1.44)}{0.81} = 16.0$$

$$P \simeq 0.07.$$

6. Decision: The χ^2 statistic is not significant at the 0.05 level. However, there is some evidence that $\sigma > 0.9$.

Now let us consider the problem of testing the equality of the variances σ^2_1 and σ^2_2 of two populations. That is, we shall test the null hypothesis H_0 that $\sigma^2_1 = \sigma^2_2$ against one of the usual alternatives $\sigma^2_1 < \sigma^2_2$, $\sigma^2_1 > \sigma^2_2$, or $\sigma^2_1 \neq \sigma^2_2$. For independent ran-

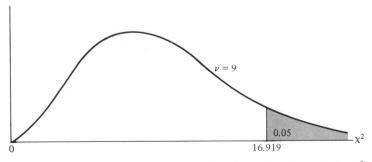

FIGURE 10.18   Critical region for the alternative hypothesis $\sigma^2 > 81$.

dom samples of size n_1 and n_2, respectively, from the two proportions, the **f-value for testing $\sigma_1^2 = \sigma_2^2$** is the ratio

$$f = \frac{s_1^2}{s_2^2},$$

where s_1^2 and s_2^2 are the variances computed from the two samples. If the two populations are approximately normally distributed and the null hypothesis is true, according to Theorem 8.8 the ratio $f = s_1^2/s_2^2$ is a value of the F-distribution with $v_1 = n_1 - 1$ and $v_2 = n_2 - 1$ degrees of freedom. Therefore, the critical regions of size α corresponding to the one-sided alternatives $\sigma_1^2 < \sigma_2^2$ and $\sigma_1^2 > \sigma_2^2$ are, respectively, $f < f_{1-\alpha}(v_1, v_2)$ and $f > f_\alpha(v_1, v_2)$. For the two-sided alternative $\sigma_1^2 \neq \sigma_2^2$, the critical region is given by $f < f_{1-\alpha/2}(v_1, v_2)$ and $f > f_{\alpha/2}(v_1, v_2)$.

EXAMPLE 10.14   In testing for the difference in the abrasive wear of the two materials in Example 10.6, we assumed that the two unknown population variances are equal. Were we justified in making this assumption? Use a 0.10 level of significance.

SOLUTION
Let σ_1^2 and σ_2^2 be the population variances for the abrasive wear of material 1 and material 2, respectively.

1. H_0: $\sigma_1^2 = \sigma_2^2$.
2. H_1: $\sigma_1^2 \neq \sigma_2^2$.
3. $\alpha = 0.10$.
4. Critical region: From Figure 10.19, we see that

$$f_{0.05}(11, 9) = 3.11$$

and, by using Theorem 8.7,

$$f_{0.95}(11, 9) = \frac{1}{f_{0.05}(9, 11)} = 0.34.$$

Therefore, the null hypothesis is rejected when $f < 0.34$ or $f > 3.11$, where $f = s_1^2/s_2^2$ with $v_1 = 11$ and $v_2 = 9$ degrees of freedom.

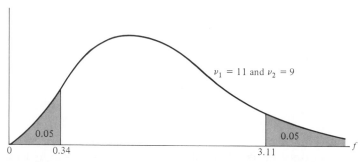

FIGURE 10.19   Critical region for the alternative $\sigma_1^2 \neq \sigma_2^2$.

5. Computations: $s_1^2 = 16$, $s_2^2 = 25$, and hence

$$f = \tfrac{16}{25} = 0.64.$$

6. Decision: Do not reject H_0. Conclude that there is insufficient evidence that the variances differ.

Exercises

1. The volume of containers of a particular lubricant is known to be normally distributed with a variance of 0.03 liter. Test the hypothesis that $\sigma^2 = 0.03$ against the alternative that $\sigma^2 \neq 0.03$ for the random sample of 10 containers in Exercise 7 on page 330. Use a 0.01 level of significance.

2. Past experience indicates that the time for high school seniors to complete a standardized test is a normal random variable with a standard deviation of 6 minutes. Test the hypothesis that $\sigma = 6$ against the alternative that $\sigma < 6$ if a random sample of 20 high school seniors has a standard deviation $s = 4.51$. Use a 0.05 level of significance.

3. The nicotine content of a certain brand of cigarettes is known to be normally distributed with a variance of 1.3 milligrams. Test the hypothesis that $\sigma^2 = 1.3$ against the alternative that $\sigma^2 \neq 1.3$ if a random sample of 8 of these cigarettes has a standard deviation $s = 1.8$. Use a 0.05 level of significance.

4. Past data indicate that the amount of money contributed by the working residents of a large city to a volunteer rescue squad is a normal random variable with a standard deviation of $1.40. It has been suggested that the contributions to the rescue squad from just the employees of the sanitation department are much more variable. If the contributions of a random sample of 12 employees from the sanitation department had a standard deviation of $1.75, can we conclude at the 0.01 level of significance that the standard deviation of the contributions of all sanitation workers is greater than that of all workers living in this city?

5. A soft-drink dispensing machine is said to be out of control if the variance of the contents exceeds 1.15 deciliters. If a random sample of 25 drinks from this machine has a variance of 2.03 deciliters, does this indicate at the 0.05 level of significance that the machine is out of control? Assume that the contents are approximately normally distributed.

6. **Large-Sample Test of $\sigma^2 = \sigma_0^2$:** When $n \geq 30$ we can test the null hypothesis that $\sigma^2 = \sigma_0^2$, or $\sigma = \sigma_0$, by computing

$$z = \frac{s - \sigma_0}{\sigma_0 / \sqrt{2n}},$$

which is a value of a random variable whose sampling distribution is approximately the standard normal distribution.

(a) With reference to Example 10.3, test at the 0.05 level of significance whether $\sigma = 7.5$ years against the alternative that $\sigma \neq 7.5$ years.

(b) It is suspected that the variance of the distribution of distances in kilometers achieved per 5 liters of fuel by a new automobile model equipped with a diesel engine is less than the variance of the distribution of distances achieved by the same model equipped with a six-cylinder gasoline engine, which is known to be $\sigma^2 = 6.25$. If 72 test runs in the diesel model have a variance of 4.41, can we conclude at the 0.02 level of significance that the variance of the distances achieved by the diesel model is less than that of the gasoline model?

7. A study is conducted to compare the length of time between men and women to assemble a certain product. Past experience indicates that the distribution of times for both men and women is approximately normal but the variance of the times for women is less than that for men. A random sample of times for 11 men and 14 women produced the following data:

Men	Women
$n_1 = 11$	$n_2 = 14$
$s_1 = 6.1$	$s_2 = 5.3$

Test the hypothesis that $\sigma_1^2 = \sigma_2^2$ against the alternative that $\sigma_1^2 > \sigma_2^2$. Use a 0.01 level of significance.

8. In Exercise 23 on page 332, test the hypothesis at the 0.02 level of significance that $\sigma_1^2 = \sigma_2^2$ against the alternative that $\sigma_1^2 \neq \sigma_2^2$, where σ_1^2 and σ_2^2 are the variances for the number of organisms per square meter at the two different locations on Cedar Run.

9. With reference to Exercise 18 on page 331, test the hypothesis that $\sigma_1 = \sigma_2$ against the alternative that $\sigma_1 < \sigma_2$, where σ_1 and σ_2 are the standard deviations of the distances obtained by brand A and brand B tires, respectively. Use a 0.05 level of significance.

10. With reference to Exercise 21 on page 331, test the hypothesis that $\sigma_1^2 = \sigma_2^2$ against the alternative that $\sigma_1^2 \neq \sigma_2^2$, where σ_1^2 and σ_2^2 are the variances for the running times of films produced by company 1 and company 2, respectively. Use a 0.10 level of significance.

11. Two types of instruments for measuring the amount of sulfur monoxide in the atmosphere are being compared in an air-pollution experiment. It is desired to determine whether the two types of instruments yield measurements having the same variability. The following readings were recorded for the two instruments:

Sulfur Monoxide	
Instrument A	Instrument B
0.86	0.87
0.82	0.74
0.75	0.63
0.61	0.55
0.89	0.76
0.64	0.70
0.81	0.69
0.68	0.57
0.65	0.53

Assuming the populations of measurements to be approximately normally distributed, test the hypothesis that $\sigma_A = \sigma_B$ against the alternative that $\sigma_A \neq \sigma_B$. Use a P-value.

12. An experiment was conducted to compare the alcohol contents in a soy sauce at two different production lines. Production was monitored eight times a day. The data are shown here.

Production line 1:
0.48   0.39   0.42   0.52   0.40   0.48   0.52   0.52

Production line 2:
0.38   0.37   0.39   0.41   0.38   0.39   0.40   0.39

It is suspected that production line 1 is not producing as consistently as production line 2 in terms of alcohol contents. Test the hypothesis that $\sigma_1 = \sigma_2$ against the alternative that $\sigma_1 \neq \sigma_2$. Use a P-value.

13. The hydrocarbon emissions are known to have decreased dramatically during the 1980s. A study was conducted to compare the hydrocarbon emissions at idling speed, in parts per million (ppm), for automobiles of 1980 and 1990. Twenty cars of each year model were randomly selected and their hydrocarbon emission levels were recorded. The data:

1980 models:
141 359 247 940 882 494 306 210 105 880 200
223 188 940 241 190 300 435 241 380

1990 models:
140 160 20 20 223 60 20 95 360 70 220
400 217 58 235 380 200 175 85 65

Test the hypothesis that $\sigma_1 = \sigma_2$ against the alternative that $\sigma_1 \neq \sigma_2$.

10.14 Goodness-of-Fit Test

Throughout this chapter we have been concerned with the testing of statistical hypotheses about single population parameters such as μ, σ^2, and p. Now we shall consider a test to determine if a population has a specified theoretical distribution. The test is based on how good a fit we have between the frequency of occurrence of observations in an observed sample and the expected frequencies obtained from the hypothesized distribution.

To illustrate, consider the tossing of a die. We hypothesize that the die is honest, which is equivalent to testing the hypothesis that the distribution of outcomes is the discrete uniform distribution

$$f(x) = \tfrac{1}{6}, \qquad x = 1, 2, \ldots, 6.$$

Suppose that the die is tossed 120 times and each outcome is recorded. Theoretically, if the die is balanced, we would expect each face to occur 20 times. The results are given in Table 10.3. By comparing the observed frequencies with the corresponding expected frequencies, we must decide whether these discrepancies are likely to occur as a result of sampling fluctuations and the die is balanced or the die is not honest and the distribution of outcomes is not uniform. It is common practice to refer to each possible outcome of an experiment as a cell. Hence, in our illustration, we have 6 cells. The appropriate statistic on which we base our decision criterion for an experiment involving k cells is defined by the following theorem.

THEOREM 10.1
A **goodness-of-fit test** *between observed and expected frequencies is based on the quantity*

$$\chi^2 = \sum_{i=1}^{k} \frac{(o_i - e_i)^2}{e_i},$$

where χ^2 is a value of a random variable whose sampling distribution is approximated very closely by the chi-squared distribution with $v = k - 1$ degrees of freedom. The symbols o_i and e_i represent the observed and expected frequencies, respectively, for the ith cell. ■

The number of degrees of freedom associated with the chi-squared distribution used here is equal to $k - 1$, since there are only $k - 1$ freely determined cell frequencies. That is, once $k - 1$ cell frequencies are determined, so is the frequency for the kth cell.

If the observed frequencies are close to the corresponding expected frequencies, the χ^2-value will be small, indicating a good fit. If the observed frequencies differ considerably from the expected frequencies, the χ^2-value will be large and the fit is poor. A good fit leads to the acceptance of H_0, whereas a poor fit leads to its rejection. The critical region will, therefore, fall in the right tail of the chi-squared distribution. For a level of significance equal to α, we find the critical value χ_α^2 from

TABLE 10.3 Observed and Expected Frequencies of 120 Tosses of a Die

	Face					
	1	2	3	4	5	6
Observed	20	22	17	18	19	24
Expected	20	20	20	20	20	20

Table A.5, and then $\chi^2 > \chi_\alpha^2$ constitutes the critical region. **The decision criterion described here should not be used unless each of the expected frequencies is at least equal to 5**. This restriction may require the combining of adjacent cells resulting in a reduction in the number of degrees of freedom.

From Table 10.3, we find the χ^2-value to be

$$\chi^2 = \frac{(20-20)^2}{20} + \frac{(22-20)^2}{20} + \frac{(17-20)^2}{20}$$
$$+ \frac{(18-20)^2}{20} + \frac{(19-20)^2}{20} + \frac{(24-20)^2}{20}$$
$$= 1.7.$$

Using Table A.5, we find $\chi_{0.05}^2 = 11.070$ for $v = 5$ degrees of freedom. Since 1.7 is less than the critical value, we fail to reject H_0. Conclude that there is insufficient evidence that the die is not balanced.

As a second illustration let us test the hypothesis that the frequency distribution of battery lives given in Table 3.4 may be approximated by a normal distribution with mean $\mu = 3.5$ and standard deviation $\sigma = 0.7$. The expected frequencies for the 7 classes (cells), listed in Table 10.4, are obtained by computing the areas under the hypothesized normal curve that fall between the various class boundaries.

For example, the z-values corresponding to the boundaries of the fourth class are

$$z_1 = \frac{2.95 - 3.5}{0.7} = -0.79$$

and

$$z_2 = \frac{3.45 - 3.5}{0.7} = -0.07.$$

From Table A.3 we find the area between $z_1 = -0.79$ and $z_2 = -0.07$ to be

$$\text{area} = P(-0.79 < Z < -0.07)$$
$$= P(Z < -0.07) - P(Z < -0.79)$$
$$= 0.4721 - 0.2148$$
$$= 0.2573.$$

Hence the expected frequency for the fourth class is

$$e_4 = (0.2573)(40) = 10.3.$$

It is customary to round these frequencies to one decimal.

The expected frequency for the first class interval is obtained by using the total area under the normal curve to the left of the boundary 1.95. For the last class interval, we use the total area to the right of the boundary 4.45. All other expected frequencies are determined by the method described for the fourth class. Note that we have combined adjacent classes in Table 10.4, where the expected frequencies

TABLE 10.4 Observed and Expected Frequencies of
Battery Lives Assuming Normality

Class Boundaries	o_i	e_i
1.45–1.95	2 ⎫	0.5 ⎫
1.95–2.45	1 ⎬ 7	2.1 ⎬ 8.5
2.45–2.95	4 ⎭	5.9 ⎭
2.95–3.45	15	10.3
3.45–3.95	10	10.7
3.95–4.45	5 ⎫ 8	7.0 ⎫ 10.5
4.45–4.95	3 ⎭	3.5 ⎭

are less than 5. Consequently, the total number of intervals is reduced from 7 to 4, resulting in $v = 3$ degrees of freedom. The χ^2-value is then given by

$$\chi^2 = \frac{(7 - 8.5)^2}{8.5} + \frac{(15 - 10.3)^2}{10.3} + \frac{(10 - 10.7)^2}{10.7} + \frac{(8 - 10.5)^2}{10.5}$$

$$= 3.05.$$

Since the computed χ^2-value is less than $\chi^2_{0.05} = 7.815$ for 3 degrees of freedom, we have no reason to reject the null hypothesis and conclude that the normal distribution with $\mu = 3.5$ and $\sigma = 0.7$ provides a good fit for the distribution of battery lives.

The chi-squared goodness-of-fit test is an extremely important tool, particularly since so many statistical procedures in practice depend, in a theoretical sense, on the assumption that the data gathered come from a specific distribution type. As we have already seen, the *normality assumption* is quite often made. In the chapters that follow we shall continue to make normality assumptions in order to provide a theoretical basis for certain tests and confidence intervals.

There are tests in the literature that are more powerful than the chi-squared test for testing normality. One such test is called **Geary's test**. This test is based on a very simple statistic which is a ratio of two estimators of the population standard deviation σ. Suppose that a random sample $X_1, X_2, \ldots, X_n$ is taken from a normal distribution, $N(\mu, \sigma)$. Consider the ratio

$$u = \frac{\sqrt{\pi/2} \sum_{i=1}^{n} |X_i - \overline{X}|/n}{\sqrt{\sum_{i=1}^{n} (X_i - \overline{X})^2/n}}.$$

The reader should recognize that the denominator is a reasonable estimator of σ whether the distribution is normal or not. The numerator is a good estimator of σ if the distribution is normal but may overestimate or underestimate σ when there are departures from normality. Thus values of U differing considerably from 1.0 represent the signal that the hypothesis of normality should be rejected.

For large samples a reasonable test is based on approximate normality of U. The test statistic is then a standardization of U. This is given by

$$Z = \frac{U - 1}{0.2661/\sqrt{n}}.$$

Of course, the test procedure involves the two-sided critical region. We compute a value of z from the data and do not reject the hypothesis of normality when

$$-z_{\alpha/2} < z < z_{\alpha/2}.$$

A reference to a paper dealing with Geary's test is given in the Bibliography.

10.15 Test for Independence

The chi-squared test procedure discussed in Section 10.14 can also be used to test the hypothesis of independence of two variables of classification. Suppose that we wish to determine whether the opinions of the voting residents of the state of Illinois concerning a new tax reform are independent of their levels of income. A random sample of 1000 registered voters from the state of Illinois are classified as to whether they are in a low, medium, or high income bracket and whether or not they favor a new tax reform. The observed frequencies are presented in Table 10.5, which is known as a **contingency table**.

A contingency table with r rows and c columns is referred to as an $r \times c$ table ("$r \times c$" is read "r by c"). The row and column totals in Table 10.5 are called **marginal frequencies**. Our decision to accept or reject the null hypothesis, H_0, of independence between a voter's opinion concerning the new tax reform and his or her level of income is based upon how good a fit we have between the observed frequencies in each of the 6 cells of Table 10.5 and the frequencies that we would expect for each cell under the assumption that H_0 is true. To find these expected frequencies, let us define the following events:

  L: A person selected is in the low-income level.

  M: A person selected is in the medium-income level.

  H: A person selected is in the high-income level.

  F: A person selected is for the new tax reform.

  A: A person selected is against the new tax reform.

By using the marginal frequencies, we can list the following probability estimates:

$$P(L) = \frac{336}{1000}, \qquad P(M) = \frac{351}{1000}, \qquad P(H) = \frac{313}{1000},$$

$$P(F) = \frac{598}{1000}, \qquad P(A) = \frac{402}{1000}.$$

TABLE 10.5 2 × 3 Contingency Table

| Tax Reform | Income Level | | | |
	Low	Medium	High	Total
For	182	213	203	598
Against	154	138	110	402
Total	336	351	313	1000

Now, if H_0 is true and the two variables are independent, we should have

$$P(L \cap F) = P(L)P(F) = \left(\frac{336}{1000}\right)\left(\frac{598}{1000}\right),$$

$$P(L \cap A) = P(L)P(A) = \left(\frac{336}{1000}\right)\left(\frac{402}{1000}\right),$$

$$P(M \cap F) = P(M)P(F) = \left(\frac{351}{1000}\right)\left(\frac{598}{1000}\right),$$

$$P(M \cap A) = P(M)P(A) = \left(\frac{351}{1000}\right)\left(\frac{402}{1000}\right),$$

$$P(H \cap F) = P(H)P(F) = \left(\frac{313}{1000}\right)\left(\frac{598}{1000}\right),$$

$$P(H \cap A) = P(H)P(A) = \left(\frac{313}{1000}\right)\left(\frac{402}{1000}\right).$$

The expected frequencies are obtained by multiplying each cell probability by the total number of observations. As before, we round these frequencies to one decimal. Thus the expected number of low-income voters in our sample who favor the new tax reform is estimated to be

$$\left(\frac{336}{1000}\right)\left(\frac{598}{1000}\right)(1000) = \frac{(336)(598)}{1000} = 200.9$$

when H_0 is true. The general rule for obtaining the **expected frequency** of any cell is given by the following formula:

$$expected\ frequency = \frac{(column\ total) \times (row\ total)}{grand\ total}.$$ ■

The expected frequency for each cell is recorded in parentheses beside the actual observed value in Table 10.6. Note that the expected frequencies in any row or column add up to the appropriate marginal total. In our example we need to compute only the two expected frequencies in the top row of Table 10.6 and then find

TABLE 10.6 Observed and Expected Frequencies

		Income Level		
Tax Reform	Low	Medium	High	Total
For	182(200.9)	213(209.9)	203(187.2)	598
Against	154(135.1)	138(141.1)	110(125.8)	402
Total	336	351	313	1000

the others by subtraction. The number of degrees of freedom associated with the chi-squared test used here is equal to the number of cell frequencies that may be filled in freely when we are given the marginal totals and the grand total, and in this illustration that number is 2. A simple formula providing the correct number of degrees of freedom is given by

$$v = (r - 1)(c - 1).$$

Hence, for our example, $v = (2 - 1)(3 - 1) = 2$ degrees of freedom.

To test the null hypothesis of independence, we use the following decision criterion:

TEST FOR
INDEPENDENCE

Calculate

$$\chi^2 = \sum_i \frac{(o_i - e_i)^2}{e_i},$$

where the summation extends over all rc cells in the r × c contingency table. If $\chi^2 > \chi_\alpha^2$ with $v = (r - 1)(c - 1)$ degrees of freedom, reject the null hypothesis of independence at the α level of significance; otherwise, accept the null hypothesis. ■

Applying this criterion to our example, we find that

$$\chi^2 = \frac{(182 - 200.9)^2}{200.9} + \frac{(213 - 209.9)^2}{209.9} + \frac{(203 - 187.2)^2}{187.2}$$

$$+ \frac{(154 - 135.1)^2}{135.1} + \frac{(138 - 141.1)^2}{141.1} + \frac{(110 - 125.8)^2}{125.8}$$

$$= 7.85,$$

$$P \simeq 0.02.$$

From Table A.5 we find that $\chi_{0.05}^2 = 5.991$ for $v = (2 - 1)(3 - 1) = 2$ degrees of freedom. The null hypothesis is rejected. We conclude that a voter's opinion concerning the new tax reform and his or her level of income are not independent.

It is important to remember that the statistic on which we base our decision has a distribution that is only approximated by the chi-squared distribution. The computed χ^2-values depend on the cell frequencies and consequently are discrete. The continuous chi-squared distribution seems to approximate the discrete sampling distribution of χ^2 very well, provided that the number of degrees of freedom is greater than 1. In a 2×2 contingency table, where we have only 1 degree of freedom, a correction called **Yates' correction for continuity** is applied. The corrected formula then becomes

$$\chi^2 \text{ (corrected)} = \sum_i \frac{(|o_i - e_i| - 0.5)^2}{e_i}.$$

If the expected cell frequencies are large, the corrected and uncorrected results are almost the same. When the expected frequencies are between 5 and 10, Yates' correction should be applied. For expected frequencies less than 5, the Fisher-Irwin exact test should be used. A discussion of this test may be found in *Basic Concepts of Probability and Statistics* by Hodges and Lehmann (see the Bibliography). The Fisher-Irwin test may be avoided, however, by choosing a larger sample.

10.16 Test for Homogeneity

When we tested for independence in Section 10.15, a random sample of 1000 voters was selected and the row and column totals for our contingency table were determined by chance. Another type of problem for which the method of Section 10.15 applies is one in which either the row or column totals are predetermined. Suppose, for example, that we decide in advance to select 200 Democrats, 150 Republicans, and 150 Independents from the voters of the state of North Carolina and record whether they are for a proposed abortion law, against it, or undecided. The observed responses are given in Table 10.7.

Now, rather than test for independence, we test the hypothesis that the population proportions within each row are the same. That is, we test the hypothesis that the proportions of Democrats, Republicans, and Independents favoring the abortion law are the same; the proportions of each political affiliation against the law are the same; and the proportions of each political affiliation that are undecided are the same. We are basically interested in determining whether the three categories of voters are **homogeneous** with respect to their opinions concerning the proposed abortion law. Such a test is called a **test for homogeneity**.

TABLE 10.7 3×3 Contingency Table

Abortion Law	Political Affiliation			Total
	Democrat	Republican	Independent	
For	82	70	62	214
Against	93	62	67	222
Undecided	25	18	21	64
Total	200	150	150	500

Assuming homogeneity, we again find the expected cell frequencies by multiplying the corresponding row and column totals and then dividing by the grand total. The analysis then proceeds using the same chi-squared statistic as before. We illustrate this process in the following example for the data of Table 10.7.

EXAMPLE 10.15   Referring to the data of Table 10.7, test the hypothesis that the opinions concerning the proposed abortion law are the same within each political affiliation. Use a 0.05 level of significance.

SOLUTION

1. H_0:   For each opinion the proportions of Democrats, Republicans, and Independents are the same.
2. H_1:   For at least one opinion the proportions of Democrats, Republicans, and Independents are not the same.
3. $\alpha = 0.05$.
4. Critical region: $\chi^2 > 9.488$ with $v = 4$ degrees of freedom.
5. Computations: Using the expected cell frequency formula on page 349 we need to compute the 4 cell frequencies. All other frequencies are found by subtraction. The observed and expected cell frequencies are displayed in Table 10.8.

Now,

$$\chi^2 = \frac{(82 - 85.6)^2}{85.6} + \frac{(70 - 64.2)^2}{64.2} + \frac{(62 - 64.2)^2}{64.2}$$
$$+ \frac{(93 - 88.8)^2}{88.8} + \frac{(62 - 66.6)^2}{66.6} + \frac{(67 - 66.6)^2}{66.6}$$
$$+ \frac{(25 - 25.6)^2}{25.6} + \frac{(18 - 19.2)^2}{19.2} + \frac{(21 - 19.2)^2}{19.2}$$
$$= 1.53.$$

6. Decision: Do not reject H_0. There is insufficient evidence to conclude that the proportion of Democrats, Republicans, and Independents differs for each stated opinion.

TABLE 10.8   Observed and Expected Frequencies

Abortion Law	Political Affiliation			Total
	Democrat	Republican	Independent	
For	82(85.6)	70(64.2)	62(64.2)	214
Against	93(88.8)	62(66.6)	67(66.6)	222
Undecided	25(25.6)	18(19.2)	21(19.2)	64
Total	200	150	150	500

10.17 Testing for Several Proportions ───────────────

The chi-squared statistic for testing for homogeneity is also applicable when testing the hypothesis that k binomial parameters have the same value. This is, therefore, an extension of the test presented in Section 10.12 for determining differences between two proportions to a test for determining differences among k proportions. Hence we are interested in testing the null hypothesis

$$H_0: \quad p_1 = p_2 = \cdots = p_k$$

against the alternative hypothesis, H_1, that the population proportions are *not all equal*. To perform this test, we first observe independent random samples of size $n_1, n_2, \ldots, n_k$ from the k populations and arrange the data as in the $2 \times k$ contingency table, Table 10.9.

TABLE 10.9 k Independent Binomial Samples

	Sample			
	1	2	$\cdots$	k
Successes	x_1	x_2	$\cdots$	x_k
Failures	$n_1 - x_1$	$n_2 - x_2$	$\cdots$	$n_k - x_k$

If we depend on whether the sizes of the random samples were predetermined or occurred at random, the test procedure is identical to the test for homogeneity or the test for independence. Therefore, the expected cell frequencies are calculated as before and substituted together with the observed frequencies into the chi-squared statistic

$$\chi^2 = \sum_i \frac{(o_i - e_i)^2}{e_i},$$

with

$$v = (2 - 1)(k - 1) = k - 1$$

degrees of freedom. By selecting the appropriate upper-tail critical region of the form $\chi^2 > \chi_\alpha^2$, one can now reach a decision concerning H_0.

EXAMPLE 10.16   In a shop study, a set of data was collected to determine whether or not the proportion of defectives produced by workers was the same for the day, evening, or night shift worked. The following data were collected:

	Shift		
	Day	*Evening*	*Night*
Defectives	45	55	70
Nondefectives	905	890	870

Use a 0.025 level of significance to determine if the proportion of defectives is the same for all three shifts.

SOLUTION

Let p_1, p_2, and p_3 represent the true proportion of defectives for the day, evening, and night shift, respectively.

1. H_0: $p_1 = p_2 = p_3$.
2. H_1: p_1, p_2, and p_3 are not all equal.
3. $\alpha = 0.025$.
4. Critical region: $\chi^2 > 7.378$ for $v = 2$ degrees of freedom.
5. Computations: Corresponding to the observed frequencies $o_1 = 45$ and $o_2 = 55$, we find

$$e_1 = \frac{(950)(170)}{2835} = 57.0 \quad \text{and} \quad e_2 = \frac{(945)(170)}{2835} = 56.7.$$

All other expected frequencies are found by subtraction and are displayed in Table 10.10. Now

TABLE 10.10   Observed and Expected Frequencies

	Shift			
	Day	Evening	Night	Total
Defectives	45(57.0)	55(56.7)	70(56.3)	170
Nondefectives	905(893.0)	890(888.3)	870(883.7)	2665
Total	950	945	940	2835

$$\chi^2 = \frac{(45 - 57.0)^2}{57.0} + \frac{(55 - 56.7)^2}{56.7} + \frac{(70 - 56.3)^2}{56.3}$$

$$+ \frac{(905 - 893.0)^2}{893.0} + \frac{(890 - 888.3)^2}{888.3} + \frac{(870 - 883.7)^2}{883.7}$$

$$= 6.29$$

$$P \simeq 0.04.$$

6. Decision: We do not reject H_0 at $\alpha = 0.025$. Nevertheless, with the above P-value computed, it would certainly be dangerous to conclude that the proportion of defectives produced is the same for all shifts.

10.18   Two-Sample Case Study

In this section we consider a study in which we show a thorough analysis using both graphical and formal analysis along with annotated computer printout and conclusions. In a data analysis study conducted by personnel at the Statistics Consulting Center at Virginia Tech, two different materials, say alloy A and alloy B, were compared in terms of breaking strength. Alloy *B* is more expensive, but it should certainly be adopted if it can be demonstrated that it is stronger than alloy *A*. The consistency of performance of the two alloys should also be taken into account.

Random samples of beams for each alloy were selected and the strength was measured in a 0.001-inch deflection as a fixed force was applied at both ends of the beam. Twenty specimens were used for each of the two alloys. The data are as follows:

Alloy A			Alloy B		
88	82	87	75	81	80
79	85	90	77	78	81
84	88	83	86	78	77
89	80	81	84	82	78
81	85		80	80	
83	87		78	76	
82	80		83	85	
79	78		76	79	

It is important that the engineer compare the two alloys. Of concern is average strength and reproducibility. It is of interest to determine if there is a severe violation of the normality assumption required of both the *t*- and *F*-tests. Figures 10.20

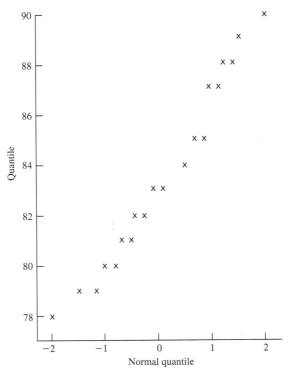

FIGURE 10.20 Normal quantile-quantile plot of data of alloy A.

and 10.21 are normal quantile–quantile plots of the samples for the two alloys. There does not appear to be any serious violation of the normality assumption. In addition, Figure 10.22 shows two box and whisker plots on the same graph. The box and whisker plots would suggest that there is no appreciable difference in the variability in deflection for the two alloys. However, it seems to suggest that the mean of alloy B is significantly smaller, suggesting at least graphically that alloy B is stronger. The sample means and standard deviations are given by

$$\bar{y}_A = 83.55 \qquad s_A = 3.663$$
$$\bar{y}_B = 79.70 \qquad s_B = 3.097.$$

The SAS printout for the PROC t-test is shown in Figure 10.23.

The F-test suggests no significant difference in variances ($P = 0.4709$) and the two-sample t statistic for testing

$$H_0: \quad \mu_A = \mu_B$$

$$H_1: \quad \mu_B > \mu_A$$

($t = 3.5895$, $P = 0.0009$) rejects H_0 in favor of H_1 and thus confirms what the graphical information suggests. Here we use the t-test that pools the two-sample variances together in light of the results of the F-test. On the basis of this analysis the adoption of alloy B would seem to be in order.

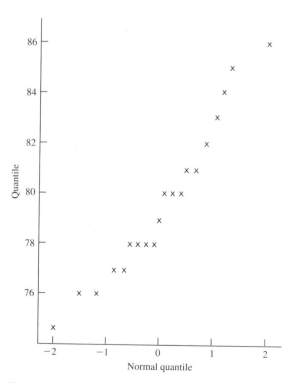

FIGURE 10.21 Normal quantile-quantile plot of data of alloy B.

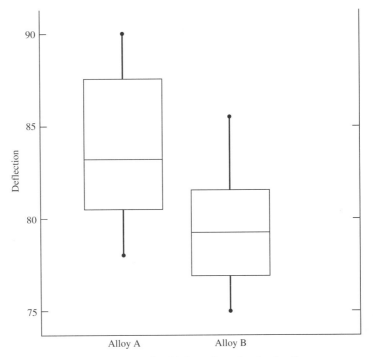

FIGURE 10.22   Box and whisker plots for both alloys.

t-test procedure

ALLOY	N	Mean	Std Dev	Std Error
1	20	83.55000000	3.66311630	0.81909771
2	20	79.70000000	3.09668753	0.69244038

Variances	T	DF	Prob > |T|
Unequal	3.5895	37.0	0.0010
Equal	3.5895	38.0	0.0009

For H0: Variances are equal, F' = 1.40 DF = (19,19)
Prob > F' = 0.4709

FIGURE 10.23   Annotated SAS printout for alloy data.

Statistical Significance and Engineering or Scientific Significance

While the statistician may feel quite comfortable with the results of the comparison between the two alloys in the case study above, a dilemma remains for the engineer. The analysis demonstrated a statistically significant improvement with the use of alloy *B*. However, is the difference found really worth it since alloy *B* is more expensive? This illustration highlights a very important issue often overlooked by statisticians and data analysts—*the distinction between statistical significance and engineering or scientific significance*. Here the average difference in deflection is $\bar{y}_A - \bar{y}_B = 0.00385$ inch. In a complete analysis the engineer must determine if the difference is sufficient to justify the extra cost in the long run.

This is an economic and engineering issue. The reader should understand that a statistically significant difference merely implies that the difference in the sample means found in the data could hardly have occurred by chance. It does not imply that the difference in the population means is profound or particularly significant in the context of the problem. For example, in Section 10.7, annotated computer printout was used to show evidence that a pH meter was, in fact, biased. That is, it did not demonstrate a mean pH of 7.00 for the material on which it was tested. But the variability among the observations in the sample is very small. The engineer may decide that the small deviations from 7.0 render the pH meter quite adequate.

Exercises

1. A die is tossed 180 times with the following results:

x	1	2	3	4	5	6
f	28	36	36	30	27	23

Is this a balanced die? Use a 0.01 level of significance.

2. In 100 tosses of a coin, 63 heads and 37 tails are observed. Is this a balanced coin? Use a 0.05 level of significance.

3. A machine is supposed to mix peanuts, hazelnuts, cashews, and pecans in the ratio $5:2:2:1$. A can containing 500 of these mixed nuts was found to have 269 peanuts, 112 hazelnuts, 74 cashews, and 45 pecans. At the 0.05 level of significance, test the hypothesis that the machine is mixing the nuts in the ratio $5:2:2:1$.

4. The grades in a statistics course for a particular semester were as follows:

Grade	A	B	C	D	F
f	14	18	32	20	16

Test the hypothesis, at the 0.05 level of significance, that the distribution of grades is uniform.

5. Three cards are drawn from an ordinary deck of playing cards, with replacement, and the number Y of spades is recorded. After repeating the experiment 64 times, the following outcomes were recorded:

y	0	1	2	3
f	21	31	12	0

Test the hypothesis of 0.01 level of significance that the recorded data may be fitted by the binomial distribution $b(y; 3, 1/4)$, $y = 0, 1, 2, 3$.

6. Three marbles are selected from an urn containing 5 red marbles and 3 green marbles. After recording the number X of red marbles, the marbles are replaced in the urn and the experiment repeated 112 times. The results obtained are as follows:

x	0	1	2	3
f	1	31	55	25

Test the hypothesis at the 0.05 level of significance that the recorded data may be fitted by the hypergeometric distribution $h(x; 8, 3, 5)$, $x = 0, 1, 2, 3$.

7. A coin is thrown until a head occurs and the number X of tosses recorded. After repeating the experiment 256 times, we obtained the following results:

x	1	2	3	4	5	6	7	8
f	136	60	34	12	9	1	3	1

Test the hypothesis at the 0.05 level of significance that the observed distribution of X may be fitted by the geometric distribution $g(x; 1/2)$, $x = 1, 2, 3, \ldots$

8. Repeat Exercise 5 using a new set of data obtained by actually carrying out the described experiment 64 times.

9. Repeat Exercise 7 using a new set of data obtained by performing the described experiment 256 times.

10. In Exercise 1 on page 67, test the goodness of fit between the observed class frequencies and the corresponding expected frequencies of a normal distribution with $\mu = 65$ and $\sigma = 21$, using a 0.05 level of significance.

11. In Exercise 5 on page 68, test the goodness of fit between the observed class frequencies and the corresponding expected frequencies of a normal distribution with $\mu = 1.8$ and $\sigma = 0.4$, using a 0.01 level of significance.

12. In an experiment to study the dependence of hypertension on smoking habits, the following data were taken on 180 individuals:

	Nonsmokers	Moderate Smokers	Heavy Smokers
Hypertension	21	36	30
No hypertension	48	26	19

Test the hypothesis that the presence or absence of hypertension is independent of smoking habits. Use a 0.05 level of significance.

13. A random sample of 90 adults are classified according to sex and the number of hours they watch television during a week:

	Sex	
	Male	Female
Over 25 hours	15	29
Under 25 hours	27	19

Use a 0.01 level of significance and test the hypothesis that the time spent watching television is independent of whether the viewer is male or female.

14. A random sample of 200 married men, all retired, were classified according to education and number of children:

Education	Number of Children		
	0–1	2–3	Over 3
Elementary	14	37	32
Secondary	19	42	17
College	12	17	10

Test the hypothesis, at the 0.05 level of significance, that the size of a family is independent of the level of education attained by the father.

15. A criminologist conducted a survey to determine whether the incidence of certain types of crime varied from one part of a large city to another. The particular crimes of interest were assault, burglary, larceny, and homicide. The following table shows the numbers of crimes committed in four areas of the city during the past year.

District	Type of Crime			
	Assault	Burglary	Larceny	Homicide
1	162	118	451	18
2	310	196	996	25
3	258	193	458	10
4	280	175	390	19

Can we conclude from these data at the 0.01 level of significance that the occurrence of these types of crime is dependent on the city district?

16. A college infirmary conducted an experiment to determine the degree of relief provided by three cough remedies. Each cough remedy was tried on 50 students and the following data recorded:

	Cough Remedy		
	NyQuil	Robitussin	Triaminic
No relief	11	13	9
Some relief	32	28	27
Total relief	7	9	14

Test the hypothesis, at the 0.05 level of significance, that the three cough remedies are equally effective.

17. To determine current attitudes about prayers in public schools, a survey was conducted in 4 Virginia counties. The following table gives the attitudes of 200 parents from Craig County, 150 parents from Giles County, 100 parents from Franklin County, and 100 parents from Montgomery County:

	County			
Attitude	Craig	Giles	Franklin	Montgomery
Favor	65	66	40	34
Oppose	42	30	33	42
No opinion	93	54	27	24

At the 0.01 level of significance, test for homogeneity of attitudes among the 4 counties concerning prayers in the public schools.

18. According to a new Johns Hopkins University study published in the *American Journal of Public Health,* widows survive longer than widowers. Consider the following survival data collected on 100 widows and 100 widowers following the death of a spouse:

Years Lived	Widow	Widower
Less than 5	25	39
5 to 10	42	40
More than 10	33	21

Can we conclude at the 0.05 level of significance that the proportions of widows and widowers are equal with respect to the different time periods that a spouse survives after the death of his or her mate?

19. The following responses concerning the standard of living at the time of an independent opinion poll of 1000 households versus one year earlier seems to be in agreement with the results of a study published in *Across the Board* (June 1981):

	Standard of Living			
Period	Somewhat Better	Same	Not as Good	Total
1980: Jan.	72	144	84	300
May	63	135	102	300
Sept.	47	100	53	200
1981: Jan.	40	105	55	200

Test the hypothesis at the 0.05 level of significance that the proportions of households within each standard of living category are the same for each of the four time periods.

20. A survey was conducted in Indiana, Kentucky, and Ohio to determine the attitude of voters concerning school busing. A poll of 200 voters from each of these states yielded the following results:

	Voter Attitude		
State	Support	Do Not Support	Undecided
Indiana	82	97	21
Kentucky	107	66	27
Ohio	93	74	33

At the 0.025 level of significance, test the null hypothesis that the proportions of voters within each attitude category are the same for each of the three states.

21. A survey was conducted in two Virginia cities to determine voter sentiment for two gubernatorial candidates in an upcoming election. Five hundred voters were randomly selected from each city and the following data were recorded:

	City	
Voter Sentiment	Richmond	Norfolk
Favor A	204	225
Favor B	211	198
Undecided	85	77

At the 0.05 level of significance, test the null hypothesis that proportions of voters favoring candidate A, candidate B, or undecided are the same for each city.

22. In a study to estimate the proportion of wives who regularly watch soap operas, it is found that 52 of 200 wives in Denver, 31 of 150 wives in Phoenix, and 37 of 150 wives in Rochester watch at least one soap opera. Use a 0.05 level of significance to test the hypothesis that there is no difference among the true proportions of wives who watch soap operas in these 3 cities.

Review Exercises

1. A geneticist is interested in the proportion of males and females in a population that have a certain minor blood disorder. In a random sample of 100 males, 31 are found to be afflicted, whereas only 24 of 100 females tested appear to have the disorder. Can we conclude at the 0.01 level of significance that the proportion of men in the population afflicted with this blood disorder is significantly greater than the proportion of women afflicted?

2. Consider the situation of Exercise 36 on p. 334. Oxygen consumption in ml/kg/min was also measured on the nine subjects.

Subject	With CO	Without CO
1	26.46	25.41
2	17.46	22.53
3	16.32	16.32
4	20.19	27.48
5	19.84	24.97
6	20.65	21.77
7	28.21	28.17
8	33.94	32.02
9	29.32	28.96

It is conjectured that the oxygen consumption should be higher in an environment relatively free of CO. Do a significance test and discuss the conjecture.

3. State the null and alternative hypotheses to be used in testing the following claims and determine generally where the critical region is located:

(a) The mean snowfall at Lake George during the month of February is 21.8 centimeters.

(b) No more than 20% of the faculty at the local university contributed to the annual giving fund.

(c) On the average, children attend schools within 6.2 kilometers of their homes in suburban St. Louis.

(d) At least 70% of next year's new cars will be in the compact and subcompact category.

(e) The proportion of voters favoring the incumbent in the upcoming election is 0.58.

(f) The average rib-eye steak at the Longhorn Steak House is at least 340 grams.

4. A study was made to determine whether more Italians than Americans prefer white champagne to pink champagne at weddings. Of the 300 Italians selected at random, 72 preferred white champagne, and of the 400 Americans selected, 70 preferred white champagne rather than pink. Can we conclude that a higher proportion of Italians than Americans prefer white champagne at weddings? Use a 0.05 level of significance.

5. In a set of data analyzed by the Statistics Consulting Center at Virginia Polytechnic Institute and State University, a group of subjects was asked to complete a certain task on the computer. The response measured was the time to completion. The purpose of the experiment was to test a set of facilitation tools developed by the Department of Computer Science at VPI & SU. There were 10 subjects involved. With a random assignment, five were given a standard procedure using Fortran language for completion of the task. The additional five were asked to do the task with the use of the facilitation tools. Following are the data on the completion times for the task.

Group 1 (Standard Procedure)	Group 2 (Facilitation Tool)
161	132
169	162
174	134
158	138
163	133

Assuming that the population variances are the same for the two groups, support or refute the conjecture that the facilitation tools increase the speed with which the task can be accomplished.

6. State the null and alternative hypotheses to be used in testing the following claims and determine generally where the critical region is located:

(a) At most, 20% of next year's wheat crop will be exported to the Soviet Union.

(b) On the average, American housewives drink 3 cups of coffee per day.

(c) The proportion of graduates in Virginia this year majoring in the social sciences is at least 0.15.

(d) The average donation to the American Lung Association is no more than $10.

(e) Residents in suburban Richmond commute, on the average, 15 kilometers to their place of employment.

7. If a can containing 500 nuts is selected at random from each of three different distributors of mixed nuts and there are, respectively, 345, 313, and 359 peanuts in each of the cans, can we conclude at the 0.01 level of significance that the mixed nuts of the three distributors contain equal proportions of peanuts?

8. **z-value for testing $p_1 - p_2 = d_0$:** To test the null hypothesis H_0 that $p_1 - p_2 = d_0$, where $d_0 \neq 0$, we base our decision on

$$z = \frac{\hat{p}_1 - \hat{p}_2 - d_0}{\sqrt{\hat{p}_1 \hat{q}_1 / n_1 + \hat{p}_2 \hat{q}_2 / n}},$$

which is a value of a random variable whose distribution approximates the standard normal distribution as long as n_1 and n_2 are both large.

(a) With reference to Example 10.12 on page 338, test the hypothesis that the percentage of town voters favoring the construction of the chemical plant will not exceed the percentage of county voters by more than 3%. Use a P-value in your conclusion.

(b) With reference to Exercise 11 on page 340, test the hypothesis at the 0.06 level of significance that brand A outsells brand B by 10% against the alternative hypothesis that the difference is less than 10%.

9. A study was made to determine whether there is a difference between the proportions of parents in the states of Maryland, Virginia, Georgia, and Alabama who favor placing Bibles in the elementary schools. The responses of 100 parents selected at random in each of these states are recorded in the following table:

	State			
Preference	*Maryland*	*Virginia*	*Georgia*	*Alabama*
Yes	65	71	78	82
No	35	29	22	18

Can we conclude that the proportions of parents who favor placing Bibles in the schools are the same for these four states? Use a 0.025 level of significance.

10. A study was conducted at the Virginia–Maryland Regional College of Veterinary Medicine Equine Center to determine if the performance of a certain type of surgery on young horses had any effect on certain kinds of blood cell types in the animal. Fluid samples were taken from each of six foals before and after surgery. The samples were analyzed for the number of postoperative WBC leukograms. A preoperative measure of WBC leukograms was also measured. Use a paired sample t-test to determine if there is a significant change in WBC leukograms with the surgery.

Foal	*Presurgery*	*Postsurgery*
1	10.80	10.60
2	12.90	16.60
3	9.59	17.20
4	8.81	14.00
5	12.00	10.60
6	6.07	8.60

All values $\times 10^{-3}$.

11. A study was conducted at the Department of Health and Physical Education at Virginia Polytechnic Institute and State University to determine if 8 weeks of lectures truly reduces the cholesterol levels of the participants. A treatment group consisting of 15 people were given lectures twice a week on how to reduce their cholesterol level. Another group of 18 people of similar age were randomly selected as a control group. All participants' cholesterol levels were recorded at the end of the 8-week program and are listed below.

Treatment:
129 131 154 172 115 126 175 191 122 238 159
156 176 175 126

Control:
151 132 196 195 188 198 187 168 115 165 137
208 133 217 191 193 140 146

Can we conclude, at the 5% level of significance, that the average cholesterol level has been reduced due to the program?

12. In a study conducted by the Department of Mechanical Engineering and analyzed by the Statistics Consulting Center at the Virginia Polytechnic Institute

and State University, the steel rods supplied by two different companies were compared. Ten sample springs were made out of the steel rods supplied by each company and the "bounciness" was studied. The data are as follows:

Company A:
9.3 8.8 6.8 8.7 8.5 6.7 8.0 6.5 9.2 7.0

Company B:
11.0 9.8 9.9 10.2 10.1 9.7 11.0 11.1 10.2 9.6

Can you conclude that there is virtually no difference between the steel rods supplied by the two companies? Use a P-value to reach your conclusion.

13. In a study conducted by the Water Resources Center and analyzed by the Statistics Consulting Center at the Virginia Polytechnic Institute and State University, two different wastewater treatment plants are compared. Plant A is located where the median household income is below \$15,000 a year, and plant B is located where the median household income is above \$40,000 a year. The amount of wastewater treated at each plant (thousand gallons/day) was ran-

domly sampled for 10 days during July 1990. The data are as follows:

Plant A:
21 19 20 23 22 28 32 19 13 18

Plant B:
20 39 24 33 30 28 30 22 33 24

Can we conclude, at the 5% level of significance, that the average amount of wastewater treated at the high-income neighborhood is more than that from the low-income area?

14. The following data show the number of defects in 100,000 lines of code in a particular type of software program made in the United States and Japan. Is there enough evidence to claim that there is a significant difference between the programs of the two countries?

United	48	39	42	52	40	48	52	52
States	54	48	52	55	43	46	48	52
Japan	50	48	42	40	43	48	50	46
	38	38	36	40	40	48	48	45

11

Simple Linear Regression and Correlation

11.1 Introduction to Linear Regression

Often, in practice, one is called upon to solve problems involving sets of variables when it is known that there exists some inherent relationship among the variables. For example, in an industrial situation it may be known that the tar content in the outlet stream in a chemical process is related to the inlet temperature. It may be of interest to develop a method of prediction, that is, a procedure for estimating the tar content for various levels of the inlet temperature from experimental information. The statistical aspect of the problem then becomes one of arriving at the best estimate of the relationship between the variables.

For this example and most applications there is a clear distinction between the variables as far as their role in the experimental process is concerned. Quite often there is a single *dependent variable* or response Y, which is uncontrolled in the experiment. This response depends on one or more *independent* **regressor variables**, say $x_1, x_2, \ldots, x_k$, which are measured with negligible error and indeed are often controlled in the experiment. Thus the independent variables $x_1, x_2, \ldots, x_k$ are *not* random variables and therefore have no distributional properties. In the example cited earlier, inlet temperature is the independent variable or regressor variable x and tar content is the response Y. The relationship fit to a set of experimental data is characterized by a prediction equation called a **regression equation**. In the case of a single Y and a single x, the situation becomes a regression of Y on x. For k independent variables, we speak in terms of a regression of Y on x_1, $x_2, \ldots, x_k$. A chemical engineer may, in fact, be concerned with the amount of hydrogen lost from samples of a particular metal when the material is placed in storage. In this case there may be two inputs, storage time x_1 in hours and storage temperature x_2 in degrees centigrade. The response would then be hydrogen loss, Y, in parts per million.

In this chapter we shall deal with the topic of **simple linear regression**, treating only the case of a single regressor variable. For the case of more than one regressor variable, the reader is referred to Chapter 12. Let us denote a random sample of size n by the set $\{(x_i, y_i); i = 1, 2, \ldots, n\}$. If additional samples were taken using exactly the same values of x, we should expect the y values to vary. Hence the value y_i in the ordered pair (x_i, y_i) is a value of some random variable Y_i. For convenience we define $Y|x$ to be the random variable Y corresponding to a fixed value x and denote its mean and variance by $\mu_{Y|x}$ and $\sigma^2_{Y|x}$, respectively. Clearly then, if $x = x_i$, the symbol $Y|x_i$ represents the random variable Y_i with mean $\mu_{Y|x_i}$ and variance $\sigma^2_{Y|x_i}$.

The term **linear regression** implies that $\mu_{Y|x}$ is linearly related to x by the **population regression equation**

$$\mu_{Y|x} = \alpha + \beta x,$$

where the **regression coefficients** α and β are parameters to be estimated from the sample data. Denoting their estimates by a and b, respectively, we can then estimate $\mu_{Y|x}$ by $\hat{y}$ from the sample regression or the **fitted regression line**

$$\hat{y} = a + bx,$$

where the estimates a and b represent the y intercept and slope, respectively. The symbol $\hat{y}$ is used here to distinguish between the estimated or predicted value given by the sample regression line and an actual observed experimental value y for some value of x.

One of the more challenging problems confronting the water pollution control field is presented by the tanning industry. Tannery wastes are chemically complex. They are characterized by high values of biochemical oxygen demand, volatile solids, and other pollution measures. Consider the experimental data of Table 11.1, which was obtained from 33 samples of chemically treated waste in the study *"Chemical Treatment of Spent Vegetable Tan Liquor"* conducted at the Virginia Polytechnic Institute and State University in 1970. Readings on x, the percent reduction in total solids, and y, the percent reduction in chemical oxygen demand for the 33 samples were recorded.

The data of Table 11.1 have been plotted in Figure 11.1 to give a **scatter diagram**. From an inspection of this scatter diagram, it is seen that the points follow closely a straight line, indicating that the assumption of linearity between the two variables appears to be reasonable.

TABLE 11.1   Measures of Solids and Chemical Oxygen
Demand

Solids Reduction, x (%)	Chemical Oxygen Demand, y (%)
3	5
7	11
11	21
15	16
18	16
27	28
29	27
30	25
30	35
31	30
31	40
32	32
33	34
33	32
34	34
36	37
36	38
36	34
37	36
38	38
39	37
39	36
39	45
40	39
41	41
42	40
42	44
43	37
44	44
45	46
46	46
47	49
50	51

The fitted regression line and a hypothetical true regression line have been drawn on the scatter diagram of Figure 11.1. One expects that the agreement between the sample line and the unknown hypothetical line will be good when a large amount of data is available.

In the following section we shall develop procedures for finding estimates of the regression coefficients, α and β, in order that the regression equation can be used for predicting or estimating the mean response or an individual response for a specific value of the independent variable x.

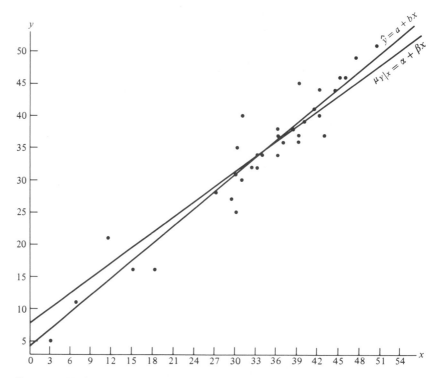

FIGURE 11.1   Scatter diagram with regression lines.

11.2 Simple Linear Regression

In the case of simple linear regression where there is a single independent regressor variable x and a single dependent random variable Y, the data may be represented by the pairs of observations $\{(x_i, y_i); i = 1, 2, \ldots, n\}$. It is informative to use the concepts from the preceding section to define each random variable $Y_i = Y|x_i$ by means of a **statistical model**. If we postulate that all means $\mu_{Y|x_i}$ fall on a straight line, each Y_i may be described by the **simple linear regression model**

$$Y_i = \mu_{Y|x_i} + E_i = \alpha + \beta x_i + E_i,$$

where the random error E_i, the model error, must necessarily have a mean of zero. Each observation (x_i, y_i) in our sample satisfies the equation

$$y_i = \alpha + \beta x_i + \varepsilon_i,$$

where ε_i is the value assumed by E_i when Y_i takes on the value y_i. *The above equation can be viewed as the model for a single observation* y_i.

Similarly, using the estimated or fitted regression line

$$\hat{y} = a + bx,$$

each pair of observations satisfies the relation

$$y_i = a + bx_i + e_i,$$

where $e_i = y_i - \hat{y}_i$ is called a **residual** and describes the error in the fit of the model at the ith data point. The difference between e_i and ε_i is clearly shown in Figure 11.2.

Figure 11.1 depicts the line fit to this set of data, namely $\hat{y} = a + bx$, and the line reflecting the model $\mu_{Y|x} = \alpha + \beta x$. Now, of course, α and β are unknown parameters. The fitted line is an estimate of the line produced by the statistical model. Keep in mind that the line $\mu_{Y|x} = \alpha + \beta x$ is not known but rather is a simple conceptual notion of how the data were generated in the scientific process. As a result, the realization of E_i, namely ε_i, is never actually observed. However, the residual e_i is observed.

The Method of Least Squares

We shall find a and b, the estimates of α and β, so that the sum of the squares of the residuals is a minimum. The residual sum of squares is often called the sum of squares of the errors about the regression line and denoted by SSE. This minimization procedure for estimating the parameters is called the **method of least squares**. Hence we shall find a and b so as to minimize

$$SSE = \sum_{i=1}^{n} e_i^2 = \sum_{i=1}^{n} (y_i - \hat{y}_i)^2 = \sum_{i=1}^{n} (y_i - a - bx_i)^2.$$

Differentiating SSE with respect to a and b, we have

$$\frac{\partial(SSE)}{\partial a} = -2 \sum_{i=1}^{n} (y_i - a - bx_i)$$

$$\frac{\partial(SSE)}{\partial b} = -2 \sum_{i=1}^{n} (y_i - a - bx_i)x_i.$$

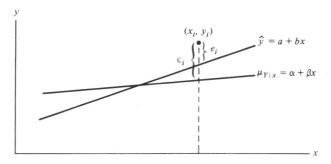

FIGURE 11.2 Comparing ε_i with the residual e_i.

Setting the partial derivatives equal to zero and rearranging the terms, we obtain the equations (called the **normal equations**)

$$na + b \sum_{i=1}^{n} x_i = \sum_{i=1}^{n} y_i$$

$$a \sum_{i=1}^{n} x_i + b \sum_{i=1}^{n} x_i^2 = \sum_{i=1}^{n} x_i y_i,$$

which may be solved simultaneously to yield computing formulas for a and b.

ESTIMATING THE REGRESSION COEFFICIENTS

Given the sample $\{(x_i, y_i); i = 1, 2, \ldots, n\}$, the least squares estimates a and b of the regression coefficients α and β are computed from the formulas

$$b = \frac{n \sum_{i=1}^{n} x_i y_i - \left(\sum_{i=1}^{n} x_i \right) \left(\sum_{i=1}^{n} y_i \right)}{n \sum_{i=1}^{n} x_i^2 - \left(\sum_{i=1}^{n} x_i \right)^2} = \frac{\sum_{i=1}^{n} (x_i - \bar{x})(y_i - \bar{y})}{\sum_{i=1}^{n} (x_i - \bar{x})^2}$$

and

$$a = \frac{\sum_{i=1}^{n} y_i - b \sum_{i=1}^{n} x_i}{n} = \bar{y} - b\bar{x}.$$ ∎

The calculations of a and b using the data of Table 11.1 are illustrated in the following example.

EXAMPLE 11.1   Estimate the regression line for the pollution data of Table 11.1.

SOLUTION

$$\sum_{i=1}^{33} x_i = 1104, \qquad \sum_{i=1}^{33} y_i = 1124, \qquad \sum_{i=1}^{33} x_i y_i = 41,355, \qquad \sum_{i=1}^{33} x_i^2 = 41,086.$$

Therefore,

$$b = \frac{(33)(41,355) - (1104)(1124)}{(33)(41,086) - (1104)^2} = 0.903643$$

and

$$a = \frac{1124 - (0.903643)(1104)}{33} = 3.829633.$$

Thus the estimated regression line is given by

$$\hat{y} = 3.8296 + 0.9036x.$$

where the coefficients are rounded to four decimals. By substituting two values of x into this equation, say $x = 3$ and $x = 50$, we obtain the predicted values $\hat{y} = 6.5$ and $\hat{y} = 49$. The sample regression line in Figure 11.1 was drawn by connecting these two points with a straight line.

Using the regression line of Example 11.1, we would predict a 31% reduction in the chemical oxygen demand when the reduction in the total solids is 30%. The 31% reduction in the chemical oxygen demand may be interpreted as an estimate of the population mean $\mu_{Y|30}$ or as an estimate of a new observation when the reduction in total solids is 30%. Such estimates, however, are subject to error. Even when the experiment is controlled so that the reduction in total solids is 30%, it is unlikely that we would measure a reduction in the chemical oxygen demand exactly equal to 31%. In fact, the original data recorded in Table 11.1 show that measurements of 25% and 35% were recorded for the reduction in oxygen demand when the reduction in total solids was kept at 30%.

Exercises

1. The study "*Development of LIFTEST, A Dynamic Technique to Assess Individual Capability to Lift Material*" was conducted at the Virginia Polytechnic Institute and State University in 1982 to determine if certain static arm strength measures have an influence on the "dynamic lift" characteristics of an individual. Twenty-five individuals were subjected to strength tests and then were asked to perform a weight-lifting test in which weight was dynamically lifted overhead. The data are as follows:

Individual	Arm Strength, x	Dynamic Lift, y
1	17.3	71.7
2	19.3	48.3
3	19.5	88.3
4	19.7	75.0
5	22.9	91.7
6	23.1	100.0
7	26.4	73.3
8	26.8	65.0
9	27.6	75.0
10	28.1	88.3
11	28.2	68.3
12	28.7	96.7
13	29.0	76.7
14	29.6	78.3
15	29.9	60.0
16	29.9	71.7
17	30.3	85.0
18	31.3	85.0
19	36.0	88.3
20	39.5	100.0
21	40.4	100.0
22	44.3	100.0
23	44.6	91.7
24	50.4	100.0
25	55.9	71.7

(a) Estimate α and β for the linear regression curve $\mu_{Y|x} = \alpha + \beta x$.

(b) Find a point estimate of $\mu_{Y|30}$.

2. The grades of a class of 9 students on a midterm report (x) and on the final examination (y) are as follows:

x	77	50	71	72	81	94	96	99	67
y	82	66	78	34	47	85	99	99	68

(a) Estimate the linear regression line.

(b) Estimate the final examination grade of a student who received a grade of 85 on the midterm report.

3. A study was made on the amount of converted sugar in a certain process at various temperatures. The data were coded and recorded as follows:

Temperature, x	Converted Sugar, y
1.0	8.1
1.1	7.8
1.2	8.5
1.3	9.8
1.4	9.5
1.5	8.9
1.6	8.6
1.7	10.2
1.8	9.3
1.9	9.2
2.0	10.5

(a) Estimate the linear regression line.

(b) Estimate the mean amount of converted sugar produced when the coded temperature is 1.75.

4. In a certain type of metal test specimen, the normal stress on a specimen is known to be functionally related to the shear resistance. The following is a set of coded experimental data on the two variables:

Normal Stress, x	Shear Resistance, y
26.8	26.5
25.4	27.3
28.9	24.2
23.6	27.1
27.7	23.6
23.9	25.9
24.7	26.3
28.1	22.5
26.9	21.7
27.4	21.4
22.6	25.8
25.6	24.9

(a) Estimate the regression line $\mu_{Y|x} = \alpha + \beta x$.
(b) Estimate the shear resistance for a normal stress of 24.5 kilograms per square centimeter.

5. The amounts of a chemical compound y, which dissolved in 100 grams of water at various temperatures, x, were recorded as follows:

x (°C)	y (grams)		
0	8	6	8
15	12	10	14
30	25	21	24
45	31	33	28
60	44	39	42
75	48	51	44

(a) Find the equation of the regression line.
(b) Graph the line on a scatter diagram.
(c) Estimate the amount of chemical that will dissolve in 100 grams of water at 50°C.

6. A mathematics placement test is given to all entering freshmen at a small college. A student who receives a grade below 35 is denied admission to the regular mathematics course and placed in a remedial class. The placement test scores and the final grades for 20 students who took the regular course were recorded as follows:

Placement Test	Course Grade	Placement Test	Course Grade
50	53	90	54
35	41	80	91
35	61	60	48
40	56	60	71
55	68	60	71
65	36	40	47
35	11	55	53
60	70	50	68
90	79	65	57
35	59	50	79

(a) Plot a scatter diagram.
(b) Find the equation of the regression line to predict course grades from placement test scores.
(c) Graph the line on the scatter diagram.
(d) If 60 is the minimum passing grade, below which placement test score should students in the future be denied admission to this course?

7. A study was made by a retail merchant to determine the relation between weekly advertising expenditures and sales. The following data were recorded:

Advertising Costs ($)	Sales ($)
40	385
20	400
25	395
20	365
30	475
50	440
40	490
20	420
50	560
40	525
25	480
50	510

(a) Plot a scatter diagram.
(b) Find the equation of the regression line to predict weekly sales from advertising expenditures.
(c) Estimate the weekly sales when advertising costs are $35.

8. The following data were collected to determine the relationship between high school rank in class and grade-point average at the end of the freshman year in college:

Grade-Point Average, y	Decile Rank, x	Grade-Point Average, y	Decile Rank, x
1.93	3	1.40	8
2.55	2	1.45	4
1.72	1	1.72	8
2.48	1	3.80	1
2.87	1	2.13	5
1.87	3	1.81	6
1.34	4	2.33	1
3.03	1	2.53	1
2.54	2	2.04	2
2.34	2	3.20	2

(a) Find the equation of the regression line to predict the grade-point average of college freshmen from high school rank in class.

(b) Predict the grade-point average for an entering freshman who ranks in the third decile of her graduating class.

9. In a study between the amount of rainfall and the quantity of air pollution removed, the following data were collected:

Daily Rainfall, x (0.01 centimeter)	Particulate Removed, y (micrograms per cubic meter)
4.3	126
4.5	121
5.9	116
5.6	118
6.1	114
5.2	118
3.8	132
2.1	141
7.5	108

(a) Find the equation of the regression line to predict the particulate removed from the amount of daily rainfall.

(b) Estimate the amount of particulate removed when the daily rainfall is $x = 4.8$ units.

10. The following data are the selling prices z of a certain make and model of used cars w years old:

w (years)	z (dollars)
1	6350
2	5695
2	5750
3	5395
5	4985
5	4895

(a) Fit a curve of the form $\mu_{z|w} = \gamma \delta^w$ by means of the nonlinear sample regression equation $\hat{z} = cd^w$. HINT: Write

$$\ln \hat{z} = \ln c + (\ln d)w$$

$$= a + bw.$$

11.3 Properties of the Least Squares Estimators _____

In addition to the assumptions that the error term in the model

$$Y_i = \alpha = \beta x_i + E_i$$

is a random variable with mean zero, suppose that we make the further assumption that each E_i has the same variance σ^2 and that $E_1, E_2, \ldots, E_n$ are independent from run to run in the experiment. With these assumptions on the E_i's, we have a procedure for finding the means and variances for the estimators of α and β.

It is important to remember that our values of a and b, based on a given sample of n observations, are only estimates of true parameters α and β. If the experiment is repeated over and over again, each time using the same fixed values of x, the resulting estimates of α and β will most likely differ from experiment to experiment. These different estimates may be viewed as values assumed by the random variables A and B.

Since the values of x remain fixed, the values of A and B depend on the variations in the values of y, or, more precisely, on the values of the random variables, Y_1, Y_2, . . . , Y_n. The distributional assumptions on the E_i's imply that the Y_i's, $i = 1, 2,$ . . . , n, are also independently distributed, with mean $\mu_{Y|x_i} = \alpha + \beta x_i$ and equal variances σ^2; that is, $\sigma^2_{Y|x_i} = \sigma^2$ for $i = 1, 2, \ldots, n$.

In what follows we show that the estimator B is unbiased for β and demonstrate the variances of both A and B. This will begin a series of developments that lead to hypothesis testing and confidence interval estimation on the intercept and slope.

Since the estimator

$$B = \frac{\sum\limits_{i=1}^{n} (x_i - \bar{x})(Y_i - \bar{Y})}{\sum\limits_{i=1}^{n} (x_i - \bar{x})^2}$$

$$= \frac{\sum\limits_{i=1}^{n} (x_i - \bar{x})Y_i}{\sum\limits_{i=1}^{n} (x_i - \bar{x})^2}$$

is of the form $\sum\limits_{i=1}^{n} a_i Y_i$, where

$$a_i = \frac{x_i - \bar{x}}{\sum\limits_{i=1}^{n} (x_i - \bar{x})^2}, \qquad i = 1, 2, \ldots, n,$$

we may conclude from Corollary 2 of Theorem 4.7 that

$$\mu_B = E(B) = \frac{\sum\limits_{i=1}^{n} (x_i - \bar{x})E(Y_i)}{\sum\limits_{i=1}^{n} (x_i - \bar{x})^2}$$

$$= \frac{\sum\limits_{i=1}^{n} (x_i - \bar{x})(\alpha + \beta x_i)}{\sum\limits_{i=1}^{n} (x_i - \bar{x})^2} = \beta$$

and then using Corollary 3 of Theorem 4.10,

$$\sigma_B^2 = \frac{\sum\limits_{i=1}^{n} (x_i - \bar{x})^2 \sigma_{Y_i}^2}{\left[\sum\limits_{i=1}^{n} (x_i - \bar{x})^2 \right]^2} = \frac{\sigma^2}{\sum\limits_{i=1}^{n} (x_i - \bar{x})^2}.$$

It can be shown (Exercise 1 on page 383) that the random variable A has the mean

$$\mu_A = \alpha$$

and variance

$$\sigma_A^2 = \frac{\sum\limits_{i=1}^{n} x_i^2}{n \sum\limits_{i=1}^{n} (x_i - \bar{x})^2} \, \sigma^2.$$

From the foregoing results, it becomes apparent that the least squares estimators for α and β are both unbiased estimators.

To be able to draw inferences on α and β, it becomes necessary to arrive at an estimate of the parameter σ^2 appearing in the two preceding variance formulas for A and B. The parameter σ^2, the model error variance, reflects random variation or experimental error variation, around the regression line. In much of what follows it is advantageous to use the notation

$$S_{xx} = \sum_{i=1}^{n} (x_i - \bar{x})^2; \; S_{yy} = \sum_{i=1}^{n} (y_i - \bar{y})^2; \; S_{xy} = \sum_{i=1}^{n} (x_i - \bar{x})(y_i - \bar{y}).$$

Now we may write the error sum of squares as follows:

$$SSE = \sum_{i=1}^{n} (y_i - a - bx_i)^2$$

$$= \sum_{i=1}^{n} [(y_i - \bar{y}) - b(x_i - \bar{x})]^2$$

$$= \sum_{i=1}^{n} (y_i - \bar{y})^2 - 2b \sum_{i=1}^{n} (x_i - \bar{x})(y_i - \bar{y}) + b^2 \sum_{i=1}^{n} (x_i - \bar{x})^2$$

$$= S_{yy} - 2bS_{xy} + b^2 S_{xx}$$

$$= S_{yy} - bS_{xy},$$

the final step following from the fact that $b = S_{xy}/S_{xx}$.

THEOREM 11.1   *An unbiased estimate of σ^2 is given by*

$$s^2 = \frac{SSE}{n-2} = \sum_{i=1}^{n} \frac{(y_i - \hat{y}_i)^2}{n-2} = \frac{S_{yy} - bS_{xy}}{n-2}. \qquad \blacksquare$$

The proof of Theorem 11.1 is left as an exercise (see Review Exercise 7).

11.4 Inferences Concerning the Regression Coefficients

Aside from merely estimating the linear relationship between x and Y for purposes of prediction, the experimenter may also be interested in drawing certain inferences about the slope and intercept. In order to allow for the testing of hypotheses and the construction of confidence intervals on α and β, one must be willing to make the further assumption that each E_i, $i = 1, 2, \ldots, n$, is normally distributed. This assumption implies that $Y_1, Y_2, \ldots, Y_n$ are also normally distributed, each with probability distribution $n(y_i; \alpha + \beta x_i, \sigma)$. Since A and B are linear functions of independent normal variables, we may now deduce from Theorem 7.11 that A and B are normally distributed with probability distributions $n(a; \alpha, \sigma_A)$ and $n(b; \beta, \sigma_B)$, respectively

It turns out that under the normality assumption, a result very much analogous to that given in Theorem 8.4 allows us to conclude that $(n-2)S^2/\sigma^2$ is a chi-squared variable with $n-2$ degrees of freedom, independent of the random variable B. Theorem 8.5 then assures us that the statistic

$$T = \frac{(B - \beta)/(\sigma/\sqrt{S_{xx}})}{S/\sigma}$$

$$= \frac{B - \beta}{S/\sqrt{S_{xx}}}$$

has a t-distribution with $n-2$ degrees of freedom. The statistic T can be used to construct a $(1 - \alpha)100\%$ confidence interval for the coefficient β.

CONFIDENCE INTERVAL FOR β   *A $(1 - \alpha)100\%$ confidence interval for the parameter β in the regression line $\mu_{Y|x} = \alpha + \beta x$ is*

$$b - \frac{t_{\alpha/2} s}{\sqrt{S_{xx}}} < \beta < b + \frac{t_{\alpha/2} s}{\sqrt{S_{xx}}},$$

where $t_{\alpha/2}$ is a value of the t-distribution with $n-2$ degrees of freedom. $\blacksquare$

EXAMPLE 11.2   Find a 95% confidence interval for β in the regression line $\mu_{Y|x} = \alpha + \beta x$ based on the pollution data in Table 11.1.

SOLUTION
In Example 11.1 we found that

$$S_{xx} = 4152.18, \qquad S_{xy} = 3752.09.$$

In addition, we find that $S_{yy} = 3713.88$. Recall that $b = 0.903643$. Hence

$$s^2 = \frac{S_{yy} - bS_{xy}}{n - 2}$$

$$= \frac{3713.88 - (0.903643)(3752.09)}{31} = 10.4299.$$

Therefore, taking the square root we obtain $s = 3.2295$. Using Table A.4, we find $t_{0.025} \approx 2.045$ for 31 degrees of freedom. Therefore, a 95% confidence interval for β is given by

$$0.903643 - \frac{(2.045)(3.2295)}{\sqrt{4152.18}} < \beta < 0.903643 + \frac{(2.045)(3.2295)}{\sqrt{4152.18}},$$

which simplifies to

$$0.8011 < \beta < 1.0061.$$

To test the null hypothesis H_0 that $\beta = \beta_0$ against a suitable alternative, we again use the t-distribution with $n - 2$ degrees of freedom to establish a critical region and then base our decision on the value of

$$t = \frac{b - \beta_0}{s/\sqrt{S_{xx}}}.$$

The method is illustrated in the following example.

EXAMPLE 11.3   Using the estimated value $b = 0.903643$ of Example 11.1, test the hypothesis that $\beta = 1.0$ against the alternative that $\beta < 1.0$.

SOLUTION

$$H_0: \quad \beta = 1.0.$$

$$H_1: \quad \beta < 1.0.$$

$$t = \frac{0.903643 - 1.0}{3.2295/\sqrt{4152.18}} = -1.92$$

with $n - 2 = 31$ degrees of freedom ($P \approx 0.03$).
Decision: The t-value is significant at the 0.03 level, suggesting strong evidence that $\beta < 1.0$.

Confidence intervals and hypothesis testing on the coefficient α may be established from the fact that A is also normally distributed. It is not difficult to show that

$$T = \frac{A - \alpha}{S \sqrt{\sum_{i=1}^{n} x_i^2 / nS_{xx}}}$$

has a t-distribution with $n - 2$ degrees of freedom from which we may construct a $(1 - \alpha)100\%$ confidence interval for α.

CONFIDENCE INTERVAL FOR α

A $(1 - \alpha)100\%$ *confidence interval for the parameter α in the regression line* $\mu_{Y|x} = \alpha + \beta x$ *is*

$$a - \frac{t_{\alpha/2} s \sqrt{\sum_{i=1}^{n} x_i^2}}{\sqrt{nS_{xx}}} < \alpha < a + \frac{t_{\alpha/2} s \sqrt{\sum_{i=1}^{n} x_i^2}}{\sqrt{nS_{xx}}},$$

where $t_{a/2}$ is a value of the t-distribution with $n - 2$ degrees of freedom. ■

Note that the symbol α is being used here in two totally unrelated ways, first as the level of significance and then as the intercept of the regression line.

EXAMPLE 11.4   Find a 95% confidence interval for α in the regression line $\mu_{Y|x} = \alpha + \beta x$, based on the data in Table 11.1.

SOLUTION
In Examples 11.1 and 11.2 we found that $S_{xx} = 4152.18$ and $s = 3.2295$. From Example 11.1 we had $\sum_{i=1}^{n} x_i^2 = 41,086$ and $a = 3.829633$. Using Table A.4, we find $t_{0.025} \simeq 2.045$ for 31 degrees of freedom. Therefore, a 95% confidence interval for α is given by

$$3.829633 - \frac{(2.045)(3.2295)\sqrt{41,086}}{\sqrt{(33)(4152.18)}} < \alpha < 3.829633$$

$$+ \frac{(2.045)(3.2295)\sqrt{41,086}}{\sqrt{(33)(4152.18)}},$$

which simplifies to

$$0.2131 < \alpha < 7.4461.$$

To test the null hypothesis H_0 that $\alpha = \alpha_0$ against a suitable alternative, we can use the t-distribution with $n - 54$ 2 degrees of freedom to establish a critical region and then base our decision on the value of

$$t = \frac{a - \alpha_0}{s \sqrt{\sum_{i=1}^{n} x_i^2 / nS_{xx}}}.$$

EXAMPLE 11.5   Using the estimated value $a = 3.829640$ in Example 11.1, test the hypothesis that $\alpha = 0$ at the 0.05 level of significance against the alternative that $\alpha \neq 0$.

SOLUTION

$$H_0: \quad \alpha = 0.$$
$$H_1: \quad \alpha \neq 0.$$

$$t = \frac{3.829633 - 0}{3.2295\sqrt{41,086/(33)(4152.18)}} = 2.17$$

with 31 degrees of freedom. Thus $P = 0.02$ and conclude that $\alpha \neq 0$.

11.5 Prediction

There are several purposes for building a linear regression. One, of course, is to predict response values at one or more values of the independent variable. In this section the focus will be on errors associated with prediction.

The equation $\hat{y} = a + bx$ may be used to predict or estimate the **mean response** $\mu_{Y|x_0}$ at $x = x_0$, where x_0 is not necessarily one of the prechosen values, or it may be used to predict a single value y_0 of the variable Y_0 when $x = x_0$. We would expect the error of prediction to be higher in the case of a single predicted value than in the case where a mean is predicted. This, then, will affect the width of our intervals for the values being predicted.

Suppose that the experimenter wishes to construct a confidence interval for $\mu_{Y|x_0}$. We shall use the point estimator $\hat{Y}_0 = A + Bx_0$ to estimate $\mu_{Y|x_0} = \alpha + \beta x_0$. It can be shown that the sampling distribution of $\hat{Y}_0$ is normal with mean

$$\mu_{\hat{Y}_0} = E(\hat{Y}_0) = E(A + Bx_0) = \alpha + \beta x_0 = \mu_{Y|x_0}$$

and variance

$$\sigma_{\hat{Y}_0}^2 = \sigma_{A+Bx_0}^2 = \sigma_{\bar{Y}+B(x_0-\bar{x})}^2$$
$$= \sigma^2 \left[\frac{1}{n} + \frac{(x_0 - \bar{x})^2}{S_{xx}} \right],$$

the latter following from the fact that $\text{Cov}(\bar{Y}, B) = 0$ (see Exercise 2 on page 383). Thus the $(1 - \alpha)100\%$ confidence interval on the mean response $\mu_{Y|x_0}$ can now be constructed from the statistic

$$T = \frac{\hat{Y}_0 - \mu_{Y|x_0}}{S\sqrt{(1/n) + [(x_0 - \bar{x})^2/S_{xx}]}},$$

which has a t-distribution with $n - 2$ degrees of freedom.

CONFIDENCE INTERVAL FOR $\mu_{Y|x_0}$

A $(1 - \alpha)100\%$ confidence interval for the mean response $\mu_{Y|x_0}$ is given by

$$\hat{y}_0 - t_{\alpha/2}s\sqrt{\frac{1}{n} + \frac{(x_0 - \bar{x})^2}{S_{xx}}} < \mu_{Y|x_0} < \hat{y}_0 + t_{\alpha/2}s\sqrt{\frac{1}{n} + \frac{(x_0 - \bar{x})^2}{S_{xx}}},$$

where $t_{\alpha/2}$ is a value of the t-distribution with $n - 2$ degrees of freedom.   ■

EXAMPLE 11.6   Using the data of Table 11.1, construct 95% confidence limits for the mean response $\mu_{Y|x}$.

SOLUTION
From the regression equation we find for $x_0 = 20$, say,

$$\hat{y}_0 = 3.829633 + (0.903643)(20) = 21.9025.$$

In addition, $\bar{x} = 33.4545$, $S_{xx} = 4152.18$, $s = 3.2295$, and $t_{0.025} \simeq 2.045$ for 31 degrees of freedom. Therefore, a 95% confidence interval for $\mu_{Y|20}$ is given by

$$21.9025 - (2.045)(3.2295)\sqrt{\frac{1}{33} + \frac{(20 - 33.4545)^2}{4152.18}} < \mu_{Y|20}$$

$$< 21.9025 + (2.045)(3.2295)\sqrt{\frac{1}{33} + \frac{(20 - 33.4545)^2}{4152.18}}$$

or simply

$$20.1071 < \mu_{Y|20} < 23.6979.$$

Repeating the previous calculations for each of several different values of x_0, one can obtain the corresponding confidence limits on each $\mu_{Y|x_0}$. Figure 11.3 displays the data points, the estimated regression line, and the upper and lower confidence limits on the mean of $Y|x$.

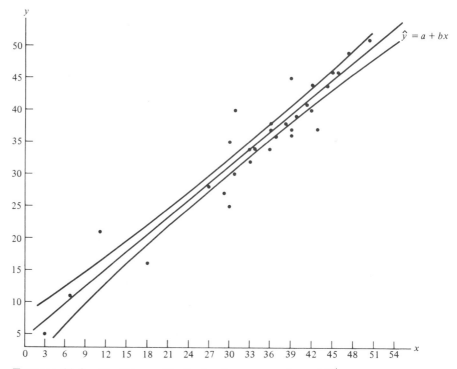

FIGURE 11.3   Confidence limits for the mean value of $Y|x$.

In Example 11.6 we are 95% confident that the population chemical oxygen demand is between 20.1071 and 23.6979%.

Prediction Interval

Another type of interval that is often misinterpreted and confused with that given for $\mu_{Y|x}$ is the prediction interval on a future observed response. Actually, in many instances the prediction interval is more relevant to the scientist or engineer than the confidence interval on the mean. In the tar content–inlet temperature example, cited in Section 11.1, there would certainly be interest not only in estimating the mean tar content at a specific temperature but also in constructing an interval that reflects the error in predicting a future observed amount of tar content at the given temperature.

To obtain a **prediction interval** for any single value y_0 of the variable Y_0, it is necessary to estimate the variance of the differences between the ordinates $\hat{y}_0$, obtained from the computed regression lines in repeated sampling when $x = y_0$, and the corresponding true ordinate y_0. We can think of the difference $\hat{y}_0 - y_0$ as a value of the random variable $\hat{Y}_0 - Y_0$, whose sampling distribution can be shown to be normal with mean

$$
\begin{aligned}
\mu_{\hat{Y}_0 - Y_0} &= E(\hat{Y}_0 - Y_0) \\
&= E[A + Bx_0 - (\alpha + \beta x_0 + E_0)] \\
&= 0
\end{aligned}
$$

and variance

$$
\begin{aligned}
\sigma^2_{\hat{Y}_0 - Y_0} &= \sigma^2_{A + Bx_0 - E_0} \\
&= \sigma^2_{\bar{Y}_0 + B(x_0 - \bar{x}) - E_0} \\
&= \sigma^2 \left[1 + \frac{1}{n} + \frac{(x_0 - \bar{x})^2}{S_{xx}} \right].
\end{aligned}
$$

Thus the $(1 - \alpha)100\%$ prediction interval for a single predicted value y_0 can be constructed from the statistic

$$
T = \frac{\hat{Y}_0 - Y_0}{S\sqrt{1 + (1/n) + [(x_0 - \bar{x})^2/S_{xx}]}},
$$

which has a t-distribution with $n - 2$ degrees of freedom.

PREDICTION INTERVAL FOR y_0

A $(1 - \alpha)100\%$ prediction interval for a single response y_0 is given by

$$
\hat{y}_0 - t_{\alpha/2}s\sqrt{1 + \frac{1}{n} + \frac{(x_0 - \bar{x})^2}{S_{xx}}} < y_0 < \hat{y}_0 + t_{\alpha/2}s\sqrt{1 + \frac{1}{n} + \frac{(x_0 - \bar{x})^2}{S_{xx}}},
$$

where $t_{\alpha/2}$ is a value of the t-distribution with $n - 2$ degrees of freedom. ∎

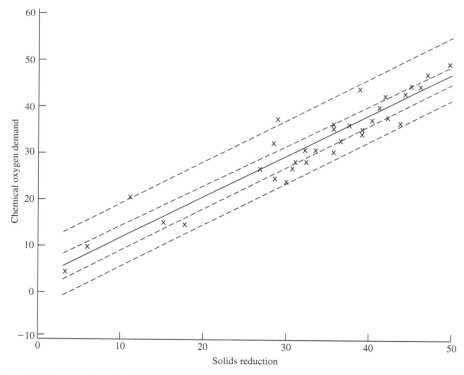

FIGURE 11.4   Confidence and prediction intervals for the chemical oxygen demand data.

Clearly, there is a distinction between the concept of a confidence interval and the prediction interval described above. The confidence interval interpretation is identical to that described for all confidence intervals on population parameters discussed throughout the book. Indeed, $\mu_{Y|x_0}$ is a population parameter. The computed prediction interval, however, represents an interval that has a probability equal to $1 - \alpha$ of containing not a parameter but a future value y_0 of the random variable Y_0.

EXAMPLE 11.7   Using the data of Table 11.1, construct a 95% prediction interval for y_0 when $x_0 = 20\%$.

SOLUTION
We have $n = 33$, $x_0 = 20$, $\bar{x} = 33.4545$, $\hat{y}_0 = 21.9025$, $S_{xx} = 4152.18$, $s = 3.2295$, and $t_{0.025} \approx 2.045$ for 31 degrees of freedom. Therefore, a 95% prediction interval for y_0 is given by

$$21.9025 - (2.045)(3.2295)\sqrt{1 + \frac{1}{33} + \frac{(20 - 33.4545)^2}{4152.18}} < y_0$$

$$< 21.9025 + (2.045)(3.2295)\sqrt{1 + \frac{1}{33} + \frac{(20 - 33.4545)^2}{4152.18}},$$

which simplifies to

$$15.0585 < y_0 < 28.7464.$$

Figure 11.4 shows another plot of the chemical oxygen demand data, with both the confidence interval on mean response and the prediction interval on an individual response plotted. The plot reflects a much tighter interval around the regression line in the case of the mean response.

Exercises

1. Assuming that the E_i's are normal, independent with zero means and common variance σ^2, show that A, the least squares estimator of α in $\mu_{Y|x} = \alpha + \beta x$ is normally distributed with mean α and variance

$$\sigma_A^2 = \frac{\displaystyle\sum_{i=1}^{n} x_i^2}{n \displaystyle\sum_{i=1}^{n} (x_i - \bar{x})^2} \, \sigma^2.$$

2. For a simple linear regression model

$$Y_i = \alpha + \beta x_i + E_i, \qquad i = 1, 2, \ldots, n,$$

where the E_i's are independent and normally distributed with zero means and equal variances σ^2, show that $\bar{Y}$ and

$$B = \frac{\displaystyle\sum_{i=1}^{n} (x_i - \bar{x}) Y_i}{\displaystyle\sum_{i=1}^{n} (x_i - \bar{x})^2}.$$

have zero covariance.

3. With reference to Exercise 1 on page 371,
   (a) evaluate s^2;
   (b) test the hypothesis that $\beta = 0$ against the alternative that $\beta \neq 0$ at the 0.05 level of significance and interpret the resulting decision.

4. With reference to Exercise 2 on page 371,
   (a) evaluate s^2;
   (b) construct a 95% confidence interval for α;
   (c) construct a 95% confidence interval for β.

5. With reference to Exercise 3 on page 371,
   (a) evaluate s^2;
   (b) construct a 95% confidence interval for α;
   (c) construct a 95% confidence interval for β.

6. With reference to Exercise 4 on page 372,
   (a) evaluate s^2;

(b) construct a 99% confidence interval for α;
(c) construct a 99% confidence interval for β.

7. With reference to Exercise 5 on page 372,
   (a) evaluate s^2;
   (b) construct a 99% confidence interval for α;
   (c) construct a 99% confidence interval for β.

8. Test the hypothesis that $\alpha = 10$ in Exercise 6 on page 372 against the alternative that $\alpha > 10$. Use a 0.05 level of significance.

9. Test the hypothesis that $\beta = 6$ in Exercise 7 on page 372 against the alternative that $\beta < 6$. Use a 0.025 level of significance.

10. Using the value of s^2 found in Exercise 4(a), construct a 95% confidence interval for $\mu_{Y|80}$ in Exercise 2 on page 371.

11. With reference to Exercise 4 on page 372, use the value of s^2 found in Exercise 6(a) to construct
    (a) a 95% confidence interval for the mean shear resistance when $x = 24.5$;
    (b) a 95% prediction interval for a single predicted value of the shear resistance when $x = 24.5$.

12. Using the value of s^2 found in Exercise 5(a), graph the regression line and the 95% confidence bands for the mean response $\mu_{Y|x}$ for the data of Exercise 3 on page 371.

13. Using the value of s^2 found in Exercise 5(a), construct a 95% confidence interval for the amount of converted sugar corresponding to $x = 1.6$ in Exercise 3 on page 371.

14. With reference to Exercise 5 on page 372, use the value of s^2 found in Exercise 7(a) to construct
    (a) a 99% confidence interval for the average amount of chemical that will dissolve in 100 grams of water at 50°C
    (b) a 99% prediction interval for the amount of chemical that will dissolve in 100 grams of water at 50°C.

11.6   Choice of a Regression Model _____

Much of what has been presented to this point on regression involving a single independent variable depends on the assumption that the model chosen is correct, the presumption that $\mu_{Y|x}$ is related to x linearly in the parameters. Certainly, one would not expect the prediction of the response to be good if there are several independent variables, not considered in the model, that are affecting the response and are varying in the system. In addition, the prediction would certainly be inadequate if the true structure relating $\mu_{Y|x}$ to x is extremely nonlinear in the range of the variables considered.

Often the simple linear regression model is used even though it is known that the model is something other than linear or that the true structure is unknown. This approach is often sound, particularly when the range of x is narrow. Thus the model used becomes an approximating function that one hopes is an adequate representation of the true picture in the region of interest. One should note, however, the effect of an inadequate model on the results presented thus far. For example, if the true model, unknown to the experimenter, is linear in more than one x, say,

$$\mu_{Y|x_1, x_2} = \alpha + \beta x_1 + \gamma x_2,$$

then the ordinary squares estimate $b = S_{xy}/S_{xx}$, calculated by only considering x_1 in the experiment, is, under general circumstances, a biased estimate of the coefficient β, the bias being a function of the additional coefficient γ (see Exercise 3 on page 371). Also, the estimate s^2 for σ^2 is biased due to the additional variable.

11.7   Analysis-of-Variance Approach _____

Often the problem of analyzing the quality of the estimated regression line is handled through an **analysis-of-variance** approach. This is merely a procedure whereby the total variation in the dependent variable is subdivided into meaningful components that are then observed and treated in a systematic fashion. The analysis of variance, discussed extensively in Chapter 13, is a powerful tool that is used in many applications.

Suppose that we have n experimental data points in the usual form (x_i, y_i) and that the regression line is estimated. In our estimation of σ^2 in Section 11.3 we established the identity

$$S_{yy} = b S_{xy} + SSE.$$

An alternative and perhaps more informative formulation is given by

$$\sum_{i=1}^{n} (y_i - \bar{y})^2 = \sum_{i=1}^{n} (\hat{y}_i - \bar{y})^2 + \sum_{i=1}^{n} (y_i - \hat{y}_i)^2.$$

So we have achieved a partitioning of the **total corrected sum of squares of y** into two components that should reflect particular meaning to the experimenter. We shall indicate this partitioning symbolically as

$$SST = SSR + SSE.$$

The first component of the right is called the **regression sum of squares** and it reflects the amount of variation in the y-values **explained by the model**, in this case the postulated straight line. The second component is just the familiar error sum of squares, which reflects variation about the regression line.

Suppose that we are interested in testing the hypothesis

$$H_0: \quad \beta = 0,$$

$$H_1: \quad \beta \neq 0,$$

where the null hypothesis essentially says that the model is $\mu_{Y|x} = \alpha$. That is, the variation in Y results from chance or random fluctuations which are independent of the values of x. Under the conditions of this null hypothesis it can be shown that SSR/σ^2 and SSE/σ^2 are values of independent chi-squared variables with 1 and $n - 2$ degrees of freedom, respectively, and then by Theorem 7.12 it follows that SST/σ^2 is also a value of a chi-squared variable with $n - 1$ degrees of freedom. To carry out a test of the hypothesis above, we compute

$$f = \frac{SSR/1}{SSE/(n - 2)} = \frac{SSR}{s^2}$$

and reject H_0 at the α-level of significance when $f > f_\alpha(1, n - 2)$.

The computations are usually summarized by means of an **analysis-of-variance table**, as indicated in Table 11.2. It is customary to refer to the various sums of squares divided by their respective degrees of freedom as the **mean squares**.

When the null hypothesis is rejected, that is, when the computed F-statistic exceeds the critical value $f_\alpha(1, n - 2)$, we conclude that there is a significant amount of variation in the response accounted for by the postulated model, the straight-line function. If the F-statistic is in the acceptance region, we conclude that the data did not reflect sufficient evidence to support the model postulated.

In Section 11.4 a procedure was given whereby the statistic

$$T = \frac{B - \beta_0}{S/\sqrt{S_{xx}}}$$

was used to test the hypothesis

$$H_0: \quad \beta = \beta_0,$$

$$H_1: \quad \beta \neq \beta_0,$$

TABLE 11.2 Analysis of Variance for Testing $\beta = 0$

Source of Variation	Sum of Squares	Degrees of Freedom	Mean Square	Computed f
Regression	SSR	1	SSR	SSR/s^2
Error	SSE	$n - 2$	$s^2 = \dfrac{SSE}{n - 2}$	
Total	SST	$n - 1$		

where T follows the t-distribution with $n - 2$ degrees of freedom. The hypothesis is rejected if $|t| > t_{\alpha/2}$ for an α-level of significance. It is interesting to note that in the special case in which we are testing

$$H_0: \quad \beta = \beta_0,$$

$$H_1: \quad \beta \neq \beta_0,$$

the value of our T-statistic becomes

$$t = \frac{b}{s/\sqrt{S_{xx}}}$$

and the hypothesis under consideration is identical to that being tested in Table 11.2. Namely, the null hypothesis states that the variation in the response is due merely to chance. The analysis of variance uses the F-distribution rather than the t-distribution. For the two-sided alternative, the two approaches are identical. This we can see by writing

$$t^2 = \frac{b^2 S_{xx}}{s^2} = \frac{b S_{xy}}{s^2} = \frac{SSR}{s^2},$$

which is identical to the f-value used in the analysis of variance. The basic relationship between the t-distribution with v degrees of freedom and the F-distribution with 1 and v degrees of freedom is given by

$$t_{\alpha/2}^2 = f_\alpha(1, v).$$

Of course, the t-test allows for testing against a one-sided alternative while the F-test is restricted to testing against a two-sided alternative.

Annotated Computer Printout for Simple Linear Regression

Consider again the chemical oxygen demand data of Table 11.1. Figure 11.5 shows an annotated computer printout. In this case the MINITAB PC software is illustrated. The t-ratio column indicates tests for null hypotheses of zero values on the parameter. The term "fit" denotes $\hat{y}$-values, often called fitted values. The term "stdev. fit" is used in computing confidence intervals on mean response. The item R^2 is computed as $(SSR/SST) \times 100$ and signifies the proportion of variation in y explained by the straight-line regression. We discuss the R^2 statistic in more detail in Chapter 12. Also shown are confidence intervals on the mean response and prediction intervals on a new observation.

11.8 Test for Linearity of Regression: Data with Repeated Observations

In certain types of experimental situations the researcher has the capability to obtain repeated observations on the response for each value of x. Although it is not necessary to have these repetitions in order to estimate α and β, nevertheless it does

The regression equation is
$y = 3.83 + 0.904 \, x$

Predictor	Coef	Stdev	t-ratio	p
Constant	3.830	1.768	2.17	0.038
x	0.90364	0.05012	18.03	0.000

$s = 3.230$ R-sq = 91.3%

Analysis of Variance

SOURCE	DF	SS	MS	F	p
Regression	1	3390.6	3390.6	325.08	0.000
Error	31	323.3	10.4		
Total	32	3713.9			

Obs.	x	y	Fit	Stdev.Fit	Residual
1	3.0	5.000	6.541	1.627	-1.541
2	7.0	11.000	10.155	1.440	0.845
3	11.0	21.000	13.770	1.258	7.230
4	15.0	16.000	17.384	1.082	-1.384
5	18.0	16.000	20.095	0.957	-4.095
6	27.0	28.000	28.228	0.649	-0.228
7	29.0	27.000	30.035	0.605	-3.035
8	30.0	25.000	30.939	0.588	-5.939
9	30.0	35.000	30.939	0.588	4.061
10	31.0	30.000	31.843	0.575	-1.843
11	31.0	40.000	31.843	0.575	8.157
12	32.0	32.000	32.746	0.567	-0.746
13	33.0	34.000	33.650	0.563	0.350
14	33.0	32.000	33.650	0.563	-1.650
15	34.0	34.000	34.554	0.563	-0.554
16	36.0	37.000	36.361	0.576	0.639
17	36.0	38.000	36.361	0.576	1.639
18	36.0	34.000	36.361	0.576	-2.361
19	37.0	36.000	37.264	0.590	-1.264
20	38.0	38.000	38.168	0.607	-0.168
21	39.0	37.000	39.072	0.627	-2.072
22	39.0	36.000	39.072	0.627	-3.072
23	39.0	45.000	39.072	0.627	5.928
24	40.0	39.000	39.975	0.651	-0.975
25	41.0	41.000	40.879	0.678	0.121
26	42.0	40.000	41.783	0.707	-1.783
27	42.0	44.000	41.783	0.707	2.217
28	43.0	37.000	42.686	0.738	-5.686
29	44.0	44.000	43.590	0.772	0.410
30	45.0	46.000	44.494	0.807	1.506
31	46.0	46.000	45.397	0.843	0.603
32	47.0	49.000	46.301	0.881	2.699
33	50.0	51.000	49.012	1.002	1.988

FIGURE 11.5 Minitab printout for simple linear regression for chemical oxygen demand data.

Fit	Stdev.Fit	95% C.I.		95% P.I.	
6.541	1.627	(3.222,	9.859)	(−0.836,	13.917)
10.155	1.440	(7.217,	13.093)	(2.942,	17.369)
13.770	1.258	(11.203,	16.336)	(6.699,	20.840)
17.384	1.082	(15.176,	19.592)	(10.436,	24.333)
20.095	0.957	(18.143,	22.048)	(13.224,	26.967)
28.228	0.649	(26.905,	29.551)	(21.508,	34.948)
30.035	0.605	(28.801,	31.269)	(23.332,	36.738)
30.939	0.588	(29.739,	32.139)	(24.242,	37.636)
30.939	0.588	(29.739,	32.139)	(24.242,	37.636)
31.843	0.575	(30.669,	33.017)	(25.151,	38.535)
31.843	0.575	(30.669,	33.017)	(25.151,	38.535)
32.746	0.567	(31.590,	33.903)	(26.057,	39.435)
33.650	0.563	(32.502,	34.798)	(26.962,	40.337)
33.650	0.563	(32.502,	34.798)	(26.962,	40.337)
34.554	0.563	(33.405,	35.702)	(27.866,	41.241)
36.361	0.576	(35.185,	37.537)	(29.668,	43.053)
36.361	0.576	(35.185,	37.537)	(29.668,	43.053)
36.361	0.576	(35.185,	37.537)	(29.668,	43.053)
37.264	0.590	(36.062,	38.467)	(30.567,	43.962)
38.168	0.607	(36.931,	39.406)	(31.465,	44.872)
39.072	0.627	(37.792,	40.351)	(32.360,	45.783)
39.072	0.627	(37.792,	40.351)	(32.360,	45.783)
39.072	0.627	(37.792,	40.351)	(32.360,	45.783)
39.975	0.651	(38.648,	41.303)	(33.255,	46.696)
40.879	0.678	(39.497,	42.261)	(34.147,	47.611)
41.783	0.707	(40.341,	43.224)	(35.039,	48.527)
41.783	0.707	(40.341,	43.224)	(35.039,	48.527)
42.686	0.738	(41.180,	44.192)	(35.928,	49.444)
43.590	0.772	(42.016,	45.164)	(36.816,	50.364)
44.494	0.807	(42.848,	46.139)	(37.703,	51.284)
45.397	0.843	(43.677,	47.118)	(38.588,	52.206)
46.301	0.881	(44.503,	48.099)	(39.472,	53.130)
49.012	1.002	(46.968,	51.056)	(42.114,	55.910)

FIGURE 11.5 Minitab printout for simple linear regression for chemical oxygen demand data. (continued)

enable the experimenter to obtain quantitative information concerning the appropriateness of the model. In fact, if repeated observations have been generated, the experimenter can make a significance test to aid in determining whether or not the model is adequate.

Let us select a random sample of n observations using k distinct values of x, say $x_1, x_2 \ldots, x_k$, such that the sample contains n_1 observed values of the random variable Y_1 corresponding to x_1, n_2 observed values of Y_2 corresponding to $x_2, \ldots,$ n_k observed values of Y_k corresponding to x_k. Of necessity, $n = \sum_{i=1}^{k} n_i$. We define

$$y_{ij} = \text{the } j\text{th value of the random variable } Y_i,$$

$$y_{i\cdot} = T_{i\cdot} = \sum_{j=1}^{n_i} y_{ij},$$

$$\bar{y}_{i\cdot} = \frac{T_i}{n_i}.$$

Hence, if $n_4 = 3$ measurements of Y are made corresponding to $x = x_4$, we would indicate these observations by y_{41}, y_{42}, and y_{43}. Then

$$T_{4\cdot} = y_{41} + y_{42} + y_{43}.$$

Concept of Lack of Fit

The error sum of squares consists of two parts: the amount due to the variation between the values of Y within given values of x and a component that is normally called the **lack of fit** contribution. The first component reflects mere random variation or **pure experimental error**, while the second component is a measure of the systematic variation brought about by higher-order terms. In our case these are terms in x other than the linear or first-order contribution. Note that in choosing a linear model we are essentially assuming that this second component does not exist and hence our error sum of squares is completely due to random errors. If this should be the case, then $s^2 = SSE/(n - 2)$ is an unbiased estimate of σ^2. However, if the model does not adequately fit the data, then the error sum of squares is inflated and produces a biased estimate of σ^2. Whether or not the model fits the data, an unbiased estimate of σ^2 can always be obtained when we have repeated observations simply by computing

$$s_i^2 = \frac{\sum_{j=1}^{n_i} (y_{ij} - \bar{y}_{i\cdot})^2}{n_i - 1}; \qquad i = 1, 2, \ldots, k,$$

for each of the k distinct values of x and then pooling these variances to give

$$s^2 = \frac{\sum_{i=1}^{k} (n_i - 1)s_i^2}{n - k} = \frac{\sum_{i=1}^{k} \sum_{j=1}^{n_i} (y_{ij} - \bar{y}_{i\cdot})^2}{n - k}.$$

The numerator of s^2 is a measure of the pure experimental error. A computational procedure for separating the error sum of squares into the two components representing pure error and lack of fit is as follows:

1. Compute the pure error sum of squares

$$\sum_{i=1}^{k} \sum_{j=1}^{n_i} (y_{ij} - \bar{y}_{i\cdot})^2.$$

This sum of squares has $n - k$ degrees of freedom associated with it and the resulting mean square is our unbiased estimate s^2 of σ^2.

2. Subtract the pure error sum of squares from the error sum of squares, SSE, thereby obtaining the sum of squares due to lack of fit. The degrees of freedom for lack of fit are also obtained by simply subtracting $(n - 2) - (n - k) = k - 2$.

The computations required for testing hypotheses in a regression problem with repeated measurements on the response may be summarized as shown in Table 11.3.

Figures 11.6 and 11.7 show a pictorial display of the sample points for the "correct model" and "incorrect model" situations. In Figure 11.6, where the $\mu_{Y|x}$ fall on a straight line, there is no lack of fit when a linear model is assumed so that the sample variation around the regression line is a pure error resulting from the variation that occurs among repeated observations. In Figure 11.7, where the $\mu_{Y|x}$ clearly do not fall on a straight line, the lack of fit from erroneously choosing a linear model accounts for a large portion of the variation around the regression line in addition to the pure error.

The concept of lack of fit is extremely important in applications of regression analysis. In fact, the need to construct or design an experiment that will account for lack of fit becomes more critical as the problem and the underlying mechanism involved become more complicated. Surely, one cannot always be certain that his or her postulated structure, in this case the linear regression model, is correct or even an adequate representation. The following example shows how the error sum of squares is partitioned into the two components representing pure error and lack of fit. The adequacy of the model is tested at the α-level of significance by comparing the lack-of-fit mean square divided by s^2 with $f_\alpha(k - 2, n - k)$.

TABLE 11.3 Analysis of Variance for Testing Linearity of Regression

Source of Variation	Sum of Squares	Degrees of Freedom	Mean Square	Computed f
Regression	SSR	1	SSR	$\dfrac{SSR}{s^2}$
Error	SSE	$n - 2$		
Lack of fit	$\left\{\begin{array}{l} SSE - SSE(\text{pure}) \end{array}\right.$	$\left\{\begin{array}{l} k - 2 \end{array}\right.$	$\dfrac{SSE - SSE(\text{pure})}{k - 2}$	$\dfrac{SSE - SSE(\text{pure})}{s^2(k - 2)}$
Pure error	$\left.\begin{array}{l} SSE(\text{pure}) \end{array}\right.$	$\left.\begin{array}{l} n - k \end{array}\right.$	$s^2 = \dfrac{SSE(\text{pure})}{n - k}$	
Total	SST	$n - 1$		

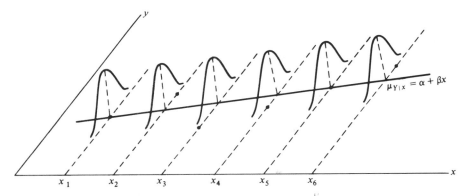

FIGURE 11.6   Correct linear model with no lack-of-fit component.

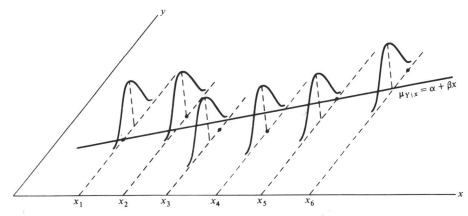

FIGURE 11.7   Incorrect linear model with lack-of-fit component.

EXAMPLE 11.8   Observations on the yield of a chemical reaction taken at various temperatures were recorded as follows:

$y(\%)$	$x(°C)$	$y(\%)$	$x(°C)$
77.4	150	88.9	250
76.7	150	89.2	250
78.2	150	89.7	250
84.1	200	94.8	300
84.5	200	94.7	300
83.7	200	95.9	300

Estimate the linear model $\mu_{Y|x} = \alpha + \beta x$ and test for lack of fit.

SOLUTION

We have $n_1 = n_2 = n_3 = n_4 = 3$. Therefore,

$$S_{yy} = \sum_{i=1}^{4} \sum_{j=1}^{3} (y_{ij} - \bar{y})^2 = 513.1167$$

$$S_{xx} = \sum_{i=1}^{4} n_i x_i^2 - \frac{\left(\sum_{i=1}^{4} n_i x_i\right)^2}{12} = \sum_{i=1}^{4} n_i(x_i - \bar{x})^2 = 37,500$$

$$S_{xy} = \sum_{i=1}^{4} \sum_{j=1}^{3} (x_i - \bar{x})(y_{ij} - \bar{y}) = 4370$$

$$\bar{y} = 86.4833 \qquad \text{and} \qquad \bar{x} = 225.$$

The regression coefficients are then given by

$$b = \frac{4370}{37,500} = 0.1165$$

and

$$a = 86.4833 - (0.1165)(225) = 60.2708.$$

Hence our estimated regression line is

$$\hat{y} = 60.2708 + 0.1165x.$$

To test for lack of fit at the $\alpha = 0.05$ level, we proceed in the usual manner:

H_0: the regression is linear in x.

H_1: the regression is nonlinear in x.

Critical region: $f > 4.46$ with 2 and 8 degrees of freedom.
Computations: We have

$$SST = S_{yy} = 513.1167$$

$$SSR = bS_{xy} = (0.1165)(4370) = 509.1050$$

$$SSE = S_{yy} - bS_{xy} = 4.0117.$$

The pure error sum of squares

$$\sum_{i} \sum_{j} (y_{ij} - \bar{y}_i)^2 = 2.660.$$

These results and the remaining computations are exhibited in Table 11.4. Conclusion: The partitioning of the total variation in this manner reveals a significant variation accounted for by the linear model and an insignificant amount of variation due to lack of fit. Thus the experimental data do not seem to suggest the need to consider terms higher than first order in the model and the null hypothesis is not rejected.

TABLE 11.4 Analysis of Variance on Yield–Temperature Data

Source of Variation	Sum of Squares	Degrees of Freedom	Mean Square	Computed f	P-Values
Regression	509.1050	1	509.1050	1531.60	<0.0001
Error	4.0117	10			
Lack of fit	1.3517 $\big\{$	2 $\big\{$	0.6758	2.03	0.19
Pure error	2.6600	8	0.3325		
Total	513.1167	11			

Exercises

1. (a) Find the least squares estimate for the parameter β in the linear equation $\mu_{Y|x} = \beta x$.
   (b) Estimate the regression line passing through the origin for the following data:

x	0.5	1.5	3.2	4.2	5.1	6.5
y	1.3	3.4	6.7	8.0	10.0	13.2

2. Suppose it is not known in Exercise 1 whether or not the true regression should pass through the origin. Estimate the linear model $\mu_{Y|x} = \alpha + \beta x$ and test the hypothesis that $\alpha = 0$ at the 0.10 level of significance against the alternative that $\alpha \neq 0$.

3. Suppose that an experimenter postulates a model of the type

$$Y_i = \alpha + \beta x_{1i} + E_i, \qquad i = 1, 2, \ldots, n,$$

when in fact, an additional variable, say x_2, also contributes linearly to the response. The true model is then given by

$$Y_i = \alpha + \beta x_{1i} + \gamma x_{2i} + E_i, \qquad i = 1, 2, \ldots, n.$$

Compute the expected value of the estimator

$$B = \frac{\sum\limits_{i=1}^{n} (x_{1i} - \bar{x}_1) Y_i}{\sum\limits_{i=1}^{n} (x_{1i} - \bar{x}_1)^2}.$$

4. Use an analysis-of-variance approach to test the hypothesis that $\beta = 0$ against the alternative hypothesis $\beta \neq 0$ in Exercise 3 on page 371 at the 0.05 level of significance.

5. Organophosphate (OP) compounds are used to a large extent as pesticides. However, it is important to study their effect on species that are exposed to them. In the laboratory study "*Some Effects of Organophosphate Pesticides on Wildlife Species*," by the Department of Fisheries and Wildlife at the Virginia Polytechnic Institute and State University (1983), an experiment was conducted in which different dosages of a particular OP pesticide were administered to 5 groups of 5 mice (peromysius leucopus). The 25 mice were female of similar age and condition. One group received no chemical. The basic response (y) was a measure of activity in the brain. It was postulated that brain activity would decrease with an increase in OP dosage. The data are as follows:

Animal	Dose, x (mg/kg body weight)	Activity, y (moles/liter/minute)
1	0.0	10.9
2	0.0	10.6
3	0.0	10.8
4	0.0	9.8
5	0.0	9.0
6	2.3	11.0
7	2.3	11.3
8	2.3	9.9
9	2.3	9.2
10	2.3	10.1
11	4.6	10.6
12	4.6	10.4
13	4.6	8.8
14	4.6	11.1

15	4.6	8.4
16	9.2	9.7
17	9.2	7.8
18	9.2	9.0
19	9.2	8.2
20	9.2	2.3
21	18.4	2.9
22	18.4	2.2
23	18.4	3.4
24	18.4	5.4
25	18.4	8.2

(a) Using the model

$$Y_i = \alpha + \beta x_i + E_i, \quad i = 1, 2, \ldots, 25,$$

find the least squares estimates of α and β.

(b) Construct an analysis-of-variance table in which the lack of fit and pure error have been separated. Determine if the lack of fit is significant at the 0.05 level. Interpret the results.

6. Test for linearity of regression in Exercise 5 on page 372. Use a 0.05 level of significance.

7. Test for linearity of regression in Exercise 6 on page 372.

11.9 Data Plots and Transformations

In this chapter we deal with building regression models in which there is one independent or regressor variable. In addition, we are assuming, through model formulation, that both x and y enter the model in a *linear fashion*. Often it is advisable to work with an alternative model in which either x and y (or both) enter in a nonlinear way. A *transformation* of the data may be indicated because of theoretical considerations inherent in the scientific study, or a simple plotting of the data may suggest the need to *reexpress* the variables in the model. The need to perform a transformation is rather simple to diagnose in the case of simple linear regression because two-dimensional plots give a true pictorial display of how each variable enters the model.

A model in which x or y is transformed should not be viewed as a *nonlinear regression model*. We normally refer to a regression model as linear when it is *linear in the parameters*. In other words, suppose the complexion of the data or other scientific information suggests that we should regress y^* against x^*, where each is a transformation on the natural variables x and y. Then the model of the form

$$y_i^* = \alpha + \beta x_i^* + \varepsilon_i$$

is a linear model since it is linear in the parameters α and β. The material given in Sections 11.2 through 11.8 remains intact with y_i^* and x_i^* replacing y_i and x_i. A simple and useful example is the log-log model given by

$$\log y_i = \alpha + \beta \log x_i + \varepsilon_i.$$

Although this model is not linear in x and y, it is linear in the parameters and is thus treated as a linear model. On the other hand, an example of a truly nonlinear model is given by

$$y_i = \beta_0 + \beta_1 x^{\beta_2} + \varepsilon_i,$$

where the parameter β_2 (as well as β_0 and β_1) is to be estimated. The model is not linear in β_2.

Transformations that may enhance the fit and predictability of the model are many in number. For a thorough discussion of transformations, the reader is referred to Myers (see the Bibliography). We choose here to indicate a few of them and show the appearance of the graph that serves as a diagnostic. Consider Table 11.5. Several functions are given describing relationships between y and x that can produce a *linear regression* through the transformation indicated. In addition, for the sake of completeness the reader is given the dependent and independent variables to use in the resulting *simple linear regression*. Figure 11.8 shows the picture depicting the situations described in Table 11.5. These serve as a guide for the analyst in choosing a transformation from the observation of the plot of y against x.

The foregoing is intended as an aid for the analyst when it is apparent that a transformation will provide an improvement. However, before we provide an example, two important points should be made. The first one revolves around the formal writing of the model when the data are transformed. Quite often the analyst does not think about this. He or she merely performs the transformation without any concern about the model form *before* and *after* the transformation. The exponential model serves as a good illustration. The model in the natural (untransformed) variables that produces an *additive error* model in the transformed variables is given by

$$y_i = \alpha e^{\beta x_i} \cdot \varepsilon_i,$$

which is a *multiplicative error model*. Clearly, taking logs produces

$$\ln y_i = \ln \alpha + \beta x_i + \ln \varepsilon_i.$$

As a result, it is on $\ln \varepsilon_i$ that the basic assumptions are made. The purpose of this presentation is merely to remind the reader that one should not view a transformation as merely an algebraic manipulation with an error added. Often a model in the transformed variables that has a proper *additive error structure* is a result of a model in the natural variables with a different type of error structure.

The second important point deals with the notion of measures of improvement. Obvious measures of comparison are, of course, R^2 and the residual mean square, s^2. (Other measures of performance in comparisons between competing models are given in Chapter 12). Now, if the response y is not transformed, then clearly s^2 and R^2 can be used in measuring the utility of the transformation. The residuals will be in the same units for both the transformed and the untransformed models. But when

TABLE 11.5 Some Useful Transformations to Linearize

Functional Form Relating y to x	*Proper Transformation*	*Form of Simple Linear Regression*
Exponential: $y = \alpha e^{\beta x}$	$y^* = \ln y$	Regress y^* against x
Power: $y = \alpha x^\beta$	$y^* = \log y$; $x^* = \log x$	Regress y^* against x^*
Reciprocal: $y = \alpha + \beta\left(\dfrac{1}{x}\right)$	$x^* = \dfrac{1}{x}$	Regress y against x^*
Hyperbolic function: $y = \dfrac{x}{\alpha + \beta x}$	$y^* = \dfrac{1}{y}$; $x^* = \dfrac{1}{x}$	Regress y^* against x^*

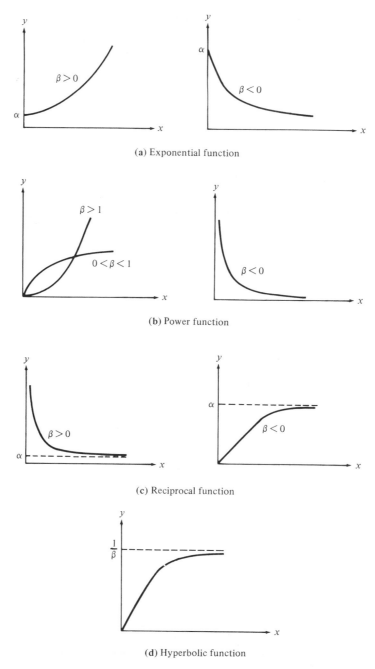

(a) Exponential function

(b) Power function

(c) Reciprocal function

(d) Hyperbolic function

FIGURE 11.8   Diagrams depicting functions described in Table 11.5.

y is transformed, performance criteria for the transformed model should be based on values of the residuals in the metric of the untransformed response. Thus comparisons that are made are proper. The example that follows provides an illustration.

EXAMPLE 11.9   The pressure P of a gas corresponding to various volumes V is recorded as follows:

V (cm^3)	50	60	70	90	100
P (kg/cm^2)	64.7	51.3	40.5	25.9	7.8

The ideal gas law is given by the functional form $PV^\gamma = C$, where γ and C are constants. Estimate the constants C and γ.

SOLUTION
Let us take natural logs of both sides of the model

$$P_i V_i^\gamma = C \cdot \varepsilon_i, \qquad i = 1, 2, 3, 4, 5.$$

As a result, a linear model can be written

$$\ln P_i = \ln C - \gamma \ln V_i + \varepsilon_i^*, \qquad i = 1, 2, 3, 4, 5,$$

where $\varepsilon_i^* = \ln \varepsilon_i$. The following represents results of the simple linear regression

Intercept: $\widehat{\ln C} = 14.7589$, $\hat{C} = 2{,}568{,}862.88$

Slope: $\hat{\gamma} = -2.65347221$.

The following represents information taken from the regression analysis.

P_i	V_i	$\ln P_i$	$\ln V_i$	$\widehat{\ln P_i}$	$\hat{P}_i$	$e_i = P_i - \hat{P}_i$
64.7	50	4.16976	3.91202	4.37853	79.7	-15.0
51.3	60	3.93769	4.09434	3.89474	49.1	2.2
40.5	70	3.70130	4.24850	3.48571	32.6	7.9
25.9	90	3.25424	4.49981	2.81885	16.8	9.1
7.8	100	2.05412	4.60517	2.53928	12.7	-4.9

It may be instructive to plot the data and the regression equation. Figure 11.9 shows a plot of the data in the untransformed pressure and volume and the curve representing the regression equation.

Diagnostic Plots of Residuals:
Graphical Detection of Violation of Assumptions

Plots of the raw data can be extremely helpful in determining the nature of the model that should be fit to the data when there is a single independent variable. We have attempted to illustrate this in the foregoing. Detection of proper model form is, however, not the only benefit gained from diagnostic plotting. As in much of the

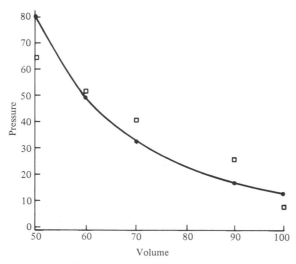

FIGURE 11.9   Pressure and volume data and fitted regression.

material associated with significance testing in Chapter 10, plotting methods can illustrate and detect violation of assumptions. The reader should recall that much of what has been illustrated in the Chapter requires assumptions made on the model errors, the E_i. In fact, we assume that the E_i are independent $N(0, \sigma)$ random variables. Now, of course, realizations of the E_i, the ε_i, are not observed. However, the $e_i = y_i - \hat{y}_i$, the *residuals,* are the error in the fit of the regression line and thus serve to mimic the ε_i. Thus the general complexion of these residuals can often highlight difficulties. Ideally, of course, the plot of the residuals are as depicted in Figure 11.10. That is, they should truly show *random fluctuations* around a value of zero.

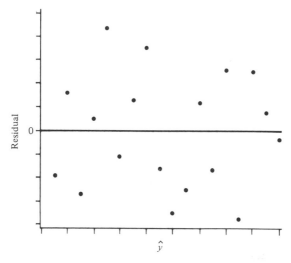

FIGURE 11.10   Ideal residual plot.

Nonhomogeneous Variance

Homogeneous variance is an important assumption made in regression analysis. Violations can often be detected through the appearance of the residual plot. Increasing error variance with an increase in the regressor variable is a common condition in scientific data. Large error variance produces large residuals, and hence a residual plot like the one in Figure 11.11 is a signal of nonhomogeneous variance. More discussion regarding these residual plots and information regarding different types of residuals will be given in Chapter 12 when we deal with multiple linear regression.

Normal Probability Plotting

The assumption that the model errors are normal is made when the data analyst deals either in hypothesis testing or confidence interval estimation. Again, the numerical counterpart to the ε_i, namely the residuals, are subjects of diagnostic plotting to detect any extreme violations. In Chapter 8 we introduced normal quantile–quantile plots and briefly discussed normal probability plots. These plots on residuals are illustrated in the case study introduced in the next section.

11.10 Simple Linear Regression Case Study

In the manufacture of commercial wood products it is important to estimate the relationship between the density of a wood product and its stiffness. A relatively new type of particleboard is being considered that can be formed with considerably more ease than the accepted commercial product. It is necessary to know at what density the stiffness compares to the well-known, well-documented commercial

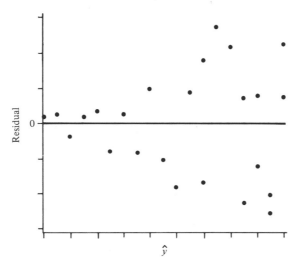

Figure 11.11 Residual plot depicting heterogeneous error variance.

product. The study was done by Terrance E. Conners, *"Investigation of Certain Mechanical Properties of a Wood-Foam Composite"* (M.S. Thesis, Department of Forestry and Wildlife Management, University of Massachusetts, 1979). Thirty particleboards were produced at densities ranging from roughly 8 to 26 pounds per cubic foot, and the stiffness was measured in pounds per square inch. Table 11.6 shows the data.

It is necessary for the data analyst to focus on an appropriate fit to the data and use inferential methods discussed in this chapter. Hypothesis testing on the slope of the regression as well as confidence or prediction interval estimation may well be appropriate. We begin by demonstrating a simple scatter plot of the raw data with a simple linear regression superimposed. Figure 11.12 shows this plot.

TABLE 11.6 Density and Stiffness for 30 Particleboards

Density, x	Stiffness, y
9.50	14,814.00
8.40	17,502.00
9.80	14,007.00
11.00	19,443.00
8.30	7,573.00
9.90	14,191.00
8.60	9,714.00
6.40	8,076.00
7.00	5,304.00
8.20	10,728.00
17.40	43,243.00
15.00	25,319.00
15.20	28,028.00
16.40	41,792.00
16.70	49,499.00
15.40	25,312.00
15.00	26,222.00
14.50	22,148.00
14.80	26,751.00
13.60	18,036.00
25.60	96,305.00
23.40	104,170.00
24.40	72,594.00
23.30	49,512.00
19.50	32,207.00
21.20	48,218.00
22.80	70,453.00
21.70	47,661.00
19.80	38,138.00
21.30	53,045.00

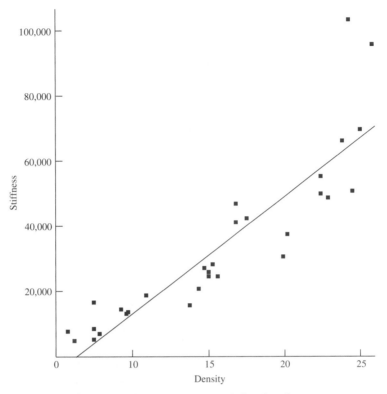

FIGURE 11.12 Scatterplot of the wood density data.

The simple linear regression fit to the data produced the fitted model

$$\hat{y} = -25{,}433.739 + 3884.976x \qquad (R^2 = 0.7975)$$

and the residuals were computed. Figure 11.13 shows the residuals plotted against the measurements of density. This is hardly an ideal or healthy set of residuals. They do not show a random scatter around a value of zero. In fact, clusters of positive and negative values might suggest that a curvilinear trend in the data should be investigated.

To gain some type of idea regarding the normal error assumption, a normal probability plot of the residuals was generated. This is the type of plot discussed in Chapter 8 in which the vertical axis represents the empirical distribution function on a scale that produces a straight-line plot when plotted against the residuals themselves. Figure 11.14 shows the normal probability plot of the residuals. Here $y_i - \hat{y}_i$ is the ith smallest residual and $f_i = \dfrac{i - 3/8}{n + 1/4}$ is plotted on the vertical axis. The normal probability plot does not reflect the straight-line appearance that one would like to see. This is another symptom of a faulty, perhaps overly simplistic choice of a regression model.

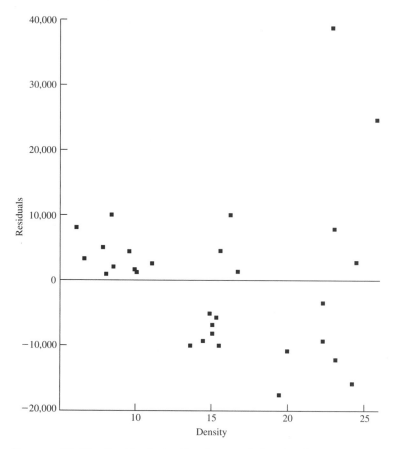

FIGURE 11.13   Residual plot for the wood density data.

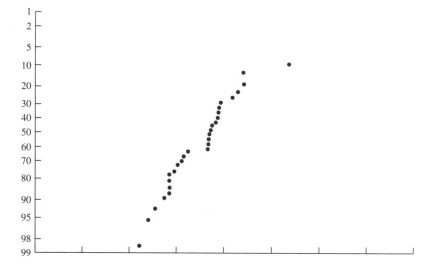

FIGURE 11.14   Normal probability plot of residuals for wood density data.

Both types of residual plots and, indeed, the scatterplot itself suggests here that a somewhat complicated model would be appropriate. One possible model is to use a natural log transformation suggested by Figure 11.8. In other words, one might choose to regress ln y against x. This produces the regression

$$\ln y = 8.257 + 0.125x_1 \qquad (R^2 = 0.9016).$$

To gain some insight on whether the transformed model is more appropriate, consider Figure 11.15 which reveals a plot of the residuals in stiffness [i.e., y_i-antilog (ln y) against density]. Figure 11.15 appears to be closer to a random pattern around zero. This in addition to the higher R^2 value would suggest that the transformed model is more appropriate.

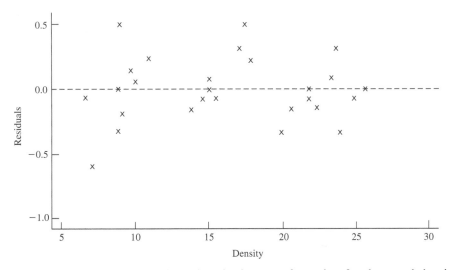

FIGURE 11.15   Residual plot using the log transformation for the wood density data.

11.11 Correlation

Up to this point we have assumed that the independent regressor variable x is a physical or scientific variable but not a random variable. In fact, in this context, x is often called a **mathematical variable**, which, in the sampling process, is measured with negligible error. In many applications of regression techniques it is more realistic to assume that both X and Y are random variables and the measurements $\{(x_i, y_i); i = 1, 2, \ldots, n\}$ are observations from a population having the joint density function $f(x, y)$. We shall consider the problem of measuring the relationship between the two variables X and Y. For example, if X and Y represent the length and circumference of a particular kind of bone in the adult body, we might conduct an anthropological study to determine whether large values of X are associated with large values of Y, and vice versa. On the other hand, if X represents the age of a used automobile and Y represents the retail book value of the automobile, we would expect large values of X to correspond to small values of Y and small values of X to correspond to large values of Y. **Correlation analysis** attempts to measure the strength of such relationships between two variables by means of a single number called a **correlation coefficient**.

In theory it is often assumed that the conditional distribution $f(y\,|\,x)$ of Y, for fixed values of X, is normal with mean $\mu_{Y|x} = \alpha + \beta x$ and variance $\sigma^2_{Y|x} = \sigma^2$ and that X is likewise normally distributed with mean μ_x and variance σ^2_X. The joint density of X and Y is then given by

$$f(x, y) = n(y\,|\,x;\ \alpha + \beta x,\ \sigma)n(x;\ \mu_X,\ \sigma_X)$$

$$= \frac{1}{2\pi\sigma_X\sigma}\exp\left\{-\left(\frac{1}{2}\right)\left[\left(\frac{y - (\alpha + \beta x)}{\sigma}\right)^2 + \left(\frac{x - \mu_X}{\sigma_X}\right)^2\right]\right\},$$

for $-\infty < x < \infty$ and $-\infty < y < \infty$.

Let us write the random variable Y in the form

$$Y = \alpha + \beta X + E,$$

where X is now a random variable independent of the random error E. Since the mean of the random error E is zero, it follows that

$$\mu_Y = \alpha + \beta\mu_X$$

and

$$\sigma^2_Y = \sigma^2 + \beta^2\sigma^2_X.$$

Substituting for α and σ^2 into the above expression for $f(x, y)$, we obtain the **bivariate normal distribution**

$$f(x, y) = \frac{1}{2\pi\sigma_X\sigma_Y\sqrt{1 - \rho^2}}$$

$$\times \exp\left\{-\frac{1}{2(1 - \rho^2)}\left[\left(\frac{x - \mu_X}{\sigma_X}\right)^2 - 2\rho\left(\frac{x - \mu_X}{\sigma_X}\right)\left(\frac{y - \mu_Y}{\sigma_Y}\right) + \left(\frac{y - \mu_Y}{\sigma_Y}\right)^2\right]\right\},$$

for $-\infty < x < \infty$ and $-\infty < y < \sigma$ where

$$\rho^2 = 1 - \frac{\sigma^2}{\sigma^2_Y} = \beta^2\frac{\sigma^2_X}{\sigma^2_Y}.$$

The constant ρ (rho) is called the **population correlation coefficient** and plays a major role in many bivariate data analysis problems. It is important for the reader to understand the physical interpretation of this correlation coefficient and the distinction between correlation and regression. The term *regression* still has meaning here. In fact, the straight line given by $\mu_{Y|x} = \alpha + \beta x$ is still called the regression line as before, and the estimates of α and β are identical to those given in Section 11.2. The value of ρ is 0 when $\beta = 0$, which results when there essentially is no linear regression; that is, the regression line is horizontal and any knowledge of X is useless in predicting Y. Since $\sigma_Y^2 \geq \sigma^2$, we must have $\rho^2 \leq 1$ and hence $-1 \leq \rho \leq 1$. Values of $\rho = \pm 1$ only occur when $\sigma^2 = 0$, in which case we have a perfect linear relationship between the two variables. Thus a value of ρ equal to $+1$ implies a perfect linear relationship with a positive slope, while a value of ρ equal to -1 results from a perfect linear relationship with a negative slope. It might be said then that sample estimates of ρ close to unity in magnitude imply good correlation or **linear association** between X and Y, while values near zero indicate little or no correlation.

To obtain a sample estimate of ρ, recall from Section 11.3 that the error sum of squares is given by

$$SSE = S_{yy} - bS_{xy}.$$

Dividing both sides of this equation by S_{yy} and replacing S_{xy} by bS_{xx}, we obtain the relation

$$b^2 \frac{S_{xx}}{S_{yy}} = 1 - \frac{SSE}{S_{yy}}.$$

The value of $b^2 S_{xx}/S_{yy}$ is zero when $b = 0$, which will occur when the sample points show no linear relationship. Since $S_{yy} \geq SSE$, we conclude that $b^2 S_{xx}/S_{yy}$ must be between 0 and 1. Consequently, $b\sqrt{S_{xx}/S_{yy}}$ must range from -1 to $+1$, negative values corresponding to lines with negative slopes and positive values to lines with positive slopes. A value of -1 or $+1$ will occur when $SSE = 0$, but this is the case where all sample points lie in a straight line. Hence a perfect linear relationship appears in the sample data when $b\sqrt{S_{xx}/S_{yy}} = \pm 1$. Clearly, the quantity $b\sqrt{S_{xx}/S_{yy}}$, which we shall henceforth designate as r, can be used as an estimate of the population correlation coefficient ρ. It is customary to refer to the estimate r as the **Pearson product-moment correlation coefficient** or simply the **sample correlation coefficient**.

CORRELATION COEFFICIENT	*The measure ρ of linear association between two variables X and Y is estimated by the* **sample correlation coefficient** r, *where*

$$r = b\sqrt{\frac{S_{xx}}{S_{yy}}} = \frac{S_{xy}}{\sqrt{S_{xx}S_{yy}}}. \qquad \blacksquare$$

For values of r between -1 and $+1$ we must be careful in our interpretation. For example, values of r equal to 0.3 and 0.6 only mean that we have two positive correlations, one somewhat stronger than the other. It is wrong to conclude that $r = 0.6$ indicates a linear relationship twice as good as that indicated by the value $r = 0.3$. On the other hand, if we write

$$r^2 = \frac{S_{xy}^2}{S_{xx}S_{yy}} = \frac{SSR}{S_{yy}},$$

then r^2, which is usually referred to as the **sample coefficient of determination**, represents the proportion of the variation of S_{yy} explained by the regression of Y on x, namely, SSR. That is, r^2 expresses the proportion of the total variation in the values of the variable Y that can be accounted for or explained by a linear relationship with the values of the random variable X. Thus a correlation of 0.6 means that 0.36 or 36% of the total variation of the values of Y in our sample is accounted for by a linear relationship with values of X.

EXAMPLE 11.10   It is important that scientific researchers in the area of forest products be able to study correlation among the anatomy and mechanical properties of trees. According to the study *"Quantitative Anatomical Characteristics of Plantation Grown Loblolly Pine (Pinus Taeda L.) and Cottonwood (Populus deltoides Bart. Ex Marsh.) and Their Relationships to Mechanical Properties"* conducted by the Department of Forestry and Forest Products at the Virginia Polytechnic Institute and State University (1983), an experiment in which 29 loblolly pines were randomly selected for investigation yielded the following table on the specific gravity in grams/cm^3 and the modulus of rupture in kilopascals (kPa):

Specific Gravity, x (g/cm^3)	Modulus of Rupture, y (kPa)
0.414	29,186
0.383	29,266
0.399	26,215
0.402	30,162
0.442	38,867
0.422	37,831
0.466	44,576
0.500	46,097
0.514	59,698
0.530	67,705
0.569	66,088
0.558	78,486
0.577	89,869
0.572	77,369
0.548	67,095
0.581	85,156
0.557	69,571
0.550	84,160
0.531	73,466
0.550	78,610
0.556	67,657
0.523	74,017
0.602	87,291
0.569	86,836
0.544	82,540
0.557	81,699
0.530	82,096
0.547	75,657
0.585	80,490

Compute and interpret the sample correlation coefficient.

SOLUTION
From the data we find that

$$S_{xx} = 0.11273, \qquad S_{yy} = 11,807,324,786, \qquad S_{xy} = 34,422.75972.$$

Therefore,

$$r = \frac{34,422.75972}{\sqrt{(0.11273)(11,807,324,786)}} = 0.9435.$$

A correlation coefficient of 0.9435 indicates a good linear relationship between X and Y. Since $r^2 = 0.8902$, we can say that approximately 89% of the variation in the values of Y is accounted for by a linear relationship with X.

A test of the special hypothesis $\rho = 0$ versus an appropriate alternative is equivalent to testing $\beta = 0$ for the simple linear regression model and therefore the procedures of Section 11.7 using either the t-distribution with $n - 2$ degrees of freedom or the F-distribution with 1 and $n - 2$ degrees of freedom are applicable. However, if one wishes to avoid the analysis-of-variance procedure and compute only the sample correlation coefficient, it can easily be verified (see Exercise 3 on page 409) that the t-value given by

$$t = \frac{b}{s/\sqrt{S_{xx}}} = \frac{\sqrt{SSR}}{s}$$

can also be written as

$$t = \frac{r\sqrt{n - 2}}{\sqrt{1 - r^2}},$$

which, as before, is a value of the statistic T having a t-distribution with $n - 2$ degrees of freedom.

EXAMPLE 11.11   For the data of Example 11.10 test the hypothesis that there is no linear association among the variables.

SOLUTION

1. H_0: $\rho = 0$.
2. H_1: $\rho \neq 0$.
3. $\alpha = 0.05$.
4. Critical region: $t < -2.052$ and $t > 2.052$.
5. Computations:

$$t = \frac{0.9435(\sqrt{27})}{\sqrt{1 - (0.9435)^2}} = 14.79$$

$$P < 0.0001.$$

6. Decision: Reject the hypothesis of no linear association.

A test of the more general hypothesis $\rho = \rho_0$ against a suitable alternative is easily conducted from the sample information. If X and Y follow the bivariate normal distribution, the quantity

$$\frac{1}{2} \ln \left(\frac{1 + r}{1 - r} \right)$$

is a value of a random variable that follows approximately the normal distribution with mean $(1/2) \ln [(1 + \rho)/1 - \rho)]$ and variance $1/(n - 3)$. Thus the test procedure is to compute

$$z = \frac{\sqrt{n - 3}}{2} \left[\ln \left(\frac{1 + r}{1 - r} \right) - \ln \left(\frac{1 + \rho_0}{1 - \rho_0} \right) \right]$$

$$= \frac{\sqrt{n - 3}}{2} \ln \left[\frac{(1 + r)(1 - \rho_0)}{(1 - r)(1 + \rho_0)} \right]$$

and compare with the critical points of the standard normal distribution.

EXAMPLE 11.12   For the data of Example 11.10 test the null hypothesis that $\rho = 0.9$ against the alternative that $\rho > 0.9$. Use a 0.05 level of significance.

SOLUTION

1. H_0: $\rho = 0.9$.
2. H_1: $\rho > 0.9$.
3. $\alpha = 0.05$.
4. Critical region: $z > 1.645$.
5. Computations:

$$z = \frac{\sqrt{26}}{2} \ln \left[\frac{(1 + 0.9435)0.1}{(1 - 0.9435)1.9} \right] = 1.51$$

$$P = 0.0655.$$

6. Decision: There is certainly some evidence that the correlation coefficient exceeds 0.9.

It should be pointed out that in correlation studies, as in linear regression problems, the results obtained are only as good as the model that is assumed. In the correlation techniques studied here, a bivariate normal density is assumed for the variables X and Y, with the mean value of Y at each x-value being linearly related to x. To observe the suitability of the linearity assumption, a preliminary plotting of the experimental data is often helpful. A value of the sample correlation coefficient close to zero will result from data that display a strictly random effect as in Figure 11.16(a), thus implying little or no causal relationship. It is important to remember that the correlation coefficient between two variables is a measure of their linear relationship, and that a value of $r = 0$ implies a lack of linearity and not a lack of association. Hence, if a strong quadratic relationship exists between X and Y as indicated in Figure 11.16(b), we can still obtain a zero correlation indicating a nonlinear relationship.

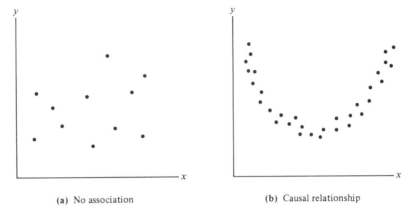

(a) No association   (b) Causal relationship

FIGURE 11.16   Scatter diagram showing zero correlation.

Exercises

1. Compute and interpret the correlation coefficient for the following grades of 6 students selected at random:

Mathematics grade	70	92	80	74	65	83
English grade	74	84	63	87	78	90

2. Test the hypothesis that $\rho = 0$ in Exercise 1 against the alternative that $\rho \neq 0$. Use a 0.05 level of significance.

3. Show the necessary steps in converting the equation

$$t = \frac{b}{s/\sqrt{S_{xx}}}$$

to the equivalent form

$$t = \frac{r\sqrt{n-2}}{\sqrt{1-r^2}}.$$

4. The following data were obtained in a study of the relationship between the weight and chest size of infants at birth:

Weight (kg)	Chest Size (cm)
2.75	29.5
2.15	26.3
4.41	32.2
5.52	36.5
3.21	27.2
4.32	27.7
2.31	28.3
4.30	30.3
3.71	28.7

(a) Calculate r.

(b) Test the null hypothesis that $\rho = 0$ against the alternative that $\rho > 0$ at the 0.01 level of significance.

(c) What percentage of the variation in the infant chest sizes is explained by difference in weight?

5. With reference to Exercise 1 on page 371,

(a) calculate r;

(b) test the hypothesis that $\rho = 0$ against the alternative that $\rho \neq 0$ at the 0.05 level of significance.

6. With reference to Exercise 9 on page 373,

(a) calculate r;

(b) test the null hypothesis that $\rho = -0.5$ against the alternative that $p < -0.5$ at the 0.025 level of significance;

(c) determine the percentage of the variation in the amount of particulate removed that is due to changes in the daily amount of rainfall.

Review Exercises

1. With reference to Exercise 6 on page 372, construct
   (a) a 95% confidence interval for the average course grade of students who make a 35 on the placement test;
   (b) a 95% prediction interval for the course grade of a student who made a 35 on the placement test.

2. The amounts of solids removed from a particular material when exposed to drying periods of different lengths are as follows:

x (hours)	y (grams)	
4.4	13.1	14.2
4.5	9.0	11.5
4.8	10.4	11.5
5.5	13.8	14.8
5.7	12.7	15.1
5.9	9.9	12.7
6.3	13.8	16.5
6.9	16.4	15.7
7.5	17.6	16.9
7.8	18.3	17.2

(a) Estimate the linear regression line.
(b) Test at the 0.05 level of significance whether the linear model is adequate.

3. The Statistics Consulting Center at Virginia Polytechnic Institute and State University analyzed data on normal woodchucks for the Department of Veterinary Medicine. The variables of interest were body weight in grams and heart weight in grams. It was also of interest to develop a linear regression equation in order to determine if there is a significant linear relationship between heart weight and total body weight. Use heart weight as the independent variable and body weight as the dependent variable and fit a simple linear regression using the following data. In addition, test the hypothesis $H_0: \beta = 0$ versus $H_1: \beta \neq 0$. Draw conclusions.

Body Weight (grams)	Heart Weight (grams)
4050	11.2
2465	12.4
3120	10.5
5700	13.2
2595	9.8
3640	11.0
2050	10.8
4235	10.4
2935	12.2
4975	11.2
3690	10.8
2800	14.2
2775	12.2
2170	10.0
2370	12.3
2055	12.5
2025	11.8
2645	16.0
2675	13.8

4. With reference to Exercise 7 on page 372, construct
   (a) a 95% confidence interval for the average weekly sales when $45 is spent on advertising;
   (b) a 95% prediction interval for the weekly sales when $45 is spent on advertising.

5. An experiment was designed for the Department of Materials Engineering at Virginia Polytechnic Institute and State University to study hydrogen embrittlement properties based on electrolytic hydrogen pressure measurements. The solution used was 0.1 N NaOH, the material being a certain type of stainless steel. The cathodic charging current density was controlled and varied at four levels. The effective hydrogen pressure was observed as the response. The data follow.

Run	Charging Current Density, x (mA/cm^2)	Effective Hydrogen Pressure, y (atm)
1	0.5	86.1
2	0.5	92.1
3	0.5	64.7
4	0.5	74.7
5	1.5	223.6
6	1.5	202.1
7	1.5	132.9
8	2.5	413.5
9	2.5	231.5
10	2.5	466.7
11	2.5	365.3
12	3.5	493.7
13	3.5	382.3
14	3.5	447.2
15	3.5	563.8

(a) Run a simple linear regression of y against x.
(b) Compute the pure error sum of squares and make a test for lack of fit.
(c) Does the information in part (b) indicate a need for a model in x beyond a first-order regression? Explain.

6. The following data represent the chemistry grades for a random sample of 12 freshmen at a certain college along with their scores on an intelligence test administered while they were still seniors in high school:

Student	Test Score, x	Chemistry Grade, y
1	65	85
2	50	74
3	55	76
4	65	90
5	55	85
6	70	87
7	65	94
8	70	98
9	55	81
10	70	91
11	50	76
12	55	74

(a) Compute and interpret the sample correlation coefficient.
(b) Test the hypothesis that $\rho = 0.5$ against the alternative that $\rho > 0.5$. Use a P-value in the conclusion.

7. For the simple linear regression model, prove that $E(s^2) = \sigma^2$.

12

Multiple Linear Regression

12.1 Introduction

In most research problems where regression analysis is applied, more than one independent variable is needed in the regression model. The complexity of most scientific mechanisms is such that in order to be able to predict an important response, a **multiple regression model** is needed. When this model is linear in the coefficients, it is called a **multiple linear regression model**. For the case of k independent variables $x_1, x_2, \ldots, x_k$, the mean of $Y | x_1, x_2, \ldots, x_k$ is given by the multiple linear regression model

$$\mu_{Y | x_1, x_2, \ldots, x_k} = \beta_0 + \beta_1 x_1 + \cdots + \beta_k x_k$$

and the estimated response is obtained from the sample regression equation

$$\hat{y} = b_0 + b_1 x_1 + \cdots + b_k x_k,$$

where each regression coefficient β_i is estimated by b_i from the sample data using the method of least squares. As in the case of a single independent variable, the multiple linear regression model can often be an adequate representation of a more complicated structure within certain ranges of the independent variables.

Similar least squares techniques can also be applied in estimating the coefficients when the linear model involves, say, powers and products of the independent variables. For example, when $k = 1$, the experimenter may feel that the means $\mu_{Y|x}$ do not fall on a straight line but are more appropriately described by the **polynomial regression model**

$$\mu_{Y|x} = \beta_0 + \beta_1 x + \beta_2 x^2 + \cdots + \beta_r x^r,$$

and the estimated response is obtained from the polynomial regression equation

$$\hat{y} = b_0 + b_1 x + b_2 x^2 + \cdots + b_r x^r.$$

Confusion arises occasionally when we speak of a polynomial model as a linear model. However, statisticians normally refer to a linear model as one in which the parameters occur linearly, regardless of how the independent variables enter the model. An example of a nonlinear model is the **exponential relationship** given by

$$\mu_{Y|x} = \alpha\beta^x,$$

which is estimated by the regression equation

$$\hat{y} = ab^x.$$

There are many phenomena in science and engineering that are inherently nonlinear in nature and, when the true structure is known, an attempt should certainly be made to fit the actual model. The literature on estimation by least squares of nonlinear models is voluminous and, except for Exercise 10 on page 373, we do not attempt to cover the subject in this book. A student who wants a good account of some aspects of this subject should consult *Classical and Modern Regression with Applications* by Myers (see the Bibliography).

12.2 Estimating the Coefficients _____

In this section we shall obtain the least squares estimators of the parameters β_0, β_1, . . . , β_k by fitting the multiple linear regression model

$$\mu_{Y|x_1, x_2, \ldots, x_k} = \beta_0 + \beta_1 x_1 + \beta_2 x_2 + \cdots + \beta_k x_k$$

to the data points

$$\{(x_{1i}, x_{2i}, \ldots, x_{ki}, y_i); \quad i = 1, 2, \ldots, n \text{ and } n > k\},$$

where y_i is the observed response to the values $x_{1i}, x_{2i}, \ldots, x_{ki}$ of the k independent variables $x_1, x_2, \ldots, x_k$. Each observation $(x_{1i}, x_{2i}, \ldots, x_{ki}, y_i)$ satisfies the equation

$$y_k = \beta_0 + \beta_1 x_{1i} + \beta_2 x_{2i} + \cdots + \beta_k x_{ki} + \varepsilon_i$$

or

$$y_i = b_0 + b_1 x_{1i} + b_2 x_{2i} + \cdots + b_k x_{ki} + e_i,$$

where ε_i and e_i are the random error and residual, respectively, associated with the response y_i. In using the concept of least squares to arrive at estimates $b_0, b_1, \ldots, b_k$, we minimize the expression

$$SSE = \sum_{i=1}^{n} e_i^2 = \sum_{i=1}^{n} (y_i - b_0 - b_1 x_{1i} - b_2 x_{2i} - \cdots - b_k x_{ki})^2.$$

Differentiating SSE in turn with respect to $b_0, b_1, b_2, \ldots, b_k$, and equating to zero, we generate the set of $k + 1$ normal equations

$$n b_0 + b_1 \sum_{i=1}^{n} x_{1i} + b_2 \sum_{i=1}^{n} x_{2i} + \cdots + b_k \sum_{i=1}^{n} x_{ki} = \sum_{i=1}^{n} y_i$$

$$b_0 \sum_{i=1}^{n} x_{1i} + b_1 \sum_{i=1}^{n} x_{1i}^2 + b_2 \sum_{i=1}^{n} x_{1i} x_{2i} + \cdots + b_k \sum_{i=1}^{n} x_{1i} x_{ki} = \sum_{i=1}^{n} x_{1i} y_i$$

$$\vdots \qquad \vdots \qquad \vdots \qquad \vdots \qquad \vdots$$

$$b_0 \sum_{i=1}^{n} x_{ki} + b_1 \sum_{i=1}^{n} x_{ki} x_{1i} + b_2 \sum_{i=1}^{n} x_{ki} x_{2i} + \cdots + b_k \sum_{i=1}^{n} x_{ki}^2 = \sum_{i=1}^{n} x_{ki} y_i.$$

These equations can be solved for $b_0, b_1, b_2, \ldots, b_k$ by any appropriate method for solving systems of linear equations.

EXAMPLE 12.1   In analytical chemistry, X-ray fluorescence analysis is a tool for estimating ingredient percentages in multicomponent mixtures. Often the estimation of concentrations depends heavily on the user's ability to fit suitable regression models. In the paper *"Corrections for Matrix Effects in X-Ray Fluorescence Analysis Using Multiple Regression Methods,"* published an *Analytical Chemistry* (Vol. 37, 1965), propellant slurries containing 4 ingredients were prepared. Concentrations of the components varied in the slurries to produce calibration-type standards. The data are as follows:

y	x_1	x_2	x_3	x_4
0.5514	1.1240	0.8980	0.8219	0.9906
0.4426	0.9285	0.8872	0.9308	0.9944
0.5631	1.1214	0.8030	0.7668	1.1221
0.5624	1.1635	0.8706	0.9272	0.9832
0.4505	0.9415	0.8064	0.9026	1.1127
0.5290	1.0712	0.8404	0.8662	1.0836
0.4702	0.9561	0.8731	0.8206	1.0290
0.5001	1.0186	0.8431	0.8346	1.0591
0.4425	0.9039	0.8314	0.7596	1.0994

The response y_i is the measured concentration of an ingredient A. The measured value x_1 is the X-ray "intensity ratio" associated with ingredient A, and the values x_2, x_3, and x_4 are intensity ratios for the additional components in the slurry. As a result of enhancement and absorption effects, the response y is best predicted after regressing it against intensity values associated with *all* components. Thus the model is given by

$$\mu_{Y|x_1, x_2, x_3, x_4} = \beta_0 + \beta_1 x_1 + \beta_2 x_2 + \beta_3 x_3 + \beta_4 x_4.$$

Fit this multiple linear regression model to the given data and then estimate the concentration of ingredient A for a mixture whose X-ray intensity ratios are, respectively, $x_1 = 1.091$, $x_2 = 0.855$, $x_3 = 0.758$, and $x_4 = 1.005$.

SOLUTION
From the given data we find $n = 9$ and

$$\sum_{i=1}^{9} x_{1i} = 9.2287 \qquad \sum_{i=1}^{9} x_{2i} = 7.6532 \qquad \sum_{i=1}^{9} x_{3i} = 7.6303$$

$$\sum_{i=1}^{9} x_{4i} = 9.4741 \qquad \sum_{i=1}^{9} x_{1i}^2 = 9.5394 \qquad \sum_{i=1}^{9} x_{2i}^2 = 6.5172$$

$$\sum_{i=1}^{9} x_{3i}^2 = 6.5015 \qquad \sum_{i=1}^{9} x_{4i}^2 = 9.9974 \qquad \sum_{i=1}^{9} x_{1i}x_{2i} = 7.8510$$

$$\sum_{i=1}^{9} x_{1i}x_{3i} = 7.8257 \qquad \sum_{i=1}^{9} x_{1i}x_{4i} = 9.7037 \qquad \sum_{i=1}^{9} x_{2i}x_{3i} = 6.4943$$

$$\sum_{i=1}^{9} x_{2i}x_{4i} = 8.0421 \qquad \sum_{i=1}^{9} x_{3i}x_{4i} = 8.0182 \qquad \sum_{i=1}^{9} y_i = 4.5118$$

$$\sum_{i=1}^{9} x_{1i}y_i = 4.6663 \qquad \sum_{i=1}^{9} x_{2i}y_i = 3.8375 \qquad \sum_{i=1}^{9} x_{3i}y_i = 3.8226$$

$$\sum_{i=1}^{9} x_{4i}y_i = 4.7456.$$

Inserting these values into the normal equations above, we obtain

$$9b_0 + 9.2287b_1 + 7.6532b_2 + 7.6303b_3 + 9.4741b_4 = 4.5118$$
$$9.2287b_0 + 9.5394b_1 + 7.8510b_2 + 7.8257b_3 + 9.7037b_4 = 4.6663$$
$$7.6532b_0 + 7.8510b_1 + 6.5172b_2 + 6.4943b_3 + 8.0421b_4 = 3.8375$$
$$7.6303b_0 + 7.8257b_1 + 6.4943b_2 + 6.5015b_3 + 8.0182b_4 = 3.8226$$
$$9.4741b_0 + 9.7037b_1 + 8.0421b_2 + 8.0182b_3 + 9.9974b_4 = 4.7456.$$

The solution of this set of equations yields the unique estimates

$$b_0 = -0.3004, \qquad b_1 = 0.5387, \qquad b_2 = 0.1770,$$
$$b_3 = -0.0704, \qquad b_4 = 0.1506.$$

Therefore, our regression equation is

$$\hat{y} = -0.3004 + 0.5387x_1 + 0.1770x_2 - 0.0704x_3 + 0.1506x_4.$$

For a mixture whose X-ray intensities are $x_1 = 1.091$, $x_2 = 0.855$, $x_3 = 0.758$, and $x_4 = 1.005$, the estimated concentration of component A is given by

$$\hat{y} = -0.3004 + (0.5387)(1.091) + (0.1770)(0.855)$$

$$-(0.0704)(0.758) + (0.1506)(1.005)$$

$$=0.5366.$$

POLYNOMIAL REGRESSION

Now suppose that we wish to fit the polynomial equation

$$\mu_{Y|x} = \beta_0 + \beta_1 x + \beta_2 x^2 + \cdots + \beta_r x^r$$

to the n pairs of observations $\{(x_i, y_i); i = 1, 2, \ldots, n\}$. Each observation, y_i, satisfies the equation

$$y_i = \beta_0 + \beta_1 x_i + \beta_2 x_i^2 + \cdots + \beta_r x_i^r + \varepsilon_i$$

or

$$y_i = b_0 + b_1 x_i + b_2 x_i^2 + \cdots + b_r x_i^r + e_i,$$

where r is the degree of the polynomial, and ε_i and e_i are again the random error and residual associated with the response y_i. Here, the number of pairs, n, must be at least as large as $r + 1$, the number of parameters to be estimated. Notice that the polynomial model can be considered a special case of the more general multiple linear regression model, where we set $x_1 = x$, $x_2 = x^2$, $\ldots$, $x_r = x^r$. The normal equations assume the form

$$nb_0 + b_1 \sum_{i=1}^{n} x_i + b_2 \sum_{i=1}^{n} x_i^2 + \cdots + b_r \sum_{i=1}^{n} x_i^r = \sum_{i=1}^{n} y_i$$

$$b_0 \sum_{i=1}^{n} x_i + b_1 \sum_{i=1}^{n} x_1^2 + b_2 \sum_{i=1}^{n} x_i^3 + \cdots + b_r \sum_{i=1}^{n} x_i^{r+1} = \sum_{i=1}^{n} x_i y_i$$

$$\vdots \qquad \vdots \qquad \vdots \qquad \vdots \qquad \vdots$$

$$b_0 \sum_{i=1}^{n} x_i^r + b_1 \sum_{i=1}^{n} x_i^{r+1} + b_2 \sum_{i=1}^{n} x_i^{r+2} + \cdots + b_r \sum_{i=1}^{n} x_i^{2r} = \sum_{i=1}^{n} x_i^r y_i,$$

which are solved as before for $b_0, b_1, \ldots, b_r$.

EXAMPLE 12.2   Given the data

x	0	1	2	3	4	5	6	7	8	9
y	9.1	7.3	3.2	4.6	4.8	2.9	5.7	7.1	8.8	10.2

fit a regression curve of the form $\mu_{Y|x} = \beta_0 + \beta_1 x + \beta_2 x^2$ and then estimate $\mu_{Y|2}$.

SOLUTION

From the data given, we find

$$\sum_{i=1}^{10} x_i = 45 \qquad \sum_{i=1}^{10} x_i^2 = 297 \qquad \sum_{i=1}^{10} x_i^3 = 2025 \qquad \sum_{i=1}^{10} x_i^4 = 15{,}332$$

$$\sum_{i=1}^{10} y_i = 53.7 \qquad \sum_{i=1}^{10} x_i y_i = 307.3 \qquad \sum_{i=1}^{10} x_i^2 y_i = 2153.3 \qquad n = 10.$$

Solving the normal equations

$$10b_0 + 45b_1 + 297b_2 = 53.7$$
$$45b_0 + 297b_1 + 2025b_2 = 307.3$$
$$297b_0 + 2025b_1 + 15{,}332b_2 = 2153.3,$$

we obtain

$$b_0 = 8.697, \qquad b_1 = -2.341, \qquad b_2 = 0.288.$$

Therefore,

$$\hat{y} = 8.697 - 2.341x + 0.288x^2.$$

When $x = 2$ our estimate of $\mu_{Y|2}$ is

$$\hat{y} = 8.697 - (2.341)(2) + (0.288)(2^2)$$
$$= 5.2.$$

12.3 Linear Regression Model Using Matrices _____

In fitting a multiple linear regression model, particularly when the number of variables exceeds two, a knowledge of matrix theory can facilitate the mathematical manipulations considerably. Suppose that the experimenter has k independent variables $x_1, x_2, \ldots, x_k$ and n observations $y_1, y_2, \ldots, y_n$, each of which can be expressed by the equation

$$y_i = \beta_0 + \beta_1 x_{1i} + \beta_2 x_{2i} + \cdots + \beta_k x_{ki} + \varepsilon_i.$$

This model essentially represents n equations describing how the response values are generated in the scientific process. Using matrix notation, we can write the equations

$$\mathbf{y} = \mathbf{X}\boldsymbol{\beta} + \boldsymbol{\varepsilon}$$

where

$$\mathbf{y} = \begin{bmatrix} y_1 \\ y_2 \\ \vdots \\ y_n \end{bmatrix}, \qquad \mathbf{X} = \begin{bmatrix} 1 & x_{11} & x_{21} & \cdots & x_{k_1} \\ 1 & x_{12} & x_{22} & \cdots & x_{k_2} \\ \vdots & \vdots & \vdots & & \vdots \\ 1 & x_{1n} & x_{2n} & \cdots & x_{kn} \end{bmatrix}, \qquad \boldsymbol{\beta} = \begin{bmatrix} \beta_0 \\ \beta_1 \\ \beta_2 \\ \vdots \\ \beta_k \end{bmatrix}$$

Then the least squares solution for estimation of $\boldsymbol{\beta}$ illustrated in Section 12.2 involves finding $\mathbf{b}$ for which

$$SSE = (\mathbf{y} - \mathbf{Xb})'(\mathbf{y} - \mathbf{Xb})$$

is minimized. This minimization process involves solving for $\mathbf{b}$ in the equation

$$\frac{\partial}{\partial \mathbf{b}} (SSE) = \mathbf{0}.$$

We will not present the details regarding solutions of the equations above. The result reduces to the solution of $\mathbf{b}$ in

$$(\mathbf{X}'\mathbf{X})\mathbf{b} = \mathbf{X}'\mathbf{y}.$$

Notice the nature of the $\mathbf{X}$ matrix. Apart from the initial element, the ith row represents the x-values that give rise to the response y_i. Writing

$$\mathbf{A} = \mathbf{X}'\mathbf{X} = \begin{bmatrix} n & \sum_{i=1}^{n} x_{1i} & \sum_{i=1}^{n} x_{2i} & \cdots & \sum_{i=1}^{n} x_{ki} \\ \sum_{i=1}^{n} x_{1i} & \sum_{i=1}^{n} x_{1i}^2 & \sum_{i=1}^{n} x_{1i}x_{2i} & \cdots & \sum_{i=1}^{n} x_{1i}x_{ki} \\ \vdots & \vdots & \vdots & & \vdots \\ \sum_{i=1}^{n} x_{ki} & \sum_{i=1}^{n} x_{ki}x_{1i} & \sum_{i=1}^{n} x_{ki}x_{2i} & \cdots & \sum_{i=1}^{n} x_{ki}^2 \end{bmatrix}$$

and

$$\mathbf{g} = \mathbf{X}'\mathbf{y} = \begin{bmatrix} g_0 = \sum_{i=1}^{n} y_i \\ g_1 = \sum_{i=1}^{n} x_{1i}y_i \\ \vdots \\ g_k = \sum_{i=1}^{n} x_{ki}y_i \end{bmatrix},$$

the normal equations can be put in the matrix form

$$\mathbf{AB} = \mathbf{g}.$$

If the matrix $\mathbf{A}$ is nonsingular, we can write the solution for the regression coefficients as

$$\mathbf{b} = \mathbf{A}^{-1}\mathbf{g} = (\mathbf{X}'\mathbf{X})^{-1}\mathbf{X}'\mathbf{y}.$$

Thus one can obtain the prediction equation or regression equation by solving a set of $k + 1$ equations in a like number of unknowns. This involves the inversion of the $k + 1$ by $k + 1$ matrix $\mathbf{X}'\mathbf{X}$. Techniques for inverting this matrix are explained in most textbooks on elementary determinants and matrices. Of course, there are many high-speed computer packages available for multiple regression problems, packages that not only print out estimates of the regression coefficients, but also provide other information relevant to making inferences concerning the regression equation.

EXAMPLE 12.3   The percent survival of a certain type of animal semen after storage was measured at various combinations of concentrations of three materials used to increase chance of survival. The data are as follows:

y (% survival)	x_1 (weight %)	x_2 (weight %)	x_3 (weight %)
25.5	1.74	5.30	10.80
31.2	6.32	5.42	9.40
25.9	6.22	8.41	7.20
38.4	10.52	4.63	8.50
18.4	1.19	11.60	9.40
26.7	1.22	5.85	9.90
26.4	4.10	6.62	8.00
25.9	6.32	8.72	9.10
32.0	4.08	4.42	8.70
25.2	4.15	7.60	9.20
39.7	10.15	4.83	9.40
35.7	1.72	3.12	7.60
26.5	1.70	5.30	8.20

Estimate the multiple linear regression model for the given data.

SOLUTION

From the experiment data we list the following sums of squares and products:

$$\sum_{i=1}^{13} y_i = 377.5 \qquad \sum_{i=1}^{13} y_i^2 = 11{,}400.15 \qquad \sum_{i=1}^{13} x_{1i} = 59.43$$

$$\sum_{i=1}^{13} x_{2i} = 81.82 \qquad \sum_{i=1}^{13} x_{3i} = 115.40 \qquad \sum_{i=1}^{13} x_{1i}^2 = 394.7255$$

$$\sum_{i=1}^{13} x_{2i}^2 = 576.7264 \qquad \sum_{i=1}^{13} x_{3i}^3 = 1035.9600 \qquad \sum_{i=1}^{13} x_{1i}y_i = 1877.567$$

$$\sum_{i=1}^{13} x_{2i}y_i = 2246.661 \qquad \sum_{i=1}^{13} x_{3i}y_i = 3337.780 \qquad \sum_{i=1}^{13} x_{1i}x_{2i} = 360.6621$$

$$\sum_{i=1}^{13} x_{1i}x_{3i} = 522.0780 \qquad \sum_{i=1}^{13} x_{2i}x_{3i} = 728.3100 \qquad n = 13.$$

The least squares estimating equations, $(\mathbf{X'X})\mathbf{b} = \mathbf{X'y}$, are given by

$$\begin{bmatrix} 13 & 59.43 & 81.82 & 115.40 \\ 59.43 & 394.7255 & 360.6621 & 522.0780 \\ 81.82 & 360.6621 & 576.7264 & 728.3100 \\ 15.40 & 522.0780 & 728.3100 & 1035.9600 \end{bmatrix} \begin{bmatrix} b_0 \\ b_1 \\ b_2 \\ b_3 \end{bmatrix} = \begin{bmatrix} 377.5 \\ 1877.567 \\ 2246.661 \\ 3337.780 \end{bmatrix}.$$

From a computer readout we obtain the elements of the inverse matrix

$$(\mathbf{X'X})^{-1} = \begin{bmatrix} 8.0648 & -0.0826 & -0.0942 & -0.7905 \\ -0.0826 & 0.0085 & 0.0017 & 0.0037 \\ -0.0942 & 0.0017 & 0.0166 & -0.0021 \\ -0.7905 & 0.0037 & -0.0021 & 0.0886 \end{bmatrix},$$

and then using the relation $\mathbf{b} = (\mathbf{X'X})^{-1}\mathbf{X'y}$, the estimated regression coefficients are given by

$$b_0 = 39.1574, \qquad b_1 = 1.0161, \qquad b_2 = -1.8616, \qquad b_3 = -0.3433.$$

Hence our estimated regression equation is

$$\hat{y} = 39.1574 + 1.0161x_1 - 1.8616x_2 - 0.3433x_3.$$

For the case of a single independent variable, the *degree* of the best-fitting polynomial can often be determined by plotting a scatter diagram of the data obtained from an experiment that yields n pairs of observations of the form $\{(x_i, y_i); i = 1, 2, \ldots, n\}$. The set of normal equations derived in Section 12.2 for estimating the coefficients of the general polynomial of degree r assumes the form

$$\begin{bmatrix} n & \sum_{i=1}^{n} x_i & \sum_{i=1}^{n} x_i^2 & \cdots & \sum_{i=1}^{n} x_i^r \\ \sum_{i=1}^{n} x_i & \sum_{i=1}^{n} x_i^2 & \sum_{i=1}^{n} x_i^3 & \cdots & \sum_{i=1}^{n} x_i^{r+1} \\ \sum_{i=1}^{n} x_i^2 & \sum_{i=1}^{n} x_i^3 & \sum_{i=1}^{n} x_i^4 & \cdots & \sum_{i=1}^{n} x_i^{r+2} \\ \vdots & \vdots & \vdots & & \vdots \\ \sum_{i=1}^{n} x_i^r & \sum_{i=1}^{n} x_i^{r+1} & \sum_{i=1}^{n} x_i^{r+2} & \cdots & \sum_{i=1}^{n} x_i^{2r} \end{bmatrix} \begin{bmatrix} b_0 \\ b_1 \\ b_2 \\ \vdots \\ b_r \end{bmatrix} = \begin{bmatrix} \sum_{i=1}^{n} y_i \\ \sum_{i=1}^{n} x_i y_i \\ \vdots \\ \sum_{i=1}^{n} x_i^r y_i \end{bmatrix}.$$

Solving these $r + 1$ equations, we obtain the estimates $b_0, b_1, \ldots, b_r$ and thereby generate the polynomial regression prediction equation

$$\hat{y} = b_0 + b_1 x + b_2 x^2 + \cdots + b_r x^r.$$

The procedure for fitting a polynomial regression model can be generalized to the case of more than one independent variable. In fact, the student of regression analysis should at this stage have the facility for fitting any linear model in, say, k independent variables. Suppose, for example, that we have a response Y with $k = 2$ independent variables and a quadratic model is postulated of the type

$$y_i = \beta_0 + \beta_1 x_{1i} + \beta_2 x_{2i} + \beta_{11} x_{1i}^2 + \beta_{22} x_{2i}^2 + \beta_{12} x_{1i} x_{2i} + \varepsilon_i,$$

where y_i, $i = 1, 2, \ldots, n$, is the response to the combination (x_{1i}, x_{2i}) of the independent variables in the experiment. In this situation n must be at least 6, since there are six parameters to estimate by the least squares procedure. In addition, since the model contains quadratic terms in both variables, at least three levels of each variable must be used. The reader should easily verify that the least squares normal equations $(\mathbf{X'X})\mathbf{b} = \mathbf{X'y}$ are given by

$$
\begin{bmatrix}
n & \sum_{i=1}^{n} x_{1i} & \sum_{i=1}^{n} x_{2i} & \sum_{i=1}^{n} x_{1i}^2 & \sum_{i=1}^{n} x_{2i}^2 & \sum_{i=1}^{n} x_{1i}x_{2i} \\[2mm]
\sum_{i=1}^{n} x_{1i} & \sum_{i=1}^{n} x_{1i}^2 & \sum_{i=1}^{n} x_{1i}x_{2i} & \sum_{i=1}^{n} x_{1i}^3 & \sum_{i=1}^{n} x_{1i}x_{2i}^2 & \sum_{i=1}^{n} x_{1i}^2 x_{2i} \\[2mm]
\sum_{i=1}^{n} x_{2i} & \sum_{i=1}^{n} x_{1i}x_{2i} & \sum_{i=1}^{n} x_{2i}^2 & \sum_{i=1}^{n} x_{1i}^2 x_{2i} & \sum_{i=1}^{n} x_{2i}^3 & \sum_{i=1}^{n} x_{1i}x_{2i}^2 \\[2mm]
\sum_{i=1}^{n} x_{1i}^2 & \sum_{i=1}^{n} x_{1i}^3 & \sum_{i=1}^{n} x_{1i}^2 x_{2i} & \sum_{i=1}^{n} x_{1i}^4 & \sum_{i=1}^{n} x_{1i}^2 x_{2i}^2 & \sum_{i=1}^{n} x_{1i}^3 x_{2i} \\[2mm]
\sum_{i=1}^{n} x_{2i}^2 & \sum_{i=1}^{n} x_{1i}x_{2i}^2 & \sum_{i=1}^{n} x_{2i}^3 & \sum_{i=1}^{n} x_{1i}^2 x_{2i}^2 & \sum_{i=1}^{n} x_{2i}^4 & \sum_{i=1}^{n} x_{1i}x_{2i}^3 \\[2mm]
\sum_{i=1}^{n} x_{1i}x_{2i} & \sum_{i=1}^{n} x_{1i}^2 x_{2i} & \sum_{i=1}^{n} x_{1i}x_{2i}^2 & \sum_{i=1}^{n} x_{1i}^3 x_{2i} & \sum_{i=1}^{n} x_{1i}x_{2i}^3 & \sum_{i=1}^{n} x_{1i}^2 x_{2i}^2
\end{bmatrix}
\begin{bmatrix}
b_0 \\[2mm] b_1 \\[2mm] b_2 \\[2mm] b_{11} \\[2mm] b_{22} \\[2mm] b_{12}
\end{bmatrix}
=
\begin{bmatrix}
\sum_{i=1}^{n} y_i \\[2mm]
\sum_{i=1}^{n} x_{1i}y_i \\[2mm]
\sum_{i=1}^{n} x_{2i}y_i \\[2mm]
\sum_{i=1}^{n} x_{1i}^2 y_i \\[2mm]
\sum_{i=1}^{n} x_{2i}^2 y_i \\[2mm]
\sum_{i=1}^{n} x_{1i}x_{2i}y_i
\end{bmatrix}
$$

EXAMPLE 12.4   The following data represent the percent of impurities that occurs at various temperatures and sterilizing times in a reaction associated with the manufacturing of a certain beverage.

Sterilizing Time, x_2 (min)	Temperature, x_1 (°C)		
	75	100	125
15	14.05	10.55	7.55
	14.93	9.48	6.59
20	16.56	13.63	9.23
	15.85	11.75	8.78
25	22.41	18.55	15.93
	21.66	17.98	16.44

Estimate the regression coefficients in the model

$$\mu_{Y|x_1, x_2} = \beta_0 + \beta_1 x_1 + \beta_2 x_2 + B_{11} x_1^2 + \beta_{22} x_2^2 + \beta_{12} x_1 x_2.$$

SOLUTION

$$b_0 = 56.4668 \qquad b_{11} = 0.00081$$

$$b_1 = -0.36235 \qquad b_{22} = 0.08171$$

$$b_2 = -2.75299 \qquad b_{12} = 0.00314,$$

and our estimated regression equation is

$$\hat{y} = 56.4668 - 0.36235x_1 - 2.75299x_2 + 0.00081x_1^2$$
$$+ 0.08171x_2^2 + 0.00314x_1 x_2.$$

Many of the principles and procedures associated with the estimation of polynomial regression functions fall into the category of **response surface methodology**, a collection of techniques that have been used quite successfully by scientists and engineers in many fields. Such problems as selecting a proper experimental design, particularly in cases where a large number of variables are in the model, and choosing "optimum" operating conditions on $x_1, x_2, \ldots, x_k$ are often approached through the use of these methods. For an extensive exposure the reader is referred to *Response Surface Methodology* by Myers (see the Bibliography).

Exercises

1. Suppose in Exercise 6 on page 411 that we are also given the number of class periods missed by the 12 students taking the chemistry course. The complete data are recorded as follows:

Student	Chemistry Grade, y	Test Score, x_1	Classes Missed, x_2
1	85	65	1
2	74	50	7
3	76	55	5
4	90	65	2
5	85	55	6
6	87	70	3
7	94	65	2
8	98	70	5
9	81	55	4
10	91	70	3
11	76	50	1
12	74	55	4

(a) Fit a multiple linear regression equation of the form $\mu_{Y|x_1, x_2} = \beta_0 + \beta_1 x_1 + \beta_2 x_2$.

(b) Estimate the chemistry grade for a student who has an intelligence test score of 60 and missed 4 classes.

2. The following data represent a set of 10 experimental runs in which two independent variables x_1 and x_2 are controlled and values of a response, y, are observed:

y	x_1	x_2
61.5	2400	54.5
61.2	2450	56.4
32.0	2500	43.2
52.5	2700	65.2
31.5	2750	45.5
22.5	2800	47.5
53.0	2900	65.0
56.8	3000	66.5
34.8	3100	57.3
52.7	3200	68.0

Estimate the multiple linear regression equation $\mu_{Y|x_1, x_2} = \beta_0 + \beta_1 x_1 + \beta_2 x_2$.

3. A set of experimental runs was made to determine a way of predicting cooking time y at various levels of oven width x_1 and flue temperature x_2. The coded data were recorded as follows:

y	x_1	x_2
6.40	1.32	1.15
15.05	2.69	3.40
18.75	3.56	4.10
30.25	4.41	8.75
44.85	5.35	14.82
48.94	6.20	15.15
51.55	7.12	15.32
61.50	8.87	18.18
100.44	9.80	35.19
111.42	10.65	40.40

Estimate the multiple linear regression equation $\mu_{Y|x_1, x_2} = \beta_0 + \beta_1 x_1 + \beta_2 x_2$.

4. An experiment was conducted to determine if the weight of an animal can be predicted after a given period of time on the basis of the initial weight of the animal and the amount of feed that was eaten. The following data, measured in kilograms, were recorded:

Final Weight, y	Initial Weight, x_1	Feed Eaten, x_2
95	42	272
77	33	226
80	33	259
100	45	292
97	39	311
70	36	183
50	32	173
80	41	236
92	40	230
84	38	235

(a) Fit a multiple regression equation of the form $\mu_{Y|x_1, x_2} = \beta_0 + \beta_1 x_1 + \beta_2 x_2$.

(b) Predict the final weight of an animal having an initial weight of 35 kilograms that is fed 250 kilograms of feed.

5. (a) Fit a multiple regression equation of the form $\mu_{Y|x} = \beta_0 + \beta_1 x + \beta_2 x^2$ to the data of Example 11.8.

(b) Estimate the yield of the chemical reaction for a temperature of 225°C.

6. An experiment was conducted on a new model of a particular make of an automobile to determine the stopping distance at various speeds. The following data were recorded.

Speed, v (km/hr)	35	50	65	80	95	110
Stopping distance, d (m)	16	26	41	62	88	119

(a) Fit a multiple regression curve of the form $\mu_{D|v} = \beta_0 + \beta_1 v + \beta_2 v^2$.

(b) Estimate the stopping distance when the car is traveling at 70 kilometers per hour.

7. An experiment was conducted in order to determine if cerebral blood flow in human beings can be predicted from arterial oxygen tension (millimeters of mercury). Fifteen patients were used in the study and the following data were observed:

Blood Flow, y	Arterial Oxygen Tension, x
84.33	603.40
87.80	582.50
82.20	556.20
78.21	594.60
78.44	558.90
80.01	575.20
83.53	580.10
79.46	451.20
75.22	404.00
76.58	484.00
77.90	452.40
78.80	448.40
80.67	334.80
86.60	320.30
78.20	350.30

Estimate the quadratic regression equation

$$\mu_{Y|x} = \beta_0 + \beta_1 x + \beta_2 x^2.$$

8. The following is a set of coded experimental data on the compressive strength of a particular alloy at various values of the concentration of some additive:

Concentration, x	Compressive Strength, y		
10.0	25.2	27.3	28.7
15.0	29.8	31.1	27.8
20.0	31.2	32.6	29.7
25.0	31.7	30.1	32.3
30.0	29.4	30.8	32.8

(a) Estimate the quadratic regression equation $\mu_{Y|x} = \beta_0 + \beta_1 x + \beta_2 x^2$.

(b) Test for lack of fit of the model.

9. The following data resulted from 15 experimental runs made on four independent variables and a single response y.

y	x_1	x_2	x_3	x_4
14.8	7.8	4.3	11.5	6.3
12.1	6.9	3.9	14.3	7.4
19.0	9.3	8.4	9.4	5.9
14.5	6.8	10.3	15.2	8.7
16.6	11.7	6.4	8.8	9.1
17.2	8.5	5.7	9.8	5.6
17.5	12.6	6.8	11.2	6.8
14.1	7.5	4.2	10.9	7.4
13.8	8.4	7.3	14.7	8.2
14.7	11.3	8.8	15.1	9.2
17.7	10.7	3.6	8.7	4.7
17.0	7.3	4.9	8.6	5.5
17.6	8.4	7.3	9.3	6.6
16.3	6.7	9.7	10.8	8.7
18.2	9.6	8.4	11.9	5.4

Estimate the multiple linear regression model relating y to x_1, x_2, x_3, and x_4.

10. Given the data

x	0	1	2	3	4	5	6
y	1	4	5	3	2	3	4

(a) Fit the cubic model $\mu_{Y|x} = \beta_0 + \beta_1 x + \beta_2 x^2 + \beta_3 x^3$.

(b) Predict Y when $x = 2$.

11. The personnel department of a certain industrial firm used 12 subjects in a study to determine the relationship between job performance rating (y) and scores of four tests. The data are as follows:

y	x_1	x_2	x_3	x_4
11.2	56.5	71.0	38.5	43.0
14.5	59.5	72.5	38.2	44.8
17.2	69.2	76.0	42.5	49.0
17.8	74.5	79.5	43.4	56.3
19.3	81.2	84.0	47.5	60.2
24.5	88.0	86.2	47.4	62.0
21.2	78.2	80.5	44.5	58.1
16.9	69.0	72.0	41.8	48.1
14.8	58.1	68.0	42.1	46.0
20.0	80.5	85.0	48.1	60.3
13.2	58.3	71.0	37.5	47.1
22.5	84.0	87.2	51.0	65.2

Estimate the regression coefficients in the model

$$\mu_{Y|x_1, x_2, x_3, x_4} = \beta_0 + \beta_1 x_1 + \beta_2 x_2 + \beta_3 x_3 + \beta_4 x_4.$$

12. The following data reflect information taken from 17 U.S. Naval hospitals at various sites around the world. The regressors are workload variables, that is, items that result in the need for personnel in a hospital installation. A brief description of the variables is as follows:

    y = monthly labor-hours

    x_1 = average daily patient load

    x_2 = monthly X-ray exposures

    x_3 = monthly occupied bed-days

    x_4 = eligible population in the area/1000

    x_5 = average length of patient's stay, in days

Site	x_1	x_2	x_3	x_4	x_5	y
1	15.57	2,463	472.92	18.0	4.45	566.52
2	44.02	2,048	1,339.75	9.5	6.92	696.82
3	20.42	3,940	620.25	12.8	4.28	1,033.15
4	18.74	6,505	568.33	36.7	3.90	1,603.62
5	49.20	5,723	1,497.60	35.7	5.50	1,611.37
6	44.92	11,520	1,365.83	24.0	4.60	1,613.27
7	55.48	5,779	1,687.00	43.3	5.62	1,854.17
8	59.28	5,969	1,639.92	46.7	5.15	2,160.55
9	94.39	8,461	2,872.33	78.7	6.18	2,305.58
10	128.02	20,106	3,655.08	180.5	6.15	3,503.93
11	96.00	13,313	2,912.00	60.9	5.88	3,571.89
12	131.42	10,771	3,921.00	103.7	4.88	3,741.40
13	127.21	15,543	3,865.67	126.8	5.50	4,026.52
14	252.90	36,194	7,684.10	157.7	7.00	10,343.81
15	409.20	34,703	12,446.33	169.4	10.78	11,732.17
16	463.70	39,204	14,098.40	331.4	7.05	15,414.94
17	510.22	86,533	15,524.00	371.6	6.35	18,854.45

The goal here is to produce an empirical equation that will estimate (or predict) personnel needs for Naval hospitals. Estimate the multiple linear regression equation

$$\mu_{y|x_1, x_2, x_3, x_4, x_5}$$
$$= \beta_0 + \beta_1 x_1 + \beta_2 x_2 + \beta_3 x_3 + \beta_4 x_4 + \beta_5 x_5.$$

13. An experiment was conducted to study the size of squid eaten by sharks and tuna. The regressor variables are characteristics of the beak or mouth of the squid. The regressor variables and response considered for the study are

    x_1 = rostral length, in inches

    x_2 = wing length, in inches

    x_3 = rostral to notch length, in inches

    x_4 = notch to wing length, in inches

    x_5 = width, in inches

    y = weight, in pounds

x_1	x_2	x_3	x_4	x_5	y
1.31	1.07	0.44	0.75	0.35	1.95
1.55	1.49	0.53	0.90	0.47	2.90
0.99	0.84	0.34	0.57	0.32	0.72
0.99	0.83	0.34	0.54	0.27	0.81
1.05	0.90	0.36	0.64	0.30	1.09
1.09	0.93	0.42	0.61	0.31	1.22
1.08	0.90	0.40	0.51	0.31	1.02
1.27	1.08	0.44	0.77	0.34	1.93
0.99	0.85	0.36	0.56	0.29	0.64
1.34	1.13	0.45	0.77	0.37	2.08
1.30	1.10	0.45	0.76	0.38	1.98
1.33	1.10	0.48	0.77	0.38	1.90
1.86	1.47	0.60	1.01	0.65	8.56
1.58	1.34	0.52	0.95	0.50	4.49
1.97	1.59	0.67	1.20	0.59	8.49
1.80	1.56	0.66	1.02	0.59	6.17
1.75	1.58	0.63	1.09	0.59	7.54
1.72	1.43	0.64	1.02	0.63	6.36
1.68	1.57	0.72	0.96	0.68	7.63
1.75	1.59	0.68	1.08	0.62	7.78
2.19	1.86	0.75	1.24	0.72	10.15
1.73	1.67	0.64	1.14	0.55	6.88

Estimate the multiple linear regression equation

$$\mu_{y|x_1, x_2, x_3, x_4, x_5}$$
$$= \beta_0 + \beta_1 x_1 + \beta_2 x_2 + \beta_3 x_3 + \beta_4 x_4 + \beta_5 x_5.$$

14. Twenty-three student teachers took part in an evaluation program designed to measure teacher effectiveness and determine what factors are important. Eleven female instructors took part. The response measure was a quantitative evaluation made on the cooperating teacher. The regressor variables were scores on four standardized tests given to each instructor. The data:

y	x_1	x_2	x_3	x_4
410	69	125	59.00	55.66
569	57	131	31.75	63.97
425	77	141	80.50	45.32
344	81	122	75.00	46.67
324	0	141	49.00	41.21
505	53	152	49.35	43.83
235	77	141	60.75	41.61
501	76	132	41.25	64.57
400	65	157	50.75	42.41
584	97	166	32.25	57.95
434	76	141	54.50	57.90

Estimate the multiple linear regression equation

$$\mu_{y|x_1, x_2, x_3, x_4} = \beta_0 + \beta_1 x_1 + \beta_2 x_2 + \beta_3 x_3 + \beta_4 x_4.$$

12.4 Properties of the Least Squares Estimators _____

The means and variances of the estimators $B_0, B_1, \ldots, B_k$ are easily obtained under certain assumptions on the random errors $E_1, E_2, \ldots, E_n$ that are identical to those made in the case of simple linear regression. Suppose that we assume these errors to be independent, each with zero mean and variance σ^2. It can then be shown that $B_0, B_1, \ldots, B_k$ are, respectively, unbiased estimators of the regression coefficients $\beta_0, \beta_1, \ldots, \beta_k$. In addition, the variances of B's are obtained through the elements of the inverse of the $\mathbf{A}$ matrix. One will note that the off-diagonal elements of $\mathbf{A} = \mathbf{X}'\mathbf{X}$ represent sums of products of elements in the columns of $\mathbf{X}$, while the diagonal elements of $\mathbf{A}$ represent sums of squares of elements in the columns of $\mathbf{X}$. The inverse matrix, $\mathbf{A}^{-1}$, apart from the multiplier σ^2, represents the **variance-covariance matrix** of the estimated regression coefficients. That is, the elements of the matrix $\mathbf{A}^{-1}\sigma^2$ display the variances of $B_0, B_1, \ldots, B_k$ on the main diagonal and covariances on the off-diagonal. For example, in a $k = 2$ multiple linear regression problem, we might write

$$(\mathbf{X'X})^{-1} = \begin{bmatrix} c_{00} & c_{01} & c_{02} \\ c_{10} & c_{11} & c_{12} \\ c_{20} & c_{21} & c_{22} \end{bmatrix}$$

with the elements below the main diagonal determined through the symmetry of the matrix. Then we can write

$$\sigma_{B_i}^2 = c_{ii}\sigma^2, \qquad\qquad i = 0,\ 1,\ 2$$

$$\sigma_{B_i B_j} = \mathrm{Cov}(B_i,\ B_j) = c_{ij}\sigma^2, \qquad i \neq j.$$

Of course, the estimates of the variances and hence the standard errors of these estimators are obtained by replacing σ^2 with the appropriate estimate obtained through experimental data. An unbiased estimate of σ^2 is once again defined in terms of the error sum of squares, which is computed using the formula established in Theorem 12.1. In the theorem we are making the assumptions on the E_i described above.

THEOREM 12.1

For the linear regression equation

$$y = X\boldsymbol{\beta} + \boldsymbol{\varepsilon},$$

an unbiased estimate of σ^2 is given by the error or residual mean square

$$s^2 = \frac{SSE}{n - k - 1},$$

where

$$SSE = \sum_{i=1}^{n} e_i^2 = \sum_{i=1}^{n} (y_i - \hat{y}_i)^2. \qquad\qquad \blacksquare$$

One can easily see that Theorem 12.1 represents a generalization of Theorem 11.1 for the simple linear regression case. The proof is left for the reader. As in the simpler linear regression case, the estimate s^2 is a measure of the variation in the prediction errors or residuals. Other important inferences regarding the fitted regression equation, based on the values of the individual residuals $e_i = y_i - \hat{y}_i$, $i = 1$, $2, \ldots, n$, will be discussed in Sections 12.9 and 12.10.

The error and regression sum of squares take on the same form and play the same role as in the simple linear regression case. In fact, the sum-of-squares identity

$$\sum_{i=1}^{n} (y_i - \bar{y})^2 = \sum_{i=1}^{n} (\hat{y}_i - \bar{y})^2 + \sum_{i=1}^{n} (y - \hat{y}_i)^2$$

continues to hold and we retain our previous notation, namely,

$$SST = SSR + SSE$$

with

$$SST = \sum_{i=1}^{n} (y_i - \bar{y})^2 = \text{total sum of squares}$$

and

$$SSR = \sum_{i=1}^{n} (\hat{y}_i - \bar{y})^2 = \text{regression sum of squares.}$$

There are k degrees of freedom associated with SSR and, as always, SST has $n - 1$ degrees of freedom. Therefore, after subtraction, SSE has $n - k - 1$ degrees of freedom. Thus our estimate of σ^2 is again given by the error sum of squares divided by its degrees of freedom. All three of these sums of squares will appear on the printout of most multiple regression computer packages.

The utility of the error mean square (or residual mean square) lies in its use in hypothesis testing and confidence interval estimation, which is discussed in Section 12.5. In addition, the error mean square plays an important role in situations where the scientist is searching for the best from a set of competing models. Many model-building criteria involve the statistic s^2. Criteria for comparing competing models are discussed in Section 12.10.

12.5   Inferences in Multiple Linear Regression ⎯⎯⎯⎯

One of the most useful inferences that can be made regarding the quality of the predicted response y_0 corresponding to the values $x_{10}, x_{20}, \ldots, x_{k0}$ is the confidence interval on the mean response $\mu_{Y|x_{10}, x_{20} \ldots, x_{k0}}$. We are interested in constructing a confidence interval on the mean response for the set of conditions given by $\mathbf{x}_0' = [1, x_{10}, x_{20}, \ldots, x_{k0}]$. We augment the conditions on the x's by the number 1 in order to facilitate using matrix notation. As in the $k = 1$ case, if we make the additional assumption that the errors are independent and normally distributed, then the B_j's are normal, with mean, variances, and covariances as indicated in Section 12.4, and hence

$$\hat{Y} = B_0 + \sum_{j=1}^{k} B_j x_{j0}$$

is likewise normally distributed and is, in fact, an unbiased estimator for the **mean response** on which we are attempting to attach confidence intervals. The variance of $\hat{Y}_0$ written in matrix notation simply as a function of σ^2, $(\mathbf{X'X})^{-1}$, and the condition vector $\mathbf{x}_0'$, is

$$\sigma_{\hat{Y}_0}^2 = \sigma^2 \mathbf{x}_0'(X'X)^{-1}\mathbf{x}_0.$$

If this expression is expanded for a given case, say $k = 2$, it is easily seen that it appropriately accounts for the variances and covariances of the B_i's. After replacing σ^2 by s^2 as given by Theorem 12.1, the $100(1 - \alpha)\%$ confidence interval on $\mu_{Y|x_{10}, x_{20}, \ldots, x_{k0}}$ can be constructed from the statistic

$$T = \frac{\hat{Y}_0 - \mu_{Y|x_{10}, x_{20}, \ldots, x_{k0}}}{S\sqrt{\mathbf{x}_0'(\mathbf{X}'\mathbf{X})^{-1}\mathbf{x}_0}},$$

which has a t-distribution with $n - k - 1$ degrees of freedom.

CONFIDENCE INTERVAL FOR $\mu_{Y|x_{10}, x_{20}, \ldots, x_{k0}}$

A $(1 - \alpha)100\%$ confidence interval for the **mean response** $\mu_{Y|x_{10}, x_{20}, \ldots, x_{k0}}$ *is given by*

$$\hat{y}_0 - t_{\alpha/2}s\sqrt{\mathbf{x}_0'(\mathbf{X}'\mathbf{X})^{-1}\mathbf{x}_0} < \mu_{Y|x_{10}, x_{20}, \ldots, x_{k0}} < \hat{y}_0 + t_{\alpha/2}s\sqrt{\mathbf{x}_0'(\mathbf{X}'\mathbf{X})^{-1}\mathbf{x}_0}$$

where $t_{\alpha/2}$ is a value of the t-distribution with $n - k - 1$ degrees of freedom. ∎

The quantity $s\sqrt{\mathbf{x}_0'(\mathbf{X}'\mathbf{X})^{-1}\mathbf{x}_0}$ is often called the **standard error of prediction** and usually appears on the printout of many regression computer packages.

EXAMPLE 12.5 Using the data of Example 12.3, construct a 95% confidence interval for the mean response when $x_1 = 3\%$, $x_2 = 8\%$, and $x_3 = 9\%$.

SOLUTION
From the regression equation of Example 12.3 the estimated percent survival when $x_1 = 3\%$, $x_2 = 8\%$, and $x_3 = 9\%$ is

$$\hat{y} = 39.1574 + (1.0161)(3) - (1.8616)(8) - (0.3433)(9) = 24.2232.$$

Next we find

$$\mathbf{x}_0'(\mathbf{X}'\mathbf{X})^{-1}\mathbf{x}_0 = [1, 3, 8, 9] \begin{bmatrix} 8.064 & -0.0826 & -0.0942 & -0.7905 \\ -0.0826 & 0.0085 & 0.0017 & 0.0037 \\ -0.0942 & 0.0017 & 0.0166 & -0.0021 \\ -0.7905 & 0.0037 & -0.0021 & 0.0886 \end{bmatrix} \begin{bmatrix} 1 \\ 3 \\ 8 \\ 9 \end{bmatrix}$$

$$= 0.1267.$$

Using the error mean square, $s^2 = 4.298$ or $s = 2.073$, and Table A.4, we see that $t_{0.025} = 2.262$ for 9 degrees of freedom. Therefore, a 95% confidence interval for the mean percent survival for $x_1 = 3\%$, $x_2 = 8\%$, and $x_3 = 9\%$ is given by

$$24.2232 - (2.262)(2.073)\sqrt{0.1267} < \mu_{Y|3, 8, 9} < 24.2232$$
$$+ (2.262)(2.073)\sqrt{0.1267}$$

or simply

$$22.5541 < \mu_{Y|3, 8, 9} < 25.8923.$$

As in the case of simple linear regression, we need to make a clear distinction between the confidence interval on mean response and the prediction interval on an *observed response*. The latter provides a bound within which we can say with a preselected degree of certainty that a new observed response will fall.

A prediction interval for a single predicted response y_0 is once again established by considering the differences $\hat{y}_0 - y_0$ of the random variable $\hat{Y}_0 - Y_0$. The sampling distribution can be shown to be normal with mean

$$\mu_{\hat{Y}_0 - Y_0} = 0$$

and variance

$$\mu^2_{\hat{Y}_0 - Y_0} = \sigma^2[1 + \mathbf{x}'_0(\mathbf{X}'\mathbf{X})^{-1}\mathbf{x}_0].$$

Thus the $(1 - \alpha)100\%$ prediction interval for a single prediction value y_0 can be constructed from the statistic

$$T = \frac{\hat{Y}_0 - Y_0}{S\sqrt{1 + \mathbf{x}'_0(\mathbf{X}'\mathbf{X})^{-1}\mathbf{x}_0}},$$

which has a t-distribution with $n - k - 1$ degrees of freedom.

PREDICTION INTERVAL FOR y_0

*A $(1 - \alpha)100\%$ prediction interval for a **single response** y_0 is given by*

$$\hat{y}_0 - t_{\alpha/2}s\sqrt{1 + \mathbf{x}'_0(\mathbf{X}'\mathbf{X})^{-1}\mathbf{x}_0} < y_0 < \hat{y}_0 + t_{\alpha/2}s\sqrt{1 + \mathbf{x}'_0(\mathbf{X}'\mathbf{X})^{-1}\mathbf{x}_0},$$

where $t_{\alpha/2}$ is a value of the t-distribution with $n - k - 1$ degrees of freedom.

EXAMPLE 12.6   Using the data of Example 12.3, construct a 95% prediction interval for an individual percent survival response when $x_1 = 3\%$, $x_2 = 8\%$, and $x_3 = 9\%$.

SOLUTION
Referring to the results of Example 12.5, we find that the 95% prediction interval for the response y_0 when $x_1 = 3\%$, $x_2 = 8\%$, and $x_3 = 9\%$ is

$$24.2232 - (2.262)(2.073)\sqrt{1.1267} < y_0 < 24.2232 + (2.262)(2.073)\sqrt{1.1267},$$

which reduces to

$$19.2459 < y_0 < 29.2005.$$

Notice, as one expects, that the prediction interval is considerably less tight than the confidence interval for mean percent survival in Example 12.5.

A knowledge of the distributions of the individual coefficient estimators enables the experimenter to construct confidence intervals for the coefficients and to test hypotheses about them. One recalls from Section 12.4 that the B_j's $(j = 0, 1, 2, \ldots, k)$ are normally distributed with mean β_j and variance $c_{jj}\sigma^2$. Thus we can use the statistic

$$T = \frac{B_j - \beta_j}{S\sqrt{c_{jj}}}$$

with $n - k - 1$ degrees of freedom to test hypotheses and construct confidence intervals on β_j. For example, if we wish to test

$$H_0: \quad \beta_j = \beta_{j0},$$
$$H_1: \quad \beta_j \neq \beta_{j0},$$

we compute the statistic

$$t = \frac{b_j - \beta_{j0}}{s\sqrt{c_{jj}}}$$

and do not reject H_0 if

$$-t_{\alpha/2} < t < t_{\alpha/2},$$

where $t_{\alpha/2}$ has $n - k - 1$ degrees of freedom.

EXAMPLE 12.7   For the model of Example 12.3, test the hypothesis that $\beta_2 = -2.5$ at the 0.05 level of significance against the alternative that $\beta_2 > -2.5$.

SOLUTION

$$H_0: \quad \beta_2 = -2.5.$$
$$H_1: \quad \beta_2 > -2.5.$$

Computations:

$$t = \frac{b_2 - \beta_{20}}{s\sqrt{c_{22}}} = \frac{-1.8616 + 2.5}{2.073\sqrt{0.0166}} = 2.391$$

$$P = P(T > 2.391) = 0.04.$$

Decision: Reject H_0 and conclude that $\beta_2 > -2.5$.

Annotated Printout for Data of Example 12.3

Figure 12.1 shows annotated computer printout for a multiple linear regression fit to the data of Example 12.3. The package used is SAS.

Note the model parameter estimates, the standard errors, and the t-statistics. The standard errors are computed from square roots of diagonal elements of $(X'X)^{-1}s^2$. In this illustration the variable x_3 is insignificant in the presence of x_1 and x_2. The terms CLM and CLI are confidence intervals on mean response and prediction limits on an individual observation, respectively.

Exercises

1. For the data of Exercise 2 on page 424, estimate σ^2.

2. For the data of Exercise 3 on page 424, estimate σ^2.

3. For the data of Exercise 9 on page 425, estimate σ^2.

4. Obtain estimates of the variances and the covariance of the estimators B_1 and B_2 of Exercise 2 on page 424.

5. Referring to Exercise 9 on page 425, find the estimate of
   (a) $\sigma_{B_2}^2$.   (b) $\mathrm{Cov}(B_1, B_4)$.

6. Using the data of Exercise 2 on page 424 and the estimate of σ^2 from Exercise 1, construct 95% confidence intervals for the predicted response and the mean response when $x_1 = 2500$ and $x_2 = 48.0$.

7. For Exercise 8 on page 425, construct a 90% confidence interval for the mean compressive strength when the concentration is $x = 19.5$ and a quadratic model is used.

8. Using the data of Exercise 9 on page 425 and the estimate of σ^2 from Exercise 3, construct 95% confidence intervals for the predicted response and the mean response when $x_1 = 8.2$, $x_2 = 6.0$, $x_3 = 10.3$, and $x_4 = 5.8$.

9. For the model of Exercise 7 on page 425, test the hypothesis that $\beta_2 = 0$ at the 0.05 level of significance against the alternative that $\beta_2 \neq 0$.

10. For the model of Exercise 2 on page 424, test the hypothesis that $\beta_1 = 0$ at the 0.05 level of significance against the alternative that $\beta_1 \neq 0$.

11. For the model of Exercise 3 on page 424, test the hypotheses that $\beta_1 = 2$ against the alternative $\beta_1 \neq 2$. Use a P-value in your conclusion.

Analysis of Variance

Source	DF	Sum of Squares	Mean Square	F Value	Prob > F
Model	3	399.45437	133.15146	30.984	0.0001
Error	9	38.67640	4.29738		
C Total	12	438.13077			

Root MSE	2.0730	R-square	0.9117	
Dep Mean	29.03846			
C.V.	7.13885			

Parameter Estimates

Variable	DF	Parameter Estimate	Standard Error	T for HO: Parameter = 0	Prob > |T|
INTERCEP	1	39.157350	5.88705963	6.651	0.0001
X1	1	1.016100	0.19089520	5.323	0.0005
X2	1	-1.861649	0.26732550	-6.964	0.0001
X3	1	-0.343260	0.61705204	-0.556	0.5916

OBS	Y	PREDICTED	RESIDUAL	CLM_L	CLM_U	CLI_L	CLI_U
1	25.5	27.3514	-1.85141	24.1500	30.5528	21.6733	33.0295
2	31.2	32.2623	-1.06232	30.4875	34.0371	27.2482	37.2764
3	25.9	27.3495	-1.44955	24.2757	30.4234	21.7424	32.9567
4	38.4	38.3096	0.09042	35.4098	41.2093	32.7960	43.8232
5	18.4	15.5447	2.85527	11.9729	19.1165	9.6499	21.4396
6	26.7	26.1081	0.59193	23.7649	28.4512	20.8657	31.3504
7	26.4	28.2532	-1.85316	26.4222	30.0841	23.2189	33.2874
8	25.9	26.2219	-0.32185	24.0204	28.4233	21.0413	31.4024
9	32.0	32.0882	-0.08818	30.3174	33.8589	27.0755	37.1009
10	25.2	26.0676	-0.86764	24.5024	27.6329	21.1238	31.0115
11	39.7	37.2524	2.44764	34.2957	40.2090	31.7086	42.7961
12	35.7	32.4879	3.21208	29.1743	35.8016	26.7458	38.2300
13	26.5	28.2032	-1.70324	25.9771	30.4294	23.0121	33.3943

FIGURE 12.1   SAS printout for data in Example 12.3.

12.6 Choice of a Fitted Model Through Hypothesis Testing

In many regression situations, individual coefficients are of importance to the experimenter. For example, in an economics application, β_1, β_2, . . . , might have some particular significance, and thus confidence intervals and tests of hypotheses on these parameters are of interest to the economist. However, consider an industrial chemical situation in which the postulated model assumes that reaction yield is dependent linearly on reaction temperature and concentration of a certain catalyst. It is probably known that this is not the true model but an adequate approximation, so the interest is likely not to be in the individual parameters but rather in the ability of the entire function to predict the true response in the range of the variables considered. Therefore, in this situation, one would put more emphasis on $\sigma_{\hat{Y}}^2$, confidence intervals on the mean response, and so forth, and likely deemphasize inferences on individual parameters.

The experimenter using regression analysis is also interested in deletion of variables when the situation dictates that, in addition to arriving at a workable prediction equation, he or she must find the "best regression" involving only variables that are useful predictors. There are a number of available computer programs that sequentially arrive at the so-called best regression equation depending on certain criteria. We shall discuss this further in Section 12.8.

One criterion that is commonly used to illustrate the adequacy of a fitted regression model is the **coefficient of multiple determination**:

$$R^2 = \frac{SSR}{SST} = \frac{\sum\limits_{i=1}^{n} (\hat{y}_i - \bar{y})^2}{\sum\limits_{i=1}^{n} (y_i - \bar{y})^2}.$$

This quantity merely indicates what proportion of the total variation in the response Y is explained by the fitted model. Often an experimenter will report $R^2 \times 100\%$ and interpret the result as percentage variation explained by the postulated model. The square root of R^2 is called the **multiple correlation coefficient** between Y and the set $x_1, x_2, \ldots, x_k$. In Example 12.3, the value of R^2 indicating the proportion of variation explained by the three independent variables x_1, x_2, and x_3 is found to be

$$R^2 = \frac{SSR}{SST} = \frac{399.45}{438.13} = 0.9117,$$

which means that 91.17% of the variation in percent survival has been explained by the linear regression model.

The regression sum of squares can be used to give some indication concerning whether or not the model is an adequate explanation of the true situation. One can test the hypothesis H_0 that the regression is not significant by merely forming the ratio

$$f = \frac{SSR/k}{SSE/(n - k - 1)} = \frac{SSR/k}{s^2}$$

and rejecting H_0 at the α-level of significance when $f > f_\alpha(k, n - k - 1)$. For the data of Example 12.3 we obtain

$$f = \frac{399.45/3}{4.298} = 30.98.$$

This value of F exceeds that tabulated critical point 6.99 of the F-distribution for 3 and 9 degrees of freedom at the $\alpha = 0.01$ level. The result here should not be misinterpreted. Although it does indicate that the regression explained by the model is significant, this does not rule out the possibility that

1. The linear regression model in this set of x's is not the only model that can be used to explain the data; indeed, there might be other models with transformations on the x's that might give a larger value of the F-statistic.
2. The model might have been more effective with the inclusion of other variables in addition to x_1, x_2, and x_3 or perhaps with the deletion of one or more of the variables in the model.

Tests on Subsets and Individual Coefficients

The addition of any single variable to a regression system *will increase the regression sum of squares* and thus *reduce the error sum of squares*. Consequently, we must decide whether the increase in regression is sufficient to warrant using it in the model. As one might expect, the use of unimportant variables can reduce the effectiveness of the prediction equation by increasing the variance of the estimated response. We shall pursue this point further by considering the importance of x_3 in Example 12.3. Initially, we can test

$$H_0: \quad \beta_3 = 0,$$

$$H_1: \quad \beta_3 \neq 0,$$

by using the t-distribution with 9 degrees of freedom. We have

$$t = \frac{b_3 - 0}{S\sqrt{c_{33}}} = \frac{-0.3433}{2.073\sqrt{0.0886}} = -0.556,$$

which indicates that β_3 does not differ significantly from zero, and hence one may very well feel justified in removing x_3 from the model. Suppose that we consider the regression of Y on the set (x_1, x_2), the least squares normal equations now reducing to

$$\begin{bmatrix} 13 & 59.43 & 81.82 \\ 59.43 & 394.7255 & 360.6621 \\ 81.82 & 360.6621 & 576.7264 \end{bmatrix} \begin{bmatrix} b_0 \\ b_1 \\ b_2 \end{bmatrix} = \begin{bmatrix} 377.75 \\ 1877.5670 \\ 2246.6610 \end{bmatrix}.$$

The estimated regression coefficients for this reduced model are given by

$$b_0 = 36.094, \quad b_1 = 1.031, \quad b_2 = -1.870,$$

and the resulting regression sum of squares with 2 degrees of freedom is as follows:

$$R(\beta_1, \beta_2) = 398.12.$$

Here we use the notation $R(\beta_1, \beta_2)$ to indicate the regression sum of squares of the restricted model and it is not to be confused with SSR, the regression sum of squares of the original model with 3 degrees of freedom. The new error sum of squares is then given by

$$SST - R(\beta_1, \beta_2) = 438.13 - 398.12$$

$$= 40.01,$$

and the resulting error mean square with 10 degrees of freedom becomes

$$s^2 = \frac{40.01}{10} = 4.001.$$

The amount of variation in the response, the percent survival, which is attributed to x_3, the weight percent of the third additive, in the presence of the variables x_1 and x_2, is given by

$$R(\beta_3 | \beta_1, \beta_2) = SSR - R(\beta_1, \beta_2)$$
$$= 399.45 - 398.12$$
$$= 1.33,$$

which represents a small proportion of the entire regression variation. This amount of added regression is statistically insignificant, as indicated by our previous test on β_3. An equivalent test involves the formation of the ratio

$$f = \frac{R(\beta_3 | \beta_1, \beta_2)}{s^2}$$
$$= \frac{1.33}{4.298} = 0.309,$$

which is a value of the F-distribution with 1 and 9 degrees of freedom. Recall that the basic relationship between the t-distribution with v degrees of freedom and the F-distribution with 1 and v degrees of freedom is given by

$$t_{\alpha/2}^2 = f_\alpha(1, v)$$

and we note that the f-value of 0.309 is indeed the square of the t-value of 0.556.

We can provide additional support for deleting x_3 from the model by considering $\sigma_{\hat{Y}}^2$ under both the full and reduced regression equation. We first note that the error mean square, s^2, was reduced from 4.298 to 4.001 by deleting x_3 from the model. In the case of the full model an estimate of the variance of the estimated response $\hat{Y}_0$ when $x_1 = 4.00$, $x_2 = 5.5$, and $x_3 = 8.90$, is

$$\hat{\sigma}_{\hat{Y}_0}^2 = s^2 \mathbf{x}_0'(\mathbf{X}'\mathbf{X})^{-1}\mathbf{x}_0$$
$$= 4.298[1, 4.00, 5.5, 8.90]$$

$$\times \begin{bmatrix} 8.064 & -0.0826 & -0.0942 & -0.7905 \\ -0.0826 & 0.0085 & 0.0017 & 0.0037 \\ -0.0942 & 0.0017 & 0.0166 & -0.0021 \\ -0.7905 & 0.0037 & -0.0021 & 0.0886 \end{bmatrix} \begin{bmatrix} 1 \\ 4.00 \\ 5.5 \\ 8.90 \end{bmatrix}$$

$$= 0.3936.$$

For the reduced model our point of interest becomes $\mathbf{x}_0' = [1, 4.00, 5.5]$, our $\mathbf{X}'\mathbf{X}$ matrix has been reduced to a 3×3 matrix with inverse given by

$$(\mathbf{X}'\mathbf{X})^{-1} = \begin{bmatrix} 1.0114 & -0.0494 & -0.1126 \\ -0.0494 & 0.0083 & 0.0018 \\ -0.1126 & 0.0018 & 0.0166 \end{bmatrix}$$

and the variance of the estimated response is now estimated to be

$$\hat{\sigma}_{\hat{Y}_0}^2 = 4.001[1,\ 4.00,\ 5.5] \begin{bmatrix} 1.0114 & -0.0494 & -0.1126 \\ -0.0494 & 0.0083 & 0.0018 \\ -0.1126 & 0.0018 & 0.0166 \end{bmatrix} \begin{bmatrix} 1 \\ 4.00 \\ 5.5 \end{bmatrix}$$

$$= 0.3668,$$

which represents a reduction over that found for the complete model.

To generalize the concepts above, one can assess the work of an independent variable x_i in the general multiple linear regression model

$$\mu_{Y|x_1, x_2, \ldots, x_k} = \beta_0 + \beta_1 x_1 + \cdots + \beta_k x_k$$

by observing the amount of regression attributed to x_i over and above that attributed to the other variables, that is, the regression on x_i *adjusted* for the other variables. This is computed by subtracting the regression sum of squares for a model with x_i removed, from *SSR*. For example, we say that x_1 is assessed by calculating

$$R(\beta_1 | \beta_2, \beta_3, \ldots, \beta_k) = SSR - R(\beta_2, \beta_3, \ldots, \beta_k),$$

where $R(\beta_2, \beta_3, \ldots, \beta_k)$ is the regression sum of squares with $\beta_1 x_1$ removed from the model. To test the hypothesis

$$H_0: \quad \beta_1 = 0,$$

$$H_1: \quad \beta_1 \neq 0,$$

compute

$$f = \frac{R(\beta_1 | \beta_2, \beta_3, \ldots, \beta_k)}{s^2}$$

and compare with $f_\alpha(1, n - k - 1)$.

In a similar manner we can test for the significance of a *set* of the variables. For example, to investigate simultaneously the importance of including x_1 and x_2 in the model, we test the hypothesis

$$H_0: \quad \beta_1 = \beta_2 = 0,$$

$$H_1: \quad \beta_1 \text{ and } \beta_2 \text{ are not both zero}$$

by computing

$$f = \frac{[R(\beta_1, \beta_2 | \beta_3, \beta_4, \ldots, \beta_k)]/2}{s^2}$$

$$= \frac{[SSR - R(\beta_3, \beta_4, \ldots, \beta_k)]/2}{s^2}$$

and comparing with $f_\alpha(2, n - k - 1)$. The number of degrees of freedom associated with the numerator, in this case 2, equals the number of variables in the set.

12.7 Special Case of Orthogonality

Prior to our original development of the general linear regression problem, the assumption was made that the independent variables were measured without error and are often controlled by the experimenter. Quite often they occur as a result of an elaborately *designed experiment*. In fact, one can increase the effectiveness of the resulting prediction equation with the use of a suitable experimental plan.

Suppose that we once again consider the **X** matrix as defined in Section 12.3. We can rewrite it to read

$$\mathbf{X} = [\mathbf{1}, \mathbf{x}_1, \mathbf{x}_2, \ldots, \mathbf{x}_k],$$

were **1** represents a column of ones and $\mathbf{x}_j$ is a column vector representing the levels of x_j. If

$$\mathbf{x}_p' \mathbf{x}_q = 0, \qquad p \neq q,$$

the variables x_p and x_q are said to be *orthogonal* to each other. There are certain obvious advantages to having a completely orthogonal situation whereby $\mathbf{x}_p' \mathbf{x}_q = 0$ for all possible p and q, $p \neq q$, and, in addition,

$$\sum_{i=1}^{n} x_{ji} = 0, \qquad j = 1, 2, \ldots, k.$$

The resulting $\mathbf{X}'\mathbf{X}$ is a diagonal matrix and the normal equations in Section 10.3 reduce to

$$nb_0 = \sum_{i=1}^{n} y_i$$

$$b_1 \sum_{i=1}^{n} x_{1i}^2 = \sum_{i=1}^{n} x_{1i} y_i$$

$$\vdots \qquad \vdots$$

$$b_k \sum_{i=1}^{n} x_{ki}^2 = \sum_{i=1}^{n} x_{ki} y_i.$$

An important advantage is that one is easily able to partition *SSR* into **single-degree-of-freedom components**, each of which corresponds to the amount of variation in Y accounted for by a given controlled variable. In the orthogonal situation we can write

$$SSR = \sum_{i=1}^{n} (\hat{y}_i - \bar{y})^2$$

$$= \sum_{i=1}^{n} (b_0 + b_1 x_{1i} + \cdots + b_k x_{ki} - b_0)^2$$

$$= b_1^2 \sum_{i=1}^{n} x_{1i}^2 + b_2^2 \sum_{i=1}^{n} x_{2i}^2 + \cdots + b_k^2 \sum_{i=1}^{n} x_{ki}^2$$

$$= R(\beta_1) + R(\beta_2) + \cdots + R(\beta_k).$$

The quantity $R(\beta_i)$ is the amount of regression sum of squares associated with a model involving a single independent variable x_i.

To test simultaneously for the significance of a set of m variables in an orthogonal situation, the regression sum of squares becomes

$$R(\beta_1, \beta_2, \ldots, \beta_m \mid \beta_{m+1}, \beta_{m+2}, \ldots, \beta_k) = R(\beta_1) + R(\beta_2) + \cdots + R(\beta_m)$$

and thus we have the further simplification

$$R(\beta_1 \mid \beta_2, \beta_3, \ldots, \beta_k) = R(\beta_1)$$

when evaluating a single independent variable. Therefore, the contribution of a given variable or set of variables is essentially found by *ignoring* the other variables in the model. Independent evaluations of the worth of the individual variables are accomplished using analysis-of-variance techniques as given in Table 12.1. The total variation in the response is partitioned into single-degree-of-freedom components plus the error term with $n - k - 1$ degrees of freedom. Each computed f-value is used to test one of the hypotheses

$$\left. \begin{array}{l} H_0: \quad \beta_i = 0 \\ H_1: \quad \beta_i \neq 0 \end{array} \right\} \quad i = 1, 2, \ldots, k$$

by comparing with the critical point $f_\alpha(1, n - k - 1)$.

TABLE 12.1 Analysis of Variance for Orthogonal Variables

Source of Variation	Sum of Squares	Degrees of Freedom	Mean Square	Computed f
β_1	$R(\beta_1) = b_1^2 \sum_{i=1}^{n} x_{1i}^2$	1	$R(\beta_1)$	$\dfrac{R(\beta_1)}{s^2}$
β_2	$R(\beta_2) = b_2^2 \sum_{i=1}^{n} x_{2i}^2$	1	$R(\beta_2)$	$\dfrac{R(\beta_2)}{s^2}$
$\vdots$	$\vdots$	$\vdots$	$\vdots$	$\vdots$
β_k	$R(\beta_k) = b_k^2 \sum_{i=1}^{n} x_{ki}^2$	1	$R(\beta_k)$	$\dfrac{R(\beta_k)}{s^2}$
Error	SSE	$n - k - 1$	$s^2 = \dfrac{SSE}{n - k - 1}$	
Total	$SST = S_{yy}$	$n - 1$		

EXAMPLE 12.8   Suppose that a scientist takes experimental data on the radius of a propellant grain Y as a function of powder temperature x_1, extrusion rate x_2, and die temperature x_3. Fit a linear regression model for predicting grain radius and determine the effectiveness of each variable in the model. The data are given as follows:

Grain Radius	Powder Temperature	Extrusion Rate	Die Temperature
82	150 (-1)	12 (-1)	220 (-1)
93	190 (1)	12 (-1)	220 (-1)
114	150 (-1)	24 (1)	220 (-1)
124	150 (-1)	12 (-1)	250 (1)
111	190 (1)	24 (1)	220 (-1)
129	190 (1)	12 (-1)	250 (1)
157	150 (-1)	24 (1)	250 (1)
164	190 (1)	24 (1)	250 (1)

SOLUTION

Note that each variable is controlled at two levels and the experiment represents each of the eight possible combinations. The data on the independent variables are coded for convenience by means of the following formulas:

$$x_1 = \frac{\text{powder temperature} - 170}{20}$$

$$x_2 = \frac{\text{extrusion rate} - 18}{6}$$

$$x_3 = \frac{\text{die temperature} - 235}{15}.$$

The resulting levels of x_1, x_2, and x_3 take on the values -1 and $+1$ as indicated in the table of data. This particular *experimental design* affords the orthogonality that we are illustrating here. A more thorough treatment of this type of experimental layout will be given in Chapter 15. The **X** matrix is given by

$$X = \begin{bmatrix} 1 & -1 & -1 & -1 \\ 1 & 1 & -1 & -1 \\ 1 & -1 & 1 & -1 \\ 1 & -1 & -1 & 1 \\ 1 & 1 & 1 & -1 \\ 1 & 1 & -1 & 1 \\ 1 & -1 & 1 & 1 \\ 1 & 1 & 1 & 1 \end{bmatrix}$$

and it is easy to verify the orthogonality conditions. One can now compute the coefficients

$$b_0 = \frac{\sum_{i=1}^{8} y_i}{8} = 121.75$$

$$b_1 = \frac{\sum_{i=1}^{8} x_{1i} y_i}{8} = \frac{20}{8} = 2.5$$

$$b_2 = \frac{\sum_{i=1}^{8} x_{2i} y_i}{\sum_{i=1}^{8} x_{2i}^2} = \frac{118}{8} = 14.75$$

$$b_3 = \frac{\sum_{i=1}^{8} x_{3i} y_i}{\sum_{i=1}^{8} x_{3i}^2} = \frac{174}{8} = 21.75,$$

so in terms of the *coded* variables, the prediction equation is given by

$$\hat{y} = 121.75 + 2.5x_1 + 14.75x_2 + 21.75x_3.$$

The analysis-of-variance table showing independent contributions to SSR for each variable is given in Table 12.2. The results, when compared to the $f_{0.05}(1, 4)$ critical point of 7.71, indicate that x_1 does not contribute significantly at the 0.05 level, while variables x_2 and x_3 are significant. In this example the estimate for σ^2 is 23.1250. As in the single-independent-variable case, it should be pointed out that this estimate does not solely contain experimental error variation unless the postulated model is correct. Otherwise, the estimate is "contaminated" by lack of fit in addition to pure error, and the lack of fit can be separated only if one obtains multiple experimental observations at the various (x_1, x_2, x_3) combinations.

TABLE 12.2 Analysis of Variance on Grain Radius Data

Source of Variation	Sum of Squares	Degrees of Freedom	Mean Square	Computed f
β_1	$(2.5)^2(8) = 50$	1	50	2.16
β_2	$(14.75)^2(8) = 1740.50$	1	1740.50	75.26
β_3	$(21.75)^2(8) = 3784.50$	1	3784.50	163.65
Error	92.5	4	23.1250	
Total	5667.50	7		

Exercises

1. Compute and interpret the coefficient of multiple determination for the variables of Exercise 3 on page 424.

2. Test whether the regression explained by the model in Exercise 3 on page 424 is significant at the 0.01 level of significance.

3. Test whether the regression explained by the model in Exercise 9 on page 425 is significant at the 0.01 level of significance.

4. For the model of Exercise 9 on page 425, test the hypothesis

$$H_0: \quad \beta_1 = \beta_2 = 0,$$

$$H_1: \quad \beta_1 \text{ and } \beta_2 \text{ are not both zero.}$$

5. Repeat Exercise 10 on page 432 using an F-statistic.

6. A small experiment is conducted to fit a multiple regression equation relating the yield y to temperature x_1, reaction time x_2, and concentration of one of the reactants x_3. Two levels of each variable were chosen and measurements corresponding to the coded independent variables were recorded as follows:

y	x_1	x_2	x_3
7.6	-1	-1	-1
8.4	1	-1	-1
9.2	-1	1	-1
10.3	-1	-1	1
9.8	1	1	-1
11.1	1	-1	1
10.2	-1	1	1
12.6	1	1	1

(a) Using the coded variables, estimate the multiple linear regression equation

$$\mu_{Y|x_1, x_2, x_3} = \beta_0 + \beta_1 x_1 + \beta_2 x_2 + \beta_3 x_3.$$

(b) Partition SSR, the regression sum of squares, into three single-degree-of-freedom components attributable to x_1, x_2, and x_3, respectively. Show an analysis-of-variance table, indicating significance tests on each variable.

12.8 Sequential Methods for Model Selection

At times the significance tests outlined in Section 12.6 are quite adequate in determining which variables should be used in the final regression model. These tests are certainly effective if the experiment can be planned and the variables are orthogonal to each other. Even if the variables are not orthogonal, the individual t-tests can be of some use in many problems where the number of variables under investigation is small. However, there are many problems in which it is necessary to use more elaborate techniques for screening variables, particularly when the experiment exhibits a substantial deviation from orthogonality. Useful measures of **multicollinearity** (linear dependency) among the independent variables are provided by the sample correlation coefficients $r_{x_i x_j}$. Since we are concerned only with linear dependency among independent variables, no confusion will result if we drop the x's from our notation and simply write $r_{x_i x_j} = r_{ij}$, where

$$r_{ij} = \frac{S_{ij}}{\sqrt{S_{ii} S_{jj}}}.$$

It should be noted that the r_{ij}'s do not give true estimates of population correlation coefficients in the strict sense, since the x's are actually not random variables in the context discussed here. Thus the term "correlation," albeit standard, is perhaps a misnomer.

When one or more of these sample correlation coefficients deviate substantially from zero, it can be quite difficult to find the most effective subset of variables for inclusion in our prediction equation. In fact, for some problems the multicollinearity will be so extreme that a suitable predictor cannot be found unless all possible subsets of the variables are investigated. Of course, the latter approach is prohibitive for a large number of variables. Informative discussions of model selection in regression by Hocking are cited in the Bibliography. Procedures for detection of multicollinearity are discussed in the textbook by Myers, also cited.

The user of multiple linear regression attempts to accomplish one of three objectives:

1. Obtain estimates of individual coefficients in a complete model.
2. Screen variables to determine which have a significant effect on the response.
3. Arrive at the most effective prediction equation.

In (1) it is known a priori that all variables are to be included in the model. In (2) prediction is secondary, while in (3) individual regression coefficients are not as important as the quality of the estimated response $\hat{y}$. In each of the situations above, multicollinearity in the experiment can have a profound effect on the success of the regression.

In this section some standard sequential procedures for selecting variables are discussed. They are based on the notion that a single variable or a collection of variables should not appear in the estimating equation unless they result in a significant increase in the regression sum of squares or, equivalently, a significant increase in R^2, the coefficient of multiple determination.

EXAMPLE 12.9   Consider the data in Table 12.3 in which measurements were taken on 9 infants. The purpose of the experiment was to arrive at a suitable estimating equation relating the length of an infant to all or a subset of the independent variables. The sample correlation coefficients, indicating the linear dependency among the independent variables, are displayed in the symmetric matrix

$$
\begin{array}{cccc}
x_1 & x_2 & x_3 & x_4
\end{array}
$$
$$
\begin{bmatrix}
1.0000 & 0.9523 & 0.5340 & 0.3900 \\
0.9523 & 1.0000 & 0.2626 & 0.1549 \\
0.5340 & 0.2626 & 1.0000 & 0.7847 \\
0.3900 & 0.1549 & 0.7847 & 1.0000
\end{bmatrix}.
$$

Note that there appears to be an appreciable amount of multicollinearity. Using the least squares technique outlined in Section 10.3, the estimated regression equation using the complete model was fitted and is given by

$$\hat{y} = 7.1475 + 0.1000x_1 + 0.7264x_2 + 3.0758x_3 + 0.0300x_4.$$

The value of s^2 with 4 degrees of freedom is 0.7414, and the value for the coefficient of determination for this model is found to be 0.9907. Regression sum of

TABLE 12.3 Data Relating to Infant Length[a]

Infant Length, y (cm)	Age, x_1 (days)	Length at Birth, x_2 (cm)	Weight at Birth, x_3 (kg)	Chest Size at Birth, x_4 (cm)
57.5	78	48.2	2.75	29.5
52.8	69	45.5	2.15	26.3
61.3	77	46.3	4.41	32.2
67.0	88	49.0	5.52	36.5
53.5	67	43.0	3.21	27.2
62.7	80	48.0	4.32	27.7
56.2	74	48.0	2.31	28.3
68.5	94	53.0	4.30	30.3
69.2	102	58.0	3.71	28.7

[a] Data analyzed by the Statistical Consulting Center, Virginia Polytechnic Institute and State University, Blacksburg, Virginia, 1976.

squares measuring the variation attributed to each individual variable in the presence of the others, and the corresponding t-values, are given in Table 12.4.

A two-tailed critical region with 4 degrees of freedom at the 0.05 level of significance is given by $|t| > 2.776$. Of the four computed t-values, only variable x_3 appears to be significant. However, it should be recalled that although the t-statistic described in Section 12.6 measures the worth of a variable adjusted for all other variables, it does not detect the potential importance of a variable in combination with a subset of the variables. For example, consider the model with only the variables x_2 and x_3 in the equation. The data analysis gives the regression function

$$\hat{y} = 2.1833 + 0.9576x_2 + 3.3253x_3,$$

with $R^2 = 0.9905$, certainly not a substantial reduction from $R^2 = 0.9907$ for the complete model. However, unless the performance characteristics of this particular combination had been observed, one would not be aware of its predictive potential. This, of course, lends support for a methodology that observes *all possible regressions* or a systematic sequential procedure designed to test several subsets.

STEPWISE
REGRESSION

One standard procedure for searching for the "optimum subset" of variables in the absence of othogonality is a technique called **stepwise regression**. It is based on the procedure of sequentially introducing the variables into the model one at a time. The description of the stepwise routine will be better understood by the reader if the methods of **forward selection** and **backward elimination** are described first.

Forward selection is based on the notion that variables should be inserted one at a time until a satisfactory regression equation is found. The procedure is as follows:

STEP 1   Choose the variable that gives the largest regression sum of squares when performing a simple linear regression with y, or equivalently, that which gives the largest value of R^2. We shall call this initial variable x_1.

TABLE 12.4 t-Values for the Regression Data of Table 12.3

Variable x_1	Variable x_2	Variable x_3	Variable x_4
$R(\beta_1 \mid \beta_2, \beta_3, \beta_4)$	$R(\beta_2 \mid \beta_1, \beta_3, \beta_4)$	$R(\beta_3 \mid \beta_1, \beta_2, \beta_4)$	$R(\beta_4 \mid \beta_1, \beta_2, \beta_3)$
$= 0.0644$	$= 0.6334$	$= 6.2523$	$= 0.0241$
$t = 0.2947$	$t = 0.9243$	$t = 2.9040$	$t = -0.1805$

STEP 2   Choose the variable that when inserted in the model gives the largest increase in R^2, in the presence of x_1, over the R^2 found in step 1. This, of course, is the variable x_j, for which

$$R(\beta_j \mid \beta_1) = R(\beta_1, \beta_j) - R(\beta_1)$$

is largest. Let us call this variable x_2. The regression model with x_1 and x_2 is then fitted and R^2 observed.

STEP 3   Choose the variable x_j that gives the largest value of

$$R(\beta_j \mid \beta_1, \beta_2) = R(\beta_1, \beta_2, \beta_j) - R(\beta_1, \beta_2),$$

again resulting in the largest increase of R^2 over that given in step 2. Calling this variable x_3, we now have a regression model involving x_1, x_2, and x_3.

This process is continued until the most recent variable inserted fails to induce a significant increase in the explained regression. Such an increase can be determined at each step by using the appropriate F-test or t-test. For example, in step 2 the value

$$f = \frac{R(\beta_2 \mid \beta_1)}{s^2}$$

can be determined to test the appropriateness of x_2 in the model. Here the value of s^2 is the error mean square for the model containing the variables x_1 and x_2. Similarly, in step 3, the ratio

$$f = \frac{R(\beta_3 \mid \beta_1, \beta_2)}{s^2}$$

tests the appropriateness of x_3 in the model. Now, however, the value for s^2 is the error mean square for the model that contains the three variables x_1, x_2, and x_3. If $f < f_\alpha(1, n - 3)$ at step 2, for a prechosen significance level, x_2 is not included and the process is terminated, resulting in a simple linear equation relating y and x_1. However, if $f > f_\alpha(1, n - 3)$, we proceed to step 3. Again, if $f < f_\alpha(1, n - 4)$ at step 3, x_3 is not included and the process is terminated with the appropriate regression equation containing the variables x_1 and x_2.

Backward elimination involves the same concepts as forward selection except that one begins with all the variables in the model. Suppose, for example, that there are five variables under consideration. The steps are as follows:

STEP 1   Fit a regression equation with all five variables included in the model. Choose the variable that gives the smallest value of the regression sum of squares

adjusted for the others. Suppose that this variable is x_2. Remove x_2 from the model if

$$g = \frac{R(\beta_2 \mid \beta_1, \beta_3, \beta_4, \beta_5)}{s^2}$$

is insignificant

STEP 2   Fit a regression equation using the remaining variables x_1, x_3, x_4, and x_5 and repeat step 1. Suppose that variable x_5 is chosen this time. Once again if

$$f = \frac{R(\beta_5 \mid \beta_1, \beta_3, \beta_4)}{s^2}$$

is insignificant, the variable x_5 is removed from the model. At each step the s^2 used in the F-test is the error mean square for the regression model at that stage.

This process is repeated until at some step the variable with the smallest adjusted regression sum of squares results in a significant f-value for some predetermined significance level.

Stepwise regression is accomplished with a slight but important modification of the forward selection procedure. The modification involves further testing at each stage to ensure the continued effectiveness of variables that had been inserted into the model at an earlier stage. This represents an improvement over forward selection, since it is quite possible that a variable entering the regression equation at an early stage might have been rendered unimportant or redundant because of relationships that exist between it and other variables entering at later stages. Therefore, at a stage in which a new variable has been entered into the regression equation through a significant increase in R^2 as determined by the F-test, all the variables already in the model are subjected to F-tests (or, equivalently to t-tests) in light of this new variable, and are deleted if they do not display a significant f-value. The procedure is continued until a stage is reached in which no additional variables can be inserted or deleted. We illustrate the stepwise procedure in the following example.

EXAMPLE 12.10   Using the techniques of stepwise regression, find an appropriate linear regression model for predicting the length of infants for the data of Table 12.3.

SOLUTION

STEP 1   In considering each variable separately, four individual simple linear regression equations are fitted. The following pertinent regression sums of squares are computed:

$$R(\beta_1) = 288.1468 \qquad R(\beta_2) = 215.3013$$
$$R(\beta_3) = 186.1065 \qquad R(\beta_4) = 100.8594.$$

Variable x_1 very clearly gives the largest regression sum of squares. The error mean square for the equation involving only x_1 is $s^2 = 4.7276$, and since

$$f = \frac{R(\beta_1)}{s^2} = \frac{288.1468}{4.7276} = 60.9500,$$

which exceeds $f_{0.05}(1, 7) = 5.59$, the variable x_1 is entered into the model.

STEP 2   Three regression equations are fitted at this stage, all containing x_1. The important results for the combinations (x_1, x_2), (x_1, x_3) and (x_1, x_4) are

$$R(\beta_2|\beta_1) = 23.8703, \qquad R(\beta_3|\beta_1) = 29.3086, \qquad R(\beta_4|\beta_1) = 13.8178.$$

Variable x_3 displays the largest regression sum of squares in the presence of x_1. The regression involving x_1 and x_3 gives a new value of $s^2 = 0.6307$, and since

$$f = \frac{R(\beta_3|\beta_1)}{s^2} = \frac{29.3086}{0.6307} = 46.47,$$

which exceeds $f_{0.05}(1, 6) = 5.99$, the variable x_3 is included along with x_1 in the model. Now we must subject x_1 in the presence of x_3 to a significance test. We find that $R(\beta_1|\beta_3) = 131.349$, and hence

$$f = \frac{R(\beta_2|\beta_3)}{s^2} = \frac{131.349}{0.6307} = 208.26,$$

which is highly significant. Therefore, x_1 is retained along with x_3.

STEP 3   With x_1 and x_3 already in the model, we now require $R(\beta_2|\beta_1, \beta_3)$ and $R(\beta_4|\beta_1, \beta_3)$ in order to determine which, if any, of the remaining two variables is entered at this stage. From the regression analysis using x_2 along with x_1 and x_3, we find $R(\beta_2|\beta_1, \beta_3) = 0.7948$, and when x_4 is used along with x_1 and x_3, we obtain $R(\beta_4|\beta_1, \beta_3) = 0.1855$. The value of s^2 is 0.5979 for the (x_1, x_2, x_3) combination and 0.7198 for the (x_1, x_2, x_4) combination. Since neither f-value is significant at the $\alpha = 0.05$ level, the final regression model includes only the variables x_1 and x_3. The estimating equation is found to be

$$\hat{y} = 20.1084 + 0.4136x_1 + 2.0253x_3$$

and the coefficient of determination for this model is $R^2 = 0.9882$.

Although (x_1, x_3) is the combination chosen by stepwise regression, it is not necessarily the combination of two variables that gives the largest value of R^2. In fact, we have already observed that the combination (x_2, x_3) gives an $R^2 = 0.9905$. Of course, the stepwise procedure never actually observed this combination. A rational argument could be made that there is actually a negligible difference in performance between these two estimating equations, at least in terms of percent variation explained. It is interesting to observe however, that the backward elimination procedure gives the combination (x_2, x_3) in the final equation (see Exercise 3 on page 459).

The main function of each of the procedures outlined in this section is to expose the variables to a systematic methodology designed to ensure the eventual inclusion of the best combinations of the variables. Obviously, there is no assurance that this will happen in all problems, and, of course, it is possible that the multicollinearity is so extensive that one has no alternative but to resort to estimation procedures other than least squares. These estimation procedures are discussed in Myers, listed in the Bibliography.

The sequential procedures discussed here represent three of many such methods that have been put forth in the literature and appear in various regression computer packages that are available. These methods are designed to be computationally efficient but, of course, do not give results for all possible subsets of the variables. As a result, the procedures are most effective in data sets that involve a **large number of variables**. In regression problems involving a relatively small number of variables, certainly when $k \leq 10$, modern regression computer packages allow for the computation and summarization of quantitative information on **all models** for every possible subset of the variables. Illustrations will be given in Section 12.10.

12.9 Study of Residuals and Violation of Assumptions

It was suggested earlier in this chapter that the residuals, or errors in the regression fit, often carry information that can be very informative to the data analyst. The $e_i = y_i - \hat{y}_i$, $i = 1, 2, \ldots, n$, which are the numerical counterpart to the ε_i's, the model errors, often shed light on the possible violation of assumptions or the presence of "suspect" data points. Suppose that we let the vector $\mathbf{x}_i$ denote the values of the regressor variables corresponding to the ith data point, supplemented by a 1 in the initial position. That is,

$$\mathbf{x}_i' = (1, x_{1i}, x_{2i}, \ldots, x_{ki}].$$

Consider the quantity

$$h_{ii} = \mathbf{x}_i'(\mathbf{X'X})^{-1}\mathbf{x}_i, \qquad i = 1, 2, \ldots, n.$$

The reader should recognize that h_{ii} was used in the computation of the confidence intervals on the mean response in Section 12.5. Apart from σ^2, h_{ii} represents the variance of the fitted value $\hat{Y}_i$. The h_{ii} values are the diagonal elements of the **HAT matrix** given by

$$\mathbf{H} = \mathbf{X}(\mathbf{X'X})^{-1}\mathbf{X'},$$

which plays an important role in any study of residuals and in other modern aspects of regression analysis (see the reference to Myers listed in the Bibliography). The term *HAT matrix* is derived from the fact that $\mathbf{H}$ generates the "y-hats" or the fitted values when multiplied by the vector $\mathbf{y}$ of observed responses. That is,

$$\hat{\mathbf{y}} = \mathbf{X}(\mathbf{X'X})^{-1}\mathbf{X'y} = \mathbf{Hy},$$

where $\hat{\mathbf{y}}$ is the vector whose ith element is $\hat{y}_i$.

If we make the usual assumptions that the E_i are independent and normally distributed with zero mean and variance σ^2, the statistical properties of the residuals are easily characterized. Then

$$E(E_i) = E(Y_i - \hat{Y}_i) = 0$$

and

$$\sigma_{E_i}^2 = (1 - h_{ii})\sigma^2$$

for $i = 1, 2, \ldots , n$. See the Myers reference for details. It can be shown that the HAT diagonal values are bounded according to the inequality

$$0 < h_{ii} < 1.$$

In addition, $\sum_{i=1}^{n} h_{ii} = k + 1$, the number of regression parameters. As a result, any data point whose HAT diagonal element is large, that is, well above the average value of $(k + 1)/n$, is in a position in the data set where the variance of $\hat{Y}_i$ is relatively large, and the variance of a residual is relatively small. As a result, the data analyst can easily gain some insight on how large a residual may become before its deviation from zero can be attributed to something other than mere chance. Many of the commercial regression computer packages produce the set of **studentized residuals**

STUDENTIZED
RESIDUAL

$$r_i = \frac{e_i}{s\sqrt{1 - h_{ii}}}, \qquad i = 1, 2, \ldots , n.$$

Here each residual has been divided by an estimate of its standard deviation, creating a *t-like* statistic that is designed to give the analyst a scale-free quantity that provides information regarding the *size* of the residual. In addition, standard computer packages often give values of another set of studentized-type residuals, called the **R-Student values** and given by

R-STUDENT VALUE

$$t_i = \frac{e_i}{s_{-i}\sqrt{1 - h_{ii}}}, \qquad i = 1, 2, \ldots , n,$$

where s_{-i} is an estimate of the error standard deviation, calculated with the ith data point deleted.

There are three types of violations of assumptions that are readily detected through use of residuals or *residual plots*. While plots of the raw residuals, the e_i, can be helpful, it is often more informative to plot the studentized residuals. The three violations are as follows:

1. Presence of outliers.
2. Heterogeneous error variance.
3. Model misspecification.

In case 1, we choose to define an *outlier* as a data point in which there is a deviation from the usual assumption $E(E_i) = 0$ for a specific value of i. If there is a reason to believe that a specific data point is an outlier exerting a large influence on the fitted model, r_i or t_i may be informative. The R-Student values can be expected to be more sensitive to outliers than the r_i values.

In fact, under the condition that $E(E_i) = 0$, t_i is a value of a random variable following a *t*-distribution with $n - 1 - (k + 1) = n - k - 2$ degrees of freedom.

Thus a two-sided t-test can be used to provide information for detecting whether or not the ith point is an outlier.

Although the R-Student statistic, t_i, produces an exact t-test for detection of an outlier at a specific data location, the t-distribution would not apply for simultaneously testing for outliers at all locations. As a result, the studentized residuals or R-Student values should be used strictly as diagnostic tools *without* formal hypothesis testing as the mechanism. The implication is that these statistics highlight data points where the error of fit is larger than what is expected by chance. Large R-Student values in magnitude suggest a need for "checking" the data with whatever resources possible. The practice of eliminating observations from regression data sets should not be done indiscriminantly. Further information regarding the use of outlier diagnostics is given in the book by Myers listed in the Bibliography.

EXAMPLE 12.11   In a biological experiment conducted at the Virginia Polytechnic Institute and State University by the Department of Entomology, n experimental runs were made with two different methods for capturing grasshoppers. The methods are: drop net catch and sweet net catch. The average number of grasshoppers caught in a set of field quadrants on a given date is recorded for each of the two methods. An additional regressor variable, the average plant height in the quadrants, was also recorded. The experimental data are as follows:

Observation	Drop Net Catch, y	Sweep Net Catch, x_1	Plant Height, x_2 (cm)
1	18.000	4.15476	52.705
2	8.8750	2.02381	42.069
3	2.0000	0.15909	34.766
4	20.0000	2.32812	27.622
5	2.3750	0.25521	45.879
6	2.7500	0.57292	97.472
7	3.3333	0.70139	102.062
8	1.0000	0.13542	97.790
9	1.3333	0.12121	88.265
10	1.7500	0.10937	58.737
11	4.1250	0.56250	42.386
12	12.8750	2.45312	31.274
13	5.3750	0.45312	31.750
14	28.0000	6.68750	35.401
15	4.7500	0.86979	64.516
16	1.7500	0.14583	25.241
17	0.1333	0.01562	36.354

The goal is to be able to estimate grasshopper catch by using only the sweep net method, which is less costly. There was some concern about the validity of the fourth data point. The observed catch that was reported using the net drop method seemed unusually high given the other conditions and, indeed, it was felt that the figure might be erroneous. Fit a model of the type

$$\mu_{Y|x_1, x_2} = \beta_0 + \beta_1 x_1 + \beta_2 x_2$$

to the 17 data points and study the residuals to determine if data point 4 is an outlier.

SOLUTION
A computer package generated the fitted regression model

$$\hat{y} = 3.6870 + 4.1050 x_1 - 0.0367 x_2$$

along with the statistics $R^2 = 0.9244$ and $s^2 = 5.580$. The residuals and other diagnostic information were also generated and recorded as follows:

Obs.	y_i	$\hat{y}_i$	$y_i - \hat{y}_i$	h_{ii}	$s\sqrt{1 - h_{ii}}$	r_i	t_i
1	18.000	18.809	−0.809	0.2291	2.074	−0.390	−0.3780
2	8.875	10.452	−1.577	0.0766	2.270	−0.695	−0.6812
3	2.000	3.065	−1.065	0.1364	2.195	−0.485	−0.4715
4	20.000	12.231	7.769	0.1256	2.209	3.517	9.9315
5	2.375	3.052	−0.677	0.0931	2.250	−0.301	−0.2909
6	2.750	2.464	0.286	0.2276	2.076	0.138	0.1329
7	3.333	2.823	0.510	0.2669	2.023	0.252	0.2437
8	1.000	0.656	0.344	0.2318	2.071	0.166	0.1601
9	1.333	0.947	0.386	0.1691	2.153	0.179	0.1729
10	1.750	1.982	−0.232	0.0852	2.260	−0.103	−0.0989
11	4.125	4.442	−0.317	0.0884	2.255	−0.140	−0.1353
12	12.875	12.610	0.265	0.1152	2.222	0.119	0.1149
13	5.375	4.383	0.992	0.1339	2.199	0.451	0.4382
14	28.000	29.841	−1.841	0.6233	1.450	−1.270	−1.3005
15	4.750	4.891	−0.141	0.0699	2.278	−0.062	−0.0598
16	1.750	3.360	−1.610	0.1891	2.127	−0.757	−0.7447
17	0.133	2.418	−2.285	0.1386	2.193	−1.042	−1.0454

As expected, the residual at the fourth location appears to be unusually high, namely 7.769. The vital issue here is whether or not this residual is larger than one would expect by chance. The residual standard error for point 4 is 2.209. The R-Student value t_4 is found to be 9.9315. Viewing this as a value of a random variable having a t-distribution with 13 degrees of freedom, one would certainly conclude that the residual of the fourth observation is estimating something greater than 0 and that the suspected measurement error is supported by the study of residuals. Notice that no other residual results in an R-Student value that produces any cause for alarm.

Plotting of Residuals

In Chapter 11 we discussed, in some detail, the usefulness of plotting residuals in regression analysis. Violation of model assumptions can often be detected through these plots. In multiple regression normal probability plotting of residuals or plots of residuals against $\hat{y}$ may be useful. However, it is often preferable to plot studentized residuals.

The reader should keep in mind that the preference of the studentized residuals over ordinary residuals for plotting purposes stems from the fact that since the variance of the ith residual depends on the ith HAT diagonal, variances of residuals will differ if there is a dispersion in the HAT diagonals. Thus the appearance of a plot of residuals may depict heterogeneity because the residuals themselves do not behave, in general, in an ideal way. The purpose of using studentized residuals is to provide a *standardization*. Clearly, if σ were known, then, under ideal conditions (i.e., a correct model and homogeneous variance), we have

$$E\left(\frac{E_i}{\sigma\sqrt{1-h_{ii}}}\right) = 0$$

and

$$\text{Var}\left(\frac{E_i}{\sigma\sqrt{1-h_{ii}}}\right) = 1.$$

So the studentized residuals produce a set of statistics that behaves in a standard way under ideal conditions.

12.10 Cross Validation, C_p, and Other Criteria for Model Selection

In many regression problems the experimenter must choose between various alternative models or model forms that are developed from the same data set. Quite often, in fact, the model that best predicts or estimates mean response is required. The experimenter should take into account the relative sizes of the s^2-values for the candidate models, and certainly the general nature of the confidence intervals on the mean response. One must also consider how well the model predicts response values that were **not used in building the candidate models**. The models should be subjected to **cross validation**. What is required, then, are cross-validation errors rather than fitting errors. Such errors in prediction are the **PRESS residuals**

$$\delta_i = y_i - \hat{y}_{i, -i}, \qquad i = 1, 2, \ldots, n,$$

where $\hat{y}_{i, -i}$ is the prediction of the ith data point by a model that did not make use of the ith point in the calculation of the coefficients. These PRESS residuals are easily calculated from the formula

$$\delta_i = \frac{e_i}{1 - h_{ii}}, \qquad i = 1, 2, \ldots, n.$$

The derivation can be found in the regression textbook by Myers.

Use of the PRESS Statistic

The motivation for PRESS and the utility of PRESS residuals is very simple to understand. The purpose of extracting or *setting aside* data points one at a time is to allow the use of separate methodologies for fitting and assessment of a specific model. For assessment of a model the "$-i$" indicates that the PRESS residual gives a prediction error where the observation being predicted is *independent of the model fit.*

Criteria that make use of the PRESS residuals are given by

$$\sum_{i=1}^{n} |\delta_i| \quad \text{and} \quad \text{PRESS} = \sum_{i=1}^{n} \delta_i^2.$$

The term *PRESS* is an acronym for the **prediction sum of squares**. We suggest that both of these criteria be used. It is possible for PRESS to be dominated by one or only a few large PRESS residuals. Clearly, the criteria $\sum_{i=1}^{n} |\delta_i|$ is less sensitive to a small number of large values.

In addition to the PRESS statistic itself, the analyst can simply compute an "R^2-like" statistic reflecting prediction performance. The statistic is often called R^2_{pred} and is given as follows:

Given a fitted model with a specific value for PRESS, R^2_{pred} is given by

$$R^2_{\text{pred}} = 1 - \frac{\text{PRESS}}{\sum_{i=1}^{n} (y_i - \bar{y})^2}. \qquad \blacksquare$$

In the following example a "case-study" illustration is provided in which many candidate models are fit to a set of data and the best model is chosen. The sequential procedures described in Section 12.8 are not used. Rather, the role of the PRESS residuals and other statistical values in selecting the best regression equation is illustrated.

EXAMPLE 12.12   Leg strength is a necessary ingredient of a successful punter in American football. One measure of the quality of a good punt is the "hang time." This is the time that the ball hangs in the air before being caught by the punt returner. To determine what leg strength factors influence hang time and to develop an empirical model for predicting this response, a study on "*The Relationship Between Selected Physical Performance Variables and Football Punting Ability*" was conducted by the Department of Health, Physical Education, and Recreation at the Virginia Polytechnic Institute and State University in 1983. Thirteen punters were chosen for the experiment and each punted a football 10 times. The average hang time, along with the strength measures used in the analysis, were recorded as follows:

Punter	Hang Time, y (sec)	RLS, x_1	LLS, x_2	RHF, x_3	LHF, x_4	Power, x_5
1	4.75	170	170	106	106	240.57
2	4.07	140	130	92	93	195.49
3	4.04	180	170	93	78	152.99
4	4.18	160	160	103	93	197.09
5	4.35	170	150	104	93	266.56
6	4.16	150	150	101	87	260.56
7	4.43	170	180	108	106	219.25
8	3.20	110	110	86	92	132.68
9	3.02	120	110	90	86	130.24
10	3.64	130	120	85	80	205.88
11	3.68	120	140	89	83	153.92
12	3.60	140	130	92	94	154.64
13	3.85	160	150	95	95	240.57

Each regressor variable is defined as follows:

1. *RLS*—right leg strength, pounds.
2. *LLS*—left leg strength, pounds.
3. *RHF*—right hamstring muscle flexibility, degrees.
4. *LHF*—left hamstring muscle flexibility, degrees.
5. *Power*—overall leg strength, foot-pounds.

Determine the most appropriate model for predicting hang time.

SOLUTION
In the search for the "best" of the candidate models for predicting hang time, the information of Table 12.5 was obtained from a regression computer package. The models are ranked in ascending order of the values of the PRESS statistic. This display provides enough information on all possible models to enable the user to eliminate from consideration all but a few models. The model x_2, x_5 (*LLS* and *Power*) appears to be superior for predicting punter hang time. Also note that all models with low PRESS, low s^2, low $\sum_{i=1}^{n} |\delta_i|$, and high R^2 values contain these two variables.

In order to gain some insight from the residuals of the fitted regression

$$\hat{y}_i = b_0 + b_2 x_{2i} + b_5 x_{5i},$$

the residuals and PRESS residuals were generated. The actual prediction model (see Exercise 1 on page 458) is given by

$$\hat{y} = 1.10765 + 0.01370 x_2 + 0.00429 x_5.$$

Residuals, hat diagonal values, and PRESS variables are listed in Table 12.6. Note the relatively good fit of the two-variable regression model to the data. The PRESS residuals reflect the capability of the regression equation to predict hang time if independent predictions were to be made. For example, for punter number 4, the

TABLE 12.5   Comparing Different Regression Models

Model	s^2	$\sum |\delta_i|$	PRESS	R^2
$x_2 x_5$	0.036907	1.93583	0.54683	0.871300
$x_1 x_2 x_5$	0.041001	2.06489	0.58998	0.871321
$x_2 x_4 x_5$	0.037708	2.18797	0.59915	0.881658
$x_2 x_3 x_5$	0.039636	2.09553	0.66182	0.875606
$x_1 x_2 x_4 x_5$	0.042265	2.42194	0.67840	0.882093
$x_1 x_2 x_3 x_5$	0.044578	2.26283	0.70958	0.875642
$x_2 x_3 x_4 x_5$	0.042421	2.55789	0.86236	0.881658
$x_1 x_3 x_5$	0.053664	2.65276	0.87325	0.831580
$x_1 x_4 x_5$	0.056279	2.75390	0.89551	0.823375
$x_1 x_5$	0.059621	2.99434	0.97483	0.792094
$x_2 x_3$	0.056153	2.95310	0.98815	0.804187
$x_1 x_3$	0.059400	3.01436	0.99697	0.792864
$x_1 x_2 x_3 x_4 x_5$	0.048302	2.87302	1.00920	0.882096
x_2	0.066894	3.22319	1.04564	0.743404
$x_3 x_5$	0.065678	3.09474	1.05708	0.770971
$x_1 x_2$	0.068402	3.09047	1.09726	0.761474
x_3	0.074518	3.06754	1.13555	0.714161
$x_1 x_3 x_4$	0.065414	3.36304	1.15043	0.794705
$x_2 x_3 x_4$	0.062082	3.32392	1.17491	0.805163
$x_2 x_4$	0.063744	3.59101	1.18531	0.777716
$x_1 x_2 x_3$	0.059670	3.41287	1.26558	0.812730
$x_3 x_4$	0.080605	3.28004	1.28314	0.718921
$x_1 x_4$	0.069965	3.64415	1.30194	0.756023
x_1	0.080208	3.31562	1.30275	0.692334
$x_1 x_3 x_4 x_5$	0.059169	3.37362	1.36867	0.834936
$x_1 x_2 x_4$	0.064143	3.89402	1.39834	0.798692
$x_3 x_4 x_5$	0.072505	3.49695	1.42036	0.772450
$x_1 x_2 x_3 x_4$	0.066088	3.95854	1.52344	0.815633
x_5	0.111779	4.17839	1.72511	0.571234
$x_4 x_5$	0.105648	4.12729	1.87734	0.631593
x_4	0.186708	4.88870	2.82207	0.283819

hang time of 4.180 would encounter a prediction error of 0.039 if the model constructed by using the remaining 12 punters were used. For this model, the average prediction error or cross-validation error is given by

$$\frac{\sum_{i=1}^{n} |\delta_i|}{13} = 0.1489 \text{ second},$$

which is quite small compared to the average hang time for the 13 punters.

We indicated in Section 12.8 that the use of all possible subset regressions is often advisable when searching for the best model. The sequential methods dis-

TABLE 12.6   PRESS Residuals

Punter	y_i	$\hat{y}_i$	$e_i = y_i - \hat{y}_i$	h_{ii}	δ_i
1	4.750	4.470	0.280	0.198	0.349
2	4.070	3.728	0.342	0.118	0.388
3	4.040	4.094	−0.054	0.444	−0.097
4	4.180	4.146	0.034	0.132	0.039
5	4.350	4.307	0.043	0.286	0.060
6	4.160	4.281	−0.121	0.250	−0.161
7	4.430	4.515	−0.085	0.298	−0.121
8	3.200	3.184	0.016	0.294	0.023
9	3.020	3.174	−0.154	0.301	−0.220
10	3.640	3.636	0.004	0.231	0.005
11	3.680	3.687	−0.007	0.152	−0.008
12	3.600	3.553	0.047	0.142	0.055
13	3.850	4.196	−0.346	0.154	−0.409

cussed in Section 12.8 should be used only when computer time prohibits the computation of all possible regressions. Most commercial statistics software packages contain an *all possible regressions* routine. These algorithms compute various criteria for all subsets of model terms. Obviously, criteria such as R^2, s^2, and PRESS are reasonable for choosing among candidate subsets. Another very popular and useful statistic, particularly in areas in the physical sciences and engineering, is the C_p statistic, which is described in the following paragraphs.

The C_p Statistic

Quite often the choice of the most appropriate model involves many considerations. Obviously, the number of model terms is important. The matter of parsimony is a consideration that cannot be ignored. On the other hand, the analyst cannot be pleased with a model that is too simple, to the point where there is serious underspecification. A single statistic that represents a nice compromise in this regard is the C_p statistic. (See the Mallows reference in the Bibliography.)

The C_p statistic appeals nicely to common sense and is developed from considerations of the proper compromise between excessive bias incurred when one underfits (chooses too few model terms) and excessive prediction variance produced when one overfits (has redundancies in the model). The C_p statistic is a simple function of the total number of parameters in the candidate model and the error mean square, s^2.

We will not present the entire development of the C_p statistic. For details the reader is referred to the textbook by Myers in the Bibliography. The C_p for a particular subset model *is an estimate* of the following:

$$\Gamma_{(p)} = \frac{\sum_{i=1}^{n} \text{Var}(\hat{y}_i)}{\sigma^2} + \frac{\sum_{i=1}^{n} (\text{Bias } \hat{y}_i)^2}{\sigma^2}.$$

It turns out that under the standard least squares assumptions indicated earlier in this chapter, and assuming that the "true" model is the model containing all candidate variables,

$$\frac{\sum_{i=1}^{n} \text{Var}(\hat{y}_i)}{\sigma^2} = p \quad \text{(number of parameters in the candidate model)}$$

(see Review Exercise 4) and an unbiased estimate of $\dfrac{\sum_{i=1}^{n} (\text{Bias } \hat{y}_i)^2}{\sigma^2}$ is given by

$$\frac{\sum_{i=1}^{n} (\widehat{\text{Bias } \hat{y}_i})^2}{\sigma^2} = \frac{(s^2 - \sigma^2)(n - p)}{\sigma^2}.$$

In the above, s^2 is the error mean square for the candidate model and σ^2 is the population error variance. Thus if we assume that some estimate $\hat{\sigma}^2$ is available for σ^2, C_p is given by

---

DEFINITION 12.1 C_p *statistic*

$$C_p = p + \frac{(s^2 - \hat{\sigma}^2)(n - p)}{\hat{\sigma}^2},$$

where p is the number of model parameters, s^2 is the error mean square for the candidate model, and $\hat{\sigma}^2$ is an estimate of σ^2. Obviously, the scientists should adopt models with small values of C_p. ∎

The reader should first note that unlike the PRESS statistic, C_p is scale free. In addition, one can gain some insight concerning adequacy of a candidate model by observing its value of C_p. For example, $C_p > p$ indicates a model that is biased due to being an underfitted model, while $C_p \approx p$ indicates a reasonable model.

There is often confusion concerning where $\hat{\sigma}^2$ comes from in the formula for C_p. Obviously, the scientist or engineer does not have access to the population quantity σ^2. In applications where replicated runs are available, say in an experimental design situation, a model-independent estimate of σ^2 is available (see Chapters 11 and 15). However, most software packages will use as $\hat{\sigma}^2$ the *error mean square from the most complete model*. Obviously, if this is not a good estimate, the bias portion of the C_p statistic can be negative. Thus C_p can be less than p.

EXAMPLE 12.13   Consider the following data set, in which a maker of asphalt shingles is interested in the relationship between sales for a particular year and factors that influence sales. The data were taken from Neter, Wassermann, and Kutner (see the Bibliography).

District	Promotional Accounts, x_1	Active Accounts, x_2	Competing Brands, x_3	Potential, x_4	Sales, y (thousands of dollars)
1	5.5	31	10	8	79.3
2	2.5	55	8	6	200.1
3	8.0	67	12	9	163.2
4	3.0	50	7	16	200.1
5	3.0	38	8	15	146.0
6	2.9	71	12	17	177.7
7	8.0	30	12	8	30.9
8	9.0	56	5	10	291.9
9	4.0	42	8	4	160.0
10	6.5	73	5	16	339.4
11	5.5	60	11	7	159.6
12	5.0	44	12	12	86.3
13	6.0	50	6	6	237.5
14	5.0	39	10	4	107.2
15	3.5	55	10	4	155.0

Of the possible subset models, three are of particular interest. These three contain the variable combinations (x_2, x_3), (x_1, x_2, x_3), and (x_1, x_2, x_3, x_4). The following represents pertinent information in comparing the three models. We have included the PRESS statistics for the three models to supplement the decision making.

Model	R^2	R^2_{pred}	s^2	PRESS	C_p
x_2, x_3	0.9940	0.9913	44.5552	782.1896	11.4013
x_1, x_2, x_3	0.9970	0.9928	24.7956	643.3578	3.4075
x_1, x_2, x_3, x_4	0.9971	0.9917	26.2073	741.7557	5.0

It seems clear from the information in the table that the model x_1, x_2, x_3 is preferable to the other two. Notice that for the full model, $C_p = 5.0$. This occurs since the *bias portion* is zero and $\hat{\sigma}^2 = 26.2073$ is the error mean square from the full model.

Exercises

1. Consider the "hang time" punting data given in Example 12.12, using only the variables x_2 and x_5,
   (a) verify the regression equation shown on page 454;
   (b) predict punter hang time for a punter with $LLS = 180$ pounds and *Power* = 260 foot-pounds;
   (c) construct a 95% confidence interval for the mean hang time of a punter with $LLS = 180$ pounds and *Power* = 260 foot-pounds.

2. For the data of Exercise 11 on page 426, use the techniques of
   (a) *forward selection* with a 0.05 level of significance to choose a linear regression model;

(b) *backward elimination* with a 0.05 level of significance to choose a linear regression model;
(c) *stepwise regression* with a 0.05 level of significance to choose a linear regression model.

3. Use the techniques of *backward elimination* with $\alpha = 0.05$ to choose a prediction equation for the data of Table 12.3.

4. For the punter data in Example 12.12, an additional response, "punting distance," was also recorded. The following are average distance values for each of the 13 punters:

Punter	Distance, y (ft)
1	162.50
2	144.00
3	147.50
4	163.50
5	192.00
6	171.75
7	162.00
8	104.93
9	105.67
10	117.59
11	140.25
12	150.17
13	165.16

(a) Using the distance data rather than the hang times, estimate a multiple linear regression model of the type

$$\mu_{Y|x_1, x_2, x_3, x_4, x_5}$$
$$= \beta_0 + \beta_1 x_1 + \beta_2 x_2 + \beta_3 x_3 + \beta_4 x_4 + \beta_5 x_5$$

for predicting punting distance.
(b) Use stepwise regression with a significance level of 0.10 to select a combination of variables.
(c) Generate values for s^2, R^2, PRESS and $\sum_{i=1}^{13} |\delta_i|$ for the entire set of 31 models. Use this information to determine the best combination of variables for predicting punting distance.

5. The following is a set of data for y, the amount of money (thousands of dollars) contributed to the

alumni association at Virginia Tech by the Class of 1960, and x, the number of years following gradua-tion:

y	x
812.52	1
822.50	2
1211.50	3
1348.00	4
1301.00	8
2567.50	9
2526.50	10
2755.00	11
4390.50	12
5581.50	13
5548.00	14
6086.00	15
5764.00	16
8903.00	17

(a) Fit a regression model of the type

$$\mu_{Y|x} = \beta_0 + \beta_1 x.$$

(b) Fit a quadratic model of the type

$$\mu_{Y|x} = \beta_0 + \beta_1 x + \beta_{11} x^2.$$

(c) Determine which of the models in (a) or (b) is preferable. Use s^2, R^2, and the PRESS residuals to support your decision.

6. For the model of Exercise 4(a), test the hypothesis

$$H_0: \quad \beta_4 = 0,$$
$$H_1: \quad \beta_4 \neq 0.$$

Use a P-value in your conclusion.

7. For the quadratic model of Exercise 5(b), give esti-mates of the variances and covariances of the esti-mates of β_1 and β_{11}.

8. A client from the Department of Mechanical Engi-neering approached the Consulting Center at Virginia Polytechnic Institute and State University for help in analyzing an experiment dealing with gas turbine engines. Voltage output of engines was measured at various combinations of blade speed and voltage measuring sensor extension. The data are as follows:

y (volts)	Speed, x_1 (in./sec)	Extension, x_2 (in.)
1.95	6336	0.000
2.50	7099	0.000
2.93	8026	0.000
1.69	6230	0.000
1.23	5369	0.000
3.13	8343	0.000
1.55	6522	0.006
1.94	7310	0.006
2.18	7974	0.006
2.70	8501	0.006
1.32	6646	0.012
1.60	7384	0.012
1.89	8000	0.012
2.15	8545	0.012
1.09	6755	0.018
1.26	7362	0.018
1.57	7934	0.018
1.92	8554	0.018

(a) Fit a multiple linear regression to the data.
(b) Compute t-tests on coefficients. Give P-values.
(c) Comment on the quality of the fitted model.

9. Rayon whiteness is an important factor for scientists dealing in fabric quality. Whiteness is affected by pulp quality and other processing variables. Some of the variables include: acid bath temperature, °C (x_1); cascade acid concentration, % (x_2); water temperature, °C (x_3); sulfide concentration, % (x_4); amount of chlorine bleach, lb/min (x_5); blanket finish temperature, °C (x_6). Following is a set of data taken on rayon specimens. The response, y, is the measure of whiteness.

y	x_1	x_2	x_3	x_4	x_5	x_6
88.7	43	0.211	85	0.243	0.606	48
89.3	42	0.604	89	0.237	0.600	55
75.5	47	0.450	87	0.198	0.527	61
92.1	46	0.641	90	0.194	0.500	65
83.4	52	0.370	93	0.198	0.485	54
44.8	50	0.526	85	0.221	0.533	60
50.9	43	0.486	83	0.203	0.510	57
78.0	49	0.504	93	0.279	0.489	49
86.8	51	0.609	90	0.220	0.462	64
47.3	51	0.702	86	0.198	0.478	63
53.7	48	0.397	92	0.231	0.411	61
92.0	46	0.488	88	0.211	0.387	88
87.9	43	0.525	85	0.199	0.437	63
90.3	45	0.486	84	0.189	0.499	58
94.2	53	0.527	87	0.245	0.530	65
89.5	47	0.601	95	0.208	0.500	67

Use the criteria MSE, C_p, and PRESS to give the best model from among all subset models.

10. In an effort to model executive compensation for the year 1979, 33 firms were selected, and data were gathered on compensation, sales, profits, and employment. The following data were gathered for the year 1979. Consider the model.

$$y_i = \beta_0 + \beta_1 \ln x_{1i} + \beta_2 \ln x_{2i} + \beta_3 \ln x_{3i} + \varepsilon_i \quad (i = 1, 2, \ldots, 33).$$

Fit the regression with the model above.

Firm	Compensation, y (thousands of dollars)	Sales, x_1 (millions of dollars)	Profits, x_2 (millions of dollars)	Employment x_3
1	450	4,600.6	128.1	48,000
2	387	9,255.4	783.9	55,900
3	368	1,526.2	136.0	13,783
4	277	1,683.2	179.0	27,765
5	676	2,752.8	231.5	34,000
6	454	2,205.8	329.5	26,500
7	507	2,384.6	381.8	30,800
8	496	2,746.0	237.9	41,000
9	487	1,434.0	222.3	25,900
10	383	470.6	63.7	8,600
11	311	1,508.0	149.5	21,075
12	271	464.4	30.0	6,874
13	524	9,329.3	577.3	39,000
14	498	2,377.5	250.7	34,300
15	343	1,174.3	82.6	19,405
16	354	409.3	61.5	3,586
17	324	724.7	90.8	3,905
18	225	578.9	63.3	4,139
19	254	966.8	42.8	6,255
20	208	591.0	48.5	10,605
21	518	4,933.1	310.6	65,392
22	406	7,613.2	491.6	89,400
23	332	3,457.4	228.0	55,200
24	340	545.3	54.6	7,800
25	698	22,862.8	3,011.3	337,119
26	306	2,361.0	203.0	52,000
27	613	2,614.1	201.0	50,500
28	302	1,013.2	121.3	18,625
29	540	4,560.3	194.6	97,937
30	293	855.7	63.4	12,300
31	528	4,211.6	352.1	71,800
32	456	5,440.4	655.2	87,700
33	417	1,229.9	97.5	14,600

Review Exercises

1. In the Department of Fisheries and Wildlife at Virginia Polytechnic Institute and State University, an experiment was conducted to study the effect of stream characteristics on fish biomass. The regressor variables are as follows: average depth (of 50 cells) (x_1); area of in-stream cover (i.e., undercut banks, logs, boulders, etc.) (x_2); percent canopy cover (average of 12) (x_3); area ≥ 25 centimeters in depth (x_4). The response is y, the fish biomass. The data are as follows:

Observation	y	x_1	x_2	x_3	x_4
1	100	14.3	15.0	12.2	48.0
2	388	19.1	29.4	26.0	152.2
3	755	54.6	58.0	24.2	469.7
4	1288	28.8	42.6	26.1	485.9
5	230	16.1	15.9	31.6	87.6
6	0	10.0	56.4	23.3	6.9
7	551	28.5	95.1	13.0	192.9
8	345	13.8	60.6	7.5	105.8
9	0	10.7	35.2	40.3	0.0
10	348	25.9	52.0	40.3	116.6

(a) Compute s^2, C_p, PRESS, and the sum of the absolute PRESS residuals for the model involving all four variables with response y.

(b) Compute s^2, C_p, PRESS, and the sum of the absolute PRESS residuals for the model x_1, x_2, x_4 with response y.

(c) Compare the appropriateness of the models in parts (a) and (b) for predicting fish biomass.

2. In a chemical engineering experiment dealing with heat transfer in a shallow fluidized bed, data are collected on the following four regressor variables: fluidizing gas flow rate, lb/hr (x_1); supernatant gas flow rate, lb/hr (x_2); supernatant gas inlet nozzle opening, millimeters (x_3); supernatant gas inlet temperature, °F (x_4). The responses measured are: heat transfer efficiency (y_1); thermal efficiency (y_2). The data are as follows:

Observation	y_1	y_2	x_1	x_2	x_3	x_4
1	41.852	38.75	69.69	170.83	45	219.74
2	155.329	51.87	113.46	230.06	25	181.22
3	99.628	53.79	113.54	228.19	65	179.06
4	49.409	53.84	118.75	117.73	65	281.30
5	72.958	49.17	119.72	117.69	25	282.20
6	107.702	47.61	168.38	173.46	45	216.14
7	97.239	64.19	169.85	169.85	45	223.88
8	105.856	52.73	169.85	170.86	45	222.80
9	99.348	51.00	170.89	173.92	80	218.84
10	111.907	47.37	171.31	173.34	25	218.12
11	100.008	43.18	171.43	171.43	45	219.20
12	175.380	71.23	171.59	263.49	45	168.62
13	117.800	49.30	171.63	171.63	45	217.58
14	217.409	50.87	171.93	170.91	10	219.92
15	41.725	54.44	173.92	71.73	45	296.60
16	151.139	47.93	221.44	217.39	65	189.14
17	220.630	42.91	222.74	221.73	25	186.08
18	131.666	66.60	228.90	114.40	25	285.80
19	80.537	64.94	231.19	113.52	65	286.34
20	152.966	43.18	236.84	167.77	45	221.72

Consider the model for predicting the heat transfer coefficient response

$$y_{1i} = \beta_0 + \sum_{j=1}^{4} \beta_j x_{ji} + \sum_{j=1}^{4} \beta_{jj} x_{ji}^2$$

$$+ \sum\sum \beta_{jl} x_{ji} x_{li} + \varepsilon_i \qquad (i = 1, 2, \ldots, 20).$$

(a) Compute PRESS and $\sum_{i=1}^{n} |y_i - \hat{y}_{i, -i}|$ for the least squares regression fit to the model above.

(b) Fit a second-order model with x_4 completely eliminated (i.e., deleting all terms involving x_4). Compute the prediction criteria for the reduced model. Comment on the appropriateness of x_4 for prediction of the heat transfer coefficient.

(c) Repeat parts (a) and (b) for thermal efficiency.

3. In exercise physiology, an objective measure of aerobic fitness is the oxygen consumption in volume per unit body weight per unit time. Thirty-one individuals were used in an experiment in order to be able to model oxygen consumption against: age in years (x_1); weight in kilograms (x_2); time to run 1½ miles (x_3); resting pulse rate (x_4); pulse rate at end of run (x_5); maximum pulse rate during run (x_6). Use the following data and do a stepwise regression with significance level 0.25. Quote the final model.

Individual	y	x_1	x_2	x_3	x_4	x_5	x_6
1	44.609	44	89.47	11.37	62	178	182
2	45.313	40	75.07	10.07	62	185	185
3	54.297	44	85.84	8.65	45	156	168
4	59.571	42	68.15	8.17	40	166	172
5	49.874	38	89.02	9.22	55	178	180
6	44.811	47	77.45	11.63	58	176	176
7	45.681	40	75.98	11.95	70	176	180
8	49.091	43	81.19	10.85	64	162	170
9	39.442	44	81.42	13.08	63	174	176
10	60.055	38	81.87	8.63	48	170	186
11	50.541	44	73.03	10.13	45	168	168
12	37.388	45	87.66	14.03	56	186	192
13	44.754	45	66.45	11.12	51	176	176
14	47.273	47	79.15	10.60	47	162	164
15	51.855	54	83.12	10.33	50	166	170
16	49.156	49	81.42	8.95	44	180	185
17	40.836	51	69.63	10.95	57	168	172
18	46.672	51	77.91	10.00	48	162	168
19	46.774	48	91.63	10.25	48	162	164
20	50.388	49	73.37	10.08	76	168	168
21	39.407	57	73.37	12.63	58	174	176
22	46.080	54	79.38	11.17	62	156	165
23	45.441	52	76.32	9.63	48	164	166
24	54.625	50	70.87	8.92	48	146	155
25	45.118	51	67.25	11.08	48	172	172
26	39.203	54	91.63	12.88	44	168	172
27	45.790	51	73.71	10.47	59	186	188
28	50.545	57	59.08	9.93	49	148	155
29	48.673	49	76.32	9.40	56	186	188
30	47.920	48	61.24	11.50	52	170	176
31	47.467	52	82.78	10.50	53	170	172

4. Show that in a multiple linear regression data set,

$$\sum_{i=1}^{n} h_{ii} = p.$$

5. A small experiment is conducted to fit a multiple regression equation relating the yield y to temperature x_1, reaction time x_2, and concentration of one of the reactants x_3. Two levels of each variable were chosen and measurements corresponding to the coded independent variables were recorded as follows:

y	x_1	x_2	x_3
7.6	−1	−1	−1
5.5	1	−1	−1
9.2	−1	1	−1
10.3	−1	−1	1
11.6	1	1	−1
11.1	1	−1	1
10.2	−1	1	1
14.0	1	1	1

(a) Using the coded variables, estimate the multiple linear regression equation

$$\mu_{Y|x_1, x_2, x_3} = \beta_0 + \beta_1 x_1 + \beta_2 x_2 + \beta_3 x_3.$$

(b) Partition SSR, the regression sum of squares, into three single-degree-of-freedom components attributable to x_1, x_2, and x_3, respectively. Show an analysis-of-variance table, indicating significance tests on each variable. Comment on the results.

13

One-Factor Experiments: General

13.1 Analysis-of-Variance Technique

In the estimation and hypotheses testing material covered in Chapters 9 and 10, we were restricted in each case to considering no more than two population parameters. Such was the case, for example, in testing for the equality of two population means using independent samples from normal populations with common but unknown variance, where it was necessary to obtain a pooled estimate of σ^2.

This material dealing in two-sample inference represents a special case of what we call the *one-factor problem*. For example, in Exercise 17 following Section 10.10, the survival time is measured for two samples of mice where one sample received a new serum for leukemia treatment and the other sample received no treatment. In this case we say that there is *one factor*, namely *treatment*, and the factor is at *two levels*. If several competing treatments were being used in the sampling process, more samples of mice would be necessary. In this case the problem would involve one factor with more than two levels and thus more than two samples.

In the $k > 2$ sample problem, it will be assumed that there are k samples from k populations. One very common procedure used to deal with testing population means is called the *analysis of variance*.

The analysis of variance is certainly not a new technique if the reader has followed the material on regression theory. We used the analysis-of-variance approach to partition the total sum of squares into a portion due to regression and a portion due to error. Further, we were able, in some cases, to conveniently partition *SSR* into meaningful components for the purpose of testing relevant hypotheses on the parameters in the model. The term *analysis of variance* describes a technique whereby the total variation is being analyzed or divided into meaningful components.

Regression problems in which the model contains *quantitative variables* (like those we have discussed to this point) are not the only type in which the analysis of variance plays an important role. In this chapter we present and study other types of models in which this technique is used. The degree of difficulty of the analysis depends on the complexity of the problem.

Suppose in an industrial experiment that an engineer is interested in how the mean absorption of moisture in concrete varies among five different concrete aggregates. The samples are exposed to moisture for 48 hours. It is decided that 6 samples are to be tested for each aggregate, requiring a total of 30 samples to be tested. The data are recorded in Table 13.1.

The model for this situation may be considered as follows. There are 6 observations taken from each of 5 populations with means $\mu_1, \mu_2, \ldots, \mu_5$, respectively. We might wish to test

$$H_0: \quad \mu_1 = \mu_2 = \cdots = \mu_5,$$

$$H_1: \quad \text{At least two of the means are not equal.}$$

In addition, we might be interested in making individual comparisons among these 5 population means.

TABLE 13.1 Absorption of Moisture in Concrete Aggregates

	\multicolumn{5}{c}{Aggregate}				
	1	2	3	4	5	
	551	595	639	417	563	
	457	580	615	449	631	
	450	508	511	517	522	
	731	583	573	438	613	
	499	633	648	415	656	
	632	517	677	555	679	
Total	3320	3416	3663	2791	3664	16,854
Mean	553.33	569.33	610.50	465.17	610.67	561.80

In the analysis-of-variance procedure, it is assumed that whatever variation exists between the aggregate averages is attributed to (1) variation in absorption among observations within aggregate types, and (2) variation due to aggregate types, that is, due to differences in the chemical composition of the aggregates. The **within-aggregate variation** is, of course, brought about by various causes. Perhaps humidity and temperature conditions were not kept entirely constant throughout the experiment. It is possible that there was a certain amount of heterogeneity in the batches of raw materials that were used. At any rate, we shall consider the within-sample variation to be **chance** or **random variation**, and part of the goal of the analysis of variance is to determine if the differences between the 5 sample means are what one would expect due to random variation alone or if indeed there is also a contribution from the systematic variation attributed to the aggregate types. The procedure essentially, then, separates the total variability into the following important two components:

1. Variability between aggregates, measuring systematic and random variation.
2. Variability within aggregates, measuring only random variation.

There remains then the task of determining if component 1 is significantly larger than component 2.

Many pointed questions appear at this stage concerning the preceding problem. For example, how many samples must be tested in each aggregate? This is a question that continually haunts the practitioner. In addition, what if the within-sample variation is so large that it is difficult for a statistical procedure to detect the systematic differences? Can we systematically control extraneous sources of variation and thus remove them from the portion we call random variation? We shall attempt to answer these and other questions in the following sections.

13.2 The Strategy of Experimental Design

In Chapters 9 and 10 the notion of estimation and testing in the two-sample case was covered under the important backdrop of the way the experiment was conducted. This falls into the broad category of design of experiments. For example, for the test that was called the pooled t-test discussed in Chapter 10, it is assumed that the factor levels (treatments in the mice exercise) were assigned randomly to the experimental units (mice). The notion of experimental units is discussed in Chapters 9 and 10 and is illustrated through examples. Simply put, experimental units are the units (mice, patients, concrete specimens, time) that provide the heterogeneity that produces experimental error in a scientific investigation. The random assignment eliminates bias that could result by systematic assignment. The goal is to distribute uniformly among the factor levels the risks brought about by the heterogeneity of the experimental units. A random assignment best simulates the conditions that are assumed by the model. In Section 13.8 we discuss *blocking* in experiments. The notion of blocking was presented in Chapters 9 and 10, when comparisons between means was accomplished with *pairing*, that is, the division of the experimental units into homogeneous pairs called blocks. The factor levels or treatments are then assigned randomly within blocks. The purpose of blocking is to reduce the effective experimental error. In this chapter we naturally extend the pairing to *larger block sizes*, with analysis of variance being the primary analytical tool.

13.3 One-Way Analysis of Variance: Completely Randomized Design

Random samples of size n are selected from each of k populations. The k different populations are classified on the basis of a single criterion such as different treatments or groups. Today the term **treatment** is used very generally to refer to the various classifications, whether they be different aggregates, different analysts, different fertilizers, or different regions of the country. It will be assumed that the k populations are independent and normally distributed with means $\mu_1, \mu_2, \ldots, \mu_k$ and common variance σ^2.

As we indicated in Section 13.2, these assumptions are made more palatable by randomization. We wish to derive appropriate methods for testing the hypothesis

$$H_0: \quad \mu_1 = \mu_2 = \cdots = \mu_k,$$

$$H_1: \quad \text{At least two of the means are not equal.}$$

Let y_{ij} denote the jth observation from the ith treatment and arrange the data as in Table 13.2. Here, T_i. is the total of all observations in the sample from the ith treatment, $\bar{y}_i$. is the mean of all observations in the sample from the ith treatment, $T_{..}$ is the total of all nk observations, and $\bar{y}_{..}$ is the mean of all nk observations. Each observation may be written in the form

$$y_{ij} = \mu_i + \varepsilon_{ij},$$

where ε_{ij} measures the deviation of the jth observation of the ith sample from the corresponding treatment mean. The ε_{ij} term represents random error and plays the

TABLE 13.2 k Random Samples

	Treatment						
	1	2	$\cdots$	i	$\cdots$	k	
	y_{11}	y_{11}	$\cdots$	y_{i1}	$\cdots$	y_{k1}	
	y_{12}	y_{22}	$\cdots$	y_{i2}	$\cdots$	y_{k2}	
	$\vdots$	$\vdots$		$\vdots$		$\vdots$	
	y_{1n}	y_{2n}	$\cdots$	y_{in}	$\cdots$	y_{kn}	
Total	$T_1.$	$T_2.$	$\cdots$	$T_i.$	$\cdots$	$T_k.$	$T_{..}$
Mean	$\bar{y}_1.$	$\bar{y}_2.$	$\cdots$	$\bar{y}_i.$	$\cdots$	$\bar{y}_k.$	$\bar{y}_{..}$

same role as the error terms in the regression models. An alternative and preferred form of this equation is obtained by substituting $\mu_i = \mu + \alpha_i$, subject to the constraint $\sum_{i=1}^{k} \alpha_i = 0$. Hence we may write

$$y_{ij} = \mu + \alpha_i + \varepsilon_{ij},$$

where μ is just the **grand mean** of all the μ_i's; that is,

$$\mu = \frac{\sum_{i=1}^{k} \mu_i}{k},$$

and α_i is called the **effect** of the ith treatment.

The null hypothesis that the k population means are equal against the alternative that at least two of the means are unequal may now be replaced by the equivalent hypothesis,

$$H_0: \quad \alpha_1 = \alpha_2 = \cdots = \alpha_k = 0,$$

$$H_1: \quad \text{At least one of the } \alpha_i\text{'s is not equal to zero.}$$

Our test will be based on a comparison of two independent estimates of the common population variance σ^2. These estimates will be obtained by partitioning the total variability of our data, designated by the double summation $\sum_{i=1}^{k} \sum_{j=1}^{n} (y_{ij} - \bar{y}_{..})^2$, into two components.

THEOREM 13.1
SUM-OF-SQUARES
IDENTITY

$$\sum_{i=1}^{k} \sum_{j=1}^{n} (y_{ij} - \bar{y}_{..})^2 = n \sum_{i=1}^{k} (\bar{y}_{i.} - \bar{y}_{..})^2 + \sum_{i=1}^{k} \sum_{j=1}^{n} (y_{ij} - \bar{y}_{i.})^2. \qquad \blacksquare$$

PROOF

$$\sum_{i=1}^{k} \sum_{j=1}^{n} (y_{ij} - \bar{y}_{..})^2 = \sum_{i=1}^{k} \sum_{j=1}^{n} [(\bar{y}_{i.}) - \bar{y}_{..}) + (y_{ij} - \bar{y}_{i.})]^2$$

$$= \sum_{i=1}^{k} \sum_{j=1}^{n} [(\bar{y}_{i.} - \bar{y}_{..})^2 + 2(\bar{y}_{i.} - \bar{y}_{..})(y_{ij} - \bar{y}_{i.})$$

$$+ (y_{ij} - \bar{y}_{i.})^2]$$

$$= \sum_{i=1}^{k} \sum_{j=1}^{n} (\bar{y}_{i.} - \bar{y}_{..})^2$$

$$+ 2 \sum_{i=1}^{k} \sum_{j=1}^{n} (\bar{y}_{i.} - \bar{y}_{..})(y_{ij} - \bar{y}_{i.})$$

$$+ \sum_{i=1}^{k} \sum_{j=1}^{n} (y_{ij} - \bar{y}_{i.})^2.$$

The middle term is zero, since

$$\sum_{j=1}^{n} (y_{ij} - \bar{y}_{i\cdot}) = \sum_{j=1}^{n} y_{ij} - n\bar{y}_{i\cdot} = \sum_{j=1}^{n} y_{ij} - n\frac{\sum_{j=1}^{n} y_{ij}}{n} = 0.$$

The first sum does not have a j as a subscript and therefore may be written as

$$\sum_{i=1}^{k} \sum_{j=1}^{n} (\bar{y}_{i\cdot} - \bar{y}_{\cdot\cdot})^2 = n \sum_{i=1}^{k} (\bar{y}_{i\cdot} - \bar{y}_{\cdot\cdot})^2.$$

Hence

$$\sum_{i=1}^{k} \sum_{j=1}^{n} (y_{ij} - \bar{y}_{\cdot\cdot})^2 = n \sum_{i=1}^{k} (\bar{y}_{i\cdot} - \bar{y}_{\cdot\cdot})^2 + \sum_{i=1}^{k} \sum_{j=1}^{n} (y_{ij} - \bar{y}_{i\cdot})^2.$$

It will be convenient in what follows to identify the terms of the sum-of-squares identify by the following notation:

$$SST = \sum_{i=1}^{k} \sum_{j=1}^{n} (y_{ij} - \bar{y}_{\cdot\cdot})^2 = \text{total sum of squares}$$

$$SSA = n \sum_{i=1}^{k} (\bar{y}_{i\cdot} - \bar{y}_{\cdot\cdot})^2 = \text{treatment sum of squares}$$

$$SSE = \sum_{i=1}^{k} \sum_{j=1}^{n} (y_{ij} - \bar{y}_{i\cdot})^2 = \text{error sum of squares.}$$

The sum-of-squares identity can then be represented symbolically by the equation

$$SST = SSA + SSE.$$

The identity above expresses how between-treatment and within-treatment variation add to the total sum of squares. However, much insight can be gained by investigating the expected value of both SSA and SSE. Eventually, we shall develop variance estimates that formulate the ratio to be used to test the equality of population means.

As implied earlier, we need to compare the appropriate measure of the between-treatment variation with the within-treatment variation in order to detect significant differences in the observations due to the treatment effects. Suppose that we look at the expected value of the treatment sum of squares.

THEOREM 13.2

$$E(SSA) = (k - 1)\sigma^2 + n \sum_{i=1}^{k} \alpha_i^2$$

The proof of the theorem is left as an exercise (see Exercise 2 on page 475).

If H_0 is true, an estimate of σ^2, based on $k - 1$ degrees of freedom, is given by the expression

TREATMENT
MEAN SQUARE

$$s_1^2 = \frac{SSA}{k - 1}.$$

If H_0 is true and thus each α_i in Theorem 13.2 is equal to zero, we see that

$$E\left(\frac{SSA}{k - 1}\right) = \sigma^2$$

and s_1^2 is an unbiased estimate of σ^2. However, if H_1 is true, we have

$$E\left(\frac{SSA}{k - 1}\right) = \sigma^2 + \frac{n \sum_{i=1}^{k} \alpha_i^2}{k - 1}$$

and s_1^2 estimates σ^2 plus an additional term, which measures variation due to the systematic effects.

A second and independent estimate of σ^2, based on $k(n - 1)$ degrees of freedom, is the familiar formula

ERROR MEAN
SQUARE

$$s^2 = \frac{SSE}{k(n - 1)}.$$

The estimate s^2 is unbiased regardless of the truth or falsity of the null hypothesis (see Exercise 1 on page 475). It is important to note that the sum-of-squares identity has not only partitioned the total variability of the data, but also the total number of degrees of freedom. That is,

$$nk - 1 = k - 1 + k(n - 1).$$

When H_0 is true, the ratio

$$f = \frac{s_1^2}{s^2}$$

is a value of the random variable F having the F-distribution with $k - 1$ and $k(n - 1)$ degrees of freedom. Since s_1^2 overestimates σ^2 when H_0 is false, we have a one-tailed test with the critical region entirely in the right tail of the distribution. The null hypothesis H_0 is rejected at the α-level of significance when

$$f > f_\alpha[k - 1, k(n - 1)].$$

Another approach, the P-value approach, suggests that the evidence in favor of or against H_0 is given by

$$P = P[F[k - 1, k(n - 1)] > f].$$

The computations in an analysis-of-variance problem are usually summarized in tabular form as shown in Table 13.3.

EXAMPLE 13.1   Test the hypothesis $\mu_1 = \mu_2 = \cdots = \mu_5$ at the 0.05 level of significance for the data of Table 13.1 on absorption of moisture by various types of cement aggregates.

SOLUTION

$$H_0:\quad \mu_1 = \mu_2 = \cdots = \mu_5.$$

H_1:   At least two of the means are not equal.

$\alpha = 0.05.$

Critical region: $f > 2.76$ with $v_1 = 4$ and $v_2 = 25$ degrees of freedom.
Computations:

$$SST = (557 - 561.8)^2 + (457 - 561.8)^2 + \cdots + (679 - 561.8)^2 = 209{,}377,$$

$$SSA = 6[(553.33 - 561.8)^2 + (569.33 - 561.8)^2 + \cdots + (610.67 - 561.8)^2]$$
$$= 85{,}356,$$

$$SSE = 209{,}377 - 85{,}356 = 124{,}021.$$

These results and the remaining computations are exhibited in Table 13.4. Decision: Reject H_0 and conclude that the aggregates do not have the same mean absorption. The P-value for $f = 4.30$ is smaller than 0.01.

In experimental work one often loses some of the desired observations. Experimental animals die, experimental material may be damaged, and human subjects

TABLE 13.3   Analysis of Variance for the One-Way Classification

Source of Variation	Sum of Squares	Degrees of Freedom	Mean Square	Computed f
Treatments	SSA	$k - 1$	$s_1^2 = \dfrac{SSA}{k-1}$	$\dfrac{s_1^2}{s^2}$
Error	SSE	$k(n - 1)$	$s^2 = \dfrac{SSE}{k(n-1)}$	
Total	SST	$nk - 1$		

TABLE 13.4   Analysis of Variance for the Data of Table 13.1

Source of Variation	Sum of Squares	Degrees of Freedom	Mean Square	Computed f
Aggregates	85,356	4	21,339	4.30
Error	124,021	25	4,961	
Total	209,377	29		

may drop out of a study. The previous analysis for equal sample size will still be valid by slightly modifying the sum of squares formulas. We now assume the k random samples to be of size $n_1, n_2, \ldots, n_k$, respectively.

SUM OF SQUARES; UNEQUAL SAMPLE SIZES

$$SST = \sum_{i=1}^{k} \sum_{j=1}^{n_i} (y_{ij} - \bar{y}_{i\cdot})^2$$

$$SSA = \sum_{i=1}^{n} n_i(\bar{y}_{i\cdot} - \bar{y}_{\cdot\cdot})^2$$

$$SSE = SST - SSA. \qquad \blacksquare$$

The degrees of freedom are then partitioned as before: $N - 1$ for SST, $k - 1$ for SSA, and $N - 1 - (k - 1) = N - k$ for SSE.

EXAMPLE 13.2   Part of the study "*Serum Inorganic Phosphorus Levels in Children with Seizure Disorders Taking Anticonvulsant Drugs,*" conducted at the Virginia Polytechnic Institute and State University in 1982, was designed to measure serum alkaline phosphatase activity levels (Bessey–Lowry Units) in children with seizure disorders who were receiving anticonvulsant therapy under the care of a private physician. Forty-five subjects were found for the study and categorized into four drug groups:

G-1:   control (not receiving anticonvulsants and having no history of seizure disorders),

G-2:   phenobarbital,

G-3:   carbamazepine,

G-4:   other anticonvulsants.

From blood samples collected on each subject the serum alkaline phosphatase activity level was determined and recorded in Table 13.5. Test the hypothesis at the 0.05 level of significance that the average serum alkaline phosphatase activity level is the same for the four drug groups.

SOLUTION

H_0: $\mu_1 = \mu_2 = \mu_3 = \mu_4$.

H_1:   At least two of the means are not equal.

$\alpha = 0.05$.

Critical region: $f > 2.836$, by interpolating in Table A.6.

Computations: $T_{1\cdot} = 1460.25$, $T_{2\cdot} = 440.36$, $T_{3\cdot} = 842.45$, $T_{4\cdot} = 707.41$, and $T_{\cdot\cdot} = 3450.47$. The analysis of variance table is given in Table 13.6.

Decision: Reject H_0 and conclude that the average serum alkaline phosphatase activity levels for the four drug groups are not all the same. The P-value is 0.02.

In concluding our discussion on the analysis of variance for the one-way classification, we state the advantages of choosing equal sample sizes over the choice of

TABLE 13.5   Serum Alkaline Phosphatase Activity Level

		Drug Group		
G-1		G-2	G-3	G-4
49.20	97.50	97.07	62.10	110.60
44.54	105.00	73.40	94.95	57.10
45.80	58.05	68.50	142.50	117.60
95.84	86.60	91.85	53.00	77.71
30.10	58.35	106.60	175.00	150.00
36.50	72.80	0.57	79.50	82.90
82.30	116.70	0.79	29.50	111.50
87.85	45.15	0.77	78.40	
105.00	70.35	0.81	127.50	
95.22	77.40			

TABLE 13.6   Analysis of Variance for the Data of Table 13.5

Source of Variation	Sum of Squares	Degrees of Freedom	Mean Square	Computed f
Drug groups	13,938.6013	3	4646.2004	3.57
Error	53,376.2972	41	1301.8609	
Total	67,314.8985	44		

unequal sample sizes. The first advantage is that the f-ratio is insensitive to slight departures from the assumption of equal variances for the k populations when the samples are of equal size. Second, the choice of equal sample size minimizes the probability of committing a type II error.

13.4   Tests for the Equality of Several Variances _____

Although the f-ratio obtained from the analysis-of-variance procedure is insensitive to departures from the assumption of equal variances for the k normal populations when the samples are of equal size, we may still prefer to exercise caution and run a preliminary test for homogeneity of variances. Such a test would certainly be advisable in the case of unequal sample sizes if there is a reasonable doubt concerning the homogeneity of the populations variances. Suppose, therefore, that we wish to test the null hypothesis

$$H_0:   \sigma_1^2 = \sigma_2^2 = \cdots = \sigma_k^2$$

against the alternative

$$H_1:   \text{The variances are not all equal.}$$

The test that we shall use, called **Bartlett's test**, is based on a statistic whose sampling distribution provides exact critical values when the sample sizes are equal. These critical values for equal sample sizes can also be used to yield highly accurate approximations to the critical values for unequal sample sizes.

First, we compute the k sample variances $s_1^2, s_2^2, \ldots, s_k^2$ from samples of size n_1, $n_2, \ldots, n_k$, with $\sum_{i=1}^{k} n_i = N$. Second, combine the sample variances to give the pooled estimate

$$s_p^2 = \frac{\sum_{i=1}^{k} (n_i - 1)s_i^2}{N - k}.$$

Now

$$b = \frac{[(s_1^2)^{n_1-1}(s_2^2)^{n_2-1} \cdots (s_k^2)^{n_k-1}]^{1/(N-k)}}{s_p^2}$$

is a value of a random variable B having the **Bartlett distribution**. For the special case when $n_1 = n_2 = \cdots = n_k = n$, we reject H_0 at the α-level of significance if

$$b < b_k(\alpha; n),$$

where $b_k(\alpha; n)$ is the critical value leaving an area of size α in the left tail of the Bartlett distribution. Table A.10 gives the critical values, $b_k(\alpha; n)$, for $\alpha = 0.01$ and 0.05; $k = 2, 3, \ldots, 10$; and selected values of n from 3 to 100.

When the sample sizes are unequal, the null hypothesis is rejected at the α-level of significance if

$$b < b_k(\alpha; n_1, n_2, \ldots, n_k),$$

where

$$b_k(\alpha; n_1, n_2, \ldots, n_k) \simeq \frac{n_1 b_k(\alpha; n_1) + n_2 b_k(\alpha; n_2) + \cdots + n_k b_k(\alpha; n_k)}{N}.$$

As before, all the $b(\alpha; n_i)$ for sample sizes $n_1, n_2, \ldots, n_k$ are obtained from Table A.10.

EXAMPLE 13.3   Use Bartlett's test to test the hypothesis at the 0.01 level of significance that the population variances of the four drug groups in Example 13.2 are equal.

SOLUTION

$$H_0: \quad \sigma_1^2 = \sigma_2^2 = \sigma_3^2 = \sigma_4^2.$$

H_1:   The variances are not all equal.

$$\alpha = 0.01.$$

Critical region: Referring to Example 13.2, we have $n_1 = 20$, $n_2 = 9$, $n_3 = 9$, $n_4 = 7$, $N = 45$, and $k = 4$. Therefore, we reject when

$$b < b_4(0.01, 20, 9, 9, 7)$$

$$\simeq \frac{(20)(0.8586) + (9)(0.6892) + (9)(0.6892) + (7)(0.6045)}{45}$$

$$= 0.7513.$$

Computations: First compute

$$s_1^2 = 662.862, \qquad s_2^2 = 2219.781, \qquad s_3^2 = 2168.434, \qquad s_4^2 = 946.032,$$

and then

$$s_p^2 = \frac{(19)(662.862) + (8)(2219.781) + (8)(2168.434) + (6)(946.032)}{41}$$

$$= 1301.861.$$

Now

$$b = \frac{[(662.862)^{19}(2219.781)^8(2168.434)^8(946.032)^6]^{1/4}}{1301.861}$$

$$= 0.8557.$$

Decision: Do not reject the hypothesis and conclude that the population variances of the four drug groups are not significantly different.

Although Bartlett's test is most often used in testing for homogeneity of variances, other methods are available. A method due to Cochran provides a computationally simple procedure, but it is restricted to situations in which the sample sizes are equal. **Cochran's test** is particularly useful in detecting if one variance is much larger than the others. The statistic that is used is given by

$$G = \frac{\text{largest } S_i^2}{\sum\limits_{i=1}^{k} S_i^2}$$

and the hypothesis of equality of variances is rejected if $g > g_\alpha$, where the value of g_α is obtained from Table A.11.

To illustrate Cochran's test, let us refer again to the data of Table 13.1 on the absorption of moisture in concrete aggregates. Were we justified in assuming equal variances when we performed the analysis of variance in Example 13.1? We find that

$$s_1^2 = 12,134, \qquad s_2^2 = 2303, \qquad s_3^2 = 3594, \qquad s_4^2 = 3319, \qquad s_5^2 = 3455.$$

Therefore,

$$g = \frac{12,134}{24,805} = 0.4892,$$

which does not exceed the tabled value $g_{0.05} = 0.5065$. Hence we conclude that the assumption of equal variances is reasonable.

Exercises

1. Show that the error mean square

$$s^2 = \frac{SSE}{k(n-1)}$$

for the analysis of variance in a one-way classification is an unbiased estimate of σ^2.

2. Prove Theorem 13.2.

3. Six different machines are being considered for use in manufacturing rubber seals. The machines are being compared with respect to tensile strength of the product. A random sample of 4 seals from each machine is used to determine whether or not the mean tensile strength varies from machine to machine. The following are the tensile-strength measurements in kilograms per square centimeter $\times\ 10^{-1}$:

		Machine			
1	2	3	4	5	6
17.5	16.4	20.3	14.6	17.5	18.3
16.9	19.2	15.7	16.7	19.2	16.2
15.8	17.7	17.8	20.8	16.5	17.5
18.6	15.4	18.9	18.9	20.5	20.1

Perform the analysis of variance at the 0.05 level of significance and indicate whether or not the mean tensile strengths differ significantly for the 6 machines.

4. The data in the following table represent the number of hours of relief provided by 5 different brands of headache tablets administered to 25 subjects experiencing fevers of 38°C or more. Perform the analysis of variance and test the hypothesis at the 0.05 level of significance that the mean number of hours of relief provided by the tablets is the same for all 5 brands.

		Tablet		
A	B	C	D	E
5	9	3	2	7
4	7	5	3	6
8	8	2	4	9
6	6	3	1	4
3	9	7	4	7

5. In the article "*Shelf-Space Strategy in Retailing,*" published in the *Proceedings: Southern Marketing Association* (1975), the effect of shelf height on the supermarket sales of canned dog food is investigated. An experiment was conducted at a small supermarket for a period of 8 days on the sales of a single brand of dog food, referred to as Arf dog food, involving three levels of shelf height: knee level, waist level, and eye level. During each day the shelf height of the canned dog food was randomly changed on three different occasions. The remaining sections of the gondola that housed the given brand were filled with a mixture of dog food brands that were both familiar and unfamiliar to customers in this particular geographic area. Sales, in hundreds of dollars, of Arf dog food per day for the three shelf heights are as follows:

	Shelf Height	
Knee Level	Waist Level	Eye Level
77	88	85
82	94	85
86	93	87
78	90	81
81	91	80
86	94	79
77	90	87
81	87	93

Is there a significant difference in the average daily sales of this dog food based on shelf height? Use a 0.01 level of significance.

6. Immobilization of free-ranging white-tailed deer by drugs allows researchers the opportunity to closely examine deer and gather valuable physiological information. In the study "*Influence of Physical Restraint and Restraint-Facilitating Drugs on Blood Measurements of White-Tailed Deer and Other Selected Mammals*" conducted at the Virginia Polytechnic Institute and State University in 1976, wildlife biologists tested the "knockdown" time (time from injection to immobilization) of three different immobilizing drugs. Immobilization, in this case, is defined as the point where the animal no longer has enough muscle control to remain standing. Thirty male white-tailed deer were randomly assigned to each of three treatments. Group A received 5 milligrams of liquid succinylcholine chloride (SCC); group B received 8 milligrams of powdered SCC; and group C received 200 milligrams of phencyclidine hydrochloride. Knockdown times, in minutes, were recorded as follows:

	Group	
A	*B*	*C*
11	10	4
5	7	4
14	16	6
7	7	3
10	7	5
7	5	6
23	10	8
4	10	3
11	6	7
11	12	3

Perform an analysis of variance at the 0.01 level of significance and determine whether or not the average knockdown time for the 3 drugs is the same.

7. It has been shown that the fertilizer magnesium ammonium phosphate, $MgNH_4Po_4$, is an effective supplier of the nutrients necessary for plant growth. The compounds supplied by this fertilizer are highly soluble in water, allowing the fertilizer to be applied directly on the soil surface or mixed with the growth substrate during the potting process. A study on the "*Effect of Magnesium Ammonium Phosphate on Height of Chrysanthemums*" was conducted at George Mason University in 1980 to determine a possible optimum level of fertilization, based on the enhanced vertical growth response of the chrysanthemums. Forty chrysanthemum seedlings were divided into 4 groups each containing 10 plants. Each was planted in a similar pot containing a uniform growth medium. To each group of plants an increasing concentration of $MgNH_4Po_4$, measured in grams per bushel, was added. The 4 groups of plants were grown under uniform conditions in a greenhouse for a period of four weeks. The treatments and the respective changes in heights, measured in centimeters, are shown in the following table:

	Treatment		
50 g/bu	100 g/bu	200 g/bu	400 g/bu
13.2	16.0	7.8	21.0
12.4	12.6	14.4	14.8
12.8	14.8	20.0	19.1
17.2	13.0	15.8	15.8
13.0	14.0	17.0	18.0
14.0	23.6	27.0	26.0
14.2	14.0	19.6	21.1
21.6	17.0	18.0	22.0
15.0	22.2	20.2	25.0
20.0	24.4	23.2	18.2

Can we conclude at the 0.05 level of significance that different concentrations of $MgNH_4Po_4$ affect the average attained height of chrysanthemums?

8. Three sections of the same elementary mathematics course are taught by three teachers. The final grades were recorded as follows:

	Teacher	
A	*B*	*C*
73	88	68
89	78	79
82	48	56
43	91	91
80	51	71
73	85	71
66	74	87
60	77	41
45	31	59
93	78	68
36	62	53
77	76	79
	96	15
	80	
	56	

Is there a significant difference in the average grades given by the three teachers? Use a 0.05 level of significance.

9. The mitochondrial enzyme NAPH:NAD transhydrogenase of the common rat tapeworm (*Hymenolegsis diminuta*) catalyzes hydrogen in transfer from NADPH to NAD, producing NADH. This enzyme is known to serve a vital role in the tapeworm's anaerobic metabolism, and it has recently been hypothesized that it may serve as a proton exchange pump, transferring protons across the mitochondrial membrane. A study on "*Effect of Various Substrate Concentrations on the Conformational Variation of the NADPH:NAD Transhydrogenase of Hymenolepsis Diminuta*" conducted in 1983 at Bowling Green State University was designed to assess the ability of this enzyme to undergo conformation or shape changes. Changes in the specific activity of the enzyme caused by variations in the concentration of NADP could be interpreted as supporting the theory of conformational change. The enzyme in question is located in the inner membrane of the tapeworm's mitochondria. These tapeworms were homogenized, and through a series of centrifugations, the enzyme was isolated. Various concentrations of NADP were then added to the isolated enzyme solution, and the mixture was then incubated in a water bath at 56°C

for 3 minutes. The enzyme was then analyzed on dual-beam spectrophotometer, and the following results were calculated in terms of the specific activity of the enzyme in nanomoles per minute per milligram of protein:

NADP Concentration (nm)				
0	80	160	360	
11.01	11.38	11.02	6.04	10.31
12.09	10.67	10.67	8.65	8.30
10.55	12.33	11.50	7.76	9.48
11.26	10.08	10.31	10.13	8.89
			9.36	

Test the hypothesis at the 0.01 level that the average specific activity is the same for the four concentrations.

13.5 Single-Degree-of-Freedom Comparisons

The analysis of variance in a one-way classification or the one-factor experiment, as it is often called, merely indicates whether or not the hypothesis of equal treatment means can be rejected. Usually, an experimenter would prefer his or her analysis to probe deeper. For instance, in Example 13.1, by rejecting the null hypothesis we concluded that the means are not all equal, but we still do not know where the differences exist among the aggregates. The engineer might have the feeling *a priori* that aggregates 1 and 2 should have similar absorption properties and that the same is true for aggregates 3 and 5. However, it is of interest to study the difference between the two groups. It would seem, then, appropriate to test the hypothesis

$$H_0: \quad \mu_1 + \mu_2 - \mu_3 - \mu_5 = 0,$$

$$H_1: \quad \mu_1 + \mu_2 - \mu_3 - \mu_5 \neq 0.$$

We notice that the hypothesis is a linear function of the population means in which the coefficients sum to zero.

DEFINITION 13.1   *Any linear function of the form*

$$\omega = \sum_{i=1}^{k} c_i u_i, \quad \text{where} \sum_{i=1}^{k} c_i = 0,$$

*is called a **comparison** or **contrast** in the treatment means.*

The experimenter can often make multiple comparisons by testing the significance of contrasts in the treatment means, that is, by testing a hypothesis of the type

$$H_0: \sum_{i=1}^{k} c_i u_i = 0,$$

$$H_1: \sum_{i=1}^{k} c_i u_i \neq 0,$$

where $\sum_{i=1}^{k} c_i = 0$. The test is conducted by first computing a similar contrast in the sample means,

$$w = \sum_{i=1}^{k} c_i \bar{y}_{i\cdot}.$$

Since $\bar{Y}_{1\cdot}, \bar{Y}_{2\cdot}, \ldots, \bar{Y}_{k\cdot}$ are independent random variables having normal distributions with means $\mu_1, \mu_2, \ldots, \mu_k$ and variances $\sigma^2/n_1, \sigma^2/n_2, \ldots, \sigma^2/n_k$, respectively, Theorem 7.11 assures us that w is a value of the normal random variable W with mean

$$\mu_W = \sum_{i=1}^{k} c_i \mu_i$$

and variance

$$\sigma_W^2 = \sigma^2 \sum_{i=1}^{k} \frac{c_i^2}{n_i}.$$

Therefore, when H_0 is true, $\mu_W = 0$ and, by Example 7.5, the statistic

$$\frac{W^2}{\sigma_W^2} = \frac{\left(\sum_{i=1}^{k} c_i \bar{Y}_{i\cdot} \right)^2}{\sigma^2 \sum_{i=1}^{k} (c_i^2/n_i)}$$

is distributed as a chi-squared random variable with 1 degree of freedom. Our hypothesis is tested at the α-level of significance by computing

$$f = \frac{\left(\sum_{i=1}^{k} c_i \bar{y}_{i\cdot} \right)^2}{s^2 \sum_{i=1}^{k} (c_i^2/n_i)} = \frac{\left[\sum_{i=1}^{k} (c_i T_{i\cdot}/n_i) \right]^2}{s^2 \sum_{i=1}^{k} (c_i^2/n_i)} = \frac{SS_w}{s^2},$$

where f is a value of the random variable F having the F-distribution with 1 and $N - k$ degrees of freedom and

$$SSw = \frac{\left[\sum\limits_{i=1}^{k} (c_i T_{i.}/n_i)\right]^2}{\sum\limits_{i=1}^{k} (c_i^2/n_i)}.$$

When the sample sizes are all equal to n,

$$SSw = \frac{\left(\sum\limits_{i=1}^{k} c_i T_{i.}\right)^2}{n \sum\limits_{i=1}^{k} c_i^2}.$$

The quantity SSw, called the **contrast sum of squares**, indicates the portion of SSA that is explained by the contrast in question.

This sum of squares will be used to test the hypothesis that the contrast $\sum\limits_{i=1}^{k} c_i \mu_i = 0$.

It is often of interest to test multiple contrasts, particularly contrasts that are linearly independent or orthogonal. As a result we need the following definition:

DEFINITION 13.2   *The two contrasts*

$$\omega_1 = \sum_{i=1}^{k} b_i \mu_i \quad and \quad \omega_2 = \sum_{i=1}^{k} c_i \mu_i$$

are said to be **orthogonal** *if* $\sum\limits_{i=1}^{k} b_i c_i / n_i = 0$ *or, when the n_i's are all equal to n, if*

$$\sum_{i=1}^{k} b_i c_i = 0. \qquad ■$$

If ω_1 and ω_2 are orthogonal, then the quantities SSw_1 and SSw_2 are components of SSA each with a single degree of freedom. The treatment sum of squares with $k - 1$ degrees of freedom can be partitioned into at most $k - 1$ independent single-degree-of-freedom contrast sum of squares satisfying the identity

$$SSA = SSw_1 + SSw_2 + \cdots + SSw_{k-1}$$

if the contrasts are orthogonal to each other.

EXAMPLE 13.4   Referring to Example 13.1, find the contrast sum of squares corresponding to the orthogonal contrasts

$$\omega_1 = \mu_1 + \mu_2 - \mu_3 - \mu_5$$
$$\omega_2 = \mu_1 + \mu_2 + \mu_3 + \mu_5 - 4\mu_4$$

and carry out appropriate tests of significance. In this case it is of interest *a priori* to compare the two groups (1, 2) and (3, 5). An important and independent contrast is the comparison between the set of aggregates (1, 2, 3, 5) and aggregate 4.

SOLUTION

It is obvious that the two contrasts are orthogonal, since $(1)(1) + (1)(1) + (-1)(1) + (0)(-4) + (-1)(1) = 0$. The second contrast indicates a comparison between aggregates (1, 2, 3, and 5) and aggregate 4. One can write down two additional contrasts orthogonal to the first two, namely

$$\omega_3 = \mu_1 - \mu_2 \quad \text{(aggregate 1 versus aggregate 2)}$$

$$\omega_4 = \mu_3 - \mu_5 \quad \text{(aggregate 3 versus aggregate 5).}$$

From the data of Table 13.1, we have

$$SSw_1 = \frac{(3320 + 3416 - 3663 - 3664)^2}{6[(1)^2 + (1)^2 + (-1)^2 + (-1)^2]} = 14{,}553$$

$$SSw_2 = \frac{[3320 + 3416 + 3663 + 3664 - 4(2791)]^2}{6[(1)^2 + (1)^2 + (1)^2 + (1)^2 + (-4)^2]} = 70{,}035.$$

A more extensive analysis-of-variance table is then given in Table 13.7. We note that the two contrast sum of squares account for nearly all the aggregate sum of squares. Although there is a significant difference between aggregates in their absorption properties, the contrast w_1 is marginally significant. The P-value is 0.08. However, the f-value of 14.12 for w_2 is quite significant and the hypothesis

$$H_0: \quad \mu_1 + \mu_2 + \mu_3 + \mu_5 = 4\mu_4$$

is rejected.

Orthogonal contrasts allow the practitioner to partition the treatment variation into independent components. There are several choices available in selecting the orthogonal contrasts except for the last one. Normally, the experimenter would have certain contrasts that are of interest to him. Such was the case in our example, where a priori considerations suggest that aggregates (1, 2) and (3, 5) constitute distinct groups with different absorption properties, a postulation that was not strongly

TABLE 13.7   Analysis of Variance Using Orthogonal Contrasts

Source of Variation	Sum of Squares	Degrees of Freedom	Mean Square	Computed f
Aggregates	85,356	4	21,339	4.30
(1, 2) vs. (3, 5)	⎰ 14,553	⎰ 1	⎰ 14,533	2.93
(1, 2, 3, 5) vs. 4	⎱ 70,035	⎱ 1	⎱ 70,035	14.12
Error	124,021	25	4,961	
Total	209,377	29		

supported by the significance test. However, the second comparison supports the conclusion that aggregate 4 seems to "stand out" from the rest. In this case the complete partitioning of SSA was not necessary, since two of the four possible independent comparisons accounted for a majority of the variation in treatments.

13.6 Multiple Comparisons

The analysis of variance is a powerful procedure for testing the homogeneity of a set of means. However, if we reject the null hypothesis and accept the stated alternative—that the means are not all equal—we still do not know which of the population means are equal and which are different.

In Section 13.5 we described the use of orthogonal contrasts to make comparisons among sets of factor levels or treatments. The notion of orthogonality allows the analyst to make tests involving *independent* contrasts. Thus the variation among treatments, SSA, can be partitioned into single-degree-of-freedom components, and then proportions of this variation can be attributed to specific contrasts. However, there are situations in which the use of contrasts is not an appropriate approach. Often it is of interest to make several (perhaps all possible) paired comparisons among the treatment. Actually, a paired comparison may be viewed as a *simple* contrast, namely a test of

$$H_0: \quad \mu_i - \mu_j = 0,$$
$$H_1: \quad \mu_i - \mu_j \neq 0$$

for all $i \neq j$. All possible paired comparisons among the means can be very beneficial when particular complex contrasts are not known a priori. For example, in the aggregate data of Table 13.1, suppose that we wish to test

$$H_0: \quad \mu_1 - \mu_5 = 0,$$
$$H_1: \quad \mu_1 - \mu_5 \neq 0.$$

The test is developed through use of an F, t, or a confidence interval approach. Using the t, we have

$$t = \frac{\bar{y}_1. - \bar{y}_5.}{s\sqrt{2/n}},$$

where s is the square root of the error mean square and $n = 6$ is the sample size per treatment. In this case

$$t = \frac{553.33 - 610.67}{\sqrt{4961}\sqrt{1/3}}$$
$$= -1.41.$$

The P-value for the t-test with 25 degrees of freedom is 0.17. Thus there is not sufficient evidence to reject H_0.

Relationship Between t and F

In the foregoing we have displayed the use of a pooled t-test along the lines of that discussed in Chapter 10. The pooled estimate comes from the error mean square in order to enjoy the degrees of freedom that are pooled across all five samples. In addition, we have tested a contrast. The reader should note that if the t-value is squared, the result is exactly of the form of the value of f for a test on a contrast discussed in the preceding section. In fact,

$$
f = \frac{(\bar{y}_{1\cdot} - \bar{y}_{5\cdot})^2}{s^2(\frac{1}{6} + \frac{1}{6})}
$$

$$
= \frac{(553.33 - 610.67)^2}{4961(\frac{1}{3})}
$$

$$
= 1.988,
$$

which, of course, is t^2.

Confidence Interval Approach to a Paired Comparison

It is quite simple to solve the same problem of a paired comparison (or a contrast) using a confidence interval approach. Clearly, if we computed a $100(1 - \alpha)\%$ confidence interval on $\mu_1 - \mu_5$, we have

$$
\bar{y}_{1\cdot} - \bar{y}_{5\cdot} \pm t_{\alpha/2}\, s\sqrt{\tfrac{2}{6}},
$$

where $t_{\alpha/2}$ is the upper $100(1 - \alpha/2)\%$ point of a t-distribution with 25 degrees of freedom (degrees of freedom coming from s^2). This simple connection between tests and confidence intervals should be obvious from discussions in Chapters 9 and 10. The test of the simple contrast $\mu_1 - \mu_5$ involves no more than observing whether or not the confidence interval above covers zero. Substituting the numbers, we have as the 95% confidence interval,

$$
(553.33 - 610.67) \pm 2.060\sqrt{4961}\,\sqrt{\tfrac{1}{3}} = -57.34 \pm 83.77.
$$

Thus, since the interval covers zero, the *contrast is not significant*. In other words, we do not find a significant difference between the means of aggregates 1 and 5.

Experiment-Wise Error Rate

We have demonstrated that a simple contrast (i.e., a comparison of two means) can be made through an F-test as demonstrated in Section 13.5, a t-test, or by computing a confidence interval on the difference between the two means. However, serious difficulties occur when the analyst attempts to make many or all possible paired comparisons. For the case of k means, there will be, of course, $r = k(k - 1)/2$ possible paired comparisons. Assuming independent comparisons, the *experiment-wise error rate* (i.e., the probability of false rejection of at least one of the hypotheses) is given by $1 - (1 - \alpha)^r$, where α is the selected probability of type I error for a specific comparison. Clearly, this measure of experiment-wise type I error can be quite large. For example, even if there are only six comparisons, say in the case of 4 means, and $\alpha = 0.05$, the experiment-wise rate is

$$1 - (0.95)^6 \simeq 0.26.$$

With the task of testing many paired comparisons there is usually the need to make the effective contrast on a single comparison more conservative. That is, using the confidence interval approach, the confidence intervals would be much wider than the $\pm t_{\alpha/2}s\sqrt{2/n}$ used in the case where only a single comparison is being made.

Tukey's Test

There are several standard methods for making paired comparisons that allow a sustaining of the credibility of the type I error rate. We shall discuss and illustrate two of them here. The first one, called **Tukey's procedure**, allows formation of simultaneous $100(1 - \alpha)\%$ confidence intervals for all paired comparisons. The method is based on the *studentized* range distribution. The appropriate percentile point is a function of α, k, and $v = $ degrees of freedom for s^2. A table of upper percentage points for $\alpha = 0.05$ is given in Table A.22. The method of paired comparisons by Tukey involves finding a significant difference between means i and j $(i \neq j)$ if $(\bar{y}_{i.} - \bar{y}_{j.})$ exceeds $q[\alpha, k, v]s\sqrt{1/n}$.

Tukey's procedure is easily illustrated. Consider a hypothetical example in which we have six treatments in a one-factor completely randomized design with 5 observations taken per treatment. Suppose that the error mean square taken from the analysis-of-variance table is $s^2 = 2.45$ (24 degrees of freedom). The sample means are given by (ascending order)

$\bar{y}_{2.}$	$\bar{y}_{5.}$	$\bar{y}_{1.}$	$\bar{y}_{3.}$	$\bar{y}_{6.}$	$\bar{y}_{4.}$
14.50	16.75	19.84	21.12	22.90	23.20

With $\alpha = 0.05$, the value of $q(0.05, 6, 24) = 4.37$. Thus all absolute differences are to be compared to

$$4.37\sqrt{\frac{2.45}{5}} = 3.059.$$

As a result, the following represent means found to be significantly different using Tukey's procedure:

4 and 1,   4 and 5,   4 and 2,   6 and 1,   6 and 5

6 and 2,   3 and 5,   3 and 2,   1 and 5,   1 and 2.

Duncan's Test

The second procedure we shall discuss is called **Duncan's procedure** or **Duncan's multiple-range test**. This procedure is also based on the general notion of studentized range. The range of any subset of p sample means must exceed a certain value before any of the p means are found to be different. This value is called the **least significant range** for the p means, and is denoted by R_p, where

$$R_p = r_p \sqrt{\frac{s^2}{n}}.$$

The values of the quantity r_p, called the **least significant studentized range**, depend on the desired level of significance and the number of degrees of freedom of the error mean square. These values may be obtained from Table A.12 for $p = 2, 3, \ldots, 10$ means.

To illustrate the multiple-range test procedure, let us consider the hypothetical example in which 6 treatments are compared with 5 observations per treatment. This is the same example as that used to illustrate Tukey's test. We obtain R_p by multiplying each r_p by 0.70. The results of these computations are summarized as follows:

p	2	3	4	5	6
r_p	2.919	3.066	3.160	3.226	3.276
R_p	2.043	2.146	2.212	2.258	2.293

Comparing these least significant ranges with the differences in ordered means, we arrive at the following conclusions:

1. Since $\bar{y}_4. - \bar{y}_2. = 8.70 > R_6 = 2.293$, we conclude that $\bar{y}_4.$ and $\bar{y}_2.$ are significantly different.
2. Comparing $\bar{y}_4. - \bar{y}_5.$ and $\bar{y}_6. - \bar{y}_2.$ with R_5, we conclude that $\bar{y}_4.$ is significantly greater than $\bar{y}_5.$ and $\bar{y}_6.$ is significantly greater than $\bar{y}_2..$
3. Comparing $\bar{y}_4. - \bar{y}_1., \bar{y}_6. - \bar{y}_5.,$ and $\bar{y}_3. - \bar{y}_2.$ with R_4, we conclude that each difference is significant.
4. Comparing $\bar{y}_4. - \bar{y}_3., \bar{y}_6. - \bar{y}_1., \bar{y}_3. - \bar{y}_5.,$ and $\bar{y}_1. - \bar{y}_2$ with R_3, we find all differences significant except for $\bar{y}_4. - \bar{y}_3..$ Therefore, $\bar{y}_4., \bar{y}_3.,$ and $\bar{y}_6.$ constitute a subset of homogeneous means.
5. Comparing $\bar{y}_3. - \bar{y}_1., \bar{y}_1. - \bar{y}_5.,$ and $\bar{y}_5. - \bar{y}_2.$ with R_2, we conclude that only $\bar{y}_3.$ with $\bar{y}_1.$ are not significantly different.

It is customary to summarize the above conclusions by drawing a line under any subset of adjacent means that are not significantly different. Thus we have

$\bar{y}_2.$	$\bar{y}_5.$	$\bar{y}_1.$	$\bar{y}_3.$	$\bar{y}_6.$	$\bar{y}_4.$
14.50	16.75	19.84	21.12	22.90	23.20

It is clear that in this case the results from Tukey's and Duncan's procedures are very similar. Tukey's procedure did not detect a difference between 2 and 5, whereas Duncan's did.

13.7   Comparing Treatments with a Control _____

In many scientific and engineering problems one is not interested in drawing inferences regarding all possible comparisons among the treatment means of the type $\mu_i - \mu_j$. Rather, the experiment often dictates the need for comparing simultaneously each *treatment* with a *control*. A test procedure for determining significant differences between each treatment mean and the control, at a single joint significance level α, has been developed by C. W. Dunnett. To illustrate Dunnett's procedure, let us consider the experimental data of Table 13.8 for the one-way classification in which the effect of three catalysts on the yield of a reaction is being studied. A fourth treatment, no catalyst, was used as a control.

In general we wish to test the k hypotheses

$$\left.\begin{array}{ll} H_0\text{:} & \mu_0 = \mu_i \\ H_1\text{:} & \mu_0 \neq \mu_i \end{array}\right\} i = 1, 2, \ldots, k,$$

where μ_0 represents the mean yield for the population of measurements in which the control is used. The usual analysis-of-variance assumptions, as outlined in Section 13.2, are expected to remain valid. To test the null hypotheses specified by H_0 against two-sided alternatives for an experimental situation in which there are k treatments, excluding the control, and n observations per treatment, we first calculate the values

$$d_i = \frac{\bar{y}_{i\cdot} - \bar{y}_{0\cdot}}{\sqrt{2s^2/n}}, \qquad i = 1, 2, \ldots, k.$$

The sample variance s^2 is obtained, as before, from the error mean square in the analysis of variance. Now, the critical regime for rejecting H_0, at the α-level of significance, is established by the inequality

$$|d_i| > d_{\alpha/2}(k, v),$$

where v is the number of degrees of freedom for the error mean square. The values of the quantity $d_{\alpha/2}(k, v)$ for a two-tailed test are given in Table A.13 (see Statistical Tables) for $\alpha = 0.05$ and $\alpha = 0.01$ for various values of k and v.

TABLE 13.8   Yield of Reaction

Control	*Catalyst* 1	*Catalyst* 2	*Catalyst* 3
50.7	54.1	52.7	51.2
51.5	53.8	53.9	50.8
49.2	53.1	57.0	49.7
53.1	52.5	54.1	48.0
52.7	54.0	52.5	47.2
$\bar{y}_{0\cdot} = 51.44$	$\bar{y}_{1\cdot} = 53.50$	$\bar{y}_{2\cdot} = 54.04$	$\bar{y}_{3\cdot} = 49.38$

EXAMPLE 13.5   For the data of Table 13.8, test hypotheses comparing each catalyst with the control, using two-sided alternatives. Choose $\alpha = 0.05$ as the joint significance level.

SOLUTION

The error sum of squares with 16 degrees of freedom is obtained from the analysis-of-variance table using all $k + 1$ treatments or by direct computation from the formula

$$SSE = \sum_{i=0}^{k} \sum_{j=1}^{n} y_{ij}^2 - \frac{\sum_{i=0}^{k} T_{i \cdot}^2}{n}$$

$$= 54{,}371.960 - 54{,}335.148$$

$$= 36.812.$$

Then the error mean square is given by

$$s^2 = \frac{36.812}{16} = 2.30075$$

and

$$\sqrt{\frac{2s^2}{n}} = \sqrt{\frac{(2)(2.30075)}{5}} = 0.9593.$$

Hence

$$d_1 = \frac{53.50 - 51.44}{0.9593} = 2.147,$$

$$d_2 = \frac{54.04 - 51.44}{0.9593} = 2.710,$$

$$d_3 = \frac{49.38 - 51.44}{0.9593} = -2.147.$$

From Table A.13 of the Appendix the critical value for $\alpha = 0.05$ is found to be $d_{0.025}(3, 16) = 2.59$. Since $|d_1| < 2.59$, and $|d_3| < 2.59$, we conclude that only the mean yield for catalyst 2 is significantly different from the mean yield of the reaction using the control.

Many practical applications dictate the need for a one-tailed test in comparing treatments with a control. Certainly, when a pharmacologist is concerned with the comparison of various dosages of a drug on the effect of reducing cholesterol level and his control is zero dosage, it is of interest to determine if each dosage produces a significantly larger reduction than that of the control. Table A.14 gives the critical values of $d_\alpha(k, v)$ for one-sided alternatives.

Exercises

1. Consider the data of Exercise 5 on page 530 to make significance tests on the following contrasts:
   (a) B versus A, C, and D;
   (b) C versus A and D;
   (c) A versus D.

2. The study "*Loss of Nitrogen Through Sweat by Pre-adolescent Boys Consuming Three Levels of Dietary Protein*" was conducted by the Department of Human Nutrition and Foods at the Virginia Polytechnic Institute and State University in 1975 to determine perspiration nitrogen loss at various dietary protein levels. Twelve pre-adolescent boys ranging in age from 7 years, 8 months to 9 years, 8 months, and judged to be clinically healthy, were used in the experiment. Each boy was subjected to one of three controlled diets in which 29, 54, or 84 grams of protein per day were consumed. The following data represent the body perspiration nitrogen loss, in milligrams, collected during the last two days of the experimental period:

	Protein Level	
29 Grams	54 Grams	84 Grams
190	318	390
266	295	321
270	271	396
	438	399
	402	

(a) Perform an analysis of variance at the 0.05 level of significance to show that the mean perspiration nitrogen losses at the three protein levels are different.

(b) Use a single-degree-of-freedom contrast with $\alpha = 0.05$ to compare the mean perspiration nitrogen loss for boys who consume 29 grams of protein per day versus boys who consume 54 and 84 grams of protein.

3. The purpose of the study "*The Incorporation of a Chelating Agent into a Flame Retardant Finish of a Cotton Flannelette and the Evaluation of Selected Fabric Properties*" conducted at the Virginia Polytechnic Institute and State University in 1974 was to evaluate the use of a chelating agent as part of the flame retardant finish of cotton flannelette by determining its effects upon flammability after the fabric is laundered under specific conditions. Two baths were prepared, one with carboxymethyl cellulose and one without. Twelve pieces of fabric were laundered 5 times in bath I, and 12 other pieces of fabric were laundered 10 times in bath I. This was repeated using 24 additional pieces of cloth in bath II. After the washings, the lengths of fabric that burned and the burn times were measured. For convenience, let us define the following treatments:

Treatment 1:   5 launderings in bath I,

Treatment 2:   5 launderings in bath II,

Treatment 3:   10 launderings in bath I,

Treatment 4:   10 launderings in bath II.

Burn times, in seconds, were recorded as follows:

	Treatment		
1	2	3	4
13.7	6.2	27.2	18.2
23.0	5.4	16.8	8.8
15.7	5.0	12.9	14.5
25.5	4.4	14.9	14.7
15.8	5.0	17.1	17.1
14.8	3.3	13.0	13.9
14.0	16.0	10.8	10.6
29.4	2.5	13.5	5.8
9.7	1.6	25.5	7.3
14.0	3.9	14.2	17.7
12.3	2.5	27.4	18.3
12.3	7.1	11.5	9.9

(a) Perform an analysis of variance using a 0.01 level of significance and determine whether there are any significant differences among the treatment means.

(b) Use single-degree-of-freedom contrasts with $\alpha = 0.01$ to compare the mean burn time of treatment 1 versus treatment 2 and also treatment 3 versus treatment 4.

4. Use Duncan's multiple-range test, with a 0.05 level of significance, to analyze the means of the 5 different brands of headache tablets in Exercise 4 on page 475.

5. For the data used in Exercise 1 on page 487, perform Duncan's multiple-range test with a 0.01 level of significance, to determine which laboratories differ, on the average, in their analysis.

6. An investigation was conducted to determine the source of reduction in yield of a certain chemical product. It was known that the loss in yield occurred in the mother liquor, that is, the material removed at the filtration stage. It was felt that different blends of the original material may result in different yield reductions at the mother liquor stage. The following are results of the percent reduction for 3 batches at each of 4 preselected blends:

Blend			
1	2	3	4
25.6	25.2	20.8	31.6
24.3	28.6	26.7	29.8
27.9	24.7	22.2	34.3

(a) Perform the analysis of variance at the $\alpha = 0.05$ level of significance.
(b) Use Duncan's multiple-range test to determine which blends differ.

7. In the study "An Evaluation of the Removal Method for Estimating Benthic Populations and Diversity" conducted by the Virginia Polytechnic Institute and State University on the Jackson River in 1983, 5 different sampling procedures were used to determine the species count. Twenty samples were selected at random and each of the 5 sampling procedures were repeated 4 times. The species counts were recorded as follows:

Sampling Procedure				
Deple-tion	Modified Hess	Surber	Substrate Removal Kicknet	Kick-net
85	75	31	43	17
55	45	20	21	10
40	35	9	15	8
77	67	37	27	15

(a) Is there a significant difference in the average species count for the different sampling procedures? Use a 0.01 level of significance.
(b) Use Duncan's multiple-range test with $\alpha = 0.05$ to find which sampling procedures differ.

8. The financial structure of a firm refers to the way the firm's assets are divided by equity and debt, and the financial leverage refers to the percentage of assets financed by debt. In the paper "The Effect of Financial Leverage on Return" (1980), Tai Ma of the Virginia Polytechnic Institute and State University claims that financial leverage can be used to increase the rate of return on equity. To say it in another way, stockholders can receive higher returns on equity with the same amount of investment by the use of financial leverage. The following data show the rates of return on equity using 3 different levels of financial leverage and a control level (zero debt) for 24 randomly selected firms:

	Financial Leverage		
Control	Low	Medium	High
2.1	6.2	9.6	10.3
5.6	4.0	8.0	6.9
3.0	8.4	5.5	7.8
7.8	2.8	12.6	5.8
5.2	4.2	7.0	7.2
2.6	5.0	7.8	12.0

Source: S & P's Machinery Industry Survey, 1975.

(a) Perform the analysis of variance at the 0.05 level of significance.
(b) Use Dunnett's test at the 0.01 level of significance to determine whether the mean rates of return on equity at the low, medium, and high levels of financial leverage are higher than at the control level.

9. In the following biological experiment 4 concentrations of a certain chemical are used to enhance the growth in centimeters of a certain type of plant over time. Five plants are used at each concentration and the growth in each plant is measured. The following growth data are taken. A control (no chemical) is also applied.

	Concentration			
Control	1	2	3	4
6.8	8.2	7.7	6.9	5.9
7.3	8.7	8.4	5.8	6.1
6.3	9.4	8.6	7.2	6.9
6.9	9.2	8.1	6.8	5.7
7.1	8.6	8.0	7.4	6.1

Use Dunnett's two-sided test at the 0.05 level of significance to simultaneously compare the concentrations with the control.

10. The following table (A. Hald, *Statistical Theory with Engineering Applications,* John Wiley & Sons, New York, 1952) gives tensile strengths of less than 340 of wires taken from nine cables to be used for a high-voltage network. Each cable is made from 12 wires. We want to know whether the mean strengths of the wires in the nine cables are the same. If the cables are different, which ones differ?

Cable												
1	5	−13	−5	−2	−10	−6	−5	0	−3	2	−7	−5
2	−11	−13	−8	8	−3	−12	−12	−10	5	−6	−12	−10
3	0	−10	−15	−12	−2	−8	−5	0	−4	−1	−5	−11
4	−12	4	2	10	−5	−8	−12	0	−5	−3	−3	0
5	7	1	5	0	10	6	5	2	0	−1	−10	−2
6	1	0	−5	−4	−1	0	2	5	1	−2	6	7
7	−1	0	2	1	−4	2	7	5	1	0	−4	2
8	−1	0	7	5	10	8	1	2	−3	6	0	5
9	2	6	7	8	15	11	−7	7	10	7	8	1

11. It is suspected that the environmental temperature in which batteries are activated affects their activated life. Thirty homogeneous batteries were tested, six at each of five temperatures, and the data shown below were obtained (activated life in seconds). Analyze and interpret the data. (C. R. Hicks, *Fundamental Concepts in the Design of Experiments*, Holt, Rinehart and Winston, New York, 1973.)

		Temperature (°C)		
0	25	50	75	100
55	60	70	72	65
55	61	72	72	66
57	60	73	72	60
54	60	68	70	64
54	60	77	68	65
56	60	77	69	65

12. The following data are values of pressure (psi) in a torsion spring for several settings of the angle between the legs of the spring in a free position:

	Angle of Legs of Spring (°)			
67	71	75	79	83
83	84	86	89	90
85	85	87	90	92
	85	87	90	
	86	87	91	
	86	88		
	87	88		
		88		
		88		
		88		
		89		
		90		

Compute a one-way analysis of variance for this experiment and state your conclusion concerning the effect of angle on the pressure in the spring. (C. R. Hicks, *Fundamental Concepts in the Design of Experiments,* Holt, Rinehart and Winston, New York, 1973.)

13.8 Comparing a Set of Treatments in Blocks _____

In Section 13.2 we discussed the idea of blocking, that is, isolating sets of experimental units that are reasonably homogeneous and randomly assigning treatments to these units. This is an extension of the "pairing" concept that was discussed in Chapters 9 and 10 and is done to reduce experimental error, since the units in blocks have characteristics that are more common than units that are in different blocks.

The reader should not view blocks as a second factor, although this is a tempting way of visualizing the design. In fact, the main factor (treatments) still carries the major thrust of the experiment. Experimental units are still the source of error, just as in the completely randomized design. We merely treat sets of these units more systematically when blocking is accomplished. In this way, we say there are restrictions in randomization. For example, in a chemical experiment designed to determine if there is a difference in mean reaction yield among four catalysts, samples of materials to be tested are drawn from the same batches of raw materials, while other conditions, such as temperature and concentration of reactants, are held constant. In this case the time of day for the experimental runs might represent the experimental units, and if the experimenter feels that there could possibly be a slight time effect, he or she would *randomize* the assignment of the catalysts to the runs to counteract the possible trend. This type of experimental strategy is the **completely randomized design**. As a second example of such a design, consider an experiment to compare four methods of measuring a particular physical property of a fluid substance. Suppose the sampling process is destructive; that is, once a sample of the substance has been measured by one method, it cannot be measured again by one of the other methods. If it is decided that five measurements are to be taken for each method, then 20 samples of the material are selected from a large batch *at random* and are used in the experiment to compare the four measuring devices. The experimental units are the randomly selected samples. Any variation from sample to sample will appear in the error variation, as measured by s^2 in the analysis.

If the variation due to heterogeneity in experimental units is so large that the sensitivity of detecting treatment differences is reduced due to an inflated value of s^2, a better plan might be to ''block off'' variation due to these units and thus reduce the extraneous variation to that accounted for by smaller or more homogeneous blocks. For example, suppose that in the previous catalyst illustration it is known *a priori* that there definitely is a significant day-to-day effect on the yield and that we can measure the yield for four catalysts on a given day. Rather than assign the four catalysts to the 20 test runs completely at random, we choose, say, five days and run each of the four catalysts on each day, randomly assigning the catalysts to the runs within days. In this way the day-to-day variation is removed in the analysis and consequently the experimental error, which still includes any time trend *within days*, more accurately represents chance variation. Each day is referred to as a **block**.

The classical example, using a randomized block design, is an agricultural experiment in which different fertilizers are being compared for their ability to increase the yield of a particular crop. Rather than assign fertilizers at random to many plots over a large area of variable soil composition, one should assign the fertilizers to smaller blocks comprised of homogeneous plots. The variation between these blocks, which is most likely significant compared to the uniformity of the plots within a block, is then removed from the experimental error in the analysis of variance.

The most straightforward of the randomized block designs is one in which we randomly assign each treatment once to every block. Such an experimental layout is called a **randomized complete block design**, each block constituting a single **replication** of the treatments.

13.9 Randomized Complete Block Designs

A typical layout for the randomized complete block design using 3 measurements in 4 blocks is as follows:

Block 1	*Block* 2	*Block* 3	*Block* 4
t_2	t_1	t_3	t_2
t_1	t_3	t_2	t_1
t_3	t_2	t_1	t_3

The t's denote the assignment to blocks of each of the 3 treatments. Of course, the true allocation of treatments to units within blocks is done at random. Once the experiment has been completed, the data can be recorded as in the following 3×4 array:

	Block			
Treatment	1	2	3	4
1	y_{11}	y_{12}	y_{13}	y_{14}
2	y_{21}	y_{22}	y_{23}	y_{24}
3	y_{31}	y_{32}	y_{33}	y_{34}

where y_{11} represents the response obtained by using treatment 1 in block 1, y_{12} represents the response obtained by using treatment 1 in block 2, . . . , and y_{34} represents the response obtained by using treatment 3 in block 4.

Let us now generalize and consider the case of k treatments assigned to b blocks. The data may be summarized as shown in the $k \times b$ rectangular array of Table 13.9. It will be assumed that the y_{ij}, $i = 1, 2, \ldots, k$ and $j = 1, 2, \ldots, b$, are values of independent random variables having normal distributions with means μ_{ij} and common variance σ^2. In Table 13.9 we define

$T_{i\cdot}$ = sum of the observations for the ith treatment

$T_{\cdot j}$ = sum of the observations in the jth block

$T_{\cdot\cdot}$ = sum of all bk observations

$\bar{y}_{i\cdot}$ = mean of the observations for the ith treatment

$\bar{y}_{\cdot j}$ = mean of the observations in the jth block

$\bar{y}_{\cdot\cdot}$ = mean of all bk observations.

TABLE 13.9 $k \times b$ Array for a Randomized Complete Block Design

Treatment	Block 1	2	$\cdots$	j	$\cdots$	b	Total	Mean
1	y_{11}	y_{12}	$\cdots$	y_{1j}	$\cdots$	y_{1b}	$T_{1\cdot}$	$\bar{y}_{1\cdot}$
2	y_{21}	y_{22}	$\cdots$	y_{2j}	$\cdots$	y_{2b}	$T_{2\cdot}$	$\bar{y}_{2\cdot}$
$\vdots$	$\vdots$	$\vdots$		$\vdots$		$\vdots$	$\vdots$	$\vdots$
i	y_{i1}	y_{i2}	$\cdots$	y_{ij}	$\cdots$	y_{ib}	$T_{i\cdot}$	$\bar{y}_{i\cdot}$
$\vdots$	$\vdots$	$\vdots$		$\vdots$		$\vdots$	$\vdots$	$\vdots$
k	y_{k1}	y_{k2}	$\cdots$	y_{kj}	$\cdots$	y_{kb}	$T_{k\cdot}$	$\bar{y}_{k\cdot}$
Total	$T_{\cdot 1}$	$T_{\cdot 2}$	$\cdots$	$T_{\cdot j}$	$\cdots$	$T_{\cdot b}$	$T_{\cdot\cdot}$	
Mean	$\bar{y}_{\cdot 1}$	$\bar{y}_{\cdot 2}$	$\cdots$	$\bar{y}_{\cdot j}$	$\cdots$	$\bar{y}_{\cdot b}$		$\bar{y}_{\cdot\cdot}$

Let $\mu_{i\cdot}$ represent the average (rather than the total) of the b population means for the ith treatment. That is,

$$\mu_{i\cdot} = \frac{\sum_{j=1}^{b} \mu_{ij}}{b}.$$

Similarly, the average of the population means for the jth block, $\mu_{\cdot j}$, is defined by

$$\mu_{\cdot j} = \frac{\sum_{i=1}^{k} \mu_{ij}}{k},$$

and the average of the bk population means, μ, is defined by

$$\mu = \frac{\sum_{i=1}^{k} \sum_{j=1}^{b} \mu_{ij}}{bk}.$$

To determine if part of the variation in our observations is due to differences among the treatments, we consider the test

$$H_0': \quad \mu_{1\cdot} = \mu_{2\cdot} = \cdots = \mu_{k\cdot} = \mu,$$

$$H_1': \quad \text{The } \mu_{i\cdot}\text{'s are not all equal.}$$

Each observation may be written in the form

$$y_{ij} = \mu_{ij} + \varepsilon_{ij},$$

where ε_{ij} measures the deviation of the observed value y_{ij} from the population mean μ_{ij}. The preferred form of this equation is obtained by substituting

$$\mu_{ij} = \mu + \alpha_i + \beta_j,$$

where α_i is, as before, the effect of the ith treatment and β_j is the effect of the jth block. It is assumed that the treatment and block effects are additive. Hence we may write

$$y_{ij} = \mu + \alpha_i + \beta_j + \varepsilon_{ij}.$$

Notice that the model resembles that of the one-way classification, the essential difference being the introduction of the block effect β_j. The basic concept is much like that of the one-way classification except that we must account in the analysis for the additional effect due to blocks, since we are now systematically controlling variation in *two directions*. If we now impose the restrictions that

$$\sum_{i=1}^{k} \alpha_i = 0 \qquad \text{and} \qquad \sum_{j=1}^{b} \beta_j = 0,$$

then

$$\mu_{i\cdot} = \frac{\sum_{j=1}^{b} (\mu + \alpha_i + \beta_j)}{b} = \mu + \alpha_i$$

and

$$\mu_{\cdot j} = \frac{\sum_{i=1}^{k} (\mu + \alpha_i + \beta_j)}{k} = \mu + \beta_j.$$

The null hypothesis that the k treatment means $\mu_{i\cdot}$ are equal, and therefore equal to μ, is now equivalent to testing the hypothesis

$$H_0': \quad \alpha_1 = \alpha_2 = \cdots = \alpha_k = 0,$$

$$H_1': \quad \text{At least one of the } \alpha_i\text{'s is not equal to zero.}$$

Each of the tests on treatments will be based on a comparison of independent estimates of the common population variance σ^2. These estimates will be obtained by splitting the total sum of squares of our data into three components by means of the following identity.

THEOREM 13.3
SUM-OF-SQUARES
IDENTITY

$$\sum_{i=1}^{k} \sum_{j=1}^{b} (y_{ij} - \bar{y}..)^2 = b \sum_{i=1}^{k} (\bar{y}_{i\cdot} - \bar{y}..)^2 + k \sum_{j=1}^{b} (\bar{y}_{\cdot j} - \bar{y}..)^2$$

$$+ \sum_{i=1}^{k} \sum_{j=1}^{b} (y_{ij} - \bar{y}_{i\cdot} - \bar{y}_{\cdot j} + \bar{y}..)^2. \blacksquare$$

PROOF

$$\sum_{i=1}^{k} \sum_{j=1}^{b} (y_{ij} - \bar{y}..)^2 = \sum_{i=1}^{k} \sum_{j=1}^{b} [(\bar{y}_{i\cdot} - \bar{y}..)$$

$$+ (\bar{y}_{\cdot j} - \bar{y}..) + (y_{ij} - \bar{y}_{i\cdot} - \bar{y}_{\cdot j} + \bar{y}..)]^2$$

$$= \sum_{i=1}^{k} \sum_{j=1}^{b} (\bar{y}_{i\cdot} - \bar{y}..)^2 + \sum_{i=1}^{k} \sum_{j=1}^{b} (\bar{y}_{\cdot j} - \bar{y}..)^2$$

$$+ \sum_{i=1}^{k} \sum_{j=1}^{b} (y_{ij} - \bar{y}_{i\cdot} - \bar{y}_{\cdot j} + \bar{y}..)^2$$

$$+ 2 \sum_{i=1}^{k} \sum_{j=1}^{b} (\bar{y}_{i\cdot} - \bar{y}..)(\bar{y}_{\cdot j} - \bar{y}..)$$

$$+ 2 \sum_{i=1}^{k} \sum_{j=1}^{b} (\bar{y}_{i\cdot} - \bar{y}..)(y_{ij} - \bar{y}_{i\cdot} - \bar{y}_{\cdot j} + \bar{y}..)$$

$$+ 2 \sum_{i=1}^{k} \sum_{j=1}^{b} (\bar{y}_{\cdot j} - \bar{y}..)(y_{ij} - \bar{y}_{i\cdot} - \bar{y}_{\cdot j} + \bar{y}..).$$

The cross-product terms are all equal to zero. Hence

$$\sum_{i=1}^{k} \sum_{j=1}^{b} (y_{ij} - \bar{y}..)^2 = b \sum_{i=1}^{k} (\bar{y}_{i\cdot} - \bar{y}..)^2 + k \sum_{j=1}^{b} (\bar{y}_{\cdot j} - \bar{y}..)^2$$

$$+ \sum_{i=1}^{k} \sum_{j=1}^{b} (y_{ij} - \bar{y}_{i\cdot} - \bar{y}_{\cdot j} + \bar{y}..)^2.$$

The sum-of-squares identity may be presented symbolically by the equation

$$SST = SSA + SSB + SSE,$$

where

$$SST = \sum_{i=1}^{k} \sum_{j=1}^{b} (y_{ij} - \bar{y}..)^2 = \text{total sum of squares}$$

$$SSA = b \sum_{i=1}^{k} (\bar{y}_{i.} - \bar{y}..)^2 = \text{treatment sum of squares}$$

$$SSB = k \sum_{j=1}^{b} (\bar{y}_{.j} - \bar{y}..)^2 = \text{block sum of squares}$$

$$SSE = \sum_{i=1}^{k} \sum_{j=1}^{b} (y_{ij} - \bar{y}_{i.} - \bar{y}_{.j} + \bar{y}..)^2 = \text{error sum of squares.}$$

Following the procedure outlined in Theorem 13.2, where we interpret the sum of squares as functions of the independent random variables, $Y_{11}, Y_{12}, \ldots, Y_{kb}$, we can show that the expected values of the treatment, block, and error sum of squares are given by

$$E(SSA) = (k - 1)\sigma^2 + b \sum_{i=1}^{k} \alpha_1^2$$

$$E(SSB) = (b - 1)\sigma^2 + k \sum_{j=1}^{b} \beta_j^2$$

$$E(SSE) = (b - 1)(k - 1)\sigma^2.$$

As in the case of the one-factor problem, we have the treatment mean square

$$s_1^2 = \frac{SSA}{k - 1}.$$

If the treatment affects $\alpha_1 = \alpha_2 = \cdots = \alpha_k = 0$, s_1^2 is an unbiased estimate of σ^2. However, if the treatment effects are not all zero, we have

$$E\left(\frac{SSA}{k - 1}\right) = \sigma^2 + \frac{b \sum_{i=1}^{k} \alpha_i^2}{k - 1}$$

and s_1^2 overestimates σ^2. A second estimate of σ^2, based on $b - 1$ degrees of freedom, is given by

$$s_2^2 = \frac{SSB}{b - 1}.$$

The estimate s_2^2 is an unbiased estimate of σ^2 when the block effects $\beta_1 = \beta_2 = \cdots = \beta_b = 0$. If the block effects are not all zero, then

$$E\left(\frac{SSB}{b-1}\right) = \sigma^2 + \frac{k\sum\limits_{j=1}^{b} \beta_j^2}{b-1}$$

and s_2^2 will overestimate σ^2. A third estimate of σ^2, based on $(k-1)(b-1)$ degrees of freedom and independent of s_1^2 and s_2^2, is given by

$$s^2 = \frac{SSE}{(k-1)(b-1)},$$

which is unbiased regardless of the truth or falsity of either null hypothesis.

To test the null hypothesis that the treatment effects are all equal to zero, we compute the ratio

$$f_1 = \frac{s_1^2}{s^2},$$

which is a value of the random variable F_1 having an F-distribution with $k-1$ and $(k-1)(b-1)$ degrees of freedom when the null hypothesis is true. The null hypothesis is rejected at the α-level of significance when $f_1 > f_\alpha[k-1, (k-1)(b-1)]$.

In practice we first compute SST, SSA, and SSB, and then, using the sum-of-squares identity, obtain SSE by subtraction. The degrees of freedom associated with SSE are also usually obtained by subtraction; that is,

$$(k-1)(b-1) = (bk-1) - (k-1) - (b-1).$$

The computations in an analysis-of-variance problem for a randomized complete block design may be summarized as shown in Table 13.10.

Table 13.10   Analysis of Variance for the Randomized Complete Block Design

Source of Variation	Sum of Squares	Degrees of Freedom	Mean Square	Computed f
Treatments	SSA	$k-1$	$s_1^2 = \dfrac{SSA}{k-1}$	$f_1 = \dfrac{s_1^2}{s^2}$
Blocks	SSB	$b-1$	$s_2^2 = \dfrac{SSB}{b-1}$	
Error	SSE	$(k-1)(b-1)$	$s^2 = \dfrac{SSE}{(b-1)(k-1)}$	
Total	SST	$bk-1$		

EXAMPLE 13.6 Four different machines, M_1, M_2, M_3, and M_4, are to be considered in the assembling of a particular product. It is decided that six different operators are to be used in a randomized block experiment to compare the machines. The machines are assigned in a random order to each operator. The operation of the machines requires a certain amount of physical dexterity, and it is anticipated that there will be a difference among the operators in the speed with which they operate the machines (Table 13.11). The times, in seconds, were recorded for assembling the given product:

TABLE 13.11 Time in Seconds to Assembly Product

	Operator						
Machine	1	2	3	4	5	6	*Total*
1	42.5	39.3	39.6	39.9	42.9	43.6	247.8
2	39.8	40.1	40.5	42.3	42.5	43.1	248.3
3	40.2	40.5	41.3	43.4	44.9	45.1	255.4
4	41.3	42.2	43.5	44.2	45.9	42.3	259.4
Total	163.8	162.1	164.9	169.8	176.2	174.1	1010.9

Test the hypothesis H_0', at the 0.05 level of significance, that the machines perform at the same mean rate of speed.

SOLUTION

H_0: $\alpha_1 = \alpha_2 = \alpha_3 = \alpha_4 = 0$ (machine effects are zero).

H_1: At least one of the α_i's is not equal to zero.

The sum of squares formulas shown on page 495 and the degrees of freedom are used to produce the analysis of variance in Table 13.12. The value $f = 3.34$ is significant at $P = 0.048$. If one uses $\alpha = 0.05$ as at least an approximate yardstick, one concludes that the machines do not perform at the same mean rate of speed.

TABLE 13.12 Analysis of Variance for the Data of Table 13.11

Source of Variation	*Sum of Squares*	*Degrees of Freedom*	*Mean Square*	*Computed f*
Machines	15.93	3	5.31	3.34
Operators	42.09	5	8.42	
Error	23.84	15	1.59	
Total	81.86	23		

Further Comments Concerning Blocking

In Chapter 10 we presented a procedure for comparing means when the observations were *paired*. The procedure involved ''subtracting out'' the effect due to the homogeneous pair and thus working with differences. This is a special case of a randomized complete block design with $k = 2$ treatments. The n homogeneous units to which the treatments were assigned take on the role of blocks.

If there is heterogeneity in the experimental units, the experimenter should not be misled into believing that it is always advantageous to reduce the experimental error through the use of small homogeneous blocks. Indeed, there may be instances where it would not be desirable to block. The purpose in reducing the error variance is to increase the *sensitivity* of the test for detecting differences in the treatment means. This is reflected in the power of the test procedure. (The power of the analysis-of-variance test procedure is discussed more extensively in Section 13.12.) The power for detecting certain differences among the treatments means increases with a decrease in the error variance. However, the power is also affected by the degrees of freedom with which this variance is estimated, and blocking reduces the degrees of freedom that are available from $k(b - 1)$ for the one-way classification to $(k - 1)(b - 1)$. So one could lose power by blocking if there is not a significant reduction in the error variance.

Interaction Between Blocks and Treatments

Another important assumption that is implicit in writing the model for a randomized complete block design is that the treatment and block effects are assumed to be additive. This is equivalent to stating that $\mu_{ij} - u_{ij'} = \mu_{i'j} - \mu_{i'j'}$ or $\mu_{ij} - \mu_{i'j} = \mu_{ij'} - \mu_{i'j'}$ for every value of i, i', j, and j'. That is, the difference between the population means for blocks j and j' is the same for every treatment and the difference between the population means for treatments i and i' is the same for every block. The parallel lines of Figure 13.1(a) illustrate a set of mean responses for which the treatment and block effects are additive, whereas the intersecting lines of Figure 13.1(b) show a situation in which treatment and block effects are said to **interact**. Referring to Example 13.6, we see that if operator 3 is 0.5 second faster on the average than operator 2 when machine 1 is used, then operator 3 will still be 0.5 second faster on the average than operator 2 when machine 2, 3, or 4 is used. In many experiments the assumption of additivity does not hold and the analysis of Section 13.8 leads to erroneous conclusions. Suppose, for instance, that operator 3 is 0.5 second faster on the average than operator 2 when machine 1 is used but is

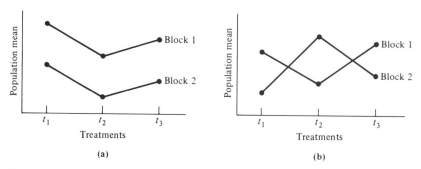

FIGURE 13.1 Population means for (a) additive effects, and (b) interacting effects.

0.2 second slower on the average than operator 2 when machine 2 is used. The operators and machines are now interacting.

An inspection of Table 13.11 suggests the presence of possible interaction. This apparent interaction may be real or it may be due to experimental error. The analysis of Example 13.6 was based on the assumption that the apparent interaction was due entirely to experimental error. If the total variability of our data was in part due to an interaction effect, this source of variation remained a part of the error sum of squares, causing the error mean square to overestimate σ^2, and thereby increasing the probability of committing a type II error. We have, in effect, assumed an incorrect model. If we let $(\alpha\beta)_{ij}$ denote the interaction effect of the ith treatment and the jth block, we can write a more appropriate model in the form

$$y_{ij} = \mu + \alpha_i + \beta_j + (\alpha\beta)_{ij} + \varepsilon_{ij},$$

on which we impose the additional restrictions $\sum_{i=1}^{k} (\alpha\beta)_{ij} = \sum_{j=1}^{b} (\alpha\beta)_{ij} = 0$. One can now very easily verify that

$$E\left[\frac{SSE}{(b-1)(k-1)}\right] = \sigma^2 + \frac{\sum_{i=1}^{k}\sum_{j=1}^{b} (\alpha\beta)_{ij}^2}{(b-1)(k-1)}.$$

Thus the error mean square is seen to be a biased estimate of σ^2 when existing interaction has been ignored. It would seem necessary at this point to arrive at a procedure for the detection of interaction for cases where there is suspicion that it exists. Such a procedure requires the availability of an unbiased and independent estimate of σ^2. Unfortunately, the randomized block design does not lend itself to such a test unless the experimental setup is altered. This is discussed extensively in Chapter 14.

13.10 Graphical Methods and Further Diagnostics _____

In several chapters we have made reference to graphical procedures displaying data and analytical results. In early chapters we used stem and leaf and box plots and visuals that aid in summarizing samples. We used similar diagnostics to better understand the data in two sample problems in Chapters 9 and 10. In Chapter 9 we introduced the notion of residual plots (ordinary and studentized residuals) to detect violations of standard assumptions. In recent years much attention in data analysis has centered on graphical methods. Like regression, analysis of variance lends itself to graphics that aid in summarizing data as well as detecting violations. For example, a simple plotting of the raw observations around each treatment mean can give the analyst a feel for variability between sample means and within samples. Figure 13.2 depicts such a plot for the aggregate data of Table 13.1. By the appearance of the plot one might even gain a graphical insight about which aggregates (if any) stand out from the others. It is clear that aggregate 4 stands out from the others. Aggregates 3 and 5 certainly form a homogeneous group, as do aggregates 1 and 2.

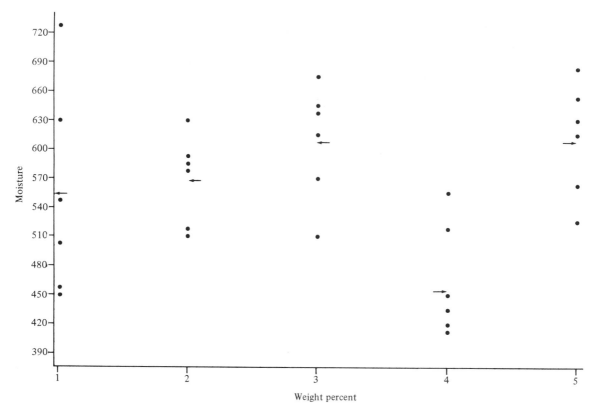

FIGURE 13.2 Plot of data around mean for the aggregate data of Table 13.1.

As in the case of regression, residuals can be helpful in analysis of variance in providing a diagnostic that may detect violations of assumptions. To form the residuals, we merely need to consider the model of the one-factor problem, namely

$$y_{ij} = \mu_i + \varepsilon_{ij}.$$

It is straightforward to determine that the estimate of μ_i is $\bar{y}_{i\cdot}$. Hence $\hat{y}_{ij} = \bar{y}_{i\cdot}$ and thus the ijth residual is $y_{ij} - \bar{y}_{i\cdot}$. This is easily extendable to the randomized complete block model. It may very well be instructive to have the residuals plotted for each aggregate in order to gain some insight regarding the homogeneous variance assumption. This plot is shown in Figure 13.3.

Trends in plots such as these may reveal difficulties in some situations, particularly when the violation of a particular assumption is very graphic. In the case of Figure 13.3 the residuals seem to indicate that the *within-treatment* variances are reasonably homogeneous apart from aggregate 1. There is some graphical evidence that the variance for aggregate 1 is larger than the rest.

The randomized complete block is another design in which graphical displays can make the analyst feel comfortable with an "ideal picture" or perhaps highlight difficulties. Recall that the model for the randomized complete block is given by

$$y_{ij} = \mu + \alpha_i + \beta_j + \varepsilon_{ij} \qquad \begin{aligned} i &= 1, 2, \ldots, a \\ j &= 1, 2, \ldots, b, \end{aligned}$$

with the imposed constraints $\sum_{i=1}^{a} \alpha_i = 0$; $\sum_{j=1}^{b} \beta_j = 0$. In order to determine what indeed constitutes a residual, consider that

$$\alpha_i = \mu_{i.} - \mu$$

$$\beta_j = \mu_{.j} - \mu$$

and that μ is estimated by $\bar{y}_{..}$, $\mu_{i.}$ is estimated by $\bar{y}_{i.}$, and μ_j is estimated by $\bar{y}_{.j}$. As a result, the predicted or *fitted value* $\hat{y}_{ij}$ is given by

$$\hat{y}_{ij} = \hat{\mu} + \hat{\alpha}_i + \hat{\beta}_j$$

$$= \bar{y}_{i.} + \bar{y}_{.j} - \bar{y}_{..}$$

and thus the residual at the (i, j) observation is given by

$$y_{ij} - \hat{y}_{ij} = y_{ij} - \bar{y}_{i.} - \bar{y}_{.j} + \bar{y}_{..}$$

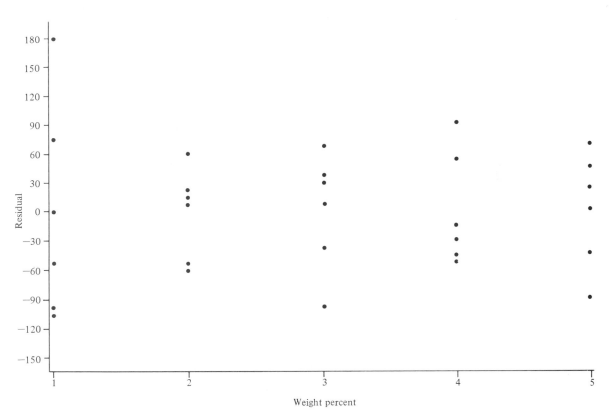

FIGURE 13.3 Plots of residuals for five aggregates using data in Table 13.1.

Note that $\hat{y}_{ij}$, the fitted value, is an estimate of the mean μ_{ij}. This is consistent with the partitioning of variability given in Theorem 13.3, in which the error sum of squares is given by

$$SSE = \sum_{j} \sum_{j} (y_{ij} - \bar{y}_{i\cdot} - \bar{y}_{\cdot j} + \bar{y}_{\cdot\cdot})^2.$$

The visual displays in the randomized complete block involve plotting the residuals separately for each treatment and for each block. The analyst should expect roughly equal variability if the homogeneous variance assumption holds. The reader should recall that in Chapter 12 we discussed plots in which the residuals are plotted for the purpose of detecting model misspecification. In the case of the randomized complete block, the serious model misspecification may be related to our assumption of additivity (i.e., no interaction). If no interaction is present, a random pattern should appear.

Consider the data of Example 13.6, in which treatments are three machines and blocks are operators. Figures 13.4 and 13.5 give the residual plots for separate

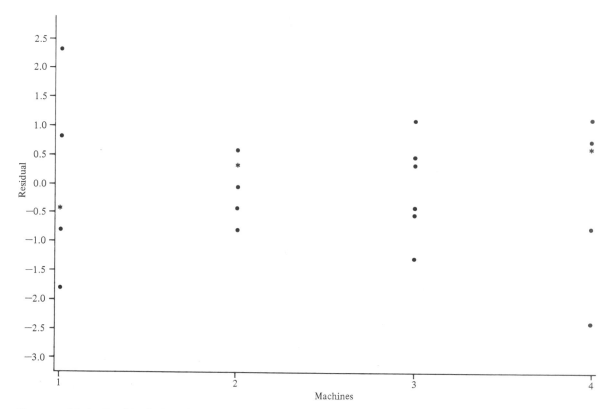

FIGURE 13.4 Residual plots for the four machines for the data of Example 13.6. The asterisks indicate multiple residuals.

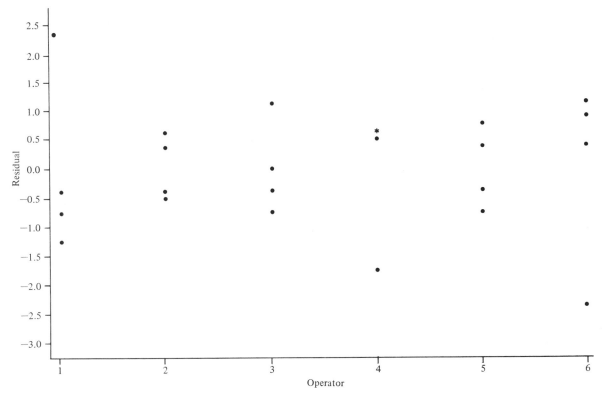

FIGURE 13.5 Residual plots for the six operators for the data of Example 13.6. The asterisk indicates multiple residuals.

treatments and separate blocks. Figure 13.6 shows a plot of the residuals against the fitted values. Figure 13.4 reveals that the error variance may not be the same for all machines. The same may be true for error variance at each of the six operators. However, two unusually large residuals appear to produce the apparent difficulty. Figure 13.6 reveals a plot of residuals that shows reasonable evidence of a random behavior. However, the two large residuals displayed earlier stand out.

13.11 Latin Squares

The randomized block design is very effective in reducing the experimental error by removing one source of variation. Another design that is particularly useful in controlling two sources of variation, while reducing the required number of treatment combinations, is called the **Latin square**. Suppose that we are interested in the yields of 4 varieties of wheat using 4 different fertilizers over a period of 4 years. The total number of treatment combinations for a completely randomized design would be 64. By selecting the same number of categories for all three criteria of classification, we may select a Latin square design and perform the analysis of

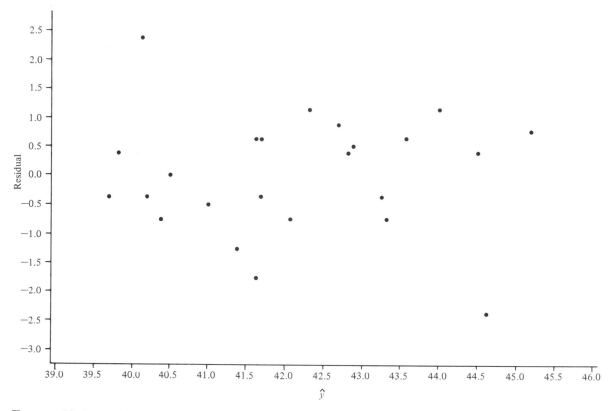

FIGURE 13.6   Residuals plotted against fitted values for the data of Example 13.6.

variance using the results of only 16 treatment combinations. A typical Latin square, selected at random from all possible 4 × 4 squares, might be the following:

	Column			
Row	1	2	3	4
1	A	B	C	D
2	D	A	B	C
3	C	D	A	B
4	B	C	D	A

The four letters, A, B, C, and D represent the 4 varieties of wheat that are referred to as the treatments. The rows and columns, represented by the 4 fertilizers and the 4 years, respectively, are the two sources of variation that we wish to control. We now see that each treatment occurs exactly once in each row and each column. With such a balanced arrangement the analysis of variance enables one to separate the variation due to the different fertilizers and different years from the error sum of squares and thereby obtain a more accurate test for differences in the yielding capabilities of the 4 varieties of wheat. When there is interaction present between any of the sources of variation, the f-values in the analysis of variance are no longer valid. In this case, the Latin square design would be inappropriate.

We shall now generalize and consider an $r \times r$ Latin square where y_{ijk} denotes an observation in the ith row and jth column corresponding to the kth letter. Note that once i and j are specified for a particular Latin square, we automatically know the letter given by k. For example, when $i = 2$ and $j = 3$ in the 4×4 Latin square above, we have $k = \text{B}$. Hence k is a function of i and j. If α_i and β_j are the effects of the ith row and jth column, τ_k the effect of the kth treatment, μ the grand mean, and ε_{ijk} the random error, then we can write

$$y_{ijk} = \mu + \alpha_i + \beta_j + \tau_k + \varepsilon_{ijk},$$

on which we impose the restrictions

$$\sum_i \alpha_i = \sum_j \beta_j = \sum_k \tau_k = 0.$$

As before, the y_{ijk} are assumed to be values of independent random variables having normal distributions with means

$$\mu_{ijk} = \mu + \alpha_i + \beta_j + \tau_k$$

and common variance σ^2.

The hypothesis to be tested is as follows:

$$H_0: \quad \alpha_1 = \alpha_2 = \cdots = \alpha_r = 0,$$

$$H_1: \quad \text{At least one of the } \alpha_i\text{'s is not equal to zero.}$$

This test will be based on a comparison of independent estimates of σ^2 provided by splitting the total sum of squares of our data into four components by means of the following identity. The reader is asked to provide the proof in Exercise 13 on page 510.

THEOREM 13.4
SUM-OF-SQUARES
IDENTITY

$$\sum_i \sum_j \sum_k (y_{ijk} - \bar{y}_{...})^2 = r \sum_i (\bar{y}_{i..} - \bar{y}_{...})^2 + r \sum_j (\bar{y}_{.j.} - \bar{y}_{...})^2$$

$$+ r \sum_k (\bar{y}_{..k} - \bar{y}_{...})^2 + \sum_i \sum_j \sum_k (y_{ijk} - \bar{y}_{i..} - \bar{y}_{.j.} - \bar{y}_{..k} + 2\bar{y}_{...})^2. \quad \blacksquare$$

Symbolically, we write the sum-of-squares identity as

$$SST = SSR + SSC + SSTr + SSE,$$

where SSR and SSC are called the row sum of squares and column sum of squares, respectively; $SSTr$ is called the treatment sum of squares; and SSE is the error sum of squares. The degrees of freedom are partitioned according to the identity

$$r^2 - 1 = (r - 1) + (r - 1) + (r - 1) + (r - 1)(r - 2).$$

Dividing each of the sum of squares on the right side of the sum-of-squares identity by their corresponding number of degrees of freedom, we obtain the four independent estimates

$$s_1^2 = \frac{SSR}{r-1}, \qquad s_2^2 = \frac{SSC}{r-1}, \qquad s_3^2 = \frac{SSTr}{r-1}, \qquad s^2 = \frac{SSE}{(r-1)(r-2)}$$

of σ^2. Interpreting the sums of squares as functions of independent random variables, it is not difficult to verify that

$$E(S_1^2) = E\left[\frac{SSR}{r-1}\right] = \sigma^2 + \frac{r\sum_i \alpha_i^2}{r-1},$$

$$E(S_2^2) = E\left[\frac{SSC}{r-1}\right] = \sigma^2 + \frac{r\sum_j \beta_j^2}{r-1},$$

$$E(S_3^2) = E\left[\frac{SSTr}{r-1}\right] = \sigma^2 + \frac{r\sum_k \tau_k^2}{r-1},$$

$$E(S^2) = E\left[\frac{SSE}{(r-1)(r-2)}\right] = \sigma^2.$$

The analysis of variance (Table 13.13) indicates the appropriate F-test for treatments.

TABLE 13.13 Analysis of Variance for an $r \times r$ Latin Square

Source of Variation	Sum of Squares	Degrees of Freedom	Mean Square	Computed f
Row	SSR	$r-1$	$s_1^2 = \dfrac{SSR}{r-1}$	
Columns	SSC	$r-1$	$s_2^2 = \dfrac{SSC}{r-1}$	
Treatments	SSTr	$r-1$	$s_3^2 = \dfrac{SSTr}{r-1}$	$f = \dfrac{s_3^2}{s^2}$
Error	SSE	$(r-1)(r-2)$	$s^2 = \dfrac{SSE}{(r-1)(t-2)}$	
Total	SST	r^2-1		

EXAMPLE 13.7   To illustrate the analysis of a Latin square design, let us return to the experiment in which the letters A, B, C, and D represent 4 varieties of wheat; the rows represent 4 different fertilizers; and the columns account for 4 different years. The data in Table 13.14 are the yields for the 4 varieties of wheat measured in kilograms per plot. It is assumed that the various sources of variation do not interact. Using a 0.05 level of significance, test the hypothesis H_0: there is no difference in the average yields of the 4 varieties of wheat.

SOLUTION

$$H_0: \quad \alpha_1 = \alpha_2 = \alpha_3 = \alpha_4 = 0,$$

$$H_1: \quad \text{At least one of the } \alpha_i\text{'s is not equal to zero.}$$

The sum of squares and degrees-of-freedom layout of Table 13.13 was used. The sum of squares formulas appear on page 505. Here, of course, the analysis-of-variance table (Table 13.15) must reflect variability accounted for due to fertilizer, years, and treatment types. The $f = 2.02$ is on 3 and 6 degrees of freedom. The P-value of approximately 0.2 is certainly too large to conclude that wheat varieties significantly affect wheat yield.

TABLE 13.14   Yields of Wheat in Kilograms per Plot

Fertilizer Treatment	1981		1982		1983		1984	
t_1	A	70	B	75	C	68	D	81
t_2	D	66	A	59	B	55	C	63
t_3	C	59	D	66	A	39	B	42
t_4	B	41	C	57	D	39	A	55

TABLE 13.15   Analysis of Variance for the Data of Table 13.14

Source of Variation	Sum of Squares	Degrees of Freedom	Mean Square	Computed f
Fertilizer	1557	3	519.000	
Year	418	3	139.333	
Treatments	264	3	88.000	2.02
Error	261	6	43.500	
Total	2500	15		

Exercises

1. Show that the computing formula for *SSB*, in the analysis of variance of the randomized complete block design, is equivalent to the corresponding term in the identity of Theorem 13.3.

2. For the randomized block design with k treatments and b blocks, show that

$$E(SSB) = (b - 1)\sigma^2 + k \sum_{j=1}^{b} \beta_j^2.$$

3. Four kinds of fertilizer f_1, f_2, f_3, and f_4 are used to study the yield of beans. The soil is divided into 3 blocks each containing 4 homogeneous plots. The yields in kilograms per plot and the corresponding treatments are as follows:

Block 1	Block 2	Block 3
$f_1 = 42.7$	$f_3 = 50.9$	$f_4 = 51.1$
$f_3 = 48.5$	$f_1 = 50.0$	$f_2 = 46.3$
$f_4 = 32.8$	$f_2 = 38.0$	$f_1 = 51.9$
$f_2 = 39.3$	$f_4 = 40.2$	$f_3 = 53.5$

(a) Conduct an analysis of variance at the 0.05 level of significance using the randomized complete block model.

(b) Use single-degree-of-freedom contrasts and a 0.01 level of significance to compare the fertilizers (f_1, f_3) versus (f_2, f_4) and f_1 versus f_3.

4. Three varieties of potatoes are being compared for yield. The experiment was conducted by assigning each variety at random to 3 equal-size plots at each of 4 different locations. The following yields for varieties A, B, and C, in 100 kilograms per plot, were recorded:

Location 1	Location 2	Location 3	Location 4
B 13	C 21	C 9	A 11
A 18	A 20	B 12	C 10
C 12	B 23	A 14	B 17

Perform a randomized block analysis of variance to test the hypothesis that there is no difference in the yielding capabilities of the 3 varieties of potatoes. Use a 0.05 level of significance.

5. The following data are the percents of foreign additives measured by 5 analysts for 3 similar brands of strawberry jam, A, B, and C:

Analyst 1	Analyst 2	Analyst 3	Analyst 4	Analyst 5
B 2.7	C 7.5	B 2.8	A 1.7	C 8.1
C 3.6	A 1.6	A 2.7	B 1.9	A 2.0
A 3.8	B 5.2	C 6.4	C 2.6	B 4.8

Perform the analysis of variance and test the hypothesis, at the 0.05 level of significance, that the percent of foreign additives is the same for all 3 brands of jam.

6. The following data represent the final grades obtained by 5 students in mathematics, English, French, and biology:

Student	Mathematics	English	French	Biology
1	68	57	73	61
2	83	94	91	86
3	72	81	63	59
4	55	73	77	66
5	92	68	75	87

Use a 0.05 level of significance to test the hypothesis that the courses are of equal difficulty.

7. In a study on "*The Periphyton of the South River, Virginia: Mercury Concentration, Productivity, and Autotropic Index Studies*" conducted by the Department of Environmental Sciences and Engineering at the Virginia Polytechnic Institute and State University in 1979, the total mercury concentration in periphyton total solids is being measured at 6 differ-

ent stations on 6 different data. The following data were recorded:

			Station			
Date	CA	CB	E1	E2	E3	E4
April 8	0.45	3.24	1.33	2.04	3.93	5.93
June 23	0.10	0.10	0.99	4.31	9.92	6.49
July 1	0.25	0.25	1.65	3.13	7.39	4.43
July 8	0.09	0.06	0.92	3.66	7.88	6.24
July 15	0.15	0.16	2.17	3.50	8.82	5.39
July 23	0.17	0.39	4.30	2.91	5.5G	4.29

Use a 0.01 level of significance to test the hypothesis that the mean mercury concentration is the same for the 6 stations.

8. A nuclear power facility produces a vast amount of heat which is usually discharged into aquatic systems. This heat raises the temperature of the aquatic system, resulting in a greater concentration of chlorophyll *a*, which in turn extends the growing season. To study this effect, water samples were collected monthly at 3 stations for a period of 12 months. Station A is located closest to a potential heated water discharge, station C is located farthest away from the discharge, while station B is located halfway between stations A and C. The following concentrations of chlorophyll *a* were recorded:

	Station		
Month	A	B	C
January	9.867	3.723	4.410
February	14.035	8.416	11.100
March	10.700	20.723	4.470
April	13.853	9.168	8.010
May	7.067	4.778	34.080
June	11.670	9.145	8.990
July	7.357	8.463	3.350
August	3.358	4.086	4.500
September	4.210	4.233	6.830
October	3.630	2.320	5.800
November	2.953	3.843	3.480
December	2.640	3.610	3.020

Perform an analysis of variance and test the hypothesis, at the 0.05 level of significance, that there is no difference in the mean concentrations of chlorophyll *a* at the 3 stations.

9. In a study conducted by the Department of Health and Physical Education at the Virginia Polytechnic Institute and State University in 1983, 3 diets were assigned for a period of 3 days to each of 6 subjects in a randomized block design. The subjects, playing the role of blocks, were assigned the following 3 diets in a random order:

Diet 1: mixed fat and carbohydrates,

Diet 2: high fat,

Diet 3: high carbohydrates.

At the end of the 3-day period each subject was put on a treadmill and the time to exhaustion, in seconds, was measured. The following data were recorded:

	Subject					
Diet	1	2	3	4	5	6
1	84	35	91	57	56	45
2	91	48	71	45	61	61
3	122	53	110	71	91	122

Perform the analysis of variance, separating out the diet, subject, and error sum of squares. Use a 0.01 level of significance to determine if there are significant differences among the diets.

10. Organic arsenicals are used by forestry personnel as silvicides. The amount of arsenic that is taken into the body when exposed to these silvicides is a major health problem. It is important that the amount of exposure be determined quickly so that a field worker with a high level of arsenic can be removed from the job. In an experiment reported in the paper "*A Rapid Method for the Determination of Arsenic Concentrations in Urine at Field Locations,*" published in the *Amer. Ind. Hyg. Assoc. J.* (Vol. 37, 1976), urine specimens from 4 forest service personnel were divided equally into 3 samples so that each individual could be analyzed for arsenic by a university laboratory, by a chemist using a portable system, and by a forest employee after a brief orientation. The following arsenic levels, in parts per million, were recorded:

		Analyst	
Individual	Employee	Chemist	Laboratory
1	0.05	0.05	0.04
2	0.05	0.05	0.04
3	0.04	0.04	0.03
4	0.15	0.17	0.10

Perform an analysis of variance and test the hypothesis, at the 0.05 level of significance, that there is no difference in the arsenic levels for the 3 methods of analysis.

11. Scientists in the Department of Plant Pathology at Virginia Tech devised an experiment in which 5 different treatments were applied to 6 different locations in an apple orchard to determine if there were significant differences in growth among the treatments. Treatments 1 through 4 represent different herbicides and treatment 5 represents a control. The growth period was from May to November in 1982, and the new growth, measured in centimeters, for samples selected from the 6 locations in the orchard were recorded as follows:

			Locations			
Treatment	1	2	3	4	5	6
1	455	72	61	215	695	501
2	622	82	444	170	437	134
3	695	56	50	443	701	373
4	607	650	493	257	490	262
5	388	263	185	103	518	622

Perform an analysis of variance, separating out the treatment, location, and error sum of squares. Use a 0.01 level of significance to test the hypothesis that there are no differences among the treatment means.

12. In the paper "*Self-Control and Therapist Control in the Behavioral Treatment of Overweight Women*" published in *Behavioral Research and Therapy* (Vol. 10, 1972), two reduction treatments and a control treatment were studied for their effects on the weight change of obese women. The two reduction treatments involved were, respectively, a self-induced weight reduction program and a therapist-controlled reduction program. Each of 10 subjects were assigned to the 3 treatment programs in a random order and measured for weight loss. The following weight changes were recorded:

		Treatment	
Subject	Control	Self-induced	Therapist
1	1.00	−2.25	−10.50
2	3.75	−6.00	−13.50
3	0.00	−2.00	0.75
4	−0.25	−1.50	−4.50
5	−2.25	−3.25	−6.00
6	−1.00	−1.50	4.00
7	−1.00	−10.75	−12.25
8	3.75	−0.75	−2.75
9	1.50	0.00	−6.75
10	−0.50	−3.75	−7.00

Perform an analysis of variance and test the hypothesis, at the 0.01 level of significance, that there is no difference in the mean weight losses for the 3 treatments.

13. Verify the sum-of-squares identity of Theorem 13.4 on page 505.

14. For the $r \times r$ Latin square design, show that

$$E(SSTr) = (r - 1)\sigma^2 + r \sum_k \tau_k^2.$$

15. The mathematics department of a large university wishes to evaluate the teaching capabilities of 4 professors. In order to eliminate any effects due to different mathematics courses and different times of the day, it was decided to conduct an experiment using a Latin square design in which the letters A, B, C, and D represent the 4 different professors. Each professor taught one section of each of 4 different courses scheduled at each of 4 different times during the day. The data in the following table show the grades assigned by these professors to 16 students of approximately equal ability.

Time Period	Course			
	Algebra	Geometry	Statistics	Calculus
1	A 84	B 79	C 63	D 97
2	B 91	C 82	D 80	A 93
3	C 59	D 70	A 77	B 80
4	D 75	A 91	B 75	C 68

Use a 0.05 level of significance to test the hypothesis that different professors have no effect on the grades.

16. A manufacturing firm wants to investigate the effects of 5 color additives on the setting time of a new concrete mix. Variations in the setting times can be expected from day-to-day changes in temperature and humidity and also from the different workers who prepare the test molds. To eliminate these extraneous sources of variation, a 5 × 5 Latin square design was used in which the letters A, B, C, D, and E represent the 5 additives. The setting times, in hours, for the 25 molds are shown in the following table.

Worker	Day				
	1	2	3	4	5
1	D 10.7	E 10.3	B 11.2	A 10.9	C 10.5
2	E 11.3	C 10.5	D 12.0	B 11.5	A 10.3
3	A 11.8	B 10.9	C 10.5	D 11.3	E 7.5
4	B 14.1	A 11.6	E 11.0	C 11.7	D 11.5
5	C 14.5	D 11.5	A 11.5	E 12.7	B 10.9

At the 0.05 level of significance, can we say that the color additives have any effect on the setting time of the concrete mix?

17. In the book *Design of Experiments for the Quality Improvement* published by the Japanese Standards Association (1989), a study on the amount of dye needed to get the best color for a certain type of a fabric was conducted. The three amounts of dye: $\frac{1}{3}$% owf ($\frac{1}{3}$% of the weight of a fabric), 1% owf, and 3% owf were each administered at two different plants. The color density of a fabric was then observed four times for each level of dye at each plant.

	Amount of dye					
	1		2		3	
Plant 1	5.2	6.0	12.3	10.5	22.4	17.8
	5.9	5.9	12.4	10.9	22.5	18.4
Plant 2	6.5	5.5	14.5	11.8	29.0	23.2
	6.4	5.9	16.0	13.6	29.7	24.0

Perform an analysis of variance to test the hypothesis, at the 0.05 level of significance, that there is no difference in the color density of a fabric for the three levels of dye. Consider plants to be blocks.

13.12 Random Effects Models

Throughout this chapter we have dealt with analysis-of-variance procedures in which the primary goal was to study the effect on some response of certain fixed or predetermined treatments. Experiments in which the treatments or treatment levels are preselected by the experimenter as opposed to being chosen randomly are called **fixed effects experiments** or **model I experiments**. For the fixed effects model, inferences were made only on those particular treatments used in the experiment.

It is often important that the experimenter be able to draw inferences about a population of treatments by means of an experiment in which the treatments used are chosen randomly from the population. For example, a biologist may be interested in whether or not there is a significant variance in some physiological characteristic due to animal type. The animal types actually used in the experiment are then chosen randomly and represent the treatment effects. A chemist may be interested in studying the effect of analytical laboratories on the chemical analysis of a substance. He is not concerned with particular laboratories but rather with a large population of laboratories. He might then select a group of laboratories at random and allocate samples to each for analysis. The statistical inference would then involve (1) testing whether or not the laboratories contribute a nonzero variance to the analytical results, and (2) estimating the variance due to laboratories and the variance within laboratories.

The one-way **random effects model**, often referred to as **model II**, is written like the fixed effects model but with the terms taking on different meanings. The response

$$y_{ij} = \mu + \alpha_i + \varepsilon_{ij}$$

is now a value of the random variable

$$Y_{ij} = \mu + A_i + E_{ij}$$

with $i = 1, 2, \ldots, k$ and $j = 1, 2, \ldots, n$, where the A_i's are normally and independently distributed with mean zero and variance σ_α^2 and are independent of the E_{ij}'s. As for the fixed effects model, the E_{ij}'s are also normally and independently distributed with mean zero and variance σ^2. Note that for a model II experiment the random variable $\sum_{i=1}^{k} A_i$ assumes the value $\sum_{i=1}^{k} \alpha_i$: and the constraint that these α_i's sum to zero no longer applies.

THEOREM 13.5   *For the random effects one-way analysis-of-variance model*

$$E(SSA) = (k - 1)\sigma^2 + n(k - 1)\sigma_\alpha^2$$

and

$$E(SSE) = k(n - 1)\sigma^2.$$

∎

PROOF. From the model

$$Y_{ij} = \mu + A_i + E_{ij}$$

we obtain

$$\overline{Y}_{i\cdot} = \mu + A_i + \overline{E}_{i\cdot},$$

$$\overline{Y}_{\cdot\cdot} = \mu + \overline{A}_\cdot + \overline{E}_{\cdot\cdot}.$$

Hence

$$SSA = n \sum_{i=1}^{k} (\overline{Y}_{i\cdot} - \overline{Y}_{\cdot\cdot})^2$$

$$= n \sum_{i=1}^{k} [(A_i - \overline{A}_\cdot) + (\overline{E}_{i\cdot} - \overline{E}_{\cdot\cdot})]^2$$

and

$$E(SSA) = n \sum_{i=1}^{k} E(A_i^2) - nkE(\overline{A}_\cdot^2) + n \sum_{i=1}^{k} E(\overline{E}_{i\cdot}^2) - nkE(\overline{E}_{\cdot\cdot}^2)$$

$$= nk\sigma_\alpha^2 - n\sigma_\alpha^2 + k\sigma^2 - \sigma^2$$

$$= (k - 1)\sigma^2 + n(k - 1)\sigma_\alpha^2.$$

Following the same steps as above, we also find that

$$SSE = \sum_{i=1}^{k} \sum_{j=1}^{n} (Y_{ij} - \overline{Y}_{i\cdot})^2$$

$$= \sum_{i=1}^{k} \sum_{j=1}^{n} (E_{ij} - \overline{E}_{i\cdot})^2$$

and therefore

$$E(SSE) = \sum_{i=1}^{k} \sum_{j=1}^{n} E(E_{ij}^2) - n \sum_{i=1}^{k} E(\overline{E}_{i\cdot}^2)$$

$$= nk\sigma^2 - k\sigma^2$$

$$= k(n - 1)\sigma^2.$$

Table 13.16 shows the expected mean squares for both a model I and a model II experiment. The computations for a model II experiment are carried out in exactly the same way as for a model I experiment. That is, the sum-of-squares, degrees-of-freedom, and mean-square columns in an analysis-of-variance table are the same for both models.

TABLE 13.16 Expected Mean Squares for the One-Factor Experiment

Source of Variation	Degrees of Freedom	Mean Squares	Expected Mean Squares Model I	Model II
Treatments	$k - 1$	s_1^2	$\sigma^2 + \dfrac{n \sum_{i=1}^{k} \alpha_i^2}{k - 1}$	$\sigma^2 + n\sigma_\alpha^2$
Error	$k(n - 1)$	s^2	σ^2	σ^2
Total	$nk - 1$			

For the random effects model, the hypothesis that the treatment effects are all zero is written

$$H_0: \quad \sigma_\alpha^2 = 0,$$

$$H_1: \quad \sigma_\alpha^2 \neq 0,$$

which says that the different treatments contribute nothing to the variability of the response. It is obvious from Table 13.16 that s_1^2 and s^2 are both estimates of σ^2 when H_0 is true and that the ratio

$$f = \frac{s_1^2}{s^2}$$

is a value of random variable F having the F-distribution with $k - 1$ and $k(n - 1)$ degrees of freedom. The null hypothesis is rejected at the α-level of significance when

$$f > f_\alpha[k - 1, \, k(n - 1)].$$

In many scientific and engineering studies, interest is not centered on the F-test. The scientist knows that the random effect does, indeed, have a significant effect. What is more important is estimation of the various variance components. This produces a sense of *ranking* in terms of what factors produce the most variability and by how much. In the present context it may be of interest to quantify how much larger the *single-factor variance component* is than that produced by chance (random variation).

Table 13.16 can also be used to estimate the **variance components** σ^2 and σ_α^2. Since s_1^2 estimates $\sigma^2 + n\sigma_\alpha^2$ and s^2 estimates σ^2,

$$\hat{\sigma}^2 = s^2$$

$$\hat{\sigma}_\alpha^2 = \frac{s_1^2 - s^2}{n}.$$

EXAMPLE 13.8   The following data are coded observations on the yield of a chemical process using 5 batches of raw material selected randomly:

	Batch				
	1	2	3	4	5
	9.7	10.4	15.9	8.6	9.7
	5.6	9.6	14.4	11.1	12.8
	8.4	7.3	8.3	10.7	8.7
	7.9	6.8	12.8	7.6	13.4
	8.2	8.8	7.9	6.4	8.3
	7.7	9.2	11.6	5.9	11.7
	8.1	7.6	9.8	8.1	10.7
Total	55.6	59.7	80.7	58.4	75.3	329.7

Show that the batch variance component is significantly greater than zero and obtain its estimate.

SOLUTION
The total, batch, and error sum of squares are given by

$$SST = 194.64$$

$$SSA = 72.60$$

$$SSE = 194.64 - 72.60 = 122.04.$$

These results, with the remaining computations, are given in Table 13.17. The f-ratio is significant at the $\alpha = 0.05$ level, indicating that the hypothesis of a zero batch component is rejected. An estimate of the batch variance component is

$$\hat{\sigma}_\alpha^2 = \frac{18.15 - 4.07}{7} = 2.01.$$

TABLE 13.17   Analysis of Variance for Example 13.8

Source of Variation	Sum of Squares	Degrees of Freedom	Mean Square	Computed f
Batches	72.60	4	18.15	4.46
Error	122.04	30	4.07	
Total	194.64	34		

Randomized Block Design with Random Blocks

In a randomized complete block experiment in which the blocks represent days, it is conceivable that the experimenter would like the results to apply not only to the actual days used in the analysis but to every day in the year. He or she would then select the days on which to run the experiment as well as the treatments at random and use the random effects model $Y_{ij} = \mu + A_i + B_j + E_{ij}$, $i = 1, 2, \ldots, k$ and $j = 1, 2, \ldots, b$, with the A_i, B_j, and E_{ij} being independent random variables with means zero and variances σ_α^2, σ_β^2, and σ^2, respectively. The expected mean squares for a model II randomized complete block design are obtained using the same procedure as for the one-factor problem and are presented along with those for a model I experiment in Table 13.18.

TABLE 13.18   Expected Mean Squares for the Randomized Complete Block Design

Source of Variation	Degrees of Freedom	Mean Square	Expected Mean Squares Model I	Model II
Treatments	$k - 1$	s_1^2	$\sigma^2 + \dfrac{b \sum\limits_{i=1}^{k} \alpha_i^2}{k - 1}$	$\sigma^2 + b\sigma_\alpha^2$
Blocks	$b - 1$	s_2^2	$\sigma^2 + \dfrac{k \sum\limits_{j=1}^{b} \beta_j^2}{b - 1}$	$\sigma^2 + k\sigma_\beta^2$
Error	$(k - 1)(b - 1)$	s^2	σ^2	σ^2
Total	$bk - 1$			

Again the computations for the individual sum of squares and degrees of freedom are identical to those of the fixed effects model. The hypothesis

$$H_0: \quad \sigma_\alpha^2 = 0,$$
$$H_1: \quad \sigma_\alpha^2 \neq 0,$$

is carried out by computing

$$f = \frac{s_1^2}{s^2}$$

and rejecting H_0 when $f > f_\alpha[k - 1, (b - 1)(k - 1)]$.

The unbiased estimates of the variance components are given by

$$\hat{\sigma}^2 = s^2$$
$$\hat{\sigma}_\alpha^2 = \frac{s_1^2 - s^2}{b}$$
$$\hat{\sigma}_\beta^2 = \frac{s_2^2 - s^2}{k}.$$

For the Latin square design, the random effects model is written

$$Y_{ijk} = \mu + A_i + B_j + T_k + E_{ijk},$$

$i = 1, 2, \ldots, r; j = 1, 2, \ldots, r;$ and $k = A, B, C, \ldots,$ with $A_i, B_j, T_k,$ and E_{ijk} being independent random variables with means zero and variances σ_α^2, σ_β^2, σ_τ^2, and σ^2, respectively. The derivation of the expected mean squares for a model II Latin

TABLE 13.19 Expected Mean Squares for a Latin Square Design

Source of Variation	Degrees of Freedom	Mean Square	Expected Mean Squares	
			Model I	Model II
Rows	$r - 1$	s_1^2	$\sigma^2 + \dfrac{r \sum_i \alpha_i^2}{r - 1}$	$\sigma^2 + r\sigma_\alpha^2$
Columns	$r - 1$	s_2^2	$\sigma^2 + \dfrac{r \sum_j \beta_j^2}{r - 1}$	$\sigma^2 + r\sigma_\beta^2$
Treatments	$r - 1$	s_3^2	$\sigma^2 + \dfrac{r \sum_k \tau_k^2}{r - 1}$	$\sigma^2 + r\sigma_\tau^2$
Error	$(r - 1)(r - 2)$	s^2	σ^2	σ^2
Total	$r^2 - 1$			

square design is straightforward and, for comparison, we present them along with those for a model I experiment in Table 13.19.

Tests of hypotheses concerning the various variance components are made by computing the ratios of appropriate mean squares as indicated in Table 13.19, and comparing with corresponding f-values from Table A.6.

13.13 Regression Approach to Analysis of Variance _____

So far, we have treated the regression models and the analysis-of-variance models as two separate and unrelated topics. Although this has become the accepted approach in dealing with these procedures on an elementary level, one can treat an analysis-of-variance model as a special case of a multiple linear regression model. In this section we show the relationship between the two models and indicate how the analysis-of-variance techniques can be developed through a regression approach.

Suppose that we consider two models, the multiple linear regression model

$$y_i = \beta_0 + \beta_1 x_{1i} + \beta_2 x_{2i} + \cdots + \beta_k x_{ki} + \varepsilon_i$$

and the one-way classification analysis-of-variance model

$$y_{ij} = \mu + \alpha_i + \varepsilon_{ij}.$$

Traditionally, the two are presented as methods for handling different practical problems, the regression model being a means of arriving at a procedure for predicting some response as a function of one or more quantitative independent variables, and the analysis-of-variance model for arriving at significance tests on multiple population means. However, any mathematical model that is linear in the parameters, such as the analysis-of-variance model, can be considered a special case of the multiple linear regression model. We can use conventional matrix notation to describe how each observation is expressed as a function of the parameters for the two models. For the regression model

$$\mathbf{y} = \mathbf{X}\boldsymbol{\beta} + \boldsymbol{\varepsilon},$$

or, more explicitly,

$$
\begin{bmatrix} y_1 \\ y_2 \\ \vdots \\ y_n \end{bmatrix} = \begin{bmatrix} 1 & x_{11} & x_{21} & \cdots & x_{k1} \\ 1 & x_{12} & x_{22} & \cdots & x_{k2} \\ \vdots & \vdots & \vdots & & \vdots \\ 1 & x_{1n} & x_{2n} & \cdots & x_{kn} \end{bmatrix} \begin{bmatrix} \beta_0 \\ \beta_1 \\ \vdots \\ \beta_k \end{bmatrix} + \begin{bmatrix} \varepsilon_1 \\ \varepsilon_2 \\ \vdots \\ \varepsilon_n \end{bmatrix},
$$

where the $\mathbf{y}$ vector on the left of the equality sign is the array of responses in the experiment. The $\mathbf{X}$ matrix has already been described and used in Section 12.3. The $\boldsymbol{\beta}$ vector is the vector of parameters appearing in the model, and the $\boldsymbol{\varepsilon}$ vector completes the model by the addition of the random error. The reader will recall that the least squares estimates $b_0, b_1, \ldots, b_k$ of the parameters $\beta_0, \beta_1, \ldots, \beta_k$ are obtained by solving the equation

$$\mathbf{Ab} = \mathbf{g},$$

where $\mathbf{A} = \mathbf{X}'\mathbf{X}$ is a nonsingular matrix and $\mathbf{g} = \mathbf{X}'\mathbf{y}$ is a vector whose elements are sums of products of elements in the columns of $\mathbf{X}$ and the elements in the vector $\mathbf{y}$. Thus the estimates are given by

$$\mathbf{b} = (\mathbf{X}'\mathbf{X})^{-1}\mathbf{X}'\mathbf{y}.$$

Consider now the analysis-of-variance model in matrix form:

$$
\begin{bmatrix} y_{11} \\ y_{12} \\ \vdots \\ y_{1n} \\ \hline y_{21} \\ y_{22} \\ \vdots \\ y_{2n} \\ \hline \vdots \\ \hline y_{k1} \\ y_{k2} \\ \vdots \\ y_{kn} \end{bmatrix} = \begin{bmatrix} 1 & 1 & 0 & \cdots & 0 \\ 1 & 1 & 0 & \cdots & 0 \\ \vdots & \vdots & \vdots & & \vdots \\ 1 & 1 & 0 & \cdots & 0 \\ \hline 1 & 0 & 1 & \cdots & 0 \\ 1 & 0 & 1 & \cdots & 0 \\ \vdots & \vdots & \vdots & & \vdots \\ 1 & 0 & 1 & \cdots & 0 \\ \hline \vdots & \vdots & \vdots & & \vdots \\ \hline 1 & 0 & 0 & \cdots & 1 \\ 1 & 0 & 0 & \cdots & 1 \\ \vdots & \vdots & \vdots & & \vdots \\ 1 & 0 & 0 & \cdots & 1 \end{bmatrix} \begin{bmatrix} \mu \\ \alpha_1 \\ \alpha_2 \\ \vdots \\ \alpha_k \end{bmatrix} + \begin{bmatrix} \varepsilon_{11} \\ \varepsilon_{12} \\ \vdots \\ \varepsilon_{1n} \\ \hline \varepsilon_{21} \\ \varepsilon_{22} \\ \vdots \\ \varepsilon_{2n} \\ \hline \vdots \\ \hline \varepsilon_{k1} \\ \varepsilon_{k2} \\ \vdots \\ \varepsilon_{kn} \end{bmatrix}
$$

Again each observation is expressed as a function of the parameters. Here the very important $\mathbf{X}$ matrix, the matrix of experimental conditions, consists of ones and zeros. Similar formulations can be written for the randomized complete block model.

Let us apply the least squares approach to the one-way analysis-of-variance model. The normal equations are given by

$$\begin{bmatrix} nk & n & n & \cdots & n \\ n & n & 0 & \cdots & 0 \\ n & 0 & n & \cdots & 0 \\ \vdots & \vdots & \vdots & & \vdots \\ n & 0 & 0 & \cdots & n \end{bmatrix} \begin{bmatrix} \hat{\mu} \\ \hat{\alpha}_1 \\ \hat{\alpha}_2 \\ \vdots \\ \hat{\alpha}_k \end{bmatrix} = \begin{bmatrix} T. \\ T_1 \\ T_2 \\ \vdots \\ T_k \end{bmatrix}.$$

At this stage it is simple to illustrate why a distinction is made in the presentation of the two models. The last k columns of the $\mathbf{A}$ matrix for the analysis-of-variance model add to the first column and thus the matrix is **singular**, implying that there is no unique solution to the estimating equations. This initially seems like a serious drawback as far as the model is concerned. In fact, we say that the parameters in the model are not **estimable**. The reader will recall that significance tests were performed on the population means, $\mu_1 = \mu + \alpha_1$, $\mu_2 = \mu + \alpha_2$, . . . , $\mu_k = \mu +$ α_k, and, in formulating the test procedure, the linear constraint $\sum\limits_{i=1}^{k} \alpha_i = 0$ was applied. Thus the α_i's take on the role of deviations (plus or minus) of the treatment or population means from the overall mean μ. Testing equality of population means then becomes equivalent to testing that the α_i's ($i = 1, 2, . . . , k$) are all zero.

With the constraint that the α_i's sum to zero, the estimating equations can be solved to yield

$$\hat{\mu} = \frac{T_{..}}{nk} = \bar{y}_{..}$$

$$\hat{\alpha}_i = \frac{T_{i.}}{n} - \frac{T_{..}}{nk} = \bar{y}_{i.} - \bar{y}_{..}, \qquad i = 1, 2, . . . , k.$$

While these estimates are not unique, since they are dependent on the constraint that was applied to the α_i's, they do give us a basis for using the general regression procedure outlined in Section 12.6 to determine if the deletion of the α_i's from the model significantly increases the error sum of squares, thereby providing us, in the regression context, with a test of hypothesis of no significant treatment effects.

If we approach the hypothesis-testing problem for the one-way analysis-of-variance model following the multiple regression procedures in Chapter 12, we might begin by computing the regression sum of squares for the parameters $\alpha_1, \alpha_2, . . . ,$ α_k. These parameters take on the same role as the coefficients $\beta_1, \beta_2, . . . , \beta_k$ in the multiple linear regression model. We would then compute the regression sum of squares

$$R(\alpha_1, \alpha_2, \ldots, \alpha_k) = SSR$$

$$= b_0 g_0 + b_1 g_1 + \cdots + b_k g_k - \frac{\left(\sum\limits_{i=1}^{k} \sum\limits_{j=1}^{n} y_{ij}^2\right)}{nk}$$

$$= b_1 g_1 + \cdots + b_k g_k$$

$$= \hat{\alpha}_1 g_1 + \cdots + \hat{\alpha}_k g_k.$$

The right side of the estimating equations gives $g_1 = T_{1\cdot}, g_2 = T_{2\cdot}, \ldots, g_k = T_{k\cdot}$. Hence

$$R(\alpha_1, \alpha_2, \ldots, \alpha_k) = \sum_{i=1}^{k} \left(\frac{T_{i\cdot}}{n} - \frac{T_{\cdot\cdot}}{nk}\right) T_{i\cdot}$$

$$= \sum_{i=1}^{k} \frac{T_{i\cdot}^2}{n} - \frac{T_{\cdot\cdot}^2}{nk}$$

$$= SSA,$$

with $k - 1$ degrees of freedom rather than k. One degree of freedom is lost because of the single linear restraint imposed on the treatment effects. The error sum of squares with $(nk - 1) - (k - 1) = k(n - 1)$ degrees of freedom is given by

$$SSE = SST - R(\alpha_1, \alpha_2, \ldots, \alpha_k)$$

$$= SST - SSA,$$

which is identical to the expression developed earlier in this chapter.

The hypothesis that the regression on the α_i's is insignificant, that is, $a_i = 0$ for all i's, is tested by forming the ratio

$$f = \frac{R(\alpha_1, \alpha_2, \ldots, \alpha_k)/(k - 1)}{SSE/k(n - 1)} = \frac{SSA/(k - 1)}{s^2}.$$

A value of $f > f_\alpha[(k - 1), k(n - 1)]$ implies that regression is significantly increased and consequently the error sum of squares is significantly decreased by including the treatment effects in the model.

The regression approach to analysis-of-variance-type models can be extended to the randomized complete block and Latin square designs discussed in Sections 13.8 and 13.9, and also to the factorial designs of Chapter 14.

13.14 Power of Analysis-of-Variance Tests

As we indicated earlier, the research worker is often plagued by the problem of not knowing how large a sample to choose. In planning a one-factor completely randomized design with n observations per treatment, the main objective is to test the hypothesis of equality of treatment means.

$$H_0: \quad \alpha_1 = \alpha_2 = \cdots = \alpha_k = 0,$$

$$H_1: \quad \text{At least one of the } \alpha_i\text{'s is not equal to zero.}$$

Quite often, however, the experimental error variance, σ^2, is so large that the test procedure will be insensitive to actual differences among the k treatment means. In Section 13.3 the expected values of the mean squares for the one-way model were given by

$$E(S_1^2) = E\left(\frac{SSA}{k-1}\right) = \sigma^2 + \frac{n \sum\limits_{i=1}^{k} \alpha_i^2}{k-1}$$

$$E(S^2) = E\left(\frac{SSE}{k(n-1)}\right) = \sigma^2.$$

Thus, for a given deviation from the null hypothesis H_0, as measured by n

$$\sum_{i=1}^{k} \alpha_i^2/(k-1),$$

large values of σ^2 decrease the chance of obtaining a value $f = s_1^2/s^2$ that is in the critical region for the test. The sensitivity of the test describes the ability of the procedure to detect differences in the population means and is measured by the power of the test (see Section 10.2), which is merely $1 - \beta$, where β is the probability of accepting a false hypothesis. We can interpret the power of our analysis-of-variance tests, then, as the probability that the F-statistic is in the critical region when, in fact, the null hypothesis is false and the treatment means do differ. For the one-way analysis-of-variance test, the power, $1 - \beta$, is given by

$$1 - \beta = P\left[\frac{S_1^2}{S^2} > f_\alpha(v_1, v_2) \text{ when } H_1 \text{ is true}\right]$$

$$= P\left[\frac{S_1^2}{S^2} > f_\alpha(v_1, v_2) \text{ when } \sum_{i=1}^{k} \alpha_i^2 \neq 0\right].$$

The term $f_\alpha(v_1, v_2)$ is, of course, the upper-tail critical point of the F-distribution with v_1 and v_2 degrees of freedom. For given values of $\sum\limits_{i=1}^{k} \alpha_i^2/(k-1)$ and σ^2, the power can be increased by using a larger sample size n. The problem becomes one of designing the experiment with a value of n so that the power requirements are met. For example, we might require that for specific values of $\sum\limits_{i=1}^{k} \alpha_i^2 \neq 0$ and σ^2, the hypothesis be rejected with probability 0.9. When the power of the test is low, it severely limits the scope of the inferences that can be drawn from the experimental data.

Fixed Effects Case

In the analysis of variance the power depends on the distribution of the F-ratio under the alternative hypothesis that the treatment means differ. Therefore, in the case of the one-way fixed effects model, we require the distribution of S_1^2/S^2 when, in fact,

$\sum_{i=1}^{k} \alpha_i^2 \neq 0$. Of course, when the hypothesis is true, $\alpha_1 = 0$ for $i = 1, 2, \ldots, k$,

and the statistic follows the F-distribution with $k - 1$ and $N - k$ degrees of free-

dom. If $\sum_{i=1}^{k} \alpha_i^2 \neq 0$, the ratio follows a **noncentral F-distribution**.

The basic random variable of the noncentral F is denoted by F'. Let $f'_\alpha(v_1, v_2, \lambda)$ be a value of F' with parameters v_1, v_2, and λ. The parameters v_1 and v_2 of the distribution are the degrees of freedom associated with S_1^2 and S^2, respectively, and λ is called the **noncentrality parameter**. When $\lambda = 0$, the noncentral F simply reduces to the ordinary F-distribution with v_1 and v_2 degrees of freedom.

For the fixed effects, one-way analysis of variance with sample sizes n_1, n_2, $\ldots$, n_k, we define

$$\lambda = \frac{\sum_{i=1}^{k} n_i \alpha_i^2}{2\sigma^2}.$$

If we have tables of the noncentral F at our disposal, the power for detecting a particular alternative is obtained by evaluating the following probability:

$$1 - \beta = P\left[\frac{S_1^2}{S^2} > f_\alpha(k - 1, N - k) \text{ when } \lambda = \frac{\sum_{i=1}^{k} n_i \alpha_i^2}{2\sigma^2}\right]$$

$$= P[F' > f_\alpha(k - 1, N - k)].$$

Although the noncentral F is normally defined in terms of λ, it is more convenient, for purposes of tabulation, to work with

$$\phi^2 = \frac{2\lambda}{v_1 + 1}.$$

Table A.15 gives graphs of the power of the analysis of variance as a function of ϕ for various values of v_1, v_2, and the significance level α. These **power charts** can be used not only for the fixed effects models discussed in this chapter, but also for the multifactor models of Chapter 14. It remains now to give a procedure whereby the noncentrality parameter λ, and thus ϕ, can be found for these fixed effects cases.

The noncentrality parameter λ can be written in terms of the **expected values of the numerator mean square** of the F-ratio in the analysis of variance. We have

$$\lambda = \frac{v_1[E(S_i^2)]}{2\sigma^2} - \frac{v_1}{2}$$

and thus

$$\phi^2 = \frac{[E(S_i^2) - \sigma^2]}{\sigma^2} \frac{v_1}{v_1 + 1}.$$

Expressions for λ and ϕ^2 for the one-way model, the randomized complete block design, and the Latin square design are given in Table 13.20.

Note from Table A.15 that for given values of v_1 and v_2, the power of the test increases with increasing values of ϕ. The value of λ depends, of course, on σ^2, and in a practical problem one may often need to substitute the error mean square as an estimate in determining ϕ^2.

TABLE 13.20 Noncentrality Parameter λ and ϕ^2 for Fixed Effects Model

	One-Way Classification	Randomized Complete Block	Latin Square
λ	$\dfrac{\sum\limits_{i} n_i \alpha_i^2}{2\sigma^2}$	$\dfrac{b \sum\limits_{i} \alpha_i^2}{2\sigma^2}$	$\dfrac{r \sum\limits_{k} \tau_k^2}{2\sigma^2}$
ϕ^2	$\dfrac{\sum\limits_{i} n_i \alpha_i^2}{k\sigma^2}$	$\dfrac{b \sum\limits_{i} \alpha_i^2}{k\sigma^2}$	$\dfrac{\sum\limits_{k} \tau_k^2}{\sigma^2}$

EXAMPLE 13.9   In a randomized block experiment 4 treatments are to be compared in 6 blocks, resulting in 15 degrees of freedom for error. Are 6 blocks sufficient if the power of our test for detecting differences among the treatment means, at the 0.05 level of significance, is to be at least 0.8 when the true means are $\mu_{1.} = 5.0$, $\mu_{2.} = 7.0$, $\mu_{3.} = 4.0$, and $\mu_{4.} = 4.0$? An estimate of σ^2 to be used in the computation of the power is given by $\hat{\sigma}^2 = 2.0$.

SOLUTION
Recall that the treatment means are given by $\mu_{i.} = \mu + \alpha_i$. If we invoke the restriction that $\sum\limits_{i=1}^{4} \alpha_i = 0$, we have

$$\mu = \frac{\sum\limits_{i=1}^{4} \mu_{i.}}{4} = 5.0,$$

and then $\alpha_1 = 0$, $\alpha_2 = 2.0$, $\alpha_3 = -1.0$, and $\alpha_4 = -1.0$. Therefore,

$$\phi^2 = \frac{b \sum\limits_{i=1}^{k} \alpha_i^2}{k\sigma^2} = \frac{(6)(6)}{(4)(2)} = 4.5,$$

from which we obtain $\phi = 2.121$. Using Table A.15, the power is found to be approximately 0.89 and thus the power requirements are met. This means that if the value of $\sum_{i=1}^{4} \alpha_i^2 = 6$ and $\sigma^2 = 2.0$, the use of 6 blocks will result in rejecting the hypothesis of equal treatment means with probability 0.89.

Random Effects Case

In the fixed effects case, the computation of power requires the use of the noncentral F-distribution. Such is not the case in the random effects model. In fact, the power is computed very simply by the use of the standard F-tables. Consider, for example, the one-way random effects model, n observations per treatment, with the hypothesis

$$H_0: \quad \sigma_\alpha^2 = 0,$$

$$H_1: \quad \sigma_\alpha^2 \neq 0.$$

When H_1 is true, the ratio

$$f = \frac{SSA/[(k-1)(\sigma^2 + n\sigma_\alpha^2)]}{SSE/k(n-1)\sigma^2} = \frac{s_1^2}{s^2(1 + n\sigma_\alpha^2/\sigma^2)}$$

is a value of the random variable F having the F-distribution with $k-1$ and $k(n-1)$ degrees of freedom. The problem becomes one, then, of determining the probability of rejecting H_0 under the condition that the true treatment variance component $\sigma_\alpha^2 \neq 0$. We have then

$$1 - \beta = P\left\{\frac{S_1^2}{S^2} > f_\alpha[(k-1), k(n-1)] \text{ when } H_1 \text{ is true}\right\}$$

$$= P\left\{\frac{S_1^2}{S^2(1 + n\sigma_\alpha^2/\sigma^2)} > \frac{f_\alpha[(k-1), k(n-1)]}{1 + n\sigma_\alpha^2/\sigma^2}\right\}$$

$$= P\left\{F > \frac{f_\alpha[(k-1), k(n-1)]}{1 + n\sigma_\alpha^2/\sigma^2}\right\}.$$

Note that as n increases, the value $f_\alpha[(k-1), k(n-1)]/(1 + n\sigma_\alpha^2/\sigma^2)$, approaches zero, resulting in an increase in the power of the test. An illustration of the power for this kind of situation is given in Figure 13.7. The lighter shaded area is the significance level α, while the entire shaded area is the power of the test.

EXAMPLE 13.10   Suppose in a one-factor problem it is of interest to test for the significance of the variance component σ_α^2. Four treatments are to be used in the experiment with 5 observations per treatment. What will be the probability of rejecting the hypothesis $\sigma_\alpha^2 = 0$, when in fact the treatment variance component is $(3/4)\sigma^2$?

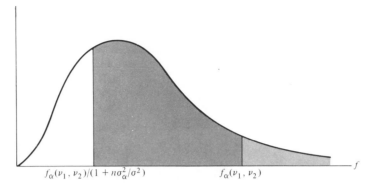

FIGURE 13.7 Power for the random effects one-way analysis of variance.

SOLUTION
Using an $\alpha = 0.05$ significance level, we have

$$1 - \beta = P\left\{ F > \frac{f_{0.05}(3,\ 16)}{1 + (5)(3)/4} \right\}$$

$$= P\left[F > \frac{f_{0.05}(3,\ 16)}{4.75} \right]$$

$$= P\left(F > \frac{3.24}{4.75} \right)$$

$$= P(F > 0.682).$$

Using Theorem 8.7 and then Table A-7c of *Introduction to Statistics*, 3rd ed., by Dixon and Massey (see the Bibliography), we see that

$$1 - \beta \approx 0.58.$$

Therefore, only about 58% of the time will the test procedure detect a variance component that is $(3/4)\sigma^2$.

13.15 Case Study

Personnel in the Chemistry Department of Virginia Tech were called upon to analyze a data set that was produced to compare four different methods of analysis of aluminum in a certain solid igniter mixture. To get a broad range of analytical laboratories involved, five laboratories were used in the experiment. These laboratories were selected because they are generally adept in doing these types of analyses. Twenty samples of igniter material containing 2.70% aluminum were assigned randomly, four to each laboratory, and directions were given on how to carry out the chemical analysis using all four methods. The data retrieved are as follows:

| Method | Laboratory | | | | | Mean |
	1	2	3	4	5	
A	2.67	2.69	2.62	2.66	2.70	2.668
B	2.71	2.74	2.69	2.70	2.77	2.722
C	2.76	2.76	2.70	2.76	2.81	2.758
D	2.65	2.69	2.60	2.64	2.73	2.662

The laboratories are not considered as random effects since they were not selected randomly from a larger population of laboratories. The data were analyzed as a randomized complete block design. Plots of these data are sought to determine if an additive model of the type

$$y_{ij} = \mu + m_i + l_i + \varepsilon_{ij}$$

is appropriate: in other words, a model with additive effects. The randomized block is not appropriate when interaction between laboratories and methods exist. Consider the plot shown in Figure 13.8. Although this plot is a bit difficult to interpret because each point is a single observation, there appears to be no appreciable interaction between methods and laboratories.

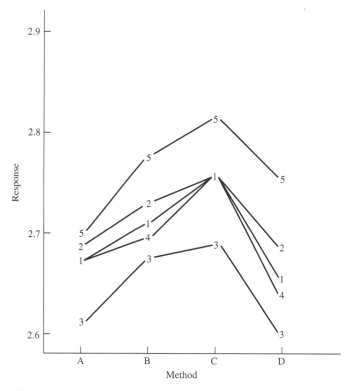

FIGURE 13.8 Interaction plot for data in Case Study 13.15.

Residual Plots

Residual plots were used as diagnostic indications regarding the homogeneous variance assumption. Figure 13.9 shows a plot of residuals against analytical methods. The variability depicted in the residuals seem to be remarkably homogeneous. To be complete, a normal probability plot of the residuals is shown in Figure 13.10.

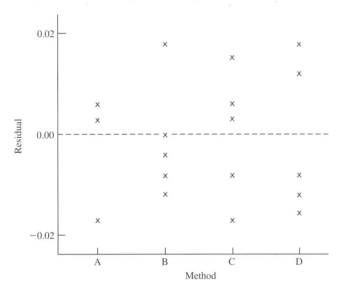

FIGURE 13.9   Plot of residuals against method for the data of Case Study 13.15. (Two observations are hidden.)

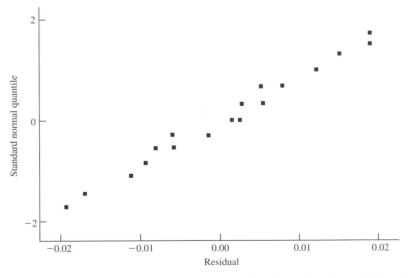

FIGURE 13.10   Normal probability plot of residuals for data of Case Study 13.15. (Two observations are hidden.)

SAS

General Linear Models Procedure

Class Level Information

Class	Levels	Values
TREATMENT	4	A B C D
BLOCK	5	1 2 3 4 5

Number of observations in data set = 20

Dependent Variable: Y

Source	DF	Sum of Squares	Mean Square	F Value	Pr > F
Model	7	0.0534050	0.0076293	42.19	0.0001
Error	12	0.0021700	0.0001808		
Corrected Total	19	0.0555750			

		Root MSE	Y Mean
		0.0134	2.702500

Source	DF	Mean Square	F Value	Pr > F
TRT	3	0.0314550 0.0104850	57.98	0.0001
BLK	4	0.0219500 0.0054875		

General Linear Models Procedure

Observation	Observed Value	Predicted Value	Residual
1	2.67000000	2.66300000	0.00700000
2	2.71000000	2.71700000	-0.00700000
3	2.76000000	2.75300000	0.00700000
4	2.65000000	2.65700000	-0.00700000
5	2.69000000	2.68550000	0.00450000
6	2.74000000	2.73950000	0.00050000
7	2.76000000	2.77550000	-0.01550000
8	2.69000000	2.67950000	0.01050000
9	2.62000000	2.61800000	0.00200000
10	2.69000000	2.67200000	0.01800000
11	2.70000000	2.70800000	-0.00800000
12	2.60000000	2.61200000	-0.01200000
13	2.66000000	2.65550000	0.00450000
14	2.70000000	2.70950000	-0.00950000
15	2.76000000	2.74550000	0.01450000
16	2.64000000	2.64950000	-0.00950000
17	2.70000000	2.71800000	-0.01800000
18	2.77000000	2.77200000	-0.00200000
19	2.81000000	2.80800000	0.00200000
20	2.73000000	2.71200000	0.01800000

FIGURE 13.11 SAS printout for data of Case Study 13.15.

The residual plots show no difficulty with either the assumption of normal errors or homogeneous variance. SAS PROC GLM was used to conduct the analysis of variance. Figure 13.11 shows the annotated computer printout. The computed f- and P-values do indicate a significant difference between analytical methods. This analysis can be followed by a multiple comparison analysis to determine where the differences are among the methods.

Exercises

1. The following data show the effect of 4 operators, chosen randomly, on the output of a particular machine:

	Operator		
1	2	3	4
175.4	168.5	170.1	175.2
171.7	162.7	173.4	175.7
173.0	165.0	175.7	180.1
170.5	164.1	170.7	183.7

(a) Perform a model II analysis of variance at the 0.05 level of significance.
(b) Compute an estimate of the operator variance component and the experimental error variance component.

2. Assuming a random effects model, show that

$$E(SSB) = (b - 1)\sigma^2 + k(b - 1)\sigma_\beta^2$$

for the randomized complete block design.

3. An experiment is conducted in which 4 treatments are to be compared in 5 blocks. The following data are generated:

	Block				
Treatment	1	2	3	4	5
1	12.8	10.6	11.7	10.7	11.0
2	11.7	14.2	11.8	9.9	13.8
3	11.5	14.7	13.6	10.7	15.9
4	12.6	16.5	15.4	9.6	17.1

(a) Assuming a random effects model, test the hypothesis at the 0.05 level of significance that there is no difference between treatment means.
(b) Compute estimates of the treatment and block variance components.

4. Assuming a random effects model, show that

$$E(SSTr) = (r - 1)(\sigma^2 + r\sigma_\tau^2)$$

for the Latin square design.

5. (a) Using a regression approach for the randomized complete block design, obtain the normal equations $\mathbf{Ab} = \mathbf{g}$ in matrix form.
(b) Show that $R(\beta_1, \beta_2, \ldots, \beta_b | \alpha_1, \alpha_2, \ldots, \alpha_k) = SSB$.

6. In Exercise 1, if we are interested in testing for the significance of the operator variance component, do we have large enough samples to ensure with a probability as large as 0.95 a significant variance component if the true σ_α^2 is $1.5\sigma^2$? If not, how many runs are necessary for each operator? Use a 0.05 level of significance.

7. If one assumes a fixed effects model in Exercise 3 and uses an $\alpha = 0.05$ level test, how many blocks are needed in order that we accept the hypothesis of equality of treatment means with probability 0.1 when, in fact,

$$\frac{\sum_{i=1}^{4} \alpha_i^2}{\sigma^2} = 2.0?$$

8. Verify the values given for λ and ϕ^2 in Table 13.20 for the randomized complete block design.

Review Exercises

1. An analysis was conducted by the Statistics Consulting Center at Virginia Polytechnic Institute and State University in conjunction with the Department of Forestry. A certain treatment was applied to a set of tree stumps. The chemical Garlon was used with the purpose of regenerating the roots of the stumps. A spray was used with four levels of Garlon concentration. After a period of time the height of the shoots was observed. Treat the following data as a one-factor analysis of variance. Test to see if the concentration of Garlon has a significant impact on the height of the shoots. Use $\alpha = 0.05$.

Garlon level	1	2.87	2.31	3.91	2.04
	2	3.27	2.66	3.15	2.00
	3	2.39	1.91	2.89	1.89
	4	3.05	0.91	2.43	0.01

2. Consider the aggregate data of Example 13.1. Perform Bartlett's test to determine if there is heterogeneity of variance among the aggregates.

3. In 1983 the Department of Dairy Science at the Virginia Polytechnic Institute and State University conducted an experiment to study the effect of feed rations, differing by source of protein, on the average daily milk production of cows. There were 5 rations used in the experiment. A 5 × 5 Latin square was used in which the rows represented different cows and the columns were different lactation periods. The following data, recorded in kilograms, were analyzed by the Statistical Consulting Center at Virginia Tech. At the 0.01 level of significance can we conclude that rations with different sources of protein have an effect on the average daily milk production of cows?

Lactation Periods	Cows 1	2	3	4	5
1	A 33.1	B 34.4	C 26.4	D 34.6	E 33.9
2	C 30.7	D 28.7	E 24.9	A 28.8	B 28.0
3	D 28.7	E 28.8	A 20.0	B 31.9	C 22.7
4	E 31.4	A 22.3	B 18.7	C 31.0	D 21.3
5	B 28.9	C 22.3	D 15.8	E 30.9	A 19.0

4. Three catalysts are used in a chemical process with a control (no catalyst) being included. The following are yield data from the process:

Control	Catalyst 1	2	3
74.5	77.5	81.5	78.1
76.1	82.0	82.3	80.2
75.9	80.6	81.4	81.5
78.1	84.9	79.5	83.0
76.2	81.0	83.0	82.1

Use Dunnett's test at the $\alpha = 0.01$ level of significance to determine if a significantly higher yield is obtained with the catalysts than with no catalyst.

5. Four laboratories are being used to perform chemical analyses. Samples of the same material are sent to the laboratories for analysis as part of the study to determine whether or not they give, on the average, the same results. The analytical results for the four laboratories are as follows:

	Laboratory		
A	B	C	D
58.7	62.7	55.9	60.7
61.4	64.5	56.1	60.3
60.9	63.1	57.3	60.9
59.1	59.2	55.2	61.4
58.2	60.3	58.1	62.3

(a) Use Bartlett's test to show that the within-laboratory variances are not significantly different at the $\alpha = 0.05$ level of significance.
(b) Perform the analysis of variance and give conclusions concerning the laboratories.

6. Use Bartlett's test at the 0.01 level of significance to test for homogeneity of variances in Exercise 9 on page 476.

7. Use Cochran's test at the 0.01 level of significance to test for homogeneity of variances in Exercise 6 on page 475.

8. Use Bartlett's test at the 0.05 level of significance to test for homogeneity of variances in Exercise 8 on page 476.

9. An experiment was designed for personnel in the Department of Animal Science at Virginia Polytechnic Institute and State University with the purpose of studying urea and aqueous ammonia treatment of wheat straw. The purpose was to improve nutrition value for male sheep. The diet treatments are: control; urea at feeding; ammonia-treated straw; urea-treated straw. Twenty-four sheep were used in the experiment, and they were separated according to relative weight. There were six sheep in each homogeneous group. Each of the six was given each of the four diets in random order. For each of the 24 sheep the percent dry matter digested was measured. The data follow.

Diet	Group by weight (block)					
	1	2	3	4	5	6
Control	32.68	36.22	36.36	40.95	34.99	33.89
Urea at feeding	35.90	38.73	37.55	34.64	37.36	34.35
Ammonia treated	49.43	53.50	52.86	45.00	47.20	49.76
Urea treated	46.58	42.82	45.41	45.08	43.81	47.40

(a) Use a randomized block type of analysis to test for differences between the diets. Use $\alpha = 0.05$.
(b) Use Dunnett's test to compare the three diets with the control. Use $\alpha = 0.05$.

10. In a data set that was analyzed for personnel in the Department of Biochemistry at Virginia Polytechnic Institute and State University, three diets were given to a group of rats in order to study the effect of each on dietary residual zinc in the bloodstream. Five pregnant rats were randomly assigned to each diet group and each was given the diet on day 22 of pregnancy. The amount of zinc in parts per million was measured. The data are as follows:

Diet					
1	0.50	0.42	0.65	0.47	0.44
2	0.42	0.40	0.73	0.47	0.69
3	1.06	0.82	0.72	0.72	0.82

Determine if there is a significant difference in residual dietary zinc among the three diets. Use $\alpha = 0.05$. Perform a one-way ANOVA.

14

Factorial Experiments

14.1 Introduction

Consider a situation in which it is of interest to study the effect of **two factors** A and B on some response. For example, in a chemical experiment we would like to simultaneously vary the reaction pressure and reaction time and study the effect of each on the yield. In a biological experiment, it is of interest to study the effect of drying time and temperature on the amount of solids (percent by weight) left in samples of yeast. As in Chapter 13, the term **factor** is used in a general sense to denote any feature of the experiment such as temperature, time, or pressure that may be varied from trial to trial. We define the **levels** of a factor to be the actual values used in the experiment.

In each of these cases it is important not only to determine if the two factors have an influence on the response, but also if there is a significant interaction between the two factors. As far as terminology is concerned, the experiment described here is a two-way classification or a two-factor experiment and the experimental design may be either a completely randomized design, in which the various treatment combinations are assigned randomly to all the experimental units, or a randomized complete block design, in which factor combinations are assigned randomly to blocks. In the case of the yeast example, the various treatment combinations of temperature and drying time would be assigned randomly to the samples of yeast if we are using a completely randomized design.

Many of the concepts studied in Chapter 13 are extended in this chapter to two and three factors. The main thrust in this material will be the use of the completely randomized design with a *factorial experiment*. A factorial experiment in two factors involves experimental trials (or a single trial) at all factor combinations. For example, in the temperature-drying time example, with, say, three levels of each and $n = 2$ runs at each of the nine combinations, we have a *two-factor-factorial in a completely randomized design*. Neither factor is a blocking factor; we are interested in how both influence percent solids in the samples and whether or not they interact. The biologist would then have available 18 physical samples of material which are experimental units. These would then be assigned randomly to the 18 combinations (nine treatment combinations, each duplicated).

Before we launch into analytical details, sums of squares, and so on, it might be of interest for the reader to observe the obvious connection between what we have described and the situation with the one-factor problem. Consider the yeast experiment. Explanation of degrees of freedom aids the reader or the analyst in visualizing the extension. One should initially view the nine treatment combinations as if they represent one factor with nine levels (eight degrees of freedom). Thus an initial look at degrees of freedom gives

Treatment combinations	8
Error	9
Total	17

Main Effects and Interaction

Actually, the experiment could be analyzed as described in this table. However, the *F*-test for combinations would probably not give the analyst the information he or she desires, namely that which considers the role of temperature and drying time. Three drying times have two associated degrees of freedom; three temperatures have two degrees of freedom. The main factors temperature and drying time are called *main effects*. The main effects represent four of the eight degrees of freedom for *factor combinations*. The additional four degrees of freedom are associated with *interaction* between the two factors. As a result, the analysis involves

Combinations		8
Temperature	2	
Drying time	2	
Interaction	4	
Error		9
Total		17

The reader should recall from Chapter 13 that factors in an analysis of variance may be viewed as fixed or random, depending on the type of inference desired and how the levels were chosen. Here we must consider fixed effects, random effects, and even cases where effects are mixed. Most attention will be drawn toward expected mean squares when we advance to these topics. In the following section we put more focus on the concept of interaction.

14.2 Interaction and the Two-Factor Experiment _____

In the randomized block model discussed previously it was assumed that one observation on each treatment is taken in each block. If the model assumption is correct, that is, if blocks and treatments are the only real effects and interaction does not exist, the expected value of the error mean square is the experimental error variance σ^2. Suppose, however, that there is interaction occurring between treatments and blocks as indicated by the model

$$y_{ij} = \mu + \alpha_i + \beta_j + (\alpha\beta)_{ij} + \varepsilon_{ij}$$

of Section 13.9. The expected value of the error mean square was then given as

$$E\left[\frac{SSE}{(b-1)(k-1)}\right] = \sigma^2 + \frac{\sum_{i=1}^{k}\sum_{j=1}^{b}(\alpha\beta)_{ij}^2}{(b-1)(k-1)}.$$

The treatment and block effects do not appear in the expected error mean square, but the interaction effects do. Thus, if there is interaction in the model, the error mean square reflects variation due to experimental error *plus* an interaction contribution and, for this experimental plan, there is no way of separating them.

From an experimenter's point of view it should seem necessary to arrive at a significance test on the existence of interaction by separating true error variation from that due to interaction. The main effects, A and B, take on a different meaning in the presence of interaction. In the previous biological example the effect that drying time has on the amount of solids left in the yeast might very well depend on the temperature to which the samples are exposed. In general, there could very well be experimental situations in which factor A has a positive effect on the response at one level of factor B, while at a different level of factor B the effect of A is negative. We use the term **positive effect** here to indicate that the yield or response increases as the levels of a given factor increase according to some defined order. In the same sense a **negative effect** corresponds to a decrease in yield for increasing levels of the factor. Consider, for example, the following hypothetical data taken on two factors each at three levels:

		B		
A	b_1	b_2	b_3	Total
a_1	4.4	8.8	5.2	18.4
a_2	7.5	8.5	2.4	18.4
a_3	9.7	7.9	0.8	18.4
Total	21.6	25.2	8.4	55.2

Clearly, the effect of A is positive at b_1 and negative at b_3. These differences in the levels of A at different levels of B are of interest to the experimenter, but an ordinary significance test on factor A would yield a value of zero for SSA, since the totals for each level of A are all of the same magnitude. We say, then, that the presence of interaction is *masking* the effect of factor A. Therefore, if we consider the average influence of A, over all three levels of B, *there is no effect*. However, this is most likely not what is pertinent to the experimenter.

Before drawing any final conclusions resulting from tests of significance on the main effects and interaction effects, the experimenter should first observe whether or not the test for interaction is significant. If interaction is not significant, then the results of the tests on the main effects are meaningful. However, if interaction should be significant, then only those tests on the main effects that turn out to be significant are meaningful. Nonsignificant main effects in the presence of interaction might well be a result of masking and dictate the need to observe the influence of each factor at fixed levels of the other.

Interaction and experimental error are separated in the two-factor experiment only if multiple observations are taken at the various treatment combinations. For maximum efficiency, there should be the same number, n, of observations at each combination. These should be true replications, not just repeated measurements. For example, in the yeast illustration, if we take $n = 2$ observations at each combination of temperature and drying time, there should be two separate samples and not merely repeated measurements on the same sample. This will allow variability due to experimental units to appear in "error," so the variation is not merely measurement error.

14.3 Two-Factor Analysis of Variance

To present general formulas for the analysis of variance of a two-factor experiment using repeated observations in a completely randomized design, we shall consider the case of n replications of the treatment combinations determined by a levels of factor A and b levels of factor B. The observations may be classified by means of a rectangular array in which the rows represent the levels of factor A and the columns represent the levels of factor B. Each treatment combination defines a cell in our array. Thus we have ab cells, each cell containing n observations. Denoting the kth observation taken at the ith level of factor A and the jth level of factor B by y_{ijk}, the abn observations are shown in Table 14.1.

The observations in the (ij)th cell constitute a random sample of size n from a population that is assumed to be normally distributed with mean μ_{ij} and variance σ^2. All ab populations are assumed to have the same variance σ^2. Let us define the following useful symbols, some of which are used in Table 14.1:

$$T_{ij\cdot} = \text{sum of the observations in the } (ij)\text{th cell}$$
$$T_{i\cdot\cdot} = \text{sum of the observations for the } i\text{th level of factor } A$$
$$T_{\cdot j\cdot} = \text{sum of the observations for the } j\text{th level of factor } B$$
$$T_{\cdots} = \text{sum of all } abn \text{ observations}$$
$$\bar{y}_{ij\cdot} = \text{mean of the observations in the } (ij)\text{th cell}$$
$$\bar{y}_{i\cdot\cdot} = \text{mean of the observations for the } i\text{th level of factor } A$$
$$\bar{y}_{\cdot j\cdot} = \text{mean of the observations for the } j\text{th level of factor } B$$
$$\bar{y}_{\cdots} = \text{mean of all } abn \text{ observations.}$$

TABLE 14.1 Two-Factor Experiment with n Replications

A	B 1	2	$\cdots$	b	Total	Mean
1	y_{111}	y_{121}	$\cdots$	y_{1b1}	$T_{1..}$	$\bar{y}_{1..}$
	y_{112}	y_{122}	$\cdots$	y_{1b2}		
	$\vdots$	$\vdots$		$\vdots$		
	y_{11n}	y_{12n}	$\cdots$	y_{1bn}		
2	y_{211}	y_{221}	$\cdots$	y_{2b1}	$T_{2..}$	$\bar{y}_{2..}$
	y_{212}	y_{222}	$\cdots$	y_{2b2}		
	$\vdots$	$\vdots$		$\vdots$		
	y_{21n}	y_{22n}	$\cdots$	y_{2bn}		
$\vdots$					$\vdots$	$\vdots$
a	y_{a11}	y_{a21}	$\cdots$	y_{ab1}	$T_{a..}$	$\bar{y}_{a..}$
	y_{a12}	y_{a22}	$\cdots$	y_{ab2}		
	$\vdots$	$\vdots$		$\vdots$		
	y_{a1n}	y_{a2n}	$\cdots$	y_{abn}		
Total	$T_{.1.}$	$T_{.2.}$	$\cdots$	$T_{.b.}$	$T_{...}$	
Mean	$\bar{y}_{.1.}$	$\bar{y}_{.2.}$	$\cdots$	$\bar{y}_{.b.}$		$\bar{y}_{...}$

Each observation in Table 14.1 may be written in the form

$$y_{ijk} = \mu_{ij} + \varepsilon_{ijk},$$

where ε_{ijk} measures the deviations of the observed y_{ijk} values in the (ij)th cell from the population mean μ_{ij}. If we let $(\alpha\beta)_{ij}$ denote the interaction effect of the ith level of factor A and the jth level of factor B, α_i the effect of the ith level of factor A, β_j the effect of the jth level of factor B, and μ the overall mean, we can write

$$\mu_{ij} = \mu + \alpha_i + \beta_j + (\alpha\beta)_{ij},$$

and then

$$y_{ijk} = \mu + \alpha_i + \beta_j + (\alpha\beta)_{ij} + \varepsilon_{ijk},$$

on which we impose the restrictions

$$\sum_{i=1}^{a} \alpha_i = 0, \qquad \sum_{j=1}^{b} \beta_j = 0, \qquad \sum_{i=1}^{a} (\alpha\beta)_{ij} = 0, \qquad \sum_{j=1}^{b} (\alpha\beta)_{ij} = 0.$$

The three hypotheses to be tested are as follows:

1. H_0': $\alpha_1 = \alpha_2 = \cdots = \alpha_a = 0,$
   H_1':   At least one of the α_i's is not equal to zero.

2. H_0'': $\beta_1 = \beta_2 = \cdots = \beta_b = 0,$
   H_1'':   At least one of the β_j's is not equal to zero.

3. H_0''': $(\alpha\beta)_{11} = (\alpha\beta)_{12} = \cdots = (\alpha\beta)_{ab} = 0,$
   H_1''':   At least one of the $(\alpha\beta)_{ij}$'s is not equal to zero.

Each of these tests will be based on a comparison of independent estimates of σ^2 provided by the splitting of the total sum of squares of our data into 4 components by means of the following identity.

THEOREM 14.1
SUM-OF-SQUARES
IDENTITY

$$\sum_{i=1}^{a} \sum_{j=1}^{b} \sum_{k=1}^{n} (y_{ijk} - \bar{y}_{...})^2 = bn \sum_{i=1}^{a} (\bar{y}_{i..} - \bar{y}_{...})^2$$

$$+ an \sum_{j=1}^{b} (\bar{y}_{\cdot j \cdot} - \bar{y}_{...})^2$$

$$+ n \sum_{i=1}^{a} \sum_{j=1}^{b} (\bar{y}_{ij\cdot} - \bar{y}_{i..} - \bar{y}_{\cdot j \cdot} + \bar{y}_{...})^2$$

$$+ \sum_{i=1}^{a} \sum_{j=1}^{b} \sum_{k=1}^{n} (y_{ijk} - \bar{y}_{ij\cdot})^2.$$

■

PROOF

$$\sum_{i=1}^{a} \sum_{j=1}^{b} \sum_{k=1}^{n} (y_{ijk} - \bar{y}_{...})^2 = \sum_{i=1}^{a} \sum_{j=1}^{b} \sum_{k=1}^{n} [(\bar{y}_{i..} - \bar{y}_{...}) + (\bar{y}_{\cdot j \cdot} - \bar{y}_{...})$$

$$+ (\bar{y}_{ij\cdot} - \bar{y}_{i..} - \bar{y}_{\cdot j \cdot} + \bar{y}_{...}) + (y_{ijk} - \bar{y}_{ij\cdot})]^2$$

$$= \sum_{i=1}^{a} \sum_{j=1}^{b} \sum_{k=1}^{n} (\bar{y}_{i..} - \bar{y}_{...})^2$$

$$+ \sum_{i=1}^{a} \sum_{j=1}^{b} \sum_{k=1}^{n} (\bar{y}_{\cdot j \cdot} - \bar{y}_{...})^2$$

$$+ \sum_{i=1}^{a} \sum_{j=1}^{b} \sum_{k=1}^{n} (\bar{y}_{ij\cdot} - \bar{y}_{i..} - \bar{y}_{\cdot j \cdot} + \bar{y}_{...})^2$$

$$+ \sum_{i=1}^{a} \sum_{j=1}^{b} \sum_{k=1}^{n} (y_{ijk} - \bar{y}_{ij\cdot})^2$$

$$+ 6 \text{ cross-product terms.}$$

The cross-product terms are all equal to zero. Hence

$$\sum_{i=1}^{a} \sum_{j=1}^{b} \sum_{k=1}^{n} (y_{ijk} - \bar{y}_{...})^2 = bn \sum_{i=1}^{a} (\bar{y}_{i..} - \bar{y}_{...})^2 + an \sum_{j=1}^{b} (\bar{y}_{\cdot j \cdot} - y_{...})^2$$

$$+ n \sum_{i=1}^{a} \sum_{j=1}^{b} (\bar{y}_{ij\cdot} - \bar{y}_{i..} - \bar{y}_{\cdot j \cdot} + \bar{y}_{...})^2$$

$$+ \sum_{i=1}^{a} \sum_{j=1}^{b} \sum_{k=1}^{n} (y_{ijk} - \bar{y}_{ij\cdot})^2.$$

Symbolically, we write the sum-of-squares identity as

$$SST = SAA + SSB + SS(AB) + SSE,$$

where SSA and SSB are called the sum of squares for the main effects A and B, respectively, $SS(AB)$ is called the interaction sum of squares for A and B, and SSE is the error sum of squares. The degrees of freedom are partitioned according to the identity

$$abn - 1 = (a - 1) + (b - 1) + (a - 1)(b - 1) + ab(n - 1).$$

Formation of Mean Squares

If we divide each of the sum of squares on the right side of the sum-of-squares identity by their corresponding number of degrees of freedom, we obtain the four statistics

$$s_1^2 = \frac{SSA}{a - 1}, \qquad s_2^2 = \frac{SSB}{b - 1}, \qquad s_3^2 = \frac{SS(AB)}{(a - 1)(b - 1)}, \qquad s^2 = \frac{SSE}{ab(n - 1)}.$$

All of these variance estimates are independent estimates of σ^2 under the condition that there are no effects α_i, β_j, and, of course, $(\alpha\beta)_{ij}$. If we interpret the sum of squares as functions of the independent random variables $Y_{111}, Y_{112}, \ldots, Y_{abn}$, it is not difficult to verify that

$$E(S_1^2) = E\left[\frac{SSA}{a - 1}\right] = \sigma^2 + \frac{nb \sum\limits_{i=1}^{a} \alpha_i^2}{a - 1}$$

$$E(S_2^2) = E\left[\frac{SSB}{b - 1}\right] = \sigma^2 + \frac{na \sum\limits_{j=1}^{b} \beta_j^2}{b - 1}$$

$$E(S_3^2) = E\left[\frac{SS(AB)}{(a - 1)(b - 1)}\right] = \sigma^2 + \frac{n \sum\limits_{i=1}^{a} \sum\limits_{j=1}^{b} (\alpha\beta)_{ij}^2}{(a - 1)(b - 1)}$$

$$E(S^2) = E\left[\frac{SSE}{ab(n - 1)}\right] = \sigma^2,$$

from which we immediately observe that all four estimates of σ^2 are unbiased when H_0', H_0'', and H_0''' are true.

To test the hypothesis H_0', that the effects of factors A are all equal to zero, we compute the ratio

$$f_1 = \frac{s_1^2}{s^2},$$

which is a value of the random variable F_1 having the F-distribution with $a - 1$ and $ab(n - 1)$ degrees of freedom when H_0' is true. The null hypothesis is rejected at the α-level of significance when $f_1 > f_\alpha[a - 1, ab(n - 1)]$. Similarly, to test the hypothesis H_0'', that the effects of factor B are all equal to zero, we compute the ratio

$$f_2 = \frac{s_2^2}{s^2},$$

which is a value of the random variable F_2 having the F-distribution with $b - 1$ and $ab(n - 1)$ degrees of freedom when H_0'' is true. This hypothesis is rejected at the α-level of significance when $f_2 > f_\alpha[b - 1, ab(n - 1)]$. Finally, to test the hypothesis H_0''', that the interaction effects are all equal to zero, we compute the ratio

$$f_3 = \frac{s_3^2}{s^2},$$

which is a value of the random variable F_3 having the F-distribution with $(a - 1)(b - 1)$ and $ab(n - 1)$ degrees of freedom when H_0''' is true. We conclude that interaction is present when $f_3 > f_\alpha[(a - 1)(b - 1), ab(n - 1)]$.

As indicated in Section 14.2, it is advisable to conduct the test for interaction before attempting to draw inferences on the main effects. If interaction is not significant, there is certainly evidence that the tests on main effects are interpretable. However, a significant interaction could very well imply that the data should be analyzed in a somewhat different manner—perhaps observing the effect of factor A at fixed levels of factor B, and so forth.

The computations in an analysis-of-variance problem, for a two-factor experiment with n replications, are usually summarized as in Table 14.2.

TABLE 14.2   Analysis of Variance for the Two-Factor Experiment with n Replications

Source of Variation	Sum of Squares	Degrees of Freedom	Mean Square	Computed f
Main effect				
A	SSA	$a - 1$	$s_1^2 = \dfrac{SSA}{a - 1}$	$f_1 = \dfrac{s_1^2}{s^2}$
B	SSB	$b - 1$	$s_2^2 = \dfrac{SSB}{b - 1}$	$f_2 = \dfrac{s_2^2}{s^2}$
Two-factor interactions				
AB	$SS(AB)$	$(a - 1)(b - 1)$	$s_3^2 = \dfrac{SS(AB)}{(a - 1)(b - 1)}$	$f_3 = \dfrac{s_3^2}{s^2}$
Error	SSE	$ab(n - 1)$	$s^2 = \dfrac{SSE}{ab(n - 1)}$	
Total	SST	$abn - 1$		

EXAMPLE 14.1   In an experiment conducted to determine which of 3 different missile systems is preferable, the propellant burning rate for 24 static firings was measured. Four different propellant types were used. The experiment yielded duplicate observations of burning rates at each combination of the treatments. The data, after coding, are given in Table 14.3. Test the following hypotheses: (a) H_0': there is no difference in the mean propellant burning rates when different missile systems are used; (b) H_0'': there is no difference in the mean propellant burning rates of the 4 propellant types; (c) H_0''': there is no interaction between the different missile systems and the different propellant types.

TABLE 14.3   Propellant Burning Rates

| Missile System | Propellant Type | | | |
	b_1	b_2	b_3	b_4
a_1	34.0	30.1	29.8	29.0
	32.7	32.8	26.7	28.9
a_2	32.0	30.2	28.7	27.6
	33.2	29.8	28.1	27.8
a_3	28.4	27.3	29.7	28.8
	29.3	28.9	27.3	29.1

SOLUTION

1. (a) H_0': $\alpha_1 = \alpha_2 = \alpha_3 = 0$.
   (b) H_0'': $\beta_1 = \beta_2 = \beta_3 = \beta_4 = 0$.
   (c) H_0''': $(\alpha\beta)_{11} = (\alpha\beta)_{12} = \cdots = (\alpha\beta)_{34} = 0$.
2. (a) H_0':   At least one of the α_i's is not equal to zero.
   (b) H_0'':   At least one of the β_j's is not equal to zero.
   (c) H_0''':   At least one of the $(\alpha\beta)_{ij}$'s is not equal to zero.

The sum of squares formulas are used as described on page 538. The analysis of variance is shown in Table 14.4.

TABLE 14.4   Analysis of Variance for the Data of Table 14.3

Source of Variation	Sum of Squares	Degrees of Freedom	Mean Square	Computed f
Missile system	14.52	2	7.26	5.85
Propellant type	40.08	3	13.36	10.77
Interaction	22.17	6	3.70	2.98
Error	14.91	12	1.24	
Total	91.68	23		

(a) Reject H_0' and conclude that different missile systems result in different mean propellant burning rates. The P-value is approximately 0.02.
(b) Reject H_0'' and conclude that the mean propellant burning rates are not the same for the four propellant types. The P-value is smaller than 0.01.
(c) Interaction is barely insignificant at the 0.05 level, but the P-value of approximately 0.055 would indicate that interaction must be taken seriously.

At this point one should draw some type of interpretation of the interaction. It should be emphasized that statistical significance of a main effect merely implies that that *marginal means are significantly different*. However, consider the following two-way table of averages.

	b_1	b_2	b_3	b_4	Average
a_1	66.7	6.29	56.5	57.9	30.5
a_2	65.2	60.0	56.8	55.4	29.7
a_3	57.7	56.2	57.0	57.9	28.6
Average	31.6	29.9	28.4	28.5	

It is quite apparent that more important information exists in the body of the table—trends that are inconsistent with the trend depicted by marginal averages. This table certainly suggests that the effect of propellant type depends on the system being used. For example, for system 3 the propellant-type effect does not appear to be important, although it does have a large effect if either system 1 or system 2 is used. This explains the "significant" interaction. More will be revealed subsequently concerning this interaction.

EXAMPLE 14.2 Referring to Example 14.1, choose two orthogonal contrasts to partition the sum of squares for the missile systems into single-degree-of-freedom components to be used in comparing systems 1 and 2 with 3 and system 1 versus system 2.

SOLUTION
The contrast for comparing systems 1 and 2 with 3 is given by

$$\omega_1 = \mu_{1.} + \mu_{2.} - 2\mu_{3..}$$

A second contrast, orthogonal to ω_1, for comparing system 1 with system 2, is given by $\omega_2 = \mu_{1.} - \mu_{2..}$. The single-degree-of-freedom sums of squares are

$$SSw_1 = \frac{[244.0 + 237.4 - (2)(228.8)]^2}{(8)[(1)^2 + (1)^2 + (-2)^2]} = 11.80$$

and

$$SSw_2 = \frac{(244.0 - 237.4)^2}{(8)[(1)^2 + (-1)^2]} = 2.72.$$

Notice that $SSw_1 + SSw_2 = SSA$, as expected. The computed f-values corresponding to w_1 and w_2 are, respectively,

$$f_1 = \frac{11.80}{1.24} = 9.5$$

and

$$f_2 = \frac{2.72}{1.24} = 2.2.$$

Compared to the critical value $f_{0.05}(1, 12) = 4.75$, we find f_1 to be significant. In fact, the P-value is less than 0.01. Thus the first contrast indicates that the hypothesis

$$H_0: \quad \frac{\mu_1. + \mu_2.}{2} = \mu_3.$$

is rejected. Since $f_2 < 4.75$, the mean burning rates of the first and second systems are not significantly different.

Impact of Significant Interaction

If the hypothesis of no interaction in Example 14.1 is true, we could make the *general* comparisons of Example 14.2 regarding our missile systems rather than separate comparisons for each propellant. Similarly, we might make general comparisons among the propellants rather than separate comparisons for each missile system. For example, we could compare propellants 1 and 2 with 3 and 4 and also propellant 1 versus propellant 2. The resulting f-ratios, each with 1 and 12 degrees of freedom, turn out to be 24.86 and 7.41, respectively, and both are quite significant at the 0.05 level.

From propellant averages there appears to be evidence that propellant 1 gives the highest mean burning rate. A prudent experimenter might be somewhat cautious in making overall conclusions in a problem such as this one, where the f-ratio for interaction is barely below the 0.05 critical value. For example, the overall evidence, 31.6 versus 29.9 on the average for the two propellants, certainly indicates that propellant 1 is superior, in terms of a higher burning rate, to propellant 2. However, if we restrict ourselves to system 3, where we have an average of 28.85 for propellant 1 as opposed to 28.10 for propellant 2, there appears to be little or no difference between these two propellants. In fact, there appears to be a stabilization of burning rates for the different propellants if we operate with system 3. There is certainly overall evidence which indicates that system 1 gives a higher burning rate than system 3, but if we restrict ourselves to propellant 4, this conclusion does not appear to hold.

The analyst can conduct a simple t-test using average burning rates at system 3 in order to display conclusive evidence that interaction is *producing considerable difficulty in allowing broad conclusions on main effects*. Consider a comparison of propellant 1 against propellant 2 only using system 3. Borrowing an estimate of σ^2

from the overall analysis, that is, using $s^2 = 1.24$ with 12 degrees of freedom, we can use

$$|t| = \frac{0.75}{\sqrt{2s^2/n}}$$

$$= \frac{0.75}{\sqrt{1.24}}$$

$$= 0.68,$$

which is not even close to being significant. This illustration suggests that one must be cautious about strict interpretation of main effects in the presence of interaction.

14.4 Graphical Analysis in the Two-Factor Problem

Many of the same types of graphical displays that were suggested in the one-factor problems certainly apply in the two-factor case. Two-dimensional plots of cell means or treatment combination means can provide an insight into the presence of interactions between the two factors. In addition, a plot of residuals against fitted values may well provide an indication of whether or not the homogeneous variance

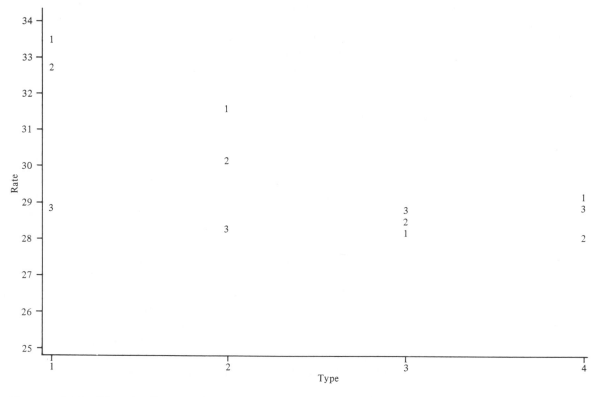

FIGURE 14.1 Plot of cell means for data in Example 14.1. Numbers represent missile systems.

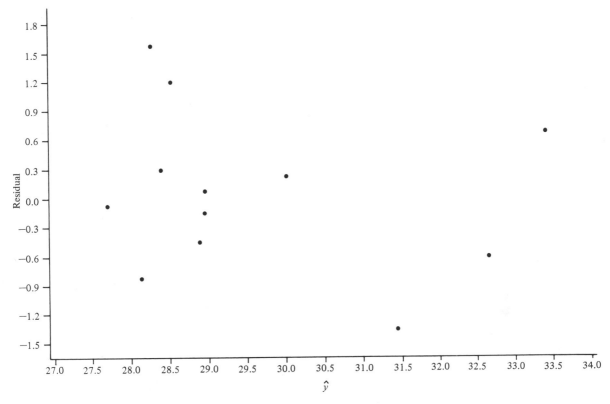

FIGURE 14.2   Residual plot of data in Example 14.1.

assumption holds. Often, of course, a violation of the homogeneous variance assumption involves an increase in the error variance as *the response numbers get larger*. As a result, this plot may point out the violation.

Figure 14.1 shows the plot of cell means in the case of the missile system-propellant illustration in Example 14.1. Notice how graphic (in this case) the lack of parallelism shows through. Note the flatness of the part of the figure showing the propellant effect at system 3. This suggests interaction among the factors. Figure 14.2 shows the plot for residuals against fitted values for the same data. There is no apparent sign of difficulty with the homogeneous variance assumption.

Exercises

1. An experiment was conducted to study the effect of temperature and type of oven on the life of a particular component being tested. Four types of ovens and 3 temperature levels were used in the experiment. Twenty-four pieces were assigned randomly, 2 to each combination of treatments, and the following results recorded:

Temperature (degrees)	Oven			
	O_1	O_2	O_3	O_4
500	227	214	225	260
	221	259	236	229
550	187	181	232	246
	208	179	198	273
600	174	198	178	206
	202	194	213	219

Using a 0.05 level of significance, test the hypothesis that

(a) different temperatures have no effect on the life of the component;

(b) different ovens have no effect on the life of the component;

(c) the type of oven and temperature do not interact.

2. To ascertain the stability of vitamin C in reconstituted frozen orange juice concentrate stored in a refrigerator for a period of up to one week, the study *"Vitamin C Retention in Reconstituted Frozen Orange Juice"* was conducted by the Department of Human Nutrition and Foods at the Virginia Polytechnic Institute and State University in 1975. Three types of frozen orange juice concentrate were tested using 3 different time periods. The time periods refer to the number of days from when the orange juice was blended until it was tested. The results, in milligrams of ascorbic acid per liter, were recorded as follows:

Brand	Time (days)					
	0		3		7	
Richfood	52.6	54.2	49.4	49.2	42.7	48.8
	49.8	46.5	42.8	53.2	40.4	47.6
Sealed-Sweet	56.0	48.0	48.8	44.0	49.2	44.0
	49.6	48.4	44.0	42.4	42.0	43.2
Minute Maid	52.5	52.0	48.0	47.0	48.5	43.3
	51.8	53.6	48.2	49.6	45.2	47.6

Use a 0.05 level of significance to test the hypothesis that

(a) there is no difference in ascorbic acid contents among the different brands of orange juice concentrate;

(b) there is no difference in ascorbic acid contents for the different time periods;

(c) the brands of orange juice concentrate and the number of days from the time the juice was blended until it is tested do not interact.

3. Three strains of rats were studied under 2 environmental conditions for their performance in a maze test. The error scores for the 48 rats were recorded as follows:

Environment	Strain					
	Bright		Mixed		Dull	
Free	28	12	33	83	101	94
	22	23	36	14	33	56
	25	10	41	76	122	83
	36	86	22	58	35	23
Restricted	72	32	60	89	136	120
	48	93	35	126	38	153
	25	31	83	110	64	128
	91	19	99	118	87	140

Using a 0.01 level of significance to test the hypothesis that

(a) there is no difference in error scores for different environments;

(b) there is no difference in error scores for different strains;

(c) the environments and strains of rats do not interact.

4. Corrosion fatigue in metals has been defined as the simultaneous action of cyclic stress and chemical attack on a metal structure. A widely used technique for minimizing corrosion-fatigue damage in aluminum involves the application of a protective coating. In the study *"Effect of Humidity and Several Surface Coatings on the Fatigue Life of 2024-T351 Aluminum Alloy"* conducted by the Department of Mechanical Engineering at the Virginia Polytechnic Institute and State University in 1979, 3 different levels of humidity

Low: 20–25% relative humidity,

Medium: 55–60% relative humidity,

High: 86–91% relative humidity,

and 3 types of surface coatings

Uncoated: no coating,

Anodized: sulfuric acid
           anodic oxide coating,

Conversion: chromate chemical
            conversion coating,

were used. The corrosion-fatigue data, expressed in thousands of cycles to failure, were recorded as follows:

| Coating | Relative Humidity | | | | | |
	Low		Medium		High	
Uncoated	361	469	314	522	1344	1216
	466	937	244	739	1027	1097
	1069	1357	261	134	663	1011
Anodized	114	1032	322	471	78	466
	1236	92	306	130	387	107
	533	211	68	398	130	327
Conversion	130	1482	252	874	586	524
	841	529	105	755	402	751
	1595	754	847	573	846	529

(a) Perform an analysis of variance with $\alpha = 0.05$ to test for significant main and interaction effects.

(b) Use Duncan's multiple-range test at the 0.05 level of significance to determine which humidity levels result in different corrosion-fatigue damage.

5. To determine which muscles need to be subjected to a conditioning program in order to improve one's performance on the flat serve used in tennis, the study *"An Electromyographic-Cinematrographic Analysis of the Tennis Serve"* was conducted by the Department of Health, Physical Education and Recreation at the Virginia Polytechnic Institute and State University in 1978. Five different muscles

1:   anterior deltoid,

2:   pectorial major,

3:   posterior deltoid,

4:   middle deltoid,

5:   triceps,

were tested on each of 3 subjects, and the experiment was carried out 3 times for each treatment combination. The electromyographic data, recorded during the serve, are as follows:

| Subject | Muscle | | | | |
	1	2	3	4	5
1	32	5	58	10	19
	59	1.5	61	10	20
	38	2	66	14	23
2	63	10	64	45	43
	60	9	78	61	61
	50	7	78	71	42
3	43	41	26	63	61
	54	43	29	46	85
	47	42	23	55	95

Use a 0.01 level of significance to test the hypothesis that

(a) different subjects have equal electromyographic measurements;

(b) different muscles have no effect on electromyographic measurements;

(c) subjects and types of muscle do not interact.

6. An experiment was conducted to increase the adhesiveness of rubber products. Sixteen products were

made with the new additive and another 16 without the new additive. The observed adhesiveness is recorded below.

	Temperature (°C)			
	50	60	70	80
Without additives	2.3	3.4	3.8	3.9
	2.9	3.7	3.9	3.2
	3.1	3.6	4.1	3.0
	3.2	3.2	3.8	2.7
With additives	4.3	3.8	3.9	3.5
	3.9	3.8	4.0	3.6
	3.9	3.9	3.7	3.8
	4.2	3.5	3.6	3.9

Perform an analysis of variance to test for significant main and interaction effects.

7. The extraction rate of a certain polymer is known to depend on the reaction temperature and the amount of catalyst used. An experiment was conducted at four levels of temperature and five levels of the catalyst, and the extraction rate was recorded in the following table.

	Amount of Catalyst				
	0.5%	0.6%	0.7%	0.8%	0.9%
50°C	38	45	57	59	57
	41	47	59	61	58
	44	56	70	73	61
60°C	43	57	69	72	58
	44	56	70	73	61
70°C	47	60	67	61	59
	49	62	70	62	53
80°C	47	65	55	69	58
	51	64	53	72	60

Perform an analysis of variance. Test for significant main and interaction effects.

14.5 Three-Factor Experiments

In this section we consider an experiment with three factors A, B, and C at a, b, and c levels, respectively, in a completely randomized experimental design. Assume again that we have n observations for each of the abc treatment combinations. We shall proceed to outline significance tests for the three main effects and interactions involved. It is hoped that the reader can then use the description given here to generalize the analysis to $k > 3$ factors.

The model for the three-factor experiment is given by

$$y_{ijkl} = \mu + \alpha_i + \beta_j + \gamma_k + (\alpha\beta)_{ij} + (\alpha\gamma)_{ik} + (\beta\gamma)_{jk} + (\alpha\beta\gamma)_{ijk} + \varepsilon_{ijkl},$$

$i = 1, 2, \ldots, a$; $j = 1, 2, \ldots, b$; $k = 1, 2, \ldots, c$; and $l = 1, 2, \ldots, n$—where α_i, β_j, and γ_k are the main effects; $(\alpha\beta)_{ij}$, $(\alpha\gamma)_{ik}$, and $(\beta\gamma)_{jk}$ are the two-factor interaction effects that have the same interpretation as in the two-factor experiment. The term $(\alpha\beta\gamma)_{ijk}$ is called the **three-factor interaction effect**, a term that represents a nonadditivity of the $(\alpha\beta)_{ij}$ over the different levels of the factor C. As before, the sum of all main effects is zero and the sum over any subscript of the two- and three-factor interaction effects is zero. In many experimental situations these higher-order interactions are insignificant and their mean squares reflect only random variation, but we shall outline the analysis in its most general detail.

Again, in order that valid significance tests can be made, we must assume that the errors are values of independent and normally distributed random variables, each with zero mean and common variance σ^2.

The general philosophy concerning the analysis is the same as that discussed for the one- and two-factor experiments. The sum of squares is partitioned into eight terms, each representing a source of variation from which we obtain independent estimates of σ^2 when all the main effects and interaction effects are zero. If the effects of any given factor or interaction are not all zero, then the mean square will estimate the error variance plus a component due to the systematic effect in question.

Although we emphasize interpretation of annotated computer printout in this section rather than being concerned with laborious computation of sum of squares, we do offer the following as the sums of squares for the three main effects and interactions. Notice the obvious extension from the two- to three-factor problem.

$$SSA = bcn \sum_{i=1}^{a} (\bar{y}_{i\cdots} - \bar{y}_{\cdots})^2$$

$$SSB = acn \sum_{j=1}^{b} (\bar{y}_{\cdot j\cdot\cdot} - \bar{y}_{\cdots})^2$$

$$SSC = abn \sum_{k=1}^{c} (\bar{y}_{\cdot\cdot k\cdot} - \bar{y}_{\cdots})^2$$

$$SS(AB) = cn \sum_{i} \sum_{j} (\bar{y}_{ij\cdot\cdot} - \bar{y}_{i\cdots} - \bar{y}_{\cdot j\cdot\cdot} + \bar{y}_{\cdots})^2$$

$$SS(AC) = bn \sum_{i} \sum_{k} (\bar{y}_{i\cdot k\cdot} - \bar{y}_{i\cdots} - \bar{y}_{\cdot\cdot k\cdot} + \bar{y}_{\cdots})^2$$

$$SS(BC) = an \sum_{j} \sum_{k} (\bar{y}_{\cdot jk\cdot} - \bar{y}_{\cdot j\cdot\cdot} - \bar{y}_{\cdot\cdot k\cdot} + \bar{y}_{\cdots})^2$$

$$SS(ABC) = n \sum_{i} \sum_{j} \sum_{k} (\bar{y}_{ijk\cdot} - \bar{y}_{ij\cdot\cdot} - \bar{y}_{i\cdot k\cdot} - \bar{y}_{\cdot jk\cdot} + \bar{y}_{i\cdots} + \bar{y}_{\cdot j\cdot\cdot} + \bar{y}_{\cdot\cdot k\cdot} - \bar{y}_{\cdots})^2$$

$$SST = \sum_{i} \sum_{j} \sum_{k} \sum_{l} (y_{ijkl} - \bar{y}_{\cdots})^2$$

$$SSE = \sum_{i} \sum_{j} \sum_{k} \sum_{l} (y_{ijkl} - \bar{y}_{ijk\cdot})^2$$

The averages in the formulas above are defined as follows:

$\bar{y}_{....}$ = average of all $abcn$ observations

$\bar{y}_{i...}$ = average of the observations for the ith level of factor A

$\bar{y}_{.j..}$ = average of the observations for the jth level of factor B

$\bar{y}_{..k.}$ = average of the observations for the kth level of factor C

$\bar{y}_{ij..}$ = average of the observations for the ith level of A and the jth level of B

$\bar{y}_{i.k.}$ = average of the observations for the ith level of A and the kth level of C

$\bar{y}_{.jk}$ = average of the observations for the jth level of B and the kth level of C

$\bar{y}_{....}$ = average of the observations for the ijkth treatment combination.

The computations in an analysis-of-variance table for a three-factor problem with n replicated runs at each factor combination is summarized in Table 14.5.

For the three-factor experiment with a single experimental run per combination we may use the analysis of Table 14.5 by setting $n = 1$ and using the ABC interac-

TABLE 14.5   Analysis of Variance for the Three-Factor Experiment with n Replications

Source of Variation	Sum of Squares	Degrees of Freedom	Mean Square	Computed f
Main effect				
A	SSA	$a - 1$	s_1^2	$f_1 = \dfrac{s_1^2}{s^2}$
B	SSB	$b - 1$	s_2^2	$f_2 = \dfrac{s_2^2}{s^2}$
C	SSC	$c - 1$	s_3^2	$f_3 = \dfrac{s_3^2}{s^2}$
Two-factor interaction				
AB	SS(AB)	$(a - 1)(b - 1)$	s_4^2	$f_4 = \dfrac{s_4^2}{s^2}$
AC	SS(AC)	$(a - 1)(c - 1)$	s_5^2	$f_5 = \dfrac{s_5^2}{s^2}$
BC	SS(BC)	$(b - 1)(c - 1)$	s_6^2	$f_6 = \dfrac{s_6^2}{s^2}$
Three-factor interaction				
ABC	SS(ABC)	$(a - 1)(b - 1)(c - 1)$	s_7^2	$f_7 = \dfrac{s_7^2}{s^2}$
Error	SSE	$abc(n - 1)$	s^2	
Total	SST	$abcn - 1$		

tion sum of squares for *SSE*. In this case we are assuming that the $(\alpha\beta\gamma)_{ijk}$ interaction effects are all equal to zero so that

$$E\left[\frac{SS(ABC)}{(a-1)(b-1)(c-1)}\right] = \sigma^2 + \frac{n\sum\limits_{i=1}^{a}\sum\limits_{j=1}^{b}\sum\limits_{k=1}^{c}(\alpha\beta\gamma)_{ijk}^2}{(a-1)(b-1)(c-1)}$$

$$= \sigma^2.$$

That is, *SS(ABC)* represents variation due only to experimental error. Its mean square thereby provides an unbiased estimate of the error variance. With $n = 1$ and $SSE = SS(ABC)$, the error sum of squares is found by subtracting the sums of squares of the main effects and two-factor interactions from the total sum of squares.

EXAMPLE 14.3   In the production of a particular material three variables are of interest: *A*, the operator effect (three operators); *B*, the catalyst used in the experiment (three catalysts); and *C*, the washing time of the product following the cooling process (15 minutes and 20 minutes). Three runs were made at each combination of factors. It was felt that all interactions among the factors should be studied. The coded yields are as follows:

	Washing Time, C					
	15 minutes			20 minutes		
	B (catalyst)			B (catalyst)		
A (operator)	1	2	3	1	2	3
1	10.7	10.3	11.2	10.9	10.5	12.2
	10.8	10.2	11.6	12.1	11.1	11.7
	11.3	10.5	12.0	11.5	10.3	11.0
2	11.4	10.2	10.7	9.8	12.6	10.8
	11.8	10.9	10.5	11.3	7.5	10.2
	11.5	10.5	10.2	10.9	9.9	11.5
3	13.6	12.0	11.1	10.7	10.2	11.9
	14.1	11.6	11.0	11.7	11.5	11.6
	14.5	11.5	11.5	12.7	10.9	12.2

Perform an analysis of variance to test for significant effects.

SOLUTION
Figure 14.3 depicts annotated computer printout for the analysis of variance of the data of Example 14.3. The SAS PROC GLM was used. None of the interactions show a significant effect at the $\alpha = 0.05$ level. However, the *P*-value for *BC* is 0.0610. Thus it should not be ignored. The operator and catalyst effects are significant, while the effect of washing time is not significant.

General Linear Models Procedure
Class Level Information

Class	Levels	Values
A	3	1 2 3
B	3	1 2 3
C	2	1 2

Number of observations in data set = 54

General Linear Models Procedure

Source	DF	Sum of Squares	Mean Square	F Value	Pr>F
Model	17	41.579259	2.445839	4.07	0.0002
Error	36	21.613333	0.600370		
Corrected Total	53	63.192593			

		Root MSE			Y Mean
		0.7748			11.22963

Source	DF	Type I SS	Mean Square	F Value	Pr>F
A	2	13.982593	6.991296	11.64	0.0001
B	2	10.182593	5.091296	8.48	0.0010
C	1	1.185185	1.185185	1.97	0.1686
A*B	4	4.774074	1.193519	1.99	0.1172
A*C	2	2.913704	1.456852	2.43	0.1027
B*C	2	3.633704	1.816852	3.03	0.0610
A*B*C	4	4.907407	1.226852	2.04	0.1089

Source	DF	Type III SS	Mean Square	F Value	Pr>F
A	2	13.982593	6.991296	11.64	0.0001
B	2	10.182593	5.091296	8.48	0.0010
C	1	1.185185	1.185185	1.97	0.1686
A*B	4	4.774074	1.193519	1.99	0.1172
A*C	2	2.913704	1.456852	2.43	0.1027
B*C	2	3.633704	1.816852	3.03	0.0610
A*B*C	4	4.907407	1.226852	2.04	0.1089

FIGURE 14.3   Computer printout for Example 14.3.

Impact of Interaction

More should be discussed and presented regarding Example 14.3, particularly in dealing with the effect that the interaction between catalyst and washing time is having on the test on the washing time main effect (factor C). Recall our discussion in Section 14.2. Illustrations were given of how the presence of interaction could change the interpretation that we make regarding main effects. In Example 14.3 the BC interaction is significant at the 0.06 level. Suppose, however, that we observe a two-way table of means.

Washing time, C

		15 min	20 min
	1	12.19	11.29
Catalyst, B	2	10.86	10.50
	3	11.09	11.46
Means		11.38	11.08

It is clear why washing time was found not to be significant. A nonthorough analyst may get the impression that washing time can be eliminated from any future study in which yield is being measured. However, it is easy to notice how the effect of washing time changes from a negative effect for the first catalyst to what appears to be a positive effect for the third catalyst. If we merely focus on the data for catalyst 1, a simple comparison between the means at the two washing times will produce a simple t-statistic:

$$t = \frac{12.19 - 11.29}{\sqrt{0.6(2/9)}}$$

$$= 2.5,$$

which is significant at a level less than 0.02. Thus an important negative effect of washing time for catalyst 1 might very well be ignored if the analyst makes the incorrect broad interpretation of the insignificant F-ratio on washing time.

14.6 Specific Multifactor Models

We have described the three-factor model and its analysis in the most general form by including all possible interactions in the model. Of course, there are many situations in which it is known *a priori* that the model should not contain certain interactions. We can then take advantage of this knowledge by combining or pooling the sums of squares corresponding to negligible interactions with the error sum of squares to form a new estimator for σ^2 with a larger number of degrees of freedom. For example, in a metallurgy experiment designed to study the effect on film thickness of three important processing variables, suppose it is known that factor A, acid concentration, does not interact with factors B and C. The sums of squares SSA, SSB, SSC, and $SS(BC)$ are computed using the methods described in Section 14.5. The mean squares for the remaining effects will now all independently estimate the error variance σ^2. Therefore, we form our new error mean square by pooling $SS(AB)$, $SS(AC)$, $SS(ABC)$, and SSE, along with the corresponding degrees of freedom. The resulting denominator for the significance tests is then the error mean square given by

$$s^2 = \frac{SS(AB) + SS(AC) + SS(ABC) + SSE}{(a-1)(b-1) + (a-1)(c-1) + (a-1)(b-1)(c-1) + abc(n-1)}.$$

Computationally, of course, one obtains the pooled sum of squares and the pooled degrees of freedom by subtraction once SST and the sums of squares for the existing effects are computed. The analysis-of-variance table would then take the form of Table 14.6.

In our analysis of the two-factor experiment in Section 14.3 a completely randomized design was used. By interpreting the levels of factor A in Table 14.6 as different blocks, we then have the analysis-of-variance procedure for a two-factor experiment in a randomized block design. For example, if we interpret the operators in Example 14.3 as blocks and assume no interaction between blocks and the other two factors, the analysis of variance takes the form of Table 14.7 rather than that

TABLE 14.6   Analysis of Variance with Factor A Noninteracting

Source of Variation	Sum of Squares	Degrees of Freedom	Mean Square	Computed f
Main effect				
A	SSA	$a - 1$	s_1^2	$f_1 = \dfrac{s_1^2}{s^2}$
B	SSB	$b - 1$	s_2^2	$f_2 = \dfrac{s_2^2}{s^2}$
C	SSC	$c - 1$	s_3^2	$f_3 = \dfrac{s_3^2}{s^2}$
Two-factor interaction				
BC	$SS(BC)$	$(b - 1)(c - 1)$	s_4^2	$f_4 = \dfrac{s_4^2}{s^2}$
Error	SSE	Subtraction	s^2	
Total	SST	$abcn - 1$		

TABLE 14.7   Analysis of Variance for a Two-Factor Experiment in a Randomized Block Design

Source of Variation	Sum of Squares	Degrees of Freedom	Mean Square	Computed f
Blocks	13.98	2	6.99	
Main effect				
B	10.18	2	5.09	6.88
C	1.18	1	1.18	1.59
Two-factor interaction				
BC	3.64	2	1.82	2.46
Error	34.21	46	0.74	
Total	63.19	53		

given in Table 14.6. The reader can easily verify that the error mean square is also given by

$$s^2 = \frac{4.78 + 2.92 + 4.89 + 21.62}{4 + 2 + 4 + 36} = 0.74,$$

which demonstrates the pooling of the sums of squares for the nonexisting interaction effects.

Exercises

1. The following data are taken in a study involving three factors A, B, and C, all fixed effects:

	C_1			C_2			C_3		
	B_1	B_2	B_3	B_1	B_2	B_3	B_1	B_2	B_3
A_1	15.0	14.8	15.9	16.8	14.2	13.2	15.8	15.5	19.2
	18.5	13.6	14.8	15.4	12.9	11.6	14.3	13.7	13.5
	22.1	12.2	13.6	14.3	13.0	10.1	13.0	12.6	11.1
A_2	11.3	17.2	16.1	18.9	15.4	12.4	12.7	17.3	7.8
	14.6	15.5	14.7	17.3	17.0	13.6	14.2	15.8	11.5
	18.2	14.2	13.4	16.1	18.6	15.2	15.9	14.6	12.2

(a) Perform tests of significance on all interactions at the $\alpha = 0.05$ level.

(b) Perform tests of significance on the main effects at the $\alpha = 0.05$ level.

(c) Give an explanation of how a significant interaction has masked the effect of factor C.

2. The method of X-ray fluorescence is an important analytical tool for determining the concentration of material in solid missile propellants. In the paper "*An X-ray Fluorescence Method for Analyzing Polybutadiene-Acrylic Acid (PBAA) Propellants,*" *Quarterly Report*, RK-TR-62-1, Army Ordinance Missile Command (1962), it is postulated that the propellant mixing process and analysis time have an influence on the homogeneity of the material and hence on the accuracy of X-ray intensity measurements. An experiment was conducted using 3 factors: A, the mixing conditions (4 levels); B, the analysis time (2 levels); and C, the method of loading propellant into sample holders (hot and room temperature). The following data, which represent the analysis in weight percent of ammonium perchlorate in a particular propellant, were recorded:

	Method of Loading, C			
	Hot		*Room Temperature*	
	B		B	
A	1	2	1	2
1	38.62	38.45	39.82	39.82
	37.20	38.64	39.15	40.26
	38.02	38.75	39.78	39.72
2	37.67	37.81	39.53	39.56
	37.57	37.75	39.76	39.25
	37.85	37.91	39.90	39.04
3	37.51	37.21	39.34	39.74
	37.74	37.42	39.60	39.49
	37.58	37.79	39.62	39.45
4	37.52	37.60	40.09	39.36
	37.15	37.55	39.63	39.38
	37.51	37.91	39.67	39.00

Perform an analysis of variance with $\alpha = 0.01$ to test for significant main and interaction effects.

3. Corrosion fatigue in metals has been defined as the simultaneous action of cyclic stress and chemical attack on a metal structure. In the study "*Effect of Humidity and Several Surface Coatings on the Fatigue Life of 2024-T351 Aluminum Alloy*" conducted by the Department of Mechanical Engineering at the Virginia Polytechnic Institute and State University in 1979, a technique involving the application of a protective chromate coating was used to minimize corrosion-fatigue damage in aluminum. Three factors were used in the investigation with 5 replicates for each treatment combination: coating, at 2 levels, and humidity and shear stress, both with 3 levels. The fatigue data, recorded in thousands of cycles to failure, are as follows:

Coating	Humidity	Shear Stress		
		13,000 psi	17,000 psi	20,000 psi
	Low (20–25% RH)	4,580	5,252	361
		10,126	897	466
		1,341	1,465	1,069
		6,414	2,694	469
		3,549	1,017	937
	Medium (50–60% RH)	2,858	799	314
Uncoated		8,829	3,471	244
		10,914	685	261
		4,067	810	522
		2,595	3,409	739
	High (86–91% RH)	6,489	1,862	1,344
		5,248	2,710	1,027
		6,816	2,632	663
		5,860	2,131	1,216
		5,901	2,470	1,097
	Low (20–25% RH)	5,395	4,035	130
		2,768	2,022	841
		1,821	914	1,595
		3,604	2,036	1,482
		4,106	3,524	529
	Medium (50–60% RH)	4,833	1,847	252
		7,414	1,684	105
Chromated		10,022	3,042	847
		7,463	4,482	874
		21,906	996	755
	High (86–91% RH)	3,287	1,319	586
		5,200	929	402
		5,493	1,263	846
		4,145	2,236	524
		3,336	1,392	751

Perform an analysis of variance with $\alpha = 0.01$ to test for significant main and interaction effects.

4. Consider an experimental situation involving factors A, B, and C, where we assume a three-way fixed-effects model of the form

$$y_{ijkl} = \mu + \alpha_i + \beta_j + \gamma_k + (\beta\gamma)_{jk} + \varepsilon_{ijkl}.$$

All other interactions are considered to be nonexistent or negligible. The data were recorded as follows:

	B_1			B_2		
	C_1	C_2	C_3	C_1	C_2	C_3
A_1	4.0	3.4	3.9	4.4	3.1	3.1
	4.9	4.1	4.3	3.4	3.5	3.7
A_2	3.6	2.8	3.1	2.7	2.9	3.7
	3.9	3.2	3.5	3.0	3.2	4.2
A_3	4.8	3.3	3.6	3.6	2.9	2.9
	3.7	3.8	4.2	3.8	3.3	3.5
A_4	3.6	3.2	3.2	2.2	2.9	3.6
	3.9	2.8	3.4	3.5	3.2	4.3

(a) Perform a test of significance on the BC interaction at the $\alpha = 0.05$ level.

(b) Perform tests of significance on the main effects A, B, and C using a pooled error mean square at the $\alpha = 0.05$ level.

5. Electronic copiers make copies by gluing black ink on paper using static electricity. Heating and gluing the ink on the paper comprise the final stage of the copying process. The gluing power in this final process determines the quality of the copy. It is postulated that temperature, surface state of gluing the roller, and hardness of the press roller influence the gluing power of the copier. An experiment was run with treatments consisting of a combination of these three factors at each of three levels. The following data show the gluing power at each treatment combination.

Temperature	Surface State of Gluing Roller	Hardness of the Press Roller (hrc)		
		20	40	60
Low	Soft	0.52	0.54	0.60
		0.44	0.52	0.55
		0.57	0.65	0.78
		0.53	0.56	0.68
	Medium	0.64	0.79	0.49
		0.59	0.73	0.48
		0.58	0.79	0.74
		0.64	0.78	0.50
	Hard	0.67	0.58	0.55
		0.77	0.68	0.65
		0.74	0.57	0.57
		0.65	0.59	0.58
Medium	Soft	0.46	0.31	0.56
		0.40	0.49	0.42
		0.58	0.48	0.49
		0.37	0.66	0.49
	Medium	0.60	0.66	0.64
		0.43	0.57	0.54
		0.62	0.72	0.74
		0.61	0.56	0.56
	Hard	0.53	0.53	0.56
		0.65	0.45	0.66
		0.66	0.59	0.71
		0.56	0.47	0.67
High	Soft	0.52	0.54	0.65
		0.44	0.52	0.49
		0.57	0.65	0.65
		0.53	0.56	0.52
	Medium	0.53	0.53	0.49
		0.65	0.45	0.48
		0.66	0.59	0.74
		0.56	0.47	0.50
	Hard	0.43	0.48	0.55
		0.43	0.31	0.65
		0.47	0.43	0.57
		0.44	0.27	0.58

Perform an analysis of variance with $\alpha = 0.05$ to test for significant main and interaction effects.

14.7 Model II Factorial Experiments _____

In a two-factor experiment with random effects we have the model

$$Y_{ijk} = \mu + A_i + B_j + (AB)_{ij} + E_{ijk},$$

$i = 1, 2, \ldots, a; j = 1, 2, \ldots, b; k = 1, 2, \ldots, n$, where the A_i, B_j, $(AB)_{ij}$, and E_{ijk} are independent random variables with zero means and variances σ_α^2, σ_β^2, $\sigma_{\alpha\beta}^2$, and σ^2, respectively. The sum of squares for the model II experiments are computed in exactly the same way as for the model I experiments. We are now interested in testing hypotheses of the form

$$H_0': \quad \sigma_\alpha^2 = 0, \qquad H_0'': \quad \sigma_\beta^2 = 0, \qquad H_0''': \quad \sigma_{\alpha\beta}^2 = 0,$$

$$H_1': \quad \sigma_\alpha^2 \neq 0, \qquad H_1'': \quad \sigma_\beta^2 \neq 0, \qquad H_1''': \quad \sigma_{\alpha\beta}^2 \neq 0,$$

where the denominator in the f-ratio is not necessarily the error mean square. The appropriate denominator can be determined by examining the expected values of the various mean squares. These are shown in Table 14.8.

From Table 14.8 we see that H_0' and H_0'' are tested by using s_3^2 in the denominator of the f-ratio, while H_0''' is tested using s^2 in the denominator. The unbiased estimates of the variance components are given by

$$\hat{\sigma}^2 = s^2$$

$$\hat{\sigma}_{\alpha\beta}^2 = \frac{s_3^2 - s^2}{n}$$

$$\hat{\sigma}_\alpha^2 = \frac{s_1^2 - s_3^2}{bn}$$

$$\hat{\sigma}_\beta^2 = \frac{s_2^2 - s_3^2}{an}.$$

The expected mean squares for the three-factor experiment with random effects in a completely randomized design are shown in Table 14.9. It is evident from the expected mean squares of Table 14.9 that one can form appropriate f-ratios for testing all two-factor and three-factor interaction variance components. However, to test a hypothesis of the form

$$H_0: \quad \sigma_\alpha^2 = 0$$

$$H_1: \quad \sigma_\alpha^2 \neq 0,$$

there appears to be no appropriate f-ratio unless we have found one or more of the two-factor interaction variance components not significant. Suppose, for example, that we have compared s_5^2 with s_7^2 and found $\sigma_{\alpha\gamma}^2$ to be negligible. We could then argue that the term $\sigma_{\alpha\gamma}^2$ should be dropped from all the expected mean squares of Table 14.9; then the ratio s_1^2/s_4^2 provides a test for the significance of the variance component σ_α^2. Therefore, if we are to test hypotheses concerning the variance

TABLE 14.8 Expected Mean Squares for a Model II Two-Factor Experiment

Source of Variation	Degrees of Freedom	Mean Square	Expected Mean Square
A	$a - 1$	s_1^2	$\sigma^2 + n\sigma_{\alpha\beta}^2 + bn\sigma_\alpha^2$
B	$b - 1$	s_2^2	$\sigma^2 + n\sigma_{\alpha\beta}^2 + an\sigma_\beta^2$
AB	$(a - 1)(b - 1)$	s_3^2	$\sigma^2 + n\sigma_{\alpha\beta}^2$
Error	$ab(n - 1)$	s^2	σ^2
Total	$abn - 1$		

TABLE 14.9 Expected Mean Squares for a Model II Three-Factor Experiment

Source of Variation	Degrees of Freedom	Mean Square	Expected Mean Square
A	$a - 1$	s_1^2	$\sigma^2 + n\sigma_{\alpha\beta\gamma}^2 + cn\sigma_{\alpha\beta}^2 + bn\sigma_{\alpha\gamma}^2 + bcn\sigma_\alpha^2$
B	$b - 1$	s_2^2	$\sigma^2 + n\sigma_{\alpha\beta\gamma}^2 + cn\sigma_{\alpha\beta}^2 + an\sigma_{\beta\gamma}^2 + acn\sigma_\beta^2$
C	$c - 1$	s_3^2	$\sigma^2 + n\sigma_{\alpha\beta\gamma}^2 + bn\sigma_{\alpha\gamma}^2 + an\sigma_{\beta\gamma}^2 + abn\sigma_\gamma^2$
AB	$(a - 1)(b - 1)$	s_4^2	$\sigma^2 + n\sigma_{\alpha\beta\gamma}^2 + cn\sigma_{\alpha\beta}^2$
AC	$(a - 1)(c - 1)$	s_5^2	$\sigma^2 + n\sigma_{\alpha\beta\gamma}^2 + bn\sigma_{\alpha\gamma}^2$
BC	$(b - 1)(c - 1)$	s_6^2	$\sigma^2 + n\sigma_{\alpha\beta\gamma}^2 + an\sigma_{\beta\gamma}^2$
ABC	$(a - 1)(b - 1)(c - 1)$	s_7^2	$\sigma^2 + n\sigma_{\alpha\beta\gamma}^2$
Error	$abc(n - 1)$	s^2	σ^2
Total	$abcn - 1$		

components of the main effects, it is necessary first to investigate the significance of the two-factor interaction components. An approximate test derived by Satterthwaite may be used when certain two-factor interaction variance components are found to be significant and hence must remain a part of the expected mean square.

EXAMPLE 14.4 In a study to determine which are the important sources of variation in an industrial process, 3 measurements are taken on yield for 3 operators chosen randomly and 4 batches of raw materials chosen randomly. It was decided that a significance test should be made at the 0.05 level of significance to determine if the variance components due to batches, operators, and interaction are significant. In addition, estimates of variance components are to be computed. The data are as follows, with the response being percent by weight:

Operator	Batch			
	1	2	3	4
1	66.9	68.3	69.0	69.3
	68.1	67.4	69.8	70.9
	67.2	67.7	67.5	71.4
2	66.3	68.1	69.7	69.4
	65.4	66.9	68.8	69.6
	65.8	67.6	69.2	70.0
3	65.6	66.0	67.1	67.9
	66.3	66.9	66.2	68.4
	65.2	67.3	67.4	68.7

SOLUTION

The sums of squares are found in the usual way, with the results given by

$$SST \text{ (total)} = 84.5564$$

$$SSA \text{ (operators)} = 18.2106$$

$$SSB \text{ (batches)} = 50.1564$$

$$SS(AB) \text{ (interaction)} = 5.5161$$

$$SSE \text{(error)} = 10.6733.$$

All other computations are carried out and exhibited in Table 14.10. Since $f_{0.05}(2, 6) = 5.14$, $f_{0.05}(3, 6) = 4.76$, and $f_{0.05}(6, 24) = 2.51$, we find the operator and batch variance components to be significant. Although the interaction variance is not significant at the $\alpha = 0.05$ level, the P-value is 0.095. Estimates of the main effect variance components are given by

$$\hat{\sigma}_\alpha^2 = \frac{9.1053 - 0.9194}{12} = 0.68$$

$$\hat{\sigma}_\beta^2 = \frac{16.7188 - 0.9144}{9} = 1.76.$$

TABLE 14.10   Analysis of Variance for Example 14.4

Source of Variation	Sum of Squares	Degrees of Freedom	Mean Square	Computed f
Operators	18.2106	2	9.1053	9.90
Batches	50.1564	3	16.7188	18.18
Interaction	5.5161	6	0.9194	2.07
Error	10.6733	24	0.4447	
Total	84.5564	35		

14.8 Choice of Sample Size

Our study of factorial experiments throughout this chapter has been restricted to the use of a completely randomized design with the exception of Section 14.6 where we demonstrated the analysis of a two-factor experiment in a randomized block design. The completely randomized design is very easy to lay out and the analysis is simple to perform; however, it should be used only when the number of treatment combinations is small and the experimental material is homogeneous. Although the randomized block design is ideal for dividing a large group of heterogeneous units into subgroups of homogeneous units, it is generally difficult to obtain uniform blocks with enough units to which a large number of treatment combinations may be assigned. This disadvantage may be overcome by choosing a design from the catalog of **incomplete block designs**. These designs allow one to investigate differences among t treatments arranged in b blocks each containing k experimental units, where $k < t$. The reader may consult Box, Hunter, and Hunter for details.

Once the experimenter has selected a completely randomized design, he must decide if the number of replications is sufficient to yield tests in the analysis of variance with high power. If not, he must add additional replications, which in turn may force him into a randomized complete block design. Had he started with a randomized block design, it would still be necessary to determine if the number of blocks is sufficient to yield powerful tests. Basically, then, we are back to the question of sample size.

The power of a fixed effects test for a given sample size is found from Table A.15 by computing the noncentrality parameter λ and the function ϕ discussed in Section 13.14. Expressions for λ and ϕ^2 for the two-factor and three-factor fixed effects experiments are given in Table 14.11.

The results of Section 13.14 for the random effects model can be extended very easily to the two- and three-factor models. Once again the general procedure is based on the values of the expected mean squares. For example, if we are testing

TABLE 14.11 Noncentrality Parameter λ and ϕ^2 for Two-Factor and Three-Factor Models

| | Two-Factor Experiments | | Three-Factor Experiments | | |
	A	B	A	B	C
λ	$\dfrac{bn\sum_{i=1}^{a}\alpha_i^2}{2\sigma^2}$	$\dfrac{an\sum_{j=1}^{b}\beta_j^2}{2\sigma^2}$	$\dfrac{bcn\sum_{i=1}^{a}\alpha_i^2}{2\sigma^2}$	$\dfrac{acn\sum_{j=1}^{b}\beta_j^2}{2\sigma^2}$	$\dfrac{abn\sum_{k=1}^{c}\gamma_k^2}{2\sigma^2}$
ϕ^2	$\dfrac{bn\sum_{i=1}^{a}\alpha_i^2}{a\sigma^2}$	$\dfrac{an\sum_{j=1}^{b}\beta_j^2}{b\sigma^2}$	$\dfrac{bcn\sum_{i=1}^{a}\alpha_i^2}{a\sigma^2}$	$\dfrac{acn\sum_{j=1}^{b}\beta_j^2}{b\sigma^2}$	$\dfrac{abn\sum_{k=1}^{c}\gamma_k^2}{c\sigma^2}$

$\sigma_\alpha^2 = 0$ in a two-factor experiment by computing the ratio s_1^2/s_3^2 (see Table 14.8), then

$$f = \frac{s_1^2/(\sigma^2 + n\sigma_{\alpha\beta}^2 + bn\sigma_\alpha^2)}{s_3^2/(\sigma^2 + n\sigma_{\alpha\beta}^2)}$$

is a value of the random variable F having the F-distribution with $a - 1$ and $(a - 1)(b - 1)$ degrees of freedom, and the power of the test is given by

$$1 - \beta = P\left\{ \frac{S_1^2}{S_3^2} > f_\alpha[(a - 1), (a - 1)(b - 1)] \text{ when } \sigma_\alpha^2 \neq 0 \right\}$$

$$= P\left\{ F > \frac{f_\alpha[(a - 1), (a - 1)(b - 1)](\sigma^2 + n\sigma_{\alpha\beta}^2)}{\sigma^2 + n\sigma_{\alpha\beta}^2 + bn\sigma_\alpha^2} \right\}.$$

Exercises

1. To estimate the various components of variability in a filtration process, the percent of material lost in the mother liquor was measured for 12 experimental conditions, 3 runs on each condition. Three filters and 4 operators were selected at random to use in the experiment, resulting in the following measurements:

Filter	Operator			
	1	2	3	4
1	16.2	15.9	15.6	14.9
	16.8	15.1	15.9	15.2
	17.1	14.5	16.1	14.9
2	16.6	16.0	16.1	15.4
	16.9	16.3	16.0	14.6
	16.8	16.5	17.2	15.9
3	16.7	16.5	16.4	16.1
	16.9	16.9	17.4	15.4
	17.1	16.8	16.9	15.6

(a) Test the hypothesis of no interaction variance component between filters and operators at the $\alpha = 0.05$ level of significance.

(b) Test the hypotheses that the operators and the filters have no effect on the variability of the filtration process at the $\alpha = 0.05$ level of significance.

(c) Estimate the components of variance due to filters, operators, and experimental error.

2. Assuming a model II experiment for Exercise 2 on page 546, estimate the variance components for brand of orange juice concentrate, for number of days from when orange juice was blended until it was tested, and for experimental error.

3. Consider the following analysis of variance for a model II experiment:

Source of Variation	Degrees of Freedom	Mean Square
A	3	140
B	1	480
C	2	325
AB	3	15
AC	6	24
BC	2	18
ABC	6	2
Error	24	5
Total	47	

Test for significant variance components among all main effects and interaction effects at the 0.01 level of significance

(a) by using a pooled estimate of error when appropriate;

(b) by not pooling sums of squares of insignificant effects.

4. Are 2 observations for each treatment combination in Exercise 4 on page 556 sufficient if the power of our

test for detecting differences among the levels of factor C at the 0.05 level of significance is to be at least 0.8 when $\gamma_1 = -0.2$, $\gamma_2 = 0.4$, and $\gamma_3 = -0.2$? Use the same pooled estimate of σ^2 that was used in the analysis of variance.

Review Exercises

1. The Statistics Consulting Center at Virginia Polytechnic Institute and State University was involved in analyzing a set of data taken by personnel in the Human Nutrition and Foods Department in which it was of interest to study the effects of flour type and percent sweetener on certain physical attributes of a type of cake. All-purpose flour and cake flour were used and the percent sweetener was varied at four levels. The following data show information on specific gravity of cake samples. Three cakes were prepared at each of the eight factor combinations.
   (a) Treat the analysis as a two-factor analysis of variance. Test for differences between flour type. Test for differences between sweetener concentration.
   (b) Discuss the effect of interaction, if any. Give P-values on all tests.

		Flour	
		All-purpose	Cake
Sweetener concentration	0	0.90 0.87 0.90	0.91 0.90 0.80
	50	0.86 0.89 0.91	0.88 0.82 0.83
	75	0.93 0.88 0.87	0.86 0.85 0.80
	100	0.79 0.82 0.80	0.86 0.85 0.85

2. An experiment was conducted in the Department of Food Science at Virginia Polytechnic Institute and State University. It was of interest to characterize the texture of certain types of fish in the herring family. The effect of sauce types used in preparing the fish was also studied. The response in the experiment was "texture value" measured with a machine that sliced the fish product. The following are data on texture values:

	Unbleached menhaden	Bleached menhaden	Herring
Sour cream	27.6 57.4 47.8 71.1 53.8	64.0 66.9 66.5 66.8 53.8	107.0 83.9 110.4 93.4 83.1
Wine sauce	49.8 31.0 11.8 35.1 16.1	48.3 62.2 54.6 43.6 41.8	88.0 95.2 108.2 86.7 105.2

   (a) Do an analysis of variance. Determine whether or not there is an interaction between sauce type and fish type.
   (b) Based on your results from part (a) and on F-tests on main effects, determine if there is a difference in texture due to sauce types, and determine whether there is a significant difference in fish types.

3. A study was made to determine if humidity conditions have an effect on the force required to pull apart pieces of glued plastic. Three types of plastic were tested using 4 different levels of humidity. The results, in kilograms, are given as follows:

	Humidity			
Plastic Type	30%	50%	70%	90%
A	39.0 42.8	33.1 37.8	33.8 30.7	33.0 32.9
B	36.9 41.0	27.2 26.8	29.7 29.1	28.5 27.9
C	27.4 30.3	29.2 29.9	26.7 32.0	30.9 31.5

5. Using the estimates of the variance components in Exercise 1, evaluate the power when we test the variance component due to filters to be zero.

(a) Assuming a model I experiment, perform an analysis of variance and test the hypothesis of no interaction between humidity and plastic type at the 0.05 level of significance.

(b) Using only plastics A and B and the value of s^2 from part (a), once again test for the presence of interaction at the 0.05 level of significance.

(c) Use a single-degree-of-freedom comparison and the value of s^2 from part (a) to compare, at the 0.05 level of significance, the force required at 30% humidity versus 50%, 70%, and 90% humidity.

(d) Using only plastic C and the value of s^2 from part (a), repeat part (c).

4. Personnel in the Materials Engineering Department at Virginia Polytechnic Institute and State University conducted an experiment to study the effects of environmental factors on the stability of a certain type of copper–nickel alloy. The basic response was the fatigue life of the material. The factors are *level of stress* and *environment*. The data are as follows:

	Stress level		
	Low	Medium	High
Dry hydrogen	11.08	13.12	14.18
	10.98	13.04	14.90
	11.24	13.37	15.10
High humidity (95%)	10.75	12.73	14.15
	10.52	12.87	14.42
	10.43	12.95	14.25

Environment

(a) Do an analysis of variance to test for interaction between the factors. Use $\alpha = 0.05$.

(b) Based on part (a), do an analysis on the two main effects and draw conclusions. Use a P-value approach in drawing conclusions.

5. In the experiment of Exercise 1, cake volume was also used as a response. The units are cubic inches. Test for interaction between factors and discuss main effects. Assume that both factors are fixed effects.

	Flour	
	All-purpose	Cake
0	4.48	4.12
	3.98	4.92
	4.42	5.10
50	3.68	5.00
	5.04	4.26
	3.72	4.34
75	3.92	4.82
	3.82	4.34
	4.06	4.40
100	3.26	4.32
	3.80	4.18
	3.40	4.30

Sweetener concentration

6. A control valve needs to be very sensitive to the input voltage, thus generating a good output voltage. An engineer turns the control bolts to change the input voltage. In the book *SN-Ratio for the Quality Evaluation* published by the Japanese Standards Association (1988), a study on how these three factors (relative position of control bolts, control range of bolts, and input voltage) affect the sensitivity of a control valve was conducted. The factors and their levels are shown below. The data show the sensitivity of a control valve.

Factor A: Relative position of control bolts: center $- 0.5$, center, and center $+ 0.5$.

Factor B: Control range of bolts: 2, 4.5, and 7 (mm).

Factor C: Input voltage: 100, 120, and 150 (V).

		Factor C		
Factor A	Factor B	C_1	C_2	C_3
	B_1	151	151	151
		135	135	138
A_1	B_2	178	180	181
		171	173	174
	B_3	204	205	206
		190	190	192
	B_1	156	158	158
		148	149	150
A_2	B_2	183	183	183
		168	170	172
	B_3	210	211	213
		204	203	204
	B_1	161	162	163
		145	148	148
A_3	B_2	189	191	192
		182	184	183
	B_3	215	216	217
		202	203	205

Perform an analysis of variance with $\alpha = 0.05$ to test for significant main and interaction effects.

7. To estimate the degree of suspension of a suspension polyethylene, we extract polyethylene by using a solvent, and compare the gel contents (gel proportion). This method is called the gel proportion estimation method. In the book *Design of Experiments for the Quality Improvement* published by the Japanese Standards Association (1989), a study was conducted on the relationship between the amount of gel generated and three factors (solvent, extraction temperature, and extraction time) that influence amount of gel. The data are as follows:

		Time		
Solvent	Temperature	4	8	24
Ethanol	120	94.0	93.8	91.1
		94.0	94.2	90.5
	80	95.3	94.9	92.5
		95.1	95.3	92.4
Toluene	120	94.6	93.6	91.1
		94.5	94.1	91.0
	80	95.4	95.6	92.1
		95.4	96.0	92.1

Perform an analysis of variance with $\alpha = 0.05$ to test for significant main and interaction effects.

8. In the book *SN-Ratio for the Quality Evaluation* published by the Japanese Standards Association (1988), a study on how tire air pressure affects the maneuverability of an automobile was conducted. Three different tire air pressures were compared on three different driving surfaces. The three air pressures were: both left- and right-side tires inflated to 6 kgf/cm^2, left-side tires inflated to 6 kgf/cm^2 and right-side tires inflated to 3 kgf/cm^2, and both left- and right-side tires inflated to 3 kgf/cm^2. The three driving surfaces were: asphalt, dry asphalt, and dry cement. The turning radius of a test vehicle was observed twice for each level of tire pressure on each of the three different driving surfaces.

	Tire air pressure					
	1		2		3	
Asphalt	44.0	25.5	34.2	37.2	27.4	42.8
Dry asphalt	31.9	33.7	31.8	27.6	43.7	38.2
Dry cement	27.3	39.5	46.6	28.1	35.5	34.6

Perform an analysis of variance of the above data. Comment on the interpretation of the main and interaction effects.

15

2^k Factorial Experiments and Fractions

15.1 Introduction

In almost any experimental study in which statistical procedures are applied to a collection of scientific data, the methods involve performing certain operations or computations on the sample information, followed by the drawing of inferences about the population or populations studied. Often there are characteristics of the experiment that are subject to the control of the experimenter, quantities such as sample size, number of levels of the factors, treatment combinations to be used, and so forth. These *experimental parameters* can often have a great effect on the precision with which hypotheses are tested or estimation is accomplished.

We have already been exposed to certain experimental design concepts. The sampling plan for the simple *t*-test on the mean of a normal population and also the analysis of variance involve randomly allocating prechosen treatments to experimental units. The randomized block design, where treatments are assigned to units within relatively homogeneous blocks, involves restricted randomization.

In this chapter we give special attention to experimental designs in which the experimental plan calls for the study of the effect on a response of k factors, each at

567

two levels. These are commonly known as 2^k factorial experiments. We often denote the levels as "high" and "low," even though this notation may be arbitrary in the case of qualitative variables. The complete factorial design requires that each level of every factor occur with each level of every other factor, giving a total of 2^k treatment combinations.

Symbolism for Factor Combinations

We denote the higher levels of the factors $A, B, C, \ldots$ by the letters $a, b, c, \ldots$ and the lower levels of each factor by the notation (1). In the presence of other letters we omit the symbol (1). For example, the treatment combination in a 2^4 experiment that contains the high levels of factors B and C and the low levels of factors A and D is written simply as bc. The treatment combination that consists of the low level of all factors in the experiment is denoted by the symbol (1). In the case of a 2^3 experiment, the eight possible treatment combinations are (1), a, b, c, ab, ac, bc, and abc.

The factorial experiment allows the effect of each and every factor to be estimated and tested independently through the usual analysis of variance. In addition, the interaction effects are easily assessed.

Often the use of these two-level designs are natural when the resulting data analysis lends itself to a *regression approach*. Here, of course, the design levels are truly continuous and the scientist is interested in building a model as well as doing hypothesis testing.

The disadvantage, of course, with the factorial experiment is the excessive amount of experimentation that may be required. For example, if it is desired to study the effect of eight variables, $2^8 = 256$ treatment combinations are required. In many instances we can obtain considerable information by using only a fraction of the experimental runs. This type of design is called a **fractional factorial design**.

The intention of these fractional designs is to reduce the size of the total experiment while retaining important information regarding the factors.

15.2 Analysis of Variance

Consider initially a 2^2 factorial plan in which there are *n experimental observations per treatment combination*. Extending our previous notation, we now interpret the symbols (1), a, b, and ab to be the *total yields* for each of the four treatment combinations. Table 15.1 gives a two-way table of these total yields.

Let us define the following contrasts among the treatment totals:

$$A \text{ contrast} = ab + a - b - (1)$$
$$B \text{ contrast} = ab - a + b - (1)$$
$$AB \text{ contrast} = ab - a - b + (1).$$

TABLE 15.1 A 2^2 Factorial Experiment

		B		Mean
		(1)	b	$\dfrac{b + (1)}{2n}$
A				
		a	ab	$\dfrac{ab + a}{n}$
	Mean	$\dfrac{a + (1)}{2n}$	$\dfrac{ab + b}{2n}$	

Clearly, there will be exactly one single-degree-of-freedom contrast for the means of each factor A and B, which we shall write as

CALCULATION OF EFFECTS

$$w_A = \frac{ab + a - b - (1)}{2n} = \frac{A \text{ contrast}}{2n}$$

and

$$w_B = \frac{ab + b - a - (1)}{2n} = \frac{B \text{ contrast}}{2n}.$$

The quantity w_A is seen to be the *difference between the mean response at the low and high levels of factor A*. In fact, we call w_A the **main effect** of A. Similarly, w_B is the main effect of factor B. Apparent interaction in the data is observed by inspecting the difference between $ab - b$ and $a - (1)$ or between $ab - a$ and $b - (1)$ in Table 15.1. If, for example, $ab - a \simeq b - (1)$ or $ab - a - b + (1) \simeq 0$, a line connecting the responses for each level of factor A at the high level of factor B will be approximately parallel to a line connecting the response for each level of factor A at the low level of factor B. The nonparallel lines of Figure 15.1 suggest the presence of interaction. To test whether this apparent interaction is significant, a third contrast in the treatment totals orthogonal to the main effect contrasts, called the **interaction effect**, is constructed by evaluating

$$w_{AB} = \frac{ab - a - b + (1)}{2n} = \frac{AB \text{ contrast}}{2n}.$$

Computation of Sums of Squares

We take advantage of the fact that in the 2^2 factorial, or for that matter in the general 2^k factorial experiment, each main effect and interaction effect has associated with it a single degree of freedom. Therefore, we can write $2^k - 1$ orthogonal single-

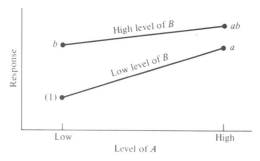

FIGURE 15.1 Response suggesting apparent interaction.

degree-of-freedom contrasts in the treatment combinations, each representing variation due to some main or interaction effect. Thus, under the usual independence and
normality assumptions in the experimental model, we can make tests to determine if
the contrast reflects systematic variation or merely chance or random variation. The
sums of squares for each contrast are found by following the procedures given in
Section 13.5. Writing $T_{1..} = b + (1)$, $T_{2..} = ab + a$, $c_1 = -1$, and $c_2 = 1$, where
$T_{1..}$ and $T_{2..}$ are the totals of $2n$ observations, we have

$$
SSA = SSw_A = \frac{\left(\sum\limits_{i=1}^{2} c_i T_{i..} \right)^2}{2n \sum\limits_{i=1}^{2} c_i^2}
$$

$$
= \frac{[ab + a - b - (1)]^2}{2^2 n} = \frac{(A \text{ contrast})^2}{2^2 n},
$$

with 1 degree of freedom. Similarly, we find that

$$
SSB = \frac{[ab + b - (1) - a]^2}{2^2 n} = \frac{(B \text{ contrast})^2}{2^2 n}
$$

and

$$
SS(AB) = \frac{[ab + (1) - a - b]^2}{2^2 n} = \frac{(AB \text{ contrast})^2}{2^2 n},
$$

each with 1 degree of freedom, while the error sum of squares, with $2^2(n - 1)$
degrees of freedom, is obtained by subtraction from the formula

$$
SSE = SST - SSA - SSB - SS(AB).
$$

In computing the sums of squares for the main effects A and B and the interaction
effect AB, it is convenient to present the total yields of the treatment combinations
along with the appropriate algebraic signs for each contrast, as in Table 15.2. The
main effects are obtained as simple comparisons between the low and high levels.
Therefore, we assign a positive sign to the treatment combination that is at the high

TABLE 15.2 Signs for Contrasts in a 2^2 Factorial Experiment

Treatment Combination	Factorial Effect		
	A	B	AB
(1)	−	−	+
a	+	−	−
b	−	+	−
ab	+	+	+

level of a given factor and a negative sign to the treatment combination at the lower level. The positive and negative signs for the interaction effect are obtained by multiplying the corresponding signs of the contrasts of the interacting factors.

The 2^3 Factorial

Let us now consider an experiment using three factors A, B, and C with levels (1), a; (1), b; and (1), c, respectively. This is a 2^3 factorial experiment giving the eight treatment combinations (1), a, b, c, ab, ac, bc, and abc. The treatment combinations and the appropriate algebraic signs for each contrast used in computing the sums of squares for the main effects and interaction effects are presented in Table 15.3.

An inspection of Table 15.3 reveals that for the 2^3 experiment any two contrasts among the seven are orthogonal and therefore the seven effects are assessed independently. The sum of squares for, say, the ABC interaction with 1 degree of freedom, is given by

$$SS(ABC) = \frac{[abc + a + b + c - (1) - ab - ac - bc]^2}{2^3 n}$$

and the ABC interaction effect is given by

$$w(ABC) = \frac{abc + a + b + c - (1) - ab - ac - bc}{4n}$$

For a 2^k factorial experiment the single-degree-of-freedom sums of squares for the main effects and interaction effects are obtained by squaring the appropriate contrasts in the treatment totals and dividing by $2^k n$, where n is the number of replications of the treatment combinations.

As before, an effect is always calculated by subtracting the average response at the "low" level from the average response at the "high" level. The high and low for main effects are quite clear. The symbolic high and low for interactions are evident from information as in Table 15.3.

The orthogonality property has the same importance here as it did in the material on comparisons discussed in Chapter 13. Orthogonality of contrasts implies that the estimated effects and thus the sums of squares are independent. This independence

TABLE 15.3   Signs for Contrasts in a 2^3 Factorial Experiment

Treatment Combination	Factorial Effect (Symbolic)						
	A	B	C	AB	AC	BC	ABC
(1)	−	−	−	+	+	+	−
a	+	−	−	−	−	+	+
b	−	+	−	−	+	−	+
c	−	−	+	+	−	−	+
ab	+	+	−	+	−	−	−
ac	+	−	+	−	+	−	−
bc	−	+	+	−	−	+	−
abc	+	+	+	+	+	+	+

is easily illustrated in a 2^3 factorial experiment if the yields, with factor A at its high level, are increased by an amount x in Table 15.3. Only the A contrast leads to a larger sum of squares, since the x effect cancels out in the formation of the six remaining contrasts as a result of the two positive and two negative signs associated with treatment combinations in which A is at the high level.

There are additional advantages produced by orthogonality. These will be pointed out subsequently when we discuss the 2^k factorial experiment in regression situations.

15.3   Yates' Technique for Computing Contrasts _____

It is laborious to write out the table of positive and negative signs for large experiments. A systematic tabular technique for deriving the factorial effects has been developed by Yates. The treatment combinations and the observations must be written down in **standard form**. For one factor the standard form is (1), a. For two factors we add b and ab, derived by multiplying the first two treatment combinations by the additional letter b. For three factors we add c, ac, bc, and abc, derived by multiplying the first four treatment combinations by the additional letter c, and so on. In the case of the three factors the standard order is then

$$(1), \quad a, \quad b, \quad ab, \quad c, \quad ac, \quad bc, \quad abc.$$

Yates' method is carried out in the following steps:

1. Place the treatment combinations and the corresponding total yields in a column in standard order.
2. Obtain the top half of a column marked (1) by adding the first two yields, then the next two, and so on. The bottom half is obtained by subtracting the first from the second of each of these same pairs.
3. Repeat the operation using the results in column (1) to obtain column (2). This operation is continued until we have k columns for a 2^k experiment.
4. The first value of the kth column will be the grand total of the yields in the experiment. Each remaining number will be a contrast in the treatment totals.

Finally, the sum of squares for the main effects and interaction effects are obtained by squaring the entries in column (k) and dividing by $2^k n$, where n is the number of replications and 2^k is the sum of the squares of the coefficients of the individual contrasts.

As an illustration we outline the procedure in Table 15.4 for the 2^3 factorial experiment.

TABLE 15.4 Yates' Technique for a 2^3 Factorial Experiment

Treatment Combination	(1)	(2)	(3)	Identification
(1)	$(1) + a$	$(1) + a + b + ab$	$(1) + a + b + ab + c + ac + bc + abc$	Total
a	$b + ab$	$b + ac + bc + abc$	$a - (1) + ab - b + ac - c + abc - bc$	A contrast
b	$c + ac$	$a - (1) + ab - b$	$b + ab - (1) - a + bc + abc - c - ac$	B contrast
ab	$bc + abc$	$ac - c + abc - bc$	$ab - b - a + (1) + abc - bc - ac + c$	AB contrast
c	$a - (1)$	$b + ab - (1) - a$	$c + ac + bc + abc - (1) - a - b - ab$	C contrast
ac	$ab - b$	$bc + abc - c - ac$	$ac - c + abc - bc - a + (1) - ab + b$	AC contrast
bc	$ac - c$	$ab - b - a + (1)$	$bc + abc - c - ac - b - ab + (1) + a$	BC contrast
abc	$abc - bc$	$abc - bc - ac + c$	$abc - bc - ac + c - ab + b + a - (1)$	ABC contrast

EXAMPLE 15.1   In a metallurgy experiment it is desired to test the effect of four factors and their interactions on the concentration (percent by weight) of a particular phosphorus compound in casting material. The variables are A, percent phosphorus in the refinement; B, percent remelted material; C, fluxing time; and D, holding time. The four factors are varied in a 2^4 factorial experiment with two castings taken at each factor combination. The 32 castings were made in random order. Using Yates' technique, calculate effects and perform the analysis of variance for the following data:

Treatment Combination	Weight % of Phosphorus Compound		
	Replication 1	Replication 2	Total
(1)	30.3	28.6	58.9
a	28.5	31.4	59.9
b	24.5	25.6	50.1
ab	25.9	27.2	53.1
c	24.8	23.4	48.2
ac	26.9	23.8	50.7
bc	24.8	27.8	52.6
abc	22.2	24.9	47.1
d	31.7	33.5	65.2
ad	24.6	26.2	50.8
bd	27.6	30.6	58.2
abd	26.3	27.8	54.1
cd	29.9	27.7	57.6
acd	26.8	24.2	51.0
bcd	26.4	24.9	51.3
$abcd$	26.9	29.3	56.2
Total	428.1	436.9	865.0

SOLUTION

Table 15.5 gives the 16 treatment totals and outlines Yates' technique for computing the individual effects and sums of squares. The total sum of squares is given by

$$SST = 30.3^2 + 28.5^2 + \cdots + 29.3^2 - \frac{865^2}{32} = 217.51.$$

We can now set up the analysis of variance as in Table 15.6. Note that all main effects and interactions BC, AD, ABD, and ACD are significant. The tests on the

TABLE 15.5 Yates' Technique for a 2^4 Example

Treatment Combination	Treatment Total	(1)	(2)	(3)	Contrast	Effect	Sum of Squares
(1)	58.9	118.8	222.0	420.6	865.0	54.06	
a	59.9	103.2	198.6	444.4	−19.2	−1.20	11.52
b	50.1	98.9	228.3	1.0	−19.6	−1.23	12.00
ab	53.1	99.7	216.1	−20.2	15.8	0.99	7.80
c	48.2	116.0	4.0	−14.8	−35.6	−2.23	39.61
ac	50.7	112.3	−3.0	−4.8	9.8	−0.61	3.00
bc	52.6	108.6	−18.5	−6.0	19.0	1.19	11.28
abc	47.1	107.5	−1.7	21.8	−8.8	−0.55	2.42
d	65.2	1.0	−15.6	−23.4	23.8	1.49	17.70
ad	50.8	3.0	0.8	−12.2	−21.2	−1.33	14.05
bd	58.2	2.5	−3.7	−7.0	10.0	0.63	3.13
abd	54.1	−5.5	−1.1	16.8	27.8	1.74	24.15
cd	57.6	−14.4	2.0	16.4	11.2	0.70	3.92
acd	51.0	−4.1	−8.0	2.6	23.8	1.49	17.70
bcd	51.3	−6.6	10.3	−10.0	−13.8	−0.86	5.95
abcd	56.2	4.9	11.5	1.2	11.2	0.70	3.92

main effects (which in the presence of interactions may be regarded as the effects *averaged* over the levels of the other factors) indicate significance in each case.

The tests and *P*-values given in Table 15.6 certainly require interpretation. A significant *P*-value suggests that the *effect differs significantly from zero*. All main effects are significant and the signs of the effects are important. An increase in the levels from low to high of A, B, and C *decreases* the response, the phosphorus content in the casting material. An increase from low to high of D increases the phosphorus content. However, because of significant interaction these main effect interpretations must be viewed as trends across the levels of the other factors. The impact of the important two-factor interactions can be assessed via two-dimensional plots of means as discussed in Chapter 14.

TABLE 15.6 Analysis of Variance for the Data of Table 15.5

Source of Variation	Effects	Sum of Squares	Degrees of Freedom	Mean Square	Computed f	P-value
Main effect						
A	−1.2000	11.52	1	11.52	4.68	0.0459
B	−1.2250	12.01	1	12.01	4.88	0.0421
C	−2.2250	39.61	1	39.61	16.10	0.0010
D	1.4875	17.70	1	17.70	7.20	0.0163
Two-factor interaction						
AB	0.9875	7.80	1	7.80	3.17	0.0939
AC	−0.6125	3.00	1	3.00	1.22	0.2857
AD	−1.3250	14.05	1	14.05	5.71	0.0295
BC	1.1875	11.28	1	11.28	4.59	0.0480
BD	0.6250	3.13	1	3.13	1.27	0.2763
CD	0.7000	3.92	1	3.92	1.59	0.2249
Three-factor interaction						
ABC	−0.5500	2.42	1	2.42	0.98	0.3360
ABD	1.7375	24.15	1	24.15	9.82	0.0064
ACD	1.4875	17.70	1	17.70	7.20	0.0163
BCD	−0.8625	5.95	1	5.95	2.42	0.1394
Four-factor interaction						
ABCD	0.7000	3.92	1	3.92	1.59	0.2249
Error		39.36	16	2.46		
Total		217.51	31			

15.4 Nonreplicated 2^k Factorial Experiment _____

The full 2^k factorial may often involve considerable experimentation, particularly when k is large. As a result, replication of each factor combination is often not allowed. If all effects, including all interactions, are included in the model of the experiment, no degrees of freedom are allowed for error. Often, when k is large, the data analyst will *pool* sums of squares and corresponding degrees of freedom for high-order interactions that are known to be or assumed to be negligible. This will produce F-tests for main effects and lower-order interactions.

Diagnostic Plotting with Unreplicated 2^k Factorial Experiments

Normal probability plotting can be a very useful methodology for determining the relative importance of effects in a reasonably large two-level factored experiment when there is no replication. This type of diagnostic plot can be particularly useful when the data analyst is hesitant to pool high-order interactions for fear that some of the effects pooled in the "error" may truly be real effects and not merely random. The reader should bear in mind that all effects that are not real (i.e., they are independent *estimates of zero*) follow a normal distribution with mean near zero and constant variance. For example, in a 2^4 factorial experiment, all effects (keep in mind $n = 1$) are of the form

$$w(AB) = \frac{contrast}{8} = \bar{y}_H - \bar{y}_L,$$

where $\bar{y}_H$ is the average of 8 independent experimental runs at the high or "+" level and $\bar{y}_L$ is the average of eight independent runs at the low or "−" level. Thus the variance of each contrast is $Var(\bar{y}_H - \bar{y}_L) = \sigma^2/4$. For any real effects, $E(\bar{y}_H - \bar{y}_L) \neq 0$. Thus normal probability plotting should reveal "significant" effects as those that fall off the straight line that depicts realizations of independent, identically distributed normal random variables.

The probability plotting can take one of many forms. The reader is referred to Chapter 8, where these plots were first presented. The empirical normal quantile–quantile plot may be used. The plotting procedure that makes use of normal probability paper may also be used. In addition, there are several other types of diagnostic normal probability plots. In summary, the diagnostic effect plots are given as follows.

PROBABILITY
EFFECT PLOTS
FOR
NONREPLICATED
2^k FACTORIAL
EXPERIMENTS

1. Calculate effects as

$$effect = \frac{contrast}{2^{k-1}}.$$

2. Construct a normal probability plot of all effects.
3. Effects that fall off the straight line should be considered real effects.

Further comments regarding probability plotting of effects are in order. First, the data analyst may feel quite frustrated if he or she uses them with a small experiment. The plotting is likely to give satisfying results when there is *effect sparsity*—many effects that are truly not real. This sparsity will be evident in large experiments where high-order interactions are not likely to be real.

15.5 Case Study

Many manufacturing companies in the United States and abroad use molded parts as components of the process. Shrinkage is often a major problem. Often, a molded die for a part will be built larger than nominal to allow for part shrinkage. In the following experimental situation a new die is being produced, and ultimately it is important to find the proper process settings to minimize shrinkage. In the following experiment, the response values are deviations from nominal (i.e., shrinkage). The factors and levels are as follows:

	Coded Levels	
	−1	+1
A, injection velocity (ft/sec)	1.0	2.0
B, mold temperature (°C)	100	150
C, mold pressure (psi)	500	1000
D, back pressure (psi)	75	120

The purpose of the experiment was to determine what effects (main effects and interaction effects) influence shrinkage. The experiment was considered a preliminary screening experiment from which the factors for a more complete analysis may be determined. Also, it was hoped that some insight on how the important factors impact shrinkage might be determined. The data from an unreplicated 2^4 factorial experiment are as follows:

Factor Combination	Response (cm $\times$ 10^4)	Factor Combination	Response (cm $\times$ 10^4)
(1)	72.68	d	73.52
a	71.74	ad	75.97
b	76.09	bd	74.28
ab	93.19	abd	92.87
c	71.25	cd	79.34
ac	70.59	acd	75.12
bc	70.92	bcd	79.67
abc	104.96	abcd	97.80

Initially, effects were calculated and placed on a normal probability plot. The calculated effects are as follows:

$$A = 10.5613$$
$$B = 12.4463$$
$$C = 2.4138$$
$$D = 2.1438$$
$$AB = 11.4038$$
$$AC = 1.2613$$
$$AD = -1.8238$$
$$BC = 1.8163$$
$$BD = -2.2787$$
$$CD = 1.4088$$
$$ABC = 2.8588$$

$$ABD = -1.7813$$
$$ACD = -3.0438$$
$$BCD = -0.4788$$
$$ABCD = -1.3063$$

The normal probability plot is shown in Figure 15.2. The plot seems to imply that effects A, B, and AB stand out as being important. The preliminary conclusions are as follows:

1. An increase in injection velocity from 1.0 to 2.0 increases shrinkage.
2. An increase in mold temperature from 100°C to 200°C increases shrinkage.
3. There is an interaction between injection velocity and mold temperature; while both main effects are important, it is crucial that we understand the impact of the two-factor interaction.

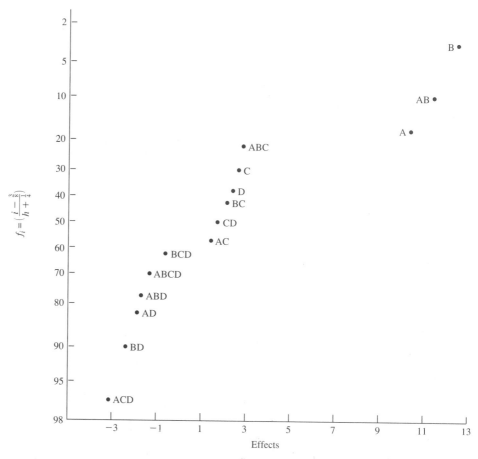

FIGURE 15.2 Normal probability plot of effects for Case Study 15.5.

Interpretation of Two-Factor Interaction

As one would expect, a two-way table of means should provide ease in interpretation of the AB interaction. Consider the following:

		100	150
A (velocity)	2	73.355	97.205
	1	74.1975	75.240

B (temperature)

Notice that the large sample mean at high velocity and high temperature created the significant interaction. The shrinkage increases in a nonadditive manner. Mold temperature appears to have a positive effect despite the velocity level. But the effect is greatest at high velocity. The velocity effect is very slight at low temperature but clearly is positive at high mold temperature. To control shrinkage at a low level *one should avoid using high injection velocity and high mold temperature simultaneously*.

Analysis with Pooled Error Mean Square: Annotated Computer Printout

It may be of interest to observe an analysis of variance of the injection molding data with high-order interactions pooled to form an error mean square. Interactions of order three and four are pooled. Figure 15.3 shows a SAS PROC GLM. The analysis of variance reveals essentially the same conclusion as that of the normal probability plot. Effects A, B, and AB are significant.

```
           General Linear Models Procedure
Dependent Variable: Y
                              Sum of        Mean
Source              DF        Squares       Square    F Value    Pr>F
Model               10       1689.2375     168.9237     9.37    0.0117
Error                5         90.1808      18.0362
Corrected Total     15       1779.4183

             R-Square          C.V.      Root MSE              Y Mean
             0.949320       5.308667      4.2469              79.999

Source              DF       Type I SS   Mean Square   F Value    Pr>F
A                    1       446.16001    446.16001     24.74    0.0042
B                    1       619.63656    619.63656     34.36    0.0020
C                    1        23.30476     23.30476      1.29    0.3072
D                    1        18.38266     18.38266      1.02    0.3590
A*B                  1       520.18206    520.18206     28.84    0.0030
A*C                  1         6.36301      6.36301      0.35    0.5784
```

FIGURE 15.3 SAS printout for Data in Case Study 15.5.

A*D	1	13.30426	13.30426	0.74	0.4297
B*C	1	13.19506	13.19506	0.73	0.4314
B*D	1	20.77081	20.77081	1.15	0.3322
C*D	1	7.93831	7.93831	0.44	0.5364

Source	DF	Type III SS	Mean Square	F Value	Pr>F
A	1	446.16001	446.16001	24.74	0.0042
B	1	619.63656	619.63656	34.36	0.0020
C	1	23.30476	23.30476	1.29	0.3072
D	1	18.38266	18.38266	1.02	0.3590
A*B	1	520.18206	520.18206	28.84	0.0030
A*C	1	6.36301	6.36301	0.35	0.5784
A*D	1	13.30426	13.30426	0.74	0.4297
B*C	1	13.19506	13.19506	0.73	0.4314
B*D	1	20.77081	20.77081	1.15	0.3322
C*D	1	7.93831	7.93831	0.44	0.5364

Parameter	Estimate	T for HO: Parameter = 0	Pr>|T|	Std Error of Estimate
INTERCEPT	79.99937500	75.35	0.0001	1.06172520
A	5.28062500	4.97	0.0042	1.06172520
B	6.22312500	5.86	0.0020	1.06172520
C	1.20687500	1.14	0.3072	1.06172520
D	1.07187500	1.01	0.3590	1.06172520
A*B	5.70187500	5.37	0.0030	1.06172520
A*C	0.63062500	0.59	0.5784	1.06172520
A*D	-0.91187500	-0.86	0.4297	1.06172520
B*C	0.90812500	0.86	0.4314	1.06172520
B*D	-1.13937500	-1.07	0.3322	1.06172520
C*D	0.70437500	0.66	0.5364	1.06172520

FIGURE 15.3   (continued)

Exercises

1. The following data were obtained from a 2^3 factorial experiment replicated three times:

Treatment Combination	Replicate 1	Replicate 2	Replicate 3
(1)	12	19	10
a	15	20	16
b	24	16	17
ab	23	17	27
c	17	25	21
ac	16	19	19
bc	24	23	29
abc	28	25	20

Evaluate the sums of squares for all factorial effects by the contrast method.

2. If an experiment conducted by the Mining Engineering Department at the Virginia Polytechnic Institute and State University in 1979 to study a particular filtering system for coal, a coagulant was added to a solution in a tank containing coal and sludge, which was then placed in a recirculation system in order that the coal could be washed. Three factors were varied in the experimental process:

Factor A: percent solids circulated initially in the overflow,

Factor B: flow rate of the polymer,

Factor C: pH of the tank.

The amount of solids in the underflow of the cleansing system determines how clean the coal has become. Two levels of each factor were used and two experimental runs were made for each of the $2^3 = 8$ combinations. The responses, percent solids by weight, in the underflow of the circulation system are as follows:

Treatment Combination	Response	
	Replicate 1	Replicate 2
(1)	4.65	5.81
a	21.42	21.35
b	12.66	12.56
ab	18.27	16.62
c	7.93	7.88
ac	13.18	12.87
bc	6.51	6.26
abc	18.23	17.83

Assuming that all interactions are potentially important, do a complete analysis of the data. Use *P*-values in your conclusion.

3. The effects of four factors on some response are to be studied. Each factor is varied at two levels in a 2^4 factorial arrangement and the following data recorded:

Treatment Combination	Response
(1)	23.8
a	19.6
b	29.9
ab	25.7
c	26.5
ac	22.6
bc	32.6
abc	28.6
d	21.6
ad	17.5
bd	27.5
abd	23.7
cd	24.6
acd	20.9
bcd	31.1
abcd	26.7

Assuming all three- and four-factor interactions to be negligible, analyze the given data by Yates' technique. Use a 0.01 level of significance.

4. A preliminary experiment is conducted to study the effects of four factors and their interactions on the output of a certain machining operation. Two runs are made at each of the treatment combinations in order to supply a measure of pure experimental error. Two levels of each factor are used, resulting in the following data:

Treatment Combination	Replicate 1	Replicate 2
(1)	7.9	9.6
a	9.1	10.2
b	8.6	5.8
c	10.4	12.0
d	7.1	8.3
ab	11.1	12.3
ac	16.4	15.5
ad	7.1	8.7
bc	12.6	15.2
bd	4.7	5.8
cd	7.4	10.9
abc	21.9	21.9
abd	9.8	7.8
acd	13.8	11.2
bcd	10.2	11.1
abcd	12.8	14.3

Make tests on all main effects and interactions at the 0.05 level of significance.

5. In the study "*An X-Ray Fluorescence Method for Analyzing Polybutadiene-Acrylic Acid (PBAA) Propellants*," Quarterly Reports, RK-TR-62-1, Army Ordnance Missile Command (1962), an experiment was conducted to determine whether or not there is a significant difference in the amount of aluminum achieved in the analysis between certain levels of the following processing variables:

> *A:* mixing time
> > level 1—2 hours
> > level 2—4 hours,
>
> *B:* Blade speed
> > level 1—36 rpm
> > level 2—78 rpm,

C: condition of nitrogen passed over
      propellant
      level 1—dry
      level 2—72% relative humidity,
D: physical state of propellant
      level 1—uncured
      level 2—cured.

The following data were recorded:

Obser- vation	Physical State	Mixing Time	Blade Speed	Nitrogen Condition	Aluminum
1	1	1	2	2	16.3
2	1	2	2	2	16.0
3	1	1	1	1	16.2
4	1	2	1	2	16.1
5	1	1	1	2	16.0
6	1	2	1	1	16.0
7	1	2	2	1	15.5
8	1	1	2	1	15.9
9	2	1	2	2	16.7
10	2	2	2	2	16.1
11	2	1	1	1	16.3
12	2	2	1	2	15.8
13	2	1	1	2	15.9
14	2	2	1	1	15.9
15	2	2	2	1	15.6
16	2	1	2	1	15.8

Assuming all three- and four-factor interactions to be negligible, analyze the data. Use a 0.05 level of significance.

15.6 Factorial Experiments in Incomplete Blocks _____

The 2^k factorial experiment lends itself to partitioning into *incomplete blocks*. For a k-factor experiment, it is often useful to use a design in 2^p blocks ($p < k$) when the entire 2^k treatment combinations cannot be applied under homogeneous conditions. The disadvantage with this experimental setup is that certain effects are completely sacrificed as a result of the blocking, the amount of sacrifice depending on the number of blocks required. For example, suppose that the eight treatment combinations in a 2^3 factorial experiment must be run in two blocks of size 4. Suppose, in addition, that one is willing to sacrifice the *ABC* interaction. Note the "contrast signs" in Table 15.3. A reasonable arrangement is

Block 1	Block 2
(1)	a
ab	b
ac	c
bc	abc

Concept of Confounding

If one assumes the usual model with the additive block effect, this effect cancels out in the formation of the contrasts on all effects except *ABC*. To illustrate, let x denote the contribution to the yield due to the difference between blocks. Writing the yields as

Block 1	Block 2
(1)	$a + x$
ab	$b + x$
ac	$c + x$
bc	$abc + x$

we see that the ABC contrast and also the contrast comparing the 2 blocks are both given by

$$ABC \text{ contrast} = (abc + x) + (c + x) + (b + x) + (a + x) - (1) - ab$$

$$- ac - bc$$

$$= abc + a + b + c - (1) - ab - ac - bc + 4x.$$

Therefore, we are measuring the ABC effect plus the block effect and there is no way of assessing the ABC interaction effect independent of blocks. We say then that the ABC interaction is **completely confounded with blocks**. By necessity, information on ABC has been sacrificed. On the other hand, the block effect cancels out in the formation of all other contrasts. For example, the A contrast is given by

$$A \text{ contrast} = (abc + x) + (a + x) + ab + ac - (b + x) - (c + x) - bc - (1)$$

$$= abc + a + ab + ac - b - c - bc - (1),$$

as in the case of a completely randomized design. We say that the effects A, B, C, AB, AC, and BC are *orthogonal to blocks*. Generally, for a 2^k factorial experiment in 2^p blocks, the number of effects confounded with blocks is 2^{p-1}, which is equivalent to the degrees of freedom for blocks.

2^k Factorial in Two Blocks

When two blocks are to be used with a 2^k factorial, one effect, usually a high-order interaction, is chosen as the **defining contrast**. This effect is to be confounded with blocks. The additional $2^k - 2$ effects are orthogonal with the defining contrast and thus with blocks.

Suppose that we represent the defining contrast as $A^{\gamma_1}B^{\gamma_2}C^{\gamma_3} \ldots$, where γ_i is either 0 or 1. This generates the expression

$$L = \gamma_1 + \gamma_2 + \cdots + \gamma_k,$$

which in turn is evaluated for each of the 2^k treatment combinations by setting γ_i equal to 0 or 1 according to whether the treatment combination contains the ith factor at its high or low level. The L values are then reduced (modulo 2) to either 0 or 1 and thereby determine to which block the treatment combinations are assigned.

In other words, the treatment combinations are divided into two blocks according to whether the L values leave a remainder of 0 or 1 when divided by 2.

EXAMPLE 15.2   Determine the values of L (modulo 2) for a 2^3 factorial experiment when the defining contrast is ABC.

SOLUTION
With ABC the defining contrast, we have

$$L = \gamma_1 + \gamma_2 + \gamma_3,$$

which is applied to each treatment combination as follows:

$$
\begin{aligned}
(1)\colon \quad & L = 0 + 0 + 0 = 0 = 0 && \text{(modulo 2)} \\
a\colon \quad & L = 1 + 0 + 0 = 1 = 1 && \text{(modulo 2)} \\
b\colon \quad & L = 0 + 1 + 0 = 1 = 1 && \text{(modulo 2)} \\
ab\colon \quad & L = 1 + 1 + 0 = 2 = 0 && \text{(modulo 2)} \\
c\colon \quad & L = 0 + 0 + 1 = 1 = 1 && \text{(modulo 2)} \\
ac\colon \quad & L = 1 + 0 + 1 = 2 = 0 && \text{(modulo 2)} \\
bc\colon \quad & L = 0 + 1 + 1 = 2 = 0 && \text{(modulo 2)} \\
abc\colon \quad & L = 1 + 1 + 1 = 3 = 1 && \text{(modulo 2).}
\end{aligned}
$$

The blocking arrangement, in which ABC is confounded, is given as before by

Block 1	Block 2
(1)	a
ab	b
ac	c
bc	abc

The A, B, C, AB, AC, and BC effects and sums of squares are computed in the usual way, ignoring blocks.

Notice that this arrangement is the same blocking scheme that would result from assigning the "+" sign factor combinations for the ABC contrast to one block and the "−" sign factor combinations for the ABC contrast to the other block.

The block containing the treatment combination (1) in Example 15.2 is called the **principal block**. This block forms an algebraic group with respect to multiplication when the exponents are reduced to the modulo 2 base. For example, the property of closure holds, since $(ab)(bc) = ab^2c = ac$, $(ab)(ab) = a^2b^2 = (1)$, and so forth.

2^k Factorial in Four Blocks

If the experimenter is required to allocate the treatment combinations to four blocks, two defining contrasts are chosen by the experimenter. A third effect, known as their **generalized interaction**, is automatically confounded with blocks, these three effects corresponding to the three degrees of freedom for blocks. The procedure for constructing the design is best explained through an example. Suppose it is decided that for a 2^4 factorial AB and CD are the defining contrasts. The third effect confounded, their generalized interaction, is formed by multiplying together the initial two modulo 2. Thus the effect

$$(AB)(CD) = ABCD$$

is also confounded with blocks. We construct the design by calculating the expressions

$$L_1 = \gamma_1 + \gamma_2 \qquad (AB)$$

$$L_2 = \gamma_3 + \gamma_4 \qquad (CD)$$

modulo 2 for each of the 16 treatment combinations to generate the following blocking scheme:

Block 1	Block 2	Block 3	Block 4
(1)	a	c	ac
ab	b	abc	bc
cd	acd	d	ad
abcd	bcd	abd	bd

$L_1 = 0$	$L_1 = 1$	$L_1 = 0$	$L_1 = 1$
$L_2 = 0$	$L_2 = 0$	$L_2 = 1$	$L_2 = 1$

A shortcut procedure can be used to construct the remaining blocks after the principal block has been generated. We begin by placing any treatment combination not in the principal block in a second block and build the block by multiplying (modulo 2) by the treatment combinations in the principal block. In the preceding example the second, third, and fourth blocks are generated as follows:

Block 2	Block 3	Block 4
$a(1) = a$	$c(1) = c$	$ac(1) = ac$
$a(ab) = b$	$c(ab) = abc$	$ac(ab) = bc$
$a(cd) = acd$	$c(cd) = d$	$ac(cd) = ad$
$a(abcd) = bcd$	$c(abcd) = abd$	$ac(abcd) = bd$

The analysis for the case of four blocks is quite simple. All effects that are orthogonal to blocks (those other than the defining contrasts) are computed in the usual fashion. In fact, Yates' technique can be used on the entire experiment, but the sums of squares for the three confounded effects are then added together to form the sum of squares due to blocks.

2^k Factorial in 2^p Blocks

The general scheme for the 2^k factorial experiment in 2^p blocks is not difficult. We select p defining contrasts such that none is the generalized interaction of any two in the group. Since there are $2^p - 1$ degrees of freedom for blocks, we have $2^p - 1 - p$ additional effects confounded with blocks. For example, in a 2^6 factorial experiment in eight blocks, we might choose ACF, $BCDE$, and $ABDF$ as the defining contrasts. Then

$$(ACF)(BCDE) = ABDEF$$

$$(ACF)(ABDF) = BCD$$

$$(BCDE)(ABDF) = ACEF$$

$$(ACF)(BCDE)(ABDF) = E$$

are the additional four effects confounded with blocks. This is not a desirable blocking scheme, since one of the confounded effects is the main effect E. The design is constructed by evaluating

$$L_1 = \gamma_1 + \gamma_3 + \gamma_6$$

$$L_2 = \gamma_2 + \gamma_3 + \gamma_4 + \gamma_5$$

$$L_3 = \gamma_1 + \gamma_2 + \gamma_4 + \gamma_6$$

and assigning treatment in combinations to blocks according to the following scheme:

Block 1: $L_1 = 0$, $L_2 = 0$, $L_3 = 0$

Block 2: $L_1 = 0$, $L_2 = 0$, $L_3 = 1$

Block 3: $L_1 = 0$, $L_2 = 1$, $L_3 = 0$

Block 4: $L_1 = 0$, $L_2 = 1$, $L_3 = 1$

Block 5: $L_1 = 1$, $L_2 = 0$, $L_3 = 0$

Block 6: $L_1 = 1$, $L_2 = 0$, $L_3 = 1$

Block 7: $L_1 = 1$, $L_2 = 1$, $L_3 = 0$

Block 8: $L_1 = 1$, $L_2 = 1$, $L_3 = 1$.

The shortcut procedure that was illustrated for the case of four blocks also applies here. Hence we can construct the remaining seven blocks from the principal block.

EXAMPLE 15.3   It is of interest to study the effect of five factors on some response with the assumption that interactions involving three, four, and five of the factors are negligible. We shall divide the 32 treatment combinations into four blocks using the defining contrasts $BCDE$ and $ABCD$. Thus $(BCDE)(ABCD) = AE$ is also confounded with blocks. The experimental design and the observations are given in Table 15.7.

The allocation of treatment combinations to experimental units within blocks is, of course, random. By pooling the unconfounded three-, four-, and five-factor

TABLE 15.7   Data for a 2^5 Experiment in Four Blocks

Block 1	Block 2	Block 3	Block 4
(1) = 30.6	a = 32.4	b = 32.6	e = 30.7
bc = 31.5	abc = 32.4	c = 31.9	bce = 31.7
bd = 32.4	abd = 32.1	d = 33.3	bde = 32.2
cd = 31.5	acd = 35.3	bcd = 33.0	cde = 31.8
abe = 32.8	be = 31.5	ae = 32.0	ab = 32.0
ace = 32.1	ce = 32.7	abce = 33.1	ac = 33.1
ade = 32.4	de = 33.4	abde = 32.9	ad = 32.2
abcde = 31.8	bcde = 32.9	acde = 35.0	abcd = 32.3

interactions to form the error term, perform the analysis of variance for the data of Table 15.7.

SOLUTION

The sums of squares for each of the 31 contrasts are computed by Yates' method and the block sum of squares is found to be

$$SS(\text{blocks}) = SS(ABCD) + SS(BCDE) + SS(AE)$$

$$= 7.538.$$

The analysis of variance is given in Table 15.8. None of the two-factor interactions

TABLE 15.8   Analysis of Variance for the Data of Table 15.7

Source of Variation	Sum of Squares	Degrees of Freedom	Mean Square	Computed f
Main effect				
A	3.251	1	3.251	6.32
B	0.320	1	0.320	0.62
C	1.361	1	1.361	2.64
D	4.061	1	4.061	7.89
E	0.005	1	0.005	0.01
Two-factor interaction				
AB	1.531	1	1.531	2.97
AC	1.125	1	1.125	2.18
AD	0.320	1	0.320	0.62
BC	1.201	1	1.201	2.33
BD	1.711	1	1.711	3.32
BE	0.020	1	0.020	0.04
CD	0.045	1	0.045	0.09
CE	0.001	1	0.001	0.002
DE	0.001	1	0.001	0.002
Blocks (ABCD, BCDE, AE)	7.538	3	2.513	
Error	7.208	14	0.515	

are significant at the $\alpha = 0.05$ level when compared to $f_{0.05}(1, 14) = 4.60$. The main effects A and D are significant and both give positive effects on the response as we go from the low to the high level.

15.7 Partial Confounding

It is possible to confound any effect with blocks by the methods described in Section 15.6. Suppose that we consider a 2^3 factorial experiment in two blocks with three complete replications. If ABC is confounded with blocks in all three replicates, we can proceed as before and determine single-degree-of-freedom sums of squares for all main effects and two-factor interaction effects. The sum of squares for blocks has 5 degrees of freedom, leaving $23 - 5 - 6 = 12$ degrees of freedom for error.

Now let us confound ABC in one replicate, AC in the second, and BC in the third. The plan for this type of experiment would be as follows:

Block				Block				Block	
1	2			1	2			1	2
abc	ab			abc	ab			abc	ab
a	ac			ac	bc			bc	ac
b	bc			b	a			a	b
c	(1)			(1)	c			(1)	c

Replicate 1	Replicate 2	Replicate 3
ABC confounded	*AC* confounded	*BC* confounded

The effects ABC, AC, and BC are said to be **partially confounded with blocks**. These three effects can be estimated from two of the three replicates. The ratio 2/3 serves as a measure of the extent of the confounding. Yates calls this ratio the **relative information** on the confounded effects. This ratio gives the amount of information available on the partially confounded effect relative to that available on an unconfounded effect.

The analysis-of-variance layout is given in Table 15.9. The sums of squares for blocks and for the unconfounded effects A, B, C, and AB are found in the usual way. The sums of squares for AC, BC, and ABC are computed from the two replicates in which the particular effect is not confounded. One must be careful to divide by 16 instead of 24 when obtaining the sums of squares for the partially confounded effects, since we are only using 16 observations. In Table 15.9 the primes are inserted with the degrees of freedom as a reminder that these effects are partially confounded and require special calculations.

TABLE 15.9 Analysis of Variance with Partial Confounding

Source of Variation	Degrees of Freedom
Blocks	5
A	1
B	1
C	1
AB	1
AC	1′
BC	1′
ABC	1′
Error	11
Total	23

Exercises

1. In a 2^3 factorial experiment with 3 replications, show the block arrangement and indicate by means of an analysis-of-variance table the effects to be tested and their degrees of freedom, when the AB interaction is confounded with blocks.

2. The following coded data represent the strength of a certain type of bread-wrapper stock produced under 16 different conditions, the latter representing two levels of each of four process variables. An operator effect was introduced into the model, since it was necessary to obtain half the experimental runs under operator 1 and half under operator 2. It was felt that operators do have an effect on the quality of the product.

Operator 1	Operator 2
(1) = 18.8	a = 14.7
ab = 16.5	b = 15.1
ac = 17.8	c = 14.7
bc = 17.3	abc = 19.0
d = 13.5	ad = 16.9
abd = 17.6	bd = 17.5
acd = 18.5	cd = 18.2
bcd = 17.6	abcd = 20.1

(a) Assuming that all interactions are negligible, make significance tests for the factors A, B, C, and D. Use a 0.05 level of significance.

(b) What interaction is confounded with operators?

3. Divide the treatment combinations of a 2^4 factorial experiment into four blocks by confounding ABC and ABD. What additional effect is also confounded with blocks?

4. An experiment was conducted to determine the breaking strength of a certain alloy containing five metals, A, B, C, D, and E. Two different percentages of each metal were used in forming the $2^5 = 32$ different alloys. Since only eight alloys could be tested on a given day, the experiment was conducted over a period of 4 days in which the $ABDE$ and the AE effects were confounded with days. The experimental data were recorded as follows:

Treatment Combination	Breaking Strength	Treatment Combination	Breaking Strength
(1)	21.4	e	29.5
a	32.5	ae	31.3
b	28.1	be	33.0
ab	25.7	abe	23.7
c	34.2	ce	26.1
ac	34.0	ace	25.9
bc	23.5	bce	35.2
abc	24.7	abce	30.4
d	32.6	de	28.5
ad	29.0	ade	36.2
bd	30.1	bde	24.7
abd	27.3	abde	29.0
cd	22.0	cde	31.3
acd	35.8	acde	34.7
bcd	26.8	bcde	26.8
abcd	36.4	abcde	23.7

(a) Set up the blocking scheme for the 4 days.
(b) What additional effect is confounded with days?
(c) Use Yates' technique to obtain the sums of squares for all main effects.

5. By confounding ABC in two replicates and AB in the third, show the block arrangement and the analysis-of-variance table for a 2^3 factorial experiment with three replicates. What is the relative information on the confounded effects?

6. The following experiment was run to study main effects and all interactions. Four factors are used at two levels each. The experiment is replicated and two blocks are necessary in each replication. The data are as follows:

Replicate I		
Block 1		**Block 2**
(1) = 17.1		a = 15.5
d = 16.8		b = 14.8
ab = 16.4		c = 16.2
ac = 17.2		ad = 17.2
bc = 16.8		bd = 18.3
abd = 18.1		cd = 17.3
abd = 19.1		abc = 17.7
bcd = 18.4		abcd = 19.2

Replicate II		
Block 3		**Block 4**
(1) = 18.7		a = 17.0
ab = 18.6		b = 17.1
ac = 18.5		c = 17.2
ad = 18.7		d = 17.6
bc = 18.9		abc = 17.5
bd = 17.0		abd = 18.3
cd = 18.7		acd = 18.4
abcd = 19.8		bcd = 18.3

(a) What effect is confounded with blocks in the first replication of the experiment? In the second replication?
(b) Conduct an appropriate analysis of variance showing tests on all main effects and interaction effects. Use a 0.05 level of significance.

15.8 Factorial Experiments in a Regression Setting

In many practical situations the scientist is provided the task of building a regression model with the regressor variables being controlled in the observational process. In Chapters 11 and 12 the regressor or independent variables in the regression analysis were not necessarily assumed to be subjected to a designed experiment. Obviously, in many chemical and biological processes and in other scientific and engineering environments, levels of regressor variables can and should be controlled. In this section we show some advantages and discuss the role of the 2^k factorial experiment in the multiple regression situation.

Recall that the regression model employed in Section 12.3 can be written in matrix notation as

$$\mathbf{y} = \mathbf{X}\boldsymbol{\beta} + \boldsymbol{\varepsilon}.$$

The $\mathbf{X}$ matrix is referred to as the model matrix. Suppose, for example, that a 2^3 factorial experiment is employed with the variables

Temperature	150°C	200°C
Humidity	15%	20%
Pressure (psi)	1000	1500

The familiar $+1$, -1 levels can be generated through the following centering and scaling to *design units*:

$$x_1 = \frac{\text{temperature} - 175}{25}$$

$$x_2 = \frac{\text{humidity} - 17.5}{2.5}$$

$$x_3 = \frac{\text{pressure} - 1250}{250}.$$

As a result the $\mathbf{X}$ matrix becomes

$$\mathbf{X} = \begin{bmatrix} 1 & -1 & -1 & -1 \\ 1 & 1 & -1 & -1 \\ 1 & -1 & 1 & -1 \\ 1 & -1 & -1 & 1 \\ 1 & 1 & 1 & -1 \\ 1 & 1 & -1 & 1 \\ 1 & -1 & 1 & 1 \\ 1 & 1 & 1 & 1 \end{bmatrix} \quad \begin{array}{c} \text{Design identification} \\ (1) \\ a \\ b \\ c \\ ab \\ ac \\ bc \\ abc \end{array}$$

with column headings x_1, x_2, x_3.

It is now easily seen that contrasts illustrated and discussed in Section 15.2 are directly related to regression coefficients. Notice that all the columns of the $\mathbf{X}$ matrix in our 2^3 example are *orthogonal*. As a result, the computation of regression coefficients as described in Section 12.3 becomes

$$\mathbf{b} = \begin{bmatrix} b_0 \\ b_1 \\ b_2 \\ b_3 \end{bmatrix}$$

$$= (\mathbf{X'X})^{-1}\mathbf{X'y}$$

$$= \left(\frac{1}{8}I\right)\mathbf{X'y}$$

$$= \frac{1}{8}\begin{bmatrix} a + ab + ac + abc + (1) + b + c + bc \\ a + ab + ac + abc - (1) - b - c - bc \\ b + ab + bc + abc - (1) - a - c - ac \\ c + ac + bc + abc - (1) - a - b - ab \end{bmatrix},$$

where a, ab, and so on, are response measures.

One can now see that the notion of *calculated main effects* that have been emphasized throughout this chapter with 2^k factorials is related to coefficients in a fitted regression model when factors are quantitative. In fact, for a 2^k with, say, n experimental runs per design point, the relationship between effects and regression coefficients are as follows:

$$\text{Effect} = \frac{\text{contrast}}{2^{k-1}(n)}$$

$$\text{Regression coefficient} = \frac{\text{contrast}}{2^k(n)}$$

$$= \frac{\text{effect}}{2}.$$

This relationship should make sense to the reader since a regression coefficient b_j is an average rate of change in response *per unit charge* in x_j. Of course, as one goes from -1 to $+1$ in x_j (low to high), the design variable has changed by 2 units.

EXAMPLE 15.4   Consider an experiment in which an engineer desires to fit a linear regression of yield (y) against holding time (x_1) and flexing time (x_2) in a certain chemical system. All other factors are held fixed. The data in the natural units are as follows:

Holding Time (hr)	Flexing Time (hr)	Yield (%)
0.5	0.10	28
0.8	0.10	39
0.5	0.20	32
0.8	0.20	46

Estimate the multiple linear regression model.

SOLUTION

As a result, the fitted regression model is given by

$$\hat{y} = b_0 + b_1 x_1 + b_2 x_2.$$

The design units are given by

$$x_1 = \frac{\text{holding time} - 0.65}{0.15}$$

$$x_2 = \frac{\text{flexing time} - 0.15}{0.05}$$

and the $\mathbf{X}$ matrix is given by

$$\mathbf{X} = \begin{array}{cc} & \begin{array}{cc} x_1 & x_2 \end{array} \\ \begin{bmatrix} 1 & -1 & -1 \\ 1 & 1 & -1 \\ 1 & -1 & 1 \\ 1 & 1 & 1 \end{bmatrix} \end{array}$$

with the regression coefficients

$$\begin{bmatrix} b_0 \\ b_1 \\ b_2 \end{bmatrix} = (\mathbf{X}'\mathbf{X})^{-1}\mathbf{X}'\mathbf{y}$$

$$= \begin{bmatrix} \dfrac{(1) + a + b + ab}{4} \\ \dfrac{a + ab - (1) - b}{4} \\ \dfrac{b + ab - (1) - a}{4} \end{bmatrix}$$

$$= \begin{bmatrix} 36.25 \\ 6.25 \\ 2.75 \end{bmatrix}.$$

Thus the least squares regression equation is given by

$$\hat{y} = 36.25 + 6.25x_1 + 2.75x_2.$$

This example provides an illustration of the use of the two-level factorial experiment in a regression setting. The four experimental runs in the 2^2 design were used to calculate a regression equation, with the obvious interpretation of the regression coefficients. The value $b_1 = 6.25$ represents the estimated increase in response (percent yield) per *design unit* change (0.15 hour) in holding time. The value $b_2 = 2.75$ represents a similar rate of change for flexing time.

Interaction in the Regression Model

The interaction contrasts discussed in Section 15.3 have definite interpretations in the regression context. In fact, interactions are accounted for in regression models by product terms. For example, in Example 15.4, the model with interaction is given by

$$\hat{y} = b_0 + b_1 x_1 + b_2 x_2 + b_{12} x_1 x_2$$

with b_0, b_1, b_2 as before and

$$b_{12} = \frac{ab + (1) - a - b}{4}$$

$$= \frac{46 + 28 - 39 - 32}{4}$$

$$= 0.75.$$

Thus the regression equation expressing two *linear main effects* and interaction is given by

$$\hat{y} = 36.25 + 6.25 x_1 + 2.75 x_2 + 0.75 x_1 x_2.$$

The regression context provides a framework in which the reader should better understand the advantage of orthogonality that is enjoyed by the 2^k factorial. In Section 15.2 the merits of orthogonality were discussed from the point of view of *analysis of variance* of the data in a 2^k experiment. It was pointed out that orthogonality among effects leads to independence among the sums of squares. Of course, the presence of regression variables certainly does not rule out the use of analysis of variance. In fact, F-tests are conducted just as they are described in Section 15.2. Of course, a distinction must be made. In the case of ANOVA, the hypotheses evolve from population means while in the regression case the hypotheses involve regression coefficients.

For instance, consider the design experiment in Exercise 2 on page 579. Each factor is continuous and suppose that the levels are

$$A(x_1): \quad 20\% \quad\quad 40\%$$
$$B(x_2): \quad 5 \text{ lb/sec} \quad 10 \text{ lb/sec}$$
$$C(x_3): \quad 5 \quad\quad 5.5$$

we have, for design levels,

$$x_1 = \frac{\text{solids} - 30}{10}$$

$$x_2 = \frac{\text{flow rate} - 7.5}{2.5}$$

$$x_3 = \frac{\text{pH} - 5.25}{0.25}.$$

Suppose that it is of interest to fit a multiple regression model in which all linear coefficients and available interactions are to be considered. In addition, it is of interest for the engineer to give some insight into what levels of the factor will *maximize* cleansing (i.e., maximize the response). This problem will be the subject of a case study in the next section.

15.9 Case Study: Coal Cleansing Experiment[*] _____

Figure 15.4 represents annotated computer printout for the regression analysis for the fitted model

$$\hat{y} = b_0 + b_1 x_1 + b_2 x_2 + b_3 x_3 + b_{12} x_1 x_2 + b_{13} x_1 x_3 + b_{23} x_2 x_3 + b_{123} x_1 x_2 x_3,$$

where x_1, x_2, and x_3 are percent solids, flow rate, and pH of the system. The computer system used is SAS PROC REG.

Note the items parameter estimates, standard error, and P-values in the printout. The parameter estimates represent coefficients in the model. Notice that all model coefficients are significant except the $x_2 x_3$ term (BC interaction). Note also that residuals, confidence intervals, and prediction intervals appear as discussed in the regression material discussed in Chapters 11 and 12.

The Orthogonal Design

In experimental situations where it is appropriate to fit models that are linear in the design variables and possibly should involve interactions or product terms, there are advantages gained from the two-level *orthogonal design*, or orthogonal array. By an orthogonal design we mean orthogonality among the columns of the **X** matrix. For example, consider the **X** matrix for the 2^2 factorial in Example 15.4. Notice that all three columns are mutually orthogonal. The **X** matrix for the 2^3 factorial on page 591 also contains orthogonal columns. The 2^3 factorial with interactions would yield an **X** matrix of the type

$$
\mathbf{X} = \begin{array}{c}
\begin{array}{cccccccc}
x_1 & x_2 & x_3 & x_1x_2 & x_1x_3 & x_2x_3 & x_1x_2x_3
\end{array} \\
\left[\begin{array}{rrrrrrrr}
1 & -1 & -1 & -1 & 1 & 1 & 1 & -1 \\
1 & 1 & -1 & -1 & -1 & -1 & 1 & 1 \\
1 & -1 & 1 & -1 & -1 & 1 & -1 & 1 \\
1 & -1 & -1 & 1 & 1 & -1 & -1 & 1 \\
1 & 1 & 1 & -1 & 1 & -1 & -1 & -1 \\
1 & 1 & -1 & 1 & -1 & 1 & -1 & -1 \\
1 & -1 & 1 & 1 & -1 & -1 & 1 & -1 \\
1 & 1 & 1 & 1 & 1 & 1 & 1 & 1
\end{array}\right]
\end{array}
$$

It is simple to verify that the 2^3 is orthogonal for the model that contains interactions. The orthogonality allows for independent sums of squares and independent

*See Exercise 2, page 579.

SAS

Dependent Variable: Y

Analysis of Variance

Source	DF	Sum of Squares	Mean Square	F Value	Prob>F
Model	7	490.23499	70.03357	254.430	0.0001
Error	8	2.20205	0.27526		
C Total	15	492.43704			

Root MSE	0.52465	R-square	0.9955	
Dep Mean	12.75188			
C.V.	4.11429			

Parameter Estimates

Variable		DF	Parameter Estimate	Standard Error	T for H0: Parameter=0	Prob>|T|
INTERCEP		1	12.751875	0.13116217	97.222	0.0001
A	(x_1)	1	4.719375	0.13116217	35.981	0.0001
B	(x_2)	1	0.865625	0.13116217	6.600	0.0002
C	(x_3)	1	-1.415625	0.13116217	-10.793	0.0001
AB	$(x_1 \cdot x_2)$	1	-0.599375	0.13116217	-4.570	0.0018
AC	$(x_1 \cdot x_3)$	1	-0.528125	0.13116217	-4.027	0.0038
BC	$(x_2 \cdot x_3)$	1	0.005625	0.13116217	0.043	0.9668
ABC	$(x_1 \cdot x_2 \cdot x_3)$	1	2.230625	0.13116217	17.007	0.0001

Obs	Dep Var Y	Predict Value	Std Err Predict	Lower 95% Mean	Upper 95% Mean	Lower 95% Predict	Upper 95% Predict
1	4.6500	5.2300	0.371	4.3745	6.0855	3.7482	6.7118
2	5.8100	5.2300	0.371	4.3745	6.0855	3.7482	6.7118
3	7.9300	7.9050	0.371	7.0495	8.7605	6.4232	9.3868
4	7.8800	7.9050	0.371	7.0495	8.7605	6.4232	9.3868
5	12.6600	12.6100	0.371	11.7545	13.4655	11.1282	14.0918
6	12.5600	12.6100	0.371	11.7545	13.4655	11.1282	14.0918
7	6.5100	6.3850	0.371	5.5295	7.2405	4.9032	7.8668
8	6.2600	6.3850	0.371	5.5295	7.2405	4.9032	7.8668
9	21.4200	21.3850	0.371	20.5295	22.2405	19.9032	22.8668
10	21.3500	21.3850	0.371	20.5295	22.2405	19.9032	22.8668
11	13.1800	13.0250	0.371	12.1695	13.8805	11.5432	14.5068
12	12.8700	13.0250	0.371	12.1695	13.8805	11.5432	14.5068
13	18.2700	17.4450	0.371	16.5895	18.3005	15.9632	18.9268
14	16.6200	17.4450	0.371	16.5895	18.3005	15.9632	18.9268
15	18.2300	18.0300	0.371	17.1745	18.8855	16.5482	19.5118
16	17.8300	18.0300	0.371	17.1745	18.8855	16.5482	19.5118

FIGURE 15.4 SAS printout for Data in Case Study 15.9.

estimates of coefficients. However, what is the impact of orthogonality on variances of these coefficients? The following theorem underscores the utility of an orthogonal design in this regard.

THEOREM 15.1   *Consider a regression model containing linear main effect terms and interaction effects among the linear terms. Also, assume that the ranges of variables and number of experimental runs are fixed. Variances of least squares estimates of coefficients are minimized for designs that allow two levels on each variable at the extremes (± 1 in design units), and columns of the $\mathbf{X}$ matrix that are orthogonal.* ■

Two-level *orthogonal designs* are often called *minimum variance* designs. Prime examples of minimum variance designs come from the 2^k factorial family. It should be clear by now that the 2^k factorial not only provides optimal estimates for linear coefficients (due to orthogonality) but also to interaction coefficients. For example, a 2^3 experiment allows for optimal (in terms of variance) estimates of β_0, β_1, β_2, β_3, β_{12}, β_{13}, β_{23}, β_{123} in a model of the type

$$E(y) = \beta_0 + \beta_1 x_1 + \beta_2 x_2 + \beta_3 x_3 + \beta_{12} x_1 x_2$$
$$+ \beta_{13} x_1 x_3 + \beta_{23} x_2 x_3 + \beta_{123} x_1 x_2 x_3.$$

Notice that there are 8 coefficients to be estimated in the model above and the design allows for 8 *design points*. In fact, each column in Table 15.3 represents one contrast or *one piece of information* available. Thus all 8 available degrees of freedom from the design points are used in estimating coefficients. As a result, *no lack-of-fit information is available*. Definition 15.1 introduces terminology used to describe such a design.

DEFINITION 15.1   *A regression design is said to be **saturated** when the number of design points (distinct factor combinations) is equal to the number of terms in the regression model.* ■

There is, of course, some danger in designing to fit a model with such a plan that no lack-of-fit degrees of freedom remain. Most prudent planners of experiments allow for some degrees of freedom to be available for both replication error (pure error) and lack of fit.

EXAMPLE 15.5   Consider the experiment in the case study discussed earlier in this section. Suppose that the engineer wishes to fit a multiple linear regression model to the data. That is, no interactions are placed in the model. Sixteen total observations on percent solids in the underflow are available. Outline the degrees of freedom for the analysis.

SOLUTION
There are 15 total degrees of freedom (apart from the mean). The model is given by

$$E(y) = \beta_0 + \beta_1 x_1 + \beta_2 x_2 + \beta_3 x_3.$$

The outline of degrees of freedom is given by

Source	d.f.	
Regression	3	
Lack of fit	4	$(x_1x_2, x_1x_3, x_2x_3, x_1x_2x_3)$
Error (pure)	8	
Total	15	

The eight degrees of freedom for pure error are obtained from the *duplicate runs* at each design point. Lack-of-fit degrees of freedom may be viewed as the difference between the number of distinct design points and the number of total model terms; in this case there are 8 points and 4 model terms.

Standard Error of Coefficients and *t*-tests

In previous sections we have shown how the designer of an experiment may exploit the notion of orthogonality to design a regression experiment with coefficients that attain minimum variance on a per cost basis. We should be able to make use of our exposure to regression in Section 12.4 to compute estimates of variances of coefficients and hence their standard errors. It is also of interest to note the relationship between the *t*-statistic on a coefficient and the *F*-statistic described and illustrated in previous chapters.

Recall from Section 12.4 that the variances and covariances of coefficients appear on A^{-1}, or in terms of present notation, the *variance–covariance matrix* of coefficients is given by

$$\sigma^2 A^{-1} = \sigma^2 (\mathbf{X}'\mathbf{X})^{-1}.$$

In the case of 2^k factorial experiment, the columns of $\mathbf{X}$ are mutually orthogonal, imposing a very special structure. In general, for the 2^k we can write

$$\mathbf{X} = [\mathbf{1} \quad \overset{x_1}{\pm 1} \quad \overset{x_2}{\pm 1} \quad \cdots \quad \overset{x_k}{\pm 1} \quad \overset{x_1x_2}{\pm 1} \quad \cdots \quad],$$

where each column contains 2^k entries or $2^k n$, where n is the number of replicate runs at each design point. Thus formation of $\mathbf{X}'\mathbf{X}$ yields

$$\mathbf{X}'\mathbf{X} = 2^k n \mathbf{I}_p,$$

where $\mathbf{I}$ is the identity matrix of dimension p, the number of model parameters.

EXAMPLE 15.6   Consider a 2^3 with duplicated runs fit to the model

$$E(y) = \beta_0 + \beta_1 x_1 + \beta_2 x_2 + \beta_3 x_3 + \beta_{12} x_1 x_2 + \beta_{13} x_1 x_3 + \beta_{23} x_2 x_3.$$

Give expressions for the standard errors of the least squares estimates of b_0, b_1, b_2, b_3, b_{12}, b_{13}, and b_{23}.

SOLUTION

$$\mathbf{X} = \begin{array}{ccccccc} x_1 & x_2 & x_3 & x_1x_2 & x_1x_3 & x_2x_3 \\ \begin{bmatrix} 1 & -1 & -1 & -1 & 1 & 1 & 1 \\ 1 & 1 & -1 & -1 & -1 & -1 & 1 \\ 1 & -1 & 1 & -1 & -1 & 1 & -1 \\ 1 & -1 & -1 & 1 & 1 & -1 & -1 \\ 1 & 1 & 1 & -1 & 1 & -1 & -1 \\ 1 & 1 & -1 & 1 & -1 & 1 & -1 \\ 1 & -1 & 1 & 1 & -1 & -1 & 1 \\ 1 & 1 & 1 & 1 & 1 & 1 & 1 \end{bmatrix} \end{array},$$

with each unit viewed as being *repeated* (i.e., each observations is duplicated). As a result,

$$\mathbf{X'X} = 16\mathbf{I}_7.$$

Thus

$$(\mathbf{X'X})^{-1} = \tfrac{1}{16}\mathbf{I}_7.$$

From the foregoing it should be clear that the variances of all coefficients for a 2^k factorial with n runs at each design point are given by

$$\text{Var}(b_j) = \frac{\sigma^2}{2^k n},$$

and, of course, all covariances are zero. As a result, standard errors of coefficients are calculated as

$$s_{b_j} = s\sqrt{\frac{1}{2^k n}},$$

where s is found from the square root of the error mean square (hopefully, obtained from adequate replication). Thus in our case with the 2^3,

$$s_{b_j} = s(\tfrac{1}{4}).$$

EXAMPLE 15.7   Consider the metallurgy experiment in Example 15.1. Suppose that the fitted model is given by

$$E(y) = \beta_0 + \beta_1 x_1 + \beta_2 x_2 + \beta_3 x_3 + \beta_4 x_4 + \beta_{12} x_1 x_2 + \beta_{13} x_1 x_3$$
$$+ \beta_{14} x_1 x_4 + \beta_{23} x_2 x_3 + \beta_{24} x_2 x_4 + \beta_{34} x_3 x_4.$$

What are the standard errors of the least squares regression coefficients?

SOLUTION
Standard errors of all coefficients for the 2^k factorial are equal and are given by

$$s_{b_j} = s\sqrt{\frac{1}{2^k n}},$$

which in this illustration is given by

$$s_{b_j} = s\sqrt{\frac{1}{(16)(2)}}.$$

In this case the pure error mean square is given by $s^2 = 2.46$ (16 degrees of freedom). Thus

$$s_{b_j} = 0.28.$$

The standard errors of coefficients can be used to construct t-statistics on all coefficients. These t-values are related to the F-statistics in the analysis of variance. We have already demonstrated that an F-statistic on a coefficient, using the 2^k factorial, is given by

$$F = \frac{(\text{contrast})^2}{(2^k)(n)s^2}.$$

These are the F-statistics, for example, in Table 15.8. It is easy to verify that if we write

$$t = \frac{b_j}{s_{b_j}}$$

where

$$b_j = \frac{\text{contrast}}{2^k n},$$

then

$$t^2 = \frac{(\text{contrast})^2}{s^2 2^k n}$$

$$= F.$$

As a result, the usual relationship holds between t-statistics on coefficients and the F-values. As one might expect, the only difference in the use of the t or F in assessing significance lies in the fact that the t-statistic indicates the sign or direction of the effect of the coefficient.

It would appear that the 2^k factorial plan would handle many practical situations in which regression models are fit. It can accommodate linear and interaction terms, providing optimal estimates of all coefficients (from a variance point of view). However, when k is large, the number of design points required is very large. Often, portions of the total design can be used and still allow orthogonality with all its advantages. These designs are discussed in Section 15.10.

15.10 Fractional Factorial Experiments ⎯⎯⎯⎯⎯⎯⎯⎯

The 2^k factorial experiment can become quite demanding, in terms of the number of experimental units required, when k is large. One of the real advantages with this experimental plan is that it allows a degree of freedom for each interaction. However, in many experimental situations, it is known that certain interactions are negligible, and thus it would be a waste of experimental effort to use the complete factorial experiment. In fact, the experimenter may have an economic constraint that disallows taking observations at all of the 2^k treatment combinations. When k is large, we can often make use of a **fractional factorial experiment** in which perhaps one-half, one-fourth, or even one-eighth of the total factorial plan is actually carried out.

Construction of 1/2 Fraction

The construction of the half-replicate design is identical to the allocation of the 2^k factorial experiment into two blocks. We begin by selecting a defining contrast that is to be completely sacrificed. We then construct the two blocks accordingly and choose either of them as the experimental plan.

Consider a 2^4 factorial experiment in which we wish to use a half-replicate. The defining contrast $ABCD$ is chosen and thus an appropriate experimental plan would be to select the principal block consisting of the following treatment combinations:

$$\{(1),\ ab,\ ac,\ ad,\ bc,\ bd,\ cd,\ abcd\}.$$

With this plan, we have contrasts on all effects except $ABCD$. Clearly,

$$A \text{ contrast} = ab + ac + ad + abcd - (1) - bc - bd - cd$$

$$AB \text{ contrast} = abcd + ab + (1) + cd - ac - ad - bc - bd,$$

with similar expressions for the contrasts of the remaining main effects and interaction effects. However, with no more than 8 of the 16 observations in our fractional design, only 7 of the 14 unconfounded contrasts are orthogonal.

Aliases in 1/2 Fraction

Consider, for example, the CD contrast given by

$$CD \text{ contrast} = abcd + cd + (1) + ab - ac - ad - bc - bd.$$

Observe that this is also the single-degree-of-freedom contrast for AB. The word **aliases** is given to two factorial effects that have the same contrast. Therefore, AB and CD are aliases. In the 2^k factorial experiments, the alias of any factorial effect is its generalized interaction with the defining contrast. For example, if $ABCD$ is the defining contrast, then the alias of A is $A(ABCD) = BCD$. It can be seen then that the complete alias structure in a half-replicate of a 2^4 factorial experiment, using *ABCD* as the defining contrast, is (the symbol $\equiv$ implies *aliased with*)

$$A \equiv BCD$$
$$B \equiv ACD$$
$$C \equiv ABD$$
$$D \equiv ABC$$
$$AB \equiv CD$$
$$AC \equiv BD$$
$$AD \equiv BC.$$

Without supplementary evidence, there is no way of explaining which of two aliased effects is actually providing the influence on the response. In a sense they *share a degree of freedom*. Herein lies the disadvantage in fractional factorial experiments. They have their greatest use when k is quite large and there is some *a priori* knowledge concerning the interactions. In the example presented, the main effects can be estimated if the three-factor interactions are known to be negligible. For testing purposes the only possible procedure, in the absence of either an outside measure of experimental error or a replication of the experiment, would be to pool the sums of squares associated with the two-factor interactions. This, of course, is desirable only if these interactions represent negligible effects.

Construction of 1/4 Fraction

The construction of the 1/4 fraction or quarter-replicate is identical to the procedure whereby one assigns 2^k treatment combinations to four blocks. This involves the sacrificing of two defining contrasts along with their generalized interaction. Any of the four resulting blocks serves as an appropriate set of experimental runs. Each effect has three aliases, which are given by the generalized interaction with the three defining contrasts. Suppose in a 1/4 fraction of a 2^6 factorial experiment, we use *ACEF* and *BDEF* as the defining contrast, resulting in

$$(ACEF)(BDEF) = ABCD$$

also being sacrificed. Using $L_1 = 0$, $L_2 = 0$ (modulo 2), where

$$L_1 = \gamma_1 + \gamma_3 + \gamma_5 + \gamma_6$$
$$L_2 = \gamma_2 + \gamma_4 + \gamma_5 + \gamma_6,$$

we have an appropriate set of experimental runs given by

$$\{(1), abcd, ef, abcdef, cde, cdf, abe, abf, acef, bdef, ac, bd, adf, ade, bcf, bce\}$$

and the alias structure for the main effects is written

$$A \equiv CEF \equiv ABDEF \equiv BCD$$

$$B \equiv ABCEF \equiv DEF \equiv ACD$$

$$C \equiv AEF \equiv BCDEF \equiv ABD$$

$$D \equiv ACDEF \equiv BEF \equiv ABC$$

$$E \equiv ACF \equiv BDF \equiv ABCDE$$

$$F \equiv ACE \equiv BDE \equiv ABCDF,$$

each with a single degree of freedom. For the two-factor interactions,

$$AB \equiv BCEF \equiv ADEF \equiv CD$$

$$AC \equiv EF \equiv ABCDEF \equiv BD$$

$$AD \equiv CDEF \equiv ABEF \equiv BC$$

$$AE \equiv CF \equiv ABDF \equiv BCDE$$

$$AF \equiv CE \equiv ABDE \equiv BCDF$$

$$BE \equiv ABCF \equiv DF \equiv ACDE$$

$$BF \equiv ABCE \equiv DE \equiv ACDF.$$

Here, of course, there is some aliasing among the two-factor interactions. The remaining 2 degrees of freedom are accounted for by the following groups:

$$ADF \equiv CDE \equiv ABE \equiv BCF$$

$$ABF \equiv BCE \equiv ADE \equiv CDF.$$

It becomes evident that one should always be aware of what the alias structure is for a fractional experiment before he or she finally recommends the experimental plan. Proper choice of defining contrasts is important, since it dictates the alias structure. For example, if one would like to study main effects and all two-factor interactions in an experiment involving eight factors and it is known that interactions involving three or more factors are negligible, a very practical design would be one in which the defining contrasts are $ACEGH$ and $BDEFGH$, resulting in a third,

$$(ACEGH)(BDEFGH) = ABCDF.$$

All main effects and two-factor interactions are not aliased with one another and are therefore estimable. The analysis of variance would contain the following:

Main effects	8 single degrees of freedom
Two-factor interactions	28 single degrees of freedom
Error	27 pooled degrees of freedom
Total	63 degrees of freedom

For the 1/8 and higher fractional factorials, the method of constructing the design generalizes. Of course, the aliasing can become quite extensive. For example, with a 1/8 fraction, each effect has seven aliases. The design is constructed by selecting three defining contrasts as if eight blocks were being constructed. Four additional effects are sacrificed and any one of the eight blocks can be properly used as the design. Additional information on higher fractions will be given in Section 15.11.

15.11   Analysis of Fractional Factorial Experiments _____

The difficulty in making formal significance tests using data from fractional factorial experiments lies in the determination of the proper error term. Unless there are data available from prior experiments, the error must come from a pooling of contrasts representing effects that are presumed to be negligible.

Sums of squares for individual effects are found using essentially the same procedures given for the complete factorial. One can form a contrast in the treatment combinations by constructing the usual table of positive and negative signs. For example, for a half-replicate of a 2^3 factorial experiment, with ABC the defining contrast, one possible set of treatment combinations and the appropriate algebraic signs for each contrast used in computing effects and the sums of squares for the various effects are presented in Table 15.10.

Note that in Table 15.10 the A and BC contrasts are identical, illustrating the aliasing. Also, $B \equiv AC$ and $C \equiv AB$. In this situation we have three orthogonal contrasts representing the 3 degrees of freedom available. If two observations are obtained for each of the four treatment combinations, we would then have an estimate of the error variance with 4 degrees of freedom. Assuming the interaction effects to be negligible, we could test all the main effects for significance.

An example effect and corresponding sum of squares is given by

$$w_A = \frac{a - b - c + abc}{2n}$$

$$SSA = \frac{(a - b - c + abc)^2}{2^2 n}.$$

TABLE 15.10   Signs for Contrasts in a Half-Replicate of a 2^3 Factorial Experiment

Treatment Combination	Factorial Effect						
	A	B	C	AB	AC	BC	ABC
a	+	−	−	−	−	+	+
b	−	+	−	−	+	−	+
c	−	−	+	+	−	−	+
abc	+	+	+	+	+	+	+

In general, the single-degree-of-freedom sum of squares for any effect in a 2^{-p} fraction of a 2^k factorial experiment $(k > p)$ is obtained by squaring contrasts in the treatment totals selected and dividing by $2^{k-p}n$, where n is the number of replications of these treatment combinations.

EXAMPLE 15.8   Suppose that we wish to use a half-replicate to study the effects of five factors, each at two levels, on some response and it is known that whatever the effect of each factor, it will be constant for each level of the other factors. In other words, there are no interactions. Let the defining contrast be $ABCDE$, causing main effects to be aliased with four factor interactions. The pooling of contrasts involving interactions provides $15 - 5 = 10$ degrees of freedom for error. Perform an analysis of variance on the following data, testing all main effects for significance at the 0.05 level:

Treatment	Response	Treatment	Response
a	11.3	bcd	14.1
b	15.6	abe	14.2
c	12.7	ace	11.7
d	10.4	ade	9.4
e	9.2	bce	16.2
abc	11.0	bde	13.9
abd	8.9	cde	14.7
acd	9.6	abcde	13.2

SOLUTION

The sums of squares and effects for the main effects are

$$SSA = \frac{(11.3 - 15.6 - \cdots - 14.7 + 13.2)^2}{2^{5-1}} = \frac{(-17.5)^2}{16} = 19.14$$

$$\text{effect of } A = -\frac{17.5}{8} = -2.19$$

$$SSB = \frac{(-11.3 + 15.6 - \cdots - 14.7 + 13.2)^2}{2^{5-1}} = \frac{(18.1)^2}{16} = 20.48$$

$$\text{effect of } B = \frac{18.1}{8} = 2.26$$

$$SSC = \frac{(-11.3 - 15.6 + \cdots + 14.7 + 13.2)^2}{2^{5-1}} = \frac{(10.3)^2}{16} = 6.63$$

$$\text{effect of } C = \frac{10.3}{8} = 1.31$$

$$SSD = \frac{(-11.3 - 15.6 - \cdots + 14.7 + 13.2)^2}{2^{5-1}} = \frac{(-7.7)^2}{16} = 3.71$$

$$\text{effect of } D = \frac{-7.7}{8} = -0.96$$

$$SS(E) = \frac{(-11.3 - 15.6 - \cdots + 14.7 + 13.2)^2}{2^{5-1}} = \frac{(8.9)^2}{16} = 4.95$$

$$\text{effect of } E = \frac{8.9}{8} = 1.11.$$

Here the E factor is enclosed in parentheses to avoid confusion with the error sum of squares. All other calculations and tests of significance are summarized in Table 15.11. The tests indicate that factor A has a significant negative effect on the response, while factor B has a significant positive effect. Factors C, D, and E are not significant at the 0.05 level.

TABLE 15.11 Analysis of Variance for the Data of a Half-Replicate of a 2^5 Factorial Experiment

Source of Variation	Sum of Squares	Degrees of Freedom	Mean Square	Computed f
Main effect				
A	19.14	1	19.14	6.21
B	20.48	1	20.48	6.65
C	6.63	1	6.63	2.15
D	3.71	1	3.71	1.20
E	4.95	1	4.95	1.61
Error	30.83	10	3.08	
Total	85.74	15		

Exercises

1. List the aliases for the various effects in a 2^5 factorial experiment when the defining contrast is $ACDE$.

2. (a) Obtain a 1/2 fraction of a 2^4 factorial design using BCD as the defining contrast.
   (b) Divide the 1/2 fraction into 2 blocks of 4 units each by confounding ABC.
   (c) Show the analysis-of-variance table (sources of variation and degrees of freedom) for testing all unconfounded main effects, assuming that all interaction effects are negligible.

3. Construct a 1/4 fraction of a 2^6 factorial design using $ABCD$ and $BDEF$ as the defining contrasts. Show what effects are aliased with the six main effects.

4. (a) Using the defining contrasts $ABCE$ and $ABDF$, obtain a 1/4 fraction of a 2^6 design.
   (b) Show the analysis-of-variation (sources of variation and degrees of freedom) for all appropriate tests assuming that E and F do not interact and all three-factor and higher interactions are negligible.

5. Seven factors are varied at two levels in an experiment involving only 16 trials. A 1/8 fraction of a 2^7 factorial experiment is used with the defining contrasts being ACD, BEF, and CEG. The data are as follows:

Treatment Combination	Response	Treatment Combination	Response
(1)	31.6	acg	31.1
ad	28.7	cdg	32.0
abce	33.1	beg	32.8
cdef	33.6	adefg	35.3
acef	33.7	efg	32.4
bcde	34.2	abdeg	35.3
abdf	32.5	bcdfg	35.6
bf	27.8	abcfg	35.1

Perform an analysis of variance on all seven main effects, assuming that interactions are negligible. Use a 0.05 level of significance.

6. An experiment was conducted so that the engineer can gain an insight into the influence of sealing temperature (A), cooling bar temperature (B), percent polyethylene additive (C), and pressure (D) on the seal strength in grams per inch of a bread-wrapper stock. A 1/2 fraction of a 2^4 factorial experiment is used with the defining contrast being ABCD. The data are as follows:

A	B	C	D	Response
−1	−1	−1	−1	6.6
1	−1	−1	1	6.9
−1	1	−1	1	7.9
1	1	−1	−1	6.1
−1	−1	1	1	9.2
1	−1	1	−1	6.8
−1	1	1	−1	10.4
1	1	1	1	7.3

Perform an analysis of variance on main effects, and two-factor interactions, assuming that all three-factor and higher interactions are negligible. Use $\alpha = 0.05$.

7. In an experiment conducted at the Department of Mechanical Engineering and analyzed by the Statistics Consulting Center at the Virginia Polytechnic Institute and State University, a sensor detects an electrical charge each time a turbine blade makes one rotation. The sensor then measures the amplitude of the electrical current. Six factors are: rpm (A), temperature (B), gap between blades (C), gap between blade and casing (D), location of input (E), and location of detection (F). A 1/4 fraction of a 2^6 factorial experiment is used, with defining contrasts being ABCE and BCDF. The data are as follows:

A	B	C	D	E	F	Response
−1	−1	−1	−1	−1	−1	3.89
1	−1	−1	−1	1	−1	10.46
−1	1	−1	−1	1	1	25.98
1	1	−1	−1	−1	1	39.88
−1	−1	1	−1	1	1	61.88
1	−1	1	−1	−1	1	3.22
−1	1	1	−1	−1	−1	8.94

A	B	C	D	E	F	Response
1	1	1	−1	1	−1	20.29
−1	−1	−1	1	−1	1	32.07
1	−1	−1	1	1	1	50.76
−1	1	−1	1	1	−1	2.80
1	1	−1	1	−1	−1	8.15
−1	−1	1	1	1	−1	16.80
1	−1	1	1	−1	−1	25.47
−1	1	1	1	−1	1	44.44
1	1	1	1	1	1	2.45

Perform an analysis of variance on main effects, and two-factor interactions, assuming that all three-factor and higher interactions are negligible. Use $\alpha = 0.05$.

8. In a study "*Durability of Rubber to Steel Adhesively Bonded Joints*" conducted at the Department of Environmental Science and Mechanics and analyzed by the Statistics Consulting Center at the Virginia Polytechnic Institute and State University, an experimenter measures the number of breakdowns in an adhesive seal. It was postulated that concentration of seawater (A), temperature (B), pH (C), voltage (D), and stress (E) influence the breakdown of an adhesive seal. A 1/2 fraction of a 2^5 factorial experiment is used, with the defining contrast being ABCDE. The data are as follows:

A	B	C	D	E	Response
−1	−1	−1	−1	1	462
1	−1	−1	−1	−1	746
−1	1	−1	−1	−1	714
1	1	−1	−1	1	1070
−1	−1	1	−1	−1	474
1	−1	1	−1	1	832
−1	1	1	−1	1	764
1	1	1	−1	−1	1087
−1	−1	−1	1	−1	522
1	−1	−1	1	1	854
−1	1	−1	1	1	773
1	1	−1	1	−1	1068
−1	−1	1	1	1	572
1	−1	1	1	−1	831
−1	1	1	1	−1	819
1	1	1	1	1	1104

Perform an analysis of variance on main effects, and two-factor interactions, assuming that all three-factor and higher interactions are negligible. Use $\alpha = 0.05$.

15.12   Higher Fractions and Screening Designs

In industrial situations the analyst is often called upon to determine which of a large number of controllable factors have an impact on some important response. The factors may be qualitative or class variables, regression variables, or a mixture of both. The analytical procedure may involve analysis of variance, regression, or both. Often the regression model used involves only linear main effects, although a few interactions may be estimated. The situation calls for variable screening and the resulting experimental designs are called *screening designs*. Clearly, two-level orthogonal designs that are saturated or nearly saturated are useful candidates.

Design Resolution

Two-level orthogonal designs are often classified according to their *resolution*, the latter determined through the following definition.

DEFINITION 15.2   *The* **resolution** *of a two-level orthogonal design is the length of the smallest (least complex) interaction among the set of defining contrasts.* ■

If the design is constructed as a full or fractional factorial [i.e., either a 2^k or 2^{k-p} ($p = 1, 2, \ldots, k - 1$) design], the notion of design resolution is an aid in categorizing the impact of the aliasing. For example, a resolution II design would have little use since there would be at least one instance of aliasing of one main effect with another. A resolution III design will have all main effects (linear effects) orthogonal to each other. However, there will be some aliasing among linear effects and two-factor interactions. Clearly, then, if the analyst is interested in studying main effects (linear effects in the case of regression) and there are no two-factor interactions, then a design of resolution at least III is required.

EXAMPLE 15.9   Suppose that an experimenter is interested in studying five factors in a regression problem. It is important that all linear effects be studied. No interactions are considered to be important. Construct a resolution III design using only eight design points.

SOLUTION
A one-fourth fraction of a 2^5 is required and 3 letters or more must be used as defining contrasts. Choose *ABC* and *BCDE* as defining contrasts. This results in *ADE* as a third defining contrast. Clearly, the design is resolution III and no main effect is aliased with any other main effect.

The use of resolution to classify two-level designs is very important. It is often desirable to construct a first-order design that is orthogonal and also is compatible with model requirements. For example, an analyst may well desire to test main effects (or linear regression coefficients) as part of a variable screening chore. If interactions are considered to be part of the system, aliasing between main effects and two-factor interactions cannot be tolerated. In this situation a resolution IV design is necessary.

15.13 Construction of Resolution III and Resolution IV Designs with 8, 16, and 32 Design Points _____

Useful designs of resolutions III and IV can be constructed for 2 to 7 variables with 8 design points. We merely begin with a 2^3 factorial that has been symbolically saturated with interactions.

$$
\begin{array}{ccccccc}
x_1 & x_2 & x_3 & x_1x_2 & x_1x_3 & x_2x_3 & x_1x_2x_3 \\
\left[\begin{array}{ccccccc}
-1 & -1 & -1 & 1 & 1 & 1 & -1 \\
1 & -1 & -1 & -1 & -1 & 1 & 1 \\
-1 & 1 & -1 & -1 & 1 & -1 & 1 \\
-1 & -1 & 1 & 1 & -1 & -1 & 1 \\
1 & 1 & -1 & 1 & -1 & -1 & -1 \\
1 & -1 & 1 & -1 & 1 & -1 & -1 \\
-1 & 1 & 1 & -1 & -1 & 1 & -1 \\
1 & 1 & 1 & 1 & 1 & 1 & 1
\end{array}\right]
\end{array}
$$

It is clear that a resolution III design can be constructed merely by replacing interaction columns by new main effects through 7 variables. For example, we may define

$$x_4 = x_1x_2 \qquad \text{(defining contrast } ABD\text{)}$$

$$x_5 = x_1x_3 \qquad \text{(defining contrast } ACE\text{)}$$

$$x_6 = x_2x_3 \qquad \text{(defining contrast } BCF\text{)}$$

$$x_7 = x_1x_2x_3 \qquad \text{(defining contrast } ABCG\text{)}$$

and obtain a 2^{-4} fraction of a 2^7 factorial. The expressions above identify the chosen defining contrasts. Eleven additional defining contrasts result and all defining contrasts contain at least three letters. Thus the design is a resolution III design. Clearly, if we begin with a *subset* of the augmented columns and conclude with a design involving fewer than seven design variables, the result is a resolution III design in fewer than 7 variables.

A similar set of possible designs can be constructed for 16 design points by beginning with a 2^4 saturated with interactions. Definitions of variables that correspond to these interactions produce resolution III designs through 15 variables. In a similar fashion designs containing 32 runs can be constructed by beginning with a 2^5.

The Foldover Technique

One can easily supplement the resolution III designs described above to produce a resolution IV design by using a *foldover technique*. Foldover involves doubling the size of the design by adding the *negative* of the design matrix constructed as described above. Table 15.12 shows a 16-run resolution IV design in 7 variables using the foldover technique. Obviously, we can construct resolution IV designs involving up to 15 variables by using the foldover technique on designs developed by the saturated 2^4 design.

TABLE 15.12 Resolution IV Two-Level Design Using the Foldover Technique

x_1	x_2	x_3	$x_4 = x_1x_2$	$x_5 = x_1x_3$	$x_6 = x_2x_3$	x_7
−1	−1	−1	1	1	1	−1
1	−1	−1	−1	−1	1	−1
−1	1	−1	−1	1	−1	−1
−1	−1	1	1	−1	−1	−1
1	1	−1	1	−1	−1	−1
1	−1	1	−1	1	−1	−1
−1	1	1	−1	−1	1	−1
1	1	1	1	1	1	−1
1	1	1	−1	−1	−1	+1
−1	1	1	1	1	−1	+1
1	−1	1	1	−1	1	+1
1	1	−1	−1	1	1	+1
−1	−1	1	−1	1	1	+1
−1	1	−1	1	−1	1	+1
1	−1	−1	1	1	−1	+1
−1	−1	−1	−1	−1	−1	+1

Foldover (bracketed group: last 8 rows)

This design was constructed by "folding over" a 1/8 fraction of a 2^6. The last column is added as a seventh factor. In practice, the last column often plays the role of a blocking variable. The foldover technique is used often in sequential experimentation where the data from the initial resolution III design are analyzed. The experimenter may then feel, based on the analysis, that a resolution IV design is needed. As a result, a blocking variable may be needed because a separation in time occurs between the two portions of the experiment. Apart from the blocking variable, the final design is a 1/4 fraction of a 2^6 experiment.

15.14 Other Two-Level Resolution III Designs; The Plackett–Burman Designs

A family of designs developed by Plackett and Burman (see the Bibliography) fills sample size voids that exist with the fractional factorials. The latter are useful with sample sizes 2^r (i.e., they involve sample sizes 4, 8, 16, 32, 64, . . .). The Plackett-Burman designs involve $2r$ design points, and thus designs of size 12, 20, 24, 28, and so on, are available. These two-level Plackett–Burman designs are resolution III designs and are very simple to construct. "Basic lines" are given for each sample size. These lines of + and − signs are $n − 1$ in number. To construct the columns of the design matrix, one begins with the basic line and does a cyclic permutation on the columns until k (the desired number of variables) columns are formed. Then fill in the last row with negative signs. The result will be a resolution III design with k variables ($k = 1, 2, . . . , N$). The basic lines are as follows:

$$
\begin{array}{llllllllllllllllllllllll}
N = 12 & + & + & - & + & + & + & - & - & - & + & - \\
N = 16 & + & + & + & + & - & + & - & + & + & - & - & + & - & - & - \\
N = 20 & + & + & - & - & + & + & + & + & - & + & - & + & - & - & - & - & + & + & - \\
N = 24 & + & + & + & + & + & - & + & - & + & + & - & - & + & + & - & - & + & - & + & - & - & - & -
\end{array}
$$

EXAMPLE 15.10   Construct a two-level screening design with 6 variables containing 12 design points.

SOLUTION

Begin with the basic line in the initial column. The second column is formed by bringing the bottom entry of the first column to the top of the second column and repeating the first column. The third column is formed in the same fashion, using entries in the second column. When there are a sufficient number of columns, simply fill in the last row with negative signs. The resulting design is given by

$$
\begin{array}{cccccc}
x_1 & x_2 & x_3 & x_4 & x_5 & x_6 \\
\left[\begin{array}{cccccc}
+ & - & + & - & - & - \\
+ & + & - & + & - & - \\
- & + & + & - & + & - \\
+ & - & + & + & - & + \\
+ & + & - & + & + & - \\
+ & + & + & - & + & + \\
- & + & + & + & - & + \\
- & - & + & + & + & - \\
- & - & - & + & + & + \\
+ & - & - & + & + & + \\
- & + & - & - & - & + \\
- & - & - & - & - & -
\end{array}\right]
\end{array}.
$$

The Plackett–Burman designs are very popular in industry for screening situations. As resolution III designs, all linear effects are orthogonal. For any sample size, the user has available a design for $k = 2, 3, \ldots, N - 1$ variables.

The alias structure for the Plackett–Burman design is very complicated and thus the user cannot construct the design with complete control over the alias structure as in the case of 2^k or 2^{k-p} designs. However, in the case of regression models the Plackett–Burman design can accommodate interactions (although they will not be orthogonal) when sufficient degrees of freedom are available.

15.15 Taguchi's Robust Parameter Design _____

In this chapter we have emphasized the notion of using design of experiments (DOE) to learn about engineering and scientific processes. In the case where the process involves a product, DOE can be used to provide product improvement or quality improvement. As we pointed out in Chapter 1, much importance has been attached to the use of statistical methods in product improvement. An important aspect of this quality improvement effort of the 1980s and 1990s is to design quality into processes and products at the research stage or the process design stage.

Much of the impetus of quality improvement methods was motivated by or, perhaps, a reaction to the apparent success that engineers and scientists in Japan have had with the use of experimental design. Design of experiments is an important engineering tool in Japan. In the early 1980s, Professor Genichi Taguchi, a Japanese engineer, introduced an approach in this country to using experimental design in the development of products and processes that have the following properties:

1. Insensitive (robust) to environmental conditions.
2. Insensitive (robust) to factors difficult to control.
3. Provide minimum variation in performance.

This philosophy and, indeed, the accompanying methods have been called *Taguchi's robust parameter design* (see Taguchi, Taguchi and Wu, and Kackar in the Bibliography). The term *design* in this context refers to the design of the process or system; *parameter* refers to the parameters in the system. These are what we have been calling *factors* or *variables*.

It is very clear that goals 1, 2, and 3 above are quite noble. It also became obvious that management personnel and engineers in the United States had not been particularly attentive to these issues. For example, a petroleum engineer may have a fine gasoline blend that performs quite well as long as conditions are ideal and stable. However, the performance may deteriorate because of changes in environmental conditions, such as type of driver, weather conditions, type of engineer, and so forth. A scientist at a food company may have a cake mix that is quite good unless the user does not exactly follow directions on the box directions that deal with oven temperature, baking time, and so forth. A product or process whose performance is consistent when exposed to these changing environmental conditions is called a **robust product** or **robust process**.

Control and Noise Variables

Taguchi emphasized the notion of using two types of design variables in a study. These factors are control factors and noise factors.

DEFINITION 15.3	**Control factors** *are variables that can be controlled in both the experiment and in the process.* **Noise factors** *are variables that may or may not be controlled in the experiment but cannot be controlled in the process (or not controlled well in the process).* ∎

Taguchi's approach (and the approach by many practicing statisticians in the United States) is to use control variables and noise variables in the same experiment as fixed effects. Orthogonal designs or orthogonal arrays are popular designs to use in this effort.

GOAL OF ROBUST
PARAMETER
DESIGN

> The goal of robust parameter design is to choose the levels of the control variables (i.e., the design of the process) that are most robust (insensitive) to changes in the noise variables.

It should be noted that *changes in the noise variables* actually imply changes during the process, changes in the field, changes in the environment, changes in handling or usage by the consumer, and so forth.

The Product Array

One approach to the design of experiments involving both control and noise variables is the use of an experimental plan that calls for an orthogonal design for both the control and the noise variables separately. The complete experiment, then, is merely the product or crossing of these two orthogonal designs. The following is a simple example of a product array with two control and two noise variables.

EXAMPLE 15.11

In an article "*The Taguchi Approach to Parameter Design*" by D. M. Byrne and S. Taguchi, in *Quality Progress,* December 1987, the authors discuss an interesting example in which a method is sought to assemble an electrometric connector to a nylon tube that delivers the required pull-off performance to be suitable for an automotive engine application. The objective is to find controllable conditions that maximize pull-off force. Among the controllable variables are A, connector wall thickness, and B, insertion depth. During routine operation there are several variables that cannot be controlled, although they will be controlled during the experiment. Among them are C, conditioning time, and D, conditioning temperature. These levels are taken for each control variable and two for each noise variable. As a result, the crossed array is as follows:

B (depth)

	Shallow	Medium	Deep
Thin	(1) c d cd	(1) c d cd	(1) c d cd
A (medium)	(1) c d cd	(1) c d cd	(1) c d cd
Thick	(1) c d cd	(1) c d cd	(1) c d cd

The control array is a 3×3 array and the noise array is a familiar 2^2 factorial with (1), c, d, and cd representing the factor combinations. The purpose of the noise factor is to create the *kind of variability in the response, pull-off force, that might be expected in day-to-day operation with the process.*

Analysis

There are several procedures for analysis of the product array. The approach advocated by Taguchi and adopted by many companies in the United States dealing in manufacturing processes involves, initially, the formation of summary statistics at each combination in the control array. This summary statistic is called a *signal-to-noise ratio*. Suppose that we call $y_1, y_2, \ldots, y_n$ a typical set of experimental runs for the noise array at a fixed control array combination. The following table describes some of the typical *SN* ratios.

Objective	SN Ratio
Maximize response	$SN_L = -10 \log \left(\dfrac{1}{n} \sum\limits_{i=1}^{n} \dfrac{1}{y_i^2} \right)$
Achieve target	$SN_T = 10 \log \dfrac{\bar{y}^2}{s^2}$
Minimize response	$SN_S = -10 \log \left(\dfrac{1}{n} \sum\limits_{i=1}^{n} y_i^2 \right)$

In each of the cases above one seeks to find the combination of the control variables that *maximizes SN.*

Marginal Means Analysis

One approach to the analysis is to simply treat the *SN* ratio as a standard measured response and do analysis of variance as we have described in this book. Although this is often done, it has been highly criticized (see, e.g., Box, 1986, in the Bibliography). Often, the control array is a highly fractionated factorial experiment and the analysis involves the use of a *marginal means* or "pick the winner" type of analysis. We shall illustrate the marginal means analysis with a case study.

EXAMPLE 15.12: CASE STUDY In an experiment described in *Understanding Industrial Designed Experiments* by Schmidt and Launsby (see the Bibliography), solder process optimization is accomplished in a printed circuit board assembly plant. Parts are inserted either manually or automatically into a bare board with a circuit printed on it. After the parts are inserted the board is put through a wave solder machine, which is used to connect all the parts into the circuit. Boards are placed on a conveyor and taken through a series of steps. They are bathed in a flux mixture to remove oxide. To minimize warpage, they are preheated before the solder is applied. Soldering takes place as the boards move across the wave of solder. The object of the experiment is to minimize the number of solder defects per million joints. The control factor and levels are as follows:

Factor	(−1)	(+1)
A, solder plot temperature (°F)	480	510
B, conveyor speed (ft/min)	7.2	10
C, flux density	0.9°	1.0°
D, preheat temperature (°F)	150	200
E, wave height (in.)	0.5	0.6

These factors are easy to control at the experimental level but not all are easy to control at the plant or process level.

Noise Factors: Tolerances on Control Factors

Often in processes such as this one the natural noise factors are tolerances in the control factors. For example, in the actual on-line process, solder pot temperature and conveyor speed are difficult to control. It is known that the control of temperature is within $\pm 5°F$ and the control of conveyor belt speed is written ± 0.2 ft/min. It is certainly conceivable that variability in the product response (soldering performance) is increased because of an inability to control these two factors at some nominal levels. The third noise factor is the type of assembly involved. In practice, one of two types of assemblies will be used. Thus we have as the noise factors:

Factor	(−1)	(+1)
A*, solder pot temperature tolerance (°F) (deviation from nominal)	−5°	5°
B*, conveyor speed tolerance (ft/min) (deviation from ideal)	−0.2	+0.2
C*, assembly type	1	2

Both the control array (inner array) and the noise array (outer array) were chosen to be fractional factorials, the former a 1/4 of a 2^5 and the latter a 1/2 of a 2^3. The *crossed array* and the response values are as follows:

Inner Array					Outer Array				
A	B	C	D	E	(1)	a*b*	a*c*	b*c*	$(SN)_S$
1	1	1	−1	−1	194	197	193	275	−46.75
1	1	−1	1	1	136	136	132	136	−42.61
1	−1	1	−1	1	185	261	264	264	−47.81
1	−1	−1	1	−1	47	125	127	42	−39.51
−1	1	1	1	−1	295	216	204	293	−48.15
−1	1	−1	−1	1	234	159	231	157	−45.97
−1	−1	1	1	1	328	326	247	322	−45.76
−1	−1	−1	−1	−1	186	187	105	104	−43.59

The first three columns of the inner array represent a 2^3. The columns $D = -AC$ and $E = -BC$. Thus the defining interactions for the inner array are ACD, BCE, and ADE. The outer array is a standard resolution III fraction of a 2^3. Notice that each inner array point contains runs from the outer array. Thus four response values are observed at each combination of the control array.

Signal-to-Noise Ratio and Analysis

As we indicated earlier, Taguchi chooses to "model" the signal-to-noise ratio rather than the natural response. In this illustration the appropriate SN ratio is

$$(SN)_S = -10 \log \left(\frac{1}{n} \sum_{i=1}^n y_i^2 \right).$$

The motivation is to use a performance criterion that reflects a "squared error loss" or average squared deviation from target, the target being zero in this case (zero faults in the solder process). With the negative sign, one chooses to determine values of A, B, C, D, and E that maximize $(SN)_S$. The analytical device, following Taguchi, is to graphically depict a "main effects only" type of analysis designed to maximize the $(SN)_S$. The plots that indicate marginal means using $(SN)_S$ as the response are shown in Figure 15.5.

The signal-to-noise ratio is intended to capture mean and variability simultaneously. This marginal means analysis would certainly suggest that temperature (A) and flux density (C) are the most important among the control factors. Similar marginal means plots can be used on the mean $\bar{y}$ (see Figure 15.6).

Clearly, temperature and flux density are the most important factors. They seem to influence both $(SN)_S$ and $\bar{y}$. Fortunately, *high temperature* and *low flux density* are preferable for both $(SN)_S$ and the mean response. Thus the "optimum" conditions are

solder temperature = 510°F

flux density = 0.9°

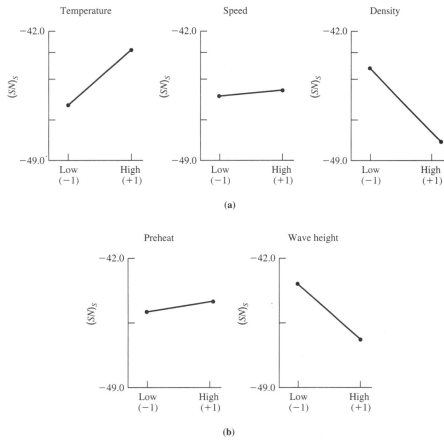

FIGURE 15.5   Plot showing the influence of factors on the signal-to-noise ratio.

If we use the analysis suggested by Taguchi, this is the *most robust* set of conditions.

Comments on Taguchi's Analytical Approach

The approach to the analysis given here is, for the most part, that advocated by Genichi Taguchi. Although his contributions are extremely important and well documented, his specific approach to analysis has been criticized. Among the forms of criticism are those that take the following lines:

· The designs he suggests are often saturated or near saturated; those leave little information (degrees of freedom) for interaction among the control variables.
· Little modeling is done. The methods discussed in Chapter 11 and 12 for building the best model among the factors are not used.

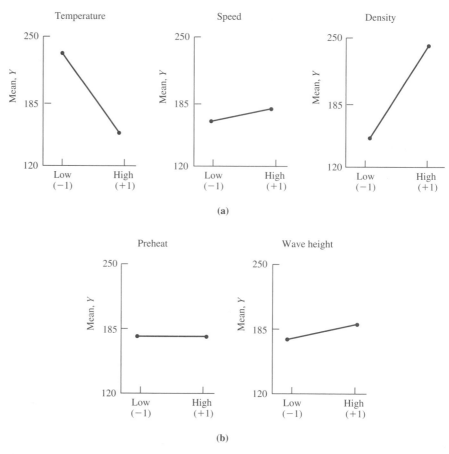

FIGURE 15.6   Plot showing the influence of factors on the mean response.

· Signal-to-noise ratios do not always capture the simultaneous sense of mean and variability.
· The crossed array approach can lead to excessive experimentation.
· Marginal means analysis uses a strictly main effects approach. This ignores interactions among control variables.

One should consult Kackar, Box, and Box and Fung for more discussion of Taguchi's approach.

Positive Aspects of Taguchi's Contribution

Taguchi's contributions to engineering design and process optimization are very important. We do not attempt to give a detailed explanation here. His work has influenced engineers and statisticians to rethink statistical methods in terms of sensitivity to environmental variables. He has prompted statisticians to resurrect old and develop new methodology for using product and process variability as a prominent part of the performance criterion. In addition to these technical contributions he has steered many engineers toward the use of statistical methods and experimental design, and he has indirectly influenced statisticians into working in the area of quality improvement.

Alternative Approaches to Robust Parameter Design

Taguchi's work has influenced many statisticians into seeking alternative approaches to solving problems in robust parameter design. Many of those follow Taguchi's principles of the use of noise variables and emphasis on dealing with response variability. One should consult Welch, Sachs, and Kang; Vining and Myers; Myers, Khuri, and Vining; and Shoemaker, Tsui, and Wu. These and others attempt to alleviate the criticisms of the Taguchi approach while holding to the robust parameter design principles.

One approach suggested by many is to model the sample mean and sample variance separately rather than combine the two separate concepts via a signal-to-noise ratio. The separate modeling will often aid the experimenter in a better understanding of the process involved. In the following example, we illustrate this approach with the solder process experiment.

EXAMPLE 15.13   Consider the data set in Example 15.12. An alternative analysis is to fit separate models for the mean $\bar{y}$ and the sample standard deviation. Suppose that we use the usual $+1$ and -1 coding for the control factors. Based on the apparent importance of solder pot temperature (x_1) and flux density (x_2) linear regression model on the response (number of errors per million joints) produces the model

$$\hat{y} = 198.175 - 28.525x_1 + 57.975x_2.$$

To find the most robust level of temperature and flux density it is important to capture a compromise between the mean response and variability. This requires a modeling of the variability. An important tool in this regard is the log transformation (see Bartlett and Kendall, or Carroll and Ruppert):

$$\ln s^2 = \gamma_0 + \gamma_1(x_1) + \gamma_2(x_2).$$

This modeling process produces the following result:

$$\ln s^2 = 7.4155 - 0.2067x_1 - 0.0309x_2.$$

The analysis that is important to the scientist or engineer is to make use of the two models simultaneously. A graphical approach can be very useful. Figure 15.7 shows a simple analysis of the mean and variance simultaneously. As one would expect, the location in temperature and flux density that minimizes the mean number of errors is the same as that which minimizes variability, namely high temperature and low flux density. The graphical multiple response approach allows the user to see trade-offs between process mean and process variability. For this example, the engineer may be dissatisfied with the extreme conditions in solder temperature and flux density. The figure offers estimation of mean and variability conditions that indicate how much is lost as one moves away from the optimum to any intermediate conditons.

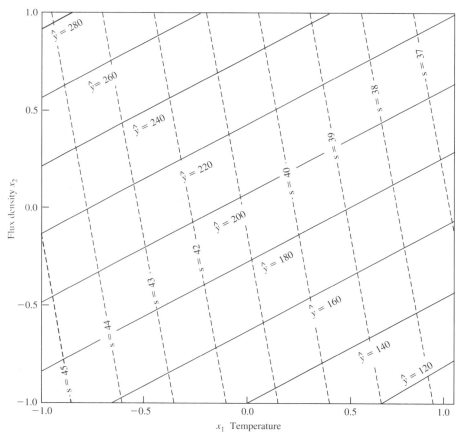

FIGURE 15.7 Mean and variance for Example 15.13.

Exercises

1. Use the coal cleansing data from Exercise 2 following Section 15.3 to fit a model of the type

$$E(y) = \beta_0 + \beta_1 x_1 + \beta_2 x_2 + \beta_3 x_3,$$

where the levels are

   x_1: percent solids: 8; 12
   x_2: flow rate: 150; 250 gal/min
   x_3: pH: 5; 6

Center and scale the variables to design units. Also conduct a test for lack of fit and comment concerning the adequacy of the linear regression model.

2. A 2^5 factorial plan is used to build a regression model containing first-order coefficients and model terms for all two-factor interactions. Duplicate runs are made for each factor. Outline the analysis-of-variance table showing degrees of freedom for regression, lack of fit, and pure error.

3. Consider the 1/16 of the 2^7 factorial discussed in Section 15.13. List the additional 11 defining contrasts.

4. Construct a Plackett–Burman design for 10 variables containing 24 experimental runs.

Review Exercises

1. A Plackett–Burman design was used for the purpose of studying the rheological properties of high-molecular-weight copolymers. Two levels of each of six variables were fixed in the experiment. The viscosity of the polymer is the response. The data were analyzed by the Statistics Consulting Center at Virginia Polytechnic Institute and State University for personnel in the Chemical Engineering Department at the University. The variables are as follows: hard block chemistry (x_1); nitrogen flow rate (x_2); heat-up time (x_3); percent compression (x_4); scans, high and low (x_5); percent strain (x_6). The following data were taken:

Observation	x_1	x_2	x_3	x_4	x_5	x_6	y
1	1	−1	1	−1	−1	−1	194,700
2	1	1	−1	1	−1	−1	588,400
3	−1	1	1	−1	1	−1	7,533
4	1	−1	1	1	−1	1	514,100
5	1	1	−1	1	1	−1	277,300
6	1	1	1	−1	1	1	493,500
7	−1	1	1	1	−1	1	8,969
8	−1	−1	1	1	1	−1	18,340
9	−1	−1	−1	1	1	1	6,793
10	1	−1	−1	−1	1	1	160,400
11	−1	1	−1	−1	−1	1	7,008
12	−1	−1	−1	−1	−1	−1	3,637

Build a regression equation relating viscosity to the levels of the six variables. Conduct t-tests for all main effects. Recommend factors that should be retained for future studies and those that should not. Use the residual mean square (5 degrees of freedom) as a measure of experimental error.

2. A large petroleum company in the Southwest regularly conducts experiments to test additives to drilling fluids. Plastic viscosity is a rheological measure reflecting the thickness of the fluid. Various polymers are added to the fluid to increase viscosity. The following is a data set in which two polymers were used at two levels each and the viscosity measured. The concentration of the polymers will be indicated as "low" and "high."

	Polymer 1	
	Low	High
Polymer 2 — Low	3 / 3.5	11.3 / 12.0
Polymer 2 — High	11.7 / 12.0	21.7 / 22.4

Conduct an analysis of the 2^2 factorial experiment. Test for effects for the two polymers and interaction.

3. A 2^2 factorial experiment was analyzed by the Statistics Consulting Center at Virginia Polytechnic Institute and State University. The client was a member of the Department of Housing, Interior Design, and Resource Management. The client was interested in comparing cold start against preheating ovens in terms of total energy being delivered to the product. In addition, the conditions of convection was being compared to regular mode. Four experimental runs were made at each of the four factor combinations. Following are the data from the experiment:

	Preheat	Cold start
Convection mode	618 / 619.3 / 629 / 611	575 / 573.7 / 574 / 572
Regular mode	581 / 585.7 / 581 / 595	558 / 562 / 562 / 566

Do an analysis of variance to study main effects and interaction. Draw conclusions.

4. Construct a design involving 12 runs in which two factors are varied at two levels each. You are further restricted in that blocks of size 2 must be used and you must be able to make significance tests on both main effects and the interaction effect.

5. In the study "*The Use of Regression Analysis for Correcting Matrix Effects in the X-Ray Fluorescence Analysis of Pyrotechnic Compositions*," published in the *Proceedings of the Tenth Conference on the Design of*

Experiments in Army Research Development and Testing, ARO-D Report 65-3 (1965), an experiment was conducted in which the concentrations of 4 components of a propellant mixture and the weights of fine and coarse particles in the slurry were each allowed to vary. Factors A, B, C, and D, each at two levels, represent the concentrations of the 4 components and factors E and F, also at two levels, represent the weights of the fine and coarse particles present in the slurry. The goal of the analysis was to determine if the X-ray intensity ratios associated with component 1 of the propellant were significantly influenced by varying the concentrations of the various components and the weights of the particle sizes in the mixture. A 1/8 fraction of a 2^6 factorial experiment was used with the defining contrasts being ADE, BCE, and ACF. The following data represent the total of a pair of intensity readings:

Batch	Treatment Combination	Intensity Ratio Total
1	abef	2.2480
2	cdef	1.8570
3	(1)	2.2428
4	ace	2.3270
5	bde	1.8830
6	abcd	1.8078
7	adf	2.1424
8	bcf	1.9122

The pooled error mean square with 8 degrees of freedom is given by 0.02005. Analyze the data using a 0.05 level of significance to determine if the concentrations of the components and the weights of the fine and coarse particles present in the slurry have a significant influence on the intensity ratios associated with component 1. Assume that no interaction exists among the 6 factors.

6. Show the blocking scheme for a 2^7 factorial experiment in eight blocks of size 16 each, using $ABCD$, $CDEFG$, and BDF as defining contrasts. Indicate what interactions are completely sacrificed in the experiment.

16

Nonparametric Statistics

16.1 Nonparametric Tests

Most of the hypothesis-testing procedures discussed in previous chapters are based on the assumption that the random samples are selected from normal populations. Fortunately, most of these tests are still reliable when we experience slight departures from normality, particularly when the sample size is large. Traditionally, these testing procedures have been referred to as **parametric methods**. In this chapter we shall consider a number of alternative test procedures, called **nonparametric** or **distribution-free methods**, that often assume no knowledge whatsoever about the distributions of the underlying populations, except perhaps that they are continuous.

Nonparametric or distribution-free procedures are being used with increasing frequency by data analysts. There are many applications in science and engineering where the data are reported not as values on a continuum but rather on an **ordinal scale** such that it is quite natural to assign ranks to the data. In fact, the reader will notice quite early in this chapter that the distribution-free methods described here involve an *analysis of ranks*. Most analysts find the computations involved in nonparametric methods to be very appealing and intuitive.

In an example where a nonparametric test is applicable, two judges might rank five brands of premium beer by assigning a rank of 1 to the brand believed to have the best overall quality, a rank of 2 to the second best, and so forth. A nonparametric test could then be used to determine where there is any agreement between the two judges.

We should also point out that there are a number of disadvantages associated with nonparametric tests. Primarily, they do not utilize all the information provided by the sample, and thus a nonparametric test will be less efficient than the corresponding parametric procedure when both methods are applicable. Consequently, to achieve the same power, a nonparametric test will require a larger sample size than will the corresponding parametric test.

As we indicated earlier, slight departures from normality result in minor deviations from the ideal for the standard parametric tests. This is particularly true for the t-test and the F-test. In the case of the t-test and the F-test, the P-value quoted may be slightly in error if there is a moderate violation of the normality assumption.

In summary, if a parametric and a nonparametric test are both applicable to the same set of data, we should carry out the more efficient parametric technique. However, we should recognize that the assumptions of normality often cannot be justified, and that we do not always have quantitative measurements. It is fortunate that statisticians have provided us with a number of useful nonparametric procedures. Armed with nonparametric techniques, the data analyst has more ammunition to accommodate a wider variety of experimental situations. It should be pointed out that even under the standard normal theory assumptions, the efficiencies of the nonparametric techniques are remarkably close to those of the corresponding parametric procedure. On the other hand, serious departures from normality will render the nonparametric method much more efficient than the parametric procedure.

16.2 Sign Test

The reader should recall that the procedures discussed in Section 10.7 for testing the null hypothesis that $\mu = \mu_0$ are valid only if the population is approximately normal or if the sample is large. However, if $n < 30$ and the population is decidedly nonnormal, we must resort to a nonparametric test.

The sign test is used to test hypotheses on a population *median*. In the case of many of the nonparametric procedures, the mean is replaced by the median as the pertinent **location parameter** under test. Recall that the sample median was defined in Chapter 8 (Definition 8.6). The population counterpart, denoted by $\tilde{\mu}$, has an analogous definition. Given a random variable X, $\tilde{\mu}$ is defined such that

$$P(X > \tilde{\mu}) = P(X < \tilde{\mu}) = 0.5.$$

Of course, if the distribution is symmetric, the population mean and median are equal. In testing the null hypothesis H_0 that $\tilde{\mu} = \tilde{\mu}_0$ against an appropriate alternative, on the basis of a random sample of size n, we replace each sample value exceeding $\tilde{\mu}_0$ with a *plus* sign and each sample value less than $\tilde{\mu}_0$ with a *minus* sign. If the null hypothesis is true and the population is symmetric, the sum of the plus signs should be approximately equal to the sum of the minus signs. When one sign appears more frequently than it should, based on chance alone, we reject the hypothesis that the population median $\tilde{\mu}$ is equal to $\tilde{\mu}_0$.

In theory the sign test is applicable only in situations where $\tilde{\mu}_0$ cannot equal the value of any of the observations. Although there is a zero probability of obtaining a sample observation exactly equal to $\tilde{\mu}_0$ when the population is continuous, nevertheless in practice a sample value equal to $\tilde{\mu}_0$ will often occur from a lack of precision in recording the data. When sample values equal to $\tilde{\mu}_0$ are observed, they are excluded from the analysis and the sample size is correspondingly reduced.

The appropriate test statistic for the sign test is the binomial random variable X, representing the number of plus signs in our random sample. If the null hypothesis that $\tilde{\mu} = \tilde{\mu}_0$ is true, the probability that a sample value results in either a plus or a minus sign is equal to $1/2$. Therefore, to test the null hypothesis that $\tilde{\mu} = \tilde{\mu}_0$, we are actually testing the null hypothesis that the number of plus signs is a value of a random variable having the binomial distribution with the parameter $p = 1/2$. P-values for both one-sided and two-sided alternatives can then be calculated using this binomial distribution. For example, in testing

$$H_0: \quad \tilde{\mu} = \tilde{\mu}_0,$$

$$H_1: \quad \tilde{\mu} < \tilde{\mu}_0,$$

we shall reject H_0 in favor of H_1 only if the proportion of plus signs is sufficiently less than $1/2$, that is, when the value x of our random variable is small. Hence, if the computed P-value

$$P = P(X \le x \text{ when } p = \tfrac{1}{2})$$

is less than or equal to some preselected significance level α, we reject H_0 in favor of H_1. For example, when $n = 15$ and $x = 3$, we find from Table A.1 that

$$P = P(X \le 3 \text{ when } p = \tfrac{1}{2})$$

$$= \sum_{x=0}^{3} b(x; 15, \tfrac{1}{2}) = 0.0176$$

so that the null hypothesis $\tilde{\mu} = \tilde{\mu}_0$ can certainly be rejected at the 0.05 level of significance but not at the 0.01 level.

To test the hypothesis

$$H_0: \quad \tilde{\mu} = \tilde{\mu}_0,$$

$$H_1: \quad \tilde{\mu} > \tilde{\mu}_0,$$

we reject H_0 in favor of H_1 only if the proportion of plus signs is sufficiently greater than $1/2$, that is, when x is large. Hence, if the computed P-value

$$P = P(X \ge x \text{ when } p = \tfrac{1}{2})$$

is less that α, we reject H_0 in favor of H_1. Finally, to test the hypothesis

$$H_0: \quad \tilde{\mu} = \tilde{\mu}_0,$$

$$H_1: \quad \tilde{\mu} \ne \tilde{\mu}_0,$$

we reject H_0 in favor of H_1 when the proportion of plus signs is significantly less than or greater than $1/2$. This, of course, is equivalent to x being sufficiently small or sufficiently large. Therefore, if $x < n/2$ and the computed P-value

$$P = 2P(X \le x \text{ when } p = \tfrac{1}{2})$$

is less than or equal to α, or if $x > n/2$ and the computed P-value

$$P = 2P(X \ge x \text{ when } p = \tfrac{1}{2})$$

is less than or equal to α, we reject H_0 in favor of H_1.

Whenever $n > 10$, binomial probabilities with $p = 1/2$ can be approximated from the normal curve, since $np = nq > 5$. Suppose, for example, that we wish to test the hypothesis

$$H_0: \quad \tilde{\mu} = \tilde{\mu}_0,$$

$$H_1: \quad \tilde{\mu} < \tilde{\mu}_0$$

at the $\alpha = 0.05$ level of significance for a random sample of size $n = 20$ that yields $x = 6$ plus signs. Using the normal-curve approximation with

$$\tilde{\mu} = np = (20)(0.5) = 10$$

and

$$\sigma = \sqrt{npq} = \sqrt{(20)(0.5)(0.5)} = 2.236,$$

we find that

$$z = \frac{6.5 - 10}{2.236} = -1.57.$$

Therefore,

$$P = P(X \le 6) \simeq P(Z < -1.57) = 0.0582,$$

which leads to the nonrejection of the null hypothesis.

EXAMPLE 16.1   The following data represent the number of hours that a rechargeable hedge trimmer operates before a recharge is required: 1.5, 2.2, 0.9, 1.3, 2.0, 1.6, 1.8, 1.5, 2.0, 1.2, and 1.7. Use the sign test to test the hypothesis at the 0.05 level of significance that this particular trimmer operates with a median of 1.8 hours before requiring a recharge.

SOLUTION

1. H_0: $\quad \tilde{\mu} = 1.8$.
2. H_1: $\quad \tilde{\mu} \ne 1.8$.
3. $\alpha = 0.05$.
4. Test statistic: Binomial variable X with $p = \tfrac{1}{2}$.
5. Computations: Replacing each value by the symbol "$+$" if it exceeds 1.8, by the symbol "$-$" if it is less than 1.8, and discarding the one measurement that equals 1.8, we obtain the sequence

$$- \quad + \quad - \quad - \quad + \quad - \quad - \quad + \quad - \quad -$$

for which $n = 10$, $x = 3$, and $n/2 = 5$. Therefore, from Table A.1 the computed P-value is

$$P = 2P(X \leq 3 \text{ when } p = \tfrac{1}{2})$$

$$= 2 \sum_{x=0}^{3} b(x; 10, \tfrac{1}{2})$$

$$= 0.3438 > 0.05.$$

6. Decision: Do not reject the null hypothesis and conclude that the median operating time is not significantly different from 1.8 hours.

One can also use the sign test to test the null hypothesis $\tilde{\mu}_1 - \tilde{\mu}_2 = d_0$ for paired observations. Here we replace each difference, d_i, with a plus or minus sign depending on whether the adjusted difference, $d_i - d_0$, is positive or negative. Throughout this section we have assumed that the populations are symmetric. However, even if populations are skewed, we can carry out the same test procedure, but the hypotheses refer to the population medians rather than the means.

EXAMPLE 16.2   A taxi company is trying to decide whether the use of radial tires instead of regular belted tires improves fuel economy. Sixteen cars were equipped with radial tires and driven over a prescribed test course. Without changing drivers, the same cars were then equipped with the regular belted tires and driven once again over the test course. The gasoline consumption, in kilometers per liter, was recorded as follows:

Car	Radial Tires	Belted Tires
1	4.2	4.1
2	4.7	4.9
3	6.6	6.2
4	7.0	6.9
5	6.7	6.8
6	4.5	4.4
7	5.7	5.7
8	6.0	5.8
9	7.4	6.9
10	4.9	4.9
11	6.1	6.0
12	5.2	4.9
13	5.7	5.3
14	6.9	6.5
15	6.8	7.1
16	4.9	4.8

Can we conclude at the 0.05 level of significance that cars equipped with radial tires give better fuel economy than those equipped with regular belted tires?

SOLUTION
Let μ_1 and μ_2 represent the mean kilometers per liter for cars equipped with radial and belted tires, respectively.

1. H_0: $\tilde{\mu}_1 - \tilde{\mu}_2 = 0$.

2. H_1: $\tilde{\mu}_1 - \tilde{\mu}_2 > 0$.

3. $\alpha = 0.05$.

4. Test statistic: Binomial variable X with $p = 1/2$.

5. Computations: After replacing each positive difference by a ''+'' symbol and each negative difference by a ''−'' symbol, and then discarding the two zero differences, we obtain the sequence

$$+ \; - \; + \; + \; - \; + \; + \; + \; + \; + \; + \; + \; - \; +$$

for which $n = 14$ and $x = 11$. Using the normal-curve approximation, we find

$$z = \frac{10.5 - 7}{\sqrt{14/2}} = 1.87,$$

and then

$$P = P(X \geq 11) \approx P(Z > 1.87) = 0.0307.$$

6. Decision: Reject H_0 and conclude that, on the average, radial tires do improve fuel economy.

Not only is the sign test one of our simplest nonparametric procedures to apply, it has the additional advantage of being applicable to dichotomous data that cannot be recorded on a numerical scale but can be represented by positive and negative responses. For example, the sign test is applicable in experiments where a qualitative response such as ''hit'' or ''miss'' is recorded, and in sensory-type experiments where a plus or minus sign is recorded depending on whether the taste tester correctly or incorrectly identifies the desired ingredient.

We shall attempt to make comparisons between many of the nonparametric procedures and the corresponding parametric tests. In the case of the sign test the competition is, of course, the t-test. If one is sampling from a normal distribution, the use of the t-test will result in the larger power of the test. If the distribution is merely symmetric, though not normal, the t-test is preferred in terms of power unless the distribution has extremely ''heavy tails'' compared to the normal distribution.

16.3 Signed-Rank Test

The reader should note that the sign test utilizes only the plus and minus signs of the differences between the observations and μ_0 in the one-sample case, or the plus and minus signs of the differences between the pairs of observations in the paired-sample case, but it does not take into consideration the magnitudes of these differences. A test utilizing both direction and magnitude was proposed in 1945 by Frank Wilcoxon and is now commonly referred to as the **Wilcoxon signed-rank test**.

The analyst can extract more information from the data in a nonparametric fashion if it is reasonable to invoke an additional restriction on the distribution from which the data were taken. The Wilcoxon signed-rank test applies in the case of a **symmetric continuous distribution**. Under this condition we can test the null hypothesis $\mu = \mu_0$. We first subtract μ_0 from each sample value, discarding all differences equal to zero. The remaining differences are then ranked without regard to sign. A rank of 1 is assigned to the smallest absolute difference (i.e., without sign), a rank of 2 to the next smallest, and so on. When the absolute value of two or more differences is the same, assign to each the average of the ranks that would have been assigned if the differences were distinguishable. For example, if the fifth and sixth smallest differences are equal in absolute value, each would be assigned a rank of 5.5. If the hypothesis $\mu = \mu_0$ is true, the total of the ranks corresponding to the positive differences should nearly equal the total of the ranks corresponding to the negative differences. Let us represent these totals by w_+ and w_-, respectively. We shall designate the smaller of the w_+ and w_- by w.

In selecting repeated samples, we would expect w_+ and w_-, and therefore w, to vary. Thus we may think of w_+, w_-, and w as values of the corresponding random variables W_+, W_-, and W. The null hypothesis $\mu = \mu_0$ can be rejected in favor of the alternative $\mu < \mu_0$ only if w_+ is small and w_- is large. Likewise, the alternative $\mu > \mu_0$ can be accepted only if w_+ is large and w_- is small. For a two-sided alternative we may reject H_0 in favor of H_1 if either w_+ or w_- and hence if w is sufficiently small. Therefore, no matter what the alternative hypothesis may be, we reject the null hypothesis when the value of the appropriate statistic W_+, W_-, or W is sufficiently small.

Two Samples with Paired Observations

To test the null hypothesis that we are sampling two continuous symmetric populations with $\mu_1 = \mu_2$ for the paired-sample case, we rank the differences of the paired observations without regard to sign and proceed as with the single-sample case. The various test procedures for both the single- and paired-sample cases are summarized in Table 16.1.

It is not difficult to show that whenever $n < 5$ and the level of significance does not exceed 0.05 for a one-tailed test or 0.10 for a two-tailed test, all possible values of w_+, w_-, or w will lead to the acceptance of the null hypothesis. However, when

TABLE 16.1 Signed-Rank Test

To Test H_0	Versus H_1	Compute
	$\mu < \mu_0$	w_+
$\mu = \mu_0$	$\mu > \mu_0$	w_-
	$\mu \neq \mu_0$	w
	$\mu_1 < \mu_2$	w_+
$\mu_1 = \mu_2$	$\mu_1 > \mu_2$	w_-
	$\mu_1 \neq \mu_2$	w

$5 \le n \le 30$, Table A.16 gives approximate critical values of W_+ and W_- for levels of significance equal to 0.01, 0.025, and 0.05 for a one-tailed test, and critical values of W for levels of significance equal to 0.02, 0.05, and 0.10 for a two-tailed test. The null hypothesis is rejected if the computed value w_+, w_-, or w is **less than or equal** to the appropriate tabled value. For example, when $n = 12$, Table A.16 shows that a value of $w_+ \le 17$ is required for the one-sided alternative $\mu < \mu_0$ to be significant at the 0.05 level.

EXAMPLE 16.3   Rework Example 16.1 by using the signed-rank test.

SOLUTION

1. H_0: $\mu = 1.8$.
2. H_1: $\mu \ne 1.8$.
3. $\alpha = 0.05$.
4. Critical region: Since $n = 10$, after discarding the one measurement that equals 1.8, Table A.16 shows the critical region to be $w \le 8$.
5. Computations: Subtracting 1.8 from each measurement and then ranking the differences without regard to sign, we have

d_i	-0.3	0.4	-0.9	-0.5	0.2	-0.2	-0.3	0.2	-0.6	-0.1
Ranks	5.5	7	10	8	3	3	5.5	3	9	1

Now $w_+ = 13$ and $w_- = 42$ so that $w = 13$, the smaller of w_+ and w_-.
6. Decision: As before, do not reject H_0 and conclude that the average operating time is not significantly different from 1.8 hours.

The signed-rank test can also be used to test the null hypothesis that $\mu_1 - \mu_2 = d_0$. In this case the populations need not be symmetric. As with the sign test we subtract d_0 from each difference, rank the adjusted differences without regard to sign, and apply the same procedure as above.

EXAMPLE 16.4   It is claimed that a college senior can increase his score in the major field area of the graduate record examination by at least 50 points if he is provided with sample problems in advance. To test this claim, 20 college seniors were divided into 10 pairs such that each matched pair had almost the same overall quality point average for their first 3 years in college. Sample problems and answers were provided at random to one member of each pair 1 week prior to the examination. The following examination scores were recorded:

	Pair									
	1	2	3	4	5	6	7	8	9	10
With sample problems	531	621	663	579	451	660	591	719	543	575
Without sample problems	509	540	688	502	424	683	568	748	530	524

Test the null hypothesis at the 0.05 level of significance that sample problems increase the scores by 50 points against the alternative hypothesis that the increase is less than 50 points.

SOLUTION
Let μ_1 and μ_2 represent the mean score of all students taking the test in question with and without sample problems, respectively.

1. H_0: $\mu_1 - \mu_2 = 50$.
2. H_1: $\mu_1 - \mu_2 < 50$.
3. $\alpha = 0.05$.
4. Critical region: Since $n = 10$, Table A.16 shows the critical region to be $w_+ \leq 11$.
5. Computations:

					Pair					
	1	2	3	4	5	6	7	8	9	10
d_i	22	81	−25	77	27	−23	23	−29	13	51
$d_i - d_0$	−28	31	−75	27	−23	−73	−27	−79	−37	1
Ranks	5	6	9	3.5	2	8	3.5	10	7	1

Now we find that $w_+ = 6 + 3.5 + 1 = 10.5$.

6. Decision: Reject H_0 and conclude that sample problems do not, on the average, increase one's graduate record score by as much as 50 points.

Normal Approximation for Large Samples

When $n \geq 15$, the sampling distribution of W_+ (or W_-) approaches the normal distribution with mean

$$\mu_{W_+} = \frac{n(n + 1)}{4}$$

and variance

$$\sigma_{W_+}^2 = \frac{n(n + 1)(2n + 1)}{24}.$$

Therefore, when n exceeds the largest value in Table A.16, the statistic

$$Z = \frac{W_+ - \mu_{W_+}}{\sigma_{W_+}}$$

can be used to determine the critical region for our test.

Exercises

1. The following data represent the time, in minutes, that a patient had to wait on 12 visits to a doctor's office before being seen by the doctor:

$$
\begin{array}{cccc}
17 & 15 & 20 & 20 \\
32 & 28 & 12 & 26 \\
25 & 25 & 35 & 24
\end{array}
$$

Use the sign test at the 0.05 level of significance to test the doctor's claim that the median waiting time for her patients is not more than 20 minutes before being admitted to the examination room.

2. The following data represent the number of hours of flight training received by 18 student pilots from a certain instructor prior to their first solo flight:

$$
\begin{array}{cccccc}
9 & 12 & 18 & 14 & 12 & 14 \\
12 & 10 & 16 & 11 & 9 & 11 \\
13 & 11 & 13 & 15 & 13 & 14
\end{array}
$$

Using binomial probabilities from Table A.1, perform a sign test at the 0.02 level of significance to test the instructor's claim that the median time required before his students solo is 12 hours of flight training.

3. A food inspector examined 16 jars of a certain brand of jam to determine the percent of foreign impurities. The following data were recorded:

$$
\begin{array}{cccc}
2.4 & 2.3 & 3.1 & 2.2 \\
2.3 & 1.2 & 1.0 & 2.4 \\
1.7 & 1.1 & 4.2 & 1.9 \\
1.7 & 3.6 & 1.6 & 2.3
\end{array}
$$

Using the normal approximation to the binomial distribution, perform a sign test at the 0.05 level of significance to test the null hypothesis that the median percent of impurities in this brand of jam is 2.5% against the alternative that the median percent of impurities is not 2.5%.

4. A paint supplier claims that a new additive will reduce the drying time of its acrylic paint. To test this claim, 12 panels of wood are painted, one-half of each panel with paint containing the regular additive and the other half with paint containing the new additive. The drying times, in hours, were recorded as follows:

| | Drying Time (hours) | |
Panel	New Additive	Regular Additive
1	6.4	6.6
2	5.8	5.8
3	7.4	7.8
4	5.5	5.7
5	6.3	6.0
6	7.8	8.4
7	8.6	8.8
8	8.2	8.4
9	7.0	7.3
10	4.9	5.8
11	5.9	5.8
12	6.5	6.5

Use the sign test at the 0.05 level to test the null hypothesis that the new additive is no better than the regular additive in reducing the drying time of this kind of paint.

5. It is claimed that a new diet will reduce a person's weight by 4.5 kilograms on the average in a period of 2 weeks. The weights of 10 women who followed this diet were recorded before and after a 2-week period, yielding the following data:

Woman	Weight Before	Weight After
1	58.5	60.0
2	60.3	54.9
3	61.7	58.1
4	69.0	62.1
5	64.0	58.5
6	62.6	59.9
7	56.7	54.4
8	63.6	60.2
9	68.2	62.3
10	59.4	58.7

Use the sign test at the 0.05 level of significance to test the hypothesis that the diet reduces the median weight by 4.5 kilograms against the alternative hypothesis that the median difference in weight is less than 4.5 kilograms.

6. Two types of instruments for measuring the amount of sulfur monoxide in the atmosphere are being compared in an air-pollution experiment. The following readings were recorded daily for a period of 2 weeks:

	Sulfur Monoxide	
Day	Instrument A	Instrument B
1	0.96	0.87
2	0.82	0.74
3	0.75	0.63
4	0.61	0.55
5	0.89	0.76
6	0.64	0.70
7	0.81	0.69
8	0.68	0.57
9	0.65	0.53
10	0.84	0.88
11	0.59	0.51
12	0.94	0.79
13	0.91	0.84
14	0.77	0.63

Using the normal approximation to the binomial distribution, perform a sign test to determine whether the different instruments lead to different results. Use a 0.05 level of significance.

7. The following figures give the systolic blood pressure of 16 joggers before and after an 8-kilometer run:

Jogger	Before	After
1	158	164
2	149	158
3	160	163
4	155	160
5	164	172
6	138	147
7	163	167
8	159	169
9	165	173
10	145	147
11	150	156
12	161	164
13	132	133
14	155	161
15	146	154
16	159	170

Use the sign test at the 0.05 level of significance to test the null hypothesis that jogging 8 kilometers increases the median systolic blood pressure by 8 points against the alternative that the increase in the median is less than 8 points.

8. Analyze the data of Exercise 1 using the signed-rank test.

9. Analyze the data of Exercise 2 using the signed-rank test.

10. The weights of 4 people before they stopped smoking and 5 weeks after they stopped smoking, in kilograms, are as follows:

	Individual				
	1	2	3	4	5
Before	66	80	69	52	75
After	71	82	68	56	73

Use the signed-rank test for paired observations to test the hypothesis, at the 0.05 level of significance, that giving up smoking has no effect on a person's weight against the alternative that one's weight increases if he or she quits smoking.

11. Rework Exercise 5 using the signed-rank test.

12. The following are the numbers of prescriptions filled by two pharmacies over a 20-day period:

Day	Pharmacy A	Pharmacy B
1	19	17
2	21	15
3	15	12
4	17	12
5	24	16
6	12	15
7	19	11
8	14	13
9	20	14
10	18	21
11	23	19
12	21	15
13	17	11
14	12	10
15	16	20
16	15	12
17	20	13
18	18	17
19	14	16
20	22	18

Use the signed-rank test at the 0.01 level of significance to determine whether the two pharmacies, on the average, fill the same number of prescriptions against the alternative that pharmacy A fills more prescriptions than pharmacy B.

13. Rework Exercise 7 using the signed-rank test.

16.4 Rank-Sum Test

As we indicated earlier, the nonparametric procedure is generally an appropriate alternative to the normal theory test when the normality assumption does not hold. When one is interested in testing equality of means of two continuous distributions that are obviously nonnormal, and samples are independent (i.e., there is no pairing of observations), the **Wilcoxon rank-sum** test or **Wilcoxon two-sample** test is an appropriate alternative to the two-sample t-test described in Chapter 10.

We shall test the null hypothesis H_0 that $\mu_1 = \mu_2$ against some suitable alternative. First we select a random sample from each of the populations. Let n_1 be the number of observations in the smaller sample, and n_2 the number of observations in the larger sample. When the samples are of equal size, n_1 and n_2 may be randomly assigned. Arrange the $n_1 + n_2$ observations of the combined samples in ascending order and substitute a rank of $1, 2, \ldots, n_1 + n_2$ for each observation. In the case of ties (identical observations), we replace the observations by the mean of the ranks that the observations would have if they were distinguishable. For example, if the seventh and eighth observations are identical, we would assign a rank of 7.5 to each of the two observations.

The sum of the ranks corresponding to the n_1 observations in the smaller sample is denoted by w_1. Similarly, the value w_2 represents the sum of the n_2 ranks corresponding to the larger sample. The total $w_1 + w_2$ depends only on the number of observations in the two samples and is in no way affected by the results of the experiment. Hence, if $n_1 = 3$ and $n_2 = 4$, then $w_1 + w_2 = 1 + 2 + \cdots + 7 = 28$, regardless of the numerical values of the observations. In general,

$$w_1 + w_2 = \frac{(n_1 + n_2)(n_1 + n_2 + 1)}{2},$$

the arithmetic sum of the integers $1, 2, \ldots, n_1 + n_2$. Once we have determined w_1, it may be easier to find w_2 by the formula

$$w_2 = \frac{(n_1 + n_2)(n_1 + n_2 + 1)}{2} - w_1.$$

In choosing repeated samples of size n_1 and n_2, we would expect w_1, and therefore w_2, to vary. Thus we may think of w_1 and w_2 as values of the random variables W_1 and W_2, respectively. The null hypothesis $\mu_1 = \mu_2$ will be rejected in favor of the alternative $\mu_1 < \mu_2$ only if w_1 is small and w_2 is large. Likewise, the alternative $\mu_1 > \mu_2$ can be accepted only if w_1 is large and w_2 is small. For a two-tailed test, we may reject H_0 in favor of H_1 if w_1 is small and w_2 is large or if w_1 is large and w_2 is small. In other words, the alternative $\mu_1 < \mu_2$ is accepted if w_1 is sufficiently small; the alternative $\mu_1 > \mu_2$ is accepted if w_2 is sufficiently small; and the alternative $\mu_1 \neq \mu_2$ is accepted if the minimum of w_1 and w_2 is sufficiently small. In actual practice we usually base our decision on the value

$$u_1 = w_1 - \frac{n_1(n_1 + 1)}{2}$$

or

$$u_2 = w_2 - \frac{n_2(n_2 + 1)}{2}$$

of the related statistic U_1 or U_2, or on the value u of the statistic U, the minimum of U_1 and U_2. These statistics simplify the construction of tables of critical values, since both U_1 and U_2 have symmetric sampling distributions and assume values in the interval from 0 to $n_1 n_2$ such that $u_1 + u_2 = n_1 n_2$.

From the formulas for u_1 and u_2 we see that u_1 will be small when w_1 is small and u_2 will be small when w_2 is small. Consequently, the null hypothesis will be rejected whenever the appropriate statistic U_1, U_2, or U assumes a value less than or equal to the desired critical value given in Table A.17. The various test procedures are summarized in Table 16.2.

Table A.17 gives critical values of U_1 and U_2 for levels of significance equal to $0.001, 0.01, 0.025$, and 0.05 for a one-tailed test, and critical values of U for levels of significance equal to $0.002, 0.02, 0.05$, and 0.10 for a two-tailed test. If the observed value of u_1, u_2, or u is **less than or equal** to the tabled critical value, the null hypothesis is rejected at the level of significance indicated by the table. Suppose, for example, that we wish to test the null hypothesis that $\mu_1 = \mu_2$ against the one-sided alternative that $\mu_1 < \mu_2$ at the 0.05 level of significance for random samples of size $n_1 = 3$ and $n_2 = 5$ that yield the value $w_1 = 8$. It follows that

$$u_1 = 8 - \frac{(3)(4)}{2} = 2.$$

Our one-tailed test is based on the statistic U_1. Using Table A.17, we reject the null hypothesis of equal means when $u_1 \leq 1$. Since $u_1 = 2$ falls in the acceptance region, the null hypothesis cannot be rejected.

EXAMPLE 16.5   The nicotine content of two brands of cigarettes, measured in milligrams, was found to be as follows:

Brand A	2.1	4.0	6.3	5.4	4.8	3.7	6.1	3.3		
Brand B	4.1	0.6	3.1	2.5	4.0	6.2	1.6	2.2	1.9	5.4

Test the hypothesis, at the 0.05 level of significance, that the average nicotine contents of the two brands are equal against the alternative that they are unequal.

TABLE 16.2   Rank-Sum Test

To Test H_0	Versus H_1	Compute
$\mu_1 = \mu_2$	$\mu_1 < \mu_2$	μ_1
	$\mu_1 > \mu_2$	μ_2
	$\mu_1 \neq \mu_2$	u

SOLUTION

1. H_0: $\mu_1 = \mu_2$.
2. H_1: $\mu_1 \neq \mu_2$.
3. $\alpha = 0.05$.
4. Critical region: $u \leq 17$ (from Table A.17).
5. Computations: The observations are arranged in ascending order and ranks from 1 to 18 assigned.

Original Data	Ranks
0.6	1
1.6	2
1.9	3
2.1	4*
2.2	5
2.5	6
3.1	7
3.3	8*
3.7	9*
4.0	10.5*
4.0	10.5
4.1	12
4.8	13*
5.4	14.5*
5.4	14.5
6.1	16
6.2	17
6.3	18*

*The starred ranks belong to sample A.

Now

$$w_1 = 4 + 8 + 9 + 10.5 + 13 + 14.5 + 18 = 93$$

and

$$w_2 = \left[\frac{(18)(19)}{2}\right] - 93 = 78.$$

Therefore,

$$u_1 = 93 - \left[\frac{(8)(9)}{2}\right] = 57$$

$$u_2 = 78 - \left[\frac{(10)(11)}{2}\right] = 23,$$

so that $u = 23$.

6. Decision: Do not reject H_0 and conclude that there is no significant difference in the average nicotine contents of the two brands of cigarettes.

Normal Theory Approximation for Two Samples

When both n_1 and n_2 exceed 8, the sampling distribution of U_1 (or U_2) approaches the normal distribution with mean

$$\mu_{U_1} = \frac{n_1 n_2}{2}$$

and variance

$$\sigma_{U_1}^2 = \frac{n_1 n_2 (n_1 + n_2 + 1)}{12}.$$

Consequently, when n_2 is greater than 20, the maximum value in Table A.17, and n_1 is at least 9, one could use the statistic

$$Z = \frac{U_1 - \mu_{U_1}}{\sigma_{U_1}}$$

for our test, with the critical region falling in either or both tails of the standard normal distribution, depending on the form of H_1.

The use of the Wilcoxon rank-sum test is not restricted to nonnormal populations. It can be used in place of the two-sample t-test when the populations are normal, although the power will be smaller. The Wilcoxon rank-sum test is always superior to the t-test for decidedly nonnormal populations.

16.5 Kruskal–Wallis Test

In Chapters 13, 14, and 15, the technique of analysis of variance was prominent as an analytical technique for testing equality of $k \geq 2$ population means. Again, however, the reader should recall that normality must be assumed in order that the F-test be theoretically correct. In this section we investigate a nonparametric alternative to analysis of variance.

The **Kruskal–Wallis test**, also called the **Kruskal–Wallis H test**, is a generalization of the rank-sum test to the case of $k > 2$ samples. It is used to test the null hypothesis H_0 that k independent samples are from identical populations. Introduced in 1952 by W. H. Kruskal and W. A. Wallis, the test is a nonparametric procedure for testing the equality of means in the one-factor analysis of variance when the experimenter wishes to avoid the assumption that the samples were selected from normal populations.

Let n_i $(i = 1, 2, \ldots, k)$ be the number of observations in the ith sample. First we combine all k samples and arrange the $n = n_1 + n_2 + \cdots + n_k$ observations in ascending order, substituting the appropriate rank from $1, 2, \ldots, n$ for each observation. In the case of ties (identical observations), we follow the usual procedure of replacing the observations by the means of the ranks that the observations would have if they were distinguishable. The sum of the ranks corresponding to the n_i observations in the ith sample is denoted by the random variable R_i. Now let us consider the statistic

$$H = \frac{12}{n(n+1)} \sum_{i=1}^{k} \frac{R_i^2}{n_i} - 3(n+1),$$

which is approximated very well by a chi-squared distribution with $k-1$ degrees of freedom when H_0 is true and if each sample consists of at least 5 observations. Note that the statistic H assumes the value h, where

$$h = \frac{12}{n(n+1)} \sum_{i=1}^{k} \frac{r_i^2}{n_i} - 3(n+1),$$

when R_1 assumes the value r_1, R_2 assumes the value r_2, and so forth. The fact that h is large when the independent samples come from populations that are not identical allows us to establish the following decision criterion for testing H_0:

KRUSKAL–WALLIS TEST

To test the null hypothesis H_0 that k independent samples are from identical populations, compute

$$h = \frac{12}{n(n+1)} \sum_{i=1}^{k} \frac{r_i^2}{n_i} - 3(n+1).$$

If h falls in the critical region $H > \chi_\alpha^2$ with $v = k-1$ degrees of freedom, reject H_0 at the α-level of significance; otherwise, accept H_0.

EXAMPLE 16.6   In an experiment to determine which of three different missile systems is preferable, the propellant burning rate was measured. The data, after coding, are given in Table 16.3. Use the Kruskal–Wallis test and a significance level of $\alpha = 0.05$ to test the hypothesis that the propellant burning rates are the same for the three missile systems.

SOLUTION

1. H_0: $\mu_1 = \mu_2 = \mu_3$.
2. H_1:   The three means are not all equal.
3. $\alpha = 0.05$.
4. Critical region: $h > \chi_{0.05}^2 = 5.991$, for $v = 2$ degrees of freedom.
5. Computations: In Table 16.4 we convert the 19 observations to ranks and sum the ranks for each missile system.

   Now, substituting $n_1 = 5$, $n_2 = 6$, $n_3 = 8$, and $r_1 = 61.0$, $r_2 = 63.5$, $r_3 = 65.5$, our test statistic H assumes the value

   $$h = \frac{12}{(19)(20)} \left(\frac{61.0^2}{5} + \frac{63.5^2}{6} + \frac{65.5^2}{8} \right) - (3)(20)$$

   $$= 1.66.$$

6. Decision: Since $h = 1.66$ does not fall in the critical region $h > 5.991$, we have insufficient evidence to reject the hypothesis that the propellant burning rates are the same for the three missile systems.

TABLE 16.3   Propellant Burning Rates

1	Missile System 2	3
24.0	23.2	18.4
16.7	19.8	19.1
22.8	18.1	17.3
19.8	17.6	17.3
18.9	20.2	19.7
	17.8	18.9
		18.8
		19.3

TABLE 16.4   Ranks for Propellant Burning Rates

1	Missile System 2	3
19	18	7
1	14.5	11
17	6	2.5
14.5	4	2.5
9.5	16	13
$r_1 = 61.0$	5	9.5
	$r_2 = 63.5$	8
		12
		$r_3 = 65.5$

Exercises

1. A cigarette manufacturer claims that the tar content of brand B cigarettes is lower than that of brand A. To test this claim, the following determinations of tar content, in milligrams, were recorded:

Brand A	12	9	13	11	14
Brand B	8	10	7		

Use the rank-sum test with $\alpha = 0.05$ to test whether the claim is valid.

2. To find out whether a new serum will arrest leukemia, 9 patients, who have all reached an advanced stage of the disease, are selected. Five patients receive the treatment and 4 do not. The survival times, in years, from the time the experiment commenced are

Treatment	2.1	5.3	1.4	4.6	0.9
No treatment	1.9	0.5	2.8	3.1	

Use the rank-sum test, at the 0.05 level of significance, to determine if the serum is effective.

3. The following data represent the number of hours that two different types of scientific pocket calculators operate before a recharge is required.

Calculator A	5.5	5.6	6.3	4.6	5.3	5.0	6.2	5.8	5.1
Calculator B	3.8	4.8	4.3	4.2	4.0	4.9	4.5	5.2	4.5

Use the rank-sum test with $\alpha = 0.01$ to determine if calculator A operates longer than calculator B on a full battery charge.

4. A fishing line is being manufactured by two processes. To determine if there is a difference in the mean breaking strength of the lines, 10 pieces by each process are selected and then tested for breaking strength. The results are as follows:

Process 1	Process 2
10.4	8.7
9.8	11.2
11.5	9.8
10.0	10.1
9.9	10.8
9.6	9.5
10.9	11.0
11.8	9.8
9.3	10.5
10.7	9.9

Use the rank-sum test with $\alpha = 0.1$ to determine if there is a difference between the mean breaking strengths of the lines manufactured by the two processes.

5. From a mathematics class of 12 equally capable students using programmed materials, 5 are selected at random and given additional instruction by the teacher. The results on the final examination were as follows:

	Grade						
Additional instruction	87	69	78	91	80		
No additional instruction	75	88	64	82	93	79	67

Use the rank-sum test with $\alpha = 0.05$ to determine if the additional instruction affects the average grade.

6. The following data represent the weights, in kilograms, of personal luggage carried on various flights by a member of a baseball team and a member of a basketball team.

Luggage Weight (kilograms)				
Baseball Player			Basketball Player	
16.3	20.0	18.6	15.4	16.3
18.1	15.0	15.4	17.7	18.1
15.9	18.6	15.6	18.6	16.8
14.1	14.5	18.3	12.7	14.1
17.7	19.1	17.4	15.0	13.6
16.3	13.6	14.8	15.9	16.3
13.2	17.2	16.5		

Use the rank-sum test with $\alpha = 0.05$ to test the null hypothesis that the two athletes carry the same amount of luggage on the average against the alternative hypothesis that the average weights of luggage for the two athletes are different.

7. The following data represent the operating times in hours for three types of scientific pocket calculators before a recharge is required:

	Calculator	
A	B	C
4.9	5.5	6.4
6.1	5.4	6.8
4.3	6.2	5.6
4.6	5.8	6.5
5.3	5.5	6.3
	5.2	6.6
	4.8	

Use the Kruskal–Wallis test, at the 0.01 level of significance, to test the hypothesis that the operating times for all three calculators are equal.

8. Random samples of four brands of cigarettes were tested for tar content. The following figures show the milligrams of tar found in the 16 cigarettes tested:

Brand A	Brand B	Brand C	Brand D
14	16	16	17
10	18	15	20
11	14	14	19
13	15	12	21

Use the Kruskal–Wallis test, at the 0.05 level of significance, to test whether there is a significant difference in tar content among the four brands of cigarettes.

9. In Exercise 8 on page 476 use the Kruskal–Wallis test, at the 0.05 level of significance to determine if the grade distributions given by the 3 teachers differ significantly.

16.6 Runs Test

In applying the many statistical concepts that were discussed throughout this book, it was always assumed that our sample data had been collected by some randomization procedure. The **runs test**, based on the order in which the sample observations are obtained, is a useful technique for testing the null hypothesis H_0 that the observations have indeed been drawn at random.

To illustrate the runs test, let us suppose that 12 people have been polled to find out if they use a certain product. One would seriously question the assumed randomness of the sample if all 12 people were of the same sex. We shall designate a male and female by the symbols M and F, respectively, and record the outcomes according to their sex in the order in which they occur. A typical sequence for the experiment might be

$$M \ M \ \ F \ F \ F \ \ M \ \ F \ F \ \ M \ M \ M \ M,$$

where we have grouped subsequences of similar symbols. Such groupings are called **runs**.

DEFINITION 16.1   *A* **run** *is a subsequence of one or more identical symbols representing a common property of the data.* ∎

Regardless of whether our sample measurements represent qualitative or quantitative data, the runs test divides the data into two mutually exclusive categories: male or female; defective or nondefective; heads or tails; above or below the median; and so forth. Consequently, a sequence will always be limited to two distinct symbols. Let n_1 be the number of symbols associated with the category that occurs the least and n_2 be the number of symbols that belong to the other category. Then the sample size $n = n_1 + n_2$.

For the $n = 12$ symbols in our poll we have five runs, with the first containing two M's, the second containing three F's, and so on. If the number of runs is larger or smaller than what we would expect by chance, the hypothesis that the sample was drawn at random should be rejected. Certainly, a sample resulting in only two runs,

$$M \ M \ M \ M \ M \ M \ M \ F \ F \ F \ F \ F,$$

or the reverse, is most unlikely to occur from a random selection process. Such a result indicates that the first seven people interviewed were all males followed by five females. Likewise, if the sample resulted in the maximum number of 12 runs, as in the alternating sequence

$$M \ F \ M \ F \ M \ F \ M \ F \ M \ F \ M \ F,$$

we would again be suspicious of the order in which the individuals were selected for the poll.

The runs test for randomness is based on the random variable V, the total number of runs that occur in the complete sequence of our experiment. In Table A.18, values of $P(V \leq v^*$ when H_0 is true) are given for $v^* = 2, 3, \ldots, 20$ runs, and values of n_1 and n_2 less than or equal to 10. The P-values for both one-tailed and two-tailed tests can be obtained using these tabled values.

In the poll taken above we exhibit a total of 5 F's and 7 M's. Hence, with $n_1 = 5$, $n_2 = 7$, and $v = 5$, we note from Table A.18 for a two-tailed test that the P-value is

$$P = 2P(V \leq 5 \text{ when } H_0 \text{ is true})$$

$$= 0.394 > 0.05.$$

That is, the value $v = 5$ is reasonable at the 0.05 level of significance when H_0 is true, and therefore we have insufficient evidence to reject the hypothesis of randomness in our sample.

When the number of runs is large, for example if $v = 11$ and $n_1 = 5$ and $n_2 = 7$, then the P-value in a two-tailed test is

$$P = 2P(V \geq 11 \text{ when } H_0 \text{ is true})$$

$$= 2[1 - P(V \leq 10) \text{ when } H_0 \text{ is true}]$$

$$= 2(1 - 0.992) = 0.016 < 0.05,$$

which leads us to reject the hypothesis that the sample values occurred at random.

The runs test can also be used to detect departures in randomness of a sequence of quantitative measurements over time, caused by trends or periodicities. Replacing each measurement in the order in which they are collected by a *plus* symbol if it falls above the median, by a *minus* symbol if it falls below the median, and omitting all measurements that are exactly equal to the median, we generate a sequence of plus and minus symbols that are tested for randomness as illustrated in the following example.

EXAMPLE 16.7   A machine is adjusted to dispense acrylic paint thinner into a container. Would you say that the amount of paint thinner being dispensed by this machine varies randomly if the contents of the next 15 containers are measured and found to be 3.6, 3.9, 4.1, 3.6, 3.8, 3.7, 3.4, 4.0, 3.8, 4.1, 3.9, 4.0, 3.8, 4.2, and 4.1 liters? Use a 0.1 level of significance.

SOLUTION

1. H_0:   Sequence is random.
2. H_1:   Sequence is not random.
3. $\alpha = 0.1$.
4. Test statistic: V, the total number of runs.
5. Computations: For the given sample we find $\bar{x} = 3.9$. Replacing each measurement by the symbol "$+$" if it falls above 3.9, by the symbol "$-$" if it falls

below 3.9, and omitting the two measurements that equal 3.9, we obtain the sequence

$$- + - - - - + + + + - + +$$

for which $n_1 = 6$, $n_2 = 7$, and $v = 6$. Therefore, from Table A.18, the computed P-value is

$$P = 2P(V \leq 6 \text{ when } H_0 \text{ is true})$$

$$= 0.592 > 0.1.$$

6. Decision: Do not reject the hypothesis that the sequence of measurements varies randomly.

The runs test, although less powerful, can also be used as an alternative to the Wilcoxon two-sample test to test the claim that two random samples come from populations having the same distributions and therefore equal means. If the populations are symmetric, rejection of the claim of equal distributions is equivalent to accepting the alternative hypothesis that the means are not equal. In performing the test, we first combine the observations from both samples and arrange them in ascending order. Now assign the letter A to each observation taken from one of the populations and the letter B to each observation from the second population, thereby generating a sequence consisting of the symbols A and B. If observations from one population are tied with observations from the other population, the sequence of A and B symbols generated will not be unique and consequently the number of runs is unlikely to be unique. Procedures for breaking ties usually result in additional tedious computations, and for this reason one might prefer to apply the Wilcoxon rank-sum test whenever these situations occur.

To illustrate the use of runs in testing for equal means, consider the survival times of the leukemia patients of Exercise 2 on page 639 for which we have

$$\begin{array}{ccccccccc} 0.5 & 0.9 & 1.4 & 1.9 & 2.1 & 2.8 & 3.1 & 4.6 & 5.3 \\ B & A & A & B & A & B & B & A & A \end{array}$$

resulting in $v = 6$ runs. If the two symmetric populations have equal means, the observations from the two samples will be intermingled, resulting in many runs. However, if the population means are significantly different, we would expect most of the observations for one of the two samples to be smaller than those for the other sample. In the extreme case where the populations do not overlap, we would obtain a sequence of the form

$$A \quad A \quad A \quad A \quad A \quad B \quad B \quad B \quad B \quad \text{or} \quad B \quad B \quad B \quad B \quad A \quad A \quad A \quad A \quad A$$

and in either case there are only two runs. Consequently, the hypothesis of equal population means will be rejected at the α-level of significance only when v is small enough so that

$$P = P(V \leq v \text{ when } H_0 \text{ is true}) \leq \alpha,$$

implying a one-tailed test.

Returning to the data of Exercise 2 on page 639 for which $n_1 = 4$, $n_2 = 5$, and $v = 6$, we find from Table A.18 that

$$P = P(V \le 6 \text{ when } H_0 \text{ is true})$$

$$= 0.786 > 0.05$$

and therefore fail to reject the null hypothesis of equal means. Hence we conclude that the new serum does not prolong life by arresting leukemia.

When n_1 and n_2 increase in size, the sampling distribution of V approaches the normal distribution with mean

$$\mu_V = \frac{2n_1n_2}{n_1 + n_2} + 1$$

and variance

$$\sigma_V^2 = \frac{2n_1n_2(2n_1n_2 - n_1 - n_2)}{(n_1 + n_2)^2(n_1 + n_2 - 1)}.$$

Consequently, when n_1 and n_2 are both greater than 10, one could use the statistic

$$Z = \frac{V - \mu_V}{\sigma_V}$$

to establish the critical region for the runs test.

16.7 Tolerance Limits

Tolerance limits for a normal distribution of measurements were discussed in Chapter 9. In this section we consider a method for constructing tolerance intervals that are independent of the shape of the underlying distribution. As one might suspect, for a reasonable degree of confidence they will be substantially longer than those constructed assuming normality, and the sample size required is generally very large. Nonparametric tolerance limits are stated in terms of the smallest and largest observations in our sample.

Two-Sided **Tolerance Limits**	*For any distribution of measurements, two-sided tolerance limits are given by the smallest and largest observations in a sample of size n, where n is determined so that one can assert with $100(1 - \gamma)\%$ confidence that **at least** the proportion $1 - \alpha$ of the distribution is included between the sample extremes.* ■

Table A.19 gives required sample sizes for selected values of γ and $1 - \alpha$. For example, when $\gamma = 0.01$ and $1 - \alpha = 0.95$, we must choose a random sample of size $n = 130$ in order to be 99% confident that at least 95% of the distribution of measurements is included between the sample extremes.

Instead of determining the sample size n such that a specified proportion of measurements are contained between the sample extremes, it is desirable in many industrial processes to determine the sample size such that a fixed proportion of the population falls below the largest (or above the smallest) observation in the sample. Such limits are called one-sided tolerance limits.

ONE-SIDED TOLERANCE LIMITS	*For any distribution of measurements, a one-sided tolerance limit is given by the smallest (largest) observation in a sample of size n, where n is determined so that one can assert with* $100(1 - \gamma)$% *confidence that* **at least** *the proportion* $1 - \alpha$ *of the distribution will exceed the smallest (be less than the largest) observation in the sample.* ∎

Table A.20 gives required sample sizes corresponding to selected values of γ and $1 - \alpha$. Hence, when $\gamma = 0.05$ and $1 - \alpha = 0.70$, we must choose a sample of size $n = 9$ in order to be 95% confident that 70% of our distribution of measurements will exceed the smallest observation in the sample.

16.8 Rank Correlation Coefficient

In Chapter 11 we used the sample correlation coefficient r to measure the linear relationship between two continuous variables X and Y. If ranks $1, 2, \ldots, n$ are assigned to the x observations in order of magnitude and similarly to the y observations, and if these ranks are then substituted for the actual numerical values into the formula for the correlation coefficient in Chapter 11, we obtain the nonparametric counterpart of the conventional correlation coefficient. A correlation coefficient calculated in this manner is known as the **Spearman rank correlation coefficient** and is denoted by r_S. When there are no ties among either set of measurements, the formula for r_S reduces to a much simpler expression involving the differences d_i between the ranks assigned to the n pairs of x's and y's, which we now state.

RANK CORRELATION COEFFICIENT	*A nonparametric measure of association between two variables X and Y is given by* the **rank correlation coefficient**

$$r_S = 1 - \frac{6 \sum_{i=1}^{n} d_i^2}{n(n^2 - 1)},$$

where d_i is the difference between the ranks assigned to x_i and y_i, and n is the number of pairs of data. ∎

In practice the preceding formula is also used when there are ties among either the x or y observations. The ranks for tied observations are assigned as in the signed-rank test by averaging the ranks that would have been assigned if the observations were distinguishable.

The value of r_S will usually be close to the value obtained by finding r based on numerical measurements and is interpreted in much the same way. As before, the value of r_S will range from -1 to $+1$. A value of $+1$ or -1 indicates perfect association between X and Y, the plus sign occurring for identical rankings and the minus sign occurring for reverse rankings. When r_S is close to zero, we would conclude that the variables are uncorrelated.

EXAMPLE 16.8   The figures listed in Table 16.5, released by the Federal Trade Commission, show the milligrams of tar and nicotine found in 10 brands of cigarettes. Calculate the rank correlation coefficient to measure the degree of relationship between tar and nicotine content in cigarettes.

SOLUTION

Let X and Y represent the tar and nicotine contents, respectively. First we assign ranks to each set of measurements, with the rank of 1 assigned to the lowest number in each set, the rank of 2 to the second lowest number in each set, and so forth, until the rank of 10 is assigned to the largest number. Table 16.6 shows the individual rankings of the measurements and the differences in ranks for the 10 pairs of observations.

Substituting into the formula for r_S, we find that

$$r_S = 1 - \frac{(6)(5.5)}{(10)(100 - 1)} = 0.97,$$

indicating a high positive correlation between the amount of tar and nicotine found in cigarettes.

Some advantages in using r_S rather than r do exist. For instance, we no longer assume the underlying relationship between X and Y to be linear and therefore, when the data possess a distinct curvilinear relationship, the rank correlation coefficient will likely be more reliable than the conventional measure. A second advantage in using the rank correlation coefficient is the fact that no assumptions of normality are made concerning the distributions of X and Y. Perhaps the greatest advantage occurs when one is unable to make meaningful numerical measurements but nevertheless can establish rankings. Such is the case, for example, when different judges rank a group of individuals according to some attribute. The rank correlation coefficient can be used in this situation as a measure of the consistency of the two judges.

To test the hypothesis that $\rho = 0$ by using a rank correlation coefficient, one needs to consider the sampling distribution of the r_S-values under the assumption of

TABLE 16.5   Tar and Nicotine Contents

Cigarette Brand	Tar Content	Nicotine Content
Viceroy	14	0.9
Marlboro	17	1.1
Chesterfield	28	1.6
Kool	17	1.3
Kent	16	1.0
Raleigh	13	0.8
Old Gold	24	1.5
Philip Morris	25	1.4
Oasis	18	1.2
Players	31	2.0

TABLE 16.6 Rankings for Tar and Nicotine Contents

Cigarette Brand	x_i	y_i	d_i
Viceroy	2	2	0
Marlboro	4.5	4	0.5
Chesterfield	9	9	0
Kool	4.5	6	−1.5
Kent	3	3	0
Raleigh	1	1	0
Old Gold	7	8	−1
Philip Morris	8	7	1
Oasis	6	5	1
Players	10	10	0

no correlation. Critical values for $\alpha = 0.05, 0.025, 0.01$, and 0.05 have been calculated and are given in Table A.21. The setup of this table is similar to the table of critical values for the t-distribution except for the left column, which now gives the number of pairs of observations rather than the degrees of freedom. Since the distribution of the r_S-values is symmetric about zero when $\rho = 0$, the r_S-value that leaves an area of α to the left is equal to the negative of the r_S-value that leaves an area of α to the right. For a two-sided alternative hypothesis, the critical region of size α falls equally in the two tails of the distribution. For a test in which the alternative hypothesis is negative, the critical region is entirely in the left tail of the distribution, and when the alternative is positive, the critical region is placed entirely in the right tail.

EXAMPLE 16.9   Refer to Example 16.8 and test the hypothesis that the correlation between the amount of tar and nicotine found in cigarettes is zero against the alternative that it is greater than zero. Use a 0.01 level of significance.

SOLUTION

1. H_0: $\rho = 0$.
2. H_1: $\rho > 0$.
3. $\alpha = 0.01$.
4. Critical region: $r_S > 0.745$, from Table A.21.
5. Computations: From Example 16.8, $r_S = 0.97$.
6. Decision: Reject H_0 and conclude that there is a significant correlation or relationship between the amount of tar and nicotine found in cigarettes.

Under the assumption of no correlation, it can be shown that the distribution of the r_S-values approaches a normal distribution with a mean of 0 and a standard deviation of $1/\sqrt{n-1}$ as n increases. Consequently, when n exceeds the values given in Table A.21, one could test for a significant correlation by computing

$$z = \frac{r_S - 0}{1/\sqrt{n-1}} = r_S\sqrt{n-1}$$

and comparing with critical values of the standard normal distribution given in Table A.3.

Exercises

1. A random sample of 15 adults living in a small town are selected to estimate the proportion of voters favoring a certain candidate for mayor. Each individual was also asked if he or she was a college graduate. By letting Y and N designate the responses of "yes" and "no" to the education question, the following sequence was obtained:

N N N N Y Y N Y Y Y N Y N N N N

   Use the runs test at the 0.1 level of significance to determine if the sequence supports the contention that the sample was selected at random.

2. A silver-plating process is being used to coat a certain type of serving tray. When the process is in control, the thickness of the silver on the trays will vary randomly following a normal distribution with a mean of 0.02 millimeter and a standard deviation of 0.005 millimeter. Suppose that the next 12 trays examined show the following thicknesses of silver: 0.019, 0.021, 0.020, 0.019, 0.020, 0.018, 0.023, 0.021, 0.024, 0.022, 0.023, 0.022. Use the runs test to determine if the fluctuations in thickness from one tray to another are random. Let $\alpha = 0.05$.

3. Use the runs test to test whether there is a difference in the average operating time for the two calculators of Exercise 3 on page 639.

4. In an industrial production line, items are inspected periodically for defectives. The following is a sequence of defective items, D, and nondefective items, N, produced by this production line:

D D N N N D N N D D N N N N

N D D D N N D N N N N D N D

   Use the large-sample theory for the runs test, with a significance level of 0.05, to determine whether the defectives are occurring at random or not.

5. Assuming that the measurements of Exercise 2 on page 67 were recorded in successive rows from left to right as they were collected, use the runs test, with $\alpha = 0.05$, to test the hypothesis that the data represent a random sequence.

6. How large a sample is required to be 95% confident that at least 85% of the distribution of measurements is included between the sample extremes?

7. What is the probability that the range of a random sample of size 24 includes at least 90% of the population?

8. How large a sample is required to be 99% confident that at least 80% of the population will be less than the largest observation in the sample?

9. What is the probability that at least 95% of a population will exceed the smallest value in a random sample of size $n = 135$?

10. The following table gives the recorded grades for 10 students on a midterm test and the final examination in a calculus course:

Student	Midterm Test	Final Examination
L.S.A.	84	73
W.P.B.	98	63
R.W.K.	91	87
J.R.L.	72	66
J.K.L.	86	78
D.L.P.	93	78
B.L.P.	80	91
D.W.M.	0	0
M.N.M.	92	88
R.H.S.	87	77

   (a) Calculate the rank correlation coefficient.
   (b) Test the null hypothesis that $\rho = 0$ against the alternative that $\rho > 0$. Use $\alpha = 0.025$.

11. With reference to the data of Exercise 1 on page 371,
   (a) calculate the rank correlation coefficient;
   (b) test the null hypothesis at the 0.05 level of significance that $\rho = 0$ against the alternative that $\rho \neq 0$. Compare your results with those obtained in Exercise 5 on page 398.

12. Calculate the rank correlation coefficient for the daily rainfall and amount of particulate removed in Exercise 9 on page 365.

13. The following data compare the rankings on November 4, 1981, of the top 15 major college football teams reported in the *Associated Press* poll with the rankings reported in the *United Press* poll:

Team	AP Poll	UPI Poll
Pittsburgh	1	1
Clemson	2	3
Southern Cal.	3	2
Georgia	4	4
Texas	5	5
Penn St.	6	6
Alabama	7	7
North Carolina	8	9
Nebraska	9	8
Michigan	10	10
Miami, Fla.	11	11
Florida St.	12	14
Mississippi St.	13	15
Washington	14	12
Oklahoma	15	13

(a) Calculate the rank correlation coefficient.

(b) Test the null hypothesis that $\rho = 0$ against the alternative that $\rho > 0$. Use a 0.01 level of significance.

14. With reference to the weights and chest sizes of infants in Exercise 4 on page 397,

   (a) calculate the rank correlation coefficient;

   (b) test the hypothesis at the 0.025 level of significance that $\rho = 0$ against the alternative that $\rho > 0$.

15. A consumer panel tested 9 makes of microwave ovens for overall quality. The ranks assigned by the panel and the suggested retail prices were as follows:

Manufacturer	Panel Rating	Suggested Price
A	6	$480
B	9	395
C	2	575
D	8	550
E	5	510
F	1	545
G	7	400
H	4	465
I	3	420

Is there a significant relationship between the quality and the price of a microwave oven? Use a 0.05 level of significance.

16. Two judges at a college homecoming parade ranked 8 floats in the following order:

		Float							
	1	2	3	4	5	6	7	8
Judge A	5	8	4	3	6	2	7	1
Judge B	7	5	4	2	8	1	6	3

(a) Calculate the rank correlation.

(b) Test the null hypothesis that $\rho = 0$ against the alternative that $\rho > 0$. Use $\alpha = 0.05$.

17. In the article called "*Risky Assumptions*" by Paul Slovic, Baruch Fischoff, and Sarah Lichtenstein, published in *Psychology Today* (June 1980), the risk of dying in the United States from 30 activities and technologies are ranked by members of the League of Women Voters and also by experts who are professionally involved in assessing risks.

The rankings are as follows:

Activity or Technology Risk	Voters	Experts
Nuclear power	1	20
Motor vehicles	2	1
Handguns	3	4
Smoking	4	2
Motorcycles	5	6
Alcoholic beverages	6	3
Private aviation	7	12
Police work	8	17
Pesticides	9	8
Surgery	10	5
Fire fighting	11	18
Large construction	12	13
Hunting	13	23
Spray cans	14	26
Mountain climbing	15	29
Bicycles	16	15
Commercial aviation	17	16
Electric power	18	9
Swimming	19	10
Contraceptives	20	11
Skiing	21	30
X-rays	22	7
Football	23	27
Railroads	24	19
Food preservatives	25	14
Food coloring	26	21
Power mowers	27	28
Antibiotics	28	24
Home appliances	29	22
Vaccinations	30	25

(a) Calculate the rank correlation coefficient.

(b) Test the null hypothesis of zero correlation between the rankings of the League of Women Voters and the experts against the alternative that the correlation is not zero. Use a 0.05 level of significance.

Review Exercises

1. The following data represent the number of hours of sleep obtained by two college students for 20 nights that precede school days:

Night	Student A	Student B
1	7.3	6.5
2	8.2	8.2
3	8.5	7.4
4	9.0	8.1
5	9.3	7.6
6	7.6	9.0
7	8.1	6.8
8	7.7	7.4
9	6.9	8.2
10	9.1	7.9
11	7.9	7.3
12	6.8	8.0
13	8.1	6.9
14	7.9	8.2
15	7.8	7.6
16	8.2	9.0
17	7.5	6.6
18	8.4	7.2
19	8.3	8.3
20	8.8	7.9

Use the signed-rank test at the 0.05 level of significance to determine whether these two students, on the average, get the same number of hours of sleep on nights preceding school days.

2. In Exercise 5 on page 530, use the Kruskal–Wallis test, at the 0.05 level of significance, to determine if the chemical analyses performed by the four laboratories give, on the average, the same results.

17

Statistical Quality Control

17.1 Introduction

The notion of using sampling and statistical analysis techniques in a production setting had its beginning in the 1920s. The objective of this highly successful concept is the systematic reduction of variability and the accompanying isolation of sources of difficulties *during production*. In 1924, Walter A. Shewhart of the Bell Telephone Laboratories developed the concept of a control chart. However, it was not until World War II that the use of control charts became widespread. This was due to the importance of maintaining quality in production processes during that period. In the 1950s and 1960s, the development of quality control and the general area of quality assurance grew rapidly, particularly with the emergence of the space program in the United States. There has been widespread and successful use of quality control in Japan thanks to the efforts of W. Edwards Deming, who served as a consultant in Japan following World War II. Quality control has been an important ingredient in the development of Japan's industry and economy.

Quality control is receiving increasing attention as a management tool in which important characteristics of a product are observed, assessed, and compared with some type of standard. The various procedures in quality control involve considerable use of sampling procedures and statistical principles that have been presented in previous chapters. The primary users of quality control are, of course, industrial corporations. It has become clear that an effective quality control program enhances the quality of the product being produced and increases profits. This is particularly true in this day and age, in which products are being produced in such high volume. Before the recent movement toward quality control methods, quality would often suffer because of lack of efficiency, which, of course, increases cost.

The Control Chart

The purpose of a control chart is to determine if the performance of a process is maintaining an acceptable level of quality. It is expected, of course, that any process will experience natural variability, that is, variability due to essentially unimportant and uncontrollable sources of variation. On the other hand, a process may experience more serious types of variability in key performance measures. These sources of variability may arise from one of several types of nonrandom "assignable causes," such as operator errors or improperly adjusted dials on a machine. A process operating in this state is called *out of control*. A process experiencing only chance variation is said to be *in statistical control*. Of course, a successful production process may operate in an in-control state for a long period. It is presumed that during this period, the process is producing an acceptable product. However, there may be either a gradual or sudden "shift" that requires detection.

A control chart is intended as a device to detect the nonrandom or out-of-control state of a process. Typically, the control chart takes the form indicated in Figure 17.1. It is important that the shift be detected quickly so that the problem can be corrected. Obviously, if detection is slow, many defective or nonconforming items are produced, resulting in considerable waste and increased cost.

Some type of quality characteristic must be under consideration and units of the process are being sampled over time, say. For example, the characteristic may be the circumference of an engine bearing. The centerline represents the average value of the characteristic when the process is in control. The points depicted in the figure may represent results of, say, sample averages of this characteristic, with the samples taken over time. The upper control limit and the lower control limit are chosen in such a way that one would expect all sample points to be covered by these boundaries if the process is in control. As a result, the general complexion of the plotted points over time determines whether or not the process is concluded to be in control. The "in control" evidence is produced by a random pattern of points, with all plotted values being inside the control limits. When a point falls outside the control limits, this is taken to be evidence of a process that is out of control, and a search for the assignable cause is suggested. In addition, a nonrandom pattern of points may be considered suspicious and certainly an indication that an investigation for the appropriate corrective action is needed.

17.2 Nature of the Control Limits _____

The fundamental ideas on which control charts are based are similar in structure to hypothesis testing. The control limits are established to control the probability of

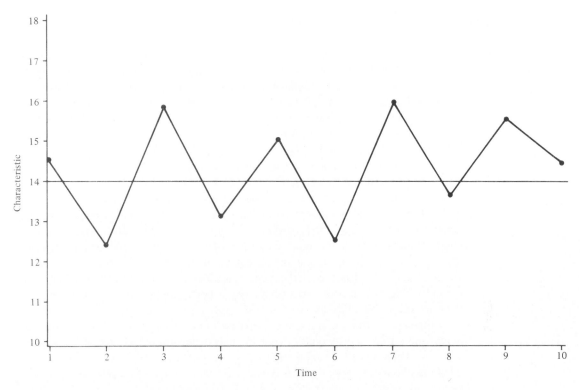

FIGURE 17.1 Typical control chart.

making the error of concluding that the process is out of control when in fact it is not. This corresponds to the probability of making a type I error if we were testing the null hypothesis that the process is in control. On the other hand, we must be attentive to the error of the second kind, namely not finding the process out of control when in fact it is (type II error). Thus the choice of control limits is similar to the choice of a critical region.

As in the case of hypothesis testing, the sample size at each point is important. The consideration of sample size depends to a large extent on the sensitivity or *power* of detection of the out-of-control state. In this application, the notion of power is very similar to that of the hypothesis-testing situation. Clearly, the larger the sample at each time period, the quicker the detection of an out-of-control process. In a sense, the control limits actually define what the user considers as being *in control*. In other words, the latitude given by the control limits obviously must depend in some sense on the process variability. As a result, the computation of the control limits will quite naturally depend on data taken from the process results. Thus any quality control must have its beginning with computation from a preliminary sample or set of samples which will establish both the centerline and the quality control limits.

17.3 Purposes of the Control Chart

One obvious purpose of the control chart is mere surveillance of the process, that is, to determine if changes need to be made. In addition, the constant systematic gathering of data often allows management to assess process capability. Clearly, if a single performance characteristic is important, continual sampling and estimation of the mean and standard deviation of the performance characteristic offer updating of what the process can do in terms of mean performance and random variation. This is valuable even if the process stays in control for long periods. The systematic and formal structure of the control chart can often prevent overreaction to changes that represent only random fluctuations. Obviously, in many situations, changes brought about by overreaction can create serious problems that are difficult to solve.

The types of quality characteristics on which control charts are constructed fall generally into *two* categories, *variables* and *attributes*. As a result, types of control charts often take the same classifications. In the case of the variables type of chart, the characteristic is usually a measurement on a continuum, such as diameter, weight, and so on. For the attribute chart, the characteristic reflects whether or not the individual product *conforms* (defective or not). Applications for these two distinct situations are obvious.

In the case of the variables chart, control must be exerted on both central tendency and variability. A quality control analyst must be concerned about whether there has been a shift in values of performance characteristic *on the average*. In addition, there will always be a concern about whether some change in process conditions results in a decrease in precision (i.e., an increase in variability). Separate control charts are essential in dealing with these two concepts. Central tendency is controlled by the $\overline{X}$-chart, where means of relatively small samples are plotted on the control chart. Variability around the mean is controlled by the *range* in the sample, or the sample *standard deviation*. In the case of attribute sampling, the *proportion defective* from a sample is often the quantity plotted on the chart. In the following section we discuss the development of control charts for the *variables* type of performance characteristic.

17.4 Control Charts for Variables

It is relatively easy to understand the rudiments of the $\overline{X}$-chart for variables through an example. Suppose that quality control charts are to be used on a process for manufacturing a certain engine part. Suppose the process mean is $\mu = 50$ mm and the standard deviation is $\sigma = 0.01$ mm. Suppose that groups of 5 are sampled every hour and the values of the *sample mean* $\overline{X}$ are recorded and plotted as in Figure 17.2. The limits for the $\overline{X}$-charts are based on the standard deviation of the random variable $\overline{X}$. We know from material in Chapter 8 that for the average of independent observations in a sample of size n,

$$\sigma_{\overline{X}} = \frac{\sigma}{\sqrt{n}},$$

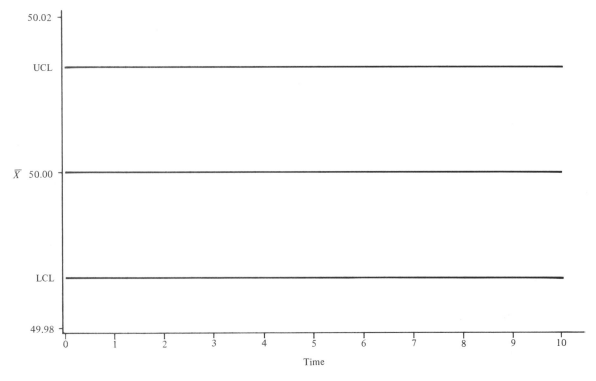

FIGURE 17.2 The 3σ control limits for the engine part example.

where σ is the standard deviation of an individual observation. The control limits are designed to result in a small probability that a given value of $\overline{X}$ is outside the limits given that, indeed, the process is in control (i.e., $\mu = 50$). If we invoke the central limit theorem, we have that under the condition that the process is in control,

$$\overline{X} \sim N\left(50, \frac{0.01}{\sqrt{5}}\right).$$

As a result, $100(1 - \alpha)\%$ of the $\overline{X}$-values fall inside the limits when the process is in control if we use the limits

$$\text{LCL} = \mu - z_{\alpha/2}\frac{\sigma}{\sqrt{n}}$$

$$= 50 - z_{\alpha/2}(0.0045)$$

$$\text{UCL} = \mu + z_{\alpha/2}\frac{\sigma}{\sqrt{n}}$$

$$= 50 + z_{\alpha/2}(0.0045).$$

Often the $\overline{X}$-charts are based on limits that are referred to as "three-sigma" limits, referring of course to $z_{\alpha/2} = 3$ and limits that become

$$\mu \pm 3\frac{\sigma}{\sqrt{n}}.$$

In our illustration the upper and lower limits become

$$UCL = 50 + 3(0.0045)$$

$$= 50.0135$$

$$LCL = 50 - 3(0.0045)$$

$$= 49.9865.$$

Thus, if we view the structure of the 3σ limits from the point of view of hypothesis testing, for a given sample point, the probability is 0.0026 that the $\overline{X}$-value falls outside control limits, given that the process is in control. This is the probability of the analyst *erroneously* determining that the process is out of control (see Table A.3).

The example above not only illustrates the $\overline{X}$-chart for variables, but also should provide the reader with an insight into the nature of control charts in general. The centerline generally reflects the ideal value of an important parameter. Control limits are established from knowledge of the sampling properties of the statistic that estimates the parameter in question. They very often involve a multiple of the standard deviation of the statistic. It has become general practice to use 3σ limits. In the case of the $\overline{X}$-chart provided here, the central limit theorem provides the user with a good approximation of the probability of falsely ruling that the process is out of control. In general, though, the user may not be able to rely on the normality of the statistic on the centerline. As a result, the exact probability of "type I error" may not be known. Despite this, it has become fairly standard to use the $k\sigma$ limits. While use of the 3σ limits is widespread, at times the user may wish to deviate from this approach. A smaller multiple of σ may be appropriate when it is important to quickly detect an out-of-control situation. Because of economic considerations, it may prove costly to allow a process to continue to run out of control for even short periods, while the cost of the search and correction of assignable causes may be relatively small. Clearly, in this case, control limits that are tighter than 3σ limits are appropriate.

Rational Subgroups

The sample values to be used in a quality control effort are divided into subgroups with a *sample* representing a subgroup. As we indicated earlier, time order of production is certainly a natural basis for selection of the subgroups. One may view the quality control effort very simply as (1) sampling, (2) detection of an out-of-control state, and (3) a search for assignable causes that may be occurring over time. The selection of the basis for these sample groups would appear to be quite straightforward. The choice of these subgroups of sampling information can have an important effect on the success of the quality control program. These subgroups are often called *rational subgroups*. Generally, if the analyst is interested in detecting a *shift in location*, it is felt that the subgroups should be chosen so that within-subgroup variability is small and that assignable causes, if they are present, can have the greatest chance of being detected. Thus we want to choose the subgroups in such a way as to maximize the between-subgroup variability. Choosing units in a subgroup that are produced close together in time, for example, is a reasonable approach. On the other hand, control charts are often used to control variability, in which case the performance statistic is *variability within the sample*. Thus it is more important to choose the rational subgroups to maximize the within-sample variability. In this case, the observations in the subgroups should behave more like a random sample and this variability within samples needs to be a depiction of the variability of the process.

It is important to note that control charts on variability should be established before the development of charts on center of location (say, $\overline{X}$-charts). Any control chart on center of location will certainly depend on variability. For example, we have seen an illustration of the central tendency chart and it depends on σ. In the sections that follow, an estimate of σ from the data will be discussed.

$\overline{X}$-Chart with Estimated Parameters

In the foregoing we have illustrated notions of the $\overline{X}$-chart that makes use of the central limit theorem and employs *known* values of the process mean and standard deviation. As we indicated earlier, the control limits

$$LCL = \mu - z_{\alpha/2}\frac{\sigma}{\sqrt{n}}$$

$$UCL = \mu + z_{\alpha/2}\frac{\sigma}{\sqrt{n}}$$

are used and an $\overline{X}$-value falling outside these limits is viewed as evidence that the mean μ has changed and thus the process may be out of control.

In many practical situations, it is unreasonable to assume that we know μ and σ. As a result, estimates must be supplied from data taken when the process is in control. Typically, the estimates are determined during a period in which *background information* or *start-up information* is gathered. A basis for rational subgroups is chosen and data are gathered with samples of size n in each subgroup. The sample sizes are usually small, say, 4, 5, or 6, and k samples are taken, with k being at least 20. During this period in which it is assumed that the process is in control, the user establishes estimates of μ and σ on which the control chart is based. The important information gathered during this period includes the sample means in the subgroup, the overall mean, and the sample range in each subgroup. In the following paragraphs we outline how this information is used to develop the control chart.

A portion of the sample information from these k samples takes the form $\overline{X}_1$, $\overline{X}_2, \ldots, \overline{X}_k$, where the random variable $\overline{X}_i$ is the average of the values in the ith sample. Obviously, the overall average is the random variable given by

$$\overline{\overline{X}} = \sum_{i=1}^{k} \frac{\overline{X}_i}{k}.$$

This is the appropriate estimator of the process mean and, as a result, is the center-line in the $\bar{X}$ control chart. In quality control applications it is often convenient to estimate σ from the information related to the *ranges* in the samples rather than sample standard deviations. Let us define for the ith sample,

$$R_i = X_{\max, i} - X_{\min, i}$$

as the range for the data in the ith sample. Here $X_{\max, i}$ and $X_{\min, i}$ are the largest and smallest observation, respectively, in the sample. The appropriate estimate of σ is a function of the average range

$$\bar{R} = \sum_{i=1}^{k} \frac{R_i}{k}.$$

An estimate of σ, say $\hat{\sigma}$, is obtained by

$$\hat{\sigma} = \frac{\bar{R}}{d_2},$$

where d_2 is a constant depending on the sample size. Values of d_2 are given in Table A.23.

Use of the range in producing an estimate of σ has roots in quality-control-type applications, particularly since the range was so easy to compute in the era prior to the period in which time of computation is considered no difficulty. The assumption of normality of the individual observations is implicit in the $\bar{X}$-chart. Of course, the existence of the central limit theorem is certainly helpful in this regard. Under the assumption of normality, we make use of a random variable called the relative range, given by

$$W = \frac{R}{\sigma}.$$

It turns out that the moments of W are simple functions of the sample size n (see the reference to Montgomery in the Bibliography). The expected value of W is often referred to as d_2. Thus by taking the expected value of W above,

$$\frac{E(R)}{\sigma} = d_2.$$

As a result, the rationale for the estimate $\hat{\sigma} = \bar{R}/d_2$ is easily seen. It is well known that the range method produces an efficient estimator of σ in relatively small samples. This makes the estimator particularly attractive in quality control applications since the sample sizes in the subgroups are generally small. Using the range method for estimation of σ results in control charts with the following parameters:

$$\text{UCL} = \bar{\bar{X}} + \frac{3\bar{R}}{d_2\sqrt{n}}$$

$$\text{centerline} = \bar{\bar{X}}$$

$$\text{LCL} = \bar{\bar{X}} - \frac{3\bar{R}}{d_2\sqrt{n}}.$$

Defining the quantity

$$A_2 = \frac{3}{d_2 \sqrt{n}},$$

we have that

$$\text{UCL} = \overline{\overline{X}} + A_2 \overline{R}$$
$$\text{LCL} = \overline{\overline{X}} - A_2 \overline{R}.$$

To simplify the structure, the user of $\overline{X}$-charts often finds values of A_2 tabulated. Tabulations of values of A_2 are given for various sample sizes in Table A.23.

R-Charts to Control Variation

Up to this point all illustrations and details have dealt with the quality control analysts' attempt at detection of out-of-control conditions produced by a *shift in the mean*. The control limits are based on the distribution of the random variable $\overline{X}$ and depend on the assumption of normality on the individual observations. It is important for control to be applied to variability as well as center of location. In fact, many experts feel as if control of variability of the performance characteristic is more important and should be established before center of location should be considered. Process variability can be controlled through the use of *plots of the sample range*. A plot over time of the sample ranges is called an *R-chart*. The same general structure can be used as in the case of the $\overline{X}$-chart, with $\overline{R}$ *being the centerline* and the control limits depending on an estimate of the standard deviation of the random variable R. Thus, as in the case of the $\overline{X}$-chart, 3σ limits will be established where "3σ" implies $3\sigma_R$. The quantity σ_R must be estimated from the data just as $\sigma_{\overline{X}}$ is estimated.

The estimate of σ_R, the standard deviation, is also based on the distribution of the relative range

$$W = \frac{R}{\sigma}.$$

The standard deviation of W is a known function of the sample size and is generally denoted by d_3. As a result,

$$\sigma_R = \sigma d_3.$$

We can now replace σ by $\hat{\sigma} = \overline{R}/d_2$, and thus the estimator of σ_R is given by

$$\hat{\sigma}_R = \frac{\overline{R} d_3}{d_2}.$$

Thus the quantities that define the R-chart are given by

$$\text{UCL} = \overline{R} D_4$$
$$\text{centerline} = \overline{R}$$
$$\text{LCL} = \overline{R} D_3,$$

where the constants D_4 and D_3 (depending only on n) are given by

$$D_4 = 1 + 3\frac{d_3}{d_2}$$

$$D_3 = 1 - 3\frac{d_3}{d_2}.$$

The constants D_4 and D_3 are tabulated in Table A.23.

$\overline{X}$- and R-Charts for Variables

A process manufacturing missile component parts is being controlled, with the performance characteristic being the tensile strength in pounds per square inch. Samples of size 5 each are taken every hour and 25 samples were reported. The data are shown in Table 17.1.

TABLE 17.1   Sample Information on Tensile Strength Data

Sample Number	Observations					$\overline{X}_i$	R_i
1	1515	1518	1512	1498	1511	1510.8	20
2	1504	1511	1507	1499	1502	1504.6	12
3	1517	1513	1504	1521	1520	1515.0	17
4	1497	1503	1510	1508	1502	1504.0	13
5	1507	1502	1497	1509	1512	1505.4	15
6	1519	1522	1523	1517	1511	1518.4	12
7	1498	1497	1507	1511	1508	1504.2	14
8	1511	1518	1507	1503	1509	1509.6	15
9	1506	1503	1498	1508	1506	1504.2	10
10	1503	1506	1511	1501	1500	1504.2	11
11	1499	1503	1507	1503	1501	1502.6	8
12	1507	1503	1502	1500	1501	1502.6	7
13	1500	1506	1501	1498	1507	1502.4	9
14	1501	1509	1503	1508	1503	1504.8	8
15	1507	1508	1502	1509	1501	1505.4	8
16	1511	1509	1503	1510	1507	1508.0	8
17	1508	1511	1513	1509	1506	1509.4	7
18	1508	1509	1512	1515	1519	1512.6	11
19	1520	1517	1519	1522	1516	1518.8	6
20	1506	1511	1517	1516	1508	1511.6	11
21	1500	1498	1503	1504	1508	1502.6	10
22	1511	1514	1509	1508	1506	1509.6	8
23	1505	1508	1500	1509	1503	1505.0	9
24	1501	1498	1505	1502	1505	1502.2	7
25	1509	1511	1507	1500	1499	1505.2	12

As we indicated earlier, it is important initially to establish "in control" conditions on variability. The calculated centerline for the R-chart is given by

$$\overline{R} = \sum_{i=1}^{25} \frac{R_i}{25}$$

$$= 10.72.$$

We find from Table A.23 that for $n = 5$, $D_3 = 0$ and $D_4 = 2.115$. As a result, the control limits for the R-chart are given by

$$\text{LCL} = \overline{R}D_3 = (10.72)(0) = 0$$

$$\text{UCL} = \overline{R}D_4 = (10.72)(2.115) = 22.6728.$$

The R-chart is shown in Figure 17.3. None of the plotted ranges fall outside the control limits. As a result, there is no indication of an out-of-control situation.

The $\overline{X}$-chart can now be constructed for the tensile strength readings. The centerline is given by

$$\overline{\overline{X}} = \frac{\sum_{i=1}^{25} \overline{X}_i}{25}$$

$$= 1507.328.$$

For samples of size 5, we find $A_2 = 0.577$ from Table A.23. Thus the control limits are

$$\text{UCL} = \overline{\overline{X}} + A_2\overline{R}$$

$$= 1507.328 + (0.577)(10.72)$$

$$= 1513.5134$$

$$\text{LCL} = \overline{\overline{X}} - A_2\overline{R}$$

$$= 1507.328 - (0.577)(10.72)$$

$$= 1501.1426.$$

The $\overline{X}$-chart is shown in Figure 17.4. As the reader can observe, three values fall outside control limits. As a result, the control limits for $\overline{X}$ should not be used for line quality control.

Further Comments About the Control Charts for Variables

A process may appear to be in control, and in fact, may stay in control for a long period. Does this necessarily mean that the process is operating successfully? A process that is operating *in control* is merely one in which the process mean and variability are stable. Apparently, no serious changes have occurred. "In control"

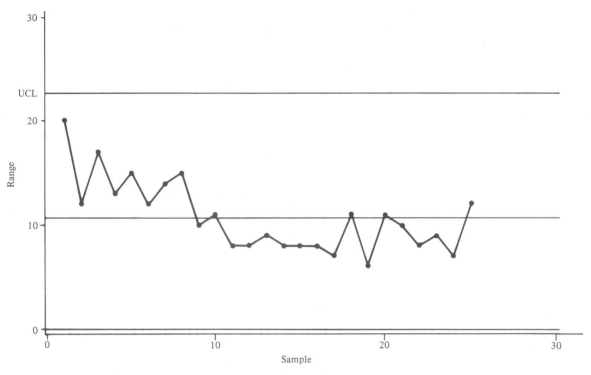

FIGURE 17.3   *R*-chart for the tensile strength example.

implies that the process remains consistent with *natural variability*. The quality control charts may be viewed as a method in which the inherent natural variability governs the width of the control limits. There is no implication, however, to what extent an in-control process satisfies predetermined *specifications* required of the process. Specifications are limits that are established by the consumer. If the current natural variability of the process is larger than that dictated by the specification, the process will not produce items that meet specifications with high frequency, even though the process is stable and in control.

We have alluded to the normality assumption on the individual observations in a variables control chart. For the $\overline{X}$-chart, if the individual observations are normal, the statistic $\overline{X}$ is normal. As a result, the quality control analyst has control over the probability of type I error in this case. If the individual X's are not normal, $\overline{X}$ is approximately normal and thus there is approximate control over the probability of type I error for the case in which σ is known. However, the use of the range method for estimating the standard deviation also depends on the normality assumption. Studies regarding the robustness of the $\overline{X}$-chart to departures from normality indicate that for samples of size $k \geq 4$, the $\overline{X}$ chart results in an α-risk close to that advertised (see the work by Montgomery, Schilling, and Nelson in the Bibliography). We earlier indicated that the $\pm k\sigma_R$ approach to the R-chart is a matter of convenience and tradition. Even if the distribution of individual observations is

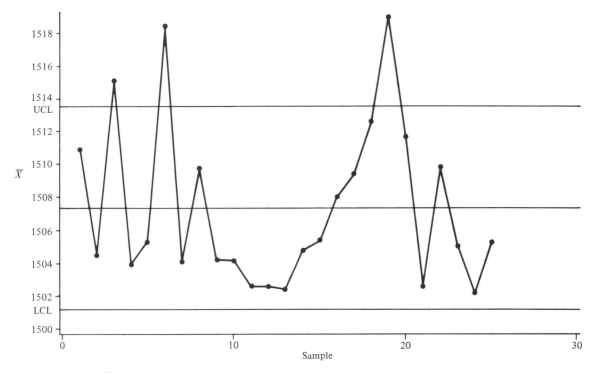

FIGURE 17.4 $\bar{X}$-chart for the tensile strength example.

normal, the distribution of R is not normal. In fact, the distribution of R is not even symmetric. The symmetric control limits of $\pm k\sigma_R$ only give an approximation to the α-risk, and in some cases the approximation is not particularly good.

Choice of Sample Size (Operating Characteristic Function) in the Case of the $\bar{X}$-Chart

Scientists and engineers dealing in quality control often refer to factors that affect the *design of the control chart*. Components that determine the design of the chart include the sample size taken in each subgroup, the width of the control limits, and the frequency of sampling. All of these factors depend to a large extent on economic and practical considerations. Frequency of sampling obviously depends on the cost of sampling and the cost incurred if the process continues out of control for a long period. These same factors affect the width of the "in-control" region. The cost that is associated with investigation and search for assignable causes has an impact on the width of the region and on frequency of sampling. A considerable amount of attention has been devoted to optimal design of control charts and extensive details will not be given here. The reader should refer to the work by Montgomery et al. cited in the Bibliography for an excellent historical account of much of this research.

Choice of sample size and frequency of sampling involve balancing available resources to these two efforts. In many cases, the analyst may need to make changes in the strategy until the proper balance is achieved. The analyst should always be aware that if the cost of producing nonconforming items is great, a high sampling frequency with relatively small sample size is a proper strategy.

Many factors must be taken into consideration in the choice of a sample size. In the illustration and discussion we have emphasized the use of $n = 4, 5$, or 6. These values are considered relatively small for general problems in statistical inference but perhaps proper sample sizes for quality control. One justification, of course, is that the quality control is a continuing process and the results produced by one sample or set of units will be followed by results from many more. Thus the "effective" sample size of the entire quality control effort is many times larger than that used in a subgroup. It is generally considered to be more effective to *sample frequently* with a small sample size.

The analyst can make use of the notion of the *power* of a test to gain some insight into the effectiveness of the sample size chosen. This is particularly important since small sample sizes are usually used in each subgroup. Refer to Chapters 10 and 13 for a discussion of the power of formal tests on means and the analysis of variance. Although formal tests of hypotheses are not actually being conducted in quality control, one can treat the sampling information as if the strategy at each subgroup is to test a hypothesis, either on the population mean μ or the standard deviation σ. Of interest is the *probability of detection* of an out-of-control condition for a given sample, and perhaps more important, the expected number of runs required for detection. The probability of detection of a specified out-of-control condition corresponds to the power of a test. It is not our intention to show development of the power for all of the types of control charts presented here, but rather, to show the development for the $\bar{X}$-chart and present power results for the R-chart.

Consider the $\bar{X}$-chart for σ known. Suppose that the in-control state has $\mu = \mu_0$. A study of the role of the subgroup sample size is tantamount to investigating the β-risk, that is, the probability that an $\bar{X}$-value remains inside the control limits, given that, indeed, a shift in the mean has occurred. Suppose that the form the shift takes is given by

$$\mu = \mu_0 + r\sigma.$$

Again, making use of the normality of $\bar{X}$, we have

$$\beta = P\{\text{LCL} \le \bar{X} \le \text{UCL} \mid \mu = \mu_0 + r\sigma\}.$$

For the case of $k\sigma$ limits, $\text{LCL} = \mu_0 = k\sigma/\sqrt{n}$ and $\text{UCL} = \mu_0 + k\sigma/\sqrt{n}$. As a result, if we denote by Z the standard normal random variable

$$\beta = P\left\{Z < \left[\frac{\mu_0 + k\sigma/\sqrt{n} - \mu}{\sigma/\sqrt{n}}\right]\right\} - P\left\{Z < \left[\frac{\mu_0 - k\sigma/\sqrt{n} - \mu}{\sigma/\sqrt{n}}\right]\right\}$$

$$= P\left\{Z < \left[\frac{\mu_0 + k\sigma/(\mu_0 + r\sigma)}{\sigma/\sqrt{n}}\right]\right\} - P\left\{Z < \left[\frac{\mu_0 - k\sigma/\sqrt{n} - (\mu_0 + r\sigma)}{\sigma/\sqrt{n}}\right]\right\}$$

$$= P\{Z < (k - r\sqrt{n})\} - P\{Z < (-k - r\sqrt{n})\}.$$

Notice the role of n, r, and k in the expression for the β-risk. The probability of not detecting a specific shift clearly increases with an increase in k, as expected. β decreases with an increase in r, the magnitude of the shift, and decreases with an increase in the sample size n.

It should be emphasized that the expression above results in the β-risk (probability of type II error) for the case of a *single sample*. For example, suppose that in the case of a sample of size 4, a shift of σ occurs in the mean. The probability of detecting the shift (power) *in the first sample following the shift* is given (assume 3σ limits) by

$$1 - \beta = 1 - [P(Z < 1) - P(Z < -5)]$$

$$= 0.159.$$

On the other hand, the probability of detecting a shift of 2σ is given by

$$1 - \beta = 1 - [P(Z < -1) - P(Z < -7)]$$

$$= 0.8413.$$

The results above illustrate a fairly modest probability of detecting a shift of magnitude σ and a fairly high probability of detecting a shift of magnitude 2σ. The complete picture of how, say, 3σ control limits perform for the $\overline{X}$-chart described here is depicted in Figure 17.5. Rather than plot power, a plot is given of β against

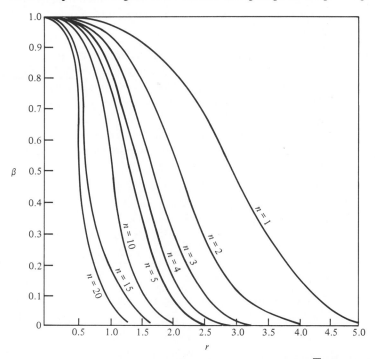

FIGURE 17.5 Operating characteristic curves for the $\overline{X}$-chart with 3σ limits. Here β is the type II probability error on the first sample after a shift in the mean of $r\sigma$.

r, where the shift in the mean is of magnitude $r\sigma$. Of course, the sample sizes of $n = 4, 5, 6$ result in a small probability of detecting a shift of 1.0σ or even 1.5σ on the first sample after the shift.

But if sampling is done frequently, the probability may not be as important as the average or expected number of runs required before detection of the shift. Quick detection is important and is certainly possible even though the probability of detection on the first sample is not high. It turns out that $\bar{X}$-charts with these small samples will result in relatively rapid detection. If β is the probability of not detecting a shift on the first sample following the shift, then the probability of detecting the shift on the sth sample after the shift is given by (assuming independent samples)

$$P_s = (1 - \beta)\beta^{s-1}.$$

The reader should recognize this as an application of the geometric distribution. The average or expected value of the number of samples required for detection is given by

$$\sum_{s=1}^{\infty} s\beta^{s-1}(1 - \beta) = \frac{1}{1 - \beta}.$$

Thus the expected number of samples required to detect the shift in the mean is the *reciprocal of the power* (i.e., the probability of detection on the first sample following the shift).

EXAMPLE 17.1   In a certain quality control effort it is important for the quality control analyst to be quickly able to detect shifts in the mean of $\pm \sigma$ while using a 3σ control chart with a sample size $n = 4$. The expected number of samples that are required following the shift for the detection of the out-of-control state can be an aid in the assessment of the quality control procedure.

From Figure 17.5, for $n = 4$ and $r = 1$, it can be seen that $\beta \approx 0.82$. If we allow s to denote the number of samples required to detect the shift, the mean of s is given by

$$E(s) = \frac{1}{1 - \beta}$$

$$= \frac{1}{0.18}$$

$$= 5.5.$$

Thus, on the average, six subgroups are required before detection of a shift of $\pm \sigma$.

Choice of Sample Size for the R-Chart

The OC curve for the R-chart is shown in Figure 17.6. Since the R-chart is used for control of the process standard deviation, the β-risk is plotted as a function of the

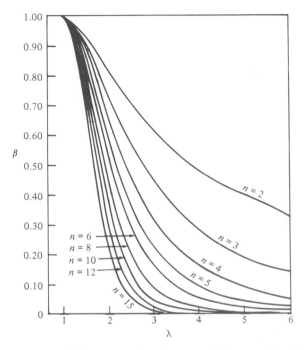

FIGURE 17.6 Operating characteristic curve for the R-charts with 3σ limits.

in-control standard deviation, σ_0, and the standard deviation after the process goes out of control. The latter standard deviation will be denoted by σ_1. Let

$$\lambda = \frac{\sigma_1}{\sigma_0}.$$

For various sample sizes, β is plotted against λ.

$\overline{X}$- and S-Charts for Variables

It is natural for the student of statistics to anticipate use of the sample variance in the $\overline{X}$-chart and in a chart to control variability. The range is efficient as an estimator for σ, but this efficiency decreases as the sample size gets larger. For n as large as 10, the familiar statistic

$$S = \sqrt{\sum_{i=1}^{n} \frac{(X_i - \overline{X})^2}{n - 1}}$$

should be used in the control chart for both the mean and variability. The reader should recall from Chapter 9 that S^2 is an unbiased estimator for σ^2 but that S is not unbiased for σ. It has become customary to correct S for bias in control chart applications. We know that, in general,

$$E(s) \neq \sigma.$$

In the case in which the X_i are independent, normally distributed with mean μ and variance σ^2,

$$E(S) = c_4\sigma,$$

where

$$c_4 = \left(\frac{2}{n-1}\right)^{1/2} \frac{\Gamma(n/2)}{\Gamma[(n-1)/2]}$$

and $\Gamma(\cdot)$ refers to the gamma function (see Chapter 6). For example, for $n = 50$, $c_4 = 3/8\sqrt{2\pi}$. In addition, the variance of the estimator S is given by

$$\text{Var}(S) = \sigma^2(1 - c_4^2).$$

We have established the properties of S that will allow us to write control limits for both $\bar{X}$ and S. To build a proper structure, we begin by assuming that σ is known. Later we discuss estimating σ from a preliminary set of samples.

If the statistic S is plotted, the obvious control chart parameters are given by

$$\text{UCL} = c_4\sigma + 3\sigma\sqrt{1 - c_4^2}$$

$$\text{centerline} = c_4\sigma$$

$$\text{LCL} = c_4\sigma - 3\sigma\sqrt{1 - c_4^2}.$$

As usual, the control limits are defined more succinctly through use of tabulated constants. Let

$$B_5 = c_4 - 3\sqrt{1 - c_4^2}$$

$$B_6 = c_4 + 3\sqrt{1 - c_4^2},$$

and thus we have

$$\text{UCL} = B_6\sigma$$

$$\text{centerline} = c_4\sigma$$

$$\text{LCL} = B_5\sigma$$

The values of B_5 and B_6 for various sample sizes are tabulated in Table A.23.

Now, of course, the control limits above serve as a basis for the development of the quality control parameters for the situation that is most often seen in practice, namely that in which σ is unknown. We must once again assume that a set of *base samples* or preliminary samples are taken to produce an estimate of σ during what is assumed to be an "in control" period. Sample standard deviations $S_1, S_2, \ldots, S_m$ are obtained from samples that are each of size n. An unbiased estimator of the type

$$\frac{\bar{S}}{c_4} = \left(\frac{1}{m}\sum_{i=1}^{m} S_i\right)\Big/ c_4$$

is often used for σ. Here, of course, $\bar{S}$, the average value of the sample standard deviation in the preliminary sample, is the logical centerline in the control chart to control variability. The upper and lower control limits are unbiased estimators of the control limits that are appropriate for the case where σ is known. Since

$$E\left(\frac{\overline{S}}{c_4}\right) = \sigma,$$

the statistic $\overline{S}$ is an appropriate centerline (as an unbiased estimator of $c_4\sigma$) and the quantities

$$\overline{S} - 3\frac{\overline{S}}{c_4}\sqrt{1 - c_4^2}$$

and

$$\overline{S} + 3\frac{\overline{S}}{c_4}\sqrt{1 - c_4^2}$$

are the appropriate lower and upper 3σ control limits, respectively. As a result, the centerline and limits for the S-chart to control variability are given by

$$\text{UCL} = B_4\overline{S}$$
$$\text{centerline} = \overline{S}$$
$$\text{LCL} = B_3\overline{S},$$

where

$$B_3 = 1 - \frac{3}{c_4}\sqrt{1 - c_4^2}$$

$$B_4 = 1 + \frac{3}{c_4}\sqrt{1 - c_4^2}.$$

The constants B_3 and B_4 are given in Table A.23.

We can now write the parameters of the corresponding $\overline{X}$-chart involving the use of the sample standard deviation. Let us assume that $\overline{S}$ and $\overline{\overline{X}}$ are available from the base preliminary sample. The centerline remains $\overline{\overline{X}}$ and the 3σ limits are merely of the form $\overline{\overline{X}} \pm 3\hat{\sigma}/\sqrt{n}$, where $\hat{\sigma}$ is an unbiased estimator. We simply supply $\overline{S}/c_4$ as an estimator for σ, and thus we have

$$\text{UCL} = \overline{\overline{X}} + A_3\overline{S}$$
$$\text{centerline} = \overline{\overline{X}}$$
$$\text{LCL} = \overline{\overline{X}} - A_3\overline{S},$$

where

$$A_3 = \frac{3}{c_4\sqrt{n}}.$$

The constant A_3 appears for various sample sizes in Table A.23.

EXAMPLE 17.2   Containers are produced in a process in which the volume of the containers is subject to a quality control. Twenty-five samples of size 10 each were used to establish the quality control parameters. Information from these samples is given in Table 17.2.

From the Appendix, $B_3 = 0$, $B_4 = 2.089$, $A_3 = 1.427$. As a result, the control limits for $\overline{X}$ are given by

$$\overline{\overline{X}} + A_3\overline{S} = 62.3771$$
$$\overline{\overline{X}} - A_3\overline{S} = 62.2740$$

and the control limits for the S-chart are given by

$$\text{UCL} = B_4\overline{S} = 0.0754$$
$$\text{LCL} = B_3\overline{S} = 0.$$

TABLE 17.2   Volume in Cubic Centimeters of Samples of Containers for 25 Samples in a Preliminary Sample

Sample	Observations					$\overline{X}_i$	S_i
1	62.255	62.301	62.289	62.289	62.311	62.269	0.0495
2	62.187	62.225	62.337	62.297	62.307	62.271	0.0622
3	62.421	62.377	62.257	62.295	62.222	62.314	0.0829
4	62.301	62.315	62.293	62.317	62.409	62.327	0.0469
5	62.400	62.375	62.295	62.272	62.372	62.343	0.0558
6	62.372	62.275	62.315	62.372	62.302	62.327	0.0434
7	62.297	62.303	62.337	62.392	62.344	62.335	0.0381
8	62.325	62.362	62.351	62.371	62.397	62.361	0.0264
9	62.327	62.297	62.318	62.342	62.318	62.320	0.0163
10	62.297	62.325	62.303	62.307	62.333	62.313	0.0153
11	62.315	62.366	62.308	62.318	62.319	62.325	0.0232
12	62.297	62.322	62.344	62.342	62.313	62.324	0.0198
13	62.375	62.287	62.362	62.319	62.382	62.345	0.0406
14	62.317	62.321	62.297	62.372	62.319	62.325	0.0279
15	62.299	62.307	62.383	62.341	62.394	62.345	0.0431
16	62.308	62.319	62.344	62.319	62.378	62.334	0.0281
17	62.319	62.357	62.277	62.315	62.295	62.313	0.0300
18	62.333	62.362	62.292	62.327	62.314	62.326	0.0257
19	62.313	62.387	62.315	62.318	62.341	62.335	0.0313
20	62.375	62.321	62.354	62.342	62.375	62.353	0.0230
21	62.399	62.308	62.292	62.372	62.299	62.334	0.0483
22	62.309	62.403	62.318	62.295	62.317	62.328	0.0427
23	62.293	62.293	62.342	62.315	62.349	62.318	0.0264
24	62.388	62.308	62.315	62.392	62.303	62.341	0.0448
25	62.328	62.318	62.317	62.295	62.319	62.314	0.0111

$$\overline{\overline{X}} = 62.3256$$
$$\overline{S} = 0.0361$$

Figures 17.7 and 17.8 show the $\overline{X}$ and S control charts, respectively, for Example 17.2. Sample information for all 25 samples in the preliminary data set is plotted on the charts. Control seems to have been established after the first few samples.

17.5 Control Charts for Attributes

As we indicated earlier in this chapter, many industrial applications of quality control require that the quality characteristic indicate no more than the statement that the item "conforms." In other words, there is no continuous measurement that is crucial to the performance of the item. An obvious illustration of this type of sampling, called *sampling for attributes*, is the performance of a light bulb—which either performs satisfactorily or does not. The item is either *defective* or *not defective*. Manufactured metal pieces may contain deformities. Containers from a production line may leak. In both of these cases a defective item negates usage by the customer. The standard control chart for this situation is the *p*-chart, or chart *for fraction defective*. As one might expect, the probability distribution involved is the *binomial* distribution. The reader is referred to Chapter 4 for background on the binomial distribution.

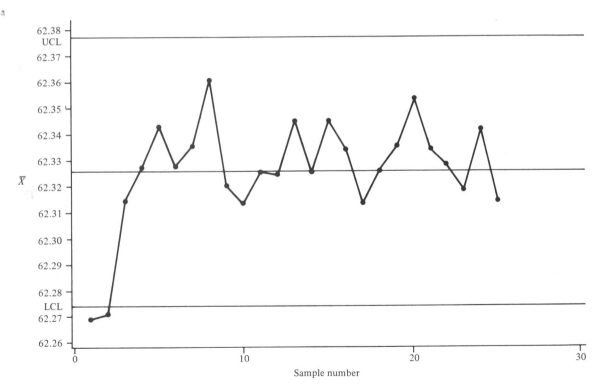

FIGURE 17.7 The $\overline{X}$-chart with control limits established by the data of Example 17.2.

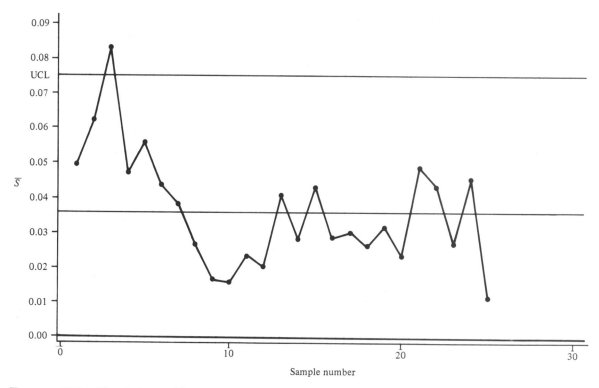

FIGURE 17.8 The *S*-chart with control limits established by the data of Example 17.2.

The *p*-Chart for Fraction Defective

Any item that is being manufactured may have several characteristics that are important and will be examined by an inspector. However, the entire development here will focus on a single characteristic. Suppose that for all items the probability of a defective item is p, and that all items are being produced independently. Then, in a random sample of n items produced, allowing X to be the number of defective items, we have

$$P(X = x) = \binom{n}{x} p^x (1 - p)^{n-x}, \qquad x = 0, 1, 2, \ldots, n.$$

As one might suspect, the mean and variance of the binomial random variable will play an important role in the development of the control chart. The reader should recall that

$$E(X) = np$$

and

$$\mathrm{Var}(X) = np(1 - p).$$

An unbiased estimator of p is the *fraction defective* or the proportion defective, $\hat{p}$, where

$$\hat{p} = \frac{\text{number of defectives in the sample of size } n}{n}.$$

As in the case of the variables control charts, the distributional properties of $\hat{p}$ are important in the development of the control chart. We know that

$$E(\hat{p}) = p$$

$$\text{Var}(\hat{p}) = \frac{p(1-p)}{n}.$$

Here we apply the same 3σ principles that we used for the variables charts. Let us assume initially that p is known. The structure, then, of the control charts involves the use of 3σ limits with

$$\sigma_{\hat{p}} = \sqrt{\frac{p(1-p)}{n}}.$$

Thus the limits are given by

$$\text{UCL} = p + 3\sqrt{\frac{p(1-p)}{n}}$$

$$\text{LCL} = p - 3\sqrt{\frac{p(1-p)}{n}},$$

with the process considered in control when the $\hat{p}$-values from the sample lie inside the control limits.

Generally, of course, the value of p is not known and must be estimated from a base set of samples very much like the case of μ and σ in the variables charts. Assume that there are m preliminary samples of size n. For a given sample, each of the n observations is reported as either "defective" or "not defective." The obvious unbiased estimator for p to use in the control chart is given by

$$\bar{p} = \sum_{i=1}^{m} \frac{\hat{p}_i}{m},$$

where $\hat{p}_i$ is the proportion defective in the ith sample. As a result, the control limits are given by

$$\text{UCL} = \bar{p} + 3\sqrt{\frac{\bar{p}(1-\bar{p})}{n}}$$

$$\text{centerline} = \bar{p}$$

$$\text{LCL} = \bar{p} - 3\sqrt{\frac{\bar{p}(1-\bar{p})}{n}}.$$

EXAMPLE 17.3   Consider the data given in Table 17.3 on the number of defective electronic components in samples of size 50. Twenty samples were taken in order to establish preliminary control chart values. The control charts determined by this preliminary period will have centerline

$$\bar{p} = 0.088$$

and control limits

$$UCL = \bar{p} + 3\sqrt{\frac{\bar{p}(1 - \bar{p})}{50}}$$

$$= 0.2082$$

$$LCL = \bar{p} - 3\sqrt{\frac{\bar{p}(1 - \bar{p})}{50}}$$

$$= -0.0322.$$

Obviously, with a computed value that is negative, the LCL will be set to zero. It is apparent from the values of the control limits that the process is in control during this preliminary period.

TABLE 17.3   Data for Example 17.3 to Establish Control Limits for p-Charts, Samples of Size 50

Sample	Number of Defective Components	Fraction Defective, $\hat{p}_i$
1	8	0.16
2	6	0.12
3	5	0.10
4	7	0.14
5	2	0.04
6	5	0.10
7	3	0.06
8	8	0.16
9	4	0.08
10	4	0.08
11	3	0.06
12	1	0.02
13	5	0.10
14	4	0.08
15	4	0.08
16	2	0.04
17	3	0.06
18	5	0.10
19	6	0.12
20	3	0.06
		$\bar{p} = 0.088$

Choice of Sample Size for the p-Chart

The choice of sample size for the p-chart for attributes involves the same general types of considerations as that of the chart for variables. A sample size is required that is sufficiently large to have a high probability of detection of an out-of-control condition when, in fact, a specified change in p has occurred. There is *no best method* for choice of sample size. However, one reasonable approach, suggested by Duncan (see the Bibliography), is to choose n so that there is probability 0.5 that we detect a shift in p of a particular amount. The resulting solution for n is quite simple. Suppose that the normal approximation to the binomial distribution applies. We wish, under the condition that p has shifted to, say, $p_1 > p_0$, that

$$P(\hat{p} \geq \text{UCL}) = 0.5$$

$$= P\left[Z \geq \frac{\text{UCL} - p_1}{\sqrt{p_1(1 - p_1)/n}}\right] = 0.5.$$

Since

$$P(Z > 0) = 0.5,$$

we set

$$\frac{\text{UCL} - p_1}{\sqrt{p_1(1 - p_1)/n}} = 0.$$

Substituting

$$p + 3\sqrt{\frac{p(1 - p)}{n}} = \text{UCL},$$

we have

$$(p - p_1) + 3\sqrt{\frac{p(1 - p)}{n}} = 0.$$

We can now solve for n, the size of each sample:

$$n = \frac{9}{\Delta^2}p(1 - p),$$

where, of course, Δ is the "shift" in the value of p, and p is the probability of a defective on which the control limits are based. Of course, if the control charts are based on $k\sigma$ limits, then

$$n = \frac{k^2}{\Delta^2}p(1 - p).$$

EXAMPLE 17.4   Suppose that an attribute quality control chart is being designed with a value of $p = 0.01$ for the in-control probability of a defective. What is the sample size per subgroup that produces a probability of 0.5 that a process shift to $p = p_1 = 0.05$ will be detected? The resulting p-chart will involve 3σ limits.

Here we have $\Delta = 0.04$. The appropriate sample size is given by

$$n = \frac{9}{(0.04)^2}(0.01)(0.99)$$

$$= 56.$$

Control Charts for Defects (Use of the Poisson Model)

In the preceding development we have assumed that the item under consideration is one that is either defective (i.e., nonfunctional) or not defective. In the latter case it is functional and thus acceptable to the consumer. In many situations this "defective or not" approach is too simplistic. Units may contain defects or nonconformities but still function quite well for the consumer. Indeed, in this case, it may be important to exert control on the *number of defects* or *number of nonconformities*. This type of quality control effort finds application when the units are either not simplistic or perhaps large. For example, the number of defects may be quite useful as the object of control when the single item or unit is, say, a personal computer. Another example is a unit defined by 50 feet of manufactured pipeline, where the number of defective welds is the object of quality control, the number of defects in 50 feet of manufactured carpeting, or the number of "bubbles" in a large manufactured sheet of glass.

It is clear from what we describe here that the binomial distribution is not appropriate. The total number of nonconformities in a unit or the average number per unit can be used as the measure for the control chart. Quite often it is assumed that the number of nonconformities in a sample of items follows the Poisson distribution. This type of chart is often called the C-chart.

Suppose that the number of defects X in one unit of product follows the Poisson distribution with parameter λ. (Here $t = 1$ for the Poisson model.) Recall that for the Poisson distribution,

$$P(X = x) = \frac{e^{-\lambda}\lambda^x}{x!}, \qquad x = 0, 1, 2, \ldots.$$

Here, the random variable X is the number of nonconformities. In Chapter 6 we learned that the mean and variance of the Poisson random variable are both λ. Thus if the quality control chart were to be structured according to the usual 3σ limits, we could have, for λ known,

$$\text{UCL} = \lambda + 3\sqrt{\lambda}$$

$$\text{centerline} = \lambda$$

$$\text{LCL} = \lambda - 3\sqrt{\lambda}.$$

As usual, λ often must come from an estimator from the data. An unbiased estimate of λ is the *average* number of nonconformities per sample. Denote this estimate by $\hat{\lambda}$. Thus the control chart has limits given by

$$UCL = \hat{\lambda} + 3\sqrt{\hat{\lambda}}$$

$$centerline = \hat{\lambda}$$

$$LCL = \hat{\lambda} - 3\sqrt{\hat{\lambda}}.$$

EXAMPLE 17.5   Table 17.4 represents the number of defects in 20 successive samples of sheet metal rolls each 100 feet long. A control chart is to be developed from these preliminary data for the purpose of controlling the number of defects in such samples. The estimate of the Poisson parameter λ is given by $\hat{\lambda} = 5.95$. As a result, the control limits suggested by these preliminary data are given by (3σ limits)

$$UCL = \hat{\lambda} + 3\sqrt{\hat{\lambda}}$$

$$= 13.2678$$

$$LCL = \hat{\lambda} - 3\sqrt{\hat{\lambda}}$$

$$= -1.3678 \qquad \text{(LCL set to zero)}.$$

Figure 17.9 shows a plot of the preliminary data with the control limits revealed.

TABLE 17.4   Data for Example 17.5; Control Involves Number of Defects in Sheet Metal Rolls

Sample Number	Number of Defects
1	8
2	7
3	5
4	4
5	4
6	7
7	6
8	4
9	5
10	6
11	3
12	7
13	5
14	9
15	7
16	7
17	8
18	6
19	7
20	4
	Ave. 5.95

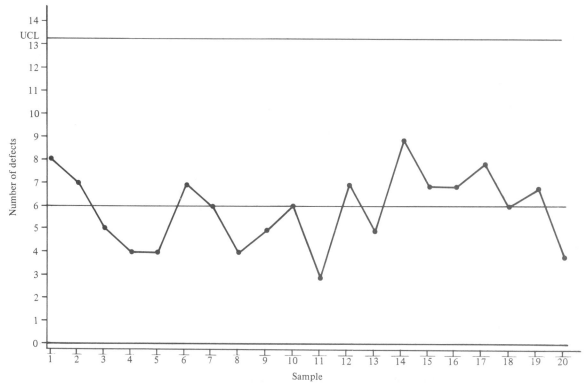

FIGURE 17.9   Preliminary data plotted on the control chart for Example 17.5.

Table 17.5 shows additional data taken from the production process. For each sample, the unit on which the chart was based, namely 100 feet of the metal, was inspected. The information on 20 samples is revealed. Figure 17.10 shows a plot of the additional production data. It is clear that the process is in control, at least through the period in which the data were taken.

In Example 17.5, we have made very clear what the sampling or inspection unit is, namely 100 feet of metal. In many cases where the item is a specific one (e.g., a personal computer or a specific type of electronic device), the inspection unit may be a *set of items*. For example, the analyst may decide to use 10 computers in each subgroup and thus observe a count of the total number of defects found. Thus the preliminary sample for construction of the control chart would involve the use of several samples, each containing 10 computers. The choice of the sample size may depend on many factors. Often, one may want a sample size that will ensure an LCL that is positive.

The analyst may wish to use the average number of defects per sampling unit as the basic measure in the control chart. For example, in the case of the personal computer, let the random variable

$$U = \frac{\text{total number of defects}}{n}$$

be measured for each sample of, say, $n = 10$. One can easily use the method of moment-generating functions to show that U is a Poisson random variable (see Exercise 1) if we assume that the number of defects per sampling unit is Poisson with parameter λ. Thus the control chart for this situation is characterized by the following:

$$\text{UCL} = \overline{U} + 3\sqrt{\frac{\overline{U}}{n}}$$

$$\text{centerline} = \overline{U}$$

$$\text{LCL} = \overline{U} - 3\sqrt{\frac{\overline{U}}{n}}.$$

Here, of course, $\overline{U}$ is the average of the U-values in the preliminary or base data set. The term $\overline{U}/n$ is derived from the result that

$$E(U) = \lambda$$

$$\text{Var}(U) = \frac{\lambda}{n},$$

and thus $\overline{U}$ is an unbiased estimate of $E(U) = \lambda$ and $\overline{U}/n$ is an unbiased estimate of $\text{Var}(U) = \lambda/n$. This type of control chart is often called a U-chart.

TABLE 17.5 Additional Data from the Production
Process of Example 17.5

Sample Number	Number of Defects
1	3
2	5
3	8
4	5
5	8
6	4
7	3
8	6
9	5
10	2
11	7
12	5
13	9
14	4
15	6
16	5
17	3
18	2
19	1
20	6

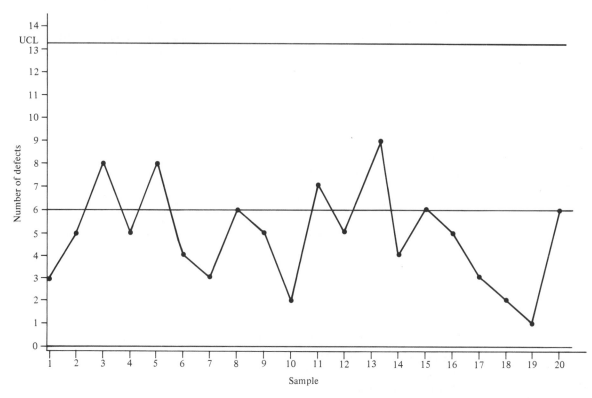

FIGURE 17.10   Additional production data for Example 17.5.

In the entire development in this section, we have based our development of control charts on the Poisson probability model. This model has been used in combination with the 3σ concept. As we have implied earlier in this chapter, the notion of 3σ limits has its roots in the normal approximation, although many users feel that the concept works well as a pragmatic tool even if normality is not even approximately correct. The difficulty, of course, is that in the absence of normality, one cannot control the probability of incorrect specification of an out-of-control state. In the case of the Poisson model, when λ is small the distribution is quite asymmetric, a condition that may result in undesirable results if one holds to the 3σ approach.

17.6   Cusum Control Charts

The disadvantage with Shewhart-type control charts, developed and illustrated in the preceding sections, lies in their inability to detect small changes in the mean. A quality control mechanism that has received considerable attention in the statistics literature and usage in industry is the cumulative sum (cusum) chart. The method for the cusum chart is very simple and its appeal is very intuitive. It should become very obvious to the reader why it is more responsive to small changes in the mean. Consider a control chart for the mean with a reference level established at value W. Consider particular observations $X_1, X_2, \ldots, X_r$.

The first r cusums are given by

$$S_1 = X_1 - W$$
$$S_2 = S_1 + (X_2 - W)$$
$$S_3 = S_2 + (X_3 - W)$$
$$\vdots$$
$$S_r = S_{r-1} + (X_r - W).$$

It becomes clear that the cusum is merely the accumulation of differences from the reference level. That is,

$$S_k = \sum_{i=1}^{k} (X_i - W), \qquad k = 1, 2, \ldots .$$

The cusum chart is then a plot of S_k against time.

Suppose that we consider the reference level w to be an acceptable value of the mean μ. Clearly, if there is no shift in μ, the cusum chart should be approximately horizontal, with some minor fluctuations balanced around zero. Now, if there is only a moderate change in the mean, a relatively large change in the *slope* of the cusum chart should result, since each new observation has a chance of contributing a shift and the measure being plotted is accumulating these shifts. Of course, the signal that the mean has shifted lies in the nature of the slope of the cusum chart. The purpose of the chart is to detect changes that are moving away from the reference level. A nonzero slope (in either direction) represents a change away from the reference level. A positive slope indicates an increase in the mean above the reference level, while a negative slope signals a decrease.

Cusum charts are often devised with a defined *acceptable quality* level (AQL) and a rejectable quality level (RQL) preestablished by the user. Both represent values of the mean. These may be viewed as playing roles somewhat similar to those of the null and alternative mean in hypothesis testing. Consider a situation in which the analyst hopes to detect an increase in the value of the process mean. We shall use the notation μ_0 for AQL and μ_1 for RQL and let $\mu_1 > \mu_0$. The reference level is now set at

$$W = \frac{\mu_0 + \mu_1}{2}.$$

The values of S_r ($r = 1, 2, \ldots$) will have a negative slope if the process is at μ_0 and a positive slope if the process mean is at μ_1.

Decision Rule for Cusum Charts

As we indicated earlier, the slope of the cusum chart provides the signal of action by the quality control analyst. The decision rule calls for action if at the rth sampling period

$$d_r > h,$$

where h is a prespecified value called the *length of the decision interval* and

$$d_r = S_r - \min_{i=1}^{r-1} S_i.$$

In other words, action is taken if the data reveal that the current cusum value exceeds the previous smallest cusum value by a specified amount.

A modification in the mechanics described above allows for ease in employing the method. We have described a procedure that plots the cusums and computes differences. A simple modification involves plotting the differences directly and allows for checking against the decision interval. The general expression for d_r is quite simple. For the cusum procedure in which one is detecting increases in the mean,

$$d_r = \max [0, d_{r-1} + (X_r - W)].$$

The choice of the value of h is, of course, very important. We do not choose in this book to provide the many details in the literature dealing with this choice. The reader is referred to Ewan and Kemp and Montgomery (see the Bibliography) for a thorough discussion. One important consideration is the expected run length. Ideally, the expected run length is quite large under $\mu = \mu_0$ and quite small when $\mu = \mu_1$.

Ewan and Kemp have produced charts that aid in the choice of the decision interval h. These charts are plotted in Figures 17.11 and 17.12. They are called nomograms and allow for choice of h so as to control average run length under the AQL $= \mu_0$ condition.

The following example will illustrate the use of the nomograms to choose a value of h. The value of n in the expressions on the chart indicates the sample size in cases where the observables $X_1, X_2, \ldots$ are *averages*.

EXAMPLE 17.6   Suppose that a quality control analyst wishes to devise a cusum chart in which the average run length at acceptable quality level is 500 when, in fact, it is of interest to detect an increase in the mean of as much as σ. What is the value of the decision interval, h? Samples of size 4 will be used with averages computed in the cusum procedure.

SOLUTION
Since

$$W = \frac{\mu_0 + \mu_1}{2}, \qquad \frac{(W - \mu_0)\sqrt{n}}{\sigma} = 1.0.$$

If we connect the point on the left vertical axis of Figure 17.11 at 1.0 with the value of 500 on the right vertical axis, we observe that

$$\frac{h\sqrt{n}}{\sigma} = 2.35.$$

Thus we use

$$h = \frac{2.35\sigma}{2} = 1.175\sigma.$$

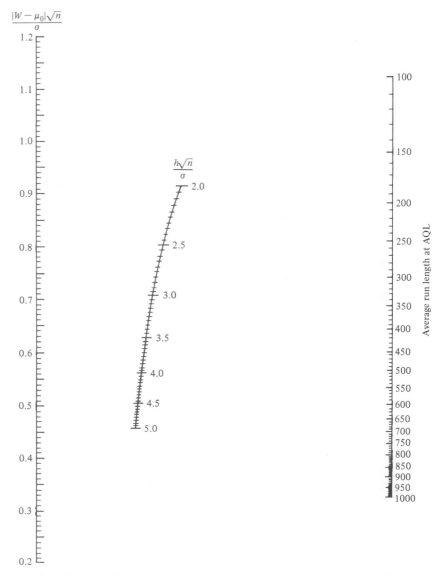

FIGURE 17.11 Nomogram for the cusum control chart in terms of average run length at AQL $= \mu_0$.

Obviously, an estimate of σ must be available in order that the decision interval be chosen in practice.

One can now easily observe by similar use of Figure 17.12 that for these conditions the average run length at RQL is approximately 3.2. This provides for a relatively quick response when, indeed, the mean has shifted by the amount σ.

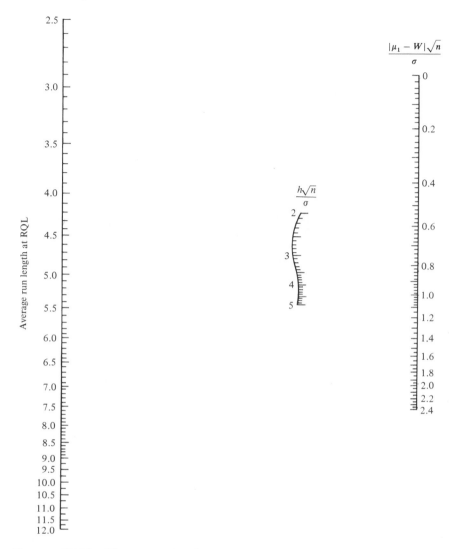

FIGURE 17.12 Nomogram for the cusum control chart in terms of average run length at RQL = μ_1.

Review Exercises

1. Consider $x_1, x_2, \ldots, x_n$ independent Poisson random variables with parameters $\mu_1, \mu_2, \ldots, \mu_n$. Use the properties of moment-generating functions to show that the random variable $\sum_{i=1}^{n} x_i$ is a Poisson random variable with mean $\sum_{i=1}^{n} \mu_i$ and variance

$$\sum_{i=1}^{n} \mu_i.$$

2. Consider the following data taken on subgroups of size 5. The data contains 20 averages and ranges on the diameter in millimeters of an important component part of an engine.

Sample	$\overline{X}$	R
1	2.3972	0.0052
2	2.4191	0.0117
3	2.4215	0.0062
4	2.3917	0.0089
5	2.4151	0.0095
6	2.4027	0.0101
7	2.3921	0.0091
8	2.4171	0.0059
9	2.3951	0.0068
10	2.4215	0.0048
11	2.3887	0.0082
12	2.4107	0.0032
13	2.4009	0.0077
14	2.3992	0.0107
15	2.3889	0.0025
16	2.4107	0.0138
17	2.4109	0.0037
18	2.3944	0.0052
19	2.3951	0.0038
20	2.4015	0.0017

Display $\overline{X}$- and R-charts. Does the process appear to be in control?

3. Suppose for Exercise 2 that the buyer has set specifications on the part. The specifications require that the diameter fall in the range covered by

$$2.4000 \pm 0.0100 \text{ mm.}$$

What proportion of units produced by this process will not conform to specifications?

4. For the situation of Exercise 2, give numerical estimates of the mean and standard deviation of the diameter for the part being manufactured in the process.

5. Consider the situation of Example 17.1. Suppose that additional samples of size 5 are taken and tensile strength recorded. The sampling produces the following results in pounds per square inch.
(a) Plot the data using the $\overline{X}$- and R-charts for the preliminary sample of Example 17.1.
(b) Does the process appear to be in control? If not, explain why not.

Sample	$\overline{X}_i$	R_i
1	1511	22
2	1508	14
3	1522	11
4	1488	18
5	1519	6
6	1524	11
7	1519	8
8	1504	7
9	1500	8
10	1519	14

6. Consider an in-control process with mean $\mu = 25$ and $\sigma = 1.0$. Suppose that subgroups of size 5 are used with control limits given by

$$\mu \pm 3\frac{\sigma}{\sqrt{n}}$$

and centerline at μ. Suppose that a shift occurs in the mean and thus the new mean is $\mu = 26.5$.
(a) What is the average number of samples required (following the shift) to detect the out-of-control situation?
(b) What is the standard deviation of the number of runs required?

7. Consider the situation of Example 17.2. The following data were taken on additional samples of size 5. Plot the $\overline{X}$- and S-values on the $\overline{X}$- and S-charts that were produced by the data in the preliminary sample. Does the process appear to be in control? Explain why or why not.

Sample	$\overline{X}_i$	S_i
1	62.280	0.062
2	62.319	0.049
3	62.297	0.077
4	62.318	0.042
5	62.315	0.038
6	62.389	0.052
7	62.401	0.059
8	62.315	0.042
9	62.298	0.036
10	62.337	0.068

8. Samples of size 50 are taken every hour from a process producing a certain type of item that is either considered defective or not defective. Twenty samples are taken.

Sample	Number of Defective Items
1	4
2	3
3	5
4	3
5	2
6	2
7	2
8	1
9	4
10	3
11	2
12	4
13	1
14	2
15	3
16	1
17	1
18	2
19	3
20	1

(a) Construct a control chart for control of proportion defective.

(b) Does the process appear to be in control? Explain.

9. For the situation of Exercise 8, suppose that additional data are collected as follows (samples of size 50):

Sample	Number of Defective Items
1	3
2	4
3	2
4	2
5	3
6	1
7	3
8	5
9	7
10	7

Does the process appear to be in control? Explain.

10. A quality control effort is being attempted on a process in which large steel plates are being manufactured and surface defects are of concern. The goal is to set up a quality control chart on the number of defects *per plate*. The data are as follows:

Sample Plot	Number of Defects
1	4
2	2
3	1
4	3
5	0
6	4
7	5
8	3
9	2
10	2
11	1
12	2
13	2
14	3
15	1
16	4
17	3
18	2
19	1
20	3

Set up the appropriate control chart using this sample information. Does the process appear to be in control?

Bibliography

BARTLETT, M. S., and D. G. KENDALL. "The Statistical Analysis of Variance Heterogeneity and the Logarithmic Transformation," *Journal of the Royal Statistical Society*, Ser. B, 8, 128–150, 1946.

BICKEL, P. J., and DOKSUM, K. A. *Mathematical Statistics*. Oakland, Calif.: Holden-Day, Inc., 1977.

BOWKER, A. H., and G. J. LIEBERMAN. *Engineering Statistics*, 2nd ed. Englewood Cliffs, N.J.: Prentice Hall, Inc., 1972.

BOX, G. E. P. "Signal-to-Noise Ratios, Performance Criteria and Transformations (with discussion)," *Technometrics*, 30, 1–40, 1988.

BOX, G. E. P., and C. A. FUNG. "Studies in Quality Improvement: Minimizing Transmitted Variation by Parameter Design," Report 8, University of Wisconsin–Madison, Center for Quality and Productivity Improvement.

BOX, G. E. P., W. G. HUNTER, and J. S. HUNTER. *Statistics for Experimenters*. New York: John Wiley & Sons, Inc., 1978.

BROWNLEE, K. A. *Statistical Theory and Methodology in Science and Engineering*, 2nd ed. New York: John Wiley & Sons, Inc., 1965.

CARROLL, R. J., and D. RUPPERT. "Transformation and Weighting in Regression," New York: Chapman and Hall, 1988, p. 348.

CHATTERJEE, S., and B. PRICE. *Regression Analysis by Example*. New York: John Wiley & Sons, Inc., 1977.

COOK, R. D., and S. WEISBERG, *Residuals and Influence in Regression*. New York: Chapman and Hall, 1982.

DANIEL, C., and F. WOOD. *Fitting Equations to Data*, 2nd ed. New York: John Wiley & Sons, Inc., 1980.

DANIEL, W. W. *Applied Nonparametric Statistics*. Boston: Houghton Mifflin Company, 1978.

DERMAN, C., L. GLASER, and I. OLKIN. *Probability Models and Applications*. New York: Macmillan Publishing Company, 1980.

DEVORE, J. L. *Probability and Statistics for Engineering and the Sciences*, 3rd ed. Monterey, Calif.: Brooks/Cole Publishing Co., 1991.

DIXON, W. J., and F. J. MASSEY, JR. *Introduction to Statistical Analysis*, 3rd ed. New York: McGraw-Hill Book Company, 1969.

DRAPER, N., and H. SMITH. *Applied Regression Analysis*, 2nd ed. New York: John Wiley & Sons, Inc., 1981.

DYER, D. D., and J. P. KEATING, "On the Determination of Critical Values for Bartlett's Test," *J. Am. Stat. Assoc.*, Vol. 75, 1980.

EWAN, W. D., and K. W. KEMP. "Sampling Inspection of Continuous Processes with No Autocorrelation Between Successive Results," *Biometrika*, Vol. 41, pp. 363–380, 1960.

GUNST, R. F., and R. L. MASON. *Regression Analysis and Its Application*: *A Data-Oriented Approach*. New York: Marcel Dekker, Inc., 1980.

GUTTMAN, I., and S. S. WILKS. *Introductory Engineering Statistics*. New York: John Wiley & Sons, Inc., 1965.

HICKS, C. R. *Fundamental Concepts in the Design of Experiments*, 2nd ed. New York: Holt, Rinehart and Winston, 1973.

HOCKING, R. R. "The Analysis and Selection of Variables in Linear Regression." *Biometrics*, Vol. 32, 1976.

HOERL, A. E., and R. W. KENNARD, "Ridge Regression: Applications to Nonorthogonal Problems," *Technometrics*, Vol. 12, No. 1, 1970.

HOGG, R. V., and A. T. CRAIG. *Introduction to Mathematical Statistics*, 4th ed. New York: Macmillan Publishing Company, 1978.

HOGG, R. V., and J. LEDOLTER. *Engineering Statistics*. New York: Macmillan Publishing Company, 1992.

HOLLANDER, M., and D. WOLFE. *Nonparametric Statistical Methods*. New York: John Wiley & Sons, Inc., 1973.

JOHNSON, N. L., and F. C. LEONE. *Statistics and Experimental Design*: *In Engineering and the Physical Sciences*, Vols. I and II, 2nd ed. New York: John Wiley & Sons, Inc., 1977.

KACKAR, R. "Off-Line Quality Control, Parameter Design, and the Taguchi Methods," *Journal of Quality Technology*, 17, 176–188, 1985.

KOOPMANS, L. H. *An Introduction to Contemporary Statistics*. Boston: Duxbury Press, 1981.

LARSEN, R. J., and M. L. MORRIS. *Introduction to Mathematical Statistics*. Englewood Cliffs, N.J.: Prentice Hall, Inc., 1981.

LEHMANN, E. I. *Nonparametrics*: *Statistical Methods Based on Ranks*. San Francisco: Holden-Day, Inc., 1975.

LENTNER, M., and T. BISHOP. *Design and Analysis of Experiments*. Blacksburg, Va.: Valley Book Co., 1986.

LI, C. C. *Introduction to Experimental Statistics*. New York: McGraw-Hill Book Company, 1964.

McLAVE, J., and F. DIETRICH. *Statistics*. San Francisco: Dellen Publishing Co., 1978.

MONTGOMERY, D. C. *Introduction to Statistical Quality Control*. New York: John Wiley & Sons, Inc., 1985.

MONTGOMERY, D. C., and E. A. PECK. *Introduction to Linear Regression Analysis*. New York: John Wiley & Sons, Inc., 1982.

MOSTELLER, F., and J. TUKEY. *Data Analysis and Regression*. Reading, Mass.: Addison-Wesley Publishing Co., Inc., 1977.

MYERS, R. H. *Classical and Modern Regression with Applications*, 2nd ed. Boston: Duxbury Press, 1990.

MYERS, R. H. *Response Surface Methodology*. Department of Statistics, Virginia Polytechnic Institute and State University, Blacksburg, Va., 1976.

MYERS, R. H., A. I. KHURI, and G. VINING. "Response Surface Alternatives to the Taguchi Robust Parameter Design Approach," *The American Statistician*, Vol. 46, No. 2, 131–139, 1992.

NETER, J., W. WASSERMAN, and M. H. KUTNER. *Applied Linear Statistical Models*, 3rd ed. Homewood, Ill.: Richard D. Irwin, Inc., 1988.

NETER, J., W. WASSERMAN, and G. A. WHITMORE. *Applied Statistics*. Boston: Allyn & Bacon, Inc., 1978.

NOETHER, G. E. *Introduction to Statistics: A Nonparametric Approach*, 2nd ed. Boston: Houghton Mifflin Company, 1976.

OTT, L. *An Introduction to Statistical Methods and Data Analysis*, 3rd ed. Boston: Duxbury Press, 1988.

PLACKETT, R. L., and J. P. BURMAN. "The Design of Multifactor Experiments," *Biometrika*, Vol. 33, pp. 305–325, 1946.

ROSS, S. *Introduction to Applied Probability Models*, 2nd ed. New York: Academic Press, Inc., 1980.

SCHILLING, E. G., and P. R. NELSON. "The Effect of Nonnormality on the Control Limits of $\overline{X}$ Charts." *J. Quality Tech.*, Vol. 8, 1976.

SCHMIDT, S. R., and R. G. LAUNSBY. "Understanding Industrial Designed Experiments," Colorado Springs, Colo., 1991.

SHOEMAKER, A. C., K. L. TSUI, and C. F. J. WU, "Economical Experimentation Methods for Robust Parameter Design," Paper presented at the Fall Technical Conference, American Society of Quality Control, Houston, Tex., 1989.

SNEDECOR, G., and W. G. COCHRAN, *Statistical Methods*, 7th ed. Ames, Iowa: The Iowa State University Press, 1980.

STEEL, R. G. D., and J. H. TORRIE. *Principles and Procedures of Statistics*, 2nd ed. New York: McGraw-Hill Book Company, 1979.

TAGUCHI, G. *Introduction to Quality Engineering*. White Plains, N.Y.: UNIPUB/Kraus International 1991.

TAGUCHI, G., and Y. WU. "Introduction to Off-Line Quality Control," Central Japan Quality Control Association (available from American Supplier Institute, 32100 Detroit Industrial Expressway, Romulus, Mich. 48174), 1980.

THOMPSON, W. O., and F. B. CADY. *Proceedings of the University of Kentucky Conference on Regression with a Large Number of Predictor Variables*, Lexington, 1973.

TUKEY, T. W. *Exploratory Data Analysis*. Reading, Mass.: Addison-Wesley Publishing Co., Inc., 1977.

VINING, G. G., and R. H. MYERS. "Combining Taguchi and Response Surface Philosophies: A Dual Response Approach," *Journal of Quality Technology*, 22, 38–45, 1990.

WELCH, W. J., T. K. YU, S. M. KANG, and J. SACKS. "Computer Experiments for Quality Control by Parameter Design," *Journal of Quality Technology*, 22, 15–22, 1990.

WINER, B. J. *Statistical Principles in Experimental Design*, 2nd ed. New York: McGraw-Hill Book Company, 1971.

YOUNGER, M. S. *A Handbook for Linear Regression*. Boston: Duxbury Press, 1979.

Appendix: Statistical Tables

TABLE A.1   Binomial Probability Sums $\sum\limits_{x=0}^{r} b(x; n, p)$

							p				
n	r	.10	.20	.25	.30	.40	.50	.60	.70	.80	.90
1	0	.9000	.8000	.7500	.7000	.6000	.5000	.4000	.3000	.2000	.1000
	1	1.0000	1.0000	1.0000	1.0000	1.0000	1.0000	1.0000	1.0000	1.0000	1.0000
2	0	.8100	.6400	.5625	.4900	.3600	.2500	.1600	.0900	.0400	.0100
	1	.9900	.9600	.9375	.9100	.8400	.7500	.6400	.5100	.3600	.1900
	2	1.0000	1.0000	1.0000	1.0000	1.0000	1.0000	1.0000	1.0000	1.0000	1.0000
3	0	.7290	.5120	.4219	.3430	.2160	.1250	.0640	.0270	.0080	.0010
	1	.9720	.8960	.8438	.7840	.6480	.5000	.3520	.2160	.1040	.0280
	2	.9990	.9920	.9844	.9730	.9360	.8750	.7840	.6570	.4880	.2710
	3	1.0000	1.0000	1.0000	1.0000	1.0000	1.0000	1.0000	1.0000	1.0000	1.0000
4	0	.6561	.4096	.3164	.2401	.1296	.0625	.0256	.0081	.0016	.0001
	1	.9477	.8192	.7383	.6517	.4752	.3125	.1792	.0837	.0272	.0037
	2	.9963	.9728	.9492	.9163	.8208	.6875	.5248	.3483	.1808	.0523
	3	.9999	.9984	.9961	.9919	.9744	.9375	.8704	.7599	.5904	.3439
	4	1.0000	1.0000	1.0000	1.0000	1.0000	1.0000	1.0000	1.0000	1.0000	1.0000
5	0	.5905	.3277	.2373	.1681	.0778	.0312	.0102	.0024	.0003	.0000
	1	.9185	.7373	.6328	.5282	.3370	.1875	.0870	.0308	.0067	.0005
	2	.9914	.9421	.8965	.8369	.6826	.5000	.3174	.1631	.0579	.0086
	3	.9995	.9933	.9844	.9692	.9130	.8125	.6630	.4718	.2627	.0815
	4	1.0000	.9997	.9990	.9976	.9898	.9688	.9222	.8319	.6723	.4095
	5		1.0000	1.0000	1.0000	1.0000	1.0000	1.0000	1.0000	1.0000	1.0000
6	0	.5314	.2621	.1780	.1176	.0467	.0156	.0041	.0007	.0001	.0000
	1	.8857	.6554	.5339	.4202	.2333	.1094	.0410	.0109	.0016	.0001
	2	.9841	.9011	.8306	.7443	.5443	.3438	.1792	.0705	.0170	.0013
	3	.9987	.9830	.9624	.9295	.8208	.6563	.4557	.2557	.0989	.0158
	4	.9999	.9984	.9954	.9891	.9590	.8906	.7667	.5798	.3447	.1143
	5	1.0000	.9999	.9998	.9993	.9959	.9844	.9533	.8824	.7379	.4686
	6		1.0000	1.0000	1.0000	1.0000	1.0000	1.0000	1.0000	1.0000	1.0000
7	0	.4783	.2097	.1335	.0824	.0280	.0078	.0016	.0002	.0000	
	1	.8503	.5767	.4449	.3294	.1586	.0625	.0188	.0038	.0004	.0000
	2	.9743	.8520	.7564	.6471	.4199	.2266	.0963	.0288	.0047	.0002
	3	.9973	.9667	.9294	.8740	.7102	.5000	.2898	.1260	.0333	.0027
	4	.9998	.9953	.9871	.9712	.9037	.7734	.5801	.3529	.1480	.0257
	5	1.0000	.9996	.9987	.9962	.9812	.9375	.8414	.6706	.4233	.1497
	6		1.0000	.9999	.9998	.9984	.9922	.9720	.9176	.7903	.5217
	7			1.0000	1.0000	1.0000	1.0000	1.0000	1.0000	1.0000	1.0000

TABLE A.1 (*continued*) Binomial Probability Sums $\sum_{x=0}^{r} b(x; n, p)$

							p				
n	r	.10	.20	.25	.30	.40	.50	.60	.70	.80	.90
8	0	.4305	.1678	.1001	.0576	.0168	.0039	.0007	.0001	.0000	
	1	.8131	.5033	.3671	.2553	.1064	.0352	.0085	.0013	.0001	
	2	.9619	.7969	.6785	.5518	.3154	.1445	.0498	.0113	.0012	.0000
	3	.9950	.9437	.8862	.8059	.5941	.3633	.1737	.0580	.0104	.0004
	4	.9996	.9896	.9727	.9420	.8263	.6367	.4059	.1941	.0563	.0050
	5	1.0000	.9988	.9958	.9887	.9502	.8555	.6846	.4482	.2031	.0381
	6		.9991	.9996	.9987	.9915	.9648	.8936	.7447	.4967	.1869
	7		1.0000	1.0000	.9999	.9993	.9961	.9832	.9424	.8322	.5695
	8				1.0000	1.0000	1.0000	1.0000	1.0000	1.0000	1.0000
9	0	.3874	.1342	.0751	.0404	.0101	.0020	.0003	.0000		
	1	.7748	.4362	.3003	.1960	.0705	.0195	.0038	.0004	.0000	
	2	.9470	.7382	.6007	.4628	.2318	.0898	.0250	.0043	.0003	.0000
	3	.9917	.9144	.8343	.7297	.4826	.2539	.0994	.0253	.0031	.0001
	4	.9991	.9804	.9511	.9012	.7334	.5000	.2666	.0988	.0196	.0009
	5	.9999	.9969	.9900	.9747	.9006	.7461	.5174	.2703	.0856	.0083
	6	1.0000	.9997	.9987	.9957	.9750	.9102	.7682	.5372	.2618	.0530
	7		1.0000	.9999	.9996	.9962	.9805	.9295	.8040	.5638	.2252
	8			1.0000	1.0000	.9997	.9980	.9899	.9596	.8658	.6126
	9					1.0000	1.0000	1.0000	1.0000	1.0000	1.0000
10	0	.3487	.1074	.0563	.0282	.0060	.0010	.0001	.0000		
	1	.7361	.3758	.2440	.1493	.0464	.0107	.0017	.0001	.0000	
	2	.9298	.6778	.5256	.3828	.1673	.0547	.0123	.0016	.0001	
	3	.9872	.8791	.7759	.6496	.3823	.1719	.0548	.0106	.0009	.0000
	4	.9984	.9672	.9219	.8497	.6331	.3770	.1662	.0474	.0064	.0002
	5	.9999	.9936	.9803	.9527	.8338	.6230	.3669	.1503	.0328	.0016
	6	1.0000	.9991	.9965	.9894	.9452	.8281	.6177	.3504	.1209	.0128
	7		.9999	.9996	.9984	.9877	.9453	.8327	.6172	.3222	.0702
	8		1.0000	1.0000	.9999	.9983	.9893	.9536	.8507	.6242	.2639
	9				1.0000	.9999	.9990	.9940	.9718	.8926	.6513
	10					1.0000	1.0000	1.0000	1.0000	1.0000	1.0000
11	0	.3138	.0859	.0422	.0198	.0036	.0005	.0000			
	1	.6974	.3221	.1971	.1130	.0302	.0059	.0007	.0000		
	2	.9104	.6174	.4552	.3127	.1189	.0327	.0059	.0006	.0000	
	3	.9815	.8369	.7133	.5696	.2963	.1133	.0293	.0043	.0002	
	4	.9972	.9496	.8854	.7897	.5328	.2744	.0994	.0216	.0020	.0000
	5	.9997	.9883	.9657	.9218	.7535	.5000	.2465	.0782	.0117	.0003
	6	1.0000	.9980	.9924	.9784	.9006	.7256	.4672	.2103	.0504	.0028
	7		.9998	.9988	.9957	.9707	.8867	.7037	.4304	.1611	.0185
	8		1.0000	.9999	.9994	.9941	.9673	.8811	.6873	.3826	.0896
	9			1.0000	1.0000	.9993	.9941	.9698	.8870	.6779	.3026
	10					1.0000	.9995	.9964	.9802	.9141	.6862
	11						1.0000	1.0000	1.0000	1.0000	1.0000

TABLE A.1 (*continued*)   Binomial Probability Sums $\sum_{x=0}^{r} b(x; n, p)$

n	r	.10	.20	.25	.30	.40	.50	.60	.70	.80	.90
12	0	.2824	.0687	.0317	.0138	.0022	.0002	.0000			
	1	.6590	.2749	.1584	.0850	.0196	.0032	.0003	.0000		
	2	.8891	.5583	.3907	.2528	.0834	.0193	.0028	.0002	.0000	
	3	.9744	.7946	.6488	.4925	.2253	.0730	.0153	.0017	.0001	
	4	.9957	.9274	.8424	.7237	.4382	.1938	.0573	.0095	.0006	.0000
	5	.9995	.9806	.9456	.8821	.6652	.3872	.1582	.0386	.0039	.0001
	6	.9999	.9961	.9857	.9614	.8418	.6128	.3348	.1178	.0194	.0005
	7	1.0000	.9994	.9972	.9905	.9427	.8062	.5618	.2763	.0726	.0043
	8		.9999	.9996	.9983	.9847	.9270	.7747	.5075	.2054	.0256
	9		1.0000	1.0000	.9998	.9972	.9807	.9166	.7472	.4417	.1109
	10				1.0000	.9997	.9968	.9804	.9150	.7251	.3410
	11					1.0000	.9998	.9978	.9862	.9313	.7176
	12						1.0000	1.0000	1.0000	1.0000	1.0000
13	0	.2542	.0550	.0238	.0097	.0013	.0001	.0000			
	1	.6213	.2336	.1267	.0637	.0126	.0017	.0001	.0000		
	2	.8661	.5017	.3326	.2025	.0579	.0112	.0013	.0001		
	3	.9658	.7473	.5843	.4206	.1686	.0461	.0078	.0007	.0000	
	4	.9935	.9009	.7940	.6543	.3530	.1334	.0321	.0040	.0002	
	5	.9991	.9700	.9198	.8346	.5744	.2905	.0977	.0182	.0012	.0000
	6	.9999	.9930	.9757	.9376	.7712	.5000	.2288	.0624	.0070	.0001
	7	1.0000	.9980	.9944	.9818	.9023	.7095	.4256	.1654	.0300	.0009
	8		.9998	.9990	.9960	.9679	.8666	.6470	.3457	.0991	.0065
	9		1.0000	.9999	.9993	.9922	.9539	.8314	.5794	.2527	.0342
	10			1.0000	.9999	.9987	.9888	.9421	.7975	.4983	.1339
	11				1.0000	.9999	.9983	.9874	.9363	.7664	.3787
	12					1.0000	.9999	.9987	.9903	.9450	.7458
	13						1.0000	1.0000	1.0000	1.0000	1.0000
14	0	.2288	.0440	.0178	.0068	.0008	.0001	.0000			
	1	.5846	.1979	.1010	.0475	.0081	.0009	.0001			
	2	.8416	.4481	.2811	.1608	.0398	.0065	.0006	.0000		
	3	.9559	.6982	.5213	.3552	.1243	.0287	.0039	.0002		
	4	.9908	.8702	.7415	.5842	.2793	.0898	.0175	.0017	.0000	
	5	.9985	.9561	.8883	.7805	.4859	.2120	.0583	.0083	.0004	
	6	.9998	.9884	.9617	.9067	.6925	.3953	.1501	.0315	.0024	.0000
	7	1.0000	.9976	.9897	.9685	.8499	.6047	.3075	.0933	.0116	.0002
	8		.9996	.9978	.9917	.9417	.7880	.5141	.2195	.0439	.0015
	9		1.0000	.9997	.9983	.9825	.9102	.7207	.4158	.1298	.0092
	10			1.0000	.9998	.9961	.9713	.8757	.6448	.3018	.0441
	11				1.0000	.9994	.9935	.9602	.8392	.5519	.1584
	12					.9999	.9991	.9919	.9525	.8021	.4154
	13					1.0000	.9999	.9992	.9932	.9560	.7712
	14						1.0000	1.0000	1.0000	1.0000	1.0000

TABLE A.1 (*continued*)   Binomial Probability Sums $\sum_{x=0}^{r} b(x; n, p)$

						p					
n	r	.10	.20	.25	.30	.40	.50	.60	.70	.80	.90
15	0	.2059	.0352	.0134	.0047	.0005	.0000				
	1	.5490	.1671	.0802	.0353	.0052	.0005	.0000			
	2	.8159	.3980	.2361	.1268	.0271	.0037	.0003	.0000		
	3	.9444	.6482	.4613	.2969	.0905	.0176	.0019	.0001		
	4	.9873	.8358	.6865	.5155	.2173	.0592	.0094	.0007	.0000	
	5	.9978	.9389	.8516	.7216	.4032	.1509	.0338	.0037	.0001	
	6	.9997	.9819	.9434	.8689	.6098	.3036	.0951	.0152	.0008	
	7	1.0000	.9958	.9827	.9500	.7869	.5000	.2131	.0500	.0042	.0000
	8		.9992	.9958	.9848	.9050	.6964	.3902	.1311	.0181	.0003
	9		.9999	.9992	.9963	.9662	.8491	.5968	.2784	.0611	.0023
	10		1.0000	.9999	.9993	.9907	.9408	.7827	.4845	.1642	.0127
	11			1.0000	.9999	.9981	.9824	.9095	.7031	.3518	.0556
	12				1.0000	.9997	.9963	.9729	.8732	.6020	.1841
	13					1.0000	.9995	.9948	.9647	.8329	.4510
	14						1.0000	.9995	.9953	.9648	.7941
	15							1.0000	1.0000	1.0000	1.0000
16	0	.1853	.0281	.0100	.0033	.0003	.0000				
	1	.5147	.1407	.0635	.0261	.0033	.0003	.0000			
	2	.7892	.3518	.1971	.0994	.0183	.0021	.0001			
	3	.9316	.5981	.4050	.2459	.0651	.0106	.0009	.0000		
	4	.9830	.7982	.6302	.4499	.1666	.0384	.0049	.0003		
	5	.9967	.9183	.8103	.6598	.3288	.1051	.0191	.0016	.0000	
	6	.9995	.9733	.9204	.8247	.5272	.2272	.0583	.0071	.0002	
	7	.9999	.9930	.9729	.9256	.7161	.4018	.1423	.0257	.0015	.0000
	8	1.0000	.9985	.9925	.9743	.8577	.5982	.2839	.0744	.0070	.0001
	9		.9998	.9984	.9929	.9417	.7728	.4728	.1753	.0267	.0005
	10		1.0000	.9997	.9984	.9809	.8949	.6712	.3402	.0817	.0033
	11			1.0000	.9997	.9951	.9616	.8334	.5501	.2018	.0170
	12				1.0000	.9991	.9894	.9349	.7541	.4019	.0684
	13					.9999	.9979	.9817	.9006	.6482	.2108
	14					1.0000	.9997	.9967	.9739	.8593	.4853
	15						1.0000	.9997	.9967	.9719	.8147
	16							1.0000	1.0000	1.0000	1.0000

TABLE A.1 (*continued*)   Binomial Probability Sums $\sum_{x=0}^{r} b(x; n, p)$

n	r	.10	.20	.25	.30	.40	.50	.60	.70	.80	.90
						p					
17	0	.1668	.0225	.0075	.0023	.0002	.0000				
	1	.4818	.1182	.0501	.0193	.0021	.0001	.0000			
	2	.7618	.3096	.1637	.0774	.0123	.0012	.0001			
	3	.9174	.5489	.3530	.2019	.0464	.0064	.0005	.0000		
	4	.9779	.7582	.5739	.3887	.1260	.0245	.0025	.0001		
	5	.9953	.8943	.7653	.5968	.2639	.0717	.0106	.0007	.0000	
	6	.9992	.9623	.8929	.7752	.4478	.1662	.0348	.0032	.0001	
	7	.9999	.9891	.9598	.8954	.6405	.3145	.0919	.0127	.0005	
	8	1.0000	.9974	.9876	.9597	.8011	.5000	.1989	.0403	.0026	.0000
	9		.9995	.9969	.9873	.9081	.6855	.3595	.1046	.0109	.0001
	10		.9999	.9994	.9968	.9652	.8338	.5522	.2248	.0377	.0008
	11		1.0000	.9999	.9993	.9894	.9283	.7361	.4032	.1057	.0047
	12			1.0000	.9999	.9975	.9755	.8740	.6113	.2418	.0221
	13				1.0000	.9995	.9936	.9536	.7981	.4511	.0826
	14					.9999	.9988	.9877	.9226	.6904	.2382
	15					1.0000	.9999	.9979	.9807	.8818	.5182
	16						1.0000	.9998	.9977	.9775	.8332
	17							1.0000	1.0000	1.0000	1,0000
18	0	.1501	.0180	.0056	.0016	.0001	.0000				
	1	.4503	.0991	.0395	.0142	.0013	.0001				
	2	.7338	.2713	.1353	.0600	.0082	.0007	.0000			
	3	.9018	.5010	.3057	.1646	.0328	.0038	.0002			
	4	.9718	.7164	.5787	.3327	.0942	.0154	.0013	.0000		
	5	.9936	.8671	.7175	.5344	.2088	.0481	.0058	.0003		
	6	.9988	.9487	.8610	.7217	.3743	.1189	.0203	.0014	.0000	
	7	.9998	.9837	.9431	.8593	.5634	.2403	.0576	.0061	.0002	
	8	1.0000	.9957	.9807	.9404	.7368	.4073	.1347	.0210	.0009	
	9		.9991	.9946	.9790	.8653	.5927	.2632	.0596	.0043	.0000
	10		.9998	.9988	.9939	.9424	.7597	.4366	.1407	.0163	.0002
	11		1.0000	.9998	.9986	.9797	.8811	.6257	.2783	.0513	.0012
	12			1.0000	.9997	.9942	.9519	.7912	.4656	.1329	.0064
	13				1.0000	.9987	.9846	.9058	.6673	.2836	.0282
	14					.9998	.9962	.9672	.8354	.4990	.0982
	15					1.0000	.9993	.9918	.9400	.7287	.2662
	16						.9999	.9987	.9858	.9009	.5497
	17						1.0000	.9999	.9984	.9820	.8499
	18							1.0000	1.0000	1.0000	1.0000

TABLE A.1 (*continued*) Binomial Probability Sums $\sum_{x=0}^{r} b(x; n, p)$

							p					
n	*r*	.10	.20	.25	.30	.40	.50	.60	.70	.80	.90
19	0	.1351	.0144	.0042	.0011	.0001					
	1	.4203	.0829	.0310	.0104	.0008	.0000				
	2	.7054	.2369	.1113	.0462	.0055	.0004	.0000			
	3	.8850	.4551	.2631	.1332	.0230	.0022	.0001			
	4	.9648	.6733	.4654	.2822	.0696	.0096	.0006	.0000		
	5	.9914	.8369	.6678	.4739	.1629	.0318	.0031	.0001		
	6	.9983	.9324	.8251	.6655	.3081	.0835	.0116	.0006		
	7	.9997	.9767	.9225	.8180	.4878	.1796	.0352	.0028	.0000	
	8	1.0000	.9933	.9713	.9161	.6675	.3238	.0885	.0105	.0003	
	9		.9984	.9911	.9674	.8139	.5000	.1861	.0326	.0016	
	10		.9997	.9977	.9895	.9115	.6762	.3325	.0839	.0067	.0000
	11		.9999	.9995	.9972	.9648	.8204	.5122	.1820	.0233	.0003
	12		1.0000	.9999	.9994	.9884	.9165	.6919	.3345	.0676	.0017
	13			1.0000	.9999	.9969	.9682	.8371	.5261	.1631	.0086
	14				1.0000	.9994	.9904	.9304	.7178	.3267	.0352
	15					.9999	.9978	.9770	.8668	.5449	.1150
	16					1.0000	.9996	.9945	.9538	.7631	.2946
	17						1.0000	.9992	.9896	.9171	.5797
	18							.9999	.9989	.9856	.8649
	19							1.0000	1.0000	1.0000	1.0000
20	0	.1216	.0115	.0032	.0008	.0000					
	1	.3917	.0692	.0243	.0076	.0005	.0000				
	2	.6769	.2061	.0913	.0355	.0036	.0002	.0000			
	3	.8670	.4114	.2252	.1071	.0160	.0013	.0001			
	4	.9568	.6296	.4148	.2375	.0510	.0059	.0003			
	5	.9887	.8042	.6172	.4164	.1256	.0207	.0016	.0000		
	6	.9976	.9133	.7858	.6080	.2500	.0577	.0065	.0003		
	7	.9996	.9679	.8982	.7723	.4159	.1316	.0210	.0013	.0000	
	8	.9999	.9900	.9591	.8867	.5956	.2517	.0565	.0051	.0001	
	9	1.0000	.9974	.9861	.9520	.7553	.4119	.1275	.0171	.0006	
	10		.9994	.9961	.9829	.8725	.5881	.2447	.0480	.0026	.0000
	11		.9999	.9991	.9949	.9435	.7483	.4044	.1133	.0100	.0001
	12		1.0000	.9998	.9987	.9790	.8684	.5841	.2277	.0321	.0004
	13			1.0000	.9997	.9935	.9423	.7500	.3920	.0867	.0024
	14				1.0000	.9984	.9793	.8744	.5836	.1958	.0113
	15					.9997	.9941	.9490	.7625	.3704	.0432
	16					1.0000	.9987	.9840	.8929	.5886	.1330
	17						.9998	.9964	.9645	.7939	.3231
	18						1.0000	.9995	.9924	.9308	.6083
	19							1.0000	.9992	.9885	.8784
	20								1.0000	1.0000	1.0000

TABLE A.2   Poisson Probability Sums $\sum_{x=0}^{r} p(x; \mu)$

	μ								
r	0.1	0.2	0.3	0.4	0.5	0.6	0.7	0.8	0.9
0	0.9048	0.8187	0.7408	0.6730	0.6065	0.5488	0.4966	0.4493	0.4066
1	0.9953	0.9825	0.9631	0.9384	0.9098	0.8781	0.8442	0.8088	0.7725
2	0.9998	0.9989	0.9964	0.9921	0.9856	0.9769	0.9659	0.9526	0.9371
3	1.0000	0.9999	0.9997	0.9992	0.9982	0.9966	0.9942	0.9909	0.9865
4		1.0000	1.0000	0.9999	0.9998	0.9996	0.9992	0.9986	0.9977
5				1.0000	1.0000	1.0000	0.9999	0.9998	0.9997
6							1.0000	1.0000	1.0000

	μ								
r	1.0	1.5	2.0	2.5	3.0	3.5	4.0	4.5	5.0
0	0.3679	0.2231	0.1353	0.0821	0.0498	0.0302	0.0183	0.0111	0.0067
1	0.7358	0.5578	0.4060	0.2873	0.1991	0.1359	0.0916	0.0611	0.0404
2	0.9197	0.8088	0.6767	0.5438	0.4232	0.3208	0.2381	0.1736	0.1247
3	0.9810	0.9344	0.8571	0.7576	0.6472	0.5366	0.4335	0.3423	0.2650
4	0.9963	0.9814	0.9473	0.8912	0.8153	0.7254	0.6288	0.5321	0.4405
5	0.9994	0.9955	0.9834	0.9580	0.9161	0.8576	0.7851	0.7029	0.6160
6	0.9999	0.9991	0.9955	0.9858	0.9665	0.9347	0.8893	0.8311	0.7622
7	1.0000	0.9998	0.9989	0.9958	0.9881	0.9733	0.9489	0.9134	0.8666
8		1.0000	0.9998	0.9989	0.9962	0.9901	0.9786	0.9597	0.9319
9			1.0000	0.9997	0.9989	0.9967	0.9919	0.9829	0.9682
10				0.9999	0.9997	0.9990	0.9972	0.9933	0.9863
11				1.0000	0.9999	0.9997	0.9991	0.9976	0.9945
12					1.0000	0.9999	0.9997	0.9992	0.9980
13						1.0000	0.9999	0.9997	0.9993
14							1.0000	0.9999	0.9998
15								1.0000	0.9999
16									1.0000

TABLE A.2 (*continued*) Poisson Probability Sums $\sum_{x=0}^{r} p(x; \mu)$

	μ								
r	5.5	6.0	6.5	7.0	7.5	8.0	8.5	9.0	9.5
0	0.0041	0.0025	0.0015	0.0009	0.0006	0.0003	0.0002	0.0001	0.0001
1	0.0266	0.0174	0.0113	0.0073	0.0047	0.0030	0.0019	0.0012	0.0008
2	0.0884	0.0620	0.0430	0.0296	0.0203	0.0138	0.0093	0.0062	0.0042
3	0.2017	0.1512	0.1118	0.0818	0.0591	0.0424	0.0301	0.0212	0.0149
4	0.3575	0.2851	0.2237	0.1730	0.1321	0.0996	0.0744	0.0550	0.0403
5	0.5289	0.4457	0.3690	0.3007	0.2414	0.1912	0.1496	0.1157	0.0885
6	0.6860	0.6063	0.5265	0.4497	0.3782	0.3134	0.2562	0.2068	0.1649
7	0.8095	0.7440	0.6728	0.5987	0.5246	0.4530	0.3856	0.3239	0.2687
8	0.8944	0.8472	0.7916	0.7291	0.6620	0.5925	0.5231	0.4557	0.3918
9	0.9462	0.9161	0.8774	0.8305	0.7764	0.7166	0.6530	0.5874	0.5218
10	0.9747	0.9574	0.9332	0.9015	0.8622	0.8159	0.7634	0.7060	0.6453
11	0.9890	0.9799	0.9661	0.9466	0.9208	0.8881	0.8487	0.8030	0.7520
12	0.9955	0.9912	0.9840	0.9730	0.9573	0.9362	0.9091	0.8758	0.8364
13	0.9983	0.9964	0.9929	0.9872	0.9784	0.9658	0.9486	0.9261	0.8981
14	0.9994	0.9986	0.9970	0.9943	0.9897	0.9827	0.9726	0.9585	0.9400
15	0.9998	0.9995	0.9988	0.9976	0.9954	0.9918	0.9862	0.9780	0.9665
16	0.9999	0.9998	0.9996	0.9990	0.9980	0.9963	0.9934	0.9889	0.9823
17	1.0000	0.9999	0.9998	0.9996	0.9992	0.9984	0.9970	0.9947	0.9911
18		1.0000	0.9999	0.9999	0.9997	0.9994	0.9987	0.9976	0.9957
19			1.0000	1.0000	0.9999	0.9997	0.9995	0.9989	0.9980
20					1.0000	0.9999	0.9998	0.9996	0.9991
21						1.0000	0.9999	0.9998	0.9996
22							1.0000	0.9999	0.9999
23								1.0000	0.9999
24									1.0000

TABLE A.2 (*continued*)   Poisson Probability Sums $\sum\limits_{x=0}^{r} p(x; \mu)$

	μ								
r	10.0	11.0	12.0	13.0	14.0	15.0	16.0	17.0	18.0
0	0.0000	0.0000	0.0000						
1	0.0005	0.0002	0.0001	0.0000	0.0000				
2	0.0028	0.0012	0.0005	0.0002	0.0001	0.0000	0.0000		
3	0.0103	0.0049	0.0023	0.0010	0.0005	0.0002	0.0001	0.0000	0.0000
4	0.0293	0.0151	0.0076	0.0037	0.0018	0.0009	0.0004	0.0002	0.0001
5	0.0671	0.0375	0.0203	0.0107	0.0055	0.0028	0.0014	0.0007	0.0003
6	0.1301	0.0786	0.0458	0.0259	0.0142	0.0076	0.0040	0.0021	0.0010
7	0.2202	0.1432	0.0895	0.0540	0.0316	0.0180	0.0100	0.0054	0.0029
8	0.3328	0.2320	0.1550	0.0998	0.0621	0.0374	0.0220	0.0126	0.0071
9	0.4579	0.3405	0.2424	0.1658	0.1094	0.0699	0.0433	0.0261	0.0154
10	0.5830	0.4599	0.3472	0.2517	0.1757	0.1185	0.0774	0.0491	0.0304
11	0.6968	0.5793	0.4616	0.3532	0.2600	0.1848	0.1270	0.0847	0.0549
12	0.7916	0.6887	0.5760	0.4631	0.3585	0.2676	0.1931	0.1350	0.0917
13	0.8645	0.7813	0.6815	0.5730	0.4644	0.3632	0.2745	0.2009	0.1426
14	0.9165	0.8540	0.7720	0.6751	0.5704	0.4657	0.3675	0.2808	0.2081
15	0.9513	0.9074	0.8444	0.7636	0.6694	0.5681	0.4667	0.3715	0.2867
16	0.9730	0.9441	0.8987	0.8355	0.7559	0.6641	0.5660	0.4677	0.3750
17	0.9857	0.9678	0.9370	0.8905	0.8272	0.7489	0.6593	0.5640	0.4686
18	0.9928	0.9823	0.9626	0.9302	0.8826	0.8195	0.7423	0.6550	0.5622
19	0.9965	0.9907	0.9787	0.9573	0.9235	0.8752	0.8122	0.7363	0.6509
20	0.9984	0.9953	0.9884	0.9750	0.9521	0.9170	0.8682	0.8055	0.7307
21	0.9993	0.9977	0.9939	0.9859	0.9712	0.9469	0.9108	0.8615	0.7991
22	0.9997	0.9990	0.9970	0.9924	0.9833	0.9673	0.9418	0.9047	0.8551
23	0.9999	0.9995	0.9985	0.9960	0.9907	0.9805	0.9633	0.9367	0.8989
24	1.0000	0.9998	0.9993	0.9980	0.9950	0.9888	0.9777	0.9594	0.9317
25		0.9999	0.9997	0.9990	0.9974	0.9938	0.9869	0.9748	0.9554
26		1.0000	0.9999	0.9995	0.9987	0.9967	0.9925	0.9848	0.9718
27			0.9999	0.9998	0.9994	0.9983	0.9959	0.9912	0.9827
28			1.0000	0.9999	0.9997	0.9991	0.9978	0.9950	0.9897
29				1.0000	0.9999	0.9996	0.9989	0.9973	0.9941
30					0.9999	0.9998	0.9994	0.9986	0.9967
31					1.0000	0.9999	0.9997	0.9993	0.9982
32						1.0000	0.9999	0.9996	0.9990
33							0.9999	0.9998	0.9995
34							1.0000	0.9999	0.9998
35								1.0000	0.9999
36									0.9999
37									1.0000

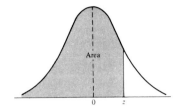

TABLE A.3 Areas Under the Normal Curve

z	.00	.01	.02	.03	.04	.05	.06	.07	.08	.09
−3.4	.0003	.0003	.0003	.0003	.0003	.0003	.0003	.0003	.0003	.0002
−3.3	.0005	.0005	.0005	.0004	.0004	.0004	.0004	.0004	.0004	.0003
−3.2	.0007	.0007	.0006	.0006	.0006	.0006	.0006	.0005	.0005	.0005
−3.1	.0010	.0009	.0009	.0009	.0008	.0008	.0008	.0008	.0007	.0007
−3.0	.0013	.0013	.0013	.0012	.0012	.0011	.0011	.0011	.0010	.0010
−2.9	.0019	.0018	.0017	.0017	.0016	.0016	.0015	.0015	.0014	.0014
−2.8	.0026	.0025	.0024	.0023	.0023	.0022	.0021	.0021	.0020	.0019
−2.7	.0035	.0034	.0033	.0032	.0031	.0030	.0029	.0028	.0027	.0026
−2.6	.0047	.0045	.0044	.0043	.0041	.0040	.0039	.0038	.0037	.0036
−2.5	.0062	.0060	.0059	.0057	.0055	.0054	.0052	.0051	.0049	.0048
−2.4	.0082	.0080	.0078	.0075	.0073	.0071	.0069	.0068	.0066	.0064
−2.3	.0107	.0104	.0102	.0099	.0096	.0094	.0091	.0089	.0087	.0084
−2.2	.0139	.0136	.0132	.0129	.0125	.0122	.0119	.0116	.0113	.0110
−2.1	.0179	.0174	.0170	.0166	.0162	.0158	.0154	.0150	.0146	.0143
−2.0	.0228	.0222	.0217	.0212	.0207	.0202	.0197	.0192	.0188	.0183
−1.9	.0287	.0281	.0274	.0268	.0262	.0256	.0250	.0244	.0239	.0233
−1.8	.0359	.0352	.0344	.0336	.0329	.0322	.0314	.0307	.0301	.0294
−1.7	.0446	.0436	.0427	.0418	.0409	.0401	.0392	.0384	.0375	.0367
−1.6	.0548	.0537	.0526	.0516	.0505	.0495	.0485	.0475	.0465	.0455
−1.5	.0668	.0655	.0643	.0630	.0618	.0606	.0594	.0582	.0571	.0559
−1.4	.0808	.0793	.0778	.0764	.0749	.0735	.0722	.0708	.0694	.0681
−1.3	.0968	.0951	.0934	.0918	.0901	.0885	.0869	.0853	.0838	.0823
−1.2	.1151	.1131	.1112	.1093	.1075	.1056	.1038	.1020	.1003	.0985
−1.1	.1357	.1335	.1314	.1292	.1271	.1251	.1230	.1210	.1190	.1170
−1.0	.1587	.1562	.1539	.1515	.1492	.1469	.1446	.1423	.1401	.1379
−0.9	.1841	.1814	.1788	.1762	.1736	.1711	.1685	.1660	.1635	.1611
−0.8	.2119	.2090	.2061	.2033	.2005	.1977	.1949	.1922	.1894	.1867
−0.7	.2420	.2389	.2358	.2327	.2296	.2266	.2236	.2206	.2177	.2148
−0.6	.2743	.2709	.2676	.2643	.2611	.2578	.2546	.2514	.2483	.2451
−0.5	.3085	.3050	.3015	.2981	.2946	.2912	.2877	.2843	.2810	.2776
−0.4	.3446	.3409	.3372	.3336	.3300	.3264	.3228	.3192	.3156	.3121
−0.3	.3821	.3783	.3745	.3707	.3669	.3632	.3594	.3557	.3520	.3483
−0.2	.4207	.4168	.4129	.4090	.4052	.4013	.3974	.3936	.3897	.3859
−0.1	.4602	.4562	.4522	.4483	.4443	.4404	.4364	.4325	.4286	.4247
−0.0	.5000	.4960	.4920	.4880	.4840	.4801	.4761	.4721	.4681	.4641

TABLE A.3 (*continued*)   Areas Under the Normal Curve

z	.00	.01	.02	.03	.04	.05	.06	.07	.08	.09
0.0	.5000	.5040	.5080	.5120	.5160	.5199	.5239	.5279	.5319	.5359
0.1	.5398	.5438	.5478	.5517	.5557	.5596	.5636	.5675	.5714	.5753
0.2	.5793	.5832	.5871	.5910	.5948	.5987	.6026	.6064	.6103	.6141
0.3	.6179	.6217	.6255	.6293	.6331	.6368	.6406	.6443	.6480	.6517
0.4	.6554	.6591	.6628	.6664	.6700	.6736	.6772	.6808	.6844	.6879
0.5	.6915	.6950	.6985	.7019	.7054	.7088	.7123	.7157	.7190	.7224
0.6	.7257	.7291	.7324	.7357	.7389	.7422	.7454	.7486	.7517	.7549
0.7	.7580	.7611	.7642	.7673	.7704	.7734	.7764	.7794	.7823	.7852
0.8	.7881	.7910	.7939	.7967	.7995	.8023	.8051	.8078	.8106	.8133
0.9	.8159	.8186	.8212	.8238	.8264	.8289	.8315	.8340	.8365	.8389
1.0	.8413	.8438	.8461	.8485	.8508	.8531	.8554	.8577	.8599	.8621
1.1	.8643	.8665	.8686	.8708	.8729	.8749	.8770	.8790	.8810	.8830
1.2	.8849	.8869	.8888	.8907	.8925	.8944	.8962	.8980	.8997	.9015
1.3	.9032	.9049	.9066	.9082	.9099	.9115	.9131	.9147	.9162	.9177
1.4	.9192	.9207	.9222	.9236	.9251	.9265	.9278	.9292	.9306	.9319
1.5	.9332	.9345	.9357	.9370	.9382	.9394	.9406	.9418	.9429	.9441
1.6	.9452	.9463	.9474	.9484	.9495	.9505	.9515	.9525	.9535	.9545
1.7	.9554	.9564	.9573	.9582	.9591	.9599	.9608	.9616	.9625	.9633
1.8	.9641	.9649	.9656	.9664	.9671	.9678	.9686	.9693	.9699	.9706
1.9	.9713	.9719	.9726	.9732	.9738	.9744	.9750	.9756	.9761	.9767
2.0	.9772	.9778	.9783	.9788	.9793	.9798	.9803	.9808	.9812	.9817
2.1	.9821	.9826	.9830	.9834	.9838	.9842	.9846	.9850	.9854	.9857
2.2	.9861	.9864	.9868	.9871	.9875	.9878	.9881	.9884	.9887	.9890
2.3	.9893	.9896	.9898	.9901	.9904	.9906	.9909	.9911	.9913	.9916
2.4	.9918	.9920	.9922	.9925	.9927	.9929	.9931	.9932	.9934	.9936
2.5	.9938	.9940	.9941	.9943	.9945	.9946	.9948	.9949	.9951	.9952
2.6	.9953	.9955	.9956	.9957	.9959	.9960	.9961	.9962	.9963	.9964
2.7	.9965	.9966	.9967	.9968	.9969	.9970	.9971	.9972	.9973	.9974
2.8	.9974	.9975	.9976	.9977	.9977	.9978	.9979	.9979	.9980	.9981
2.9	.9981	.9982	.9982	.9983	.9984	.9984	.9985	.9985	.9986	.9986
3.0	.9987	.9987	.9987	.9988	.9988	.9989	.9989	.9989	.9990	.9990
3.1	.9990	.9991	.9991	.9991	.9992	.9992	.9992	.9992	.9993	.9993
3.2	.9993	.9993	.9994	.9994	.9994	.9994	.9994	.9995	.9995	.9995
3.3	.9995	.9995	.9995	.9996	.9996	.9996	.9996	.9996	.9996	.9997
3.4	.9997	.9997	.9997	.9997	.9997	.9997	.9997	.9997	.9997	.9998

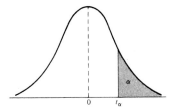

TABLE A.4 Critical Values of the *t*-Distribution

	α						
v	0.40	0.30	0.20	0.15	0.10	0.05	0.025
1	0.325	0.727	1.376	1.963	3.078	6.314	12.706
2	0.289	0.617	1.061	1.386	1.886	2.920	4.303
3	0.277	0.584	0.978	1.250	1.638	2.353	3.182
4	0.271	0.569	0.941	1.190	1.533	2.132	2.776
5	0.267	0.559	0.920	1.156	1.476	2.015	2.571
6	0.265	0.553	0.906	1.134	1.440	1.943	2.447
7	0.263	0.549	0.896	1.119	1.415	1.895	2.365
8	0.262	0.546	0.889	1.108	1.397	1.860	2.306
9	0.261	0.543	0.883	1.100	1.383	1.833	2.262
10	0.260	0.542	0.879	1.093	1.372	1.812	2.228
11	0.260	0.540	0.876	1.088	1.363	1.796	2.201
12	0.259	0.539	0.873	1.083	1.356	1.782	2.179
13	0.259	0.537	0.870	1.079	1.350	1.771	2.160
14	0.258	0.537	0.868	1.076	1.345	1.761	2.145
15	0.258	0.536	0.866	1.074	1.341	1.753	2.131
16	0.258	0.535	0.865	1.071	1.337	1.746	2.120
17	0.257	0.534	0.863	1.069	1.333	1.740	2.110
18	0.257	0.534	0.862	1.067	1.330	1.734	2.101
19	0.257	0.533	0.861	1.066	1.328	1.729	2.093
20	0.257	0.533	0.860	1.064	1.325	1.725	2.086
21	0.257	0.532	0.859	1.063	1.323	1.721	2.080
22	0.256	0.532	0.858	1.061	1.321	1.717	2.074
23	0.256	0.532	0.858	1.060	1.319	1.714	2.069
24	0.256	0.531	0.857	1.059	1.318	1.711	2.064
25	0.256	0.531	0.856	1.058	1.316	1.708	2.060
26	0.256	0.531	0.856	1.058	1.315	1.706	2.056
27	0.256	0.531	0.855	1.057	1.314	1.703	2.052
28	0.256	0.530	0.855	1.056	1.313	1.701	2.048
29	0.256	0.530	0.854	1.055	1.311	1.699	2.045
30	0.256	0.530	0.854	1.055	1.310	1.697	2.042
40	0.255	0.529	0.851	1.050	1.303	1.684	2.021
60	0.254	0.527	0.848	1.045	1.296	1.671	2.000
120	0.254	0.526	0.845	1.041	1.289	1.658	1.980
∞	0.253	0.524	0.842	1.036	1.282	1.645	1.960

TABLE A.4 (*continued*)   Critical Values of the *t*-Distribution

v	α						
	0.02	0.015	0.01	0.0075	0.005	0.0025	0.0005
1	15.895	21.205	31.821	42.434	63.657	127.322	636.590
2	4.849	5.643	6.965	8.073	9.925	14.089	31.598
3	3.482	3.896	4.541	5.047	5.841	7.453	12.924
4	2.999	3.298	3.747	4.088	4.604	5.598	8.610
5	2.757	3.003	3.365	3.634	4.032	4.773	6.869
6	2.612	2.829	3.143	3.372	3.707	4.317	5.959
7	2.517	2.715	2.998	3.203	3.499	4.029	5.408
8	2.449	2.634	2.896	3.085	3.355	3.833	5.041
9	2.398	2.574	2.821	2.998	3.250	3.690	4.781
10	2.359	2.527	2.764	2.932	3.169	3.581	4.587
11	2.328	2.491	2.718	2.879	3.106	3.497	4.437
12	2.303	2.461	2.681	2.836	3.055	3.428	4.318
13	2.282	2.436	2.650	2.801	3.012	3.372	4.221
14	2.264	2.415	2.624	2.771	2.977	3.326	4.140
15	2.249	2.397	2.602	2.746	2.947	3.286	4.073
16	2.235	2.382	2.583	2.724	2.921	3.252	4.015
17	2.224	2.368	2.567	2.706	2.898	3.222	3.965
18	2.214	2.356	2.552	2.689	2.878	3.197	3.922
19	2.205	2.346	2.539	2.674	2.861	3.174	3.883
20	2.197	2.336	2.528	2.661	2.845	3.153	3.849
21	2.189	2.328	2.518	2.649	2.831	3.135	3.819
22	2.183	2.320	2.508	2.639	2.819	3.119	3.792
23	2.177	2.313	2.500	2.629	2.807	3.104	3.768
24	2.172	2.307	2.492	2.620	2.797	3.091	3.745
25	2.167	2.301	2.485	2.612	2.787	3.078	3.725
26	2.162	2.296	2.479	2.605	2.779	3.067	3.707
27	2.158	2.291	2.473	2.598	2.771	3.057	3.690
28	2.154	2.286	2.467	2.592	2.763	3.047	3.674
29	2.150	2.282	2.462	2.586	2.756	3.038	3.659
30	2.147	2.278	2.457	2.581	2.750	3.030	3.646
40	2.125	2.250	2.423	2.542	2.704	2.971	3.551
60	2.099	2.223	2.390	2.504	2.660	2.915	3.460
120	2.076	2.196	2.358	2.468	2.617	2.860	3.373
∞	2.054	2.170	2.326	2.432	2.576	2.807	3.291

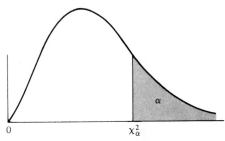

TABLE A.5 Critical Values of the Chi-Squared Distribution

	α									
ν	.995	.99	.98	.975	.95	.90	.80	.75	.70	.50
1	.0⁴393	.0³157	.0³628	.0³982	.00393	.0158	.0642	.102	.148	.455
2	.0100	.0201	.0404	.0506	.103	.211	.446	.575	.713	1.386
3	.0717	.115	.185	.216	.352	.584	1.005	1.213	1.424	2.366
4	.207	.297	.429	.484	.711	1.064	1.649	1.923	2.195	3.357
5	.412	.554	.752	.831	1.145	1.610	2.343	2.675	3.000	4.351
6	.676	.872	1.134	1.237	1.635	2.204	3.070	3.455	3.828	5.348
7	.989	1.239	1.564	1.690	2.167	2.833	3.822	4.255	4.671	6.346
8	1.344	1.646	2.032	2.180	2.733	3.490	4.594	5.071	5.527	7.344
9	1.735	2.088	2.532	2.700	3.325	4.168	5.380	5.899	6.393	8.343
10	2.156	2.558	3.059	3.247	3.940	4.865	6.179	6.737	7.267	9.342
11	2.603	3.053	3.609	3.816	4.575	5.578	6.989	7.584	8.148	10.341
12	3.074	3.571	4.178	4.404	5.226	6.304	7.807	8.438	9.034	11.340
13	3.565	4.107	4.765	5.009	5.892	7.042	8.634	9.299	9.926	12.340
14	4.075	4.660	5.368	5.629	6.571	7.790	9.467	10.165	10.821	13.339
15	4.601	5.229	5.985	6.262	7.261	8.547	10.307	11.036	11.721	14.339
16	5.142	5.812	6.614	6.908	7.962	9.312	11.152	11.912	12.624	15.338
17	5.697	6.408	7.255	7.564	8.672	10.085	12.002	12.792	13.531	16.338
18	6.265	7.015	7.906	8.231	9.390	10.865	12.857	13.675	14.440	17.338
19	6.844	7.633	8.567	8.907	10.117	11.651	13.716	14.562	15.352	18.338
20	7.434	8.260	9.237	9.591	10.851	12.443	14.578	15.452	16.266	19.337
21	8.034	8.897	9.915	10.283	11.591	13.240	15.445	16.344	17.182	20.337
22	8.643	9.542	10.600	10.982	12.338	14.041	16.314	17.240	18.101	21.337
23	9.260	10.196	11.293	11.688	13.091	14.848	17.187	18.137	19.021	22.337
24	9.886	10.856	11.992	12.401	13.848	15.659	18.062	19.037	19.943	23.337
25	10.520	11.524	12.697	13.120	14.611	16.473	18.940	19.939	20.867	24.337
26	11.160	12.198	13.409	13.844	15.379	17.292	19.820	20.843	21.792	25.336
27	11.808	12.879	14.125	14.573	16.151	18.114	20.703	21.749	22.719	26.336
28	12.461	13.565	14.847	15.308	16.928	18.939	21.588	22.657	23.647	27.336
29	13.121	14.256	15.574	16.047	17.708	19.768	22.475	23.567	24.577	28.336
30	13.787	14.953	16.306	16.791	18.493	20.599	23.364	24.478	25.508	29.336

TABLE A.5 (*continued*)   Critical Values of the Chi-Squared Distribution

	α									
v	.30	.25	.20	.10	.05	.025	.02	.01	.005	.001
1	1.074	1.323	1.642	2.706	3.841	5.024	5.412	6.635	7.879	10.827
2	2.408	2.773	3.219	4.605	5.991	7.378	7.824	9.210	10.597	13.815
3	3.665	4.108	4.642	6.251	7.815	9.348	9.837	11.345	12.838	16.268
4	4.878	5.385	5.989	7.779	9.488	11.143	11.668	13.277	14.860	18.465
5	6.064	6.626	7.289	9.236	11.070	12.832	13.388	15.086	16.750	20.517
6	7.231	7.841	8.558	10.645	12.592	14.449	15.033	16.812	18.548	22.457
7	8.383	9.037	9.803	12.017	14.067	16.013	16.622	18.475	20.278	24.322
8	9.524	10.219	11.030	13.362	15.507	17.535	18.168	20.090	21.955	26.125
9	10.656	11.389	12.242	14.684	16.919	19.023	19.679	21.666	23.589	27.877
10	11.781	12.549	13.442	15.987	18.307	20.483	21.161	23.209	25.188	29.588
11	12.899	13.701	14.631	17.275	19.675	21.920	22.618	24.725	26.757	31.264
12	14.011	14.845	15.812	18.549	21.026	23.337	24.054	26.217	28.300	32.909
13	15.119	15.984	16.985	19.812	22.362	24.736	25.472	27.688	29.819	34.528
14	16.222	17.117	18.151	21.064	23.685	26.119	26.873	29.141	31.319	36.123
15	17.322	18.245	19.311	22.307	24.996	27.488	28.259	30.578	32.801	37.697
16	18.418	19.369	20.465	23.542	26.296	28.845	29.633	32.000	34.267	39.252
17	19.511	20.489	21.615	24.769	27.587	30.191	30.995	33.409	35.718	40.790
18	20.601	21.605	22.760	25.989	28.869	31.526	32.346	34.805	37.156	42.312
19	21.689	22.718	23.900	27.204	30.144	32.852	33.687	36.191	38.582	43.820
20	22.775	23.828	25.038	28.412	31.410	34.170	35.020	37.566	39.997	45.315
21	23.858	24.935	26.171	29.615	32.671	35.479	36.343	38.932	41.401	46.797
22	24.939	26.039	27.301	30.813	33.924	36.781	37.659	40.289	42.796	48.268
23	26.018	27.141	28.429	32.007	35.172	38.076	38.968	41.638	44.181	49.728
24	27.096	28.241	29.553	33.196	36.415	39.364	40.270	42.980	45.558	51.179
25	28.172	29.339	30.675	34.382	37.652	40.646	41.566	44.314	46.928	52.620
26	29.246	30.434	31.795	35.563	38.885	41.923	42.856	45.642	48.290	54.052
27	30.319	31.528	32.912	36.741	40.113	43.194	44.140	46.963	49.645	55.476
28	31.391	32.620	34.027	37.916	41.337	44.461	45.419	48.278	50.993	56.893
29	32.461	33.711	35.139	39.087	42.557	45.722	46.693	49.588	52.336	58.302
30	33.530	34.800	36.250	40.256	43.773	46.979	47.962	50.892	53.672	59.703

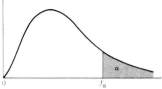

TABLE A.6*   Critical Values of the F-Distribution

$$f_{0.05}(v_1, v_2)$$

v_2	v_1								
	1	2	3	4	5	6	7	8	9
1	161.4	199.5	215.7	224.6	230.2	234.0	236.8	238.9	240.5
2	18.51	19.00	19.16	19.25	19.30	19.33	19.35	19.37	19.38
3	10.13	9.55	9.28	9.12	9.01	8.94	8.89	8.85	8.81
4	7.71	6.94	6.59	6.39	6.26	6.16	6.09	6.04	6.00
5	6.61	5.79	5.41	5.19	5.05	4.95	4.88	4.82	4.77
6	5.99	5.14	4.76	4.53	4.39	4.28	4.21	4.15	4.10
7	5.59	4.74	4.35	4.12	3.97	3.87	3.79	3.73	3.68
8	5.32	4.46	4.07	3.84	3.69	3.58	3.50	3.44	3.39
9	5.12	4.26	3.86	3.63	3.48	3.37	3.29	3.23	3.18
10	4.96	4.10	3.71	3.48	3.33	3.22	3.14	3.07	3.02
11	4.84	3.98	3.59	3.36	3.20	3.09	3.01	2.95	2.90
12	4.75	3.89	3.49	3.26	3.11	3.00	2.91	2.85	2.80
13	4.67	3.81	3.41	3.18	3.03	2.92	2.83	2.77	2.71
14	4.60	3.74	3.34	3.11	2.96	2.85	2.76	2.70	2.65
15	4.54	3.68	3.29	3.06	2.90	2.79	2.71	2.64	2.59
16	4.49	3.63	3.24	3.01	2.85	2.74	2.66	2.59	2.54
17	4.45	3.59	3.20	2.96	2.81	2.70	2.61	2.55	2.49
18	4.41	3.55	3.16	2.93	2.77	2.66	2.58	2.51	2.46
19	4.38	3.52	3.13	2.90	2.74	2.63	2.54	2.48	2.42
20	4.35	3.49	3.10	2.87	2.71	2.60	2.51	2.45	2.39
21	4.32	3.47	3.07	2.84	2.68	2.57	2.49	2.42	2.37
22	4.30	3.44	3.05	2.82	2.66	2.55	2.46	2.40	2.34
23	4.28	3.42	3.03	2.80	2.64	2.53	2.44	2.37	2.32
24	4.26	3.40	3.01	2.78	2.62	2.51	2.42	2.36	2.30
25	4.24	3.39	2.99	2.76	2.60	2.49	2.40	2.34	2.28
26	4.23	3.37	2.98	2.74	2.59	2.47	2.39	2.32	2.27
27	4.21	3.35	2.96	2.73	2.57	2.46	2.37	2.31	2.25
28	4.20	3.34	2.95	2.71	2.56	2.45	2.36	2.29	2.24
29	4.18	3.33	2.93	2.70	2.55	2.43	2.35	2.28	2.22
30	4.17	3.32	2.92	2.69	2.53	2.42	2.33	2.27	2.21
40	4.08	3.23	2.84	2.61	2.45	2.34	2.25	2.18	2.12
60	4.00	3.15	2.76	2.53	2.37	2.25	2.17	2.10	2.04
120	3.92	3.07	2.68	2.45	2.29	2.17	2.09	2.02	1.96
∞	3.84	3.00	2.60	2.37	2.21	2.10	2.01	1.94	1.88

TABLE A.6 *(continued)*   Critical Values of the *F*-Distribution

$$f_{0.05}(\nu_1, \nu_2)$$

ν_2	\multicolumn{10}{c}{ν_1}
	10	12	15	20	24	30	40	60	120	∞
1	241.9	243.9	245.9	248.0	249.1	250.1	251.1	252.2	253.3	254.3
2	19.40	19.41	19.43	19.45	19.45	19.46	19.47	19.48	19.49	19.50
3	8.79	8.74	8.70	8.66	8.64	8.62	8.59	8.57	8.55	8.53
4	5.96	5.91	5.86	5.80	5.77	5.75	5.72	5.69	5.66	5.63
5	4.74	4.68	4.62	4.56	4.53	4.50	4.46	4.43	4.40	4.36
6	4.06	4.00	3.94	3.87	3.84	3.81	3.77	3.74	3.70	3.67
7	3.64	3.57	3.51	3.44	3.41	3.38	3.34	3.30	3.27	3.23
8	3.35	3.28	3.22	3.15	3.12	3.08	3.04	3.01	2.97	2.93
9	3.14	3.07	3.01	2.94	2.90	2.86	2.83	2.79	2.75	2.71
10	2.98	2.91	2.85	2.77	2.74	2.70	2.66	2.62	2.58	2.54
11	2.85	2.79	2.72	2.65	2.61	2.57	2.53	2.49	2.45	2.40
12	2.75	2.69	2.62	2.54	2.51	2.47	2.43	2.38	2.34	2.30
13	2.67	2.60	2.53	2.46	2.42	2.38	2.34	2.30	2.25	2.21
14	2.60	2.53	2.46	2.39	2.35	2.31	2.27	2.22	2.18	2.13
15	2.54	2.48	2.40	2.33	2.29	2.25	2.20	2.16	2.11	2.07
16	2.49	2.42	2.35	2.28	2.24	2.19	2.15	2.11	2.06	2.01
17	2.45	2.38	2.31	2.23	2.19	2.15	2.10	2.06	2.01	1.96
18	2.41	2.34	2.27	2.19	2.15	2.11	2.06	2.02	1.97	1.92
19	2.38	2.31	2.23	2.16	2.11	2.07	2.03	1.98	1.93	1.88
20	2.35	2.28	2.20	2.12	2.08	2.04	1.99	1.95	1.90	1.84
21	2.32	2.25	2.18	2.10	2.05	2.01	1.96	1.92	1.87	1.81
22	2.30	2.23	2.15	2.07	2.03	1.98	1.94	1.89	1.84	1.78
23	2.27	2.20	2.13	2.05	2.01	1.96	1.91	1.86	1.81	1.76
24	2.25	2.18	2.11	2.03	1.98	1.94	1.89	1.84	1.79	1.73
25	2.24	2.16	2.09	2.01	1.96	1.92	1.87	1.82	1.77	1.71
26	2.22	2.15	2.07	1.99	1.95	1.90	1.85	1.80	1.75	1.69
27	2.20	2.13	2.06	1.97	1.93	1.88	1.84	1.79	1.73	1.67
28	2.19	2.12	2.04	1.96	1.91	1.87	1.82	1.77	1.71	1.65
29	2.18	2.10	2.03	1.94	1.90	1.85	1.81	1.75	1.70	1.64
30	2.16	2.09	2.01	1.93	1.89	1.84	1.79	1.74	1.68	1.62
40	2.08	2.00	1.92	1.84	1.79	1.74	1.69	1.64	1.58	1.51
60	1.99	1.92	1.84	1.75	1.70	1.65	1.59	1.53	1.47	1.39
120	1.91	1.83	1.75	1.66	1.61	1.55	1.50	1.43	1.35	1.25
∞	1.83	1.75	1.67	1.57	1.52	1.46	1.39	1.32	1.22	1.00

TABLE A.6 *(continued)*   Critical Values of the *F*-Distribution

$$f_{0.01}(\nu_1, \nu_2)$$

ν_2	ν_1								
	1	2	3	4	5	6	7	8	9
1	4052	4999.5	5403	5625	5764	5859	5928	5981	6022
2	98.50	99.00	99.17	99.25	99.30	99.33	99.36	99.37	99.39
3	34.12	30.82	29.46	28.71	28.24	27.91	27.67	27.49	27.35
4	21.20	18.00	16.69	15.98	15.52	15.21	14.98	14.80	14.66
5	16.26	13.27	12.06	11.39	10.97	10.67	10.46	10.29	10.16
6	13.75	10.92	9.78	9.15	8.75	8.47	8.26	8.10	7.98
7	12.25	9.55	8.45	7.85	7.46	7.19	6.99	6.84	6.72
8	11.26	8.65	7.59	7.01	6.63	6.37	6.18	6.03	5.91
9	10.56	8.02	6.99	6.42	6.06	5.80	5.61	5.47	5.35
10	10.04	7.56	6.55	5.99	5.64	5.39	5.20	5.06	4.94
11	9.65	7.21	6.22	5.67	5.32	5.07	4.89	4.74	4.63
12	9.33	6.93	5.95	5.41	5.06	4.82	4.64	4.50	4.39
13	9.07	6.70	5.74	5.21	4.86	4.62	4.44	4.30	4.19
14	8.86	6.51	5.56	5.04	4.69	4.46	4.28	4.14	4.03
15	8.68	6.36	5.42	4.89	4.56	4.32	4.14	4.00	3.89
16	8.53	6.23	5.29	4.77	4.44	4.20	4.03	3.89	3.78
17	8.40	6.11	5.18	4.67	4.34	4.10	3.93	3.79	3.68
18	8.29	6.01	5.09	4.58	4.25	4.01	3.84	3.71	3.60
19	8.18	5.93	5.01	4.50	4.17	3.94	3.77	3.63	3.52
20	8.10	5.85	4.94	4.43	4.10	3.87	3.70	3.56	3.46
21	8.02	5.78	4.87	4.37	4.04	3.81	3.64	3.51	3.40
22	7.95	5.72	4.82	4.31	3.99	3.76	3.59	3.45	3.35
23	7.88	5.66	4.76	4.26	3.94	3.71	3.54	3.41	3.30
24	7.82	5.61	4.72	4.22	3.90	3.67	3.50	3.36	3.26
25	7.77	5.57	4.68	4.18	3.85	3.63	3.46	3.32	3.22
26	7.72	5.53	4.64	4.14	3.82	3.59	3.42	3.29	3.18
27	7.68	5.49	4.60	4.11	3.78	3.56	3.39	3.26	3.15
28	7.64	5.45	4.57	4.07	3.75	3.53	3.36	3.23	3.12
29	7.60	5.42	4.54	4.04	3.73	3.50	3.33	3.20	3.09
30	7.56	5.39	4.51	4.02	3.70	3.47	3.30	3.17	3.07
40	7.31	5.18	4.31	3.83	3.51	3.29	3.12	2.99	2.89
60	7.08	4.98	4.13	3.65	3.34	3.12	2.95	2.82	2.72
120	6.85	4.79	3.95	3.48	3.17	2.96	2.79	2.66	2.56
∞	6.63	4.61	3.78	3.32	3.02	2.80	2.64	2.51	2.41

TABLE A.6 *(continued)*   Critical Values of the *F*-Distribution

$$f_{0.01}(\nu_1, \nu_2)$$

ν_2	ν_1									
	10	12	15	20	24	30	40	60	120	∞
1	6056	6106	6157	6209	6235	6261	6287	6313	6339	6366
2	99.40	99.42	99.43	99.45	99.46	99.47	99.47	99.48	99.49	99.50
3	27.23	27.05	26.87	26.69	26.60	26.50	26.41	26.32	26.22	26.13
4	14.55	14.37	14.20	14.02	13.93	13.84	13.75	13.65	13.56	13.46
5	10.05	9.89	9.72	9.55	9.47	9.38	9.29	9.20	9.11	9.02
6	7.87	7.72	7.56	7.40	7.31	7.23	7.14	7.06	6.97	6.88
7	6.62	6.47	6.31	6.16	6.07	5.99	5.91	5.82	5.74	5.65
8	5.81	5.67	5.52	5.36	5.28	5.20	5.12	5.03	4.95	4.86
9	5.26	5.11	4.96	4.81	4.73	4.65	4.57	4.48	4.40	4.31
10	4.85	4.71	4.56	4.41	4.33	4.25	4.17	4.08	4.00	3.91
11	4.54	4.40	4.25	4.10	4.02	3.94	3.86	3.78	3.69	3.60
12	4.30	4.16	4.01	3.86	3.78	3.70	3.62	3.54	3.45	3.36
13	4.10	3.96	3.82	3.66	3.59	3.51	3.43	3.34	3.25	3.17
14	3.94	3.80	3.66	3.51	3.43	3.35	3.27	3.18	3.09	3.00
15	3.80	3.67	3.52	3.37	3.29	3.21	3.13	3.05	2.96	2.87
16	3.69	3.55	3.41	3.26	3.18	3.10	3.02	2.93	2.84	2.75
17	3.59	3.46	3.31	3.16	3.08	3.00	2.92	2.83	2.75	2.65
18	3.51	3.37	3.23	3.08	3.00	2.92	2.84	2.75	2.66	2.57
19	3.43	3.30	3.15	3.00	2.92	2.84	2.76	2.67	2.58	2.49
20	3.37	3.23	3.09	2.94	2.86	2.78	2.69	2.61	2.52	2.42
21	3.31	3.17	3.03	2.88	2.80	2.72	2.64	2.55	2.46	2.36
22	3.26	3.12	2.98	2.83	2.75	2.67	2.58	2.50	2.40	2.31
23	3.21	3.07	2.93	2.78	2.70	2.62	2.54	2.45	2.35	2.26
24	3.17	3.03	2.89	2.74	2.66	2.58	2.49	2.40	2.31	2.21
25	3.13	2.99	2.85	2.70	2.62	2.54	2.45	2.36	2.27	2.17
26	3.09	2.96	2.81	2.66	2.58	2.50	2.42	2.33	2.23	2.13
27	3.06	2.93	2.78	2.63	2.55	2.47	2.38	2.29	2.20	2.10
28	3.03	2.90	2.75	2.60	2.52	2.44	2.35	2.26	2.17	2.06
29	3.00	2.87	2.73	2.57	2.49	2.41	2.33	2.23	2.14	2.03
30	2.98	2.84	2.70	2.55	2.47	2.39	2.30	2.21	2.11	2.01
40	2.80	2.66	2.52	2.37	2.29	2.20	2.11	2.02	1.92	1.80
60	2.63	2.50	2.35	2.20	2.12	2.03	1.94	1.84	1.73	1.60
120	2.47	2.34	2.19	2.03	1.95	1.86	1.76	1.66	1.53	1.38
∞	2.32	2.18	2.04	1.88	1.79	1.70	1.59	1.47	1.32	1.00

TABLE A.7*   Tolerance Factors for Normal Distributions

	$\gamma = 0.05$				$\gamma = 0.01$		
	$1 - \alpha$				$1 - \alpha$		
n	0.90	0.95	0.99	n	0.90	0.95	0.99
2	32.019	37.674	48.430	2	160.193	188.491	242.300
3	8.380	9.916	12.861	3	18.930	22.401	29.055
4	5.369	6.370	8.299	4	9.398	11.150	14.527
5	4.275	5.079	6.634	5	6.612	7.855	10.260
6	3.712	4.414	5.775	6	5.337	6.345	8.301
7	3.369	4.007	5.248	7	4.613	5.488	7.187
8	3.136	3.732	4.891	8	4.147	4.936	6.468
9	2.967	3.532	4.631	9	3.822	4.550	5.966
10	2.839	3.379	4.433	10	3.582	4.265	5.594
11	2.737	3.259	4.277	11	3.397	4.045	5.308
12	2.655	3.162	4.150	12	3.250	3.870	5.079
13	2.587	3.081	4.044	13	3.130	3.727	4.893
14	2.529	3.012	3.955	14	3.029	3.608	4.737
15	2.480	2.954	3.878	15	2.945	3.507	4.605
16	2.437	2.903	3.812	16	2.872	3.421	4.492
17	2.400	2.858	3.754	17	2.808	3.345	4.393
18	2.366	2.819	3.702	18	2.753	3.279	4.307
19	2.337	2.784	3.656	19	2.703	3.221	4.230
20	2.310	2.752	3.615	20	2.659	3.168	4.161
25	2.208	2.631	3.457	25	2.494	2.972	3.904
30	2.140	2.549	3.350	30	2.385	2.841	3.733
35	2.090	2.490	3.272	35	2.306	2.748	3.611
40	2.052	2.445	3.213	40	2.247	2.677	3.518

*Adapted from C. Eisenhart, M. W. Hastay, and W. A. Wallis, *Techniques of Statistical Analysis*, Chapter 2, McGraw-Hill Book Company, New York, 1947. Used with permission of McGraw-Hill Book Company.

TABLE A.7 (*continued*)   Tolerance Factors for Normal Distributions

	$\gamma = 0.05$				$\gamma = 0.01$		
	$1 - \alpha$				$1 - \alpha$		
n	0.90	0.95	0.99	n	0.90	0.95	0.99
45	2.021	2.408	3.165	45	2.200	2.621	3.444
50	1.996	2.379	3.126	50	2.162	2.576	3.385
55	1.976	2.354	3.094	55	2.130	2.538	3.335
60	1.958	2.333	3.066	60	2.103	2.506	3.293
65	1.943	2.315	3.042	65	2.080	2.478	3.257
70	1.929	2.299	3.021	70	2.060	2.454	3.225
75	1.917	2.285	3.002	75	2.042	2.433	3.197
80	1.907	2.272	2.986	80	2.026	2.414	3.173
85	1.897	2.261	2.971	85	2.012	2.397	3.150
90	1.889	2.251	2.958	90	1.999	2.382	3.130
95	1.881	2.241	2.945	95	1.987	2.368	3.112
100	1.874	2.233	2.934	100	1.977	2.355	3.096
150	1.825	2.175	2.859	150	1.905	2.270	2.983
200	1.798	2.143	2.816	200	1.865	2.222	2.921
250	1.780	2.121	2.788	250	1.839	2.191	2.880
300	1.767	2.106	2.767	300	1.820	2.169	2.850
400	1.749	2.084	2.739	400	1.794	2.138	2.809
500	1.737	2.070	2.721	500	1.777	2.117	2.783
600	1.729	2.060	2.707	600	1.764	2.102	2.763
700	1.722	2.052	2.697	700	1.755	2.091	2.748
800	1.717	2.046	2.688	800	1.747	2.082	2.736
900	1.712	2.040	2.682	900	1.741	2.075	2.726
1000	1.709	2.036	2.676	1000	1.736	2.068	2.718
∞	1.645	1.960	2.576	∞	1.645	1.960	2.576

TABLE A.8* Sample Size for the *t*-Test of the Mean

Column groups under **Level of t-test** (each with sub-columns for β = 0.01, 0.05, 0.1, 0.2, 0.5):

- Group 1 — Single-sided test $\alpha = 0.005$; Double-sided test $\alpha = 0.01$
- Group 2 — Single-sided test $\alpha = 0.01$; Double-sided test $\alpha = 0.02$
- Group 3 — Single-sided test $\alpha = 0.025$; Double-sided test $\alpha = 0.05$
- Group 4 — Single-sided test $\alpha = 0.05$; Double-sided test $\alpha = 0.1$

Row label: Value of $\Delta = \dfrac{|\delta|}{\sigma}$

Δ	0.01	0.05	0.1	0.2	0.5	0.01	0.05	0.1	0.2	0.5	0.01	0.05	0.1	0.2	0.5	0.01	0.05	0.1	0.2	0.5	Δ
0.05																					0.05
0.10																					0.10
0.15																				122	0.15
0.20										139					99					70	0.20
0.25					110					90				128	64			139	101	45	0.25
0.30				134	78				115	63			119	90	45		122	97	71	32	0.30
0.35			125	99	58			109	85	47		109	88	67	34		90	72	52	24	0.35
0.40		115	97	77	45		101	85	66	37	117	84	68	51	26	101	70	55	40	19	0.40
0.45		92	77	62	37	110	81	68	53	30	93	67	54	41	21	80	55	44	33	15	0.45
0.50	100	75	63	51	30	90	66	55	43	25	76	54	44	34	18	65	45	36	27	13	0.50
0.55	83	63	53	42	26	75	55	46	36	21	63	45	37	28	15	54	38	30	22	11	0.55
0.60	71	53	45	36	22	63	47	39	31	18	53	38	32	24	13	46	32	26	19	9	0.60
0.65	61	46	39	31	20	55	41	34	27	16	46	33	27	21	12	39	28	22	17	8	0.65
0.70	53	40	34	28	17	47	35	30	24	14	40	29	24	19	10	34	24	19	15	8	0.70
0.75	47	36	30	25	16	42	31	27	21	13	35	26	21	16	9	30	21	17	13	7	0.75
0.80	41	32	27	22	14	37	28	24	19	12	31	22	19	15	9	27	19	15	12	6	0.80
0.85	37	29	24	20	13	33	25	21	17	11	28	21	17	13	8	24	17	14	11	6	0.85
0.90	34	26	22	18	12	29	23	19	16	10	25	19	16	12	7	21	15	13	10	5	0.90
0.95	31	24	20	17	11	27	21	18	14	9	23	17	14	11	7	19	14	11	9	5	0.95
1.00	28	22	19	16	10	25	19	16	13	9	21	16	13	10	6	18	13	11	8	5	1.00
1.1	24	19	16	14	9	21	16	14	12	8	18	13	11	9	6	15	11	9	7		1.1
1.2	21	16	14	12	8	18	14	12	10	7	15	12	10	8	5	13	10	8	6		1.2
1.3	18	15	13	11	8	16	13	11	9	6	14	10	9	7		11	8	7	6		1.3
1.4	16	13	12	10	7	14	11	10	9	6	12	9	8	7		10	8	7	5		1.4
1.5	15	12	11	9	7	13	10	9	8	6	11	8	7	6		9	7	6			1.5
1.6	13	11	10	8	6	12	10	9	7	5	10	8	7	6		8	6	6			1.6
1.7	12	10	9	8	6	11	9	8	7		9	7	6	5		8	6	5			1.7
1.8	12	10	9	8	6	10	8	7	7		8	7	6			7	6				1.8
1.9	11	9	8	7	6	10	8	7	6		8	6	6			7	5				1.9
2.0	10	8	8	7	5	9	7	7	6		7	6	5			6					2.0
2.1	10	8	7	7		8	7	6	6		7	6				6					2.1
2.2	9	8	7	6		8	7	6	5		7	6				6					2.2
2.3	9	7	7	6		8	6	6			6	5				5					2.3
2.4	8	7	7	6		7	6	6			6										2.4
2.5	8	7	6	6		7	6	6			6										2.5
3.0	7	6	6	5		6	5	5			5										3.0
3.5	6	5	5			5															3.5
4.0	6																				4.0

*Reproduced with permission from O. L. Davies, ed., *Design and Analysis of Industrial Experiments*, Oliver & Boyd, Edinburgh, 1956.

TABLE A.9* Sample Size for the *t*-Test of the Difference Between Two Means

	Level of t-test																				
Single-sided test / Double-sided test	α = 0.005 / α = 0.01					α = 0.01 / α = 0.02					α = 0.025 / α = 0.05					α = 0.05 / α = 0.1					
β =	0.01	0.05	0.1	0.2	0.5	0.01	0.05	0.1	0.2	0.5	0.01	0.05	0.1	0.2	0.5	0.01	0.05	0.1	0.2	0.5	
0.05																					0.05
0.10																					0.10
0.15																					0.15
0.20																				137	0.20
0.25															124					88	0.25
0.30										123					87					61	0.30
0.35					110					90					64				102	45	0.35
0.40					85					70				100	50			108	78	35	0.40
0.45				118	68				101	55			105	79	39		108	86	62	28	0.45
0.50				96	55			106	82	45		106	86	64	32		88	70	51	23	0.50
0.55			101	79	46		106	88	68	38		87	71	53	27	112	73	58	42	19	0.55
0.60		101	85	67	39		90	74	58	32	104	74	60	45	23	89	61	49	36	16	0.60
0.65		87	73	57	34	104	77	64	49	27	88	63	51	39	20	76	52	42	30	14	0.65
0.70	100	75	63	50	29	90	66	55	43	24	76	55	44	34	17	66	45	36	26	12	0.70
0.75	88	66	55	44	26	79	58	48	38	21	67	48	39	29	15	57	40	32	23	11	0.75
0.80	77	58	49	39	23	70	51	43	33	19	59	42	34	26	14	50	35	28	21	10	0.80
0.85	69	51	43	35	21	62	46	38	30	17	52	37	31	23	12	45	31	25	18	9	0.85
0.90	62	46	39	31	19	55	41	34	27	15	47	34	27	21	11	40	28	22	16	8	0.90
0.95	55	42	35	28	17	50	37	31	24	14	42	30	25	19	10	36	25	20	15	7	0.95
1.00	50	38	32	26	15	45	33	28	22	13	38	27	23	17	9	33	23	18	14	7	1.00
1.1	42	32	27	22	13	38	28	23	19	11	32	23	19	14	8	27	19	15	12	6	1.1
1.2	36	27	23	18	11	32	24	20	16	9	27	20	16	12	7	23	16	13	10	5	1.2
1.3	31	23	20	16	10	28	21	17	14	8	23	17	14	11	6	20	14	11	9	5	1.3
1.4	27	20	17	14	9	24	18	15	12	8	20	15	12	10	6	17	12	10	8	4	1.4
1.5	24	18	15	13	8	21	16	14	11	7	18	13	11	9	5	15	11	9	7	4	1.5
1.6	21	16	14	11	7	19	14	12	10	6	16	12	10	8	5	14	10	8	6	4	1.6
1.7	19	15	13	10	7	17	13	11	9	6	14	11	9	7	4	12	9	7	6	3	1.7
1.8	17	13	11	10	6	15	12	10	8	5	13	10	8	6	4	11	8	7	5		1.8
1.9	16	12	11	9	6	14	11	9	8	5	12	9	7	6	4	10	7	6	5		1.9
2.0	14	11	10	8	6	13	10	9	7	5	11	8	7	6	4	9	7	6	4		2.0
2.1	13	10	9	8	5	12	9	8	7	5	10	8	6	5	3	8	6	5	4		2.1
2.2	12	10	8	7	5	11	9	7	6	4	9	7	6	5		8	6	5	4		2.2
2.3	11	9	8	7	5	10	8	7	6	4	9	7	6	5		7	5	5	4		2.3
2.4	11	9	8	6	5	10	8	7	6	4	8	6	5	4		7	5	4	4		2.4
2.5	10	8	7	6	4	9	7	6	5	4	8	6	5	4		6	5	4	3		2.5
3.0	8	6	6	5	4	7	6	5	4	3	6	5	4	4		5	4	3			3.0
3.5	6	5	5	4	3	6	5	4	4		5	4	4	3		4	3				3.5
4.0	6	5	4	4		5	4	4	3		4	4	3			4					4.0

Value of Δ = |δ|/σ

*Reproduced with permission from O. L. Davies, ed., *Design and Analysis of Industrial Experiments*, Oliver & Boyd, Edinburgh, 1956.

TABLE A.10* Critical Values for Bartlett's Test

$$b_k(0.01; n)$$

	Number of Populations, k								
n	2	3	4	5	6	7	8	9	10
3	.1411	.1672	*	*	*	*	*	*	*
4	.2843	.3165	.3475	.3729	.3937	.4110	*	*	*
5	.3984	.4304	.4607	.4850	.5046	.5207	.5343	.5458	.5558
6	.4850	.5149	.5430	.5653	.5832	.5978	.6100	.6204	.6293
7	.5512	.5787	.6045	.6248	.6410	.6542	.6652	.6744	.6824
8	.6031	.6282	.6518	.6704	.6851	.6970	.7069	.7153	.7225
9	.6445	.6676	.6892	.7062	.7197	.7305	.7395	.7471	.7536
10	.6783	.6996	.7195	.7352	.7475	.7575	.7657	.7726	.7786
11	.7063	.7260	.7445	.7590	.7703	.7795	.7871	.7935	.7990
12	.7299	.7483	.7654	.7789	.7894	.7980	.8050	.8109	.8160
13	.7501	.7672	.7832	.7958	.8056	.8135	.8201	.8256	.8303
14	.7674	.7835	.7985	.8103	.8195	.8269	.8330	.8382	.8426
15	.7825	.7977	.8118	.8229	.8315	.8385	.8443	.8491	.8532
16	.7958	.8101	.8235	.8339	.8421	.8486	.8541	.8586	.8625
17	.8076	.8211	.8338	.8436	.8514	.8576	.8627	.8670	.8707
18	.8181	.8309	.8429	.8523	.8596	.8655	.8704	.8745	.8780
19	.8275	.8397	.8512	.8601	.8670	.8727	.8773	.8811	.8845
20	.8360	.8476	.8586	.8671	.8737	.8791	.8835	.8871	.8903
21	.8437	.8548	.8653	.8734	.8797	.8848	.8890	.8926	.8956
22	.8507	.8614	.8714	.8791	.8852	.8901	.8941	.8975	.9004
23	.8571	.8673	.8769	.8844	.8902	.8949	.8988	.9020	.9047
24	.8630	.8728	.8820	.8892	.8948	.8993	.9030	.9061	.9087
25	.8684	.8779	.8867	.8936	.8990	.9034	.9069	.9099	.9124
26	.8734	.8825	.8911	.8977	.9029	.9071	.9105	.9134	.9158
27	.8781	.8869	.8951	.9015	.9065	.9105	.9138	.9166	.9190
28	.8824	.8909	.8988	.9050	.9099	.9138	.9169	.9196	.9219
29	.8864	.8946	.9023	.9083	.9130	.9167	.9198	.9224	.9246
30	.8902	.8981	.9056	.9114	.9159	.9195	.9225	.9250	.9271
40	.9175	.9235	.9291	.9335	.9370	.9397	.9420	.9439	.9455
50	.9339	.9387	.9433	.9468	.9496	.9518	.9536	.9551	.9564
60	.9449	.9489	.9527	.9557	.9580	.9599	.9614	.9626	.9637
80	.9586	.9617	.9646	.9668	.9685	.9699	.9711	.9720	.9728
100	9669	.9693	.9716	.9734	.9748	.9759	.9769	.9776	.9783

*Reproduced from D. D. Dyer and J. P. Keating, "On the determination of critical values for Bartlett's test," *J. Am. Stat. Assoc.*, Vol. 75, 1980, by permission of the Board of Directors.

TABLE A.10 (*continued*)   Critical Values for Bartlett's Test

$$b_k(0.05; n)$$

n	Number of Populations, k								
	2	3	4	5	6	7	8	9	10
3	.3123	.3058	.3173	.3299	*	*	*	*	*
4	.4780	.4699	.4803	.4921	.5028	.5122	.5204	.5277	.5341
5	.5845	.5762	.5850	.5952	.6045	.6126	.6197	.6260	.6315
6	.6563	.6483	.6559	.6646	.6727	.6798	.6860	.6914	.6961
7	.7075	.7000	.7065	.7142	.7213	.7275	.7329	.7376	.7418
8	.7456	.7387	.7444	.7512	.7574	.7629	.7677	.7719	.7757
9	.7751	.7686	.7737	.7798	.7854	.7903	.7946	.7984	.8017
10	.7984	.7924	.7970	.8025	.8076	.8121	.8160	.8194	.8224
11	.8175	.8118	.8160	.8210	.8257	.8298	.8333	.8365	.8392
12	.8332	.8280	.8317	.8364	.8407	.8444	.8477	.8506	.8531
13	.8465	.8415	.8450	.8493	.8533	.8568	.8598	.8625	.8648
14	.8578	.8532	.8564	.8604	.8641	.8673	.8701	.8726	.8748
15	.8676	.8632	.8662	.8699	.8734	.8764	.8790	.8814	.8834
16	.8761	.8719	.8747	.8782	.8815	.8843	.8868	.8890	.8909
17	.8836	.8796	.8823	.8856	.8886	.8913	.8936	.8957	.8975
18	.8902	.8865	.8890	.8921	.8949	.8975	.8997	.9016	.9033
19	.8961	.8926	.8949	.8979	.9006	.9030	.9051	.9069	.9086
20	.9015	.8980	.9003	.9031	.9057	.9080	.9100	.9117	.9132
21	.9063	.9030	.9051	.9078	.9103	.9124	.9143	.9160	.9175
22	.9106	.9075	.9095	.9120	.9144	.9165	.9183	.9199	.9213
23	.9146	.9116	.9135	.9159	.9182	.9202	.9219	.9235	.9248
24	.9182	.9153	.9172	.9195	.9217	.9236	.9253	.9267	.9280
25	.9216	.9187	.9205	.9228	.9249	.9267	.9283	.9297	.9309
26	.9246	.9219	.9236	.9258	.9278	.9296	.9311	.9325	.9336
27	.9275	.9249	.9265	.9286	.9305	.9322	.9337	.9350	.9361
28	.9301	.9276	.9292	.9312	.9330	.9347	.9361	.9374	.9385
29	.9326	.9301	.9316	.9336	.9354	.9370	.9383	.9396	.9406
30	.9348	.9325	.9340	.9358	.9376	.9391	.9404	.9416	.9426
40	.9513	.9495	.9506	.9520	.9533	.9545	.9555	.9564	.9572
50	.9612	.9597	.9606	.9617	.9628	.9637	.9645	.9652	.9658
60	.9677	.9665	.9672	.9681	.9690	.9698	.9705	.9710	.9716
80	.9758	.9749	.9754	.9761	.9768	.9774	.9779	.9783	.9787
100	.9807	.9799	.9804	.9809	.9815	.9819	.9823	.9827	.9830

TABLE A.11* Critical Values for Cochran's Test

$$\alpha = 0.01$$

k \ n	2	3	4	5	6	7	8	9	10	11	17	37	145	∞
2	0.9999	0.9950	0.9794	0.9586	0.9373	0.9172	0.8988	0.8823	0.8674	0.8539	0.7949	0.7067	0.6062	0.5000
3	0.9933	0.9423	0.8831	0.8335	0.7933	0.7606	0.7335	0.7107	0.6912	0.6743	0.6059	0.5153	0.4230	0.3333
4	0.9676	0.8643	0.7814	0.7212	0.6761	0.6410	0.6129	0.5897	0.5702	0.5536	0.4884	0.4057	0.3251	0.2500
5	0.9279	0.7885	0.6957	0.6329	0.5875	0.5531	0.5259	0.5037	0.4854	0.4697	0.4094	0.3351	0.2644	0.2000
6	0.8828	0.7218	0.6258	0.5635	0.5195	0.4866	0.4608	0.4401	0.4229	0.4084	0.3529	0.2858	0.2229	0.1667
7	0.8376	0.6644	0.5685	0.5080	0.4659	0.4347	0.4105	0.3911	0.3751	0.3616	0.3105	0.2494	0.1929	0.1429
8	0.7945	0.6152	0.5209	0.4627	0.4226	0.3932	0.3704	0.3522	0.3373	0.3248	0.2779	0.2214	0.1700	0.1250
9	0.7544	0.5727	0.4810	0.4251	0.3870	0.3592	0.3378	0.3207	0.3067	0.2950	0.2514	0.1992	0.1521	0.1111
10	0.7175	0.5353	0.4469	0.3934	0.3572	0.3308	0.3106	0.2945	0.2813	0.2704	0.2297	0.1811	0.1376	0.1000
12	0.6528	0.4751	0.3919	0.3428	0.3099	0.2861	0.2680	0.2535	0.2419	0.2320	0.1961	0.1535	0.1157	0.0833
15	0.5747	0.4069	0.3317	0.2882	0.2593	0.2386	0.2228	0.2104	0.2002	0.1918	0.1612	0.1251	0.0934	0.0667
20	0.4799	0.3297	0.2654	0.2288	0.2048	0.1877	0.1748	0.1646	0.1567	0.1501	0.1248	0.0960	0.0709	0.0500
24	0.4217	0.2871	0.2295	0.1970	0.1759	0.1608	0.1495	0.1406	0.1338	0.1283	0.1060	0.0810	0.0595	0.0417
30	0.3632	0.2412	0.1913	0.1635	0.1454	0.1327	0.1232	0.1157	0.1100	0.1054	0.0867	0.0658	0.0480	0.0333
40	0.2940	0.1915	0.1508	0.1281	0.1135	0.1033	0.0957	0.0898	0.0853	0.0816	0.0668	0.0503	0.0363	0.0250
60	0.2151	0.1371	0.1069	0.0902	0.0796	0.0722	0.0668	0.0625	0.0594	0.0567	0.0461	0.0344	0.0245	0.0167
120	0.1225	0.0759	0.0585	0.0489	0.0429	0.0387	0.0357	0.0334	0.0316	0.0302	0.0242	0.0178	0.0125	0.0083
∞	0	0	0	0	0	0	0	0	0	0	0	0	0	0

*Reproduced from C. Eisenhart, M. W. Hastay, and W. A. Wallis, *Techniques of Statistical Analysis*, Chapter 15, McGraw-Hill Book Company, New York, 1947. Used with permission of McGraw-Hill Book Company.

TABLE A.11 (*continued*)   Critical Values for Cochran's Test

$\alpha = 0.05$

k \ n	2	3	4	5	6	7	8	9	10	11	17	37	145	∞
2	0.9985	0.9750	0.9392	0.9057	0.8772	0.8534	0.8332	0.8159	0.8010	0.7880	0.7341	0.6602	0.5813	0.5000
3	0.9669	0.8709	0.7977	0.7457	0.7071	0.6771	0.6530	0.6333	0.6167	0.6025	0.5466	0.4748	0.4031	0.3333
4	0.9065	0.7679	0.6841	0.6287	0.5895	0.5598	0.5365	0.5175	0.5017	0.4884	0.4366	0.3720	0.3093	0.2500
5	0.8412	0.6838	0.5981	0.5441	0.5065	0.4783	0.4564	0.4387	0.4241	0.4118	0.3645	0.3066	0.2513	0.2000
6	0.7808	0.6161	0.5321	0.4803	0.4447	0.4184	0.3980	0.3817	0.3682	0.3568	0.3135	0.2612	0.2119	0.1667
7	0.7271	0.5612	0.4800	0.4307	0.3974	0.3726	0.3535	0.3384	0.3259	0.3154	0.2756	0.2278	0.1833	0.1429
8	0.6798	0.5157	0.4377	0.3910	0.3595	0.3362	0.3185	0.3043	0.2926	0.2829	0.2462	0.2022	0.1616	0.1250
9	0.6385	0.4775	0.4027	0.3584	0.3286	0.3067	0.2901	0.2768	0.2659	0.2568	0.2226	0.1820	0.1446	0.1111
10	6.6020	0.4450	0.3733	0.3311	0.3029	0.2823	0.2666	0.2541	0.2439	0.2353	0.2032	0.1655	0.1308	0.1000
12	0.5410	0.3924	0.3264	0.2880	0.2624	0.2439	0.2299	0.2187	0.2098	0.2020	0.1737	0.1403	0.1100	0.0833
15	0.4709	0.3346	0.2758	0.2419	0.2195	0.2034	0.1911	0.1815	0.1736	0.1671	0.1429	0.1144	0.0889	0.0667
20	0.3894	0.2705	0.2205	0.1921	0.1735	0.1602	0.1501	0.1422	0.1357	0.1303	0.1108	0.0879	0.0675	0.0500
24	0.3434	0.2354	0.1907	0.1656	0.1493	0.1374	0.1286	0.1216	0.1160	0.1113	0.0942	0.0743	0.0567	0.0417
30	0.2929	0.1980	0.1593	0.1377	0.1237	0.1137	0.1061	0.1002	0.0958	0.0921	0.0771	0.0604	0.0457	0.0333
40	0.2370	0.1576	0.1259	0.1082	0.0968	0.0887	0.0827	0.0780	0.0745	0.0713	0.0595	0.0462	0.0347	0.0250
60	0.1737	0.1131	0.0895	0.0765	0.0682	0.0623	0.0583	0.0552	0.0520	0.0497	0.0411	0.0316	0.0234	0.0167
120	0.0998	0.0632	0.0495	0.0419	0.0371	0.0337	0.0312	0.0292	0.0279	0.0266	0.0218	0.0165	0.0120	0.0083
∞	0	0	0	0	0	0	0	0	0	0	0	0	0	0

TABLE A.12*   Least Significant Studentized Ranges r_p

$$\alpha = 0.05$$

ν	\multicolumn{9}{c}{p}								
	2	3	4	5	6	7	8	9	10
1	17.97	17.97	17.97	17.97	17.97	17.97	17.97	17.97	17.97
2	6.085	6.085	6.085	6.085	6.085	6.085	6.085	6.085	6.085
3	4.501	4.516	4.516	4.516	4.516	4.516	4.516	4.516	4.516
4	3.927	4.013	4.033	4.033	4.033	4.033	4.033	4.033	4.033
5	3.635	3.749	3.797	3.814	3.814	3.814	3.814	3.814	3.814
6	3.461	3.587	3.649	3.680	3.694	3.697	3.697	3.697	3.697
7	3.344	3.477	3.548	3.588	3.611	3.622	3.626	3.626	3.626
8	3.261	3.399	3.475	3.521	3.549	3.566	3.575	3.579	3.579
9	3.199	3.339	3.420	3.470	3.502	3.523	3.536	3.544	3.547
10	3.151	3.293	3.376	3.430	3.465	3.489	3.505	3.516	3.522
11	3.113	3.256	3.342	3.397	3.435	3.462	3.480	3.493	3.501
12	3.082	3.225	3.313	3.370	3.410	3.439	3.459	3.474	3.484
13	3.055	3.200	3.289	3.348	3.389	3.419	3.442	3.458	3.470
14	3.033	3.178	3.268	3.329	3.372	3.403	3.426	3.444	3.457
15	3.014	3.160	3.250	3.312	3.356	3.389	3.413	3.432	3.446
16	2.998	3.144	3.235	3.298	3.343	3.376	3.402	3.422	3.437
17	2.984	3.130	3.222	3.285	3.331	3.366	3.392	3.412	3.429
18	2.971	3.118	3.210	3.274	3.321	3.356	3.383	3.405	3.421
19	2.960	3.107	3.199	3.264	3.311	3.347	3.375	3.397	3.415
20	2.950	3.097	3.190	3.255	3.303	3.339	3.368	3.391	3.409
24	2.919	3.066	3.160	3.226	3.276	3.315	3.345	3.370	3.390
30	2.888	3.035	3.131	3.199	3.250	3.290	3.322	3.349	3.371
40	2.858	3.006	3.102	3.171	3.224	3.266	3.300	3.328	3.352
60	2.829	2.976	3.073	3.143	3.198	3.241	3.277	3.307	3.333
120	2.800	2.947	3.045	3.116	3.172	3.217	3.254	3.287	3.314
∞	2.772	2.918	3.017	3.089	3.146	3.193	3.232	3.265	3.294

* Abridged from H. Leon Harter, "†Critical Values for Duncan's New Multiple Range Test," *Biometrics*, Vol. 16, No. 4, 1960, by permission of the author and the editor.

TABLE A.12 (*continued*)   Least Significant Studentized Ranges r_p

$\alpha = 0.01$

ν	2	3	4	5	6	7	8	9	10
1	90.03	90.03	90.03	90.03	90.03	90.03	90.03	90.03	90.03
2	14.04	14.04	14.04	14.04	14.04	14.04	14.04	14.04	14.04
3	8.261	8.321	8.321	8.321	8.321	8.321	8.321	8.321	8.321
4	6.512	6.677	6.740	6.756	6.756	6.756	6.756	6.756	6.756
5	5.702	5.893	5.989	6.040	6.065	6.074	6.074	6.074	6.074
6	5.243	5.439	5.549	5.614	5.655	5.680	5.694	5.701	5.703
7	4.949	5.145	5.260	5.334	5.383	5.416	5.439	5.454	5.464
8	4.746	4.939	5.057	5.135	5.189	5.227	5.256	5.276	5.291
9	4.596	4.787	4.906	4.986	5.043	5.086	5.118	5.142	5.160
10	4.482	4.671	4.790	4.871	4.931	4.975	5.010	5.037	5.058
11	4.392	4.579	4.697	4.780	4.841	4.887	4.924	4.952	4.975
12	4.320	4.504	4.622	4.706	4.767	4.815	4.852	4.883	4.907
13	4.260	4.442	4.560	4.644	4.706	4.755	4.793	4.824	4.850
14	4.210	4.391	4.508	4.591	4.654	4.704	4.743	4.775	4.802
15	4.168	4.347	4.463	4.547	4.610	4.660	4.700	4.733	4.760
16	4.131	4.309	4.425	4.509	4.572	4.622	4.663	4.696	4.724
17	4.099	4.275	4.391	4.475	4.539	4.589	4.630	4.664	4.693
18	4.071	4.246	4.362	4.445	4.509	4.560	4.601	4.635	4.664
19	4.046	4.220	4.335	4.419	4.483	4.534	4.575	4.610	4.639
20	4.024	4.197	4.312	4.395	4.459	4.510	4.552	4.587	4.617
24	3.956	4.126	4.239	4.322	4.386	4.437	4.480	4.516	4.546
30	3.889	4.056	4.168	4.250	4.314	4.366	4.409	4.445	4.477
40	3.825	3.988	4.098	4.180	4.244	4.296	4.339	4.376	4.408
60	3.762	3.922	4.031	4.111	4.174	4.226	4.270	4.307	4.340
120	3.702	3.858	3.965	4.044	4.107	4.158	4.202	4.239	4.272
∞	3.643	3.796	3.900	3.978	4.040	4.091	4.135	4.172	4.205

TABLE A.13*   Values of $d_{\alpha/2}(k, v)$ for Two-Sided Comparisons Between k Treatments and a Control

$$\alpha = 0.05$$

v	\multicolumn{9}{c}{k = number of treatment means (excluding control)}
	1	2	3	4	5	6	7	8	9
5	2.57	3.03	3.29	3.48	3.62	3.73	3.82	3.90	3.97
6	2.45	2.86	3.10	3.26	3.39	3.49	3.57	3.64	3.71
7	2.36	2.75	2.97	3.12	3.24	3.33	3.41	3.47	3.53
8	2.31	2.67	2.88	3.02	3.13	3.22	3.29	3.35	3.41
9	2.26	2.61	2.81	2.95	3.05	3.14	3.20	3.26	3.32
10	2.23	2.57	2.76	2.89	2.99	3.07	3.14	3.19	3.24
11	2.20	2.53	2.72	2.84	2.94	3.02	3.08	3.14	3.19
12	2.18	2.50	2.68	2.81	2.90	2.98	3.04	3.09	3.14
13	2.16	2.48	2.65	2.78	2.87	2.94	3.00	3.06	3.10
14	2.14	2.46	2.63	2.75	2.84	2.91	2.97	3.02	3.07
15	2.13	2.44	2.61	2.73	2.82	2.89	2.95	3.00	3.04
16	2.12	2.42	2.59	2.71	2.80	2.87	2.92	2.97	3.02
17	2.11	2.41	2.58	2.69	2.78	2.85	2.90	2.95	3.00
18	2.10	2.40	2.56	2.68	2.76	2.83	2.89	2.94	2.98
19	2.09	2.39	2.55	2.66	2.75	2.81	2.87	2.92	2.96
20	2.09	2.38	2.54	2.65	2.73	2.80	2.86	2.90	2.95
24	2.06	2.35	2.51	2.61	2.70	2.76	2.81	2.86	2.90
30	2.04	2.32	2.47	2.58	2.66	2.72	2.77	2.82	2.86
40	2.02	2.29	2.44	2.54	2.62	2.68	2.73	2.77	2.81
60	2.00	2.27	2.41	2.51	2.58	2.64	2.69	2.73	2.77
120	1.98	2.24	2.38	2.47	2.55	2.60	2.65	2.69	2.73
∞	1.96	2.21	2.35	2.44	2.51	2.57	2.61	2.65	2.69

*Reproduced from Charles W. Dunnett, ''New Tables for Multiple Comparison with a Control,'' *Biometrics*, Vol. 20, No. 3, 1964, by permission of the author and the editor.

TABLE A.13 (*continued*)   Values of $d_{\alpha/2}(k, v)$ for Two-Sided Comparisons Between k Treatments and a Control

$$\alpha = 0.01$$

v	\multicolumn{9}{c}{k = number of treatment means (excluding control)}
	1	2	3	4	5	6	7	8	9
5	4.03	4.63	4.98	5.22	5.41	5.56	5.69	5.80	5.89
6	3.71	4.21	4.51	4.71	4.87	5.00	5.10	5.20	5.28
7	3.50	3.95	4.21	4.39	4.53	4.64	4.74	4.82	4.89
8	3.36	3.77	4.00	4.17	4.29	4.40	4.48	4.56	4.62
9	3.25	3.63	3.85	4.01	4.12	4.22	4.30	4.37	4.43
10	3.17	3.53	3.74	3.88	3.99	4.08	4.16	4.22	4.28
11	3.11	3.45	3.65	3.79	3.89	3.98	4.05	4.11	4.16
12	3.05	3.39	3.58	3.71	3.81	3.89	3.96	4.02	4.07
13	3.01	3.33	3.52	3.65	3.74	3.82	3.89	3.94	3.99
14	2.98	3.29	3.47	3.59	3.69	3.76	3.83	3.88	3.93
15	2.95	3.25	3.43	3.55	3.64	3.71	3.78	3.83	3.88
16	2.92	3.22	3.39	3.51	3.60	3.67	3.73	3.78	3.83
17	2.90	3.19	3.36	3.47	3.56	3.63	3.69	3.74	3.79
18	2.88	3.17	3.33	3.44	3.53	3.60	3.66	3.71	3.75
19	2.86	3.15	3.31	3.42	3.50	3.57	3.63	3.68	3.72
20	2.85	3.13	3.29	3.40	3.48	3.55	3.60	3.65	3.69
24	2.80	3.07	3.22	3.32	3.40	3.47	3.52	3.57	3.61
30	2.75	3.01	3.15	3.25	3.33	3.39	3.44	3.49	3.52
40	2.70	2.95	3.09	3.19	3.26	3.32	3.37	3.41	3.44
60	2.66	2.90	3.03	3.12	3.19	3.25	3.29	3.33	3.37
120	2.62	2.85	2.97	3.06	3.12	3.18	3.22	3.26	3.29
∞	2.58	2.79	2.92	3.00	3.06	3.11	3.15	3.19	3.22

TABLE A.14*   Values of $d_\alpha(k, v)$ for One-Sided Comparisons Between k Treatments and a Control

$$\alpha = 0.05$$

v	\multicolumn{9}{c}{k = number of treatment means (excluding control)}
	1	2	3	4	5	6	7	8	9
5	2.02	2.44	2.68	2.85	2.98	3.08	3.16	3.24	3.30
6	1.94	2.34	2.56	2.71	2.83	2.92	3.00	3.07	3.12
7	1.89	2.27	2.48	2.62	2.73	2.82	2.89	2.95	3.01
8	1.86	2.22	2.42	2.55	2.66	2.74	2.81	2.87	2.92
9	1.83	2.18	2.37	2.50	2.60	2.68	2.75	2.81	2.86
10	1.81	2.15	2.34	2.47	2.56	2.64	2.70	2.76	2.81
11	1.80	2.13	2.31	2.44	2.53	2.60	2.67	2.72	2.77
12	1.78	2.11	2.29	2.41	2.50	2.58	2.64	2.69	2.74
13	1.77	2.09	2.27	2.39	2.48	2.55	2.61	2.66	2.71
14	1.76	2.08	2.25	2.37	2.46	2.53	2.59	2.64	2.69
15	1.75	2.07	2.24	2.36	2.44	2.51	2.57	2.62	2.67
16	1.75	2.06	2.23	2.34	2.43	2.50	2.56	2.61	2.65
17	1.74	2.05	2.22	2.33	2.42	2.49	2.54	2.59	2.64
18	1.73	2.04	2.21	2.32	2.41	2.48	2.53	2.58	2.62
19	1.73	2.03	2.20	2.31	2.40	2.47	2.52	2.57	2.61
20	1.72	2.03	2.19	2.30	2.39	2.46	2.51	2.56	2.60
24	1.71	2.01	2.17	2.28	2.36	2.43	2.48	2.53	2.57
30	1.70	1.99	2.15	2.25	2.33	2.40	2.45	2.50	2.54
40	1.68	1.97	2.13	2.23	2.31	2.37	2.42	2.47	2.51
60	1.67	1.95	2.10	2.21	2.28	2.35	2.39	2.44	2.48
120	1.66	1.93	2.08	2.18	2.26	2.32	2.37	2.41	2.45
∞	1.64	1.92	2.06	2.16	2.23	2.29	2.34	2.38	2.42

*Reproduced from Charles W. Dunnett, "A Multiple Comparison Procedure for Comparing Several Treatments with a Control," *J. Am. Stat. Assoc.*, Vol. 50, 1955, 1096–1121, by permission of the author and the editor.

TABLE A.14 (*continued*)   Values of $d_\alpha(k, v)$ for One-Sided Comparisons Between k Treatments and a Control

$$\alpha = 0.01$$

v	k = number of treatment means (excluding control)								
	1	2	3	4	5	6	7	8	9
5	3.37	3.90	4.21	4.43	4.60	4.73	4.85	4.94	5.03
6	3.14	3.61	3.88	4.07	4.21	4.33	4.43	4.51	4.59
7	3.00	3.42	3.66	3.83	3.96	4.07	4.15	4.23	4.30
8	2.90	3.29	3.51	3.67	3.79	3.88	3.96	4.03	4.09
9	2.82	3.19	3.40	3.55	3.66	3.75	3.82	3.89	3.94
10	2.76	3.11	3.31	3.45	3.56	3.64	3.71	3.78	3.83
11	2.72	3.06	3.25	3.38	3.48	3.56	3.63	3.69	3.74
12	2.68	3.01	3.19	3.32	3.42	3.50	3.56	3.62	3.67
13	2.65	2.97	3.15	3.27	3.37	3.44	3.51	3.56	3.61
14	2.62	2.94	3.11	3.23	3.32	3.40	3.46	3.51	3.56
15	2.60	2.91	3.08	3.20	3.29	3.36	3.42	3.47	3.52
16	2.58	2.88	3.05	3.17	3.26	3.33	3.39	3.44	3.48
17	2.57	2.86	3.03	3.14	3.23	3.30	3.36	3.41	3.45
18	2.55	2.84	3.01	3.12	3.21	3.27	3.33	3.38	3.42
19	2.54	2.83	2.99	3.10	3.18	3.25	3.31	3.36	3.40
20	2.53	2.81	2.97	3.08	3.17	3.23	3.29	3.34	3.38
24	2.49	2.77	2.92	3.03	3.11	3.17	3.22	3.27	3.31
30	2.46	2.72	2.87	2.97	3.05	3.11	3.16	3.21	3.24
40	2.42	2.68	2.82	2.92	2.99	3.05	3.10	3.14	3.18
60	2.39	2.64	2.78	2.87	2.94	3.00	3.04	3.08	3.12
120	2.36	2.60	2.73	2.82	2.89	2.94	2.99	3.03	3.06
∞	2.33	2.56	2.68	2.77	2.84	2.89	2.93	2.97	3.00

TABLE A.15* Power of the Analysis-of-Variance Test

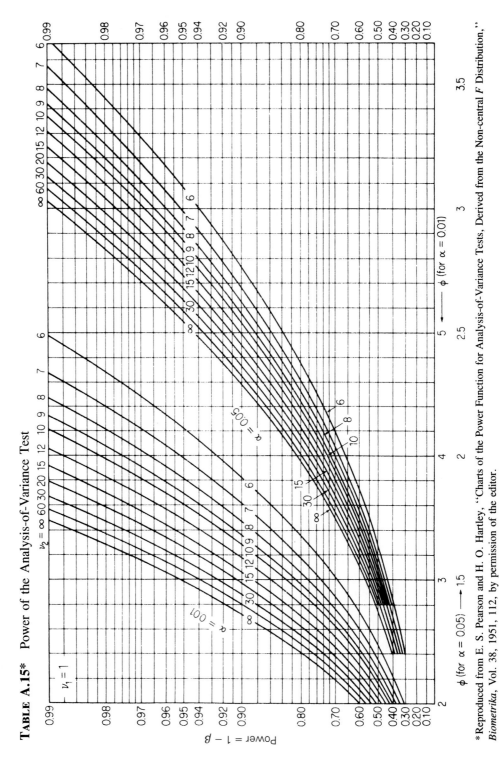

* Reproduced from E. S. Pearson and H. O. Hartley, "Charts of the Power Function for Analysis-of-Variance Tests, Derived from the Non-central F Distribution," *Biometrika*, Vol. 38, 1951, 112, by permission of the editor.

TABLE A.15 (*continued*) Power of the Analysis-of-Variance Test

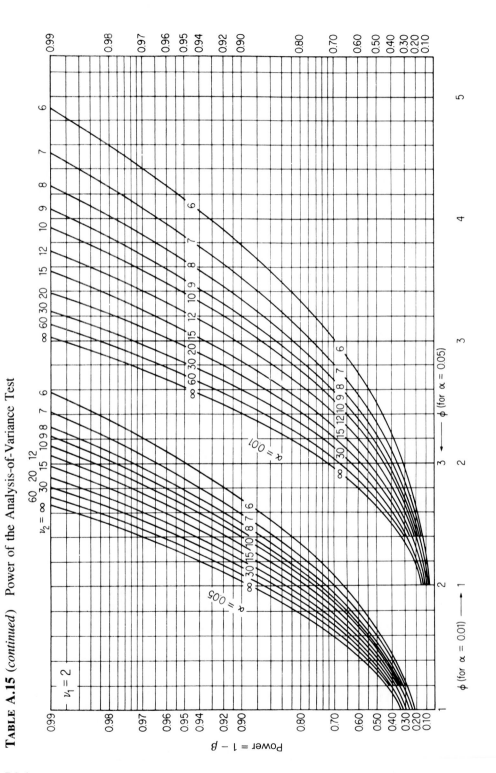

726

TABLE A.15 (*continued*) Power of the Analysis-of-Variance Test

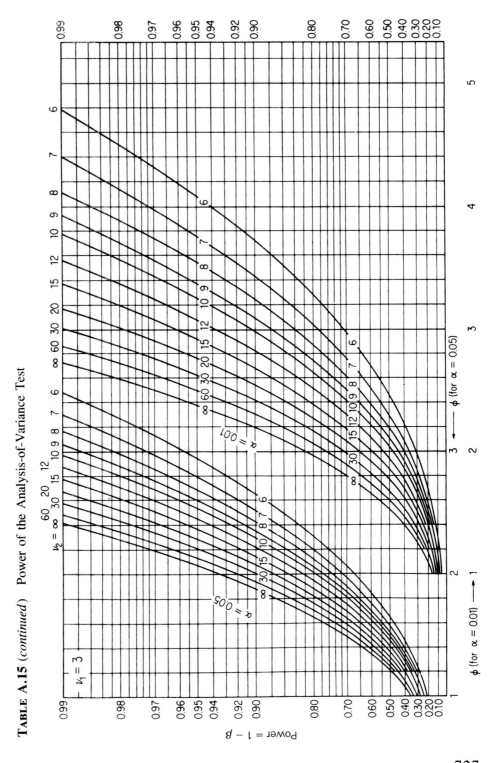

727

TABLE A.15 (*continued*) Power of the Analysis-of-Variance Test

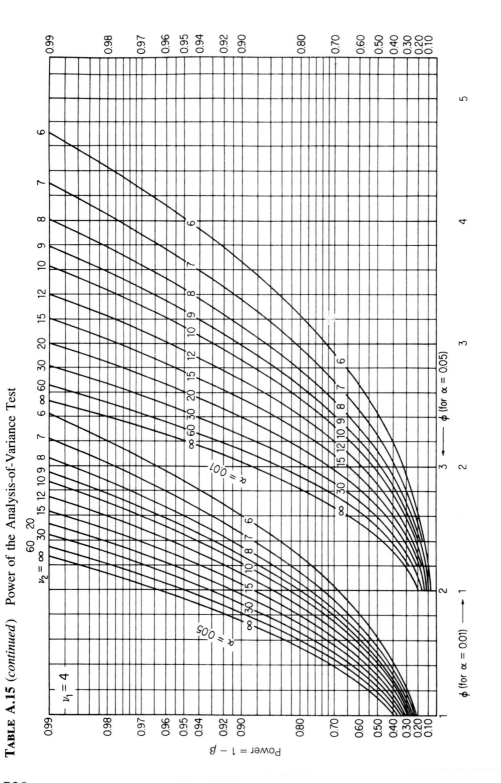

728

TABLE A.15 (*continued*) Power of the Analysis-of-Variance Test

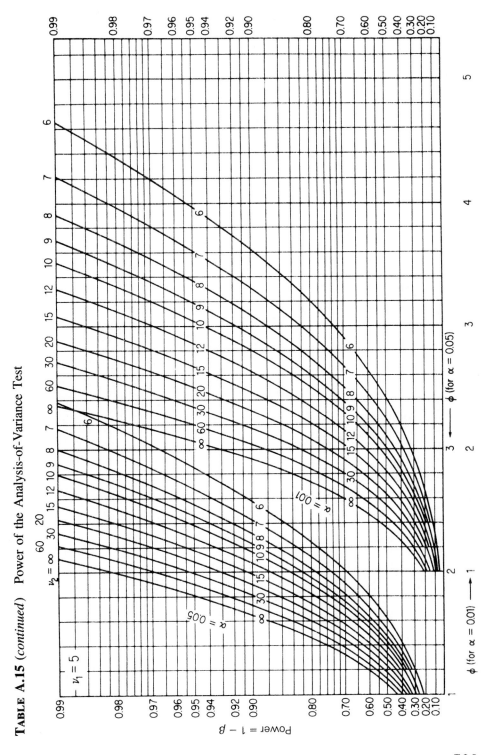

729

TABLE A.15 (*continued*) Power of the Analysis-of-Variance Test

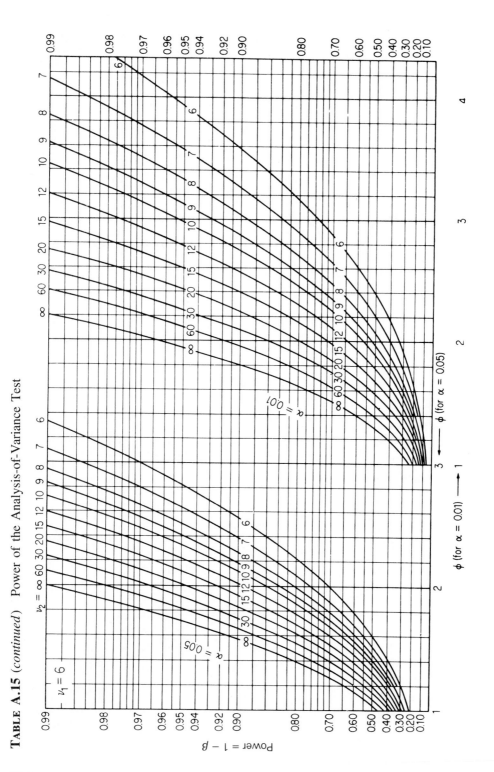

TABLE A.15 (*continued*) Power of the Analysis-of-Variance Test

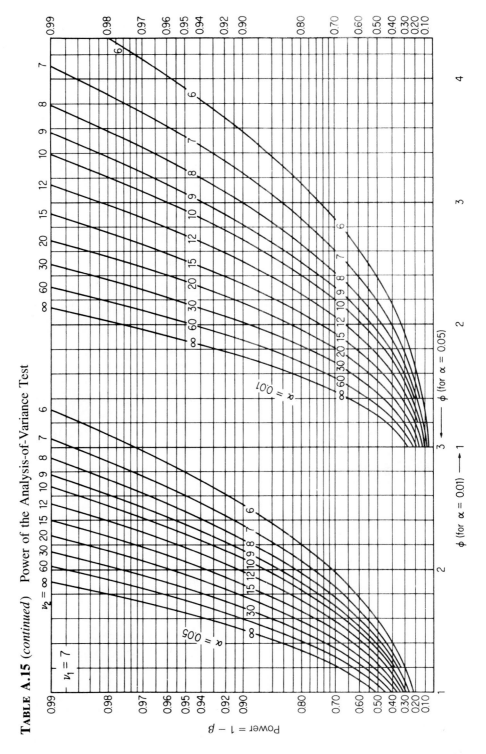

731

TABLE A.15 (*continued*) Power of the Analysis-of-Variance Test

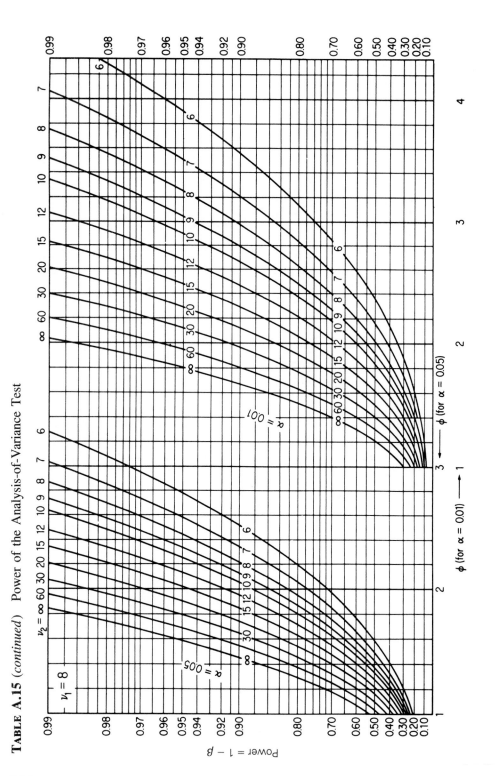

732

TABLE A.16* Critical Values for the Signed-Rank Test

n	One-sided $\alpha = 0.01$ Two-sided $\alpha = 0.02$	One-sided $\alpha = 0.025$ Two-sided $\alpha = 0.05$	One-sided $\alpha = 0.05$ Two-sided $\alpha = 0.10$
5			1
6		1	2
7	0	2	4
8	2	4	6
9	3	6	8
10	5	8	11
11	7	11	14
12	10	14	17
13	13	17	21
14	16	21	26
15	20	25	30
16	24	30	36
17	28	35	41
18	33	40	47
19	38	46	54
20	43	52	60
21	49	59	68
22	56	66	75
23	62	73	83
24	69	81	92
25	77	90	101
26	85	98	110
27	93	107	120
28	102	117	130
29	111	127	141
30	120	137	152

*Reproduced from F. Wilcoxon and R. A. Wilcox, *Some Rapid Approximate Statistical Procedures*, American Cyanamid Company, Pearl River, N.Y., 1964, by permission of the American Cyanamid Company.

TABLE A.17* Critical Values for the Rank-Sum Test

One-Tailed Test at $\alpha = 0.001$ or Two-Tailed Test at $\alpha = 0.002$

n_1	n_2=6	7	8	9	10	11	12	13	14	15	16	17	18	19	20
1															
2															
3												0	0	0	0
4					0	0	0	1	1	1	2	2	3	3	3
5		0	0	1	1	2	2	3	3	4	5	5	6	7	7
6	0	1	2	2	3	4	4	5	6	7	8	9	10	11	12
7		2	3	3	5	6	7	8	9	10	11	13	14	15	16
8			5	5	6	8	9	11	12	14	15	17	18	20	21
9				7	8	10	12	14	15	17	19	21	23	25	26
10					10	12	14	17	19	21	23	25	27	29	32
11						15	17	20	22	24	27	29	32	34	37
12							20	23	25	28	31	34	37	40	42
13								26	29	32	35	38	42	45	48
14									32	36	39	43	46	50	54
15										40	43	47	51	55	59
16											48	52	56	60	65
17												57	61	66	70
18													66	71	76
19														77	82
20															88

One-Tailed Test at $\alpha = 0.01$ or Two-Tailed Test at $\alpha = 0.02$

n_1	n_2=5	6	7	8	9	10	11	12	13	14	15	16	17	18	19	20
1																
2									0	0	0	0	0	0	1	1
3				0	1	1	1	2	2	2	3	3	4	4	4	5
4	0	1	1	2	3	3	4	5	5	6	7	7	8	9	9	10
5	1	2	3	4	5	6	7	8	9	10	11	12	13	14	15	16
6		3	4	6	7	8	9	11	12	13	15	16	18	19	20	22
7			6	8	9	11	12	14	16	17	19	21	23	24	26	28
8				10	11	13	15	17	20	22	24	26	28	30	32	34
9					14	16	18	21	23	26	28	31	33	36	38	40
10						19	22	24	27	30	33	36	38	41	44	47
11							25	28	31	34	37	41	44	47	50	53
12								31	35	38	42	46	49	53	56	60
13									39	43	47	51	55	59	63	67
14										47	51	56	60	65	69	73
15											56	61	66	70	75	80
16												66	71	76	82	87
17													77	82	88	93
18														88	94	100
19															101	107
20																114

*Based in part on Tables 1, 3, 5, and 7 of D. Auble, "Extended tables for the Mann–Whitney statistic," *Bulletin of the Institute of Educational Research at Indiana University*, Vol. 1, No. 2, 1953, by permission of the director.

TABLE A.17 (*continued*)   Critical Values for the Rank-Sum Test

One-Tailed Test at α = 0.025 or Two-Tailed Test at α = 0.05

n_1									n_2								
	4	5	6	7	8	9	10	11	12	13	14	15	16	17	18	19	20
1																	
2					0	0	0	0	1	1	1	1	1	2	2	2	2
3		0	1	1	2	2	3	3	4	4	5	5	6	6	7	7	8
4	0	1	2	3	4	4	5	6	7	8	9	10	11	11	12	13	13
5		2	3	5	6	7	8	9	11	12	13	14	15	17	18	19	20
6			5	6	8	10	11	13	14	16	17	19	21	22	24	25	27
7				8	10	12	14	16	18	20	22	24	26	28	30	32	34
8					13	15	17	19	22	24	26	29	31	34	36	38	41
9						17	20	23	26	28	31	34	37	39	42	45	48
10							23	26	29	33	36	39	42	45	48	52	55
11								30	33	37	40	44	47	51	55	58	62
12									37	41	45	49	53	57	61	65	69
13										45	50	54	59	63	67	72	76
14											55	59	64	67	74	78	83
15												64	70	75	80	85	90
16													75	81	86	92	98
17														87	93	99	105
18															99	106	112
19																113	119
20																	127

One-Tailed Test at α = 0.05 or Two-Tailed Test at α = 0.10

n_1										n_2								
	3	4	5	6	7	8	9	10	11	12	13	14	15	16	17	18	19	20
1																	0	0
2			0	0	0	1	1	1	1	2	2	3	3	3	3	4	4	4
3	0	0	1	2	2	3	4	4	5	5	6	7	7	8	9	9	10	11
4		1	2	3	4	5	6	7	8	9	10	11	12	14	15	16	17	18
5			4	5	6	8	9	11	12	13	15	16	18	19	20	22	23	25
6				7	8	10	12	14	16	17	19	21	23	25	26	28	30	32
7					11	13	15	17	19	21	24	26	28	30	33	35	37	39
8						15	18	20	23	26	28	31	33	36	39	41	44	47
9							21	24	27	30	33	36	39	42	45	48	51	54
10								27	31	34	37	41	44	48	51	55	58	62
11									34	38	42	46	50	54	57	61	65	69
12										42	47	51	55	60	64	68	72	77
13											51	56	61	65	70	75	80	84
14												61	66	71	77	82	87	92
15													72	77	83	88	94	100
16														83	89	95	101	107
17															96	102	109	115
18																109	116	123
19																	123	130
20																		138

TABLE A.18* $P(V \leq v^*$ when H_0 is true) in the Runs Test

(n_1, n_2)	v^*								
	2	3	4	5	6	7	8	9	10
(2, 3)	0.200	0.500	0.900	1.000					
(2, 4)	0.133	0.400	0.800	1.000					
(2, 5)	0.095	0.333	0.714	1.000					
(2, 6)	0.071	0.286	0.643	1.000					
(2, 7)	0.056	0.250	0.583	1.000					
(2, 8)	0.044	0.222	0.533	1.000					
(2, 9)	0.036	0.200	0.491	1.000					
(2, 10)	0.030	0.182	0.455	1.000					
(3, 3)	0.100	0.300	0.700	0.900	1.000				
(3, 4)	0.057	0.200	0.543	0.800	0.971	1.000			
(3, 5)	0.036	0.143	0.429	0.714	0.929	1.000			
(3, 6)	0.024	0.107	0.345	0.643	0.881	1.000			
(3, 7)	0.017	0.083	0.283	0.583	0.833	1.000			
(3, 8)	0.012	0.067	0.236	0.533	0.788	1.000			
(3, 9)	0.009	0.055	0.200	0.491	0.745	1.000			
(3, 10)	0.007	0.045	0.171	0.455	0.706	1.000			
(4, 4)	0.029	0.114	0.371	0.629	0.886	0.971	1.000		
(4, 5)	0.016	0.071	0.262	0.500	0.786	0.929	0.992	1.000	
(4, 6)	0.010	0.048	0.190	0.405	0.690	0.881	0.976	1.000	
(4, 7)	0.006	0.033	0.142	0.333	0.606	0.833	0.954	1.000	
(4, 8)	0.004	0.024	0.109	0.279	0.533	0.788	0.929	1.000	
(4, 9)	0.003	0.018	0.085	0.236	0.471	0.745	0.902	1.000	
(4, 10)	0.002	0.014	0.068	0.203	0.419	0.706	0.874	1.000	
(5, 5)	0.008	0.040	0.167	0.357	0.643	0.833	0.960	0.992	1.000
(5, 6)	0.004	0.024	0.110	0.262	0.522	0.738	0.911	0.976	0.998
(5, 7)	0.003	0.015	0.076	0.197	0.424	0.652	0.854	0.955	0.992
(5, 8)	0.002	0.010	0.054	0.152	0.347	0.576	0.793	0.929	0.984
(5, 9)	0.001	0.007	0.039	0.119	0.287	0.510	0.734	0.902	0.972
(5, 10)	0.001	0.005	0.029	0.095	0.239	0.455	0.678	0.874	0.958
(6, 6)	0.002	0.013	0.067	0.175	0.392	0.608	0.825	0.933	0.987
(6, 7)	0.001	0.008	0.043	0.121	0.296	0.500	0.733	0.879	0.966
(6, 8)	0.001	0.005	0.028	0.086	0.226	0.413	0.646	0.821	0.937
(6, 9)	0.000	0.003	0.019	0.063	0.175	0.343	0.566	0.762	0.902
(6, 10)	0.000	0.002	0.013	0.047	0.137	0.288	0.497	0.706	0.864
(7, 7)	0.001	0.004	0.025	0.078	0.209	0.383	0.617	0.791	0.922
(7, 8)	0.000	0.002	0.015	0.051	0.149	0.296	0.514	0.704	0.867
(7, 9)	0.000	0.001	0.010	0.035	0.108	0.231	0.427	0.622	0.806
(7, 10)	0.000	0.001	0.006	0.024	0.080	0.182	0.355	0.549	0.743
(8, 8)	0.000	0.001	0.009	0.032	0.100	0.214	0.405	0.595	0.786
(8, 9)	0.000	0.001	0.005	0.020	0.069	0.157	0.319	0.500	0.702
(8, 10)	0.000	0.000	0.003	0.013	0.048	0.117	0.251	0.419	0.621
(9, 9)	0.000	0.000	0.003	0.012	0.044	0.109	0.238	0.399	0.601
(9, 10)	0.000	0.000	0.002	0.008	0.029	0.077	0.179	0.319	0.510
(10, 10)	0.000	0.000	0.001	0.004	0.019	0.051	0.128	0.242	0.414

*Reproduced from C. Eisenhart and F. Swed, "Tables for Testing Randomness of Grouping in a Sequence of Alternatives," *Ann. Math. Stat.*, Vol. 14, 1943, by permission of the editor.

TABLE A.18 (*continued*) $P(V \leq v^*$ when H_0 is true) in the Runs Test

(n_1, n_2)	v^*									
	11	12	13	14	15	16	17	18	19	20
(2, 3)										
(2, 4)										
(2, 5)										
(2, 6)										
(2, 7)										
(2, 8)										
(2, 9)										
(2, 10)										
(3, 3)										
(3, 4)										
(3, 5)										
(3, 6)										
(3, 7)										
(3, 8)										
(3, 9)										
(3, 10)										
(4, 4)										
(4, 5)										
(4, 6)										
(4, 7)										
(4, 8)										
(4, 9)										
(4, 10)										
(5, 5)										
(5, 6)	1.000									
(5, 7)	1.000									
(5, 8)	1.000									
(5, 9)	1.000									
(5, 10)	1.000									
(6, 6)	0.998	1.000								
(6, 7)	0.992	0.999	1.000							
(6, 8)	0.984	0.998	1.000							
(6, 9)	0.972	0.994	1.000							
(6, 10)	0.958	0.990	1.000							
(7, 7)	0.975	0.996	0.999	1.000						
(7, 8)	0.949	0.988	0.998	1.000	1.000					
(7, 9)	0.916	0.975	0.994	0.999	1.000					
(7, 10)	0.879	0.957	0.990	0.998	1.000					
(8, 8)	0.900	0.968	0.991	0.999	1.000	1.000				
(8, 9)	0.843	0.939	0.980	0.996	0.999	1.000	1.000			
(8, 10)	0.782	0.903	0.964	0.990	0.998	1.000	1.000			
(9, 9)	0.762	0.891	0.956	0.988	0.997	1.000	1.000	1.000		
(9, 10)	0.681	0.834	0.923	0.974	0.992	0.999	1.000	1.000	1.000	
(10, 10)	0.586	0.758	0.872	0.949	0.981	0.996	0.999	1.000	1.000	1.000

TABLE A.19*   Sample Size for Two-Sided Nonparametric Tolerance Limits

1 − α	\multicolumn{6}{c}{1 − γ}					
	0.50	0.70	0.90	0.95	0.99	0.995
0.995	336	488	777	947	1,325	1,483
0.99	168	244	388	473	662	740
0.95	34	49	77	93	130	146
0.90	17	24	38	46	64	72
0.85	11	16	25	30	42	47
0.80	9	12	18	22	31	34
0.75	7	10	15	18	24	27
0.70	6	8	12	14	20	22
0.60	4	6	9	10	14	16
0.50	3	5	7	8	11	12

*Reproduced from Tables A-25d of Wilfrid J. Dixon and Frank J. Massey, Jr., *Introduction to Statistical Analysis*, 3rd ed., McGraw-Hill Book Company, New York, 1969. Used with permission of McGraw-Hill Book Company.

TABLE A.20*   Sample Size for One-Sided Nonparametric Tolerance Limits

1 − α	\multicolumn{5}{c}{1 − γ}				
	0.50	0.70	0.95	0.99	0.995
0.995	139	241	598	919	1,379
0.99	69	120	299	459	688
0.95	14	24	59	90	135
0.90	7	12	29	44	66
0.85	5	8	19	29	43
0.80	4	6	14	21	31
0.75	3	5	11	17	25
0.70	2	4	9	13	20
0.60	2	3	6	10	14
0.50	1	2	5	7	10

*Reproduced from Tables A-25e of Wilfrid J. Dixon and Frank J. Massey, Jr., *Introduction to Statistical Analysis*, 3rd ed., McGraw-Hill Book Company, New York, 1969. Used with permission of McGraw-Hill Book Company.

TABLE A.21* Critical Values of Spearman's Rank
Correlation Coefficient

n	$\alpha = 0.05$	$\alpha = 0.025$	$\alpha = 0.01$	$\alpha = 0.005$
5	0.900	—	—	—
6	0.829	0.886	0.943	—
7	0.714	0.786	0.893	—
8	0.643	0.738	0.833	0.881
9	0.600	0.683	0.783	0.833
10	0.564	0.648	0.745	0.794
11	0.523	0.623	0.736	0.818
12	0.497	0.591	0.703	0.780
13	0.475	0.566	0.673	0.745
14	0.457	0.545	0.646	0.716
15	0.441	0.525	0.623	0.689
16	0.425	0.507	0.601	0.666
17	0.412	0.490	0.582	0.645
18	0.399	0.476	0.564	0.625
19	0.388	0.462	0.549	0.608
20	0.377	0.450	0.534	0.591
21	0.368	0.438	0.521	0.576
22	0.359	0.428	0.508	0.562
23	0.351	0.418	0.496	0.549
24	0.343	0.409	0.485	0.537
25	0.336	0.400	0.475	0.526
26	0.329	0.392	0.465	0.515
27	0.323	0.385	0.456	0.505
28	0.317	0.377	0.448	0.496
29	0.311	0.370	0.440	0.487
30	0.305	0.364	0.432	0.478

*Reproduced from E. G. Olds, "Distribution of Sums of Squares of Rank Differences for Small Samples," *Ann. Math. Stat.*, Vol. 9, 1938, by permission of the editor.

TABLE A.22   Upper Percentage Points of the Studentized Range Distribution: Values of $q(0.05; k, v)$

Degrees of Freedom v	Number of Treatments k								
	2	3	4	5	6	7	8	9	10
1	18.0	27.0	32.8	37.2	40.5	43.1	45.4	47.3	49.1
2	6.09	8.33	9.80	10.89	11.73	12.43	13.03	13.54	13.99
3	4.50	5.91	6.83	7.51	8.04	8.47	8.85	9.18	9.46
4	3.93	5.04	5.76	6.29	6.71	7.06	7.35	7.60	7.83
5	3.64	4.60	5.22	5.67	6.03	6.33	6.58	6.80	6.99
6	3.46	4.34	4.90	5.31	5.63	5.89	6.12	6.32	6.49
7	3.34	4.16	4.68	5.06	5.35	5.59	5.80	5.99	6.15
8	3.26	4.04	4.53	4.89	5.17	5.40	5.60	5.77	5.92
9	3.20	3.95	4.42	4.76	5.02	5.24	5.43	5.60	5.74
10	3.15	3.88	4.33	4.66	4.91	5.12	5.30	5.46	5.60
11	3.11	3.82	4.26	4.58	4.82	5.03	5.20	5.35	5.49
12	3.08	3.77	4.20	4.51	4.75	4.95	5.12	5.27	5.40
13	3.06	3.73	4.15	4.46	4.69	4.88	5.05	5.19	5.32
14	3.03	3.70	4.11	4.41	4.64	4.83	4.99	5.13	5.25
15	3.01	3.67	4.08	4.37	4.59	4.78	4.94	5.08	5.20
16	3.00	3.65	4.05	4.34	4.56	4.74	4.90	5.03	5.15
17	2.98	3.62	4.02	4.31	4.52	4.70	4.86	4.99	5.11
18	2.97	3.61	4.00	4.28	4.49	4.67	4.83	4.96	5.07
19	2.96	3.59	3.98	4.26	4.47	4.64	4.79	4.92	5.04
20	2.95	3.58	3.96	4.24	4.45	4.62	4.77	4.90	5.01
24	2.92	3.53	3.90	4.17	4.37	4.54	4.68	4.81	4.92
30	2.89	3.48	3.84	4.11	4.30	4.46	4.60	4.72	4.83
40	2.86	3.44	3.79	4.04	4.23	4.39	4.52	4.63	4.74
60	2.83	3.40	3.74	3.98	4.16	4.31	4.44	4.55	4.65
120	2.80	3.36	3.69	3.92	4.10	4.24	4.36	4.47	4.56
∞	2.77	3.32	3.63	3.86	4.03	4.17	4.29	4.39	4.47

TABLE A.23 Factors for Constructing Control Charts

Observations in Sample, n	Chart for Averages — Factors for Control Limits		Chart for Standard Deviations — Factors for Central Line		Chart for Standard Deviations — Factors for Control Limits				Chart for Ranges — Factors for Central Line		Chart for Ranges — Factors for Control Limits		
	A_2	A_3	c_4	$1/c_4$	B_3	B_4	B_5	B_6	d_2	$1/d_2$	d_3	D_3	D_4
2	1.880	2.659	0.7979	1.2533	0	3.267	0	2.606	1.128	0.8865	0.853	0	3.267
3	1.023	1.954	0.8862	1.1284	0	2.568	0	2.276	1.693	0.5907	0.888	0	2.574
4	0.729	1.628	0.9213	1.0854	0	2.266	0	2.088	2.059	0.4857	0.880	0	2.282
5	0.577	1.427	0.9400	1.0638	0	2.089	0	1.964	2.326	0.4299	0.864	0	2.114
6	0.483	1.287	0.9515	1.0510	0.030	1.970	0.029	1.874	2.534	0.3946	0.848	0	2.004
7	0.419	1.182	0.9594	1.04230	0.118	1.882	0.113	1.806	2.704	0.3698	0.833	0.076	1.924
8	0.373	1.099	0.9650	1.0363	0.185	1.815	0.179	1.751	2.847	0.3512	0.820	0.136	1.864
9	0.337	1.032	0.9693	1.0317	0.239	1.761	0.232	1.707	2.970	0.3367	0.808	0.184	1.816
10	0.308	0.975	0.9727	1.0281	0.284	1.716	0.276	1.669	3.078	0.3249	0.797	0.223	1.777
11	0.285	0.927	0.9754	1.0252	0.321	1.679	0.313	1.637	3.173	0.3152	0.787	0.256	1.744
12	0.266	0.886	0.9776	1.0229	0.354	1.646	0.346	1.610	3.258	0.3069	0.778	0.283	1.717
13	0.249	0.850	0.9794	1.0210	0.382	1.618	0.374	1.585	3.336	0.2998	0.770	0.307	1.693
14	0.235	0.817	0.9810	1.0194	0.406	1.594	0.399	1.563	3.407	0.2935	0.763	0.328	1.672
15	0.223	0.789	0.9823	1.0180	0.428	1.572	0.421	1.544	3.472	0.2880	0.756	0.347	1.653
16	0.212	0.763	0.9835	1.0168	0.448	1.552	0.440	1.526	3.532	0.2831	0.750	0.363	1.637
17	0.203	0.739	0.9845	1.0157	0.466	1.534	0.458	1.511	3.588	0.2787	0.744	0.378	1.622
18	0.194	0.718	0.9854	1.0148	0.482	1.518	0.475	1.496	3.640	0.2747	0.739	0.391	1.608
19	0.187	0.698	0.9862	1.0140	0.497	1.503	0.490	1.483	3.689	0.2711	0.734	0.403	1.597
20	0.180	0.680	0.9869	1.0133	0.510	1.490	0.504	1.470	3.735	0.2677	0.729	0.415	1.585
21	0.173	0.663	0.9876	1.0126	0.523	1.477	0.516	1.459	3.778	0.2647	0.724	0.425	1.575
22	0.167	0.647	0.9882	1.0119	0.534	1.466	0.528	1.448	3.819	0.2618	0.720	0.434	1.566
23	0.162	0.633	0.9887	1.0114	0.545	1.455	0.539	1.438	3.858	0.2592	0.716	0.443	1.557
24	0.157	0.619	0.9892	1.0109	0.555	1.445	0.549	1.429	3.895	0.2567	0.712	0.451	1.548
25	0.153	0.606	0.9896	1.0105	0.565	1.435	0.559	1.420	3.931	0.2544	0.708	0.459	4.541

Answers to Exercises

CHAPTER 2

PAGE 16

1. (a) $\{8, 16, 24, 32, 40, 48\}$.
 (b) $\{-5, 1\}$.
 (c) $\{T, HT, HHT, HHH\}$.
 (d) {North America, South America, Europe, Asia, Africa, Australia, Antarctica}.
 (e) $\varnothing$.

2. $\{(x, y) \,|\, x^2 + y^2 < 9;\; x > 0,\, y > 0\}$.

3. $A = C$.

4. (a)

Green	Red					
	1	2	3	4	5	6
1	(1, 1)	(1, 2)	(1, 3)	(1, 4)	(1, 5)	(1, 6)
2	(2, 1)	(2, 2)	(2, 3)	(2, 4)	(2, 5)	(2, 6)
3	(3, 1)	(3, 2)	(3, 3)	(3, 4)	(3, 5)	(3, 6)
4	(4, 1)	(4, 2)	(4, 3)	(4, 4)	(4, 5)	(4, 6)
5	(5, 1)	(5, 2)	(5, 3)	(5, 4)	(5, 5)	(5, 6)
6	(6, 1)	(6, 2)	(6, 3)	(6, 4)	(6, 5)	(6, 6)

 (b) $S = \{(x, y) \,|\, x = 1, 2, \ldots, 6;\; y = 1, 2, \ldots, 6\}$.

5. $S = \{1HH, 1HT, 1TH, 1TT, 2H, 2T, 3HH, 3HT, 3TH, 3TT, 4H, 4T, 5HH, 5HT, 5TH, 5TT, 6H, 6T\}$.

6. $S = \{A_1A_2, A_1A_3, A_1A_4, A_2A_3, A_2A_4, A_3A_4\}$.

7. $S_1 = \{MMMM, MMMF, MMFM, MFMM, FMMM, MMFF, MFMF, MFFM, FMFM, FFMM, FMMF, MFFF, FMFF, FFMF, FFFM, FFFF\}$; $S_2 = \{0, 1, 2, 3, 4\}$.

8. (a) $A = \{(3, 6), (4, 5), (4, 6), (5, 4), (5, 5), (5, 6), (6, 3), (6, 4), (6, 5), (6, 6)\}$.
 (b) $B = \{(1, 2), (2, 2), (3, 2), (4, 2), (5, 2), (6, 2), (2, 1), (2, 3), (2, 4), (2, 5), (2, 6)\}$.
 (c) $C = \{(5, 1), (5, 2), (5, 3), (5, 4), (5, 5), (5, 6), (6, 1), (6, 2), (6, 3), (6, 4), (6, 5), (6, 6)\}$.
 (d) $A \cap C = \{(5, 4), (5, 5), (5, 6), (6, 3), (6, 4), (6, 5), (6, 6)\}$.
 (e) $A \cap B = \varnothing$.
 (f) $B \cap C = \{(5, 2), (6, 2)\}$.

(g)

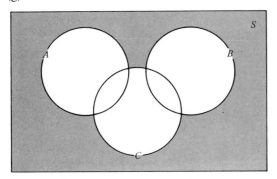

9. (a) $A = \{1HH, 1HT, 1TH, 1TT, 2H, 2T\}$.
(b) $B = \{1TT, 3TT, 5TT\}$.
(c) $A' = \{3HH, 3HT, 3TH, 3TT, 4H, 4T, 5HH, 5HT, 5TH, 5TT, 6H, 6T\}$.
(d) $A' \cap B = \{3TT, 5TT\}$.
(e) $A \cup B = \{1HH, 1HT, 1TH, 1TT, 2H, 2T, 3TT, 5TT\}$.

10. (a) $S = \{YYY, YYN, YNY, NYY, YNN, NYN, NNY, NNN\}$.
(b) $E = \{YYY, YYN, YNY, NYY\}$.
(c) One possible event: "The second woman interviewed uses brand X."

11. (a) $S = \{M_1M_2, M_1F_1, M_1F_2, M_2M_1, M_2F_1, M_2F_2, F_1M_1, F_1M_2, F_1F_2, F_2M_1, F_2M_2, F_2F_1\}$.
(b) $A = \{M_1M_2, M_1F_1, M_1F_2, M_2M_1, M_2F_1, M_2F_2\}$.
(c) $B = \{M_1F_1, M_1F_2, M_2F_1, M_2F_2, F_1M_1, F_1M_2, F_2M_1, F_2M_2\}$.
(d) $C = \{F_1F_2, F_2F_1\}$.
(e) $A \cap B = \{M_1F_1, M_1F_2, M_2F_1, M_2F_2\}$.
(f) $A \cup C = \{M_1M_2, M_1F_1, M_1F_2, M_2M_1, M_2F_1, M_2F_2, F_1F_2, F_2F_1\}$.
(g)

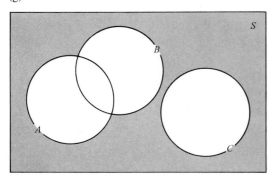

12. $S = \{Vhr, Vhb, Vmr, Vmb, Vcr, Vcb, Nhr, Nhb, Nmr, Nmb, Ncr, Ncb, Chr, Chb, Cmr, Cmb, Ccr, Ccb, Mhr, Mhb, Mmr, Mmb, Mcr, Mcb\}$.

13.

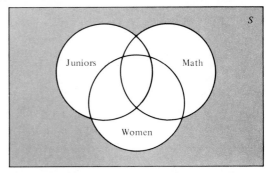

14. (a) $\{0, 2, 3, 4, 5, 6, 8\}$.
(b) $\varnothing$.   (c) $\{0, 1, 6, 7, 8, 9\}$.
(d) $\{1, 3, 5, 6, 7, 9\}$.
(e) $\{0, 1, 6, 7, 8, 9\}$.   (f) $\{2, 4\}$.

15. (a) {nitrogen, potassium, uranium, oxygen}.
(b) {copper, sodium, zinc, oxygen}.
(c) {copper, sodium, nitrogen, potassium, uranium, zinc}.
(d) {copper, uranium, zinc}.
(e) $\varnothing$.   (f) {oxygen}.

16. (a) $M \cup N = \{x \mid 0 < x < 9\}$.
(b) $M \cap N = \{x \mid 1 < x < 5\}$.
(c) $M' \cap N' = \{x \mid 9 < x < 12\}$.

17. (a)

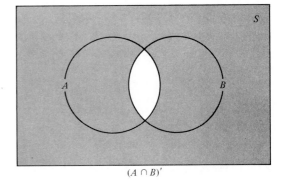

$(A \cap B)'$

(b)

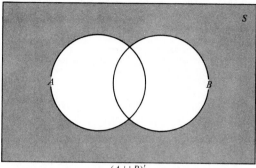

$(A \cup B)'$

(c)

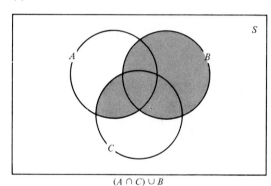

$(A \cap C) \cup B$

18. (a) Not mutually exclusive.
(b) Mutually exclusive.
(c) Not mutually exclusive.
(d) Mutually exclusive.

19. (a) The family will experience mechanical problems but will receive no ticket for a traffic violation and will not arrive at a campsite that has no vacancies.
(b) The family will receive a traffic ticket and arrive at a campsite that has no vacancies but will not experience mechanical problems.
(c) The family will experience mechanical problems and will arrive at a campsite that has no vacancies.
(d) The family will receive a traffic ticket but will not arrive at a campsite that has no vacancies.
(e) The family will not experience mechanical problems.

20. (a) 6.   (b) 2.   (c) 2, 5, 6.   (d) 4, 5, 7, 8.

PAGE 23

1. 18.

2. 24.

3. 156.

4. 8.

5. 20.

6. (a) 21.   (b) 10.

7. 48.

8. 30.

9. 210.

10. 512.

11. (a) 1024.   (b) 243.

12. (a) 5040.   (b) 720.

13. 72.

14. (a) 720.   (b) 144.   (c) 480.

15. 362,880.

16. (a) 180.   (b) 75.   (c) 105.

17. 2880.

18. (a) 40,320.   (b) 384.   (c) 576.

19. (a) 40,320.   (b) 336.

20. 6720.

21. 360.

22. 59,280.

23. 24.

24. 5040.

25. 3360.

26. 1260.

27. 7920.

28. 4410.

29. 56.

PAGE 31

1. (a) Sum of the probabilities exceeds 1.
(b) Sum of the probabilities is less than 1.
(c) A negative probability.
(d) Probability of both a heart and a black card is zero.

2. (a) 5/18.   (b) 1/3.   (c) 7/36.

3. $S = \{\$10, \$25, \$100\}$; 17/20.

4. (a) 88/500.   (b) 31/500.   (c) 171/500.

5. (a) 0.3.   (b) 0.2.

6. (a) 0.75.   (b) 0.25.

7. (a) 5/26.   (b) 9/26.   (c) 19/26.

8. (a) 3/5.   (b) 2/5.   (c) 1/10.

9. 10/117.

10. (a) 5/36.   (b) 5/18.

11. 95/663.

12. (a) 1/3.   (b) 5/14.

13. (a) 94/54,145.   (b) 143/39,984.

14. 25/2592.

15. (a) 22/25.   (b) 3/25.   (c) 17/50.

PAGE 39

1. (a) The probability that a convict who pushed dope also committed armed robbery.
(b) The probability that a convict who committed armed robbery did not push dope.
(c) The probability that a convict who did not push dope also did not commit armed robbery.

2. 5/9.

3. (a) 14/39.   (b) 95/112.

4. (a) 30/49.   (b) 16/31.

5. (a) 5/34.   (b) 3/8.

6. (a) 2/11.   (b) 5/11.

7. 6/13.

8. (a) 0.56.   (b) 0.35.

9. (a) 0.35.   (b) 0.875.   (c) 0.55.

10. (a) 0.34.   (b) 5/7.   (c) 1/12.

11. (a) 9/28.   (b) 3/4.   (c) 0.91.

12. 0.24.

13. 0.27.

14. 35/64.

15. 5/8.

16. (a) 25/102.   (b) 20/221.

17. (a) 0.0016.   (b) 0.9984.

18. 0.03.

19. (a) 1/5.   (b) 4/15.   (c) 3/5.

20. 0.018.

21. (a) 91/323.   (b) 91/323.

PAGE 45

1. 0.0744.

2. 0.27.

3. 0.2097.

4. 1/9.

5. 0.1124.

6. 0.2632.

CHAPTER 3

PAGE 60

1. Discrete; continuous; continuous; discrete, discrete; continuous.

2.

Sample Space	x
NNN	0
NNB	1
NBN	1
BNN	1
NBB	2
BNB	2
BBN	2

3.

Sample Space	w
HHH	3
HHT	1
HTH	1
THH	1
HTT	−1
THT	−1
TTH	−1
TTT	−3

4. $S = \{HHH, THHH, HTHHH, TTHHH, TTTHHH, HTTHHH, THTHHH, HHTHHH, \ldots\}$; discrete.

5. (a) 1/30.   (b) 1/10.

6. (a) 1/9.   (b) 0.1020.

7. (a) 0.68.   (b) 0.375.

8.

w	-3	-1	1	3
$P(W = w)$	$\frac{1}{27}$	$\frac{2}{9}$	$\frac{4}{9}$	$\frac{8}{27}$

9. (b) 19/80.

10. $f(x) = 1/6$, $x = 1, 2, \ldots, 6$.

11.

x	0	1	2
$f(x)$	$\frac{2}{7}$	$\frac{4}{7}$	$\frac{1}{7}$

12. (a) 1/4.   (b) 1/2.   (c) 1/2.

13.
$$F(x) = \begin{cases} 0 & \text{for } x < 0 \\ 0.41 & \text{for } 0 \le x < 1 \\ 0.78 & \text{for } 1 \le x < 2 \\ 0.94 & \text{for } 2 \le x < 3 \\ 0.99 & \text{for } 3 \le x < 4 \\ 1 & \text{for } x \ge 4. \end{cases}$$

14. (a) 0.7981.   (b) 0.7981.

15.
$$F(x) = \begin{cases} 0 & \text{for } x < 0 \\ \frac{2}{7} & \text{for } 0 \le x < 1 \\ \frac{6}{7} & \text{for } 1 \le x < 2 \\ 1 & \text{for } x \ge 2. \end{cases}$$

16. (a) 4/7.   (b) 5/7.

17. (b) 1/4.   (c) 0.3.

18. (a) 16/27.   (b) 1/3.

19. $F(x) = (x - 1)/2$; 1/4.

20. $F(x) = (x + 4)(x - 2)/27$; 1/3.

21. (a) 3/2.   (b) $F(x) = x^{3/2}$; 0.3004.

22.

x	0	1	2	3
$f(x)$	$\frac{703}{1700}$	$\frac{741}{1700}$	$\frac{117}{850}$	$\frac{11}{850}$

23.
$$F(w) = \begin{cases} 0 & \text{for } w < -3 \\ \frac{1}{27} & \text{for } -3 \le w < -1 \\ \frac{7}{27} & \text{for } -1 \le w < 1 \\ \frac{19}{27} & \text{for } 1 \le w < 3 \\ 1 & \text{for } w \ge 3. \end{cases}$$
(a) 20/27.   (b) 2/3.

24. $f(x) = \dfrac{\binom{5}{x}\binom{5}{4 - x}}{\binom{10}{4}}$, for $x = 0, 1, 2, 3, 4$.

25.

t	20	25	30
$P(T = t)$	$\frac{1}{5}$	$\frac{3}{5}$	$\frac{1}{5}$

26.

x	0	1	2	3
$f(x)$	$\frac{8}{27}$	$\frac{4}{9}$	$\frac{2}{9}$	$\frac{1}{27}$

PAGE 67

1. (a) Stems are 1, 2, 3, 4, 5, 6, 7, 8, 9, with frequencies 3, 2, 3, 4, 5, 11, 14, 14, 4.
   (b) Class intervals are 10–19, 20–29, 30–39, 40–49, 50–59, 60–69, 70–79, 80–89, 90–99 with relative frequencies 0.05, 0.03, 0.05, 0.07, 0.08, 0.18, 0.23, 0.23, 0.07.
   (c) Skewed to the left.
   (d) Relative cumulative frequencies are 0, 0.05, 0.08, 0.13, 0.20, 0.28, 0.46, 0.69, 0.92, 0.99.
   (f) 1st quartile $\approx$ 55.5; 7th decile $\approx$ 80.

2. (a) Stems are 0, 1, 2, 3, 4, 5, 6 with frequencies 8, 6, 3, 2, 3, 4, 4.
   (b) Relative frequencies are 0.267, 0.200, 0.100, 0.067, 0.100, 0.133, 0.133.
   (c) Relative cumulative frequencies are 0, 0.267, 0.467, 0.567, 0.634, 0.734, 0.867, 1.000.
   (e) Approximately 4.28 years.

3. (a) Stems are $0*, 0\cdot, 1*, 1\cdot, 2*, 2\cdot, 3*$ with frequencies 2, 17, 16, 10, 3, 1, 1.
   (b) Relative frequencies are 0.04, 0.34, 0.32, 0.20, 0.06, 0.02, 0.02.
   (c) Skewed to the right.
   (d) Relative cumulative frequencies are 0, 0.04, 0.38, 0.70, 0.90, 0.96, 0.98, 1.00.
   (f) 75th percentile $\approx$ 15.9.

4. Stems are $1d, 1e, 2a, 2b, 2c, 2d, 2e, 3a, 3b, 3c, 3d, 3e, 4a, 4b, 4c, 4d$ with frequencies 1, 1, 0, 1, 1, 2, 1, 6, 6, 5, 4, 4, 2, 2, 2, 2.

5. (a) Stems are $0d, 0e, 1a, 1b, 1c, 1d, 1e, 2a, 2b, 2c$ with frequencies 1, 1, 1, 2, 4, 13, 8, 5, 3, 2.
   (b) Relative frequencies are 0.025, 0.025, 0.025, 0.050, 0.100, 0.325, 0.200, 0.125, 0.075, 0.050.

PAGE 78

1. (a) 1/36.   (b) 1/15.

2. (a) 1/5.   (b) 7/30.   (c) 3/5.   (d) 4/15.

3. (a)

$f(x, y)$	0	1	2	3
0		$\frac{3}{70}$	$\frac{9}{70}$	$\frac{3}{70}$
y 1	$\frac{2}{70}$	$\frac{18}{70}$	$\frac{18}{70}$	$\frac{2}{70}$
2	$\frac{3}{70}$	$\frac{9}{70}$	$\frac{3}{70}$	

(b) 1/2.

4. (a) $g(x) = \begin{cases} \dfrac{2(x + 1)}{3}, & 0 \le x \le 1 \\ 0, & \text{elsewhere.} \end{cases}$

(b) $h(y) = \begin{cases} \dfrac{(1 + 4y)}{3}, & 0 \le y \le 1 \\ 0, & \text{elsewhere.} \end{cases}$

(c) 5/12.

5. (a) 1/16.
(b) $g(x) = 12x(1 - x)^2, \ 0 \le x \le 1.$
(c) 1/4.

6. 0.6321.

7. (a) 3/64.   (b) 1/2.

8. (a) $\dfrac{3}{392} \cdot 10^{-4}.$

(b) 0.25.
(c) 0.1888.

9. 0.6534.

10. (a)

x	0	1	2	3
$g(x)$	$\frac{1}{10}$	$\frac{1}{5}$	$\frac{3}{10}$	$\frac{2}{5}$

(b)

y	0	1	2
$h(y)$	$\frac{1}{5}$	$\frac{1}{3}$	$\frac{7}{15}$

11. (a) Dependent.   (b) 1/3.

12. (a)

y	0	1	2
$f(y \mid 2)$	$\frac{3}{10}$	$\frac{3}{5}$	$\frac{1}{10}$

(b) 3/10.

13. (a)

x	1	2	3
$g(x)$	0.10	0.35	0.55

(b)

y	1	2	3
$h(y)$	0.20	0.50	0.30

(c) 0.2.

14. (a)

x	2	4
$g(x)$	0.40	0.60

(b)

y	1	3	5
$h(y)$	0.25	0.50	0.25

15. (a)

$f(x, y)$	0	1	2
0	$\frac{16}{36}$	$\frac{8}{36}$	$\frac{1}{36}$
y 1	$\frac{8}{36}$	$\frac{2}{36}$	
2	$\frac{1}{36}$		

(b) 11/12.

16.

$f(x, y)$	0	1	2	3
-3	$\frac{1}{8}$			
y -1		$\frac{3}{8}$		
1			$\frac{3}{8}$	
3				$\frac{1}{8}$

17. (a)

$f(x, y)$	0	1	2	3
0	$\frac{1}{55}$	$\frac{6}{55}$	$\frac{6}{55}$	$\frac{1}{55}$
y 1	$\frac{6}{55}$	$\frac{16}{55}$	$\frac{6}{55}$	
2	$\frac{6}{55}$	$\frac{6}{55}$		
3	$\frac{1}{55}$			

(b) 42/55.

18. (a)

$f(w, z)$	0	1	2
z 0	0.36	0.24	
1		0.24	0.16

(b)

w	0	1	2
$g(w)$	0.36	0.48	0.16

(c)

z	0	1
$h(z)$	0.60	0.40

(d) 0.64.

19. 3/4.

20. Dependent.

21. Independent.

22. (b) 0.64.

23. (a) 3.   (b) 21/512.

24. Independent.

25. Dependent.

26. (a) $g(y, z) = 2yz^2/9$,
   $0 < y < 1, \ 0 < z < 3$.
   (b) $h(y) = 2y, \ 0 < y < 1$.
   (c) 7/162.
   (d) 1/4.

CHAPTER 4

PAGE 89

1. 0.

2. 3/4.

3. 25 cents.

4. 1/2.

5. 0.88.

6. $12.67.

7. $500.

8. $88.

9. $1.23.

10. $u_x = 2.16$.
   $u_y = 2.04$.

11. $2100.

12. $333.

13. $(\ln 4)/\pi$.

14. 8/15.

15. 100 hours.

16. 0.0392.

17. 209.

18. 9/8.

19. $1855.

20. 3.

21. $167.

22. 8.

23. (a) 35.2.   (b) $\mu_X = 3.20; \ \mu_Y = 3.00$.

24. (a) $-3/7$.   (b) $\mu_X = 3/2; \ \mu_Y = 1$.

25. 2.

26. 0.9752.

PAGE 99

1. $5,250.000.

2. 3.041.

3. 0.74.

4. $\mu = 1; \ \sigma^2 = 1$.

5. 1/18.

6. 37/450.

7. 1/6.

8. 21,6845.

9. 118.9.

10. 7/180.

11. $\mu_Y = 10; \ \sigma_Y^2 = 144$.

12. $-3/14$.

13. $\sigma_{XY} = 0.005$.

14. -0.6642.

15. -0.0062.

PAGE 109

1. 10.33; 6.66.

2. 8, 25.

3. 80 cents.

4. $\mu_Y = 10; \ \sigma_Y^2 = 144$.

5. 209.

6. 109 kilowatt hours.

7. $\mu = 7/2$; $\sigma^2 = 15/4$.

8. (a) -2.60.   (b) 9.60.

9. 3/14.

10. Yes.

11. 0.03125.

12. 0.0556.

13. 0.9340.

14. 68.

15. 52.

16. (a) At least 3/4.   (b) At least 8/9.

17. (a) At most 4/9.   (b) At least 5/9.
     (c) At least 21/25.   (d) 10.

18. 0.9839.

19. (a) 7.   (b) 0.   (c) 12.25.

20. 8/3.

21. 1.

22. (a) 175/12.   (b) 175/6.

CHAPTER 5

PAGE 121

1. 3/10.

2. 0.0537.

3. $\mu = 5.5$; $\sigma^2 = 8.25$.

4. (a) 0.0879.   (b) 0.3672.

5. (a) 16/81.   (b) 64/81.

6. (a) 0.647.   (b) 0.680.

7. (a) 0.0474.   (b) 0.0171.

8. (a) 0.1239.   (b) 0.5941.

9. (a) 0.7073.   (b) 0.4613.   (c) 0.1484.

10. (a) 0.6294.   (b) 0.0386.   (c) 0.7237.

11. 0.1240.

12. 0.8343.

13. 0.8369.

14. 0.0006.

15. (a) 0.0778.   (b) 0.3370.   (c) 0.0870.

16. 0.8208 and 0.8400; 2-engine plane.

17. $\mu \pm 2\sigma = 3.5 \pm 2.05$.

18. (a) 3.75.   (b) From 0.396 to 7.104.

19. $f(x_1, x_2, x_3) = \begin{pmatrix} n \\ x_1, x_2, x_3 \end{pmatrix} \begin{array}{ccc} x_1 & x_2 & x_3 \\ 0.35 & 0.05 & 0.60 \end{array}$

20. 15/128.

21. 0.0095.

22. 21/256.

23. 0.0077.

24. 0.1382.

25. 0.8670.

26. (a) 0.2090.   (b) 0.2090.

27. (a) 0.2852.   (b) 0.9887.   (c) 0.6083.

28. (a) $x = 4$.   (b) $y = 14$.

PAGE 130

1. (a) 0.3246.   (b) 0.44964.

2. 53/65.

3. 5/14.

4. (a) 1/6.   (b) 29/30.

5. $h(x; 6, 3, 4) = \dfrac{\begin{pmatrix} 4 \\ x \end{pmatrix}\begin{pmatrix} 2 \\ 2 - x \end{pmatrix}}{\begin{pmatrix} 6 \\ 3 \end{pmatrix}}$, $x = 1, 2, 3$;

   $P(2 \le X \le 3) = 4/5$.

6. 10/21.

7. 0.9517.

8. (a) 77/115.   (b) 3/25.

9. (a) 0.6815.   (b) 0.1153.

10. $\mu = 1.2$.

11. 3.25; from 0.52 to 5.98.

12. 0.2131.

13. 0.9453.

14. 0.3222.

15. 0.60776.

16. 0.0129.

17. (a) 4/33.   (b) 8/165.

18. 17/63.

PAGE 138

1. 0.0515.

2. 0.0651.

3. (a) 0.3840.   (b) 0.0067.

4. (a) 0.1172.   (b) 1/16.

5. 63/64.

6. (a) 2/243.   (b) 16/81.

7. (a) 0.0630.   (b) 0.9730.

8. (a) 0.1008.   (b) 0.4232.   (c) 0.8009.

9. (a) 0.1429.   (b) 0.1353.

10. (a) 0.1512.   (b) 0.4015.

11. (a) 0.1638.   (b) 0.032.

12. (a) 0.0458.   (b) 0.0060.

13. (a) 0.3840.   (b) 0.1395.   (c) 0.0553.

14. 0.6288.

15. 0.2657.

16. (a) 0.1321.   (b) 0.3376.

17. (a) $\mu = 4$; $\sigma^2 = 4$.   (b) From 0 to 8.

18. (a) $\mu = 10$; $\sigma^2 = 10$.
(b) From 0.51 to 19.49.

CHAPTER 6

PAGE 156

1. (a) 0.9236.   (b) 0.8133.
(c) 0.2424.   (d) 0.0823.
(e) 0.0250.   (f) 0.6435.

2. (a) 0.35.   (b) −1.21.
(c) 2.14.   (d) 1.96.

3. (a) −1.72.   (b) 0.54.   (c) 1.28.

4. (a) 0.9850.   (b) 0.0918.
(c) 0.3371.   (d) 35.04.
(e) 23.1 and 36.9.

5. (a) 0.1151.   (b) 16.1.
(c) 20.275.   (d) 0.5403.

6. 0.9974.

7. (a) 0.8980.   (b) 0.0287.   (c) 0.6080.

8. (a) 19.77%.   (b) 59.67%.   (c) 1.22%.

9. (a) 0.0548.   (b) 0.4514.
(c) 23.   (d) 189.95 milliliters.

10. (a) 0.0062.   (b) 0.6826.
(c) 9.969 centimeters.

11. (a) 0.0571.   (b) 99.11%.
(c) 0.3974.   (d) 27.952 minutes.
(e) 0.0092.

12. (a) 64.   (b) 86.   (c) 78.

13. 6.24 years.

14. (a) 16.   (b) 549.   (c) 28.   (d) 27.

15. (a) 56.99%.   (b) $10.23.

16. (a) 0.0427.   (b) 0.7642.   (c) 0.6964.

17. (a) 0.0401.   (b) 0.0244.

18. (a) 19.36%.   (b) 39.70%.

19. 26.

20. (a) 0.0045.   (b) 0.1496.   (c) 0.0526.

21. 62.

PAGE 165

1. (a) 0.8006.   (b) 0.7803.

2. (a) 0.7925.   (b) 0.0352.   (c) 0.0101.

3. (a) 0.9048.   (b) 0.6058. Normal approx. invalid.

4. (a) 0.1210.   (b) 0.2033.

5. (a) 0.9514.   (b) 0.0668.

6. (a) 0.9966.   (b) 0.1841.

7. (a) 0.1171.   (b) 0.2049.

8. (a) 0.0838.   (b) 0.1635.

9. 0.1357.

10. 0.4364.

11. (a) 0.0778.   (b) 0.0571.   (c) 0.6811.

12. (a) 0.8643.   (b) 0.2978.   (c) 0.0796.

PAGE 175

1. $2.8e^{-1.8} - 3.4e^{-2.4} = 0.1545.$

2. $4e^{-3} = 0.1992.$

4. (a) $1 - 3e^{-2} = 0.5940.$
(b) $5e^{-4} = 0.0916.$

5. (a) $\mu = 6$; $\sigma^2 = 18.$
(b) From 0 to 14.485 million liters.

6. (a) $\alpha = 3$; $\beta = 2.$
(b) $25e^{-6} = 0.0620.$

7. $\displaystyle\sum_{x=4}^{6} \binom{6}{x}(1 - e^{-3/4})^x (e^{-3/4})^{6-x} = 0.3968.$

8. 0.0352.

9. (a) 3/5.   (b) 1/2.

10. (a) 0.6.   (b) 0.7.   (c) 0.5.

12. (a) $\sqrt{\pi/2} = 1.2533.$ (b) $e^{-2} = 0.1353.$

14. $e^{-4} = 0.0183.$

15. 0.1808.

16. (a) $\mu = \alpha B = 50.$ (b) $\sigma^2 = \alpha B^2 = 500\sigma = \sqrt{500}.$ (c) 0.8155.

17. (a) 0.9995.   (b) 0.5940.

18. (a) 0.1889.   (b) 0.357.

CHAPTER 7

PAGE 195

1. $g(y) = 1/3$, $y = 1, 3, 5.$

2. $g(y) = \left(\dfrac{3}{\sqrt{y}}\right)\left(\dfrac{2}{5}\right)^{\sqrt{y}}\left(\dfrac{3}{5}\right)^{3-\sqrt{y}}$, $y = 0, 1, 4, 9.$

3. $g(y_1, y_2) = \left(\dfrac{y_1 + y_2}{2}, \dfrac{y_1 - y_2}{2}, 2 - y_1\right)$
$\times \left(\dfrac{1}{4}\right)^{(y_1+y_2)/2}\left(\dfrac{1}{3}\right)^{(y_1-y_2)/2}\left(\dfrac{5}{12}\right)^{2-y_1}$;
$y_1 = 0, 1, 2$; $y_2 = -2, -1, 0, 1, 2$;
$y_2 \le y_1$; $y_1 + y_2 = 0, 2, 4.$

4.

y	1	2	3	4	6
$h(y)$	$\frac{1}{18}$	$\frac{2}{9}$	$\frac{1}{6}$	$\frac{2}{9}$	$\frac{1}{3}$

6. $g(y) = 1/6y^{1/3}$, $0 < y < 8.$

7. Gamma distribution with $\alpha = 3/2$ and $\beta = m/2b.$

8. (a) $g(y) = y^{-1/2} - 1$, $0 < y < 1.$
(b) 3/4.

9. (a) $g(y) = 32/y^3$, $y > 4.$
(b) 1/4.

10. (a) $g(z) = 4z^3$, $0 < z < 1.$
(b) 65/256.

11. $h(z) = 2(1 - z)$, $0 < z < 1.$

13. $h(w) = 6 + 6w - 12w^{1/2}$, $0 < w < 1.$

14. $g(y) = 1/2\sqrt{y}$, $0 < y < 1.$

15.
$$g(y) = \begin{cases} \dfrac{2}{9\sqrt{y}}, & 0 < y < 1 \\[2mm] \dfrac{(\sqrt{y} + 1)}{9\sqrt{y}}, & 1 < y < 4. \end{cases}$$

18. $\mu = 1/p$; $\sigma^2 = q/p^2.$

19. Both equal $\mu.$

20. 0.9306.

CHAPTER 8

PAGE 206

1. (a) Responses of all people in Richmond who have a telephone.
(b) Outcomes for a large or infinite number of tosses of a coin.
(c) Length of life of such tennis shoes when worn on the professional tour.
(d) All possible time intervals for this lawyer to drive from her home to her office.

2. (a) Number of tickets issued by all state troopers in Montgomery County during the Memorial Day weekend.
(b) Number of tickets issued by all state troopers in South Carolina during the Memorial Day weekend.

3. (a) $\bar{x} = 2.4.$ (b) $\tilde{x} = 2.$ (c) $m = 3.$

4. (a) $\bar{x} = 8.6$ minutes.   (b) $\tilde{x} = 9.5$ minutes.
(c) Modes are 5 and 10 minutes.

5. (a) $\bar{x} = 3.2$ seconds. (b) $\tilde{x} = 3.1$ seconds.

6. (a) $\bar{x} = 35.7$ grams. (b) $\tilde{x} = 32.5$ grams.
(c) $n = 29$ grams.

7. (a) $\bar{x} = \$22.50$. (b) Modes: \$10 and \$25.

8. $\bar{x} = 22.2$ days; $\tilde{x} = 14$ days; $m = 8$ days.

9. (a) Range is 10. (b) $s = 3.307$.

10. (a) Range is 2. (b) $s^2 = 0.498$.

11. (a) 2.971. (b) 2.971.

12. (a) 11.69 milligrams. (b) $s^2 = 10.776$.

13. $s = 0.585$.

15. (a) 45.9. (b) 5.1.

PAGE 222

1. 0.3159.

2. 0.1912.

3. (a) Reduced from 0.7 to 0.4.
(b) Increased from 0.2 to 0.8.

4. 100.

5. Yes.

6. (a) $\mu_{\bar{x}} + 174.5$; $\sigma_{\bar{x}} = 1.38$.
(b) Approximately 154.
(c) Approximately 6.

7. (a) $\mu = 5.3$; $\sigma^2 = 0.81$.
(b) $\mu_{\bar{x}} = 5.3$; $\sigma_{\bar{x}}^2 = 0.0225$.
(c) 0.9082.

8. 0.0668.

9. (a) 0.6898. (b) 5.35 years.

10. (a) 0.0062. (b) 0.0668. (c) 0.3413.

12. 0.7070.

13. 0.5596.

14. (a) 0.0768. (b) 0.2812.

15. 0.9052.

PAGE 235

1. (a) 27.488. (b) 18.475. (c) 36.415.

2. (a) 16.750. (b) 30.144. (c) 26.217.

3. (a) 13.277. (b) 32.852. (c) 46.928.

4. (a) 38.932. (b) 12.592. (c) 20.483.

5. (a) 0.05. (b) 0.94.

6. Not valid.

8. (a) 2.145. (b) -1.372. (c) -3.499.

9. (a) 0.975. (b) 0.10.
(c) 0.875. (d) 0.99.

10. (a) 0.985. (b) 0.975.

11. (a) 2.500 (b) 1.319. (c) 1.714.

12. $t = -2.000$; valid claim.

13. No; $\mu > 20$.

14. $t = 1.64$; yes.

15. (a) 2.71. (b) 3.51. (c) 2.92.
(d) 0.47. (e) 0.34.

16. The F-ratio is 5.66. The variances appear to be different.

17. The F-ratio is 1.44. The variances can be considered equal.

CHAPTER 9

PAGE 252

4. $765 < \mu < 795$.

5. $2.20 < \mu < 2.30$.

6. (a) $172.23 < \mu < 176.77$. (b) Error ≤ 2.27.

7. (a) $22{,}496 < \mu < 24{,}504$. (b) Error ≤ 1004.

8. 68.

9. 11.

10. 28.

11. 56.

12. $10.15 < \mu < 12.45$.

13. $0.978 < \mu < 1.033$.

14. $1.49 < \mu < 3.71$.

15. $47.722 < \mu < 49.278$.

16. $74.34 < \mu < 84.26$.

17. 0.925 to 1.675.

18. 0.382 to 7.192.

19. 11,426 to 35,574.

20. 44.52 to 52.48.

PAGE 263

1. $2.9 < \mu_1 - \mu_2 < 7.1$.

2. $6.56 < \mu_B - \mu_A < 11.24$.

3. $2.80 < \mu_1 - \mu_2 < 3.40$.

4. $0.69 < \mu_1 - \mu_2 < 7.31$.

5. $1.5 < \mu_1 - \mu_2 < 12.5$.

6. $0.033 < \mu_2 - \mu_1 < 0.299$.

7. $0.70 < \mu_2 - \mu_1 < 3.30$.

8. $4.3 < \mu_1 - \mu_2 < 5.7$.

9. $-6522 < \mu_1 - \mu_2 < 2922$.

10. $-11.9 < \mu_{II} - \mu_I < 36.5$.

11. $(-0.74, 6.29)$.

12. $-2912 < \mu_D < 687$.

PAGE 270

1. (a) $0.498 < p < 0.642$.   (b) Error ≤ 0.072.

2. (a) $0.1422 < p < 0.1998$.
(b) Error ≤ 0.0278.

3. $0.194 < p < 0.262$.

4. $0.017 < p < 0.143$.

5. (a) $0.739 < p < 0.961$.   (b) No.

6. (a) $0.130 < p < 0.350$.   (b) Error ≤ 0.110.

7. (a) $0.644 < p < 0.690$.   (b) Error ≤ 0.023.

8. 2090.

9. 2576.

10. 467.

11. 160.

12. 9604.

13. 16,577.

14. 601.

15. $-0.0136 < p_F - p_M < 0.0636$.

16. $0.016 < p_A - p_B < 0.164$; valid claim.

17. $0.0011 < p_1 - p_2 < 0.0869$.

18. $(-1.1265, 0.6265)$. No significant difference.

19. $(-0.0817, -0.0019)$. Significantly different.

20. $(-0.0187, 0.0593)$. No significant difference.

PAGE 276

1. $0.293 < \sigma^2 < 6.736$; valid claim.

2. $8.400 < \sigma^2 < 39.827$.

3. $1.863 < \sigma < 3.578$.

4. $0.00022 < \sigma^2 < 0.00357$.

5. $1.410 < \sigma < 6.385$.

6. $1.258 < \sigma^2 < 5.410$.

7. $0.549 < \sigma_1/\sigma_2 < 2.690$.

8. $0.238 < \sigma_1^2/\sigma_2^2 < 1.895$; yes.

9. $0.016 < \sigma_1^2/\sigma_2^2 < 0.454$; no.

PAGE 284

1. $p^* = 0.173$.

2. (a)

p	0.05	0.10	0.15
$f(p \mid x = 2)$	0.12	0.55	0.33

(b) $p^* = 0.111$.

3. (a) $f(p \mid x = 1) = 40p(1 - p)^3/0.2844$.
(b) $p^* = 0.106$.

4. (a)

p	0.6	0.7
$f(p \mid x = 12)$	0.228	0.772

(b) $p^* = 0.677$.

5. $8.077 < \mu < 8.692$.

6. (a) \$7.04.   (b) $\$6.65 < \mu < \$7.43$.
(c) 0.6532.

7. (a) 0.2509.   (b) $68.71 < \mu < 71.69$.
(c) 0.0174.

8. $f(\mu \mid x_1, x_2, \ldots, x_{25}) = \dfrac{1}{\sqrt{2\pi}13.706}$
$\times\, e^{-1/2[(\mu-780)/20]^2}$, $770 < \mu < 830$.

10. $R(\hat{P}; p) = pq/n$.

11.
$$R(\Theta; \theta) = \begin{cases} 0 & \text{for } \theta = 0 \\ \frac{2}{3} & \text{for } \theta = 1 \\ \frac{2}{3} & \text{for } \theta = 2 \\ 0 & \text{for } \theta = 3. \end{cases}$$

12.
$$R(\Theta_2; \theta) = \begin{cases} 0 & \text{for } \theta = 0 \\ \frac{1}{3} & \text{for } \theta = 1 \\ 1 & \text{for } \theta = 2 \\ 0 & \text{for } \theta = 3. \end{cases}$$

13. $\hat{\Theta}_1$.

14. $\hat{\Theta}_2$.

PAGE 290

1. (a) $L(x_1, x_2, \ldots, x_n, \beta) = \left(\dfrac{1}{\beta^n}\right) e^{-\left(\sum_i \dfrac{x_i}{\beta}\right)}$.

   (b) $\displaystyle\sum_{i=1}^{n} \frac{x_i}{n} = \bar{x}$.

2. $\displaystyle\sum_{i=1}^{n} \frac{x_i}{n} = \bar{x}$.

CHAPTER 10

PAGE 308

1. (a) Conclude that fewer than 30% of the public are allergic to some cheese products when in fact 30% or more are allergic.
   (b) Conclude that at least 30% of the public are allergic to some cheese products when in fact fewer than 30% are allergic.

2. (a) The training course is effective.
   (b) The training course is effective.

3. (a) The firm is not guilty.
   (b) The firm is guilty.

4. (a) $\alpha = 0.0853$.
   (b) $\beta = 0.8287; \beta = 0.7817$.
   (c) No.

5. (a) $\alpha = 0.0536$.
   (b) $\beta = 0.0918; \beta = 0.1401$.
   (c) Fair.

6. (a) $\alpha = 0.0548$.
   (b) $\beta = 0.3504; \beta = 0.6177; \beta = 0.8281$.

7. (a) $\alpha = 0.0559$.
   (b) $\beta = 0.0017; \beta = 0.0968; \beta = 0.5557$.

8. (a) $\alpha = 0.0850$.   (b) $\beta = 0.3409$.

9. (a) $\alpha = 0.0032$.   (b) $\beta = 0.0062$.

10. (a) $\alpha = 0.4199$.   (b) $\beta = 0.3529$.

11. (a) $\alpha = 0.1357$.   (b) $\beta = 0.2578$.

12. (a) $\alpha = 0.0466$.   (b) $\beta = 0.0022$.

13. $\alpha = 0.0094; \beta = 0.0122$.

14. (a) $\alpha = 0.0793$.   (b) $\beta = 0.0793; \beta = 0.5$.

15. (a) $\alpha = 0.0718$.   (b) $\beta = 0.1151$.

16. (a) $\alpha = 0.0026$.   (b) $\beta = 0.0228$.

17. (a) $\alpha = 0.0384$.   (b) $\beta = 0.5; \beta = 0.2776$.

18.

Value of μ	Probability of Accepting H_0
184	0.08
188	0.27
192	0.58
196	0.84
200	0.93
204	0.84
208	0.58
212	0.27
216	0.08

PAGE 329

1. $z = -1.64$; do not reject $\mu = 800$ hours.

2. $z = -1.27$; do not reject $\mu = 22.2$ deciliters.

3. $z = -2.76$; yes, $\mu < 40$ months.

4. $z = 2.77$; yes, $\mu > 162.5$ centimeters. $P = 0.0056$.

5. $z = 8.97$; yes, $\mu > 20,000$ kilometers. $P < 0.000$.

6. $z = 1.58$; no, do not reject at $\alpha = 0.05$. However, $P = 0.06$.

7. $t = 0.77$; do not reject H_0.

8. $t = 4.38$; yes, $\mu > 220$ milligrams.

9. $t = 1.41$; do not reject H_0. $P = 0.90$.

10. $t = 1.78$; do not reject H_0.

11. $t = -1.98$; do not reject H_0.

12. $z = 4.22$; reject H_0, $\mu_1 \neq \mu_2$. $P < 0.000$.

13. $z = -2.60$; reject H_0, $\mu_A - \mu_B < 12$ kilograms.

14. $z = 2.45$; reject H_0, $\mu_1 - \mu_2 > \$500$.

15. $t = 1.50$; no.

16. $t = 0.92$; no. $P = 0.18$.

17. $t = -0.70$; not effective.

18. $t = -0.84$; do not reject $\mu_1 = \mu_2$.

19. $t = 2.55$; reject H_0, $\mu_1 - \mu_2 > 4$ kilometers.

20. $t = -2.07$; reject H_0, $\mu_1 - \mu_2 < 8$ months.

21. $t' = 0.22$; do not reject $\mu_{II} - \mu_I = 10$ minutes.

22. $t' = -0.42$; do not reject $\mu_S - \mu_N$. $P = 0.68$.

23. $t' = 2.76$; no, $\mu_1 > \mu_2$.

24. $t = -1.58$; Do not reject that, $\mu_1 = \mu_2$.

25. $t = 2.45$; yes.

26. $t = -0.90$; do not reject $\mu_1 - \mu_2 = 4.5$. $P = 0.20$.

27. $t = -2.53$; valid claim.

28. $t = 2.64$; yes.

29. 22.

30. 21.

31. 79.

32. 48.

33. 10.

34. 68.

35. (a) $H_0 : M_{hot} - M_{cold} = 0$,
$H_1 : M_{hot} - M_{cold} \neq 0$.
(b) Paired t. $t = 0.99$; do not reject. $P = 0.36$.

36. $t = 2.4$; $P = 0.023$. Breathing frequency significantly higher in the presence of CO.

PAGE 338

1. $P = 0.3916$; do not reject $p = 0.2$.

2. $P = 0.2131$; do not reject $p = 0.4$.

3. $P = 0.0207$; yes, the coin is not balanced.

4. $z = -1.44$; do not reject $p = 0.6$.

5. $z = 2.85$; reject H_0, $p > 1/5$.

6. $z = 1.34$; valid estimate.

7. $z = 1.44$; valid claim.

8. $z = 1.33$; no increase.

9. $z = 2.40$; yes. $P = 0.01$.

10. $z = 2.42$; yes. $P = 0.01$.

11. $z = 1.88$; yes.

12. $H_0 : p_1 = p_2; H_1 : p_1 > p_2; p_1 =$ proportion cured on medicine; $p_2 =$ proportion cured not on medicine; $z = 0.9344$. Do not reject H_0.

PAGE 343

1. $\chi^2 = 18.12$; do not reject $\sigma^2 = 0.03$.

2. $\chi^2 = 10.74$; do not reject $\sigma = 6$.

3. $\chi^2 = 17.45$; reject H_0, $\sigma^2 > 1.3$.

4. $\chi^2 = 17.19$; do not reject $\sigma = 1.40$.

5. $\chi^2 = 42.37$; machine is out of control.

6. (a) $z = 2.64$; reject H_0, $\sigma > 7.5$.
(b) $z = -1.92$; no.

7. $f = 1.33$; do not reject $\sigma_1^2 = \sigma_2^2$.

8. $f = 10.09$; reject H_0, $\sigma_1^2 > \sigma_2^2$.

9. $f = 0.75$; do not reject $\sigma_1 = \sigma_2$.

10. $f = 0.086$; reject H_0, $\sigma_1^2 < \sigma_2^2$.

11. $f = 1.18$; do not reject $\sigma_A = \sigma_B$. $P = 0.82$.

12. $F = 19.67$; $P[F > 19.67] = 0.0008$. Variances are not equal.

13. $F = 5.54$; $P[F > 5.54] = 0.0005$. Variances are not equal.

PAGE 358

1. $\chi^2 = 4.47$; yes.

2. $\chi^2 = 6.76$; no.

3. $\chi^2 = 10.14$; reject H_0, ratio is not $5:2:2:1$.

4. $\chi^2 = 10.00$; reject H_0, distribution is not uniform.

5. $\chi^2 = 2.33$; do not reject H_0, binomial distribution.

6. $\chi^2 = 1.67$; do not reject H_0, hypergeometric distribution.

7. $\chi^2 = 2.57$; do not reject H_0, geometric distribution.

10. $\chi^2 = 12.78$; reject H_0, not normal.

11. $\chi^2 = 5.19$; do not reject H_0, normal distribution.

12. $\chi^2 = 14.60$; not independent.

13. $\chi^2 = 5.40$; do not reject H_0.

14. $\chi^2 = 7.54$; do not reject H_0.

15. $\chi^2 = 124.59$; yes.

16. $\chi^2 = 3.81$; do not reject H_0.

17. $\chi^2 = 31.17$; attitudes are not homogeneous.

18. $\chi^2 = 5.78$; no.

19. $\chi^2 = 5.92$; do not reject H_0.

20. $\chi^2 = 12.56$; proportions are different.

21. $\chi^2 = 1.84$; do not reject H_0.

22. $\chi^2 = 1.39$; do not reject H_0.

CHAPTER 11

PAGE 371

1. (a) $\alpha = 64.52916$, $b = 0.56090$.
(b) $\hat{y} = 81.4$.

2. (a) $\hat{y} = 12.0623 + 0.7771x$.
(b) $\hat{y} = 78$.

3. (a) $6.4136 + 1.8091x$.
(b) $\hat{y} = 9.580$.

4. (a) $\hat{y} = 42.5818 - 0.6861x$.
(b) $\hat{y} = 25.7724$.

5. (a) $\hat{y} = 5.8254 + 0.5676x$.
(c) $\hat{y} = 34.205$.

6. (b) $\hat{y} = 32.5059 + 0.47711x$.
(d) $x = 59$.

7. (b) $\hat{y} = 343.706 + 3.221x$.
(c) $\hat{y} = \$456$.

8. (a) $\hat{y} = 2.776 - 0.180x$.
(b) $\hat{y} = 2.24$.

9. (a) $\hat{y} = 153.175 - 6.324x$.
(b) $\hat{y} = 123$.

10. (a) $\hat{z} = 6461.392 \times 0.947^w$.
(b) $\hat{z} = \$5197$.

PAGE 383

3. (a) $s^2 = 176.362$.
(b) $t = 2.04$; do not reject $\beta = 0$.

4. (a) $s^2 = 379.150$.
(b) $-69.913 < \alpha < 94.038$.
(c) $-0.248 < \beta < 1.802$.

5. (a) $s^2 = 0.40$.
(b) $4.324 < \alpha < 8.503$.
(c) $0.446 < \beta < 3.172$.

6. (a) $s^2 = 2.69$.
(b) $21.958 < \alpha < 63.205$.
(c) $-1.478 < \beta < 0.106$.

7. (a) $s^2 = 6.626$.
(b) $2.684 < \alpha < 8.968$.
(c) $0.498 < \beta < 0.637$.

8. $t = 1.78$; reject H_0, $\alpha > 10$.

9. $t = -2.24$; reject H_0, $\beta < 6$.

10. $58.808 < \mu_{Y|80} < 89.652$.

11. (a) $24.444 < \mu_{Y|24.5} < 27.112$.
(b) $21.889 < y_0 < 29.668$.

13. $7.815 < y_0 < 10.801$.

14. (a) $32.233 < \mu_{Y|50} < 36.177$.
(b) $26.432 < y_0 < 41.978$.

PAGE 393

1. (a) $b = \sum_{i=1}^{n} x_i y_i \bigg/ \sum_{i=1}^{n} x_i^2$.   (b) $\hat{y} = 2.003x$.

2. $\hat{y} = 0.349 + 1.929x$; $t = 1.40$; do not reject H_0.

3. $E(B) = \beta + \gamma \sum_{i=1}^{n} (x_{1i} - \bar{x}_1)x_{2i} \bigg/ \sum_{i=1}^{n} (x_{1i} - \bar{x}_1)^2$.

4. $f = 9.00$; reject H_0.

5. (a) $a = 10.81153$, $b = -0.34370$.
(b) $f = 0.43$; regression is linear.

6. $f = 1.58$; regression is linear.

7. $f = 1.12$; regression is linear.

PAGE 409

1. $r = 0.240$.

2. $t = 0.51$; do not reject $\rho = 0$.

4. (a) $r = 0.784$.
   (b) $t = 3.34$; reject H_0; $\rho > 0$.
   (c) 61.5%.

5. (a) $r = 0.392$.
   (b) $t = 2.04$; do not reject $\rho = 0$.

6. (a) $r = -0.979$.
   (b) $z = -4.22$; reject H_0, $\rho < -0.5$.
   (c) 95.8%.

CHAPTER 12

PAGE 424

1. (a) $\hat{y} = 27.547 + 0.922x_1 + 0.284x_2$.
   (b) $\hat{y} = 84$.

2. $\hat{y} = 55.2266 - 0.0378x_1 + 1.6816x^2$.

3. $\hat{y} = 0.5800 + 2.7122x_1 + 2.0497x_2$.

4. (a) $\hat{y} = -22.9932 + 1.3957x_1 + 0.2176x_2$.
   (b) $\hat{y} = 80$ kg.

5. (a) $\hat{y} = 56.4633 + 0.1525x - 0.00008x^2$.
   (b) $\hat{y} = 86.7\%$.

6. (a) $\hat{d} = 13.3587 - 0.3394v + 0.011825v^2$.
   (b) $\hat{d} = 47.54$.

7. $\hat{y} = 141.6118 - 0.2819x + 0.0003x^2$.

8. (a) $\hat{y} = 19.033333 + 1.0085714x - 0.020380952x^2$.
   (b) $f = 0.018$; model is adequate.

9. $\hat{y} = 19.98519 + 0.30363x_1 + 0.59635x_2 - 0.49706x_3 - 0.70378x_4$.

10. (a) $\hat{y} = 1.0714 + 4.6032x - 1.8452x^2 + 0.1944x^3$.
    (b) $\hat{y} = 4.45$.

11. $\hat{y} = 3.3205 + 0.4210x_1 - 0.2958x_2 + 0.0164x_3 + 0.1247x_4$.

12. $\hat{y} = 1962.9481 - 15.8517x_1 + 0.0559x_2 + 1.5896x_3 - 4.2187x_4 - 394.3141x_5$.

13. $\hat{y} = -6.5122 + 1.9994x_1 - 3.6751x_2 + 2.5245x_3 + 5.1581x_4 + 14.4012x_5$.

14. $\hat{y} = -884.6670 - 0.8381x_1 + 4.9066x_2 + 1.3311x_3 + 11.9313x_4$.

PAGE 432

1. 34.3699.

2. 0.4316.

3. 0.000995.

4. $\hat{\sigma}_{B_1}^2 = 0.000071$; $\hat{\sigma}_{B_2}^2 = 0.063523$;
   $\hat{\sigma}_{B_1 B_2} = -0.001134$.

5. (a) $\hat{\sigma}_{B_2}^2 = 0.00002$.   (b) $\hat{\sigma}_{B_1 B_4} = -0.000003$.

6. $26.2352 < y_0 < 57.1516$;
   $34.8580 < \mu_{Y|2500, 48.0} < 48.5288$.

7. $29.93 < \mu_{Y|19.5} < 31.97$.

8. $16.78 < y_0 < 16.93$;
   $16.83 < \mu_{Y|8.2, 6.0, 10.3, 5.8} < 16.88$.

9. $t = 2.86$; reject H_0, $\beta_2 > 0$.

10. $t = -4.48$; reject H_0, $\beta_1 < 0$.

11. $t = 3.55$; reject H_0, $\beta_1 > 2$. $P = 0.01$.

PAGE 442

1. $R^2 = 0.9997$.

2. $f = 12{,}689$; regression is significant.

3. $f = 13{,}409$; regression is significant.

4. $f = 11{,}759$; reject H_0.

5. $f = 20.07$; reject H_0, $\beta_1 < 0$.

6. (a) $\hat{x} = 9.9 + 0.575x_1 + 0.550x_2 + 1.150x_3$.
   (b) $f = 7.69$ for β_1; $f = 7.04$ for β_2;
   $f = 30.77$ for β_3.

PAGE 458

1. (b) $\hat{y} = 4.690$ seconds.
   (c) $4.450 > \mu_{Y|180, 260} < 4.930$.

2. (a) $\hat{y} = -6.33592 + 0.33738x_1$.
   (b) Same as (a).
   (c) Same as (a).

3. $\hat{y} = 2.1833 + 0.9576x_2 + 3.3253x_3$.

4. (a) $y = -29.5805 + 0.27877x_1 + 0.06971x_2 + 1.24146x_3 - 0.39535x_4 - 0.22369x_5$.
   (b) $\hat{y} = -56.93515 + 1.63432x_3 + 0.24862x_5$.
   (c) (x_2, x_5), (x_1, x_5), and (x_1, x_3, x_5).

5. (a) $\hat{y} = -587.211 + 428.433x$.
 (b) $\hat{y} = 1180 - 191.691x + 35.20945x^2$.
 (c) Quadratic model.

6. $t = -0.53$; do not reject $\beta_4 = 0$.

7. $\hat{\sigma}_{B_1}^2 = 20{,}588.04$; $\hat{\sigma}_{B_{11}}^2 = 62.6502$.

8. (a) VOLTS $= -1.641287 + 0.000556$
     SPEED $- 67.395890$ EXT.
 (b) $t = -6.64$ (intercept) $P < 0.0001$
     $t = 16.12$ (speed) $P < 0.0001$
     $t = -15.04$ (EXT) $P < 0.0001$
 (c) Both variables are significant.
             $R^2 = 0.96$.

9. Intercept model is best.

10.

Variables	PRESS	CP
ln (x_2), ln (x_3)	282194	2.0337
ln (x_2)	282276	1.5422
ln (x_1) ln (x_2)	289651	3.1039
ln (x_1), ln (x_2), ln (x_3)	294621	4.0000
ln (x_3)	297243	2.8181
ln (x_1)	304664	3.3489
ln (x_1) ln (x_3)	306820	3.9645

CHAPTER 13

PAGE 475

3. $f = 0.31$; no significant difference.

4. $f = 6.90$; mean number of hours of relief differ.

5. $f = 14.52$; yes, significant.

6. $f = 5.46$; do not reject H_0.

7. $f = 2.25$; do not reject H_0.

8. $f = 0.46$; do not reject H_0.

9. $f = 8.38$; average specific activities differ.

PAGE 487

1. (a) $f = 14.27$; reject H_0.
 (b) $f = 23.23$; reject H_0.
 (c) $f = 2.48$; do not reject H_0.

2. (a) $f = 5.15$; significant.
 (b) $f = 9.86$; significant.

3. (a) $f = 13.50$; treatment means differ.
 (b) f (1 versus 2) $= 29.35$; significant.
     f (3 versus 4) $= 3.59$; not significant.

4.

$\bar{x}_4$	$\bar{x}_3$	$\bar{x}_1$	$\bar{x}_5$	$\bar{x}_2$
2.8	4.0	5.2	6.6	7.8

5.

$\bar{x}_3$	$\bar{x}_1$	$\bar{x}_4$	$\bar{x}_2$
56.52	59.66	61.12	61.96

6. (a) $f = 7.10$; reject H_0.
 (b) Blend 4 differs significantly from all others.

7. (a) $f = 9.01$; yes, significant.
 (b) Depletion and modified Hess are significantly different from the other three procedures.

8. (a) $f = 5.34$; significant.
 (b) $d_1 = 0.5553$; not significant.
     $d_2 = 3.1158$; significant.
     $d_3 = 3.0464$; significant.

9. $d_1 = 9.0878$; significant.
   $d_2 = 6.8498$; significant.
   $d_3 = 2.3059$; significant.
   $d_4 = 2.5093$; significant.

10. $f = 9.07$ P-value ≤ 0.001. Reject H_0. Cable 9 is clearly stronger than all other cables except cable 8. Cable 8 is stronger than cables 1–4.

11. $f = 70.27$ P-value ≤ 0.0001. Reject H_0.

$\bar{x}_0$	$\bar{x}_{25}$	$\bar{x}_{100}$	$\bar{x}_{75}$	$\bar{x}_{50}$
55.167	60.167	64.167	70.500	72.833

Temperatures are important. Both 75 and 50°(C) yielded batteries with significantly longest activated life.

PAGE 508

3. (a) f(fertilizers) $= 6.11$; significant.
 (b) $f = 17.37$; significant.
     $f = 0.96$; not significant.

4. f(varieties) $= 1.74$; no significant difference in the yielding capabilities of the different varieties.

5. $f = 5.99$; percent of foreign additives is not the same for all three brands of jam.

6. f(subjects) $= 0.15$; not significant.

7. f(stations) $= 26.14$; significant.

8. f(stations) $= 0.15$; not significant.

9. f(diet) $= 11.86$; significant.

10. $f = 3.00$; not significant.

11. $f = 0.58$; not significant.

12. f(treatments) $= 8.60$; mean weight losses are different.

15. $f = 5.03$; grades are affected by different professors.

16. $f = 1.29$; color additives have no effect on setting time.

17. $p \leq 0.001$.
$f = 122.37$: the amount of dye has an effect on the color of the fabric.

PAGE 529

1. (a) $f = 14.9$; operators differ significantly.
(b) $\hat{\sigma}_\alpha^2 = 28.91$; $s^2 = 8.32$.

3. (a) $f = 3.33$; no significant difference.
(b) $\hat{\sigma}_\alpha^2 = 1.08$; $\hat{\sigma}_\beta^2 = 2.25$.

6. No; 16.

7. 9.

CHAPTER 14

PAGE 546

1. (a) $f = 8.13$; significant.
(b) $f = 5.18$; significant.
(c) $f = 1.63$; not significant.

2. (a) $f = 2.36$; not significant.
(b) $f = 16.59$; significant.
(c) $f = 0.41$; not significant.

3. (a) $f = 14.81$; significant.
(b) $f = 9.04$; significant.
(c) $f = 0.61$; not significant.

4. (a) f(humidity) $= 4.57$; significant.
f(coating) $= 6.87$; significant.
f(humidity $\times$ coating) $= 2.44$; not significant.
(b) Corrosion damage is different for medium humidity than for low or high humidity.

5. (a) $f = 34.40$; significant.
(b) $f = 26.95$; significant.
(c) $f = 20.30$; significant.

PAGE 555

1. (a) AB: $f = 3.83$; significant.
AC: $f = 3.79$; significant.
BC: $f = 1.31$; not significant.
ABC: $f = 1.63$; not significant.

(b) A: $f = 0.54$; not significant.
B: $f = 6.85$; significant.
C: $f = 2.15$; not significant.

2. Significant effects:
A: $f = 8.28$ C: $f = 507.57$
Insignificant effects:
B: $f = 0.29$ BC: $f = 5.92$
AB: $f = 3.85$ (Significant at 0.05)
AC: $f = 2.33$ ABC: $f = 1.84$

3. Significant effects:
Stress: $f = 45.96$
Insignificant effects:
Coating: $f = 0.05$
Humidity: $f = 2.13$
Coating $\times$ humidity: $f = 3.41$
Coating $\times$ stress: $f = 0.08$
Humidity $\times$ stress: $f = 3.15$
Coating $\times$ humidity $\times$ stress: $f = 1.93$

4. (a) $f = 2.69$; no significant interaction.
(b) A: $f = 3.37$; significant.
B: $f = 5.63$; significant.
C: $f = 4.82$; significant.
The pooled error includes BC.

5.

Effect	f	P-value
Temperature	14.22	≤ 0.0001
Surface	6.70	0.0020
HRC	1.67	0.1954
TXS	5.50	0.0006
T * HRC	2.69	0.0369
S * HRC	5.41	0.0007
T * S * HRC	3.02	0.0051

PAGE 562

1. (a) $f = 1.49$; no significant interaction.
(b) f(operators) $= 12.45$; significant.
f(filters) $= 8.39$; significant.
(c) $\hat{\sigma}_\alpha^2 = 0.1701$ (filters);
$\hat{\sigma}_\beta^2 = 0.3514$ (operators);
$s^2 = 0.1867$.

2. $\hat{\sigma}_\alpha^2 = 1.4678$ (brand); $\hat{\sigma}_\beta^2 = 12.1642$ (time);
$s^2 = 0.9237$.

3. (a) $\hat{\sigma}_\beta^2$, $\hat{\sigma}_\gamma^2$, $\hat{\sigma}_{\alpha\gamma}^2$ are significant.
(b) $\hat{\sigma}_\gamma^2$, $\hat{\sigma}_{\alpha\gamma}^2$ are significant.

4. Yes.

5. 0.59.

CHAPTER 15

1. $SSA = 2.6667$, $SSB = 170.6667$, $SSC = 104.1667$, $SS(AB) = 1.5000$, $SS(AC) = 42.6667$, $SS(BC) = 0.0000$, $SS(ABC) = 1.5000$.

2. Significant effects:

A: $f = 1294.65$	AB: $f = 20.88$
B: $f = 43.56$	AC: $f = 16.21$
C: $f = 116.49$	ABC: $f = 289.23$

Insignificant effect:

BC: $f = 0.00$

3. Significant effects:

A: $f = 1940.64$
B: $f = 4411.62$
C: $f = 1098.38$
D: $f = 458.50$

4. Significant effects:

A: $f = 57.85$	AC: $f = 7.08$
B: $f = 7.52$	AD: $f = 4.85$
C: $f = 127.87$	BC: $f = 10.96$
D: $f = 44.72$	BD: $f = 4.85$
AB: $f = 6.94$	CD: $f = 6.52$

Insignificant effects:

ABC: $f = 1.26$	BCD: $f = 1.20$
ABD: $f = 1.14$	$ABCD$: $f = 0.87$
ACD: $f = 1.72$	

5. Significant effects:

A: $f = 9.98$ BC: $f = 19.03$

Insignificant effects:

B: $f = 0.20$	AC: $f = 0.20$
C: $f = 6.54$	AD: $f = 0.57$
D: $f = 0.02$	BD: $f = 1.83$
AB: $f = 1.83$	CD: $f = 0.02$

1. A, B, C, AC, BC, and ABC each with 1 degree of freedom can be tested using an error mean square with 12 degrees of freedom.

2. (a) A: $f = 1.55$; not significant.
       B: $f = 1.27$; not significant.
       C: $f = 3.49$; not significant.
       D: $f = 0.79$; not significant.
   (b) ABC.

3.

Block 1	Block 2	Block 3	Block 4
(1)	c	d	a
ab	abc	ac	b
acd	ad	bc	cd
bcd	bd	abd	$abcd$

CD is also confounded.

4. (a)

Block 1	Block 2	Block 3	Block 4
(1)	a	b	ab
c	ac	bc	abc
ae	e	abe	be
bd	abd	d	ad
ace	ce	$abce$	bce
bcd	$abcd$	cd	acd
$abde$	bde	ade	de
$abcde$	$bcde$	$acde$	cde

(b) BD.
(c) $SSA = 21.9453$; $SSC = 2.4753$; $SS(E) = 1.0878$; $SSB = 40.2753$; $SSD = 7.7028$.

5.

Block		Block		Block	
1	2	1	2	1	2
abc	ab	abc	ab	(1)	a
a	ac	a	ac	c	b
b	bc	b	bc	ab	ac
c	(1)	c	(1)	abc	bc

Replicate 1	Replicate 2	Replicate 3
ABC	ABC	AB
confounded	confounded	confounded

6. (a) ABD; $ABCD$.
   (b) Significant effects:
       C: $f = 7.53$ D: $f = 13.38$
   Insignificant effects:

A: $f = 3.35$	CD: $f = 0.30$
B: $f = 0.84$	AC: $f = 1.94$
AB: $f = 0.84$	ABD: $f = 0.54$
AC: $f = 0.54$	ACD: $f = 0.03$
AD: $f = 1.20$	BCD: $f = 0.30$
BC: $f = 0.84$	$ABCD$: $f = 0.89$
BD: $f = 0.54$	

PAGE 606

1.

$A \equiv CDE$	$AE \equiv CD$
$B \equiv ABCDE$	$BC \equiv ABDE$
$C \equiv ADE$	$BD \equiv ABCE$
$D \equiv ACE$	$BE \equiv ABCD$
$E \equiv ACD$	$ABC \equiv BDE$
$AB \equiv BCDE$	$ABD \equiv BCE$
$AC \equiv DE$	$ABE \equiv BCD$
$AD \equiv CE$	

2. (a) Principal block = $\{(1), a, bc, abc, bd, abd, cd, acd\}$.

(b)

Block 1	Block 2
(1)	a
bc	abc
abd	bd
acd	cd

(c)

Source of Variation	Degrees of Freedom
A	1
B	1
C	1
D	1
Blocks	1
Error	2
Total	7

3. Principal block = $\{(1), ac, bd, abcd, abe, bce, ade, cde, abf, bcf, adf, cdf, ef, acef, bdef, abcdef\}$.

$$A \equiv BCD \equiv ABDEF \equiv CEF$$
$$B \equiv ACD \equiv DEF \equiv ABCEF$$
$$C \equiv ABD \equiv BCDEF \equiv AEF$$
$$D \equiv ABC \equiv BEF \equiv ACDEF$$
$$E \equiv ABCDE \equiv BDF \equiv ACF$$
$$F \equiv ABCDF \equiv BDE \equiv ACE.$$

4. (a) Principal block = $\{(1), ab, acd, bcd, ce, abce, ade, bde, acf, bcf, df, abdf, aef, bef, cdef, abcdef\}$.

(b)

Source of Variation	Degrees of Freedom
A	1
B	1
C	1
D	1
E	1
F	1
AB	1
AC	1
AD	1
BC	1
BD	1
CD	1
Error	3
Total	15

5. Significant effect:
   E: $f = 5.39$
Insignificant effects:

A:	$f = 0.48$	D:	$f = 1.94$
B:	$f = 1.35$	F:	$f = 1.09$
C:	$f = 3.03$	G:	$f = 4.36$

PAGE 620

1. $\hat{y} = 12.7519 + 4.7194x_1 + 0.8656x_2 - 1.4156x_3$.
Units are centered and scaled. Lack of fit:
$$F = 81.58 \qquad P < 0.0001$$

2.

Coefficients	D.F.
Intercept	1
β_1	1
β_2	1
β_3	1
β_4	1
β_5	1
Two-factor interactions	10
Lack of fit	16
Pure error	32
Total	63

3.

AFG	CEFG	ACDF
BEG	BDFG	ABEF
CDG	BCDE	ABCDEFG
DEF	ADEG	

4. Begin with basic line for $n = 24$; permute as described in Section 13.10 until 18 columns are formed.

CHAPTER 16

PAGE 632

1. $x = 7$; $P = 0.1719$, do not reject H_0.

2. $x = 10$; $P = 0.4544$, do not reject H_0.

3. $x = 3$; $P = 0.0244$, reject H_0.

4. $x = 2$; $P = 0.0547$, do not reject H_0.

5. $x = 4$; $P = 0.3770$, do not reject H_0.

6. $x = 12$; $P = 0.0160$, reject H_0.

7. $x = 4$; $P = 0.1335$, do not reject H_0.

8. $w_- = 12.5$; do not reject H_0.

9. $w = 43$; do not reject H_0.

10. $w_+ = 3.5$; do not reject H_0.

11. $w_+ = 17.5$; do not reject H_0.

12. $z = 2.80$; reject H_0.

13. $z = -2.13$; reject H_0, $\mu_1 - \mu_2 < 8$.

PAGE 639

1. $u_1 = 1$; claim is valid.

2. $u_1 = 8$; serum is not effective.

3. $u_2 = 5$; A operates longer.

4. $u = 43.5$; do not reject H_0.

5. $u = 15$; no, do not reject H_0.

6. $z = -0.84$; do not reject H_0.

7. $h = 10.47$; operating times are different.

8. $h = 11.27$; tar contents are different.

9. $h = 1.07$; no significant difference.

PAGE 648

1. $v = 7$; $P = 0.910$, random sample.

2. $v = 2$; $P = 0.016$, reject randomness.

3. $v = 6$; $P = 0.044$, do not reject H_0.

4. $z = -0.55$; defectives occur at random.

5. $z = 1.11$; random sample.

6. 30.

7. 0.70.

8. 21.

9. 0.995.

10. (a) 0.24.   (b) Do not reject H_0.

11. (a) $r_s = 0.39$.   (b) do not reject H_0.

12. $r_s = -0.99$.

13. (a) $r_s = 0.96$ (b) Reject H_0, $\rho > 0$.

14. (a) $r_s = 0.72$.   (b) Reject H_0, $\rho > 0$.

15. $r = -0.47$; no significant relationship.

16. (a) $r_s = 0.71$.   (b) Reject H_0, $\rho > 0$.

17. (a) $r_s = 0.59$.   (b) Reject H_0, $\rho > 0$.

Index